REVOLUTIONS OF GEOMETRY

PURE AND APPLIED MATHEMATICS

A Wiley-Interscience Series of Texts, Monographs, and Tracts

A complete list of the titles in this series appears at the end of this volume.

REVOLUTIONS OF GEOMETRY

MICHAEL O'LEARY
College of DuPage
Glen Ellyn, Illinois

WILEY

A JOHN WILEY & SONS, INC., PUBLICATION

Published by John Wiley & Sons, Inc., Hoboken, New Jersey.
Published simultaneously in Canada.

Library of Congress Cataloging-in-Publication Data:

O'Leary, Michael, 1966–
Revolutions of geometry / Michael O'Leary.
p. cm.
Includes bibliographical references and index.
ISBN 978-0-470-16755-7 (cloth)
1. Geometry—History. 2. Geometry—Foundations. I. Title.
QA443.5.044 2009
516.009—dc22 2009034631

Printed in the United States of America.

10 9 8 7 6 5 4 3 2 1

For my wife, Barb

CONTENTS

Preface xi

Acknowledgments xiii

PART I FOUNDATIONS

1 The First Geometers 3

1.1 Egypt 6
1.2 Babylon 13
1.3 China 20

2 Thales 27

2.1 The Axiomatic System 29
2.2 Deductive Logic 35
2.3 Proof Writing 43

3 Plato and Aristotle 53

3.1 Form 56
3.2 Categorical Propositions 62

3.3 Categorical Syllogisms 72
3.4 Figures 77

PART II THE GOLDEN AGE

4 Pythagoras 87
4.1 Number Theory 91
4.2 The Pythagorean Theorem 98
4.3 Archytas 102
4.4 The Golden Ratio 110

5 Euclid 123
5.1 The Elements 124
5.2 Constructions 130
5.3 Triangles 138
5.4 Parallel Lines 147
5.5 Circles 159
5.6 The Pythagorean Theorem Revisited 167

6 Archimedes 173
6.1 The Archimedean Library 174
6.2 The Method of Exhaustion 182
6.3 The Method 193
6.4 Preliminaries to the Proof 204
6.5 The Volume of a Sphere 214

PART III ENLIGHTENMENT

7 François Viète 227
7.1 The Analytic Art 229
7.2 Three Problems 236
7.3 Conic Sections 246
7.4 The Analytic Art in Two Variables 257

8 René Descartes 267
8.1 Compasses 269
8.2 Method 274
8.3 Analytic Geometry 279

9 Gérard Desargues **293**

9.1 Projections 294
9.2 Points at Infinity 298
9.3 Theorems of Desargues and Menelaus 306
9.4 Involutions 312

PART IV A STRANGE NEW WORLD

10 Giovanni Saccheri **323**

10.1 The Question of Parallels 324
10.2 The Three Hypotheses 330
10.3 Conclusions for Two Hypotheses 337
10.4 Properties of Parallel Lines 340
10.5 Parallelism Redefined 349

11 Johann Lambert **353**

11.1 The Three Hypotheses Revisited 355
11.2 Polygons 360
11.3 Omega Triangles 373
11.4 Pure Reason 383

12 Nicolai Lobachevski and János Bolyai **393**

12.1 Parallel Fundamentals 397
12.2 Horocycles 404
12.3 The Surface of a Sphere 414
12.4 Horospheres 424
12.5 Evaluating the Pi Function 431

PART V NEW DIRECTIONS

13 Bernhard Riemann **443**

13.1 Metric Spaces 445
13.2 Topological Spaces 457
13.3 Stereographic Projection 464
13.4 Consistency of Non-Euclidean Geometry 471

14 Jean-Victor Poncelet **483**

14.1 The Projective Plane 486
14.2 Duality 492

14.3 Perspectivity 501
14.4 Homogeneous Coordinates 507

15 Felix Klein 519

15.1 Group Theory 520
15.2 Transformation Groups 529
15.3 The Principal Group 535
15.4 Isometries of the Plane 543
15.5 Consistency of Euclidean Geometry 553

References 565

Index 573

PREFACE

One of the courses that had a strong influence on me as an undergraduate student was a class on the history of mathematics. It was a serious course with a calculus prerequisite. We used David M. Burton's *The History of Mathematics: An Introduction* and Morris Kline's *Mathematics: The Loss of Certainty*. I found the problems challenging and the introduction to the history and philosophy of mathematics interesting. Based on this experience from too many years ago, I believe that studying mathematics within a historical context can be very motivating for students. Although I certainly believe that *Revolutions of Geometry* is a text that any junior or senior mathematics student would find stimulating, my main concern is for those who plan to teach the subject as a profession. I believe that it is crucial that the mathematics teachers of the future are not only competent in the subject but also enjoy it.

For these reasons, this book is a chronological introduction to the history of geometry presenting the mathematics alongside the story of those who made the discoveries. With few exceptions, the text focuses on the original work of the mathematicians encountered. It strives to clarify and organize the arguments to make them more accessible. Since it is geometry, the work emphasizes proofs, so it will not be without its challenges.

Because the history of geometry dictated the topics that are included in the text, it has a wide range of material from which to choose for many different kinds of geometry courses. The text is divided into five parts.

- *Foundations.* After encountering pre-Greek geometry, we meet Thales, Plato, and Aristotle. We learn their philosophy and logic. Methods of proof that are needed later in the book are introduced here.

- *The Golden Age.* This is the mathematics of the classical Greeks. We read of the teachings of Pythagoras and his followers, review high school geometry through Euclid's *Elements*, and see firsthand how close Archimedes actually was to the calculus.

- *Enlightenment.* The focus is the work of three great French mathematicians: Viète's revolutionary contributions to algebra, Descartes' merging of algebra and geometry as he solves the Pappus Problem, and Desargues' development of projective geometry.

- *A Strange New World.* Saccheri and Lambert were close, but it took Bolyai and Lobachevski to finally answer the question of parallels. Their work is based on a new definition of what it means for lines to be parallel. Here we encounter Saccheri's three hypotheses, results concerning triangles and rectangles, horocycles, horospheres, and some trigonometry. The intellectual giants Gauss and Kant also make appearances.

- *New Directions.* As we approach the twentieth century, the redefinition of geometry continues. Riemann defined it as the study of manifolds. Poncelet returned to projective geometry, but now it is more abstract. Klein used group theory to characterize different geometries. Along the way questions of consistency become important. Through the work of Beltrami, Poincaré, Klein again, and Hilbert we learn that each of Saccheri's three hypotheses is without contradiction.

There are occasions when topics in the text use results in set theory, function theory, and linear algebra. This is due to the requirements of some of the geometry studied. These prerequisites did not prevent these topics from being included because they provide a very different view of geometry and a good review of the material.

The exercise sets at the end of each section are designed both to reinforce the material and to push the student beyond the contents presented. They are organized to coincide approximately with the order of topics in their section. No solutions for these problems are given in the text, but a student's solution manual is available, and instructors may contact Wiley for an instructor's manual. The text also has a web page that contains supplementary and review material. Its address is

www.revolutionsofgeometry.com.

Finally, this book was typeset in LaTeX and the illustrations were created with Adobe Illustrator, both by the author.

Michael O'Leary
College of DuPage
December 2009

ACKNOWLEDGMENTS

Thanks are due to Susanne Steitz-Filler, the mathematics editor at Wiley, and my production editor, Angioline Loredo, for their help and patience with this project. I would also like to thank some of my colleagues at College of DuPage who read and commented on some early sections of the text: James Africh, Patrick Bradley, Robert Cappetta, and Zia Mahmood. Thanks also go to Dale Brant for his assistance with Kant and Lee Kesselman for his help with the Pythagorean scale. Lastly I send my appreciation to Kenneth Mangels and Bret Taylor (Concordia University Irvine) along with Astrida Cirulis and Norman E. Young (Concordia University Chicago) who supported this project with their initial reviews.

Thanks go the Princeton University Press for their permission to use quotations from Glen R. Morrow's translation of Proclus's *A Commentary on the First Book of Euclid's* Elements.

On a personal note, I would like to express my gratitude to my parents for their continued support and understanding; to my brother and his wife, who will teach my niece mathematics; to my dissertation advisor, Paul Eklof, who guided me through graduate school; to Robert Meyer, who introduced me to the history of mathematics; to Kenneth Mangels, who taught me non-Euclidean geometry; to David Elfman, who taught me about science and computers; to my new family, who have not seen me much lately; and to my wife, Barb, whose love and patience (and some proofreading) supported me as I played the hermit to finish this book.

PART I

FOUNDATIONS

CHAPTER 1

THE FIRST GEOMETERS

The exact origin of the ancient civilization of **Egypt** is unknown, but it definitely dates to sometime before the year 4000 B.C. There were two kingdoms along the Nile, an upper kingdom and a lower. It is said that somewhere between the years 3500 to 3000 B.C. the legendary Menes unified the two kingdoms and so formed the First Dynasty. He is further claimed to have founded the city of Memphis and dedicated a temple there to the creator-god called Ptah. For the next 500 to 1000 years Egypt grew in strength as a state and a society. The Third Dynasty, which lasted from 2700 to 2625 B.C, saw the reign of King Djoser. With the assistance of his chief architect and physician, Imhotep, they directed the construction of the first great stone building, the Step Pyramid at Saqqara. It sat in the middle of a great mortuary complex and was intended to see Djoser safely into the afterlife. During the Fourth Dynasty, approximately a half century later, Khufu (Cheops) reigned. Little is known about this pharaoh. The few texts that speak of Khufu range from portraying him as a amicable monarch interested in magic to a cruel slave master driven to build a magnificent monument to himself. This monument was the Great Pyramid. It currently stands the tallest of a group of three pyramids on the Giza plateau and is the only surviving member of the Seven Wonders of the Ancient World. It originally stood 481 feet tall but has lost 30 feet off its height. Its square base has sides that measure about 755 feet each with the difference between the longest and shortest sides being only 8 inches. Scholars estimate that it was built using 2.3 million stone blocks. The average weight of each is around 2.5 tons. In July 1798, Napoleon calculated that the stones from the three pyramids at Giza could be used to build a wall around France 1 foot wide and 12 feet high (Grimal 1992, 63–67, 389; Boardman 1982, 14, 145; Clayton 1995, 37, 46).

According to the ancient Greek historian Herodotus, there was a great king of Egypt whose name was Sesostris. It was claimed that he conquered vast portions of Asia and Europe and erected pillars on which were inscribed his name and country. After he had conquered the world, Sesostris returned home and focused on domestic issues. One of his major achievements was supposedly an extensive canal system that brought water from the Nile to his people. It is unlikely that this Sesostris ever existed. There were pharaohs who shared his name. Sesostris I (Senusret I) reigned at the beginning of the 12th Dynasty (*c.* 1990 B.C.), and he was succeed by his son, Amenemhat II, who was followed by his son, Sesostris II, and then his grandson, Sesostris III. However, neither of these can be the Sesostris of Herodotus. Instead, the pharaoh described by Herodotus was probably a legend, based on the reigns of Rameses II and Sety I, that later Egyptians created at a time when the power of Egyptian civilization began to wane (Herodotus 1987, II; Shaw 2000, 160–166, 295–302).

To the northeast of Egypt between the Tigris and Euphrates rivers was the land of Mesopotamia. This was the location of the fabled cities of Babylon, Ur, Susa, and Kish. About 4000 B.C. the **Sumerians** migrated to Mesopotamia and settled in its southern regions. Their land became known as Sumer and its capital was Ur. Around 2200 B.C. they lost their independence to the **Akkadians**, who were Semites from northern Mesopotamia. At the time the Akkadians were ruled by Sargon. His political life began in the court of King Ur-Zababa of Kish. Sargon rose to power and eventually gained control of Sumer. Many cuneiform tablets laud Sargon as a great ruler and warrior who supposedly conquered Asia Minor and the Mediterranean. Although the Sumerians and the Akkadians had there own identities, it is common to refer to them under the single term **Babylonian** (Leick 2003, 6, 102, 113; Crawford 2004, 29, 33).

Probably the best known ruler of Mesopotamia was Hammurabi. He reigned from 1792 to 1750 B.C. as king of Babylon during the First Dynasty. Beginning with a small region around the city, Hammurabi gradually extended his domain. This was accomplished by a combination of military conquests and temporary diplomatic alliances that were ended in betrayal by the king. By 1755 B.C., after 30 years of strengthening his position, Hammurabi had gained control of all of Mesopotamia, and Babylon would become the leading power in western Asia. As he accomplished this, Hammurabi established a strong central government, invested in irrigation, and strengthened the walls of the city of Babylon. In order to establish justice throughout his realm, a sequence of laws was published in all of his major cities. Known as the Code of Hammurabi, it contained 282 separate laws, covering everything from serious crimes such as murder to everyday business transactions. Several copies survive, including one found in 1902 at Susa on an 8-foot-tall stone monument of polished black diorite. The death of Hammurabi saw the beginning of the decline of the First Dynasty, which fell when Mursili, king of the Hittites, attacked Babylon by surprise in 1595 B.C. or 1499 B.C. (Leick 2003, 47–48, 57–58).

Although there was some contact between the Egyptian and Mesopotamian civilizations, on the other side of the continent in **China** there developed a society independent of Western contacts. The Chinese civilization began as a collection of small city-states along the banks of the Yellow River. Near the end of the twelfth century B.C., an ancient tribe began to grow in strength. It later became known as the Zhou dynasty and was led by King Wen, a well-respected leader who had ambitions to conquer the Shang dynasty, which was then dominant. King Wen died before he could finish his plan, but his son, Prince Wu, would later join forces with neighboring tribes to completely defeat the Shang (Shaughnessy 1999, 307–309).

The now King Wu established the Zhou dynasty with Gaojing as its capital. This dynasty lasted from 1122 to 256 B.C. This long time period is divided into two periods, the first being the Western Zhou (1122–771 B.C). The Western Zhou society was feudal in the sense that the king appointed rulers as his representatives over local jurisdictions. However, the relationship between subject and king was modeled on that of kinship as opposed to the impersonal, contractual relationship that arose in Europe. This system allowed the land over which the dynasty exercised control to expand, but it expanded too quickly and led to a collapse. This resulted in the Zhou capital's moving to Luoyi. This is the beginning of the Eastern Zhou period. This second period is divided further into the Spring and Autumn Period (771–481 B.C) and the Warring States Period (402–221 B.C). Although both periods saw many wars where states vied for power, significant cultural progress was made. For example, iron was introduced, new farming and military techniques were developed which included the use of horses, and writing was further developed (Roberts 1999, 9–12).

During the Warring States Period the Qin state, which had existed since the 8th century B.C., developed into a power that could rival the Zhou. The Qin were led by Qin Shi Huangdi who wished to unify all of China under his leadership. At the time there were six other states that were in a seemingly constant state of war. They were all defeated by Qin Shi Huangdi, ushering in the Qin dynasty and bringing order and unification to China. The dynasty only lasted from 221 B.C. to 206 B.C., but it was a significant period in Chinese history during which Chinese society was restructured so that the peasants now owned the land but were taxed by the government and a new code of laws was established which applied to all equally. In addition, Qin Shi Huangdi standardized weights, measures, and the font used in writing, and he led the construction of many civil projects, including a network of walls which led to construction of the Great Wall. However, to control knowledge the emperor ordered the destruction of many books and ordered the execution of any scholar who dissented. By the end of his reign Qin Shi Huangdi had became a tyrant. At the time of his death in 210 B.C., the people had grown dissatisfied living under the harsh conditions imposed by the emperor's construction projects. Millions had been drafted to build the projects, and this led to a rebelion. Qin Shi Huangdi's successor was killed, after which the Qin quickly surrendered. The year was 206 B.C., and civil war soon followed. The winner was a leader of the peasants named Liu Bang. His victory led to the establishment in 205 B.C. of the Han dynasty which lasted for 426 years. During this time a strong central government was established, yet many of the harsh policies of the Qin were removed and taxes reduced. This was also a time when the teachings of Confucius were made the state religion, the arts and literature flourished, and Chinese philosophy was at its peak (Roberts 1999, 19–29).

1.1 EGYPT

It is believed by most historians of mathematics that the Egyptians began studying geometry due to a need for a reliable method of measuring areas of land. This view is supported by the testimony of Herodotus.

> The priests also say that it was this king [Sesostris] who divided the land among all the Egyptians, giving to each man as an allotment a square, equal in size; for the king derived his revenues, as he appointed the payment therefore of a yearly tax. If the river should carry off a portion of the allotment, the man would come to the king himself and signify what had happened, whereupon the king sent men to inspect and remeasure by how much the allotment had grown less, so that for the future it should pay proportionally less of the assigned tax. I think it was from this that geometry was discovered and came to Greece. (Herodotus 1987, 2.109, 175)

The ancient Egyptians used lengths of rope with knots at regular intervals to measure land. By stretching the ropes along the needed dimensions and applying the appropriate formula the area would be found. From this practice the subject of geometry received its name. It is a combination of two Greek words: *geo*, meaning "earth," and *metron*, meaning "measure" (Burton 1985, 56–57).

What we know of Egyptian geometry can be traced to a surprising source. In 1798, Napoleon invaded Egypt in an attempt to protect French trade interests in the region and to weaken British ties to India. Napoleon brought a group of scholars along with his army. This was evidently a propaganda ploy to cast a favorable light on his intentions. Whatever his true motivation, his decision led to the discovery of the **Rosetta Stone** in 1799. This is a stone slab on which is carved a message in Greek, in Demotic, and in the Egyptian writing called **hieroglyphics**, a picture-based system that was used in formal settings. The meaning behind the Egyptian hieroglyphics had long eluded scholars. Finding the Rosetta Stone unlocked their meaning since they were now able to compare the hieroglyphics with ancient Greek (Boyer and Merzbach 1991, 10–11; Burton 1985, 36).

In 1858, Scotland native A. Henry Rhind made another discovery. In the tombs of Thebes in Luxor he found a scroll and purchased it for his collection. Scholars estimate that it was written about 1650 B.C., placing its creation during the reign of King Apophis I of the 17th Dynasty. The author of the scroll, Ahmes, claimed that his work was a copy of a document from the Third Dynasty, possibly originating from Imhotep. The scroll material was **papyrus**, sheets made by pressing the pith of the papyrus plant and then slicing off the pieces that were needed. Since it is organic, papyrus is not a good material for preservation purposes. Although the climate in Egypt is dry, which helped to preserve the scrolls, most finds have deteriorated to some degree. This was the case with Rhind's discovery. The scroll was 1 foot wide and 18 feet long but missing its middle portions. The content of the scroll that remained was seen to be written in the ancient Egyptian writing called **hieratic**. Beginning as a simple modification of hieroglyphics, hieratic later became a language in which each syllable is represented as a symbol and each symbol represents a particular idea. Because of this connection to hieroglyphics, the Rosetta Stone made possible the translation of the scroll that we know as the **Rhind Mathematical Papyrus** (Kline 1972, 15–16).

About four years after Rhind found his papyrus, the American Edwin Smith was studying in Egypt when he purchased what he believed to be a scroll on ancient Egyptian medicine. However, it was a fraud. Someone had taken pieces from a papyrus and created what looked like a scroll, but only the outside layers were ancient Egyptian. The inside was fake. Smith was an expert on Egypt, and it seems that it would be unlikely for someone in his position to be fooled by a forgery. We must remember, however, that the condition of any found papyri will be very fragile. It takes time and care to open an ancient scroll. Fortunately for posterity, Smith kept his find, and it went with the rest of his collection to the New York Historical Society after his death. In 1922 experts at the museum determined that the fraudulent scroll was assembled using the lost pieces of Rhind's papyrus! The pieces were shipped to the British Museum, where experts were able to fit together most of the missing section and complete its translation (Burton 1985, 36).

Other papyri discovered are the Golenischev Papyrus, usually called the **Moscow Papyrus** because it is now a part of the collection of the Museum of Fine Arts in Moscow, and a second purchase of Rhind's, which we know as the **Egyptian Mathematical Leather Scroll**. All of these finds have greatly aided our understanding of ancient Egyptian mathematics. The ancient Egyptian's wrote their mathematics in the form of questions followed by answers. The goal was to calculate values associated with government and commerce, specifically the computation of areas and volumes. They used specific numbers to solve problems, and they did not generalize their results. Any formulas that they did use were not deduced from first principles but were discovered empirically by trial and error. Because of this, some of their formulas were not accurate but only approximations, which, considering their needs, often sufficed. The greatest mystery regarding the mathematics of the ancient Egyptians is that for about two millennia they made very little progress beyond their initial work (Burton 1985, 36, 52, 57, 64).

Approximations

The mathematical scrolls show that the Egyptians did not distinguish between algebra and geometry. They viewed mathematics simply as a tool by which to obtain needed values. They had formulas for basic areas and volumes. This included the volumes of a cube and a cylinder. There is some debate in academic circles as to whether the Egyptians ever gave any justifications for any of the steps in their solutions. None have been discovered, but it is clear that the Egyptians did not consider geometry to be a logical system (Kline 1972, 19–20).

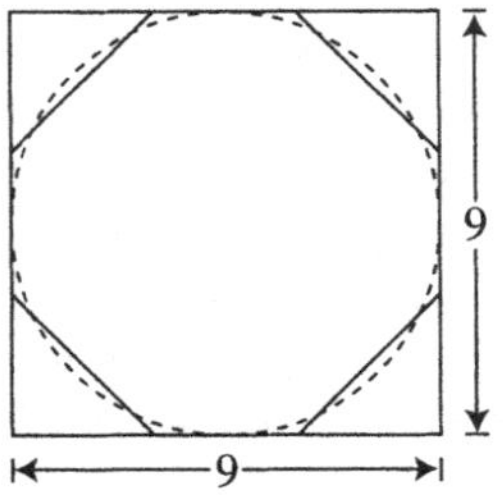

Figure 1.1 Using a 9×9 square to estimate the area of a circle.

The Rhind Mathematical Papyrus provides an example of an ancient Egyptian formula that did satisfactorily approximate its intended value. Take a square with sides of length 9. Trisect each of the sides, and then by joining consecutive points at the corners, create a regular octagon. This is illustrated in Figure 1.1, but we must note that the diagram found in the papyrus was a simple sketch of this diagram without the circle. The scribe of the papyrus considered the octagon a good estimate for the area of the circle. To find the octagon's area, we simply remove the four equilateral triangles from the square. Therefore, its area is

$$9^2 - 4 \cdot \frac{9}{2} = 63.$$

It is conjectured that this provided the ancient Egyptians a reason to believe that for a circle of diameter d,

$$A \approx \left(\frac{8d}{9}\right)^2. \tag{1.1}$$

The Egyptians could have relied on the octagonal estimate in support of Equation 1.1 because it yields an area of 64 square units when $d = 9$, which is close to the Rhind approximation. Furthermore, notice what Equation 1.1 implies about the Egyptian value of π. Since $A = \pi d^2/4$, we substitute Equation 1.1 to find

$$\pi \approx 3\frac{13}{81} = 3.16049\ldots. \tag{1.2}$$

This is very close to the common approximation of $3\frac{1}{7}$ (Burton 1985, 58).

An example of an equation from ancient Egypt that did not provide very good estimates is one that was found at the temple of Horus at Edfu. The priests of the temple received gifts in the form of patches of land. Each plot was in the form of a quadrilateral but was not necessarily rectangular. If we let a, b, c, and d represent the lengths of consecutive sides of the quadrilateral, their formula is

$$A = \frac{1}{4}(a + c)(b + d).$$

Notice that this amounts to multiplying the average of opposite sides together. It is not a bad estimate for regions that are nearly rectangular, but for most quadrilaterals it is an example of an instance when the Egyptians did try to generalize a result but did it poorly (Burton 1985, 57).

Pyramids

Herodotus recounts an encounter that he had with one of the priests (Burton 1985, 62–63). He claims that he was told that the Great Pyramid was designed so that the area of a square with side equal to the height of the pyramid is equal to the area of one of the faces of the pyramid. Using Figure 1.2, this is equivalent to writing

$$h^2 = \frac{1}{2}(2ba) = ab,$$

where $2b$ is the length of one of the sides of its square base, a is the altitude of one of the faces, and h is the height of the pyramid. From this we can calculate the ratio b/a by first noting that since

$$h^2 + b^2 = a^2,$$

we have

$$a^2 - b^2 = ab.$$

Dividing through by a^2 and a little rearranging yields

$$\left(\frac{b}{a}\right)^2 + \frac{b}{a} - 1 = 0.$$

This is a quadratic equation with b/a serving as the unknown. Thus, since both a and b are positive,

$$\frac{b}{a} = \frac{\sqrt{5} - 1}{2}. \tag{1.3}$$

Geometrically, a **pyramid** is a solid consisting of a collection of polygons joined at their edges. Recall that a **polygon** is a closed plane figure consisting of line segments called **sides**. An endpoint of a side is called a **vertex**. A polygon with n sides is often called an ***n*-gon**. If all the sides of the polygon are of the same magnitude and if the same can be said of the interior angles, the polygon is **regular**. In a pyramid one

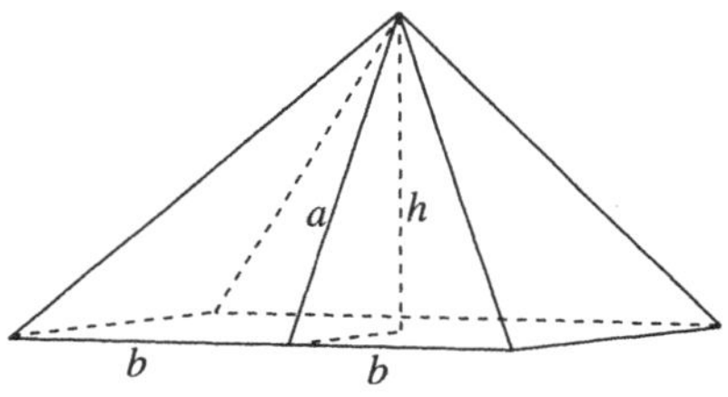

Figure 1.2 The Great Pyramid.

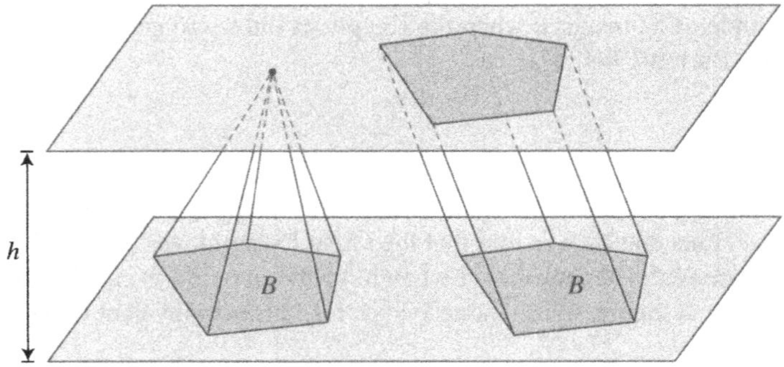

Figure 1.3 A pyramid and a prism of height h.

polygon serves as the **base** and the others are **faces**. The faces all share a common vertex called the **apex** of the pyramid. The distance between that apex and the plane containing the base is the **height** of the pyramid (Figure 1.3).

Similar in definition to a pyramid is the prism. A **prism** is a solid consisting of a collection of polygons joined at their edges. Two of the polygons are the **bases** of the prism. They are identical in shape and size (in other words, they are **congruent**) and lie in parallel planes. The other polygons are called **faces** and are formed by joining corresponding vertices of the bases (Figure 1.3). The **height** of a prism is the distance between the two parallel planes. In general, if B is the area of the base of a prism and h is its height,

$$\text{area(prism)} = Bh,$$

and the volume of a pyramid is one third the volume of a prism with the same base and height of the pyramid. This means that

$$\text{area(pyramid)} = \frac{Bh}{3}.$$

The formula for the volume of a **square pyramid** (one with a square base) quickly follows from this.

The Moscow Papyrus contains what is one of the most significant mathematical discoveries of the ancient Egyptians. The translation of Gillings (1972, 188) reads as follows:

> Method of calculating a truncated pyramid.
> If it is said to thee, a truncated pyramid of 6 [cubits] in height,
> Of 4 [cubits] of the base, by 2 on the top,
> Reckon thou with this 4, squaring. Result 16.
> Double thou this 4. Result 8.
> Reckon thou with this 2, squaring. Result 4.
> Add together this 16, with this 8, and with this 4. Result 28.
> Calculate thou [one third] of 6. Result 2.
> Calculate thou with 28 twice. Result 56.
> Lo! It is 56! Thou has found rightly.

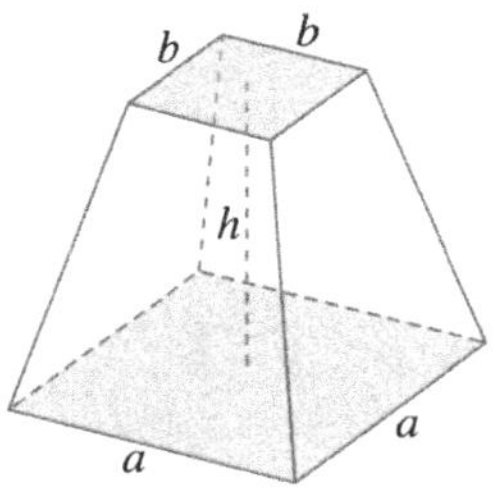

Figure 1.4 A frustum or truncated pyramid.

The Egyptians found an accurate formula for the area of a truncated square pyramid. Such a volume is called a **frustum** and is formed by taking the volume of a solid, such as a pyramid or a cone, that lies between two parallel planes. If a is the length of a side of the base of the truncated pyramid and b the length of the square at its top (Figure 1.4), then

$$V = \frac{h}{3}(a^2 + ab + b^2). \tag{1.4}$$

How Equation 1.4 was obtained is only conjecture, but a reasonable guess is that the ancient Egyptians checked such cases as when b is half of a or a third of a, and from these results concluded a formula that they could then check further in other cases (Gillings 1972, 190–191; Burton 1985, 59–60).

We can prove Equation 1.4, however, by setting up a coordinate system on a cross-section of the frustum with the origin at O (Figure 1.5). The slope of $\overline{BA}$ is $2h/(b - a)$, so the equation for the line through B and A is

$$y = \frac{2h}{b-a}x + \frac{ah}{a-b}.$$

Therefore, the height of the pyramid without the top removed is $ah/(a - b)$. The formula for the frustum is simply the volume of the large pyramid minus the volume

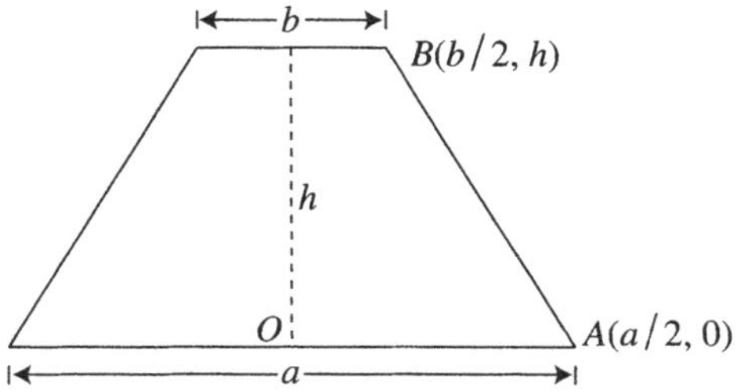

Figure 1.5 Deriving the formula for the volume of the truncated pyramid.

of the small pyramid at the tip. We calculate as follows:

$$
\begin{aligned}
V &= \frac{1}{3}\left(\frac{ha}{a-b}\right)a^2 - \frac{1}{3}\left(\frac{ha}{a-b} - h\right)b^2 \\
&= \frac{h}{3}\left(\frac{a^3}{a-b} - \frac{ab^2}{a-b} + b^2\right) \\
&= \frac{h}{3(a-b)}(a^3 - ab^2 + b^2[a-b]) \\
&= \frac{h}{3(a-b)}(a^3 - b^3) \\
&= \frac{h}{3}(a^2 + ab + b^2).
\end{aligned}
$$

Exercises

1. The Egyptians estimated the area of a quadrilateral that had consecutive sides of length a, b, c, and d with the formula $A = \frac{1}{4}(a+c)(b+d)$.
   (a) Show that if the quadrilateral is a rectangle, its area equals A.
   (b) Explain why A is equal to the product of the average of opposite sides of the quadrilateral.
   (c) Find the dimensions of a quadrilateral where the absolute value of the error between its true area and the estimate given by A is 25 percent.

2. Confirm that the ancient Egyptian approximation for π is as given in Equation 1.2.

3. Assuming $\pi = 3$, the ancients inaccurately used the formula

$$V = \frac{h}{12}\left(\frac{3}{2}[D+d]\right)^2$$

for the volume of a truncated cone where D is the diameter of the base, d is the diameter of the top, and h is the height. Derive the correct formula.

4. The **lateral surface area** of a solid is the total area of the sides of the solid excluding any bases. Find the lateral surface area of a truncated square pyramid of height h and bases with sides of lengths a and b.

5. From the Rhind Mathematical Papyrus we know that the Egyptians were able to calculate the slopes of the sides of a pyramid. This slope is called a **seked**. Its value is equal to the horizontal distance for every unit rise (Gillings 1972, 185–187).
   (a) Find the seked of a square pyramid of height 429 feet and with a base that is 618 feet long.
   (b) If the seked of a square pyramid is 5 feet per unit and its base is 400 feet long, what is the height of the pyramid?
   (c) Find an equation that expresses the seked in terms of the base and height of the pyramid.

6. Since the volume of a pyramid is equal to one third of the volume of a prism with base and height equal to that of the pyramid, it would have been simple for the ancient Egyptians to determine an empirical formula for a pyramid.
   (a) Describe an experiment that the Egyptians could reasonably have used to determine the formula for the volume of a pyramid.
   (b) Gillings (1972, 190) gives two conjectures concerning a mathematical approach to determining the formula. The first involves taking a right pyramid with a square base. Divide it into four congruent oblique pyramids by passing two planes through the vertex of the pyramid such that the planes are perpendicular to each other and the pyramid's base and divide the base into four equal squares. Show that the resulting pyramids can be arranged into a square from which the formula can be derived.
   (c) The second method is to start with a cube. Draw six pyramids such that each has as its base a side of the cube and each has a height that is equal to half the length of a side of the cube. It is best to draw these pyramids so that they share a common vertex on the interior of the cube. Each of these is called a **Juel's pyramid**. Use this construction to confirm the formula for the volume of a pyramid.

7. A **regular** pyramid has a regular base.
   (a) Find the lateral surface area of a regular hexagonal pyramid with height 10 and base with sides of length 4.
   (b) Find the pyramid's total area and volume.

8. Use calculus to confirm the formula for the volume of a square pyramid.

1.2 BABYLON

As the Rosetta Stone aided in the translation of the hieroglyphics of ancient Egypt, the 1870 discovery at the Behistun Cliff in Iran did the same for the Babylonian language. An inscription in the side of the cliff announced the ascension to power of Darius. It was written in the Persian, Elamitic, and Babylonian languages. The symbols were all various forms of **cuneiform**. Typically, cuneiform was written on clay tablets when they were still soft using a stylus with a triangular tip to impress wedges into the clay. This is how cuneiform received its name, for this word is from the Latin for "wedge." The clay was then dried. As opposed to papyrus, clay survives well over time, so there were many museums with large collections of tablets with cuneiform writing on them. However, they were not easily deciphered. In particular, there were large collections at the British Museum, the Louvre, Yale, Columbia, and the University of Pennsylvania, with many needing translation. Because of the Behistun Cliff and the later work of Otto Neugebauer and Paul Thureau-Dangin, the tablets were discovered to contain ancient Babylonian mathematics (Boyer and Merzbach 1991, 9–10; Kline 1972, 5).

Approximately two thirds of the clay tablets dated from 1800 to 1600 B.C, and they originated from the Sumerian and Akkadian civilizations. But for a few exceptions, the Babylonian mathematics proved to be significantly ahead of the Egyptians. Both civilizations used empirical methods to find their results, but the Babylonians were more theoretical and able to solve algebraic equations. They used symbols for unknown quantities and their equations would include one or more variables. They even knew a form of the Quadratic Formula. Despite this, their solutions were on a case-by-case basis. They had no concept of letting coefficients be represented as unknown parameters. More importantly, they had no concept of a proof. The closest they came would be a step-by-step explanation of how to solve an equation, yet this would be given without any justification (Burton 1985, 67, 69; Kline 1972, 13–14).

Pythagorean Triples

Although geometry as a subject on its own right played little role in Babylonian mathematics, there is one particular clay tablet that is of interest. Deciphered by Neugebauer and Abraham Sachs in 1945, it gave clear evidence that the Babylonians around the years 1900 to 1600 B.C. knew the Pythagorean Theorem. The tablet is called Plimpton 322 since it is housed in the G. A. Plimpton collection at Columbia University. The tablet contains a table displaying four columns of numbers, three of which are translated into decimal numbers as follows:

x	z	
119	169	1
3367	*4825	2
4601	6649	3
12709	18541	4
65	97	5
319	481	6
2291	3541	7
799	1249	8
*481	769	9
4961	8161	10
45	75	11
1679	2929	12
*161	289	13
1771	3229	14
56	*106	15

If we check some of the numbers, we find that the difference between the the square of a value from the z column and the square from the x column yields another perfect square: for instance,

$$169^2 - 119^2 = 120^2.$$

Four of the lines of the table did not follow this pattern (indicated by a *). These discrepancies have been shown to be due to errors on the part of the scribe. The right-most column served simply to enumerate the lines of the table (Burton 1985, 77–78; Kline 1972, 4–5).

The Babylonians wrote the numbers in the table using the **sexagesimal** system (Exercises 3 to 5). This type of numeration originated with the Sumerians and dates to at least 2000 B.C. Numbers written in sexagesimal are in base 60. This means that the Sumerians had symbols for the numbers 1 through 59. When 60 was reached, they would begin again at 1. They originally had no notation for zero. For example, since 119 = 60 + 59, they would write it in base 60 as

$$119 = 1, 59.$$

The number 120 was represented as 2. Because of the lack of a zero, their numbers were ambiguous and required some context for clarification. However, around the year 350 B.C. the Babylonians did introduce a symbol that would serve as a placeholder for missing digits. Their sexagesimal system was very efficient and remains with us today when we measure both time and degrees in minutes and seconds (Boyer and Merzbach 1991, 23, 27, 37).

Returning to the table we note that there is no extant explanation as to how the Babylonians obtained their result, but Burton (1985, 77–82) details a conjecture on their method. First we must note that the numbers in the table are too large for simple trial and error. If the Babylonians were guessing, we would expect to find simpler examples in addition to those already in the table. Instead, we can make sense of the numbers if we examine the fourth column of the table. The column is incomplete due to the condition of the tablet, but it can be seen that it enumerates the values of z^2/x^2. This suggests that the Babylonians took

$$x^2 + y^2 = z^2$$

and transformed it into

$$\left(\frac{z}{x}\right)^2 - \left(\frac{y}{x}\right)^2 = 1.$$

Letting $a = z/x$ and $b = y/x$, we have

$$a^2 - b^2 = 1.$$

To generate the table we simply construct triangles with sides of integral length 1, a, and b such that $a^2 - b^2 = 1$. However, we know that

$$a^2 - b^2 = (a+b)(a-b). \tag{1.5}$$

Hence, $a + b$ and $a - b$ are reciprocals, so we may write

$$a + b = \frac{m}{n} \text{ and } a - b = \frac{n}{m},$$

where m and n are positive integers. Adding the two equations, we find that

$$a = \frac{1}{2}\left(\frac{m}{n} + \frac{n}{m}\right),$$

and

$$b = \frac{1}{2}\left(\frac{m}{n} - \frac{n}{m}\right).$$

These equations simplify to

$$a = \frac{m^2 + n^2}{2mn}$$

and

$$b = \frac{m^2 - n^2}{2mn}.$$

Letting $x = 2mn$ and remembering that $y = bx$ and $z = ax$, we obtain

$$x = 2mn,$$
$$y = m^2 - n^2,$$
$$z = m^2 + n^2,$$

for positive integers m and n. These are well-known equations for generating **Pythagorean triples**, sets of integers that satisfy the Pythagorean Theorem.

Areas

It is unlikely that we will ever know with reasonable certainty whether the given derivation for the numbers in the table is what the Babylonians had in mind. However, it is quite certain that they knew the algebraic identity given in Equation 1.5, and how they discovered it is probably a geometric question. The expression $a^2 - b^2$ refers to the area that remains when a square with sides of length b is subtracted from one with sides of length a. The shaded regions in Figure 1.6.a represents this difference. Notice that the top shaded rectangle has dimensions a by $a - b$, and the one on the right has dimensions b by $a - b$. When we place the two rectangles side by side, forming a longer rectangle as in Figure 1.6.b, the resulting rectangle has dimensions $a + b$ by $a - b$. This confirms the identity (Burton 1985, 80).

In the 1930s a group of archeologists discovered a cache of mathematical tablets at Susa. The tablets contained one of the relatively few examples of Babylonian geometry. Not only did they include some of the oldest applications of the Pythagorean Theorem (Burton 1985, 82), but they also contained tables of ratios involving various magnitudes related to regular polygons, including the ratios between the area of

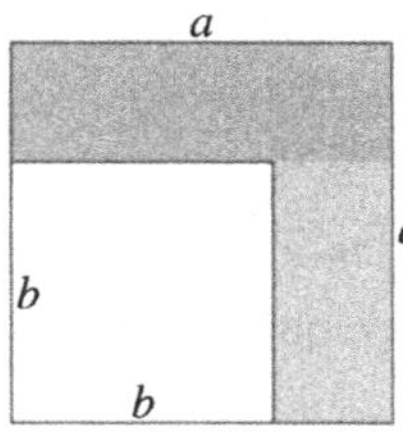

(a) $a^2 - b^2$

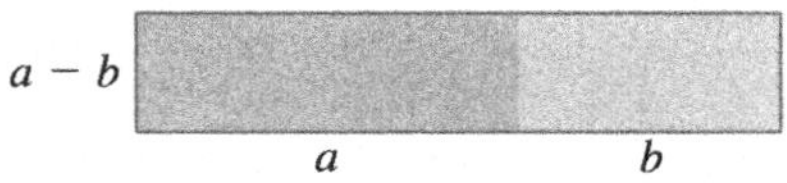

(b) $(a + b)(a - b)$

Figure 1.6 $a^2 - b^2 = (a + b)(a - b)$.

certain regular polygons and the square on one of its sides. The Babylonian ratios for the **pentagon** (5-gon), **hexagon** (6-gon), and **heptagon** (7-gon) are summarized in the next table (Boyer and Merzbach 1991, 37–38):

sides	area : square of side
5	1 2/3
6	2 5/8
7	3 41/60

Let us check the accuracy of their work for the pentagon. Figure 1.7 represents a portion of a pentagon. There is a point in the interior of every regular polygon called its **center** that has the property that it is equidistant from each of the endpoints of the sides. Let B be the center of the pentagon in question. Let M be the midpoint of side $\overline{AC}$. Point M has the property that $\overline{BM}$ is perpendicular to $\overline{AC}$. Every segment that joins the center to the midpoint of a side is called an **apothem** of the regular polygon. To find the area of the pentagon, we first find the area of $\triangle ABC$. The apothem a of the pentagon is the height of the triangle. The measure of the **central angle** in radians is $2\pi/5$, so $m\measuredangle MBC = \pi/5$. Hence, letting $r = AB$, the **radius** of the polygon, we have

$$a = r\cos\frac{\pi}{5}$$

and

$$AC = 2r\sin\frac{\pi}{5}.$$

We may then conclude that

$$\text{area}(\triangle ABC) = r^2\cos\frac{\pi}{5}\sin\frac{\pi}{5} = \frac{r^2\sin 2\pi/5}{2}.$$

Denote the area of a regular n-gon with radius r by A_n. Let s_n be the length of its side. Then $s_5 = AC$ and

$$A_5 = \frac{5r^2\sin 2\pi/5}{2}.$$

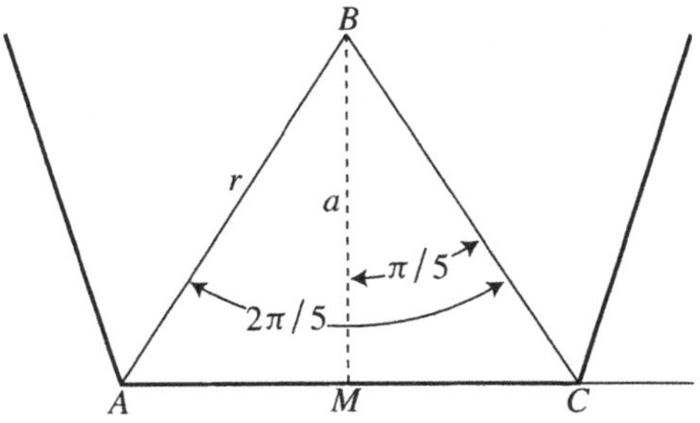

Figure 1.7 Finding the area of a regular pentagon.

This generalizes to

$$A_n = \frac{nr^2 \sin 2\pi/n}{2} \tag{1.6}$$

and

$$s_n = 2r \sin \frac{\pi}{n}. \tag{1.7}$$

Using these equations, we see that the ratio of the area of a pentagon to the square on its side is

$$\frac{A_5}{s_5^2} = \frac{5}{4 \tan \pi/5} \approx 1.72048.$$

The difference between our calculation and the value from the table is 0.05381. This amounts to a 3.12% error for the Babylonian calculation.

The clay tablets found at Susa also yielded ratios between the perimeter of a regular polygon and the circle circumscribed about it. As we did with the Egyptians, this should allow us to find the Babylonian value for π. Let us take the regular hexagon as an example. The ratio of the perimeter of a regular hexagon to the circumference of its circumscribed circle was calculated by the ancient Babylonians to be 24/25 (Boyer and Merzbach 1991, 38). Let P_n represent the perimeter of a regular polygon. By Equation 1.7,

$$P_n = 2nr \sin \frac{\pi}{n}.$$

Therefore,

$$\frac{P_6}{2\pi r} = \frac{6 \sin \pi/6}{\pi}.$$

Setting this equal to the ancient estimate, we find that

$$\pi \approx \frac{25 \cdot 6 \sin \pi/6}{24} = \frac{75}{24} = 3\frac{1}{8}.$$

The is an improvement over the early Babylonian (1800–1600 B.C) assumption of $\pi \approx 3$ and compares very favorably to the Egyptian value calculated in Equation 1.2.

Exercises

1. Show that each of the following sets of numbers are Pythagorean triples. Assume that a and b are integers.
   (a) $2ab, a^2 - b^2, a^2 + b^2$
   (b) $a, (a^2 + 1)/2, (a^2 - 1)/2$ (provided that a is odd)

2. Burton (1985, 84–85) cites a number of problems found on Babylonian tablets.
   (a) The circumference of a circle is 60 and the length of a perpendicular from the center of a chord to the circumference is 2. Find the length of the chord assuming that $\pi = 3\frac{1}{8}$.
   (b) One leg of a right triangle has length 50. Parallel to the other leg at a distance of 20, a segment is drawn that cuts off a right trapezoid of area 320. Find the lengths of the bases of the trapezoid.

(c) Given a right triangle with base of length 30 and a segment parallel to the base of length c that divides the triangle into a right triangle of height a and a right trapezoid of height b. The area of the right trapezoid is 420 more than the area of the small right triangle. Given that this information yields the equations $a - b = 20$ and

$$\frac{b(c+30)}{2} = \frac{ac+840}{2},$$

use similar triangles to find the values of a, b, and c.

3. We will use the numbers 0, 1, 2, ..., 59 separated by commas to represent integers written in sexagesimal. The number farthest right will be considered in the units (60^0) position, the next one to the left is in the 60^1 position, and so on. For example,

$$14, 43, 3 = 14 \cdot 60^2 + 43 \cdot 60^1 + 3.$$

Use this notation to convert the decimal (base 10) numbers 60, 56,861, 4,968,011, and 315,839,601 to sexagesimal.

4. Since sexagesimal is a positional numeral system, it is easy to calculate with it. (Think of the problems that arise even with simple arithmetic in the Roman numeral system.) All of the operations behave as with the decimal system, but we have to remember that it is base 60. When we "carry" or "borrow," we do it with 60s and not 10s. For example, $45 + 35 = 1, 20$, and $3, 21 - 50 = 2, 31$. Multiplication is worked similarly; for example, $2 \cdot 16 = 32$ and $30 \cdot 2 = 1, 0$. Compute the following:

(a) $3, 20 + 45, 54$
(b) $34, 19, 56 + 26, 0, 4$
(c) $34, 19, 56 - 26, 0, 4$
(d) $44, 0, 0 - 37, 37, 37$
(e) $39, 13 \times 2, 0$
(f) $3, 20 \times 45, 54$

5. In sexagesimal the notation $0; 13, 54 = 13 \cdot 60^{-1} + 54 \cdot 60^{-2}$.
   (a) Convert .365 to sexagesimal.
   (b) The ratio 1 : 2 is the number 0; 30 in sexagesimal, and 1 : 8 is 0; 7, 30. Explain how this is computed and use the found algorithm to calculate 1 : 15 and 1 : 50, and 1 : 21.
   (c) Find the sexagesimal numeral for 3 : 8.

6. Explain why the area of a regular polygon equals one half the product of its apothem and perimeter.

7. Find the area of a square whose apothem is 10.

8. Confirm that the ratio of the area of a regular hexagon and the square of one of its sides is approximately $2\frac{5}{8}$.

9. Find the ratio of the perimeter of a regular pentagon to the circumference of its circumscribed circle.

1.3 CHINA

The Han dynasty witnessed a rapid increase in science and technology. Agricultural production also saw an increase during this time. To handle the logistics of this larger endeavor and to maximize the chances of successful crops, the Chinese believed it was crucial to have accurate calendars and reliable weather forecasts. This led to improved methods in astronomy and the computational skills that went along with it, yet the mathematics upon which their science rested was the culmination of some 800 years of development dating back to the Zhou and Qin dynasties. Despite the long lineage, it was during the middle of the Former Han dynasty that the earliest extant Chinese text containing mathematics was written. Its title is the *Arithmetical Classic of the Gnomon and the Circular Paths of Heaven*, and it is primarily a book on astronomy. At the beginning of the work the author writes that the astronomers of the Zhou period would erect vertical stakes in the ground to track the motion of the sun. We call such a stake a **gnomon** (Figure 1.8). It enabled the astronomers to perform various celestial calculations. For this reason the gnomon was also known in the West. The Babylonians used it, and from there it was introduced to the Greeks (Yǎn and Shírán 1987, 25, 27; Heath 1921a, 78).

The dating of the mathematics of China is very difficult. A number of the emperors issued book-burning decrees to try to eliminate the memory of past rivals. The first of these was by the tyrant Shih Huang-ti in 213 B.C. Scholars of later dynasties would then attempt to recreate the old works by memory. Another contributing factor to the difficulty of dating is due to scribes attributing new results to past famous scholars, so it is difficult to attribute any given result accurately. In addition the texts were written originally on bamboo and later on silk, making the documents susceptible to decay (Swetz and Kao 1977, 17; Burton 1985, 27). Therefore, it was difficult for a work to survive for long periods of time. The earliest text dedicated solely to mathematics that did endure goes by the name *Nine Chapters on the Mathematical Art*. It presents a complete system of mathematics that served as the foundational work for all later Chinese mathematics. The text is written as a sequence of problems each followed by its solution. As its name implies, it is comprised of nine chapters:

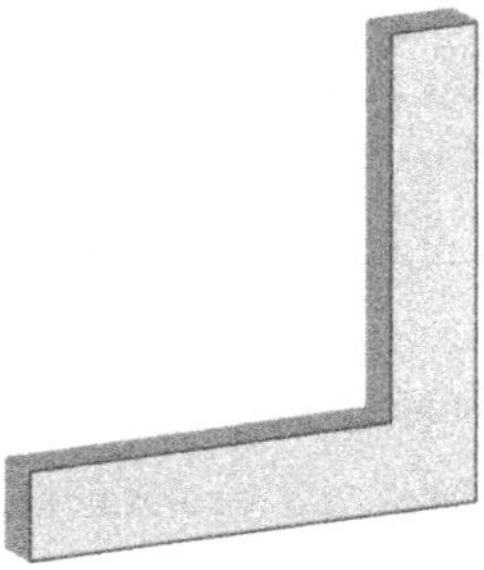

Figure 1.8 A gnomon

1. "Field measurement" shows how to calculate areas.
2. "Cereals" gives problems on proportion.
3. "Distribution by proportion" explains how to allocate goods according to a given proportional distribution.
4. In "What width?" solutions are found for problems where the length of a side is sought when an area or a volume is given. This chapter includes methods on computing square and cube roots.
5. "Construction consultations" focuses on calculating the volume of various solids.
6. "Fair taxes" describes how best to distribute grain and labor according to population and distance.
7. "Excess and deficiency" deals with solving a specific type of system of two equations in two unknowns.
8. A more general presentation of systems of linear equations is given in "Rectangular arrays." How to work with positive and negative numbers is also discussed.
9. "Gōugŭ" discusses right triangles and includes problems involving similar right triangles. An introduction to quadratic equations is also given.

The *Nine Chapters* was probably the work of several mathematicians covering centuries of work (Yăn and Shírán 1987, 33–35).

Formulas

Although the *Nine Chapters*' main emphasis is algebra, it did include some geometry including an extensive list of formulas for area and volume. Like the Egyptians and Babylonians, the Chinese discovered their formulas empirically. They did not establish them based on reasoning from first principles (Swetz and Kao 1977, 62). Some of their formulas include:

- The area of a circle,

$$A = \frac{C}{2} \cdot \frac{d}{2},$$

  where C is the circumference and d is the diameter.

- The area of a ring [Figure 1.9(a)],

$$A = \frac{1}{2}(C_1 + C_2)\Delta r,$$

  where C_1 is the outer circumference, C_2 is the inner circumference, and Δr is the difference of the radii.

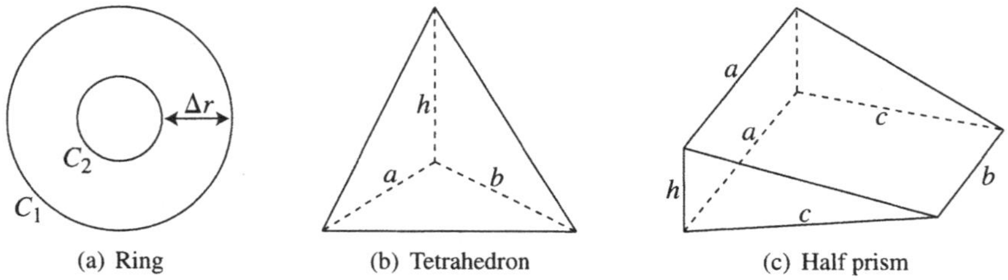

Figure 1.9 An area and two volumes.

- The volume of a tetrahedron with a base consisting of a right triangle [Figure 1.9(b)],

$$V = \frac{1}{6}abh,$$

where its height is h and the legs of its base have length a and b.

- A half-prism with an isosceles trapezoidal base [Figure 1.9(c)],

$$V = \frac{1}{6}(2b + a)ch,$$

where the bases of the trapezoidal base are a and b, the sides of the base have length c, and the height of the half-prism is h.

- The volume of a right circular cylinder,

$$V = \frac{1}{12}C^2h,$$

where C is the circumference of its base, h is its height, and $\pi = 3$.

- The volume of a sphere of diameter d,

$$V = \frac{9}{16}d^3,$$

where again $\pi = 3$.

The accuracy of these formulas is checked in Exercise 1.

Gōugǔ Theorem

The *Arithmetical Classic* emphasizes that mastery in calculation is only the first step in learning mathematics. Taking these basic tools and applying them in other situations is the more important step. In a dialogue that is recorded in the *Arithmetical Classic*, the master tells his student that "by asking one question one can reach ten thousand things." An example of a basic result that leads to many applications is

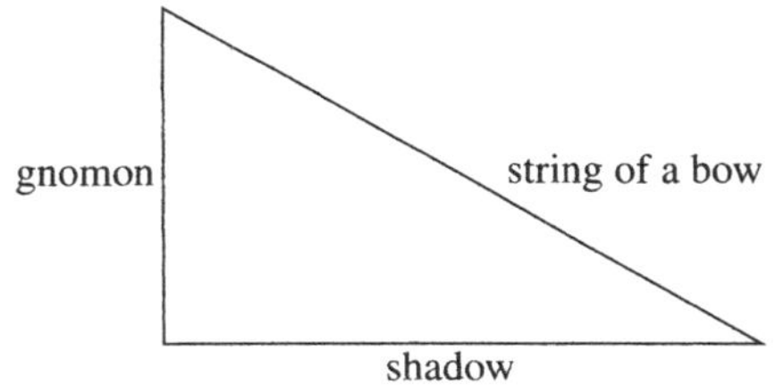

Figure 1.10 The triangle for the Gōugǔ Theorem.

the **Gōugǔ Theorem**. Take a right triangle. Call the vertical leg the *gnomon*, the horizontal leg the *shadow*, and the hypotenuse the *string of a bow* (Figure 1.10). The Gōugǔ Theorem allows us to conclude that

$$(\text{gnomon})^2 + (\text{shadow})^2 = (\text{string of a bow})^2,$$

and the *Arithmetical Classic* notes the 3-4-5 triangle as a special case. The Chinese probably mastered the Gōugǔ Theorem early because it gained almost legendary status. As with many of the significant mathematical works in ancient times, people would write **commentaries** in which they would clarify and expand on what was written. Sometimes the commentator would include tales that emphasized the importance of a result. An example appears in the *Arithmetical Classic* concerning the Gōugǔ Theorem. The author writes:

> Emperor Yǔ quells floods, he deepens rivers and streams, observes the shape of mountains and valleys, surveys the high and low places, relieves the greatest calamities and saves the people from danger. He leads the floods east into the sea and ensures no flooding or drowning. This is made possible because of the Gōugǔ theorem. . . . (Yǎn and Shírán 1987, 29–30)

A well-known problem that requires the Gōugǔ Theorem we will call the **broken bamboo problem** (Swetz and Kao 1977, 44–45). It is illustrated in Figure 1.11.

> A bamboo shoot stands 10 feet tall and perpendicular to the ground. There is a break in the shoot near the top that causes a bend in the shoot that allows the top to touch the ground 3 feet from the base of the bamboo. What is the length of the stem that is left standing erect?

This problem is also found in an Indian work, *Compendium of Calculation*. It was written by Mahavira in the ninth century. It also appeared in Philippi Calandri's *Arithmetic*. To solve it, let x be the height of the standing shoot, and then $10 - x$ is the length of the bent portion. We shall follow the traditional solution. From the figure we see that

$$9 = (10 - x)^2 - x^2 = (10 - x + x)(10 - x - x) = 10(10 - 2x).$$

Therefore,

$$\tfrac{9}{10} = 10 - 2x,$$

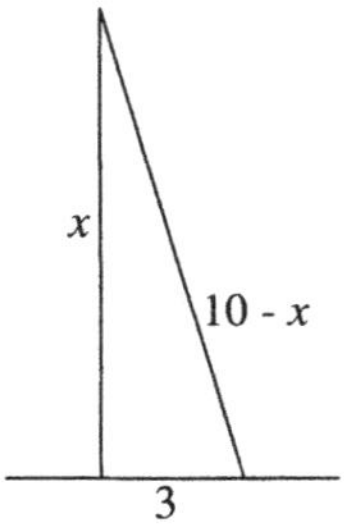

Figure 1.11 The broken bamboo problem.

from which we find that

$$x = 4\tfrac{11}{20}.$$

The fame of the Gōugŭ Theorem extended to its proof, which had its own name. It was called ***hsuan-thu***, and it is the oldest known proof of what we know as the Pythagorean Theorem. It is illustrated in Figure 1.12. Begin by taking a 3×4 rectangle like the one in the lower left-hand corner of the diagram. Its diagonal is 5 units long, so the area of the square constructed on this diagonal is 25 square units. Construct three more rectangles identical in dimension to the original rectangle and place them adjacent to each other, forming a 7×7 square. Its area is

$$(3+4)^2 = 49 \text{ square units.}$$

The shaded square cuts each of the four rectangles in half. These four halves can be joined to form two 3×4 rectangles that have total area of

$$2 \cdot 3 \cdot 4 = 24 \text{ square units.}$$

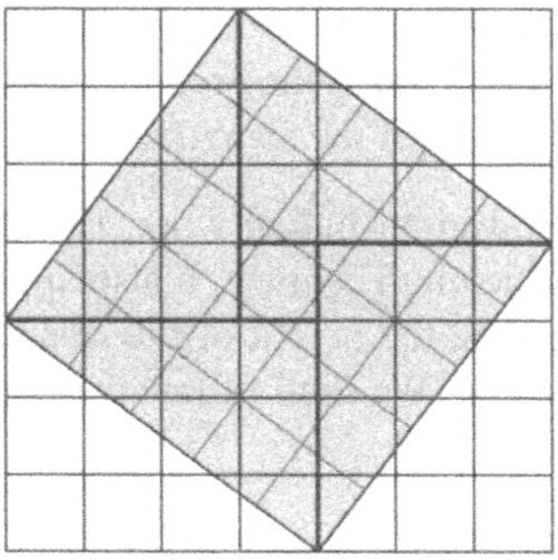

Figure 1.12 The *hsuan-thu*.

When the area of these two rectangles is subtracted from the area of the large square, the remaining region is the shaded square. This illustrates that

$$(3+4)^2 - 2 \cdot 3 \cdot 4 = 5^2,$$

that is,

$$3^2 + 2 \cdot 3 \cdot 4 + 4^2 - 2 \cdot 3 \cdot 4 = 5^2,$$

and then

$$3^2 + 4^2 = 5^2.$$

The *hsuan-thu* inspired the Hindu mathematician Bhaskara. He was born in India in A.D. 1114. For his proof he took a right triangle with legs of length a and b and hypotenuse c. He made three copies of the triangle and arranged all four into the shape of a square. The triangles are arranged such that in the middle of the large square is a smaller one with sides of length $a - b$ [Figure 1.13(a)]. What Bhaskara did next was take the five pieces of the large square and rearrange them so that he formed a shape that had area equal to $a^2 + b^2$ [Figure 1.13(b)]. We can imagine that he was so delighted with his proof that the only thing left for him to say was "Behold!" (Swetz and Kao 1977, 13; Burton 1985, 114).

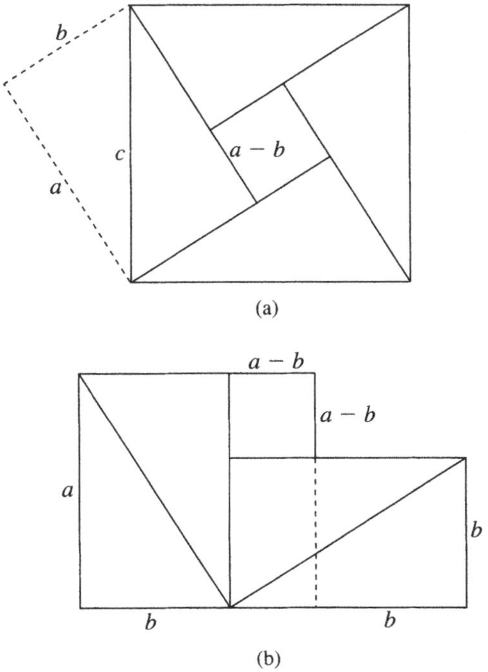

Figure 1.13 Behold!

Exercises

1. Check the accuracy of the area and volume formulas that start on page 21.

2. The following is a sample of right-triangle problems from the *Nine Chapters* (Swetz and Kao 1977, 26f).

   (a) A 7-inch-thick board is to be cut from a wooden log 2-feet, 5-inches in diameter. What is the maximum width of the board?

   (b) A 20-foot-tall tree has a circumference of 3 feet. A vine winds around the tree seven times until it reaches the top, forming a perfect helix. What is the length of the vine?

   (c) In the center of a square pond with sides of length 10 feet grows a reed. The top of the reed is 1-foot above the surface of the water, and if it is pulled toward the edge of the pond, its top will be even with the water's surface. Find the total height of the reed and the depth of the pond.

   (d) The height of a wall is 10 feet, and a pole leans against it so that the pole reaches the top of the wall. If the bottom of the pole is moved 1 foot farther from the base of the wall, the pole will fall. What is the height of the pole?

   (e) When we use a rod to measure a doorway, we find that the rod is 4 feet longer than the width of the door, 2 feet longer than the height, and the same length as the diagonal. What are the dimensions of the door?

   (f) What is the radius of a circle inscribed within a right triangle with legs of lengths 8 and 15 feet?

   (g) There is a square walled city with a door at the center of each of its four sides. Find the dimensions of the city if there is a tree that is 30 yards due north of the northern gate that can just be seen beyond the corner of the city if one stands 750 yards due west of the western gate.

3. Use dots aligned in the shape of gnomons to show that

$$1 + 3 + 5 + \cdots + (2n - 1) = n^2.$$

4. The *Nine Chapters* illustrated a "patchwork" method for calculating volumes (Yǎn and Shírán 1987, 74–75). Use any method to find the volume of the *chú méng* with height h and an $a + 2b$ by $2c$ base.

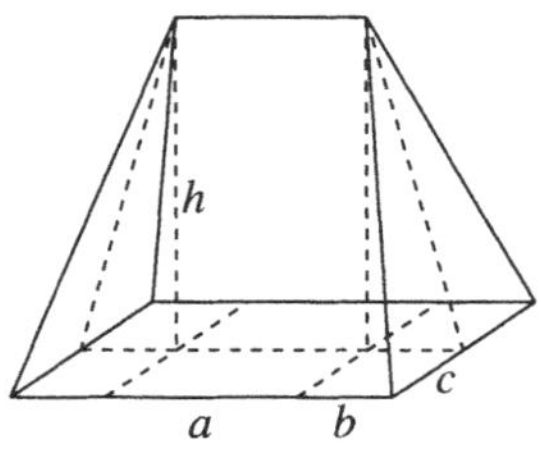

CHAPTER 2

THALES

In ancient Greece there arose people who became legendary for their work in politics, heroism on the battlefield, and skill as poets. Although they should not necessarily be considered scholars, these men were said to possess the type of wisdom that was characteristic of the first Greek philosophers (Bodéüs 2000, 137–138). They lived around 600 B.C. and were known as the Seven Sages. At least 17 different men were regarded as belonging to this group, but six from the traditional list are: Bias from Priene, Chilon from Sparta, Cleobulus from Lindos, Periander from Corinth, Pittacus from Mytilene, and Solon from Athens (Diogenes 1966, 31, 41-45).

Of all of those considered wise enough to be included among the Sages, there was one who stood out. His name was Thales. Like the others, he was an able politician and a skilled poet. As Plutarch (1932, *Lycurgus*, 51) wrote:

> Among the persons there the most renowned for their learning and their wisdom in state matters was one Thales... where, though by his outward appearance and his own profession he seemed to be no other than a lyric poet, in reality he performed the part of one of the ablest lawgivers in the world. The very songs which he composed were exhortations to obedience and concord, and the very measure and cadence of the verse, conveying impressions of order and tranquillity, had so great an influence on the minds of the listeners, that they were insensibly softened and civilized, insomuch that they renounced their

> private feuds and animosities, and were reunited in a common admiration of virtue.

We do not know much about Thales. He lived from around 624 to 548 B.C. and was a citizen of Miletus, a city of Ionia. This was a Greek region located on the western coast of what is now Turkey. As a young man he spent his time traveling on business. Tradition has it that he was a very savvy entrepreneur. One story has Thales buying all of the presses from the local olive growers. It was the end of the season, and Thales had been studying weather patterns. Based on his observations, he expected that next season would be especially suitable to olive growing, so he offered to buy the presses from the orchard owners. They were happy to make some extra money, so the deal was done. As it turned out, Thales was right. The next year saw a large harvest of olives, and although he charged them a reasonable rate, Thales rented the presses back to the former owners and made a nice profit (Guthrie 1962, 50, 52; Diogenes 1966, 23).

The story illustrates not only Thales' willingness to take a risk for profit but also that he had the intelligence to draw accurate (albeit lucky) conclusions. It is claimed that his knowledge and wisdom were acquired during his travels. These included a trip to Babylon probably during the time of Nebuchadnezzar II, who reigned from about 605 to 562 B.C. (Leick 2003, 86–87). There he would have learned astronomy using the Babylonians' astronomical tables and instruments. According to tradition, Thales predicted that an eclipse would happen during a particular year, and to the amazement of the Ionians, it did. The year is often cited as 585 B.C., because Herodotus writes that in that year a solar eclipse occurred during a battle between the Lydians and the Medes. The two kings, Alyattes of Lydia and Cyaxares of Media, were amazed by the event and were inspired to stop fighting and made peace with each other. Unfortunately, it is quite unlikely that Thales did make such a successful prediction, because the Babylonian tables that he would have used were not capable of such precise predictions at that time (Boyer and Merzbach 1991, 45; Guthrie 1962, 46).

Thales also sojourned to Egypt. There he studied with the priests, and at some point in his visit he earned their respect. This admiration reached the highest levels of society, even to the royal court. The Egyptian king grew particularly fond of Thales because of his method of easily finding the height of the Great Pyramid. He simply measured the length of its shadow at the time of day when he knew that the length of his shadow was equal to his height (Diogenes 1966, 29; Burton 1985, 94).

Thales' travels resulted in two significant developments. As he traveled abroad, he encountered various belief systems that must have significantly influenced his own. Since he was a respected public figure, it would be natural for him to want to share his thoughts and experiences. The result is his first contribution. Thales is credited with the founding of the Ionian school of philosophy. This was the first of the great Greek centers of learning, but as the first it did not generate any elaborate system of thought like those of its successors. The Ionian philosophers did, nonetheless, make a significant initial step. The population of Greece at the time was highly superstitious as is witnessed by their highly developed mythology. They believed that nature was something to be feared because the gods controlled the world arbitrarily, bringing fortune or catastrophe on a whim. Instead, Thales and his students believed that nature was orderly, and they searched for this order (Guthrie 1962, 56; Kline 1972, 27, 146-147).

2.1 THE AXIOMATIC SYSTEM

Thales' second major contribution is believed to be the introduction of geometry to Greece. According to the fifth century A.D. philosopher and mathematical commentator Proclus (1970, 65):

> Thales, who had traveled to Egypt, was the first to introduce this science [geometry] into Greece. He made many discoveries himself and taught the principles for many others to his successors, attacking some problems in a general way and others more empirically.

Others dispute Proclus's claim and believe that Greek mathematics has a stronger Babylonian influence (Waerden 1971, 36). Whichever is the case, the extent of Thales' knowledge is unknown. Tradition has it that Thales knew and could prove six basic propositions from geometry. The first is known as **Thales' Theorem** (Proclus 1970, 157, 250, 299, 352; Burton 1985, 94).

- Every angle inscribed in a semicircle is a right angle.
- A circle is bisected by its diameter.
- The base angles of an isosceles triangle are congruent.
- If two straight lines intersect, the opposite angles are congruent.
- The corresponding sides of equiangular triangles are proportional.
- Two triangles are congruent if two pairs of adjacent angles are congruent and the included sides are also congruent.

Admittedly, some of his demonstrations were "empirical," which meant that like the Egyptians and the Babylonians, he used experiment or observation to confirm the result. Others, however, were solved "in a general way." What we would expect this phrase to mean is that Thales used an **argument** or a **proof** that was supposed to guarantee that the sentence was true, not that it was simply probably true. Take, for example, the following proof of Thales' Theorem. Begin by inscribing $\sphericalangle ABC$ in a semicircle as in Figure 2.1. This gives $\triangle ABC$, with $\overline{AC}$ being the diameter of the circle. Remember that

$$\mathrm{m}\sphericalangle A = \frac{1}{2}\mathrm{m}\widehat{BC}$$

and

$$\mathrm{m}\sphericalangle C = \frac{1}{2}\mathrm{m}\widehat{AB}.$$

Therefore, using degree measure we have

$$\mathrm{m}\sphericalangle A + \mathrm{m}\sphericalangle C = \frac{1}{2}(\mathrm{m}\widehat{AB} + \mathrm{m}\widehat{BC}) = \frac{1}{2}(180) = 90.$$

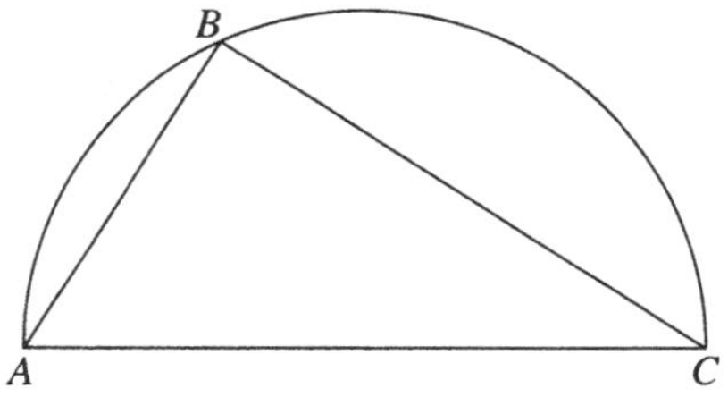

Figure 2.1 Thales' Theorem.

Since the sum of the measures of the angles of a triangle equals 180°, it must be the case that $\sphericalangle B$ is a right angle.

When we examine the parts of this proof, we find the following:

- A list of statements.
- The list includes a sentence describing what was given. In this case we were given an angle inscribed within a semicircle. Each given sentence is called a **premise**.
- Other statements in the list follow from earlier ones by using a definition (the angle joined with the diameter form a triangle), a rule of geometry (the measure of an inscribed angle is half its intercepted arc), or a rule of logic (from the arbitrary angle that we inscribed within the circle, we concluded that the result holds for all such angles).
- The list concludes with what we wanted to prove. This sentence is called the **conclusion** of the proof.

Is this Thales' "general way," and was he actually the one to introduce it to Greece? No one knows. Possibly a more interesting question is: Why introduce this type of thinking to mathematics in the first place? Was it merely a natural outgrowth of the Greek fascination with philosophical speculation, or was it based on a desire for a level of certainty that experiment and observation could not give (Boyer and Merzbach 1991, 77, 94; Kline 1972, 45–46)? Whatever is the answer to these questions, Thales represents the beginning of Greek geometry as a mixture of observation and reason. The Greeks would soon begin a complete abstraction of geometry based on the method of geometric proof given above.

To fully understand these proofs, we will analyze each of their parts using the proof of Thales' Theorem as our guide. We start with the statements themselves. Take the mathematical sentence

a circle can be inscribed within a triangle.

Since this sentence is true, we say that it has a **truth value** of **true**. The truth value of the next sentence is **false**:

there is a largest integer.

In mathematics we are only interested in statements that can be either true or false. This leads us to our first definition.

> A **proposition** is a sentence that is either true or false, but not both. Propositions are also known as **statements**.

Although no sentence can be true and false at the same time, there are many sentences that are neither true nor false. These sentences are not propositions:

- *Oh no!*—This is simply an exclamation.
- *Study geometry.*—This is a command. It is not stating that we do study geometry; it is giving an order.
- *This sentence is false.*—If it is true, then it must be false. Conversely, if it is false, then the sentence is true. Hence, it cannot be a proposition, for both cases are impossible.
- *The rectangle runs faster than roundness.*—Some may think that this is poetic, but how is it possible to assign a truth value to it?

Sometimes sentences can be neither true nor false, based on factors such as imprecision or poor structure. Take the sentence: *it is a triangle*. Is this true or false? It is impossible to know because we do not know to what *it* refers. In this sentence the word *it* is acting like the variable l as in *l is a line*. Since it is not known what l is, it cannot be determined whether *l is a line* is true or false. However, if we knew to what l refers, we could make a determination. Similarly, if we knew to what *it* referred, the sentence would be a proposition. However, this does not mean that we would necessarily know its truth value. We would simply know that it had one. This suggests that a proposition is actually an idea to which the sentence on a page refers. If each of the words of the sentence has meaning and the sentence as a whole has meaning from its words, we can identify the idea to which the sentence points and determine whether that idea can be either true or false.

To add one of these propositions to the proof, we follow very specific rules. These rules fall under one of two categories. We first consider the **rules of geometry**. These have a mathematical content that allows us to draw conclusions if certain conditions are satisfied. There are three types of these rules. Here is the first.

> A **postulate** is a proposition that is assumed to be true and is not proven.

Think of postulates as the rules of the game. If we do not set down initial rules, there would be no place to start. When we choose these rules, we want to be as confident as possible that what we prove from them will be true. For this reason we want the number of postulates as few as possible, and we want to be as sure as we can about their truth. Sometimes the term **self-evident** is used about their truth value. For example, the sentence *between any two points there exists exactly one line.*

This is one of the five initial assumptions that we will make giving us a small set of postulates. It is also a statement that is reasonable to accept as true without geometric proof. If we have concluded in a proof that we are writing that there are two points, this postulate allows us to conclude that there is a line joining the points. We may then enter this new proposition into the proof.

Usually, the term *postulate* applies only to assumptions made for geometry. Otherwise, the term **axiom** is used. For example, the basic assumptions concerning the real numbers are axioms. They are:

- $a + b$ and ab are real numbers. (**Closure**)
- $a + b = b + a$, $ab = ba$. (**Commutative**)
- $a + (b + c) = (a + b) + c$, $a(bc) = (ab)c$. (**Associative**)
- $0 + a = a + 0 = a$, $1a = a \cdot 1 = a$. (**Identity**)
- $a + (-a) = (-a) + a = 0$; $a \cdot a^{-1} = a^{-1}a = 1$, where $a \neq 0$. (**Inverse**)
- $a(b + c) = ab + ac$. (**Distributive**)
- If $a = b$, then $a + c = b + c$. (**Addition Property**)
- If $a = b$, then $ac = bc$. (**Multiplication Property**)

All of these propositions are assumed to hold for all real numbers a, b, and c. We use that standard notation of $a - b$ to represent $a + (-b)$ and $a^2 = aa$, $a^3 = aaa$, and so on.

Once the postulates give us our ground rules, we can then identify the objects that we are interested in studying.

A **definition** is a statement that identifies the meaning of a word.

For example,

a triangle is a polygon with exactly three sides

is a geometric definition. It tells us what is meant by the term *triangle*. If we have concluded in our proof that we have a polygon and that the polygon has exactly three sides, we may conclude that we have a triangle and enter the proposition saying so into our proof. Conversely, if we know that we have a triangle, we may write that our figure is a polygon that has three sides.

Two definitions based on the axioms for real numbers are these:

An **additive identity** is a real number a such that $a + b = b + a = b$ for every real number b.

An **additive inverse** of the real number a is a real number b with the property that $a + b = b + a =$ an additive identity.

For example, 0 is an additive identity, and -5 is an additive inverse of 5.

With rules and terminology we are ready to discover truths about geometry.

> A **theorem** is a proposition that has been proven using the postulates (axioms), definitions, or other theorems.

These propositions are used in exactly the same way as the postulates except that they are not assumed to be true. If the theorem is a technical result written specifically to prove another theorem, it is called a **lemma**, and a theorem is called a **corollary** when it is closely related to another theorem and proven from it quickly.

An example of a theorem is

$$a \cdot 0 = 0, \tag{2.1}$$

where a denotes a real number. The proof of this proposition will be a sequence of statements, each a consequence of one of the axioms:

1.	$0 + 0 = 0$	Identity
2.	$a(0 + 0) = a \cdot 0$	Multiplication Property
3.	$a \cdot 0 + a \cdot 0 = a \cdot 0$	Distributive
4.	$(a \cdot 0 + a \cdot 0) - a \cdot 0 = a \cdot 0 - a \cdot 0$	Addition Property
5.	$a \cdot 0 + (a \cdot 0 - a \cdot 0) = a \cdot 0 - a \cdot 0$	Associative
6.	$a \cdot 0 + 0 = 0$	Inverse
7.	$a \cdot 0 = 0$	Identity

A corollary that follows from the theorem by the Commutative Axiom is

$$0 \cdot a = 0.$$

We may also use the theorem to prove that

$$\text{an additive inverse of } ab \text{ is } a(-b). \tag{2.2}$$

This is because

$$ab + a(-b) = a(b + [-b]) = a \cdot 0 = 0.$$

The first equation follows by the Distributive Axiom, the second by Inverse, the third by Equation 2.1. The reason that this sequence of equalities proves Proposition 2.2 is because it shows that $a(-b)$ satisfies what it means to be an additive inverse.

A set of postulates and the theorems that logically follow from them is known as an **axiomatic system**. The proofs within this system are fundamentally different from the geometric proofs of the cultures before the Greeks. Those older arguments were **inductive**, designed in such a way that the conclusion could only be possibly true. If such an argument provides enough evidence that the conclusion follows with a high degree of probability, the argument is **cogent**. However, the Greeks' proofs were deductive. A **deductive argument** is an argument that claims to prove that the conclusion necessarily follows from the premises. Thus, if the reasoning is correct, we know that the conclusion is always true when the premises are. Such a deductive argument is called **valid**. If there is a time when the premises are true yet the conclusion is false, the argument is **invalid**. The proofs we have seen in this section have been valid.

Exercises

1. Identify each sentence as either a proposition or not a proposition. If it is not a proposition, explain why.
   (a) *Hello.*
   (b) *Geometers use a straightedge and compass.*
   (c) $3 + 8 = 18$.
   (d) $3 + x = 18$.
   (e) *Define the term so that it is meaningful.*
   (f) *Parallel lines do not intersect.*
   (g) *He solved the problem.*
   (h) *This statement is true.*
   (i) *This statement is not true.*

2. Identify each argument as deductive or inductive. Explain your reasoning.
   (a) *All geometers are attractive. Eudoxus was a geometer. Thus, Eudoxus was attractive.*
   (b) *Some triangles have right angles. Some squares have right angles. Therefore, some triangles are squares.*
   (c) *The high temperature in Eureka on June 1 has been rising for the last 10 years. Hence, the high temperature in Eureka on June 1 next year will be greater still.*
   (d) *Thales was born before Plato, and Plato was born before Aristotle. Hence, we know that Thales was born before Aristotle.*
   (e) *Gauss measured the angle sum of a very large triangle and found it to be less than two right angles. We conclude that there exists a triangle that has angle sum less than two right angles.*
   (f) *3 is prime, 5 is prime, and 7 is prime. Therefore, all odd integers greater than 1 are prime.*
   (g) *We are confident that Newton's law of universal gravitation is true because it is verified every time that it is tested.*
   (h) *Either Euclid discovered geometry or George Washington did. We know that Washington did not study geometry, so Euclid must have discovered geometry.*
   (i) *All geometers on Jupiter know the Pythagorean theorem. Thus, there are people on Jupiter, and they know the Pythagorean theorem.*

3. Which of the deductive arguments in Exercise 1 are valid, and which of the inductive arguments are cogent?

4. Give an example of a proposition from algebra that is self-evident and one that is not. What criteria did you use to make your choice?

5. Make a conjecture as to what method Thales used to prove that a circle is bisected by its diameter. How would you prove it?

6. Write definitions for the following terms.

(a) *triangle*
(b) *right triangle*
(c) *circle*
(d) *sphere*
(e) *point*
(f) *line*

7. Let a, b, and c represent real numbers. Prove each of the following.

(a) $(a+b)\cdot c = ac + bc$.
(b) $(a+b)^2 = a^2 + 2ab + b^2$.
(c) $a + (b+c) = (c+a) + b$.
(d) $(a-b) + (c-a) + (b-c) = 0$.
(e) $(-1)\cdot a = -a$.
(f) $(-a)\cdot(-b) = ab$.
(g) $-(a+b) = (-a) + (-b)$.

8. It is often said that a deductive argument is one that "reasons from the general to the particular." How is this similar to the definition given on page 33, and how does it differ?

2.2 DEDUCTIVE LOGIC

The second category of rules that allow us to enter propositions into proofs are the **rules of logic**. While the postulates of geometry have meaning to us and give us information about the objects that we are studying, the rules of logic that we will examine focus on the forms of sentences. These rules allow us to draw conclusions based only on the structure of the sentences.

Conjunctions, Disjunctions, and Negations

Consider the following two statements:

the base angles of an isosceles triangle are congruent (2.3)

and

a square has no right angles. (2.4)

Using the word *and*, we may create a new proposition:

the base angles of an isosceles triangle are congruent,
and a square has no right angles.

We know that this is false because Proposition 2.4 is false. For it to have been true, we would have needed both propositions to be true.

■ DEFINITION 2.2.1

A proposition of the form *p and q* is called a **conjunction**. It is true when both p and q are true and false otherwise.

There are many ways to write a conjunction. To illustrate, consider the sentence

Archimedes worked, and Rome attacked.

This can be written as:

Archimedes worked while Rome attacked.
Archimedes worked, but Rome attacked.
Archimedes worked, yet Rome attacked.
Archimedes worked; however, Rome attacked.

Each of these has its own nuances, but they are all conjunctions. This is because each sentence asserts that both propositions are true.

Just as we did with the word *and*, we may use the word *or* to create a new proposition from Propositions 2.3 and 2.4:

the base angles of an isosceles triangle are congruent,
or a square has no right angles.

We know that this statement is true because Proposition 2.3 is true. However, the sentence

$3 + 7 = 9$, *or all even integers are divisible by 3,*

is false since both $3 + 7 = 9$ and *all even integers are divisible by 3* are false.

■ DEFINITION 2.2.2

A proposition of the form *p or q* is called a **disjunction**. It is true when at least one of the original propositions is true and false when both p and q are false.

We must remember that only one of the propositions needs to be true for the entire disjunction to be true. For this reason, the logical disjunction is sometimes called an **inclusive or**. If an **exclusive or** is needed, make it explicit in the translation. For example, take the proposition

the number is greater than five,
or the number is less than five.

To make it exclusive, write

the number is greater than five,
or the number is less than five, but not both.

Both *and* and *or* are called **connectives** because they connect propositions together to make new ones. Although *and* and *or* require two propositions, there is a connective that needs only one. It is the word *not*.

■ **DEFINITION 2.2.3**

A **negation** is a proposition of the form *not p*. This negation is true when p is false, and it is false when p is true.

For example, the negation of $3 + 8 = 5$ is $3 + 8 \neq 5$. In this case we say that the proposition $3 + 8 = 5$ has been **negated**. Negating the proposition *a square has no right angles* yields

it is not the case that a square has no right angles.

A translation without the *not* is

it is false that a square has no right angles.

However, both of these are usually written as

a square has right angles.

The main inference associated with negations also involves the disjunction. The rule states that if we know that a disjunction is true but one of its propositions is false, it must be the case that the other proposition is true. This rule is called the **Disjunctive Syllogism**. A **syllogism** is a deduction with exactly two premises. In the notation that follows, the premises come before the *therefore* symbol ($\therefore$), and the conclusion is after it.

■ **LOGIC RULE 2.2.4 [Disjunctive Syllogism]**

p or q, not p / $\therefore$ *q*

An example of this rule is the following argument:

The angle is acute, obtuse, or right.
The angle is not acute.
———
$\therefore$ *The angle is obtuse or right.*

Conditionals

A common type of mathematical proposition is exemplified by the following sentence. It states that the second proposition is true when the first one is true:

if the polygon is a square, then the polygon is a rectangle. (2.5)

Two propositions have been combined using the words *if* and *then*. We consider this pair a connective since the if-then structure connects the two propositions.

■ **DEFINITION 2.2.5**

A proposition of the form *if p then q* is called a **conditional**, where p is the **antecedent** and q is the **consequent**. A conditional statement is also known as an **implication** or an **if-then statement**.

Let us examine the truth value of *if p then q*. If the proposition p is false, the conditional allows us to draw no conclusion. For instance, when the antecedent of Proposition 2.5 is false, the polygon could nonetheless be a rectangle or it could be a pentagon. That is, the truth value of the consequent is still undetermined. However, suppose that p is true. If the implication is true, we know q, but if q is false, there is no way that the conditional is true. This is because the antecedent did in fact not yield a true consequent. It is then evidently the case that the only time that *if p then q* describes a false implication is when p is true, yet q is false. This means that a conditional like

$$\textit{if } 2 + 3 = 5, \textit{ then } 4 + 5 = 19$$

is false, but

$$\textit{if } 2 + 3 = 8, \textit{ then } 4 + 5 = 9,$$

$$\textit{if } 2 + 3 = 8, \textit{ then } 4 + 5 = 19,$$

and

$$\textit{if } 2 + 3 = 5, \textit{ then } 4 + 5 = 9$$

are true. Therefore, *if p then q* is true exactly when *not p or q* is true.

Since conditionals are so common, it is not surprising that they can be written in many different ways. The conditional *if the polygon is a square, then the polygon is a rectangle* can be rewritten as follows. This list is definitely not exhaustive.

- The word *implies* is sometimes used in place of *if-then*. With this our example conditional becomes

  the polygon is a square implies that the polygon is a rectangle.

- Another way to write this is to use *only if*:

  the polygon is a square only if the polygon is a rectangle.

  The *only if* phrase means that the proposition after *only if* must be true in order for the first proposition to be true.

- There are times when *then* does not accompany *if*, as when the conclusion precedes the antecedent:

  the polygon is a rectangle if the polygon is a square.

  Other times it is simply left out, as in

  if the polygon is a square, the polygon is a rectangle.

- Sometimes the words *all* or *every* can be used:

  all squares are rectangles,

  or

  every square is a rectangle.

- The words *necessary* and *sufficient* are sometimes used. *Necessary* means "needed" or "required," as in

  it being a rectangle is necessary for the polygon to be a square.

  The sentence states that the polygon must be a rectangle in order for it also to be a square. The word *sufficient* means "adequate" or "enough." Thus, the sentence

  the polygon being a square is sufficient for it to be a rectangle

  translates our original implication. The fact that it is a square is enough for us to know that it is a rectangle. Remember that the antecedent is sufficient for the consequent, and the consequent is necessary for the antecedent.

Now that we have seen many of the different ways that an implication can be written, we are ready to use them to reason. There are two very important rules of inference related to the conditional.

■ **LOGIC RULE 2.2.6 [*Modus Ponens*]**

if p then q, p / ∴ q

■ **LOGIC RULE 2.2.7 [*Modus Tolens*]**

if p then q, not q / ∴ not p

In Latin *modus* means "standard" or "measure," and *ponere* means "to affirm." Hence, *Modus Ponens* is a rule that allows us to deduce the consequent when we "affirm" the antecedent. In Latin *tollere* means "to deny," so *Modus Tolens* allows us to conclude that the negation of the antecedent is true when we "deny" the consequent (Copi and Cohen 2002, 281). For example, this argument is an example of *Modus Ponens*:

If the corresponding angles are congruent, then the lines are parallel.
The corresponding angles are congruent.
___
∴ The lines are parallel.

This one is an example of *Modus Tolens*:

If the corresponding angles are congruent, then the lines are parallel.
The lines are not parallel.
___
∴ The corresponding angles are not congruent.

Both of these rules are commonly used when citing theorems during proof writing. In the example above, the statement *if the corresponding angles are congruent, then the lines are parallel* is a theorem. To use this as a reason to enter a line into a proof, the sentence *the corresponding angles are congruent* should already have been proven. If so, *the lines are parallel* can be entered because of *Modus Ponens*. If *the lines are not parallel*, then by *Modus Tolens* we have *the corresponding angles are not congruent.*

We must be careful when we apply *Modus Ponens* and *Modus Tolens* to make sure that we apply them in the "right direction." For instance, suppose we have proven that *if the polygon is a square, then the polygon is a rectangle*. If we know that *the polygon is a rectangle*, we cannot conclude that *the polygon is a square*. If we do, we have committed a **fallacy**, an error in logic. This error is so common that it has a name. It is called **affirming the consequent**. Similarly, if we know that *the polygon is not a square*, we cannot conclude without further information that *the polygon is not a rectangle*. If we do, we have committed the **denying the antecedent** fallacy.

Another rule of inference that we will need is the following.

■ **LOGIC RULE 2.2.8 [Hypothetical Syllogism]**

if p then q, if q then r / ∴ *if p then r*

The rule allows us to conclude *if the polygon is a square, then it is a quadrilateral* from the premises *if the polygon is a square, then it is a rectangle* and *if the polygon is a rectangle, then the polygon is a quadrilateral.*

Biconditionals

The **converse** of an implication is obtained by switching the places of the antecedent and the consequent. For example, the converse of

if the polygon is a square, then it is a rectangle

is

if the polygon is a rectangle, then it is a square.

Notice that the converse and the original implication may not have the same true value. Sometimes, though, they do coincide. Consider the following sentence:

if the alternate exterior angles are congruent, then the lines are parallel. (2.6)

We can show that the converse of this sentence,

if the lines are parallel, then the alternate exterior angles are congruent, (2.7)

also holds. Therefore, we conclude that

if the alternate exterior angles are congruent, then the lines are parallel, and if the lines are parallel, then the alternate exterior angles are congruent.

However, this sentence has an unnecessary redundancy. To remedy this, rewrite Proposition 2.6 as

the lines are parallel if the alternate exterior angles are congruent

and Proposition 2.7 as

the lines are parallel only if the alternate exterior angles are congruent.

We now combine these two versions of the original implication and its converse into one sentence and write

the lines are parallel if and only if the alternate exterior angles are congruent.

This is an example of our last proposition type.

■ **DEFINITION 2.2.9**

A proposition that is the conjunction of a conditional and its converse is called a **biconditional**. It is often written in the form p *if and only if* q.

When p *if and only if* q is true, we say that p is **equivalent** to q. Since p *if and only if* q means p *if* q, *and* p *only if* q, the propositions p and q are equivalent implies that both p and q have the same truth values. To see this, remember that p *if* q is true only when *not* q *or* p is true and p *only if* q is true exactly when *not* p *or* q is (page 38). Therefore, p *if and only if* q is true exactly when

(not p or q) and (not q or p)

is true. Suppose that p and q are both true. We then see that *not* p *or* q is true and so is *not* q *or* p. Thus, their conjunction is true. Similarly, when p and q are both false, then *not* p and *not* q are both true, which yields both *not* p *or* q and *not* q *or* p. However, the conjunction cannot be true when p and q have opposite truth values (Exercise 6).

A definition serves as an example of a biconditional. For example, to define a parallelogram we may use the sentence

a parallelogram is a quadrilateral
that has parallel opposite sides.

This is shorthand for the biconditional

a quadrilateral is a parallelogram
if and only if its opposite sides are parallel.

Typically, the first half of the biconditional contains the word that is being defined, and the second half has the conditions that must be satisfied for an object to be designated by the term. Despite definitions being biconditionals, it is traditional to write a definition as an if-then with the biconditional being understood. For example,

a quadrilateral is a parallelogram
if its opposite sides are parallel.

Exercises

1. Give the truth value of each of the propositions listed.
   (a) *A system of linear equations always has a solution, or a quadratic equation always has a real solution.*
   (b) *It is false that all right angles are congruent.*
   (c) *Vertical lines have no slope, and lines through the origin have a positive y-intercept.*
   (d) *Every integer is odd, or zero is even.*
   (e) *If every parabola has a vertex, then an ellipse must have a horizontal major axis.*

(f) *The sine function is periodic if and only if cosine is not continuous.*

(g) *The triangle has a right angle, but the square of its hypotenuse is not equal to the sum of the squares of its legs.*

(h) *It is not the case that* $2 + 4 \neq 6$.

(i) *The distance between two points is always positive if every line segment is horizontal.*

(j) *The angles being adjacent is necessary for them being supplementary.*

(k) *The derivative of the sine function being cosine is sufficient for the derivative of the cosine function being sine.*

(l) *Any real number is negative or positive, but not both.*

2. For each sentence, fill in the blank using as many of the words *and*, *or*, *if*, and *if and only if* as possible to make the sentence true.

(a) *Triangles have three sides* _______ $3 + 5 = 6$.

(b) $3 + 5 = 6$ _______ *triangles have three sides.*

(c) *Ten is the largest integer* _______ *zero is the smallest.*

(d) *The derivative of a function that is a horizontal line is zero* _______ *tangent lines for increasing functions have positive slope.*

3. Identify the antecedent and the consequent for each of the following conditionals.

(a) *If the triangle has two congruent sides, it is isosceles.*

(b) *The polynomial has at most two roots if it is a quadratic.*

(c) *The data is widely spread only if the standard deviation is large.*

(d) *The function being constant implies that its derivative is always zero.*

(e) *The system of equations is consistent is necessary for it to have a solution.*

(f) *A function is even is sufficient for its square to be even.*

4. Write the converse for each conditional in Exercise 3.

5. Rewrite the implication *if the integer is divisible by 4, then it is even* using the following words:

(a) *implies*

(b) *only if*

(c) *sufficient*

(d) *necessary*

6. Explain why *p if and only if q* is false when the truth values of *p* and *q* are opposite.

7. Identify the following arguments as valid or invalid. Explain.

(a) *The angles are supplementary if a line segment intersects another line segment forming two angles. The angles are supplementary. We conclude that the angles are formed by intersecting two line segments.*

(b) *A square exists if a rectangle exists. The angle sum of the triangle equals two right angles only if a rectangle exists, so a square exists only if a rectangle exists.*

(c) *If a quadrilateral is inscribed in a circle, the opposite angles are supplementary. The quadrilateral is inscribed in a circle. Therefore, the opposite angles are supplementary.*

(d) *Either triangles are in the same parallels or they are not congruent. The two triangles are not congruent, so they are not in the same parallels.*

(e) *Corresponding angles are congruent implies that the lines are parallel, and if the lines are parallel, the corresponding angles are congruent. Thus, lines are parallel if and only if corresponding angles are congruent.*

(f) *If the polygons are congruent, they are similar. The polygons are similar. Hence, the polygons are congruent.*

(g) *The opposite sides of a parallelogram are congruent. Opposite sides are not congruent. Thus, it is not a parallelogram.*

8. Fill in the blanks to make the arguments valid.
   (a) p, _______ / $\therefore q$
   (b) *if* p *then* q, *not* q / $\therefore$ _______
   (c) p *or* q, _______ / $\therefore q$
   (d) *if* (p *or* q) *then* r, _______ / $\therefore r$
   (e) *an integer is even if it is divisible by 4, the integer is not even* / $\therefore$ _______
   (f) _______, *the value fell outside the interval* / $\therefore$ *the hypothesis should be rejected*
   (g) *the polygon is concave or convex, it is not convex* / $\therefore$ _______
   (h) *the integer is divisible by* 8 *only if it is divisible by* 4, _______ / $\therefore$ *if the integer is divisible by* 8, *then it is divisible by* 2

2.3 PROOF WRITING

In the first section of this chapter we wrote some small proofs using the axioms for the real numbers. In this section we expand on this by writing more complicated proofs, but we continue to focus on the real numbers. We are doing this because these concepts are straightforward and serve as a good illustration of the techniques that we will need later when we expand our study of geometry.

Direct Proof

The proof of Thales' Theorem began on page 29. Rewrite the statement of the theorem in if-then form:

if an angle is inscribed in a semicircle, then it is a right angle. (2.8)

Reexamining the proof, we see that we began by taking an angle that is inscribed within a semicircle and showing that it had to be a right angle. That is, the proof that demonstrated Thales' Theorem was an argument that had the antecedent of the conditional as the given premise and had the consequent as the conclusion. This is the most common method used to prove an implication, and it is called the method of **direct proof**.

■ **LOGIC RULE 2.3.1 [Direct Proof]**

To prove *if* p *then* q, assume p and show q.

This method works because if p is false, the implication is immediately true. Hence, the only case that concerns the proof of *if* p *then* q is when p is true. If we can show that this always gives q in this case, we will have proven *if* p *then* q.

Recall that if A and B are points in a plane, we represent the line segment joining A and B by $\overline{AB}$. The length of $\overline{AB}$ is designated by AB. We define

$$\overline{AB} \text{ is } \textbf{congruent} \text{ to } \overline{CD} \text{ if } AB = CD$$

and denote this by $\overline{AB} \cong \overline{CD}$. We wish to prove that congruence of line segments is **transitive**:

$$\textit{if } \overline{AB} \cong \overline{CD} \textit{ and } \overline{CD} \cong \overline{EF}, \textit{ then } \overline{AB} \cong \overline{EF}. \tag{2.9}$$

To achieve this, we introduce three more axioms. For all real numbers a, b, and c:

- $a = a$. (**Reflexive**)
- if $a = b$, then $b = a$. (**Symmetric**)
- if $a = b$ and $b = c$, then $a = c$. (**Transitive**)

The proof of Proposition 2.9 using direct proof goes like this:

> Assume that $\overline{AB} \cong \overline{CD}$ and $\overline{CD} \cong \overline{EF}$. By definition of congruence of line segments, $AB = CD$ and $CD = EF$. Since equality of real numbers is transitive, $AB = EF$. Therefore, $\overline{AB} \cong \overline{EF}$.

It is important to see the structure of the proof. We began with the assumption of the antecedent of Proposition 2.9. From there we argued using an axiom and the definition of congruence until we were able to conclude the consequent of Proposition 2.9. This is exactly how we proved Thales' Theorem.

When studying an implication, we sometimes need to investigate the different ways that its antecedent and consequent relate to each other. For example, to determine the truth value of Thales' Theorem, we could have examined the statement

if it is not a right angle, then the angle is not inscribed in a semicircle.

This sentence is called the **contrapositive** of the original conditional, and it is true at exactly the same times that the original conditional is true. It is formed by exchanging and negating both the antecedent and the consequent.

However, in the case of Thales' Theorem, there is no advantage in trying to prove the contrapositive since the original conditional can easily be proven using direct proof. Sometimes this is not the case, and the proof of the contrapositive is either easier or simply requires fewer propositions. To see an example of this, let us show that

if the square of an integer n is odd, then n is odd. (2.10)

This theorem is from **number theory**. This field of mathematics is the study of the integers, and the ancient Greeks knew many of the basic results but only for the positive integers. For instance, they defined even and odd integers similarly to the modern definition (Euclid 1925, vol. 2, 277, 281):

n is **even** if $n = 2k$

and

n is **odd** if $n = 2k + 1$,

where k is an integer.

We need these definitions to prove Proposition 2.10. Because proving it directly will be a problem, we use direct proof instead to demonstrate its contrapositive:

if the integer n is not odd, then the square of n is not odd.

In other words, we will prove:

if n is even, then n^2 is even.

The proof is as follows:

> Let n be an even integer. This means that $n = 2k$, with k being an integer. To see that n^2 is even, calculate
>
> $$n^2 = (2k)^2 = 2(2k^2).$$

Since the contrapositive holds, the original implication is known to be true.

Proving Equivalence

Since a biconditional is the conjunction of two conditionals, there is a clear approach to its proof. We call it the method of **biconditional proof**.

■ **LOGIC RULE 2.3.2 [Biconditional Proof]**

To prove *p if and only if q*, prove both *if p then q* and *if q then p*.

Let us use this method to show:

an integer n is even if and only if its cube is even.

We will prove both "directions":

- Assume that n is even. Then $n = 2k$ for some integer k. To show that n^3 is even, we calculate

$$n^3 = (2k)^3 = 2(4k^3),$$

which means that n^3 is even.

- Now suppose that n is odd. This means that $n = 2k + 1$ for some integer k. To show that n^3 is odd, we again calculate

$$n^3 = (2k+1)^3 = 8k^3 + 12k^2 + 6k + 1 = 2(4k^3 + 6k^2 + 3k) + 1.$$

Hence, n^3 is odd.

Notice that we proved that *an integer n is even if its cube is even* in the second part by proving its contrapositive. Furthermore, since *p only if q* can be written as *p is sufficient for q*, and since *p if q* can be written as *p is necessary for q*, any proof of a biconditional may have its two parts introduced with the words *sufficient* and *necessary*. The first step could have been introduced with a phrase like

to show sufficiency,

and the second could have opened with

as for necessity.

There will be times when we need to prove a sequence of biconditionals. We say that $p_1, p_2, \ldots, p_n$ are **equivalent** if for all i, j,

p_i *if and only if* p_j.

To prove all of these biconditionals, we use direct proof to show that

if p_1 *then* p_2, *if* p_2 *then* $p_3, \ldots,$ *if* p_{n-1} *then* p_n, *if* p_n *then* p_1.

Each equivalence then follows by the Hypothetical Syllogism (Logic Rule 2.2.8) and biconditional proof. For example, supposing that $n \geqslant 3$, we know that p_1 *if and only if* p_3 follows from these conditionals because we have *if* p_1 *then* p_2 and *if* p_2 *then* p_3, so *if* p_1 *then* p_3 follows by the Hypothetical Syllogism. Continuing through the cycle and ending with *if* p_n *then* p_1, successive applications of the Hypothetical Syllogism yields the converse *if* p_3 *then* p_1. The proposition then follows by biconditional proof.

Indirect Proof

There is another method by which we could have proven Proposition 2.10. Start with direct proof by assuming that n^2 is odd. Now suppose that n is even, so we can write $n = 2k$, where k is an integer. But as above this implies that n^2 is even. Since n^2 cannot be both odd and even, we have reached a contradiction. What we mean by **contradiction** is a statement that cannot be true under any circumstances. Therefore, the assumption that led to the contradiction cannot be true, and we conclude that n is not even. This method, in which we assume the opposite of what we want to prove and then deduce a contradiction, is called the method of **indirect proof**. It is sometimes called **proof by contradiction** or ***reductio ad absurdum***. Although it is similar to proving the contrapositive, the advantage of indirect proof is that it can also be used to prove propositions that are not implications.

LOGIC RULE 2.3.3 [Indirect Proof]

To prove q, show that *not* q implies a contradiction.

Here is a nice application of indirect proof. Given integers a and b with $a \neq 0$, we say that a **divides** b if $b = ak$, for some integer k. An integer greater than one is **prime** if it is only divisible by 1 and itself; otherwise, it is **composite**. Furthermore, 1 is neither prime nor composite, and every integer greater than 1 is divisible by a prime. We will now show that

there are infinitely many prime numbers.

This is an old result (Euclid 1925, *Elements* IX.20). Here is its proof.

> Suppose that there are only finitely many primes. Let $p_1, p_2, \ldots, p_n$ be those primes. The number $p_1 p_2 \cdots p_n + 1$ is divisible by some p_i from the list because it must be divisible by a prime. Hence,
>
> $$p_1 p_2 \cdots p_n + 1 = p_i k,$$
>
> where k is an integer. Therefore,
>
> $$1 = p_i k - p_1 p_2 \cdots p_n = p_i(k - p_1 p_2 \cdots p_{i-1} p_{i+1} \cdots p_n).$$
>
> This says that 1 is divisible by a prime number, which is incorrect. Therefore, we conclude that there are infinitely many primes.

Another application of indirect proof is the demonstration of the claim that

there exists exactly one additive identity.

In other words, the additive identity is **unique**. To prove this, we will demonstrate that

$$\textit{there exists at most one additive identity} \tag{2.11}$$

and

$$\textit{there exists at least one additive identity.} \tag{2.12}$$

- We prove Proposition 2.11 by assuming that a and b are two different additive identities. This means that for all real c,

  $$a + c = c \text{ and } b + c = c.$$

  Since equality is both symmetric and transitive,

  $$a + c = b + c.$$

  Therefore, by the Addition Property

  $$(a + c) + -c = (b + c) + -c,$$

  and then by the Associative Axiom,

  $$a + (c + -c) = b + (c + -c).$$

Therefore,

$$a + 0 = b + 0$$

by the Inverse Axiom, and by the Identity Axiom we conclude that

$$a = b.$$

Therefore, the two additive identities are actually this same. This means that our initial choice of two different identities is impossible.

- Since we know that 0 is an additive identity, we know that we have at least one. This proves Proposition 2.12

Proof of Disjunctions

If we need to prove *p or q* from some hypotheses, it is not reasonable to believe that we can simply prove p and then conclude *p or q*. Indeed, if we could prove simply p, we would expect the conclusion of the argument to be p and not *p or q*. Hence, we need somehow to incorporate both p and q into the proof. We do this by assuming the negation of p. If we can prove q, the disjunction must follow. The reason this method works is that in order for a disjunction to be true, one of its propositions must be true. If p is true, we have *p or q*. However, if p is false but we have proven *if not p then q*, we may conclude q, which also shows that the disjunction is true. This is the intuition behind the method of **disjunction proof**.

■ **LOGIC RULE 2.3.4 [Disjunction Proof]**

To prove *p or q*, use direct proof to show *if not p then q*.

Our work with conditionals (page 38) confirms the method. We know that *if not p then q* is equivalent to *p or q*. Therefore, to prove *p or q*, we follow the strategy of disjunction proof and prove *if not p then q* directly.

To illustrate, let us prove

$$\textit{if } ab = 0, \textit{ then } a = 0 \textit{ or } b = 0,$$

where a and b are real numbers. We start by assuming that $ab = 0$. To prove the disjunction, suppose that $a \neq 0$. Then a^{-1} exists, and we can multiply both sides of the equation by a^{-1},

$$a^{-1}(ab) = a^{-1} \cdot 0.$$

Hence,

$$(a^{-1}a)b = 0,$$

so $b = 0$. Thus, we have proven that $a = 0$ or $b = 0$.

Proof by Cases

Geometric proofs often involve cases that have to be handled separately. For instance, suppose that there is a line and two points. There are multiple possibilities:

- Both points are on the line.
- One point is on the line and the other is not.
- Neither point is on the line, but they are on opposite sides of the line.
- Both points are on the same side of the line.

Notice that all of these cases exhaust all possible positions of the points with respect to the line. We can now assume each of the four cases separately and under each of these different assumptions try to prove a theorem. If we succeed in finding a proof in all the cases, we have found a proof for the theorem. This leads to our last method.

LOGIC RULE 2.3.5 [Proof by Cases]

Assume p_1 *or* p_2 *or* ... *or* p_n, where $p_1, p_2, \ldots, p_n$ exhaust all possible cases of a particular condition. To prove q, use direct proof to demonstrate *if* p_1 *then* q, *if* p_2 *then* $q, \ldots,$ *if* p_n *then* q,

To exemplify the method, let a and b represent integers. We will show that

if $a = b$ or $a = -b$, then a divides b and b divides a.

With $a = b$ and $a = -b$ as the two cases, we have to show both

if $a = b$, then a divides b and b divides a

and

if $a = -b$, then a divides b and b divides a.

We will use direct proof twice:

Let a and b be integers.

- We first assume that $a = b$. Then $a = b \cdot 1$ and $b = a \cdot 1$.
- Next suppose that $a = -b$. This means that $a = b \cdot (-1)$ and $b = a \cdot (-1)$.

In both cases we have proven that a divides b and b divides a.

Exercises

Assume that all lowercase variables represent integers.

1. Assume that $a \neq 0$ and $c \neq 0$. Use direct proof to prove the following results.
   (a) If a divides b, then a divides bd.
   (b) If a divides b and a divides d, then a^2 divides bd.
   (c) If a divides b and c divides d, then ac divides bd.

2. Write each of the following as an if-then sentence and prove.
   (a) The sum of two odd integers is even.
   (b) The sum of an even and an odd is odd.
   (c) The product of two even integers is even.
   (d) The product of two odd integers is odd.

3. Demonstrate the following. The contrapositive or indirect proof may be needed.
   (a) If a and b are even, then $a^4 + b^4 + 32$ is divisible by 16.
   (b) If a and b are odd, then 4 divides $a^3 + b^3 + 6$.
   (c) If a^4 is even, then a is even.
   (d) If $a^3 + a^2$ is odd, then a is odd.

4. Write proofs for the following.
   (a) The equation $x - 10 = 23$ has a unique solution.
   (b) The equation $\sqrt{2x - 5} = 2$ has a unique solution.
   (c) The equation $x^2 + 5x + 6 = 0$ has at most two solutions.

5. Demonstrate using biconditional proof.
   (a) a is even if and only if a^2 is even.
   (b) a is odd if and only if $a + 1$ is even.
   (c) $a^3 + a^2 + a$ is even if and only if a is even.
   (d) If $c \neq 0$, then a divides b if and only if ac divides bc.

6. Prove that if ab is even, then a is even or b is even using the following methods:
   (a) Direct proof
   (b) Indirect proof
   (c) Prove the contrapositive

7. Prove that the following are equivalent:
   - a is divisible by 3.
   - $3a$ is divisible by 9.
   - $a + 3$ is divisible by 3.

8. Prove by cases:
   (a) If $a = 0$ or $b = 0$, then $ab = 0$.
   (b) The square of every odd integer is of the form $8k + 1$ for some integer k. (*Hint:* Square $2\ell + 1$. Then consider two cases: ℓ is even and ℓ is odd.)
   (c) If a is an integer, then $a^2 + a + 1$ is odd.
   (d) The fourth power of every odd integer is of the form $16k + 1$.
   (e) 2 divides $a(a + 1)$.
   (f) 3 divides $a(a + 1)(a + 2)$.

9. Recall that we may define the **absolute value** of c by

$$|c| = \begin{cases} c & \text{if } c \geqslant 0 \\ -c & \text{if } c < 0. \end{cases}$$

Let $a > 0$ and prove the following about absolute value.

(a) $|x| = |-x|$.
(b) $x \leqslant |x|$.
(c) $|x| < a$ if and only if $-a < x < a$.
(d) $|x| > a$ if and only if $x > a$ or $x < -a$.

10. Prove:

(a) a divides 1 if and only if $a = \pm 1$.
(b) If $a = \pm b$, then $|a| = |b|$.
(c) If a divides b and b divides a, then $a = \pm b$.

11. Prove:

(a) $\overline{AB} \cong \overline{AB}$.
(b) If $\overline{AB} \cong \overline{CD}$, then $\overline{CD} \cong \overline{AB}$.

12. Let $n > 0$ and define

$$a \equiv b \pmod{n} \text{ if and only if } n \text{ divides } a - b.$$

Read $a \equiv b \pmod{n}$ as "a is **congruent** to b **modulo** n." As on page 44, the following properties are called **reflexive**, **symmetric**, and **transitive**, respectively. A relation such as congruence that is reflexive, symmetric, and transitive is called an **equivalence relation**. Demonstrate the following:

(a) $a \equiv b \pmod{n}$.
(b) If $a \equiv b \pmod{n}$, then $b \equiv a \pmod{n}$.
(c) If $a \equiv b \pmod{n}$ and $b \equiv c \pmod{n}$, then $a \equiv c \pmod{n}$.

13. Assume that $n > 0$. Demonstrate that if $a \equiv b \pmod{n}$ and $c \equiv d \pmod{n}$, we have:

(a) $a + c \equiv b + d \pmod{n}$.
(b) $a - c \equiv b - d \pmod{n}$.
(c) $ac \equiv bd \pmod{n}$.

14. Let $n > 0$ and prove:

(a) n divides a if and only if $a \equiv 0 \pmod{n}$.
(b) a is even if and only if $a^2 \equiv b \pmod{4}$.
(c) If $a \equiv b \pmod{n}$ and m divides n, then $a \equiv b \pmod{m}$.

CHAPTER 3

PLATO AND ARISTOTLE

Plato was born in 427 B.C. to a wealthy family, with a lineage that can be traced back to Solon on his mother's side; they had political connections. Plato was brilliant, and he believed that he should take advantage of his family's position and enter into politics. His timing was not very good. In April 404 B.C., the Spartans had just defeated the Athenians in the Peloponnesian War. This resulted in Athens being led by a small group of pro-Spartans oppressors, known as the Thirty Tyrants. Two of Plato's uncles, Critias and Charmides, became involved in this government and asked Plato if he would join them. Plato refused. Not only did Plato not want to become associated with the corruption of the Thirty, but those same relatives had disrespected Socrates by seeking his participation in their schemes. This angered Plato because he was a student of Socrates (Gottlieb 2000, 175–176; Annas 2000, 672).

After Socrates was executed by drinking hemlock, Plato decided to leave Athens and travel abroad. This led him to southern Italy. What he found was not very different from Athens. For the most part, the people he encountered were hedonistic. They enjoyed staying up all night eating and drinking, and they thought that this should be done as often as possible. This was not Plato's style. Furthermore, he observed that these cities tended to be governed badly, and this often led to civil unrest. Not all of southern Italy was like this, though. The exceptions were those communities where the people valued mathematics and philosophy. The societies that formed in these areas were internally stable and peaceful.

Plato continued his sojourns, and they took him to the city of Syracuse on the island of Sicily. At the time the city was ruled ruthlessly by Dionysius I. His territory extended beyond the city proper to other regions of Sicily. Dionysius had two wives: one, Doris, the mother of the king's heir, Dionysius II; the other, Aristomache, bore him two daughters, one married to Dionysius II and the other married to Dion, the brother of Aristomache. As it turns out, Plato and Dion became acquainted, and Dion would become one of his best students. As both son-in-law and brother-in-law to the king, Dion held a high position in the king's court. This connection brought Plato back to politics, a place where he was not comfortable. At the king's court he discussed with Dionysius matters of philosophy. This was arranged by Dion, for he hoped that if the king would become interested in philosophy, all of Syracuse would benefit. However, when the king saw that his court was enamored with Plato, he became jealous and ended all discussions. Plato realized that it would be wise to leave Sicily, so he did. It is rumored that Dionysius sent agents to try to capture Plato on his voyage home and sell him into slavery. Fortunately, no such plan ever succeeded (Gottlieb 2000, 177; Plutarch 1932, *Dion*, 1156–1157).

Finished with traveling, Plato decided that it was time to return to Athens. The year was about 387 B.C. His return would be significant because this would be the year that Plato would found the Academy. It was located in a grove outside Athens named after the hero Hekademos. It was a public park where people could come to walk and exercise. It seems that Plato purchased property either on or near park grounds on which he had built a villa where he lived, entertained guests, and housed a library. It included a shrine to the Muses and a lecture hall, but the students of the Academy usually met in the park, where passersby could watch them debate (Dillon 2003, 2, 5–6, 9; Lévy 2000, 799).

Although there were some who attended the Academy for practical reasons, this was not Plato's aim for the school. He believed that the pursuit of knowledge for its own sake is the true calling for someone who wishes to be considered an intellectual. The curriculum at the Academy reflected this. The school had a solid reputation in both mathematics and astronomy. Other subjects included geography, botany, and politics. Plato would never promise success for any of his students, for he believed that talent combined with hard work were both necessities for the scholarly life (Gottlieb 2000, 178–179). There is a long-standing myth that written above the door to the Academy were the words, "let no one destitute of geometry enter my doors." There is no direct evidence that such an inscription ever existed. The belief probably arose because of Plato's expectations of his students. The Academy that Plato founded lasted until 88 B.C. Other philosophers started their own versions of the Academy, but they cannot be traced directly to Plato's institution. The last version of the school was closed by the Christian emperor Justinian in A.D. 529 (Lévy 2000, 800; Kline 1972, 43; Heath 1921a, 284).

Possibly the most famous student of the Academy was born in northern Greece in 384 B.C. This student was Aristotle. His father was a court physician to King Amyntas II of Macedonia. When Aristotle was 17, he left his home and traveled to Athens. There he joined the Academy and studied under Plato for 20 years. He was a good student. He listened attentively to everything that Plato taught. At some point, however, he began to doubt what he was hearing. This led Aristotle to rethink everything. Over time he rejected much of what Plato taught and created his own philosophy. Some of Aristotle's Academic colleagues did not understand how he could have seemingly turned against Plato. Aristotle replied, "Plato is dear to me, but truth is dearer still" (Gottlieb 2000, 227).

When Plato died in 347 B.C., his nephew Speusippus was chosen to replace him as leader of the Academy. Either in reaction to not being chosen himself or simply because his long-time friend and teacher had died, Aristotle decided to leave Athens. He was accompanied by Xenocrates, a well-respected citizen of Athens and a true believer in Plato's theory of forms (*Oxford Classical Dictionary*, 3rd ed., s.v. "Xenocrates"). The two traveled throughout Asia Minor and Macedonia. Along the way, Aristotle would seek out students of Plato and converse with them. At the various cities he would found schools, so his philosophy began to spread throughout Greece. Aristotle's influence was found not only among his students at the schools, but also reached to the ruling class. Through some family connections in 342 B.C., Aristotle parted ways with Xenocrates and became the tutor of Alexander the Great. Alexander was only 13 at the time (Jaeger 1948, 111, 120, 124–125).

Seven years later, in 335 B.C., one year after Alexander was crowned king of the Macedonians, Aristotle returned to Athens. Four years later, Xenocrates was elected head of the Academy. Realizing that he and his friend had significant philosophical differences, Aristotle decided against a return to the Academy and, instead, founded another school. Like the Academy, it was housed at a gymnasium outside Athens. It was called the Lyceum. Soon the Lyceum would be the foremost school of philosophy in Greece. Its environment was similar to that of the Academy's. There was an extensive library and garden. The common life of the students was well regulated, and there were monthly banquets. During his years at the Lyceum, Aristotle devoted most of his time to teaching. His morning lectures focused on difficult questions, while the afternoon was reserved for more general discussions (Gottlieb 2000, 220, 226–228; Jaeger 1948, 311–316).

In 323 B.C. Alexander the Great died, prompting an anti-Macedonian revolution in Athens. There had been other revolts, but Alexander dealt with them quickly and severely. This time the opposition was able to gain control of Athens. Because Aristotle was not a citizen of Athens and, worse, was the former tutor of Alexander, the Athenians soon charged him with "impiety." Sounding all too similar to the situation that led to the death of Socrates, Aristotle decided it was wisest to leave Athens. As he put it, he did not wish the Athenians to "sin twice against philosophy." He fled to Chalcis in Euboea, the home of his mother's family. Aristotle died a year later, due to a disease of the stomach (Gottlieb 2000, 228; Jaeger 1948, 311–312, 319).

3.1 FORM

When we first learn geometry, we are usually taught that only a straightedge and compass are allowed for geometric constructions, but as we shall see, the Greeks did not always limit themselves to those instruments. They sometimes used other devices. Plato was not, however, very fond of such mechanical procedures. As Plutarch (1932, *Marcellus*, 376) wrote,

> [T]o solve the problem, so often required in constructing geometrical figures, given the two extremes, to find the two mean lines of a proportion, . . . [Archytas] had recourse to the aid of instruments, adapting to their purpose certain curves and sections of lines. But what with Plato's indignation at it, and his invectives against it as the mere corruption and annihilation of the one good of geometry, which was thus shamefully turning its back upon the unembodied objects of pure intelligence to recur to sensation, and to ask help (not to be obtained without base supervisions and depravation) from matter; so it was that mechanics came to be separated from geometry.

Why did Plato object to using devices other than the standard two when doing geometry? Why did he separate mechanics from geometry? The answer, according to Plutarch, was that **mechanics**, here defined as the use of tools to solve mathematical problems, perverted the true purpose of geometry by using matter to answer abstract questions. The source of the issue can be traced to an after-dinner speech given by Socrates in Plato's *Symposium*. The topic is the love of beauty. After the other guests had, in turn, given talks on their beliefs regarding the subject, Socrates stands up and gives a presentation that overshadows all of the others. He recounts several encounters with a priestess named Diotima. During each of these meetings the two discussed in depth the nature of love (Gottlieb 2000, 171–173). As he tells it, the conversations tended to be one-sided, with Diotima leading the discussion. She begins by observing that among people and animals there is a strong desire to procreate, and this results in a love for the beauty of an individual. Once born, the parents will diligently nurture and protect their offspring. They will do this unconditionally whether it comes with great suffering or even death. This is similar to the action of a warrior. He, too, will risk much for the same reward. What is it they both seek? It is immortality. They cannot be immutable like the gods, but they can both leave something behind once they are dead, whether it be children or fame. This is the mortal's version of immortality (Plato 1961, *Symposium* 207a, 208ac).

But such a level of love is only the first step. At some point one must notice that the physical beauty that is seen in their spouse is a beauty shared by others and thus realize that this is not a quality possessed by a single individual. One should now love all physical beauty. Soon, however, one must move beyond the love of the external and acknowledge the splendor of each soul, whether or not it inhabits an attractive body. This is the beginning of a more noble path. As the focus turns to the spiritual, one begins to recognize the fruits of the soul. One contemplates the beauty of society and its laws and institutions. One sees the glories of mathematics and science. Soon one realizes that beauty is reflected in various ways throughout creation (Plato 1961, *Symposium* 210a–210d). Socrates concludes in the words of Diotima:

> Nor will his vision of the beautiful take the form of a face, or of hands, or of anything that is of the flesh. It will be neither words, nor knowledge, nor a something that exists in something else, such as a living creature, or the earth, or the heavens, or anything that is—but subsisting of itself and by itself in an eternal oneness, while every lovely thing partakes of it in such sort that, however much the parts may wax and wane, it will be neither more nor less, but still the same inviolable whole. (Plato 1961, *Symposium* 211ab)

The point of this discourse is that there is an entity that we call Beauty. It exists independent of the various objects that are beautiful. If all of those objects would disappear, Beauty would still exist. Beauty is an example of a **form** or an **idealseeform** or an **essence**. The forms have a spiritual existence. They are eternal, unchanging, and uncreated. They cannot be seen because they exist in a realm different from ours, but we can learn about them through reason. What we see in this world are representations of the forms. We see, for example, that a waterfall is beautiful because it **partakes** of the form of Beauty. All things that are beautiful partake of this form. It is as if Beauty is a perfect prototype and all beautiful things are copies of it (Wedberg 1955, 29–30, 34).

The Cave

But these copies are imperfect. Plato explains this by way of allegory in the *Republic*. This dialogue is a conversation between Socrates and two other men, Glaucon and Adimantus. Socrates claimed that it is only the just man who is happy; the unjust are miserable. This was unbelievable to Glaucon and Adimantus, and they challenge Socrates. Do not the unjust lie and cheat so as to fool people into thinking that they are just? Will they not be popular and rich? And what about the just? Will they not be misunderstood and despised because they will treat all people fairly? Will they not be persecuted and possibly killed? Socrates is up for the challenge, but he wants to prove his point on a grander scale. He will mentally construct a utopia that is ruled not by laws that will eventually be misused by the unjust but by individuals who have become known as philosopher kings. These people have been educated from the time that they were young to recognize what is good and apply it in the daily lives of the citizens of the republic. That is, they have been taught to understand the forms. The training is not easy because forms are difficult to grasp. The world we see hides the true reality, and we must overcome this. Plato, through the words of Socrates, explained the situation:

> Picture men dwelling in a sort of subterranean cavern with a long entrance open to the light on its entire width. Conceive them as having their legs and necks fettered from childhood, so that they remain in the same spot, able to look forward only, and prevented by the fetters from turning their heads. Picture further the light from the fire burning higher up and at a distance behind them, and between the fire and the prisoners and above them a road along which a low wall has been built, as the exhibitors of puppet shows have partitions before the men themselves, above which they show the puppets. . . . See also, then, men carrying past the wall implements of all kinds that rise above the wall, and

human images and shapes of animals as well, wrought in stone and wood and every material, some of these bearers presumably speaking and others silent. (Plato 1961, *Republic* 7.514a–c)

The fire casts shadows of the shapes on the wall. The prisoners can only see the shadows, so they regard the images as reality. This is all that they have ever known. Suddenly, one of the prisoners is chosen and released. He is shown the true source of the shadows, and he can hardly believe it. He fights against the new reality, believing that the shadows are the truth and what he now sees is some kind of hoax. Worse still is that the light is no longer viewed as a reflection but directly. It blinds him, and it hurts. He wants to go back to his prison, but his release will be permanent. He is led out of the cave, up through a steep and rocky tunnel, and into the day. The sun burns his eyes and hurts his skin. This is worse than the fire in the cave. He is definitely not pleased with his new circumstances. Eventually, the released prisoner adjusts to his new surroundings. His eyes adapt to the bright light, and he sees clearly the higher world above the cave. This is the true reality. The freed prisoner always had the ability to see it. He simply needed to be released and shown the direction to look. Similarly, those who are to learn the nature of reality must be freed from their preconceptions and dedicate themselves wholeheartedly to that pursuit of truth (Plato 1961, *Republic* VII, 515c–516c, 518cd).

Geometry

This pursuit starts with a study of geometry. Although mathematics does have application to various earthly endeavors, these pale in comparison to the value that the study of mathematics has for the mind. Plato taught that the study of mathematics, particularly geometry, is the first step to knowledge of eternal truth because it teaches logic and disciplined thinking. Moreover, as we study geometry, we learn about the **intermediate forms**. These constitute a middle category of existence between the forms and the sensory world. For example, there is the form Triangle. However, we encounter many different types of triangles reflected in the world around us. We see the isosceles triangle that we draw on paper or the equilateral triangle on a billiards table. Because of their imperfections, we recognize them only as representations of perfect triangles. These perfect triangles are examples of intermediate forms. The sensory triangles **correspond** to perfect ones, and both the sensory and the perfect triangles partake of the form Triangle (Figure 3.1). The triangles in the ideal world partake perfectly, and the sensory ones partake imperfectly (Wedberg 1955, 61–62; Plato 1961, *Republic* 7.527bc).

It is because of his belief in these ideal realities that Plato rejected mechanical constructions. We cannot use the material to study the immaterial because then we would be using shadows to study reality. We can only come to a true knowledge of the forms through rational thought. The restriction to the straightedge and compass gives us a way to express these thoughts and represent the eternal mathematical forms (Plato 1961, *Republic* 6.510d, 746). The straightedge gives a representation of a line segment. It is not marked because distance is a property of creation. The compass produces a representation of a circle. This ancient compass was not like ours. It

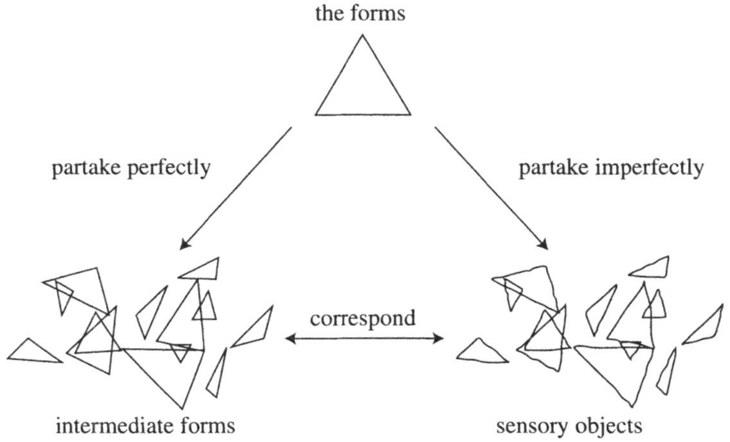

Figure 3.1 Plato's hierarchy of reality.

was **collapsible**. This means that once a circle was drawn, the compass could not be moved to produce another circle congruent to the original. Motion occurs only in the sensory world, not in the world of forms.

Rebuttal

There is no motion in the world of forms because they are are immutable, and they are immutable because the forms are perfect. They are perfect in the sense that the forms are the highest representatives of the class of objects that partake in them. For example, the form called the Good is perfectly good, and the Just is perfectly just. However, to be good means to partake in the Good, and to be just means to partake in the Just. This leads to a problem. To illustrate, take the class of all men. Each of these men partakes of the form Man because that is what it means to be a man. However, this form must itself be a man, yet to be a man requires participation in the appropriate form. Since the forms exist separately from those that partake of them, there must be another form, distinct from Man, of which the form Man partakes. Now we have a question. If we randomly choose a man from the world that we see, of which form does he actually partake? Because there are now two forms that are perfections of man, does the first man partake of the original Man, the second in our list, or the third Man? Plato taught that there is only one form for any given perfection, but we have found at least two forms of which a man can partake. This objection to Plato's system is known as the **Third Man Argument**. It is interesting to note that Plato himself wrote about this problem in the *Parmenides*, and he was very critical of himself. He was not alone. Aristotle believed that this is a fatal objection to the Platonic theory of forms (Plato 1961, *Parmenides* 132B; Wedberg 1955, 30–37; Aristotle 1984, *Sophistical Refutations* 178^b37).

This is not to say that Aristotle did not have his own theory of form. He taught that form is an abstraction from objects that already exist in the sensory world. Form is an **accidental** quality of an object. For example, a piece of paper has a rectangular shape. We see this and generalize the concept of a rectangle from it. We now have a form that we can study on its own because it now exists separate from the paper. It, nonetheless, originated from our experience of the paper. Now rip the paper into pieces. The paper's form is no longer that of a rectangle, but it is still paper. Its **substance**, the fundamental material that makes the paper paper, has not changed. As opposed to Plato, who first established his system based on his philosophical speculations and then drew conclusions about the universe, Aristotle argued that we learn about the world by observation, and once we have identified some basic properties, we can deduce new conclusions by carefully using logic.

Infinity

Since Plato and Aristotle differed radically on what they meant by form, their conceptions of mathematics were naturally at odds. First of all, Aristotle rejected Plato's use of the infinite. The problem for Aristotle was that the intermediate forms were infinite in number. This is an example of what is known as an **actual infinite**, an endless amount of objects existing all at once. There are two types. The first is when there are infinitely many **by addition**. This is the situation with the intermediate forms. When one tries to count them, Plato would say that one would soon realize that they are infinite in number. Aristotle thought that such a concept was inconceivable because this would mean that infinity was a number. Since all numbers (positive integers) can be reached by counting and since it is impossible to transverse all of the numbers to reach an infinite value, the actual infinite is impossible (Moore 1990, 36–37; Wedberg 1955, 59).

The other type of actual infinite is **by division**. Take a line segment and divide it in half. Choosing one of the halves, divide that in two. Pick again and cut again. Based on our intuitions about geometry, this can continue indefinitely, so that we define an infinite sequence a_n of segment lengths such that

$$a_n = 1/2^n,$$

where n ranges over all positive integers. The question is whether this fits reality. Can we, for instance, take some lumber and do the same thing? Is it infinitely divisible? Aristotle answered that we cannot. He relied on the line of reasoning given by Zeno to refute it (Moore 1990, 37). Zeno argued that motion does not exist. It is only an illusion. He believed this because his teacher Parmenides of Elea taught that we cannot speak coherently about things that do not exist. This means that an object that can only be defined by expressing what it is not, cannot be. For example, change cannot exist because its description must explain how the object had a particular state and now it does not have that state (*Oxford Classical Dictionary*, 3rd ed., s.v. "Parmenides"). Hence, motion, since it is a type of change, cannot exist. Supporting his master's views, Zeno gives four further arguments against the existence of motion. They are recorded by Aristotle (1984, *Physics* $239^{b}15$). Here is the second:

> The second is the so-called "Achilles," and it amounts to this, that in a race the quickest runner can never overtake the slowest, since the pursuer must first reach the point whence the pursued started, so that the slower must always hold a lead.

Aristotle agreed that the slower runner would always lead the faster if the faster runner was required to cover infinitely many intervals to catch the slower. However, Aristotle trusted his eyes. There is motion. His solution was to introduce a new understanding of the infinite known as the **potential infinite**. The positive integers are again the example. When we count, we do not expect that at some time in the future we will be done. At any given time we have only counted finitely many numbers. However, in theory this process need never end. It is potentially infinite, but that infinity is never actualized. This is Aristotle's solution to the Zeno paradox. The runner does not span infinitely many small intervals. This would yield an actually infinite set. Instead, the distance covered can be subdivided continually, yielding a set of intervals which are only potentially infinite. Therefore, no contradiction is reached, and the faster runner can now win the race in both reality and appearance (Moore 1990, 35–36, 41–42).

Number and Magnitude

Aristotle and Plato also disagreed concerning the very definition of mathematics. Plato viewed the subject as the study of form. Aristotle thought that the subject matter of mathematics is **quantity**. Aristotle will leave this term undefined, but it does have certain known properties:

- A quantity is finite.
- Quantities are potentially infinitely divisible into other quantities.
- A quantity does not have a **contrary**. In other words, if x is not the quantity 5, we cannot conclude what quantity x represents.
- One quantity is not more than another, like an army of 100,000 has more soldiers than an army of 10,000. It is true that 10 is greater than 4, but 10 is not more than 4, as if 10 contains more things than 4 does.
- Quantities have no visible characteristics.
- Quantities do not exist on their own. They are accidental.

There are two types of quantity: number and magnitude. The term **number** refers to discrete quantities such as the integers or the rational numbers. The study of number is called **arithmetic**. The term **magnitude** refers to continuous values that are assigned to objects as a form of measurement. For example, the magnitude of a line segment is its length, the magnitude of a triangle is its area, and the magnitude of a cylinder is its volume. Aristotle taught that there are no other types of magnitudes because there are only three dimensions. The study of magnitude is called **geometry**

(Apostle 1952, 96, 105); we may therefore define **mathematics** as the combined study of arithmetic and geometry.

These quantities do exist and they can only be apprehended by rational thought. Both Aristotle and Plato agreed on this, but they differed significantly on what they meant by existence. They both believed that quantity exists independent of people, but Plato further believed that mathematical objects exist independent of the sensory world. In opposition to this, Aristotle taught that quantity is a property of the world we see. With respect to geometry, the quantity of an object is what is left when all of its physical properties are mentally removed. What remains is pure form, a continuous figure without motion, color, texture, and so on. We learn about magnitude by combining reason with experience (Apostle 1952, 106; Wedberg 1955, 59).

To fully understand the concept of a magnitude or any object of study, one must identify its causes (Gottlieb 2000, 221). According to Aristotle, there are four:

- The abstraction or idea that describes the object is known as its **formal cause**. By analogy, the blueprint of a house is the formal cause of the house.
- The **material cause** refers to the substance out of which the object is made. For example, wood, bricks, and shingles are the material causes of the house.
- That which creates or causes change in an object is the **efficient cause**. The efficient cause of a house are the tradespeople who construct it.
- The **final cause** of an object answers the question: What is the purpose of the object? The purpose of a house is habitation.

According to Aristotle, philosophers err when they fail to take into account all four causes. For example, Plato was mistaken when he focused on the formal cause of the universe. His belief in the existence of the eternal forms was a confusion of the formal with the material (Gottlieb 2000, 229). He was correct in that mathematical objects do have causes. Their formal cause are the definitions. These give mathematics its structure. Their material cause varies depending on the object. For positive integers it is the unit. For a cube its material cause is the continuous width that its edges surround. The objects of mathematics are immutable, so they have no efficient cause, and these objects have no purpose, so they have no final cause (Apostle 1952, 106).

3.2 CATEGORICAL PROPOSITIONS

Plato taught that the reason that only rational thought could lead to a knowledge of the forms even though there appeared to be a great gulf between the sensory world and the world of forms is because each soul originated from among the forms. When we learn geometry, what is actually happening is that long-forgotten knowledge is being revived in our minds. Plato thought that before birth every soul existed and was conscious, but when the soul joined to matter, it forgot all that it had previously known (Wedberg 1955, 60). To illustrate this, Plato describes an encounter between Socrates and Meno. Socrates attempts to convince Meno that since the soul is immortal and has been born many times, all learning is simple recollection. He does this by leading one of Meno's young servants through a geometric argument by a sequence of questions. Socrates' claim is that if the servant, who has no geometric training, can come to recognize a geometric truth without being told that truth, the servant is remembering and not learning (Plato 1961, *Meno* 84d–85e).

The process that Socrates used to help Meno's servant remember a little of his geometry is known as **dialectic**. By using definitions and images and carefully chosen questions, one can help someone remember knowledge they had of the forms before they were born (Gottlieb 2000, 173–174; Plato 1961, *Meno* 81e). In the case of Meno's servant, Socrates knew the answer to the problem that he was posing, and this helped him frame his questions. This need not be the case. Dialectics can be used to identity the nature of a form even if no one involved in the discussion initially knows the answer. For example, in the *Republic*, Socrates and Cephalus are trying to come to agreement on a definition of what is just. Their attempt is a conversation in which various possibilities are examined. Cephalus claims that the act of returning borrowed items is just. Socrates replies:

> [S]peaking of this very thing, justice, are we to affirm thus without qualification that... paying back what one has received from anyone [is just], or may these very actions sometimes be just and sometimes unjust? I mean, for example, as everyone I presume would admit, if one took over weapons from a friend who was in his right mind and then the lender should go mad and demand them back, that we ought not to return them in that case and that he who did so return them would not be acting justly.

Socrates concludes that this is not the definition of justice because he has given a case where returning an item would not be the right thing to do (Plato 1961, *Republic* 331cd). Socrates and his associates come closer to identifying what is just by enumerating what is not necessarily just. Socrates views it as ludicrous that one would give a weapon to someone who is known to be mad, so he is able to conclude that Cephalus's definition is not sufficient for every case.

Aristotle took this rudimentary logic and systematized it, and for this reason he is considered the founder of formal logic. After his death Aristotle's students collected his logical works and bundled them in a collection that we call the ***Organon***, the *"Instrument."* It is called this because Aristotle viewed logic as a tool for mathematics and the other sciences and not as a subject of study on its own. The collection consists of five innovative pieces:

- ***Categories***. The classification of different types of descriptions.
- ***De Interpretatione***. The philosophy of logic and the study of contraries.
- ***Prior Analytics***. History's first systematic study of formal logic.
- ***Posterior Analytics***. An analysis of mathematical reasoning.
- ***Topics***. The study of dialectical reasoning.

These books were not originally written by Aristotle as a unified piece. They should be regarded as notes or study guides written for his classes at the Lyceum (Kneale and Kneale 1964, 16, 23–24; Gottlieb 2000, 220–221).

For our purposes here, we are most interested in Aristotle's teachings regarding propositions. He studied a very specific type of statement that we will call a **categorical proposition** (Copi and Cohen 2002, 181–215; Hurley 2000, 199–252; Aristotle 1984, *Prior Analytics* 25^{a}1–5). The verbs of these sentences will be various forms of *to be* and will relate one class, or **category**, of objects to another. These sentences will contain two **terms** that identify two different classes.

■ **DEFINITION 3.2.1**

- The **subject term** of a categorical proposition identifies a class of objects about which something will be asserted.
- The sentence will maintain that the objects identified by the subject term will belong to another category of things, which will be identified by the **predicate term**.

For example, the categorical proposition

all projections are involution-preserving functions

has *projections* as its subject term and *involution-preserving functions* as its predicate term. The sentence asserts that everything that is a *projection* is also an *involution-preserving function*.

The number of objects of which something is predicated is known as the **quantity** of the proposition.

■ **DEFINITION 3.2.2**

- If every object in the subject class is included in the predicate class, the quantity is **universal**.
- If the sentence only claims that at least one element of the subject class is involved, the quantity is **particular**.

We use words known as **quantifiers** to identify the quantity of a sentence. The words *all* and *no* are **universal quantifiers**, and the word *some* is an **existential quantifier**.

In addition to quantity, every categorical proposition has a **quality**.

■ **DEFINITION 3.2.3**

If objects from the subject class are said to also be members of the predicate class, the quality is **affirmative**; otherwise, the quality is **negative**.

Take, for example, the categorical proposition

every square is a rectangle. (3.1)

The subject term is *square*, and the predicate term is *rectangle*. The sentence claims that every object that is a square is also a rectangle. This means that the proposition is universal and affirmative. This statement is particular but negative:

there is a rectangle that is not a square. (3.2)

Standard Form

Proposition 3.1 does not include the word *all* even though it is clearly a universal proposition. To aid in the study of categorical propositions, we will convert every categorical proposition into a **standard form**. Let B denote the subject term and A the predicate term. Categorical propositions come in one of the following four standard forms:

Standard Form	**Quantity**	**Quality**	**Example**
All B are A.	Universal	Affirmative	All squares are rectangles.
No B are A.	Universal	Negative	No circles are rectangles.
Some B are A.	Particular	Affirmative	Some rectangles are squares.
Some B are not A.	Particular	Negative	Some rectangles are not squares.

The standard form for Proposition 3.1 is

all squares are rectangles.

If we let B represent *squares* and A represent *rectangles*, the proposition can be written as

all B are A.

The standard form for Proposition 3.2 is

some rectangles are not squares.

If B is the term *rectangles* and A the term *squares*, the standard form that represents this statement is

some B are not A.

To facilitate our understanding of the categorical propositions, we use the diagrams developed by John Venn in the late nineteenth century. These **Venn diagrams** are modern conveniences that illustrate the meaning of these sentences. A rectangle represents the class of all objects, circles represent general classes, and dots represent objects within classes. Originally, Venn used shading to mean that there are no objects in that portion of the class. Over time, however, mathematicians began to use solid shading to represent the possibility of objects existing within a class. Therefore, since we still need to use Venn's method, we will fill areas with parallel lines when we need to exclude the possibility of objects existing in a particular portion of a class. In addition to this, we will draw our diagrams so that all possible intersections of classes are represented. If we have two classes, the circles will overlap, allowing for the possibility of an object in both classes at the same time (Venn 1894, 114–115, 122–125).

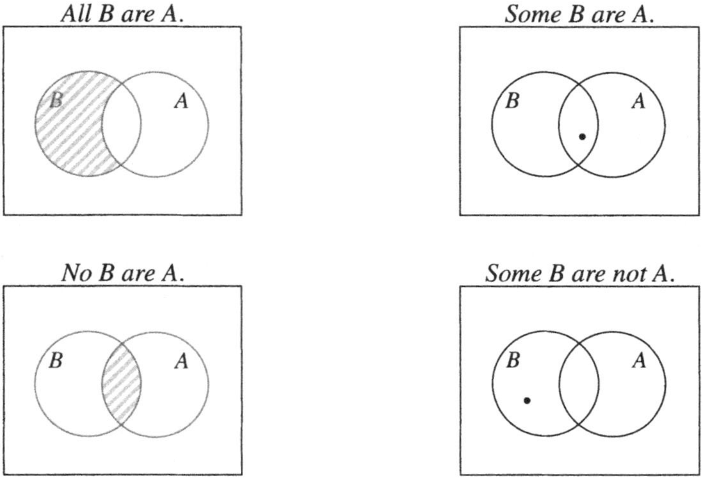

Figure 3.2 The standard forms and their Venn diagrams.

Figure 3.2 represents the four possible Venn diagrams for categorical propositions in standard form.

- In *all B are A*, the overlap between B and A is not shaded because the only objects in B are those that are also in A.
- The opposite occurs in *no B are A*. The intersection of B and A is shaded, making it impossible that there is an element of B that is also in A.
- The first particular form, *some B are A*, has a dot in the overlap between the two categories. This represents the object that is both in B and in A.
- The dot that lies in B but outside A illustrates that there is at least one element that is a B but not an A. In other words, *some B are not A*.

The Converse

With this collection of four categorical propositional forms, we may now begin to write some basic arguments. Let us begin with an argument that has a single premise and determine which categorical propositions follow from it in a single step. For example, consider

all rectangles are parallelograms. (3.3)

From Section 2.3 we know that this sentence can be translated as

if the quadrilateral is a rectangle, then it is a parallelogram.

When its antecedent and consequent exchange positions to form its converse (page 40), the result is

if the quadrilateral is a parallelogram, then it is a rectangle.

This sentence can then written as a categorical proposition in standard form:

$$\textit{all parallelograms are rectangles,} \tag{3.4}$$

which does not have the same truth value as Proposition 3.3. Because of this relationship to the conditional statement, we generalize the definition of a converse to categorical propositions.

■ DEFINITION 3.2.4

The **converse** of a categorical proposition is formed by exchanging the roles of the subject and predicate terms.

Aristotle studied the converse only in terms of syllogisms and examined when the converse follows logically from the original categorical proposition.

> First then take a universal negative with the terms A and B. Now if A belongs to no B, B will not belong to any A; for if it does belong to some B (say to C), it will not be true that A belongs to no B—for C is one of the Bs. . . . Similarly if the proposition is particular: if A belongs to some B, it is necessary for B to belong to some A; for if it belongs to none, A will belong to no B (Aristotle 1984, *Prior Analytics* $25^{a}9$–16, 20–22).

Using the $\equiv$ symbol to represent equivalence, Aristotle is saying that

$$\textit{no } B \textit{ are } A \equiv \textit{no } A \textit{ are } B$$

and

$$\textit{some } B \textit{ are } A \equiv \textit{some } A \textit{ are } B.$$

Aristotle (1984, *Prior Analytics* $25^{a}20$–25) notes, however, that not all converses are equivalent.

> But if A does not belong to some B, it is not necessary that B should not belong to some A: e.g., if B is animal and A man; for man does not belong to every animal, but animal belongs to every man.

He based his demonstration on finding a **counterexample**, an object which proves that a deduction is invalid or a universal proposition is false (Exercise 8). In this case, Aristotle needed to find B and A such that *some B is not A* is true, yet *some A is not B* is false. It could have been any number of things. He chose to let B = *animal* and A = *man*. We can now summarize the four categorical propositions with their converses. Those that are logically equivalent are noted.

Proposition		**Converse**
All B are A		All B are A
No B are A	$\equiv$	No A are B
Some B are A	$\equiv$	Some A are B
Some B are not A		Some A are not B

There is a final argument related to the converse that Aristotle mentioned that is worth comment. He showed that *all B are A* implies that *some B are A* by supposing that *A* is predicated of every *B* and then concluding that a *B* that is an *A* exists (Aristotle 1984, *Prior Analytics* 25^a9–10). This is called **existential import** and is invalid in modern logic. This is because if *B* is empty, then *all B are A* is true but *some B are A* is false. The reason that Aristotle can draw his inference is because he did not allow empty categories. For example, the term *cyclops* is not a legal term in a categorical proposition because it names a category that has no objects. Thus, when we work with Aristotle's logic, we will assume that terms identify nonempty classes.

The Contrapositive

To generalize the contrapositive to categorical propositions, we need a definition. The **complement** of a term *A* is denoted by non-*A* and represents all objects that are not described by *A*. For example, if *A* is *lines*, then a non-*A* is not a line.

■ **DEFINITION 3.2.5**

The **contrapositive** of a categorical proposition is formed by exchanging the roles of the subject and predicate terms and finding the complement of each.

As with the converse, we need to identify when the contrapositive is logically equivalent to the original proposition. We may argue as Aristotle did, or we may draw Venn diagrams. If two propositions yield identical diagrams, they are equivalent. The Venn diagrams for *all non-A are non-B* and *some non-A are not non-B* are:

All non-A are non-B

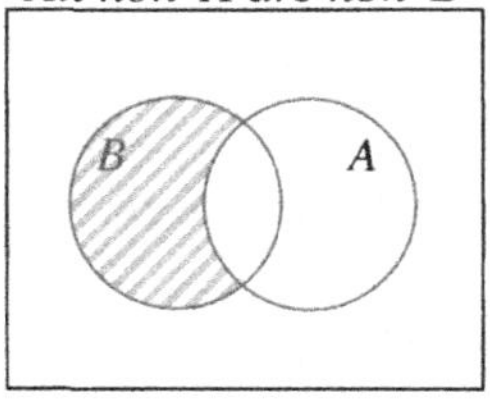

Some non-A are not non-B

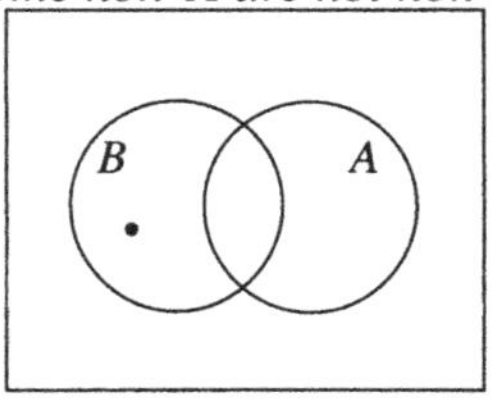

When we compare these to the Venn diagrams in Figure 3.2, we see that *all non-A are non-B* is equivalent to *all B are A* and *some non-A are not non-B* to *some B are not A* because their Venn diagrams match. However, these are not found in Figure 3.2:

No non-A are non-B

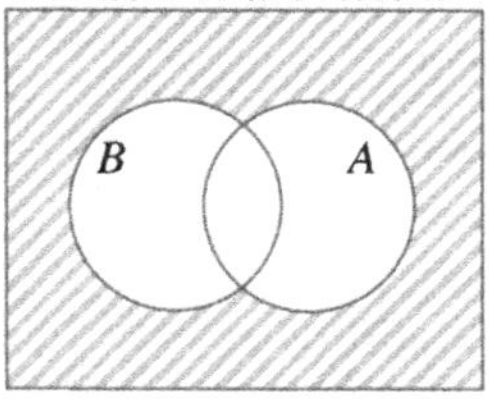

Some non-A are non-B

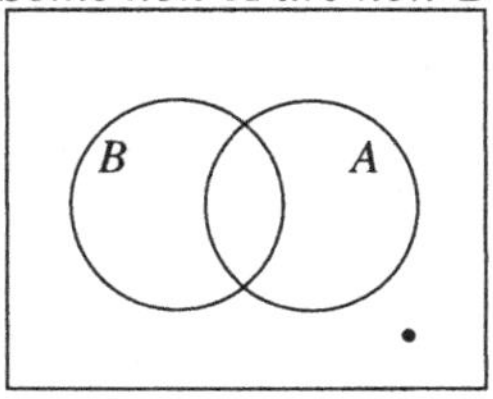

This means that neither *no non-A are non-B* nor *some non-A are non-B* are equivalent to any of those propositional forms, and we may then summarize:

Proposition		Contrapositive
All B are A	$\equiv$	All non-A are non-B
No B are A		No non-A are non-B
Some B are A		Some non-A are non-B
Some B are not A	$\equiv$	Some non-A are not non-B

The Square of Opposition

The proposition *all B are A* claims that every object that is a B is also an A. If this is true, it cannot be the case that *no B are A* is true because such a sentence states that there is no object that is both a B and an A. Similarly, if *no B are A* is true, then *all B are A* is false. However, it is possible that both are false. This happens when some objects that are Bs are also As, while some Bs are not As.

■ DEFINITION 3.2.6

Two categorical propositions that can both be false but cannot both be true are called **contrary**.

Since we are assuming that every term identifies at least one object, there exists an object that is a B. This object is either an A or not an A. For this reason the sentences *some B are A* and *some B are not A* cannot both be false.

■ DEFINITION 3.2.7

Two categorical propositions that can both be true but cannot both be false are called **subcontrary**.

Suppose that *all B are A* is true. This means that every single thing that is a B is also an A. Hence, *some B are not A* is false. On the other hand, if *all B are A* is false, there is at least one B that is not an A, so *some B are not A* is true. Similarly, *some B are A* is true exactly when *no B are A* is false.

■ DEFINITION 3.2.8

Two categorical propositions that always have opposite truth values are called **contradictory**.

Next consider the forms *all B are A* and *some B are A*. Assume that *all B are A* is true. Since there are objects that are Bs and all of these objects are also As, there is at least one element that is a B and an A by existential import. Thus, *some B are A* is true. However, if *all B are A* is false, there exists a B that is not an A. This has no impact on the truth value of *some B are A*. There can still be a B that is also an A so that *some B are A* is true, and there could be no such element, making *some B are A* false. All of this is illustrated by the Venn diagrams.

All B are A is true.
∴ *Some B are A* is true.

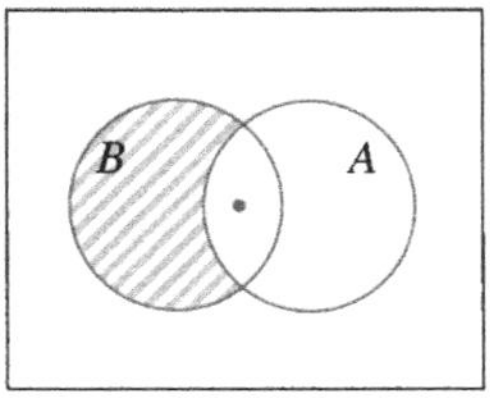

All B are A is false.
Some B are A is true.

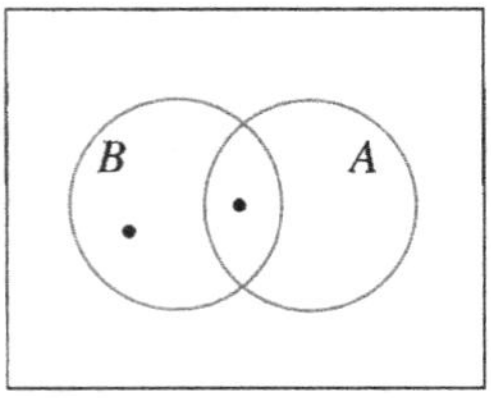

All B are A is false.
Some B are A is false.

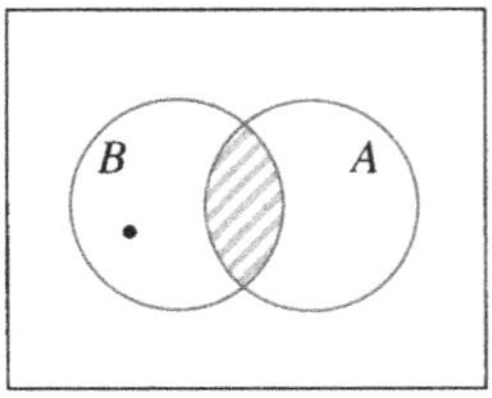

Finally, suppose that *some B are A* is false. This means that there is no object that is both a B and an A. It is therefore impossible for each B to be an A. Hence, *all B are A* is false. If *some B are A* is instead true, the truth value of *all B are A* can be either true or false.

Some B are A is false.
∴ *All B are A* is false.

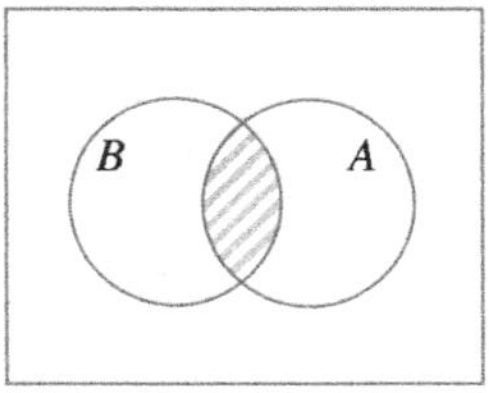

Some B are A is true.
All B are A is true.

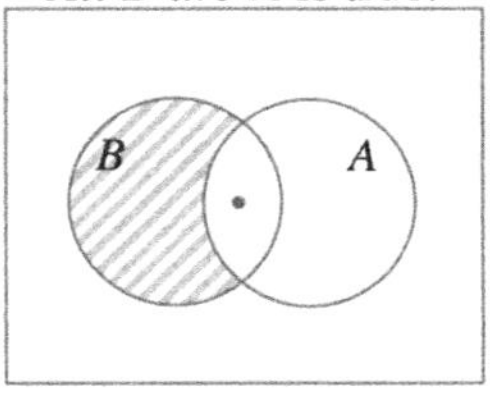

Some B are A is true.
All B are A is false.

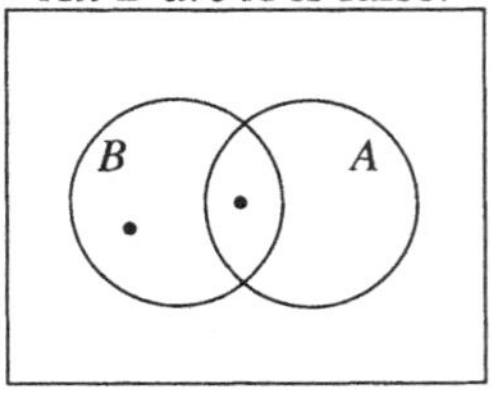

We can show similarly that if *no B are A* is true, then *some B are not A* is true, and if *some B are not A* is false, *no B are A* is false. We shall now put a name on this immediate inference.

■ **DEFINITION 3.2.9**

Subalternation is the relationship between two propositional forms p and q where the truth of p implies the truth of q and the falsity of q implies the falsity of p.

All of these results are summarized in the **square of opposition** (Figure 3.3). There is another square of opposition that is similar but does not include contraries, subcontraries, or subalternation. This modern version has the contradictories only because it is now customary to allow for empty classes.

Exercises

1. Identify the subject and predicate terms of each proposition.
   (a) *No one destitute of geometry was a student of Plato.*
   (b) *Some triangles are not equilateral.*
   (c) *All geometers are philosophers who excel at mathematics.*
   (d) *Some trapezoids that do not have congruent base angles are inscribed within a circle.*

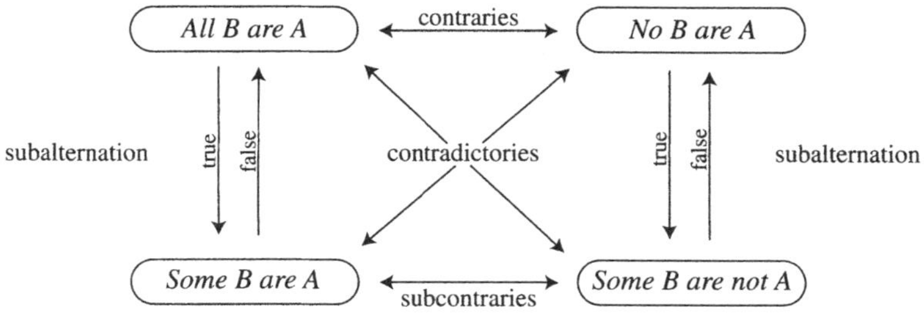

Figure 3.3 The square of opposition

(e) *No line through point P is parallel to l.*
(f) *All friends of Aristotle were people who enjoyed debate and controversy.*
(g) *Some people who listened to Thales were Egyptians who preferred to do geometry the old-fashioned way.*
(h) *All loci formed by cutting a cone with a plane are either a parabola, a hyperbola, or an ellipse.*
(i) *Every Babylonian mathematician was capable of making astronomical observations.*
(j) *No pair of distinct circles are able to share more than two points.*

2. Name the quantity and quality of each of the propositions in Exercise 1.

3. Rewrite each of the propositions in Exercise 1 by changing:
   (a) the quantity.
   (b) the quality.

4. Write the converse and contrapositive for the propositions in Exercise 1.

5. Draw Venn diagrams for the propositions in Exercise 1.

6. Using B for the subject term and A for the predicate term, write each categorical sentence in standard form. Once this is done, determine whether each pair of propositions are contradictory. If they are not, write the contradictory of each.
   (a) *Every real number has a square root.*
   Every real number does not have a square root.
   (b) *All derivatives are continuous.*
   Some derivatives are continuous.
   (c) *Every multiple of 4 is a multiple of 2.*
   Some multiples of 2 are multiples of 4.
   (d) *There is a function that has a derivative.*
   All functions do not have a derivative.
   (e) *There exists an even function.*
   There is no even function.

7. Write the contradictory of the following sentences.

(a) *For all real numbers x, there exists a real y such that $y/x = 9$.*
(b) *There exists a real number x so that $xy = 1$ for all real numbers y.*
(c) *There is a function with y-intercept equal to -3.*
(d) *Every multiple of 10 is a multiple of 5.*
(e) *No interval contains a rational number.*
(f) *All functions have a derivative at $x = 0$.*

8. Provide counterexamples for each of the following false propositions.

(a) *Every integer is a solution to $x + 1 = 0$.*
(b) *The product of any two integers is even.*
(c) *For every integer n, if n is even, n^2 is a multiple of 8.*
(d) *If a function has a minimum, it also has a maximum.*
(e) *Every function has a derivative at $x = 0$.*

9. Determine whether each of the following arguments is valid. If the argument is valid, determine whether the conclusion was obtained by finding the converse, contrapositive, contrary, subcontrary, contradictory, or using subalternation.

(a) *All B are A. / $\therefore$ Some B are A.*
(b) *Some B are not A. / $\therefore$ Some B are not A.*
(c) *No B are A. / $\therefore$ No non-A are non-B.*
(d) *Some B are A. / $\therefore$ No B are A.*
(e) *All B are A. / $\therefore$ No B are A.*
(f) *Some B are not A. / $\therefore$ Some non-A are not non-B.*
(g) *Some B are A. / $\therefore$ Some B are not A.*
(h) *Some B are not A. / $\therefore$ No B are A.*
(i) *No B are A. / $\therefore$ No A are B.*

10. For each statement, identify all categorical propositions in standard form that follow if the statement is true and identify all that follow if the statement is false.

(a) *All triangles are polygons.*
(b) *Some rectangles are square.*
(c) *No circles are squares.*
(d) *Some pentagons are not regular polygons.*

3.3 CATEGORICAL SYLLOGISMS

Theodorus, Theaetetus, and Socrates the Younger met up with Socrates. They brought with them a stranger from Elea in Italy. The man was a disciple of both Parmenides and Zeno, and he was happy to make their acquaintance. Socrates at first believed that the stranger was a god in disguise who had come down to test the group. After Theodorus convinces Socrates that the stranger was not a god but a philosopher, Socrates waxed eloquent about some similarities between the two: that both gods and philosophers sometimes appear as statesmen and other times as sophists. He then asked the stranger his opinion about the meaning of these terms and would like to know how the stranger distinguishes among sophists, statesmen, and philosophers. The modest stranger was willing but at first would rather listen since he was only a visitor. At last he was convinced, so he began:

> We had better, I think, begin by studying the Sophist and try to bring his nature to light in a clear formula. At present, you see, all that you and I possess in common is the name. The thing to which each of us gives that name we may perhaps have privately before our mind, but it is always desirable to have reached an agreement about the thing itself by means of explicit statements, rather than be content to use the same word without formulating what it means. It is not so easy to comprehend this group we intend to examine or to say what it means to be a Sophist. However, when some great task is to be properly carried through, everyone has long since found it a good rule to take something comparatively small and easy and practice on that, before attempting the big thing itself. (Plato 1961, *Sophist* 218b–d)

The term on which they will practice is *fisherman*. The stranger leads the others to begin their investigation by agreeing initially that a fisherman is a type of *artist*. This artist does not produce or create works of art, but instead, he skillfully acquires things. He is, therefore, an *acquisitive* artist, not by trade but by force or *conquest*. By a sequence of choices, they determine that the conquest is done by *hunting*, not fighting. The hunting is that of *live* animals, not dead ones. The live animals *swim*, and they do this in the water, not in the air, so the fisherman *fishes*. The fishing is done by *striking* during the *day* with a *hook*. Their inquiry is summarized in Figure 3.4,

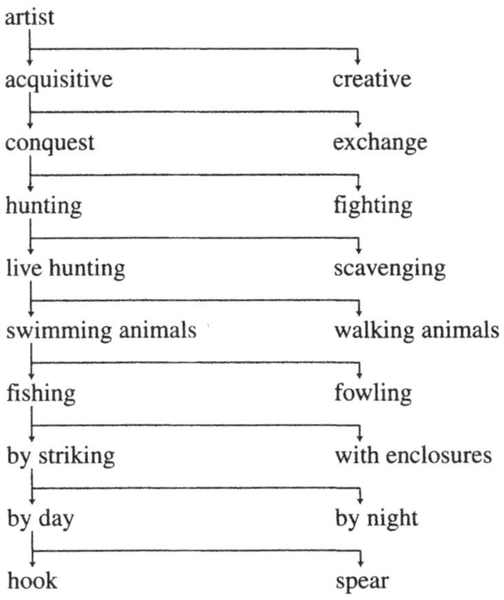

Figure 3.4 Using the method of division to define the term *fisherman*.

and they conclude that a fisherman is an artist who acquires live fish by hunting with a hook during the day (Plato 1961, *Sophist* 219a–221a).

This procedure is known as Plato's **method of division**, or ***diairesis***. It is an important dialectic technique by which a precise definition of a given term can be identified. Its later proponents, however, went further. Division became a method that some Academics used to determine properties of an object. For example, it would be in line with their thinking to modify the previous work led by the stranger into an argument from which properties of a fisherman could be deduced. That a fisherman hunts live animals is determined at the fourth stage of the division. Since

all live hunting is either the hunting of swimming or land animals,

the conclusion that the division is supposed to drive one to is that the fisherman hunts swimming animals. This argument is:

All fishermen are hunters of live animals.
All live animals are creatures that swim in water or live on land.

All fisherman are hunters of animals that swim in the water.

But why should the conclusion follow? Why cannot we conclude that fishermen also hunt land animals? The answer is because we know that fishermen do not do this. The problem is that this is never included as a premise. When the stranger asked them to decide between swimming and land animals, Theaetetus, for the group, chose swimming animals. He did this because he knew what a fisherman is, and a fisherman's target is fish. As a conclusion to the argument, he has **begged the question**. He has assumed a characteristic of the fisherman that the method is claimed to demonstrate.

Aristotle was willing to accept *diairesis* as a means to create definitions, but he made it clear that he did not like it as a deductive method (Cherniss 1962, 28–29; Aristotle 1984, *Prior Analytics* 46^b26–46^b37). However, upon investigation of the results of the division, one can find valid arguments. Here is one:

All fishing is hunting of swimming animals.
All hunting of swimming animals is live hunting.

∴ *All fishing is live hunting.*

To check the validity of this syllogism, let F equal *fishing*, H equal *hunting of swimming animals*, and L equal *live hunting*. The standard forms for these three sentences form this argument:

$$\begin{array}{l} \text{All } F \text{ are } H. \\ \text{All } H \text{ are } L. \\ \hline \therefore \text{All } F \text{ are } L. \end{array}$$

Next we use the two premises to draw a Venn diagram with three circles overlapping so that all possible intersections are included. We use shading that is upward to the right to represent the result of assuming the first premise and strokes downward to the right to represent the assumption of the second premise. The result is the following Venn diagram:

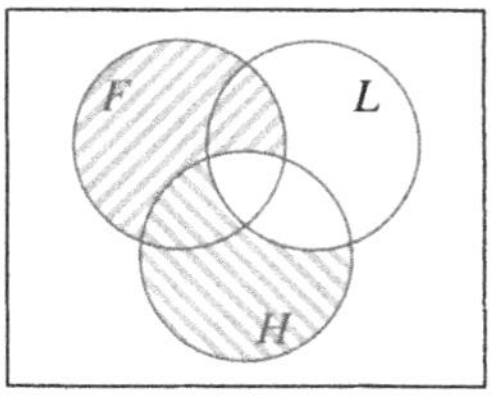

In this case we first shaded inside F with upward strokes and then downward strokes inside H. What remains of F shows us that every object in F must be in L.

We will limit ourselves to arguments that contain only categorical propositions with two premises. Such arguments are called **categorical syllogisms** (Copi and Cohen 2002, 217–236; Hurley 2000, 253–282).

■ **DEFINITION 3.3.1**

A **categorical syllogism** is a syllogism in which the premises and conclusion are categorial propositions.

Following Aristotle, our categorical syllogisms will have only two premises. They are called the **major** and **minor premises**. There will be a total of three terms in the syllogism. The **major term** is the predicate of the conclusion and is also found in the major premise. The **minor term** is the subject of the conclusion, and it can be found in the minor premise. Aristotle called these terms the **extremes**. To be able to draw a conclusion, the two premises must share a term. It is called the **middle term**.

When we study categorical syllogisms, we want the argument to be in a consistent form.

■ **DEFINITION 3.3.2**

A syllogism is in **standard form** if:

- When listing the premises, the major premise is first and the minor premise is second.
- The occurrences of each term must be identical.
- Each term of the argument must be used in the same sense throughout.

For example, take the syllogism

All triangles are two-dimensional.
Some figures are triangles.
∴ *Some figures are two-dimensional.*

The major term A is *two-dimensional*, the minor term B is *figures*, and the middle term M is *triangles*. Notice that the middle term is used to make a connection between the extremes so that a conclusion can be drawn. This can be seen in the Venn diagram for the syllogism:

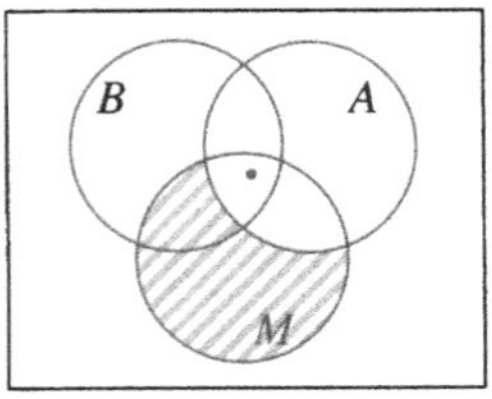

In drawing this diagram, the particular premise is represented by the dot, which is placed after the shading due to the universal premise. The Venn diagrams are always drawn in this order because the type of shading that we are using here removes possibilities from the diagram, and we do not want to place a dot where it is impossible for it to be. Finally, since the conclusion must be true in the diagram, we conclude that the syllogism is valid. The next argument is invalid:

Some polygons are triangles.
Some polygons are squares.
$\therefore$ *Some squares are triangles.*

When we let *A* be *triangles*, *B* be *squares*, and *M* be *polygons*, the syllogism's Venn diagram becomes

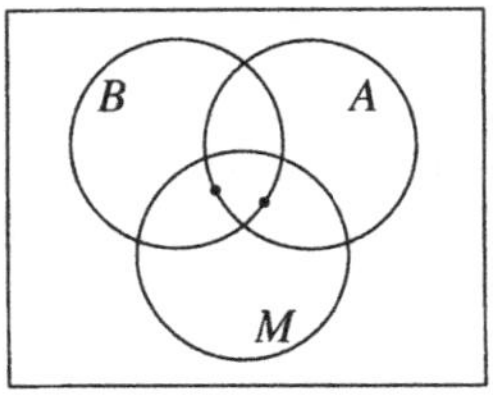

The dot on the right is placed because of the major premise, and the one on the left is placed because of the minor premise. The dots are placed on the borders since the premises do not give enough information to determine on which side they should be placed. For example, the polygon that exists because of the major premise may or may not be a square. The fact that we know that this polygon cannot be a square in reality plays no role in the decision on where to place the dots. We may only base our conclusions on the premises, and they do not state that squares and triangles are different types of polygons.

Exercises

1. Use a Venn diagram to determine if each syllogism is valid.
   (a) *All A are M. All B are M.* / $\therefore$ *All B are A.*
   (b) *No A are M. No M are B.* / $\therefore$ *No B are A.*
   (c) *All M are A. Some B are M.* / $\therefore$ *Some B are A.*
   (d) *Some A are not M. Some M are not B.* / $\therefore$ *Some B are not A.*

(e) *No A are M. No B are M. / ∴ All B are A.*
(f) *No M are A. No B are M. / ∴ No B are A.*
(g) *All M are A. Some M are B . / ∴ Some B are A.*
(h) *No M are A. Some B are M. / ∴ Some B are not A.*

2. Identify the major, minor, and middle terms of each syllogism.
   (a) Some mathematicians are geometers. Therefore, some engineers are mathematicians since all engineers are geometers.
   (b) Some transformations are not rotations, but all functions are rotations. It follows that some transformations are not functions.
   (c) No ellipses are circles, for all ellipses are loci with two foci, and no circles are loci with two foci.
   (d) Some groups are not abelian, for some groups are not permutation groups, and some permutation groups are not abelian.
   (e) No great circles are parallel lines, because no great circles are straight, and all parallel lines are straight.
   (f) Some polygons are hexagons, so some convex figures are hexagons since all polygons are convex figures.
   (g) Some points are not interior points of the circle, for no interior points of the circle are exterior points and all exterior points are points.
   (h) All squares are rectangles. Therefore, no rectangles are triangles because no triangles are squares.
   (i) No curves are line segments. All curves are graphs of a function. Hence, no line segments are graphs of a function.
   (j) No rationalists are empiricists, and all Kantians are rationalists, which means that no empiricists are Kantians.

3. Put each syllogism in Exercise 2 into standard form and draw a Venn diagram to determine whether the argument is valid.

3.4 FIGURES

Prior Analytics is a systematic analysis of syllogism forms. Its goal is to determine which patterns lead to valid arguments and which do not. Aristotle began his investigation by grouping the syllogism forms into three types based on the locations of the major and minor terms in the premises. He called these types **figures**. Let A represent the major term, B the minor term, and M the middle term. The three figures are these:

Figure 1	**Figure 2**	**Figure 3**
MA	AM	MA
BM	BM	MB
$\therefore BA$	$\therefore BA$	$\therefore BA$

The figures represent substitution patterns into which propositional forms are placed. Denote the four types of categorical proposition by the traditional letters:

$$A = \text{universal affirmative,}$$
$$E = \text{universal negative,}$$
$$I = \text{existential affirmative,}$$
$$O = \text{existential negative.}$$

Any combination of three of the letters with repeats allowed is called a **mood**. If the figure and the mood are known, the form of the syllogism is known. For example, the mood AIO of the second figure represents the syllogism

$$\begin{array}{l} \textit{All A are M.} \\ \textit{Some B are M.} \\ \hline \therefore \textit{Some B are not A.} \end{array}$$

First Figure

Aristotle called a syllogism **perfect** if no other propositions need to be added to see that the consequence follows from the premises. A syllogism is **imperfect** if it requires additional propositions that follow necessarily from the premises, yet are needed to demonstrate the consequence (Aristotle 1984, *Prior Analytics* $24^{b}23$–26). The syllogisms of the first figure are perfect. To see this, Aristotle began by noting that AAA and EAE are valid. The argument forms are seen to be valid by studying their Venn diagrams:

$$\begin{array}{l} \textit{All M are A.} \\ \textit{All B are M.} \\ \hline \therefore \textit{All B are A.} \end{array} \qquad \begin{array}{l} \textit{No M are A.} \\ \textit{All B are M.} \\ \hline \therefore \textit{No B are A.} \end{array}$$

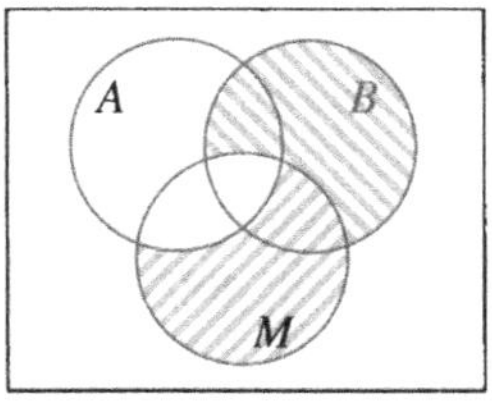

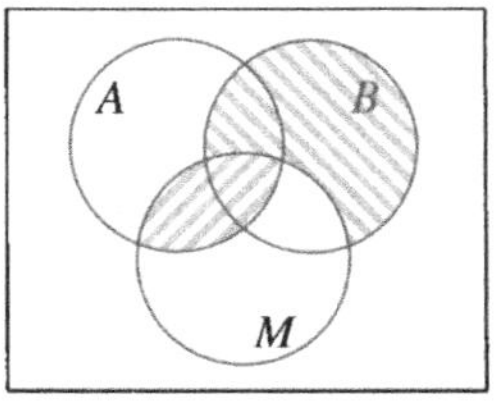

Since classes are nonempty, we may conclude that *some B are A* from *all B are A* and *some B are not A* from *no B are A*. Therefore, AAI and EAO are valid for the first figure:

$$\begin{array}{l} \textit{All M are A.} \\ \textit{All B are M.} \\ \hline \therefore \textit{Some B are A.} \end{array} \qquad \begin{array}{l} \textit{No M are A.} \\ \textit{All B are M.} \\ \hline \therefore \textit{Some B are not A.} \end{array}$$

Aristotle next examined AII and EIO, which are also valid syllogisms for this figure:

All M are A.	*No M are A.*
Some B are M.	*Some B are M.*
∴ Some B are A.	*∴ Some B are not A.*

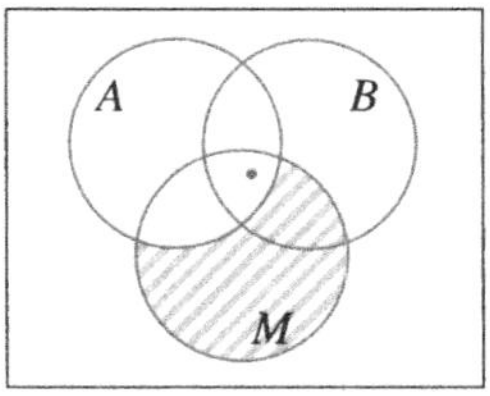

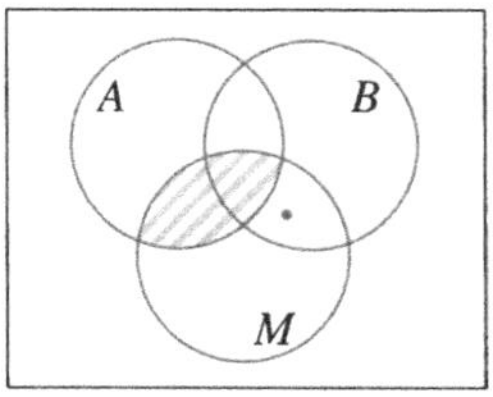

All other moods do not produce valid syllogisms for this figure. Aristotle showed this by assigning meaning to the terms in such a way that the premises are true but the conclusions are not necessarily true. He first considered a universal affirmative conclusion:

All men are animals.
No horses are men.
All horses are animals.

Since there are horses, *some horses are animals* follows from the last sentence, so we could have written instead

All men are animals.
No horses are men.
Some horses are animals.

Although all of these sentences are true, Aristotle did want to say that these conclusions followed from the premises. Indeed, consider the next argument, with *horses* replaced with *stones*:

All men are animals.
No stones are men.
All stones are animals.

Here the premises are true but the conclusion does not follow. This is Aristotle's point. He is studying the form, and in the form the conclusion does not necessarily follow. He illustrated this further with the case of a universal negative conclusion:

All men are animals.
No stones are men.
No stones are animals.

Similar to the first example, since *stones* denotes a class and classes are nonempty, *some stones are not animals* follows from *no stones are animals*. Again, all of the propositions are true, but can we truly say that the premises led us to the conclusion? The answer is no, and this is confirmed by the Venn diagram of the premises. The major premise has the form *all M are A*, and the minor premise is *no B are M*. From this we draw

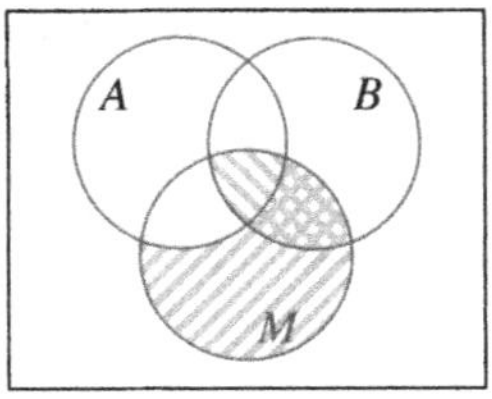

None of the categorical propositions,

all B are A,
no B are A,
some B are A, or
some B are not A

must be true or must be false in the diagram, so we cannot draw a conclusion from these premises.

We may then summarize:

THEOREM 3.4.1

The only moods from the first figure that form valid syllogisms are

AAA, AAI, AII, EAE, EAO, EIO.

Second Figure

The syllogisms of the second figure are imperfect, so Aristotle (1984, *Prior Analytics* $26^{b}34$–$28^{a}9$) will examine them differently. He will first attempt to reduce the second figure syllogism to that of the first figure by converting one of the premises to an equivalent form. Since it is now in the first figure, Aristotle may draw the appropriate conclusion. Sometimes it is necessary to convert the conclusion so that it matches the pattern of the second figure. If conversion would not help, Aristotle's next favorite method was *reductio ad absurdum* (Bocheński 1968, 46).

Consider the premises

No A are M,
All B are M.

Since *no M are A* is the converse of *no A are M* and these are logically equivalent (page 67), the premises are equivalent to

No M are A,
All B are M,

which is a EA premise set from the first figure. Thus, we know that its conclusion is either *no B are A* or *some B are not A* (Theorem 3.4.1), so the moods EAE and EAO yield valid syllogisms in the second figure:

No A are M. *All B are M.* $\therefore$ *No B are A.*	*No A are M.* *All B are M.* $\therefore$ *Some B are not A.*

A little more work is needed to show

All A are M.
No B are M.
$\therefore$ *No B are A.*

Convert the second premise and switch the order to obtain the premise set

No M are B,
All A are M.

We are now in the first figure with premise set EA, so by Theorem 3.4.1 we may conclude that

No A are B.

This converts to *no B are A*, the conclusion that we wanted. Therefore, AEE gives a valid syllogism in the second figure.

Notice that this method cannot be used to show that AEO forms a valid syllogism in the second figure because a particular negative proposition does not convert into an equivalent proposition. However, that AEO is valid in the second figure is made evident by the Venn diagram:

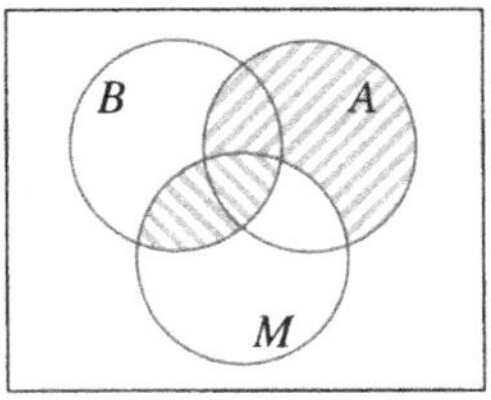

The only unshaded region of *A* must contain an element since classes are nonempty. Because this element is an *A* that is not a *B*, *some A are not B* is the necessary conclusion from these premises.

Similarly, we have

No A are M.
Some B are M.
$\therefore$ *Some B are not A.*

because this argument can be translated to the first figure by converting the first premise to *no M are A*. Hence, we include EIO among the moods of the second figure. However, the proof of the AOO cannot be completed by conversion. Instead, Aristotle used *reductio ad absurdum* to show that the next syllogism is valid:

All A are M.
Some B are not M.
$\therefore$ *Some B are not A.*

Suppose that *some B are not A* is false. This means that *all B are A* is true, so by the first premise, *all B are M*. This contradicts the second premise that states that there is at least one thing that is B but not M. Therefore, we conclude that the proposition *some B are not A* is true.

None of the other premise sets lead to valid syllogisms. Aristotle shows this by counterexamples. For example,

All A are M,
All B are M,

will not yield the conclusion *all B are A* as the substitution $A = mrn$, $B = animals$, and $M = substances$ shows. This leads to the next summary concerning the second figure.

THEOREM 3.4.2

The only moods from the second figure that form syllogisms are

AEE, AEO, AOO, EAE, EAO, EIO.

Third and Fourth Figures

In like manner, Aristotle finds the valid moods of the third figure.

THEOREM 3.4.3

The only moods from the third figure that form syllogisms are

AAI, AII, EAO, EIO, IAI, OAO.

It appears that the figures presented on page 77 are missing a pattern. Should not have Aristotle included a **fourth figure**?

$$\begin{array}{c} AM \\ MB \\ \hline \therefore BA \end{array}$$

Aristotle did not consider this last pattern a distinct figure; instead, he analyzed it as if it belonged to the first figure. He did this because if we switch the positions of the premises so that B becomes the major term and A becomes the minor term, we have the premise pattern of the first figure (Bocheński 1968, 45; Aristotle 1984, *Prior Analytics* $29^{a}19$–27, $53^{a}9$–12).

THEOREM 3.4.4

The only moods from the fourth figure that form syllogisms are

AAI, AEE, AEO, EAO, EIO, IAI.

Exercises

1. Put each syllogism into standard form, and identify the figure and mood of each argument.

   (a) *No circles are squares, but all squares are polygons. Therefore, no circles are polygons.*

   (b) *Some magnitudes are rational numbers, because all integers are magnitudes, and some rational numbers are integers.*

   (c) *All geometers are humans, but no humans are martians. It follows that no geometers are martians.*

   (d) *All conic sections are loci. Therefore, some loci are not parabolas since some conic sections are not parabolas.*

   (e) *No Aristotelians are Platonists, so because some philosophers are Aristotelians, some Platonists are not philosophers.*

   (f) *Some mathematicians are not geometers. This is because all geometers are Kantians, and some Kantians are not mathematicians.*

   (g) *All horocycles are Euclidean surfaces. Some horocycles are planes. Hence, some Euclidean surfaces are planes.*

   (h) *All friends of Mersenne are scientists. Hence, all friends of Fermat are friends of Mersenne since all friends of Fermat are scientists.*

   (i) *No pentagons are rectangles, but all squares are rectangles. It follows that no pentagons are squares.*

2. Use the theorems of the section to determine whether each of the syllogisms of Exercise 1 are valid, and confirm your decision with a Venn diagram.

3. Use A as the major term, B as the minor term, and M as the middle term to write the syllogism that arises from the given mood and figure.

(a) AAA, 1	(e) AII, 3	(i) EAE, 1
(b) AEE, 2	(f) AEE, 1	(j) AOI, 3
(c) EAO, 3	(g) EAE, 2	(k) EAO, 2
(d) AAO, 2	(h) OAO, 3	(l) IAI, 1

4. Use Venn diagrams to determine whether each of the syllogisms in Exercise 3 is valid.

5. Prove Theorem 3.4.3 using Venn diagrams.

PART II

THE GOLDEN AGE

CHAPTER 4

PYTHAGORAS

There are no surviving writings of Pythagoras or from his early followers, known as the Pythagoreans, and his teachings survived only as an oral tradition. Similarly, there is no early information regarding the life of Pythagoras. As with Thales, much must be considered legend. The biographies that are available were written about 750 years after his death. Gathered together by Guthrie (1988) in a single volume, these works were written or preserved by four men: Diogenes Laertius (*c.* 250), Porphyry (*c.* 275), Iamblichus (*c.* 300), and Photius (*c.* 850). What follows is just one possible combination of events claimed for the life of Pythagoras.

By most accounts Pythagoras was born in 570 B.C. on Samos in the Aegean Sea. His parents were Mnesarchus, a gem engraver, originally from Tyre, and his wife Pythais. Iamblichus recorded how Mnesarchus learned that he would have a son:

> Mnesarchus had gone to Delphi on a business trip, leaving his wife without any signs of pregnancy. He enquired of the oracle about the event of his return voyage to Syria, and he was informed that his trip would be lucrative, . . . [and] that his wife was new with child, and would present him with a son who would surpass all others who had ever lived in beauty and wisdom, and that he would be of the greatest benefit to the human race in everything pertaining to human achievements. (Guthrie 1988, 58)

Mnesarchus named his son Pythagoras after Pythia, the oracle of Apollo. Soon thereafter the family moved to Samos, which was the home of Pythais's family. Whether born on Samos or not, Pythagoras would grow up on the island and would not disappoint his parents' expectations. While a youth he is said to have had a quiet serenity and a profound knowledge. He was known to have reached out for the wisdom of the elders, and in this he earned their respect. Together with his mental traits, it was claimed that he had a dignified appearance. The inhabitants of Samos concluded that he was descended directly from Apollo himself.

At some time during these years, Pythagoras left Samos for Greece to be initiated in the sacred mysteries, and then he traveled to Ionia to study under Thales. He would later journey to Egypt. There the priests taught him their language and their religion, and he was allowed into the inner chambers of the temples. As with Thales the Egyptians taught Pythagoras geometry. Later journeys brought him to Babylon and Persia. He learned about astronomy and religion from the Chaldeans and about numbers and proportions from the Phoenicians.

When Pythagoras returned to Ionia, he was about 40 years old. Inspired by all that he had learned, he decided to open his own school. It was known as the Pythagorean Semicircle. He taught his students that within each person was all of the powers of the universe. There are various levels of reason from the highest among the gods to the lowest among the animals. People have reason. In nature things move, grow, and reproduce. People move, grow, and reproduce. Whatever power one can find in the world, that power can also be found within a single person. People, Pythagoras taught, were microcosms of the universe.

One day Pythagoras decided to bring his teachings back to his homeland of Samos, but Polycrates ruled the island. Polycrates had usurped power around 535 B.C. and soon made Samos a regional power. With a strong navy, he dominated the local islands, including Delos and Rheneia, which he consecrated to Apollo. Polycrates was famous for his love of the arts, but he was a tyrant (*Oxford Classical Dictionary*, 3rd ed., s.v. "Polycrates"). Believing that someone like himself would not be welcome, Pythagoras fled to Croton in southern Italy. When the residents of that area learned that Pythagoras had moved to their region, great crowds sought after him to learn his teachings. As when he was young, many of the Italians regarded him as a god. Signifying his divinity, many believed that he had a golden thigh, could be two places at once, and had control over animals. Porphyry recounts two tales (Guthrie 1988, 127). The first involves an ox.

> At Tarentum, in a pasture, seeing an ox cropping beans, he went to the herdsman, and advised him to tell the ox to abstain from beans. The countryman mocked him, proclaiming his ignorance of the ox-language. So Pythagoras himself went and whispered in the ox's ear. Not only did the bovine at once desist from his diet of beans, but would never touch any thenceforward, though he survived many years near Hera's temple at Tarentum, until very old, being called the sacred ox, and eating any food given him.

The second regards a more serious matter with a bear.

> The Daunian bear, who had committed extensive depredations in the neighborhood, he [Pythagoras] seized; and after having patted her for awhile, and given

her barley and fruits, he made her swear never again to touch a living creature, and then released her. She immediately took herself into the woods and the hills, and from that time on never attacked any irrational animal.

Pythagoras traveled throughout Italy and Sicily. He extolled the people to liberty, and many cities that were under the control of tyrants became free because of his teachings. He would tell them to live justly and with wisdom, using sayings such as

eat not the heart,

meaning that they should not burden themselves with sorrows,

help a man take up a burden, but not to lay it down,

meaning that they should encourage hard work, and

on starting a journey, do not turn back,

meaning that they should have no regrets at life's end. He told them that they should purify themselves by not sacrificing animals and never eating certain foods, such as anemones, brains, and beans. They should follow his teachings so that they would be ready to live again after they die, for he also taught them that this life was not the end. They should expect to be reincarnated many times in the future as they had in the past. They may not remember those other lives, but he could help them recall. Pythagoras could do this because he had the gift of remembering previous lives, including his own. This was granted him by Hermes. Pythagoras recollected once existing as Euphorbus, Aethalides, Hermotimus, and Pyrrhus, and he had also suffered in Hades.

As the teachings of Pythagoras spread, so did the political influence of his followers, reaching many throughout southern Italy, including Archytas of Tarentum and Hippocrates of Chios (Boyer and Merzbach 1991, 65, 70). However, the Pythagoreans were not always well received. One day Pythagoras was with some of his companions in a house when a mob decided to burn it down. It is unclear why they did this. It may have been because of petty reasons such as envy of the fame of Pythagoras or anger that they were not part of the inner circle. Some suggest that they were citizens of Croton who feared that Pythagoras would become too powerful and become a tyrant over their city. Whatever the reason, the mob burned the house, and Pythagoras and his companions scattered. Many were killed, but Pythagoras escaped. As he fled he came across a field of beans, and not wanting to trample them, he stopped and was killed. Other accounts state that he fled to the temple of the muses at Metapontum, where he died 40 days later of starvation. This occurred around 475 B.C.

After his death the disciples of Pythagoras continued to follow his ways. These not only included his moral precepts but also his view of the world. Thales believed that the fundamental substance of the world is water. He saw that things that are alive are moist to one degree or another, so moisture is a sign of life. Since he also believed that life permeated all matter, so must water. The Pythagoreans replaced water with number. They viewed the boundlessness of the universe as dark, chaotic, and without meaning (Moore 1990, 19; Kahn 2001, 15, 27). In contrast, it is among the finite where one finds structure, order, relationships, and life. This system was brought forth by the **One**, which is sometimes called the **Monad**. The One's primary importance to the world is that it is the generator of all numbers:

$$1,\ 1+1,\ 1+1+1,\ 1+1+1+1,\ \ldots$$

Following the Monad is the **Dyad**, 2. As the Monad conveys equality, unity, and measure, the Dyad conveys excess and defect. Next comes the **Triad**, 3. It is the actualization of harmony. As Porphyry notes in his *The Life of Pythagoras*, it peacefully brings together the conflicting powers of the Monad and Dyad to make them a perfect whole. Since initially the Dyad was not considered a number but the generator of the even numbers (Heath 1921a, 71), the Triad is the first number and the first odd number. The **Tetrad**, 4, is the first even number (Kahn 2001, 28–29; Guthrie 1962, 243–244; Guthrie 1988, 40, 133, 137).

The numbers continued until they reached the **Decad**, 10. It was viewed as the perfect number because it is the sum of 1, 2, 3, and 4. For this reason it is represented as a triangle, so it also goes by the name ***Tetraktys***:

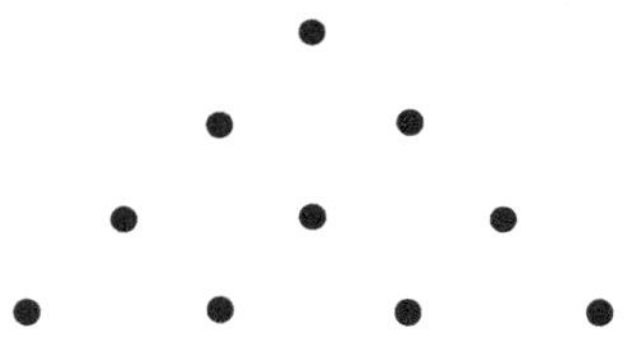

In one number the *Tetraktys* combined the different dimensions of the world. The Monad is represented by a single point, the Dyad has two points and forms a segment, the Triad can form a triangle, and the Tetrad can have the shape of a tetrahedron. The Pythagoreans would swear by the *Tetraktys* as their greatest oath, for in it they saw their guiding principles summarized (Heath 1921a, 75–76).

The One did not simply create the numbers, it also produced the universe in the same orderly fashion. The One gave order to the chaotic universe. The result was the world, and it was a perfect sphere. At the center of this sphere is located the ***Hestia*** or the *Hearth* of the universe. It is a central ball of fire that is a physical manifestation of the One. At the outer edges of the sphere are the fixed stars. About the *Hestia* orbit the other heavenly bodies. Like number they are in sequence. Moving along circles we find Saturn at the outermost position, and then working toward the center there is Jupiter, Mars, Venus, Mercury, the Sun, the Moon, and finally the Earth. Counting the sphere of fixed stars as a unit, this totals nine. The author of this system, a philosopher named Philolaus, believed that since 10 embodied perfection, there must be a tenth object. The problem was that none was to be found. Philolaus concluded that there is an **anti-earth**, which would bring the count to 10 (Figure 4.1). Exactly what is found on this planet is not known, but it is located on the earth's orbit on the exact opposite side of the *Hestia* (Kahn 2001, 27–28; Guthrie 1962, 333).

Based on the lack of source material written relatively close to the life of Pythagoras, it is difficult to determine with any certainty how much of this account or any

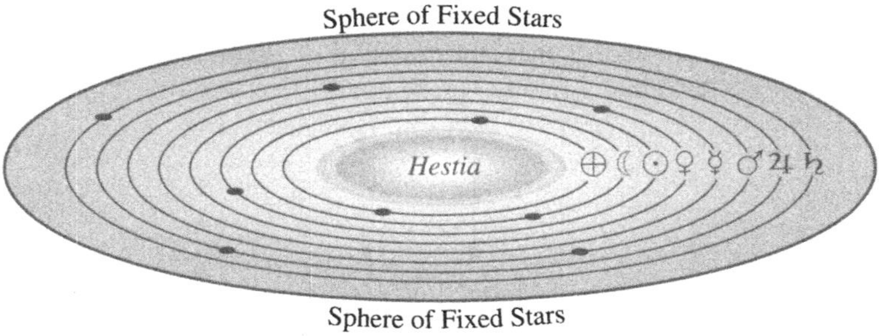

Figure 4.1 The universe according to Philolaus, with the anti-earth.

other is based on reality. We might be willing to dispense with the story of the Daunian bear but accept that he died at Metapontum. What is more difficult is the determination of what Pythagoras and his early followers actually believed. Their view of number was probably close to that described, but later Greeks often associated the views of their teachers with Pythagoras to bolster their own reputations. What does seem certain is that Pythagoras offered the Greeks a new manner of living, with a strict religious and moral code, and that he taught them a new and hopeful view of life after death. As for mathematics, it is unlikely that Pythagoras was the great mathematician of legend, but it is likely that he and his earlier followers had some interest in the subject besides their mystical pursuits (Kahn 2001, 32–32, 68–70).

4.1 NUMBER THEORY

The ordering of the universe by number led the Pythagoreans to believe that the very essence of the objects found in the world is number. It is a living quantity, not a sterile measurement. Possibly inspired by this, later Greeks furthered the study of number, not as mystics but as mathematicians. This led to many different characterizations of number, including that of being even or odd and that of prime or composite (pages 45, 47). A prime number was sometimes referred to as **rectilinear** or simply **linear** because such numbers can naturally be represented by a sequence of dots in a line, as the Pythagoreans liked to do. These dots cannot comprise a rectangle because to do so would mean that the number was the product of two numbers besides 1 and itself. The Triad, however, can also be viewed as a triangle:

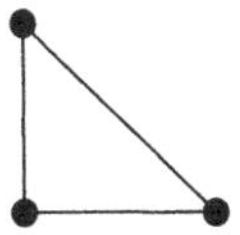

This is an example of a **triangular number** because "when it is analyzed into units, it shapes into triangular form with the equilateral placement of its parts in a plane," as was defined by the later Pythagorean Nicomachus (1926, II.8, 241). Included among the triangular numbers are 6 (the **Hexad**) and 10. Since triangular numbers can be represented by a plane figure, they are examples of **plane numbers** (Heath 1921a, 72–73, 76–77). The Tetrad is another plane number:

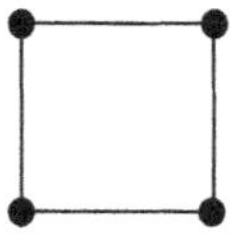

It is also a square number, for obvious reasons. According to Euclid (1925, VII, Definition 18, 278):

■ **DEFINITION 4.1.1**

> A **square number** is equal multiplied by equal, or a number that is contained by two equal numbers.

It is contained by two equal numbers in the sense that two consecutive sides of a square contain the square. This is equivalent to our modern definition since an integer n is a (perfect) square if there is an integer k so that

$$n = k^2.$$

The Tetrad is an example of a square number. It is also a **solid number** because the dots can be arranged to represent a solid. In this case it is a tetrahedron:

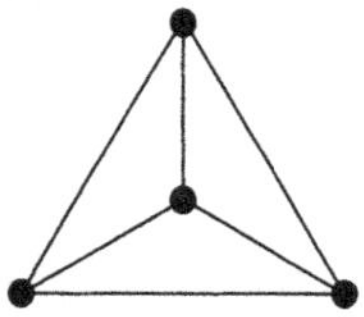

Use dots to generalize this concept by drawing pyramids that have other bases such as squares, pentagons, hexagons, and heptagons (Nicomachus 1926, II.13, 249–250).

Ratios

Supposedly, one day Pythagoras was walking by a blacksmith shop and noticed that there was a harmonious quality to the striking of the hammers. When he investigated he found that there was a predictable relationship between the weights of different hammers and the sounds that they produced. This inspired him to experiment on his own. It is claimed that he used a device known as a **monochord**. This is a simple box with a stretched string attached, something like a one-stringed guitar. The obvious first experiment is to pluck the string and listen to the resulting **tone**. Let f designate its frequency. Next, halve the string and pluck again. What happens is that the frequency at which the string vibrates doubles to $2f$. Pythagoras could not measure this precisely, of course, but he could hear the change in frequency. In other words, he could hear an increase in **pitch**. The relationship between the different frequencies can then be represented by a ratio. To the classical Greeks, a **ratio** was simply a relationship between numbers. The ratio of the frequency of the tone played on the half-string to the frequency of the original note is 2 : 1 and is called an **octave**. In general, the ratio between the frequencies of two tones is called an **interval**.

What would Pythagoras do next? Shorten the length to a third of the original and pluck again. He would hear another increase in pitch, one with frequency $3f$. Therefore, the interval between the note played on a third length and a note played on a half length is 3 : 2. If we take the note that is an octave below the note just played, its frequency will be $\frac{3}{2}f$. We now have a set of three notes, the start of a **scale**:

$$f < \frac{3}{2}f < 2f.$$

Pythagoras's objective was to fill notes between f and $2f$ in a manner that was pleasing to the ear. The interval 3 : 2 is called a **perfect fifth**. Second to the octave, it is the most pleasing musical interval. The interval 4 : 3 is called a **perfect fourth**. To the ancient Greek ear, the fourth is almost as pleasing as the fifth (Guthrie 1988, 327–328).

These ratios fit in perfectly with the Pythagorean numerology. The music appeared to originate because of the positive integers. The ratios that described the most pleasing intervals were formed by the simplest of ratios, and the fundamental octave was the ratio of the Dyad to the Monad:

Octave	2 : 1
Fifth	3 : 2
Fourth	4 : 3

Even the *Tetraktys* reflected this discovery with its single dot followed by a row of two dots, then a row of three, and finally a row of four (Heath 1921a, 76). The discovery of these ratios so impressed Pythagoras and his early followers that they concluded that there should be music in nature. Since they saw number and geometry when they gazed at the night sky, they believed that the celestial spheres that carried the planets and the stars made music. The Pythagoreans did not hear it, however, because the music was continual and they had grown accustomed to it. However, Pythagoras could hear it. According to Porphyry (Guthrie 1988, 129):

> He himself could hear the Harmony of the Universe, and understood the universal music of the spheres, and of the stars which move in concert with them, and which we cannot hear because of the limitations of our weak nature.

As number brings order to the infinite, so number brings structure to sound (Bibby 2006, 14–17; Guthrie 1988, 24–28).

Proportions

We say that two equal ratios are in **proportion**. For example,

$$3 : 2 = 6 : 4$$

is a proportion. This means that 3 : 2 and 6 : 4 represent the same relationship. The Pythagoreans considered the study of proportion as necessary to a full understanding of the universe (Nicomachus 1926, II.21, 264–265). As we will see, ratios and proportions were crucial to Greek geometry, but we would have a difficult time understanding their proofs when the ratios are written using the notation above. We think of ratios as fractions, and these fractions to us are numbers. For this reason, we will write ratios as fractions.

In the proportion

$$\frac{a}{b} = \frac{c}{d}, \tag{4.1}$$

both a and d are called the **extremes** and b and c are called the **means**. The Greeks knew that for this proportion to hold, it was necessary and sufficient for the product of the extremes to equal the product of the means. That is,

$$\frac{a}{b} = \frac{c}{d} \text{ if and only if } ad = bc.$$

As a result, they also knew that given Equation 4.1, the following five proportions would hold.

- ***permutando***

  $$\frac{a}{c} = \frac{b}{d}$$

- ***invertendo***

  $$\frac{b}{a} = \frac{d}{c}$$

- ***componendo***

  $$\frac{a+b}{b} = \frac{c+d}{d}$$

- ***separando***

  $$\frac{a-b}{b} = \frac{c-d}{d}$$

- ***convertendo***

  $$\frac{a}{a-b} = \frac{c}{c-d}$$

Since the Greeks did not recognize negative numbers, both *separando* and *convertendo* require that $a > b$ and $c > d$.

Related to the terminology above, a **mean** is a value that expresses a relationship between two numbers. There were three means known to the early Pythagoreans, and they are related to their system of music. Let a and b represent numbers and m the mean.

- The **arithmetic mean** is defined by the equation

  $$\frac{a-m}{m-b} = 1.$$

  Solving for m yields the familiar $m = (a + b)/2$. Notice that if we find the arithmetic mean of a frequency and its double,

  $$\frac{f + 2f}{2} = \frac{3}{2}f,$$

the resulting mean is equal to the perfect fifth.

- Next there is the **harmonic mean**. It is defined by

$$\frac{a-m}{m-b} = \frac{a}{b},$$

which is equivalent to

$$\frac{1}{a} + \frac{1}{b} = \frac{2}{m}.$$

Hence, $m = 2ab/(a+b)$. The harmonic mean of a frequency and its double yields the perfect fourth:

$$\frac{2 \cdot f \cdot 2f}{f + 2f} = \frac{4}{3}f.$$

- The characteristic equation for the **geometric mean** is

$$\frac{a-m}{m-b} = \frac{a}{m} = \frac{m}{b}.$$

From this proportion we can conclude that $m = \sqrt{ab}$. The ancient Greeks did not have a notation for square roots but, instead, would have described $\sqrt{ab}$ as the length of the side of a square with area ab. The geometric mean is used to fill in the other notes of what is known as the **Pythagorean scale** (Guthrie 1988, 27–28). See Exercise 8.

What is distinct about the Pythagorean theory of ratios and proportions is that the Pythagoreans' philosophy regarding the relationship between number and the world affected the manner in which they viewed their ratios. For example, consider lengths of line segments. Since everything is number, ratios of lengths should be able to be described by numbers, and number to a Pythagorean meant positive integer (Euclid 1925, *Elements* X, 10).

■ DEFINITION 4.1.2

Two line segments $\overline{AB}$ and $\overline{CD}$ are **commensurable** if there exists a line segment $\overline{UV}$ such that $AB = a \cdot UV$ and $CD = b \cdot UV$ for some positive integers a and b.

From this we conclude that segments being commensurable means that

$$\frac{AB}{CD} = \frac{a}{b},$$

where a and b are integers. The first Pythagoreans believed that all pairs of line segments were commensurable. The upshot of this is that all line segments would have **rational** length. This is because if $CD = 1$, then $AB = a/b$ from the equation above.

Exercises

1. Let t_n represent the nth triangular number, that is, the number of dots required to fill an isosceles triangle with legs containing n dots:

   (a) Find the formula for t_n.
   (b) Show both by a diagram and algebraically that $(n+1)^2 = t_n + t_{n+1}$.
   (c) Find a formula for the sum $T_n = t_1 + t_2 + \cdots + t_n$.
   (d) The value T_n is know as a **tetrahedral number**. Use dots to explain the terminology.

2. A **pyramid number** is defined as $p_n = 1 + 4 + 9 + \cdots + n^2$.
   (a) Use dots to explain why p_n is called a pyramid number.
   (b) Find a another formula for p_n.

3. An **oblong number** o_n is the number of dots required to fill a rectangle with length one dot more than its width. In other words, $o_n = n^2 + n$. Draw diagrams for the first four oblong numbers.

4. Prove each of the following equations involving oblong numbers both by a diagram and algebraically.
   (a) $o_n = 2 + 4 + 6 + \cdots + 2n$
   (b) $o_n = t_{2n} - n^2$
   (c) $o_n = \frac{1}{2}([2n+1]^2 - [n+1]^2 - n^2)$
   (d) $o_n = 2t_n$

5. The **pentagonal number** p_n is the number of dots required to outline and fill a regular pentagon as follows:

   (a) Show both by a diagram and algebraically that $p_n = n^2 + t_{n-1}$ $(n > 1)$.
   (b) Find another formula for p_n.

6. Given that $a/b = c/d$, prove that the five proportions given on page 94 (*permutando, invertendo, componendo, separando, convertendo*) hold.

7. Check the following from Plato's *Parmenides* (Heath 1921a, 294):

$$\frac{a}{b} > \frac{a+c}{b+c}$$

when $a > b$ and $c > 0$.

8. We want to identify the frequency of notes on a scale. We begin at C, F will be a perfect fourth, G a perfect fifth, and C* will be an octave higher than our initial note C:

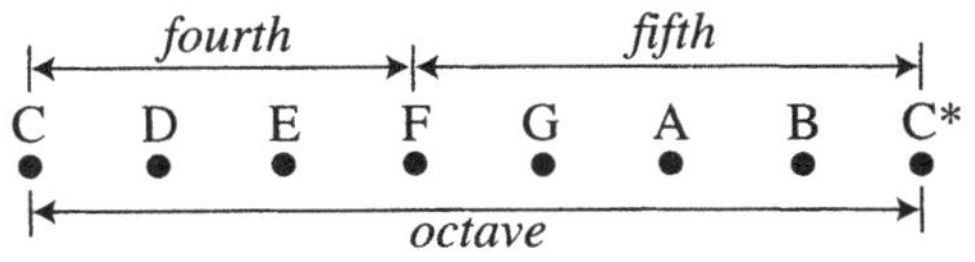

Set the note C to have a frequency of 350 **hertz** (Hz, cycles per second). We wish to find the frequencies of the other notes according to the Pythagorean scale.

(a) Find the frequencies of F, G, and C*.

(b) We know that F : C is 4 : 3 and G : C is 3 : 2. Find G : F. This ratio will be our tone.

(c) Find the frequencies of D and E so that D : C and E : D equals the tone G : F. Similarly, find the frequencies of A and B so that A : G and B : A equal the tone.

(d) Confirm that the frequency of D is the geometric mean of the frequencies of C and E and that A is the geometric mean of G and B.

9. Many believe that the proportion

$$\frac{a}{(a+b)/2} = \frac{2ab/(a+b)}{b}$$

was discovered by the Babylonians and introduced to the Greeks by Pythagoras. It has been called the "most perfect" and the "musical" proportion. Identify the two means that appear in the proportion and check that the equation is true.

10. A **geometric sequence** is a list of numbers $a_1, a_2, \ldots, a_n, \ldots$ such that for all $n > 1$, $a_n = a_1 r^{n-1}$. Each a_n is called a **term** of the sequence, and r is the **common ratio**.

(a) Write the first five terms of the sequences given by: $a_1 = 3, r = 2; a_1 = 4, r = -1; a_3 = 4, r = -1/2$.

(b) Show that $a_1 + a_2 + \cdots + a_n = \dfrac{a_1(1-r^n)}{1-r}$.

11. Let x, y, z be the first three terms of a geometric sequence. Write a, b, and c as linear functions in terms of x, y, and z:

$$a = 2x + 3y + z,$$
$$b = 2y + z,$$
$$c = y + z.$$

(a) Show that $\dfrac{a-b}{b-c} = \dfrac{a}{c}$.

(b) Find the least positive solution to the previous proportion.

12. As in Exercise 11, write a, b, and c as linear functions in the variables x, y, and z, where x, y, and z are the first three terms of a geometric sequence. Prove each of the following proportions. Also, find the least positive solution of each proportion.

(a) $\dfrac{a-b}{b-c} = \dfrac{a}{b} = \dfrac{b}{c}$

(b) $\dfrac{a-b}{b-c} = \dfrac{c}{a}$

(c) $\dfrac{a-b}{b-c} = \dfrac{b}{a}$

13. Let $\overline{AB}$ be commensurable with $\overline{CD}$ and let $\overline{A'B'}$ be commensurable with $\overline{C'D'}$. Must $\overline{EF}$ be commensurable with $\overline{GH}$ where $EF = AB + A'B'$ and $GH = CD + C'D'$? Explain.

14. Show that the sum, difference, product, and quotient of rational numbers is a rational number.

4.2 THE PYTHAGOREAN THEOREM

Not all of the Pythagoreans were interested in mathematical questions. After the death of Pythagoras, his followers split into two groups. The first, known as the ***akousmatikoi***, strove to be faithful to the rituals and prohibitions established by their master, but that is as far as they went. The second group, the ***mathêmatikoi***, followed the Pythagorean way of life as did the *akousmatikoi*, but they claimed to be truer to the master because they pursued a study of mathematics. It is because of them that it is difficult to identify whether or not Pythagoras himself was a mathematician since the early *mathêmatikoi* would attribute all of their discoveries to him (Kahn 2001, 15, 27).

One of these finds was the **Pythagorean Theorem**. Did Pythagoras discover it himself, or was it the work of one of the *mathêmatikoi*? Did one of them learn it on a journey to Babylon, or did Pythagoras one day make some patterns with stones and discover that $9 + 16 = 25$, as in Figure 4.2. After checking a few more cases, did he simply conclude that for any three integers a, b, and c,

$$a^2 + b^2 = c^2,$$

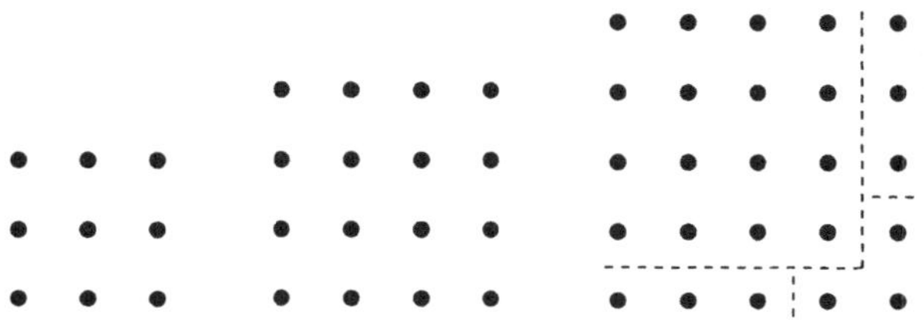

Figure 4.2 The Pythagorean triple 3-4-5 confirmed with dots.

or did Pythagoras actually give a deductive proof? The answers to these questions are unknown. It is not even known how the result came to be called the Pythagorean Theorem, unless, of course, Pythagoras did prove it (Kahn 2001, 32).

Two Proofs

It has been disputed whether the early Pythagoreans had the mathematical sophistication to prove the theorem. At best they may have been able to handle special cases, like the one involving the 45-45-90 triangle (Maor 2007, 25). To see how this would work, take $\square ABCD$ in Figure 4.3. Using diagonal $\overline{BD}$ as the hypotenuse, $\triangle ABD$ is an isosceles right triangle. On the sides of the original square, construct $\square LMBA$, $\square BEFC$, $\square DCHI$, and $\square KADJ$. These squares have the same area as $\square ABCD$. Produce $\overline{EF}$ and $\overline{IH}$ to G, forming $\square CFGH$, which also has the same area as $\square ABCD$. Since the diagonal of a square divides the square into two figures of equal area,

$$\text{area}(\square AEGI) = 2\text{area}(\square BFHD).$$

Because $\text{area}(\square AEGI) = 2[\text{area}(\square LMBA) + \text{area}(\square KADJ)]$, we may conclude that

$$\text{area}(\square LMBA) + \text{area}(\square KADJ) = \text{area}(\square BFHD).$$

We now have that

$$AB^2 + AD^2 = BD^2$$

since $\text{area}(\square LMBA) = AB^2$, $\text{area}(\square KADJ) = AD^2$, and $\text{area}(\square BFHD) = BD^2$.

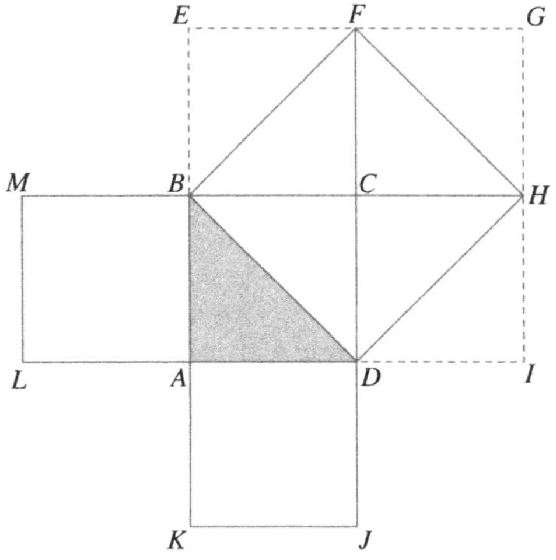

Figure 4.3 The Pythagorean Theorem for a 45-45-90 triangle.

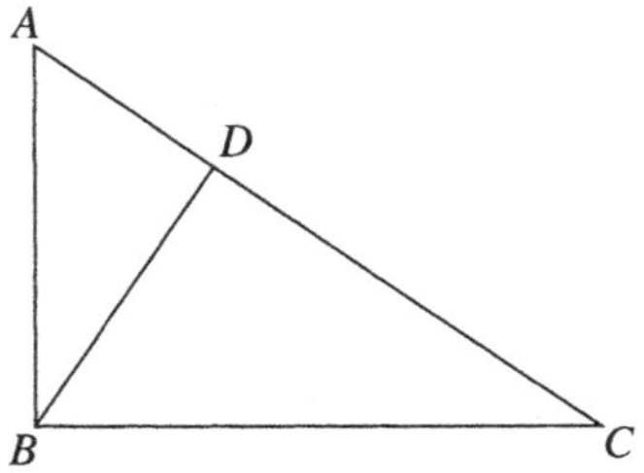

Figure 4.4 The Pythagorean Theorem proven using proportions.

If they did have the ability to prove the theorem, it is reasonable to believe that they used their theory of proportion (Heath 1921a, 148). Suppose that $\triangle ABC$ is a right triangle with $\sphericalangle B$ the right angle (Figure 4.4). Drop a perpendicular from B to the hypotenuse at D. In our diagram, all pairs of triangles are similar. In particular, since $\sphericalangle ABC \cong \sphericalangle ADB$ and $\sphericalangle BAC \cong \sphericalangle BAD$, we have $\triangle ABC \sim \triangle ADB$. Therefore,

$$\frac{AB}{AD} = \frac{AC}{AB}$$

yielding $AB^2 = AC \cdot AD$. Similarly, $\triangle ABC \sim \triangle BDC$, and we have

$$\frac{BC}{DC} = \frac{AC}{BC},$$

giving $BC^2 = AC \cdot DC$. Adding these two equations, we obtain

$$AB^2 + BC^2 = AC \cdot AD + AC \cdot DC = AC(AD + DC) = AC^2.$$

The last equality holds since $AD + DC = AC$.

The Incommensurable

It is ironic that the theorem that gave the name *Pythagorean* its fame also destroyed the Pythagorean view of number. The positive integers were supposed to be the building blocks of the universe. This is reflected in their belief that all lengths are rational. To disprove this, consider $\sqrt{2}$. We know that $\sqrt{2}$ is the length of the hypotenuse of an isosceles right triangle with legs of unit length. This follows from the Pythagorean Theorem. However, $\sqrt{2}$ is not rational. To prove this, suppose that $\sqrt{2}$ is rational. We may then write $\sqrt{2} = a/b$, where a and b are positive integers and the fraction is in lowest terms. So

$$\sqrt{2}\,b = a,$$

and then squaring both sides, we find that

$$2b^2 = a^2.$$

This means that a^2 is even which implies that a is even (Exercise 5(a) of Section 2.3). Thus, $a = 2k$ for some number k. Now substitute:

$$2b^2 = (2k)^2 = 4k^2.$$

Therefore, b^2 is even, and as before, b is even. This is impossible because both a and b are divisible by 2, but the original fraction was reduced. Hence, $\sqrt{2}$ is **irrational**. It cannot be written in the form a/b, where a and b are integers and $b \neq 0$. Furthermore, the diagonal of a square is not commensurable with its sides because if it were, $\sqrt{2}$ would be rational. When two segments are not commensurable, we say that they are **incommensurable**.

It is generally accepted that the Pythagoreans were responsible for the discovery of irrational lengths. Specifically, it is sometimes a former Pythagorean by the name of Hippasus of Croton who receives credit. He lived around the time of Philolaus and was a member of the *mathêmatikoi*. Supposedly, the Pythagoreans wanted the discovery kept private, but Hippasus made public what he had found. Not happy that their secret was leaked, the Pythagoreans, expelled Hippasus from the society. Some report that as was their custom, the Pythagoreans planted a tombstone with his name on it to indicate that they regarded Hippasus as one who had died. Others claim that the gods punished his infidelity by drowning him at sea, but another account claims that his drowning was not quite the work of the gods (Boyer and Merzbach 1991, 71–72; Kline 1972, 32).

Exercises

1. Use the two given diagrams to justify the Pythagorean Theorem.

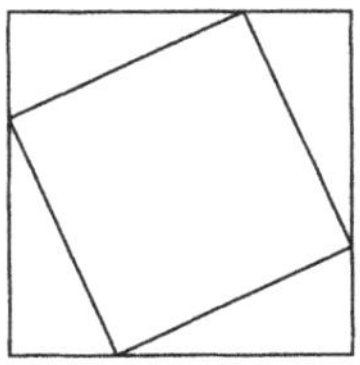

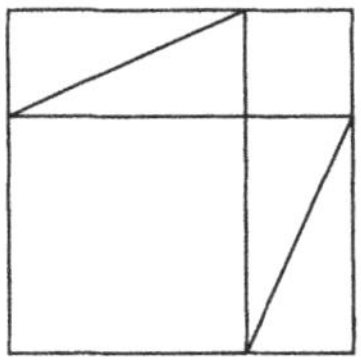

2. Let $\triangle ABC$ have a right angle at C. Label the lengths of the legs as a and b and the hypotenuse as c, as in the figure. Let D be the point such that it lies on the extension of $\overline{BC}$ and $\sphericalangle DAB$ is right.
   (a) Show that $DA = bc/a$ and $DC = b^2/a$.
   (b) Use these expressions for DA and DC to prove that $a^2 + b^2 = c^2$.

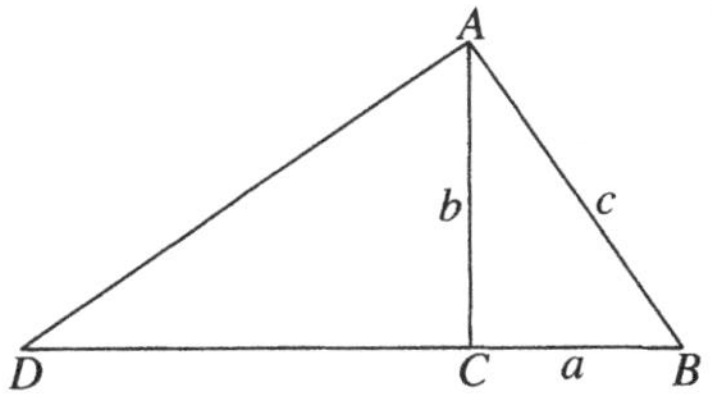

3. Use the Pythagorean Theorem to show that the distance between (x_1, y_1) and (x_2, y_2) is given by

$$\sqrt{(x_1 - x_2)^2 + (y_1 - y_2)^2},$$

while the distance between (x_1, y_1, z_1) and (x_2, y_2, z_2) is

$$\sqrt{(x_1 - x_2)^2 + (y_1 - y_2)^2 + (z_1 - z_2)^2}.$$

4. The following proof of the Pythagorean Theorem is due to James Garfield (Burton 1985, 127): Let $\triangle ABC$ be a right triangle with hypotenuse $\overline{AB}$. Let $a = CB$, $b = AC$, and $c = AB$. Suppose that $\triangle BED \cong \triangle ABC$ and join $\overline{AD}$ to form a quadrilateral. Use this to prove that $a^2 + b^2 = c^2$.

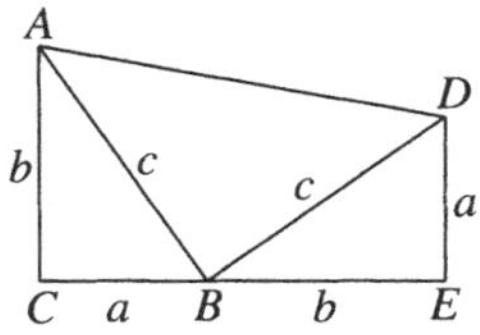

5. Show that $\sqrt{3}$ is irrational.

6. Prove:
   (a) The sum of a rational number with an irrational number is an irrational number.
   (b) The product of a rational number with an irrational number is irrational.

7. Define $\overline{AB}$ and $\overline{CD}$ to be **commensurable in square** if there exists a line segment $\overline{UV}$ such that $AB^2 = a \cdot UV$ and $CD^2 = b \cdot UV$ for positive integers a and b. Two segments are **commensurable in square only** if they are commensurable in square but not commensurable. Show that $\sqrt{3}$ and $\sqrt{5}$ are commensurable in square only.

4.3 ARCHYTAS

For seven consecutive years around 375 B.C., Archytas was the commander of the army for the city of Tarentum, a large coastal city in southern Italy. He was their leading citizen in both matters of politics and of war. He was so beloved by the townsfolk that they ignored their law that no one may serve consecutive terms. Archytas was also a Pythagorean. He believed in number and reason and applied those principles to the benefit of the people. It was said that he loved music and children. As a result, he considered education important and made sure that both mathematics and music were included among the subjects taught. In fact, it is in his *On Music*, where we have the arithmetic, harmonic, and geometric means defined, and to him is credited the curriculum known as the **quadrivium**: arithmetic, geometry, music, and astronomy (Boyer and Merzbach 1991, 28, 70–71; Heath 1921a, 21, 85, 213).

Archytas was also a long-time friend of Plato. They probably first met during the philosopher's travels in Italy (page 53), and it was this association that probably saved Plato's life. In 367 B.C. Dionysius I died and his son became king. While Dionysius II was a child, he received very little education because his father wanted to discourage any thoughts of premature ascension to the throne. At this time Dion was still an honored member of the court, and he would recall the teachings of Plato for Dionysius whenever he could, hoping that the new king would embrace philosophy. Eventually, the tyrant's interest peaked, and he sent for Plato. Needless to say, Plato did not want to return to Syracuse, but for his friend and for the opportunity to test his political philosophy, he would do so. The king warmly welcomed Plato to his court, and the studies began. Dionysius was passionate about geometry and philosophy, but politics would soon intrude. Many of the members of the king's court were fearful that if Dionysius would begin to follow Plato's teachings, their jobs would be in jeopardy. They decided to rid themselves of both Plato and Dion by starting a rumor campaign. It worked in part. The king became suspicious of Dion, so he was exiled. Dionysius still wanted Plato around but feared he would flee when he learned of Dion's fate, so he had him locked up in the palace. Soon, though, war broke out, and Dionysius sent Plato back to Athens, where he would be safe (Gottlieb 2000, 179; Plutarch 1932, *Dion*, 1159–1161; Heath 1921a, 213).

On Plato's voyage home, he located Dion, and the two returned to the Academy together. Dionysius, however, wanted his philosopher back, so he sent word to Athens that Plato should return. Plato disagreed. Dionysius was insistent and told him that Dion may return to Syracuse if Plato would come and resume his teaching. Reluctantly, Plato agreed and set sail for Sicily. For the third time Plato was warmly welcomed, and for the third time this would soon change. Dionysius recanted on his promise to let Dion return. Instead, the king sold all of Dion's property and gave his wife in marriage to another man. When Dion heard this news, he gathered forces and returned to Syracuse to battle Dionysius to regain what was his. Plato for his part escaped the island with the help of Archytas, who sent a ship for him. He would never return to Syracuse again (Kahn 2001, 39–40; Plutarch 1932, *Dion*, 1163–1165).

Not only did Plato and Archytas share some adventures in Italy, but these two friends also shared an interest in geometry. Photius writes that this began when Archytas taught Plato geometry on his first trip to Italy (Guthrie 1988, 137). In one instance it is said that they worked on the same problem. It is known as the **Delian problem**. There are various versions of the story. One begins on the island of Delos, the supposed birthplace of Apollo, when a plague had befallen the people. They approached the oracle to ascertain how they should placate the gods so that they would be saved from certain death. The oracle responded that they must direct their artisans to replace the current altar with one double in size. Since the altar was a cube, this meant that they were required to construct another cube with sides that were $\sqrt[3]{2}$ times as long as the original. They did not know how to proceed. Their only recourse was to head to Athens and search for Plato. Upon hearing their story, Plato realized that the people had misinterpreted the message behind the oracle's command. He told them that it was not the gods' intention for the people to glorify them with a larger altar; instead, the gods wanted to punish them for their disdain of geometry! Nonetheless, Plato pitied them and sought out one of his students to work on the problem (Heath 1921a, 245–246, 255).

The Delian problem is also known as **duplication of the cube**. Despite the legend, it is a natural problem for the Greeks to have considered. Since a square can be easily duplicated by constructing another square on the diagonal of the original, it is reasonable to investigate whether the same can be done with a cube. The solution to the problem was aided by a discovery of Hippocrates of Chios. Not to be confused with his namesake famous in medical circles, Hippocrates was the greatest mathematician of the fifth century B.C. His studies involved the following two definitions.

■ DEFINITION 4.3.1

Let a, b, m, and n represent positive integers.

- We say that m is the **mean proportional** between a and b if
$$\frac{a}{m} = \frac{m}{b}.$$
The term *mean proportional* is simply another name for the geometric mean (page 95).

- Call m and n **continued mean proportionals** between a and b if
$$\frac{a}{m} = \frac{m}{n} = \frac{n}{b}.$$

Hippocrates showed that the Delian problem was equivalent to finding two continued mean proportionals between two magnitudes where the greater magnitude is twice the lessor (Heath 1921a, 200, 245; Kline 1972, 38, 41). To see how this works, let a denote the length of one of the sides of the initial cube. Suppose that we can construct the two mean proportionals. Call them b and c with $b < c$. We then have

$$\frac{a}{b} = \frac{b}{c} = \frac{c}{2a}.$$

Therefore, we see that $2a^2 = bc$ and $ac = b^2$. This gives

$$2a^3 = abc = b^3.$$

A cube with sides of length b will have twice the volume of the original cube.

Archytas's Solution

Archytas showed that the continued mean proportionals between a and b can be constructed using a very clever method (Heath 1921a, 247–249; Waerden 1971, 150–151). All solutions that duplicate the cube that we will encounter are two-dimensional except for this one. Take $\overline{AB}$ and $\overline{AC}$ such that $AB = a$ and $AC = b$. Draw a circle with $\overline{AC}$ as its diameter and $\overline{AB}$ a chord. We will construct three solids that will intersect in a point, and this point will give us the solution. See Figure 4.5.

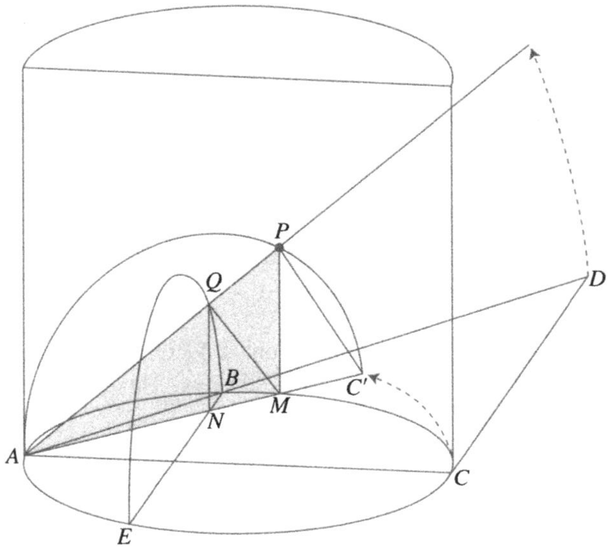

Figure 4.5 Archytas's construction of two continued mean proportionals.

- On $\overline{AC}$, construct a semicircle perpendicular to plane ABC and then rotate it about the perpendicular to the plane through A. The result is the upper half of a torus that has no hole. (A **torus** is the solid obtained by revolving a circle about an axis that does not intersect the circle in more than one point.)
- On semicircle ABC, build half a right circular cylinder high enough to intersect the torus.
- For the last, let $\overline{CD}$ be tangent to $\odot ABC$ in plane ABC and extend $\overline{AB}$ to D. Revolve $\triangle ADC$ about $\overline{AC}$ to generate a right circular cone. Observe that this causes B to trace out a semicircle with diameter $\overline{EB}$, with E on $\odot ABC$. The semicircle is perpendicular to $\odot ABC$ and $\overline{EB}$ is perpendicular to $\overline{AC}$.

Note that the cone, the torus, and the half-cylinder all meet at a point that we will label P.

Now let $\overline{AC'}$ be in the location of the diameter of the revolving semicircle when $\triangle APC'$ is perpendicular to the plane ABC. Designate the intersection of $\overline{AC'}$ and the circumference of $\odot ABC$ by M. Join $\overline{PM}$. Let $\overline{AP}$ meet the semicircle generated by B at Q, and let $\overline{AC'}$ intersect $\overline{EB}$ at N. Join $\overline{PC'}$, $\overline{QM}$, and $\overline{QN}$. Since $\overline{QN}$ is the intersection of semicircle EQB and $\triangle APM$, we conclude that $\overline{QN}$ meets $\overline{EB}$ in a right angle. We know that

$$EN \cdot NB = AN \cdot NM,$$

and

$$QN^2 = EN \cdot NB.$$

Both follow by the Chord Theorem (Theorem 5.5.3). Hence,

$$QN^2 = AN \cdot NM,$$

and since $\overline{QN}$ is perpendicular to $\overline{AM}$, we see by the Pythagorean Theorem that

$$\begin{aligned} AQ^2 + QM^2 &= AN^2 + QN^2 + QN^2 + NM^2 \\ &= AN^2 + 2(AN \cdot NM) + NM^2 \\ &= (AN + NM)^2 \\ &= AM^2. \end{aligned}$$

Therefore, $\measuredangle AQM$ is a right angle. In addition, $\measuredangle APC'$ is right for it is inscribed within the semicircle that generated the torus (Thales' Theorem, page 29). Thus, $\triangle APC'$ is similar to $\triangle AQM$. Therefore,

$$\frac{AC'}{AP} = \frac{AM}{AQ}, \tag{4.2}$$

and since $\triangle APC' \sim \triangle AMP$,

$$\frac{AC'}{AP} = \frac{AP}{AM}. \tag{4.3}$$

Because $AC' = AC$ and $AQ = AB$, we may conclude by Equations 4.2 and 4.3 that

$$\frac{AC}{AP} = \frac{AP}{AM} = \frac{AM}{AB}.$$

In other words, AP and AM are the desired mean proportionals between a and b.

A Solution Attributed to Plato

Another strategy for duplicating the cube is one that is attributed to Plato (Heath 1921a, 255–258). It is a solution that requires a device in addition to straightedge and compass. Although the device is simple, it is unlikely that Plato endorsed such a method, considering his beliefs regarding the straightedge and compass (Section 3.1). Fasten two perfectly straight sticks together to form a right angle. Call these sticks S_1 and S_2, as in Figure 4.6. Attach to S_1 another perfectly straight stick so that it may slide, yet stay parallel to S_2. To double the cube, take segments $\overline{OA}$ and $\overline{OB}$ such that $OA = 2OB$ and place them perpendicular to each other (Figure 4.7). Place the device so that B lies on side S_2. Turn it so that $\overline{OA}$ produced intersects the fixed corner. Next move the slide so that it touches A. If the $\overline{OB}$ produced intersects the vertex C, the corner where the slide is attached to S_1, the point of intersection M will yield the solution.

To prove this, suppose that the corner does not meet the $\overline{OB}$ produced; drop a perpendicular from C to R on $\overline{OA}$ extended. Let G be on the vertex of the fixed corner and extend both $\overline{OM}$ and $\overline{GC}$ so that they intersects at S. Therefore,

$$AR \cdot RG = CR^2$$

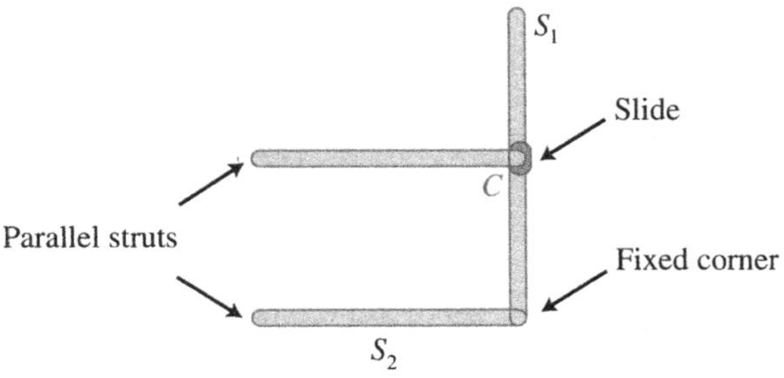

Figure 4.6 The doubling of the cube device attributed to Plato.

by similar triangles, and since $AR = OA + OR$ and $RG = OG - OR$,

$$(OA + OR)(OG - OR) = CR^2. \tag{4.4}$$

Because $\Delta CRG \sim \Delta SOG \sim \Delta GOB$,

$$\frac{CR}{RG} = \frac{SO}{OG} = \frac{OG}{OB},$$

which yields

$$\frac{CR}{OG - OR} = \frac{OG}{OB}. \tag{4.5}$$

Multiplying Equations 4.4 and 4.5 together gives

$$OB(OA + OR) = CR \cdot OG.$$

However, by Equation 4.4,

$$OG = \frac{CR^2 + OR(OA + OR)}{OA + OR},$$

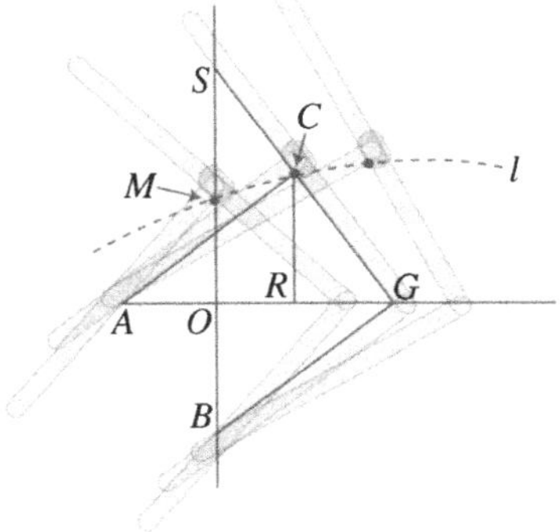

Figure 4.7 Plato's device in motion.

so substituting, we obtain

$$OB(OA + OR) = CR \cdot \frac{CR^2 + OR(OA + OR)}{OA + OR},$$

which simplifies to

$$OB(OA + OR)^2 = CR^3 + CR \cdot OR(OA + OR). \tag{4.6}$$

The goal is to continue turning the device until vertex C is on $\overline{OB}$ produced. As this is done, C traces out a path of points that is labeled l in Figure 4.7. Any such set of points that are described by a given condition the ancient Greeks called a **locus**. We see that the locus will intersect the segment, so in theory it is possible to find M. At the point in time when the device is in the desired position, perpendicular $\overline{CR}$ and $\overline{OM}$ are identical, and Equation 4.6 becomes

$$OB \cdot OA^2 = OM^3.$$

Because $OB = 2OA$, we have

$$2OA^3 = OM^3,$$

and this means that OM is the length required to duplicate the cube.

Exercises

1. We know that we can construct a segment of length $\sqrt{2}$ because it is the length of the hypothenuse of an isosceles triangle with legs of length 1.
   (a) Show that $a\sqrt{2}$ can be constructed using the Pythagorean Theorem.
   (b) Find m and n so that $a\sqrt{2}$ is the mean proportional between m and n.

2. Use three continued mean proportionals to triple a cube.

3. A **lune** is the region bounded by two intersecting circles with unequal radii. The shaded areas of the figure are lunes. Hippocrates studied these figures (Heath 1921a, 183–200).

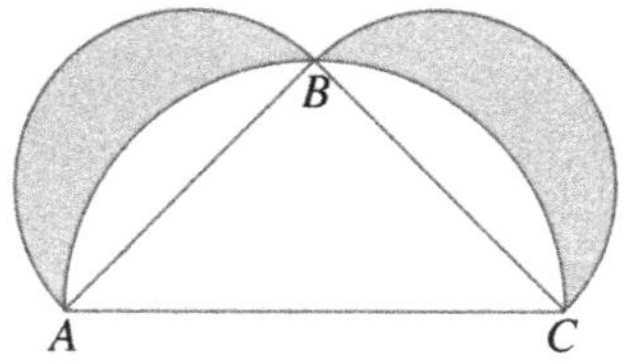

   (a) Let $\overline{AB} \cong \overline{BC}$ in $\triangle ABC$. Demonstrate that the sum of the areas of the semicircles on $\overline{AB}$ and $\overline{BC}$ equals the area of the semicircle on $\overline{AC}$.
   (b) Use the previous result to show that the total area of the two lunes equals the area of $\triangle ABC$, and then find a square that has the same area as one of the lunes.

4. Hippocrates continued his work with lunes by inscribing half a regular hexagon inside a semicircle. On each of the three sides he drew a semicircle forming three lunes, as in the diagram. Let O be the midpoint of $\overline{AD}$.

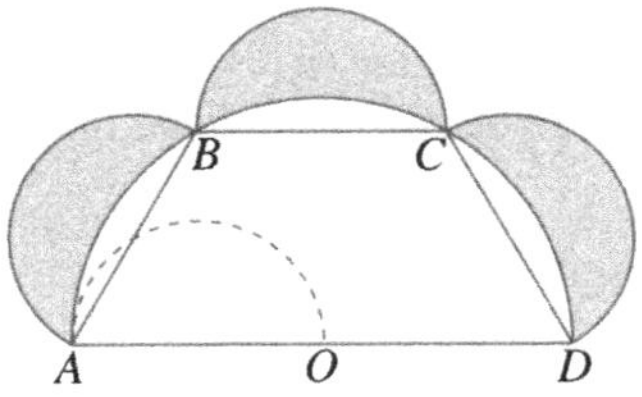

(a) Confirm that the ratio of the area of the semicircle on $\overline{AO}$ to the area of the semicircle on $\overline{AD}$ is $1/4$.
(b) Show that the area of isosceles trapezoid $ABCD$ equals the sum of the areas of the three lunes plus the area of a circle with diameter $\overline{AO}$.
(c) Suppose that for every lune there exists a square that has area equal to the given lune (as in the special case of Exercise 3). Use this assumption and the previous result to show that a rectangle can be constructed that has area equal to the semicircle on $\overline{AO}$.

5. Sketch or describe each locus in the plane.
(a) All points that are equidistant from $\overline{BA}$ and $\overline{BC}$ in $\measuredangle ABC$.
(b) All points that are equidistant from one fixed point.
(c) All points that are equidistant from two fixed points.
(d) All points that are equidistant from a fixed line and a fixed point.
(e) All points that are the vertex of the right angle of a right triangle constructed on a given hypotenuse.

6. Sketch or describe each locus in space.
(a) All points that are equidistant from one fixed point.
(b) All points that are equidistant from two parallel planes.
(c) All points that are equidistant from three noncollinear points.
(d) All points that are on both the surface of a sphere and a plane that intersects the sphere in more than one point.
(e) All points that are equidistant from the points on the surface of a sphere.

7. We are given $\overline{AB}$ and two points P and Q on A. Beginning at A, suppose that Q moves uniformly along $\overline{AB}$. At the same time P also moves uniformly so that $\overline{PQ}$ is perpendicular to $\overline{AB}$ and PQ is directly proportional to AQ. Sketch the locus of points described by P.

8. Derive the equation that describes the locus l in Figure 4.7, where O is taken to be the origin, $\overline{OG}$ the x-axis, and $\overline{OS}$ the y-axis.

9. Find the volume of the cone constructed in Archytas's solution of the Delian problem (Figure 4.5).

10. Set up three-dimensional coordinates on Figure 4.5 by choosing A to be the origin $(0, 0, 0)$, the tangent to $\odot ABC$ at A in plane ABC as the x-axis (not in the figure), $\overline{AC}$ the y-axis, and the perpendicular to the plane ABC at A as the z-axis. Find the coordinates of P.

11. Take a circle of radius r with center $(a, 0)$ with $a \geqslant r$. Rotate the circle about the y-axis to form a torus. Find both the volume and the surface area of the torus.

4.4 THE GOLDEN RATIO

Hippasus's crime (page 101) may not have directly involved the irrational. The secret that he revealed may, instead, have involved the mystical symbol of the Pythagoreans. The figure is the **regular pentagram** (Figure 4.8). It is formed by joining opposite vertices of the regular pentagon. The pentagram became the symbol for the Pythagoreans, and they used it as a means of recognizing fellow members. They named the pentagram Health (Boyer and Merzbach 1991, 72; Heath 1921a, 161).

The construction of the pentagram follows easily once the regular pentagon is constructed. To do this we begin with a definition (Euclid 1925, VI, 188).

DEFINITION 4.4.1

A straight line is said to have been **cut in extreme and mean ratio** when, as the whole line is to the greater segment, so the greater is to the less.

What this means is that if we are given $\overline{AC}$ and want to find B on $\overline{AC}$ such that it cuts the segment in extreme and mean ratio, point B must satisfy the proportion,

$$\frac{AC}{AB} = \frac{AB}{BC}. \tag{4.7}$$

This means that AB is the mean proportional (geometric mean) of AC and BC. The ratio AB/BC is known as the **golden ratio** and will be denoted by τ. This is from the Greek word $\tau o\mu\acute{\eta}$, which means "the cut" or "the section" (Livio 2002, 5).

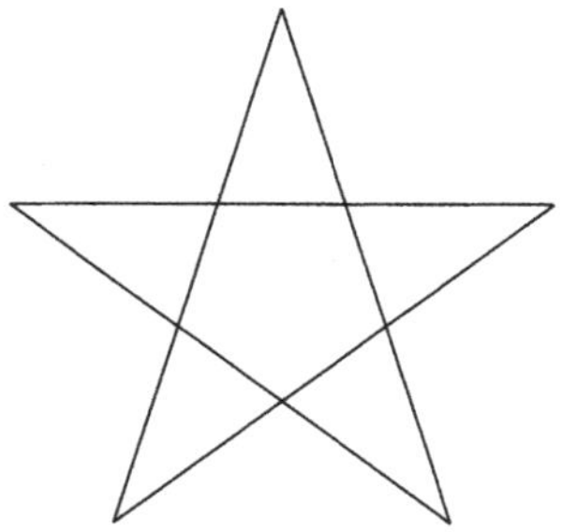

Figure 4.8 The regular pentagram, the symbol of the Pythagoreans.

To find the value of τ, let $a = AB$ and $b = BC$ and solve Equation 4.7,

$$\frac{a+b}{a} = \frac{a}{b}.$$

Since $a/b = \tau$, substitute $a = b\tau$ and write

$$\frac{b\tau + b}{b\tau} = \frac{b\tau}{b}.$$

Therefore,

$$\frac{\tau + 1}{\tau} = \tau.$$

This yields the quadratic equation

$$\tau^2 - \tau - 1 = 0, \tag{4.8}$$

which has

$$\tau = \frac{\sqrt{5}+1}{2} \approx 1.618$$

as its only positive solution. Its negative solution is $-1/\tau$, where we note that

$$\frac{1}{\tau} = \frac{\sqrt{5}-1}{2},$$

as confirmed in Exercise 1.

Now from Equation 4.8, write $\tau^2 = \tau + 1$. Therefore, for $n \geqslant 0$ we may multiply both sides of the equation by t^n to obtain

$$\tau^{n+2} = \tau^{n+1} + \tau^n.$$

We may use this last equation to generate the following sequence of equations that express powers of τ in the form $a\tau + b$:

$$\begin{aligned}
\tau^1 &= \tau, \\
\tau^2 &= \tau + 1, \\
\tau^3 &= 2\tau + 1, \\
\tau^4 &= 3\tau + 2, \\
\tau^5 &= 5\tau + 3.
\end{aligned}$$

Continuing this pattern, we see that

$$\tau^n = a_n\tau + a_{n-1}.$$

The sequence a_n is defined recursively as $a_0 = 0$, $a_1 = a_2 = 1$, and for all $n > 2$,

$$a_n = a_{n-1} + a_{n-2}. \tag{4.9}$$

This sequence is

$$1,\ 1,\ 2,\ 3,\ 5,\ 8,\ 13,\ 21,\ 34,\ \ldots,$$

where the zero has been deleted (Walser 2001, 67–70).

This list is called the **Fibonacci sequence** after the medieval mathematician Leonardo of Pisa (*c.* 1180–1250). He is sometimes called Fibonacci, meaning "son of Bonaccio." Leonardo's father was a merchant who conducted business in northern Africa. He would take his son with him on some of his trips. As a result, Leonardo learned the mathematical methods of the Muslims and traveled throughout Egypt, Syria, and Greece. No doubt this laid the foundations for his *Liber abaci* (*Book of the Abacus*). It was not a treatise on the abacus but on algebraic methods and the Hindu-Arabic numeral system. It was one of the earliest European texts that used the digits 0 through 9 to represent numbers. One of the many problems found in the *Liber abaci* is a famous one about bunnies (Boyer and Merzbach 1991, 254–256):

> How many pairs of rabbits will be produced in a year, beginning with a single pair, if in every month each pair bears a new pair which becomes productive from the second month on?

We will call a pair that has just been born a baby pair while the others are adult pairs. The population then grows according to the following table:

month	adult pairs	baby pairs
1	1	1
2	1	2
3	2	3
4	3	5
5	5	8
6	8	13

Let a_n represent the number of adult pairs during month n. During the first month there is one adult pair ($a_1 = 1$), which have one pair of offspring. In the second month there is still one adult pair of rabbits ($a_2 = 1$), which give birth to another pair. The pair born during the first month are now only a month old, so they do not reproduce. In the third month there are now two pairs of adults because the pairs of rabbits born during the first month have now grown up ($a_3 = 1 + 1$). In the fourth month there are three adult pairs, two from the previous month plus the baby pair from the second month ($a_4 = 2 + 1$). This continues so that we may conclude that the number of pairs of adults during any given month equals the number of adult pairs from the previous month plus the new adults. Since the number of new adult pairs are those born two months ago, this number equals the number of adults from that month. Hence, the current adult pairs equals the sum of the adult pairs living during the previous two months. This population grows according to Equation 4.9.

The golden ratio led to the Fibonacci sequence, and from the Fibonacci sequence we can recover the golden ratio. The first ten terms of the sequence a_{n+1}/a_n are:

$$1,\ 2,\ 1.5,\ 1.667,\ 1.6,\ 1.625,\ 1.615,\ 1.619,\ 1.618,\ 1.618.$$

We see that the ratios appear to be approaching a limit that either equals or approximates $(\sqrt{5} + 1)/2$. To find the sequence's precise limit, recall that $-1/\tau$ is a zero of

$x^2 - x - 1$, as τ is. Therefore,

$$\left(\frac{-1}{\tau}\right)^n = \frac{-a_n}{\tau} + a_{n-1}.$$

With this and the equation for τ^n, we can prove that

$$a_n = \frac{1}{\sqrt{5}}\left[\tau^n - \left(\frac{-1}{\tau}\right)^n\right]. \tag{4.10}$$

(See Exercise 3.) Since $1/\tau < 1$,

$$\lim_{n\to\infty}\left(\frac{-1}{\tau}\right)^n = 0.$$

Therefore,

$$\lim_{n\to\infty} a_n = \frac{\tau^n}{\sqrt{5}},$$

and then

$$\lim_{n\to\infty}\frac{a_n}{a_{n-1}} = \tau.$$

The rabbit problem has at once the feel of realism because it deals with populations of rabbits and the feeling of being contrived. It seems like the problem is rigged to produce the Fibonacci sequence and thus produce the golden ratio. In nature, however, the sequence and the ratio do appear. Here are just two examples from Livio (2002, 100, 111–112, 113):

- A drone bee is a male (M) that was born from an unfertilized egg, so we will say that it has only one parent. Female bees (F) have two parents because they are born from fertilized eggs. Figure 4.9 shows the family tree of a particular drone bee. The number of bees in the past generations of the drone form the Fibonacci sequence.

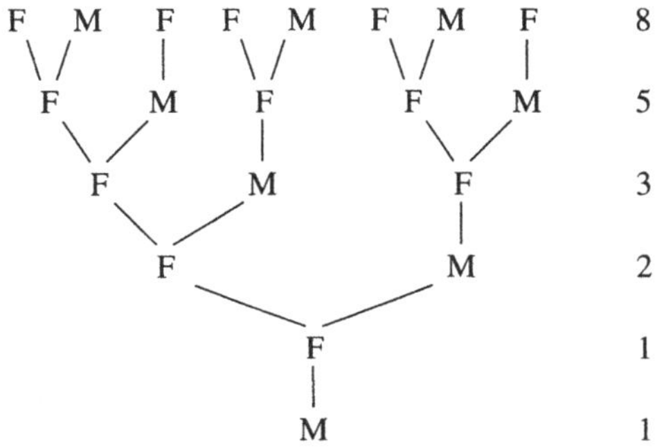

Figure 4.9 The family tree of a drone bee.

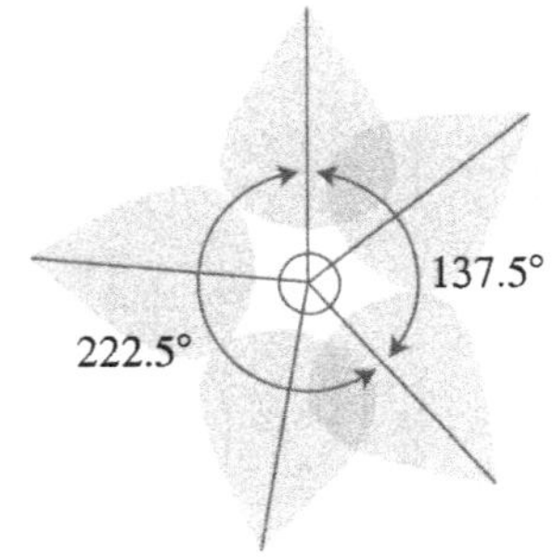

Figure 4.10 Leaves arranged around a stem.

- In 1904 the botanist A. H. Church wrote about this discovery that the leaves on a stem appear along a spiral in the order of their appearance. He found that the angle between successive leaves is approximately constant and is usually about 137.5° or 222.5° (Figure 4.10). Now for a calculation:

$$\frac{360}{\tau} \approx 222.50.$$

The Construction

The Greeks knew how to find the point that cuts a segment in extreme and mean ratio by using only a straightedge and compass. Let $\overline{AC}$ be the segment that will be cut and construct $\square ACDE$. Bisect $\overline{AE}$ at M and join $\overline{MC}$. Extend $\overline{EA}$ to F such that $FM = MC$. Draw $\square AFGB$ and extend $\overline{GB}$ to H on $\overline{ED}$ (see Figure 4.11). We claim that B cuts $\overline{AC}$ in extreme and mean ratio. To see this, first note that

$$FE = FA + 2AM.$$

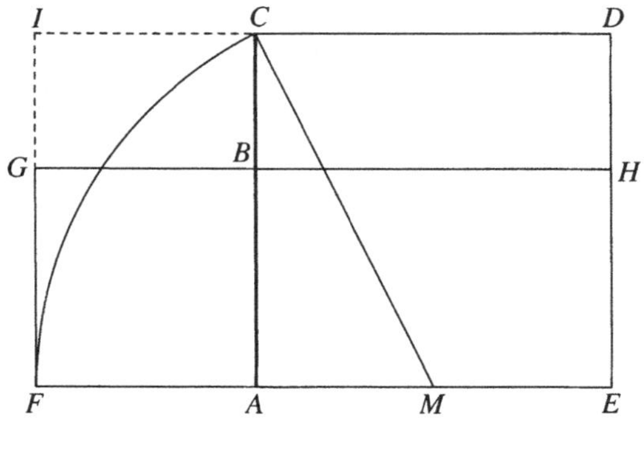

Figure 4.11 Construction of the extreme and mean ratio.

Hence,

$$\begin{aligned} FA \cdot FE + AM^2 &= FA(FA + 2AM) + AM^2 \\ &= FA^2 + 2FA \cdot AM + AM^2 \\ &= (FA + AM)^2 \\ &= FM^2. \end{aligned}$$

Since $FM = MC$,

$$FA \cdot FE + AM^2 = MC^2.$$

Because $\sphericalangle MAC$ is right, we have by the Pythagorean Theorem,

$$AC^2 + AM^2 = MC^2,$$

so

$$FA \cdot FE + AM^2 = AC^2 + AM^2,$$

and then

$$FA \cdot FE = AC^2.$$

We notice that area($\sqsubset\!\sqsupset FGHE$) $= FA \cdot FE$ since $FA = FG$, and the area of $\square ACDE$ is AC^2. Thus,

$$\text{area}(\sqsubset\!\sqsupset FGHE) = \text{area}(\square ACDE).$$

When $\sqsubset\!\sqsupset ABHE$ is taken away from both figures, the result is

$$\text{area}(\square FGBA) = \text{area}(\sqsubset\!\sqsupset BCDH).$$

In other words, $AB^2 = BC \cdot AC$, which is what we needed to prove from Equation 4.7 (Euclid 1925, II.11, VI.30).

The rectangles $FEDI$, $AEHB$, $FACI$, and $GBCI$ (Figure 4.11) all have the property that the ratio of their longer sides to their shorter ones equals the golden ratio (Exercise 8). Such a rectangle is called a **golden rectangle**. There is a long history behind this rectangle. Many claim that it was used extensively by artists and architects down through history, but Livio (2002, 9, 74, 131–135) argues that this is mostly untrue. Most of the legend can be traced to the treatise *The Divine Proportion* by Luca Pacioli. Published in Venice in 1509, it consisted of three volumes, and its subject matter was the golden ratio. The first volume was a study of the properties of the golden ratio. The second focused on proportions found in architecture and the structure of the human body. The last volume was a translation (some say was plagiarized) from the Latin of Piero della Francesca's *Five Regular Solids*. To add to the lure of the *The Divine Proportion*, it was illustrated by none other than Leonardo da Vinci. Despite Pacioli's near adoration of τ, there is no direct evidence that golden rectangles were used by the great painters of the past. Golden rectangles superimposed upon paintings of interest by later authors appear as either placed arbitrarily or have the ratio of their sides as close but not close enough

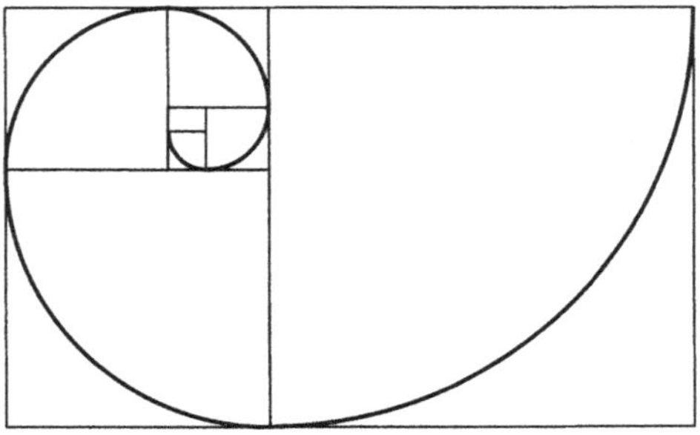

Figure 4.12 A logarithmic spiral constructed using golden rectangles.

approximations to τ. The same appears to be true for architecture. For example, the same mistake seems to be behind the claim that the golden rectangle was used to design the Parthenon. This is not to say that the golden rectangle was never deliberately used in art, but it was quite late. The first known deliberate use of the golden rectangle was not until the late nineteenth century with possibly the most famous being the *Sacrament of the Last Supper* by Salvador Dali.

Despite the lack of evidence for extensive human use, the the golden rectangle does have a connection to nature. It is found in the **logarithmic spiral** of René Descartes, which is given by the equation

$$r = ae^{b\theta}.$$

This curve is found in nature in the shell of the chambered nautilus, the horns of rams, and the tusks of elephants. It also has the property that it looks the same under all magnifications. To construct a logarithmic spiral, begin with a golden rectangle and partition off a square on the right-hand side using the construction for cutting a segment in mean and extreme ratio (Figure 4.12). The rectangle on the left-hand side is also a golden rectangle. At the bottom of this rectangle, draw a square. The rectangle at the top is a golden rectangle, and we can then continue this process for as long as we want, forming a type of spiral of squares. Starting in the upper right-hand corner of the original rectangle, join opposite corners of the first square with an appropriate curve, and then the second, and so on, so as to draw the shape of a spiral. The result is a logarithmic spiral (Livio 2002, 116–117; Burton 1985, 360).

A Regular Pentagon

We can now construct a regular pentagon. We do this by producing an isosceles triangle with a vertex of measure 72°. We can then build a regular pentagon by constructing five of these triangles around a center point. To get our first isosceles triangle, we must take $\overline{AC}$ and cut it into mean and extreme ratio with B, so we have by Equation 4.7

$$\frac{AC}{AB} = \frac{AB}{BC}.$$

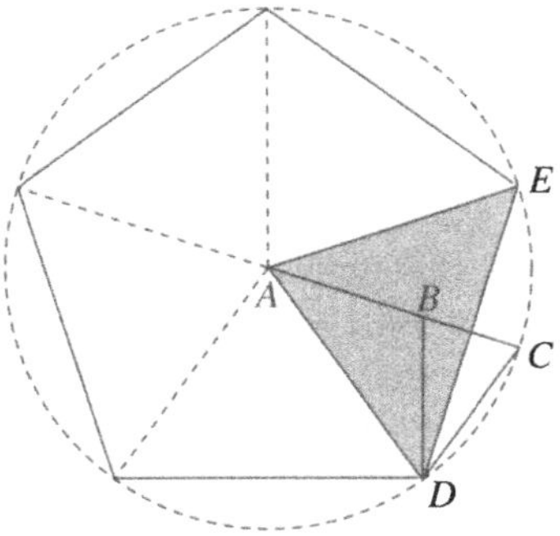

Figure 4.13 The regular pentagon.

Draw $\odot A$ with radius $\overline{AC}$ as in Figure 4.13. Find D on $\odot A$ such that $\overline{DC} \cong \overline{AB}$ and then join both $\overline{AD}$ and $\overline{BD}$. Thus,

$$\frac{AD}{DC} = \frac{DC}{BC}. \tag{4.11}$$

Because $\triangle ACD$ is isosceles, $\measuredangle ADC \cong \measuredangle DCB$. This congruence and Equation 4.11 allow us to conclude that $\triangle ADC$ and $\triangle DCB$ are similar. Since corresponding angles of similar triangles are congruent,

$$\measuredangle CDB \cong \measuredangle DAC$$

and

$$\measuredangle ACD \cong \measuredangle DBC.$$

We then see that $\triangle DCB$ is isosceles, so $\overline{BD} \cong \overline{DC}$. Therefore, $\overline{AB} \cong \overline{BD}$, which means that $\triangle ABD$ is isosceles. This leads to

$$\measuredangle BDA \cong \measuredangle DAC.$$

Hence,

$$m\measuredangle ADC = m\measuredangle BDA + m\measuredangle CDB = 2m\measuredangle DAC.$$

Therefore, since $\measuredangle ACD \cong \measuredangle ADC$ and $\measuredangle BDA \cong \measuredangle CDB$,

$$180° = m\measuredangle DAC + m\measuredangle ACD + m\measuredangle ADC = 5m\measuredangle DAC.$$

We may then calculate the measure of $\measuredangle DAC$ to be 36°. Now find E on the other side of $\overline{AC}$ such that $\measuredangle ACD \cong \measuredangle ACE$. Join $\overline{DE}$, forming $\triangle AED$. The measure of $\measuredangle DAE$ is 72°, so we may finish our construction (Burton 1985, 180-181; Euclid 1925, IV.10, 11).

Platonic Solids

When we use dialectic to come to a knowledge of the forms, we encounter the very essence of creation. In the *Timaeus* written near the end of his life, Plato hypothesized through the words of the main speaker that the world was created by a supreme being that he called the **demiurge**. This term refers to the god's action as a craftsman (Vlastos 1975, 26). Because of his goodness the demiurge decided to create the universe, and he would use the eternal forms to do it (Plato 1961, *Timaeus* 27d–30c). In the beginning he found the four elements of fire, water, earth, and air. They were in a state of chaos, and from that state the demiurge "fashioned them by form and number" (Plato 1961, *Timaeus* 53b). His actions resulted in the elements taking form as extremely small bodies, so small that they cannot be seen. Plato can only guess, but his conjecture is based on years of contemplation. Since the elements formed are bodies, they are bounded by surfaces that were probably polygons. Since these can be viewed as consisting of triangles, Plato hypothesized that the demiurge crafted the four basic elements into shapes based on the form of Triangle. He used only two kinds, and they were chosen because they were the most beautiful. The first was the isosceles right triangle and the other that "having the square of the longer side equal to three times the square of the lesser side" (Plato 1961, *Timaeus* 54c–55b). These are the 45-45-90 and 30-60-90 triangles.

The demiurge used these triangles to shape each element as a geometric solid known as **polyhedron**. These solids consist of flat faces joined together at straight edges. As with the original triangles, these should be beautiful since they were designed by the creator. Hence, each element should be regular and convex. A polyhedron is **regular** if all of the faces are regular polygons and are arranged in the same manner around each vertex. It is **convex** if the line segment joining any two points within the polyhedron is contained completely within it. A regular convex polyhedron is known as a **Platonic solid**. There are only five. They are the **tetrahedron** (four triangular faces), the cube or **hexahedron**, the **octahedron** (eight triangular faces), the **icosahedron** (20 triangular faces) in Figure 4.14, and the **dodecahedron** (12 pentagonal faces) in Figure 4.15. The ability to construct these polyhedra evidently predates Plato. Some sources state that the earlier geometers could construct all five, but others say that they only knew the first three and it was Theaetetus who first showed how to construct the more complicated dodecahedron and the icosahedron. Theaetetus was a friend of Plato who lived from 414 to 369 B.C. Plato wrote the dialogue *Theaetetus* to honor him (Boyer and Merzbach 1991, 85).

We can use the golden ration and the Pythagorean Theorem to assemble a dodecahedron. Begin with a cube with each side of unit length. Construct multiple copies of a given regular pentagon with sides of length $1/\tau$. Since for any regular pentagon,

$$\text{the length of an edge} \times \tau = \text{the length of a diagonal} \qquad (4.12)$$

by Exercise 10, the pentagon constructed has diagonals of unit length. Take two of the pentagons and place them like a roof on top of the cube so that their common edge is $\overline{HE}$ and their diagonals lie along two of the cube's edges (Figure 4.15). Drop perpendiculars from H and E that meet the surface of the cube at B and C,

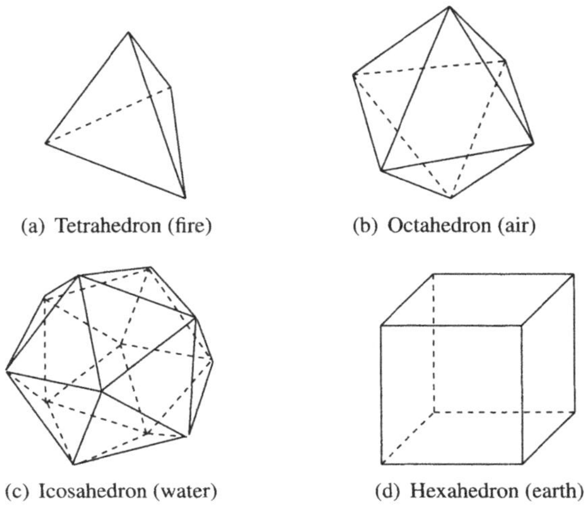

(a) Tetrahedron (fire) (b) Octahedron (air)

(c) Icosahedron (water) (d) Hexahedron (earth)

Figure 4.14 Four Platonic solids represent the four elements.

respectively. Since the pentagons are congruent, $\overline{BC}$ produced to $\overline{AD}$ divides the top side of the cube in half. Join $\overline{ED}$. Now place another congruent pentagon so that one of its vertices is at E and one side is $\overline{EI}$, with I on the corner of the cube. This pentagon fits to complete the top roof on the right-hand side of the cube. Extend $\overline{ED}$ to a point F on the side of the pentagon that is opposite E. Find the point G on the right side of the cube so that $\overline{GF}$ is perpendicular to the cube. So that we know that we may complete the construction, it must be the case that $EC = GF$ and $DG = 1/2$. Because $AD = AB + BC + CD$ and $AB = CD$,

$$CD = \frac{\tau - 1}{2\tau} = \frac{3 - \sqrt{5}}{4}.$$

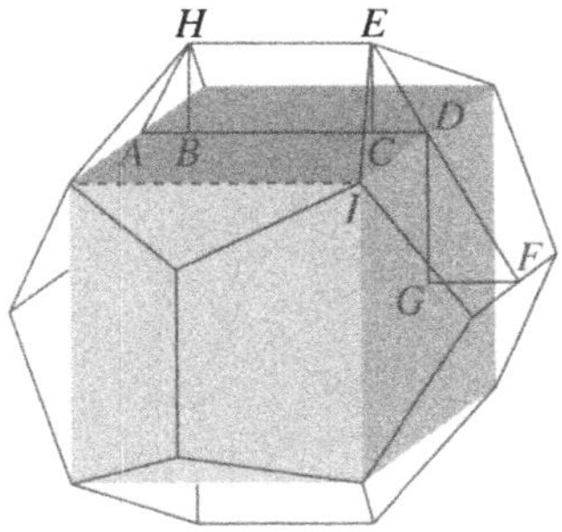

Figure 4.15 The construction of a dodecahedron.

Hence,

$$CD^2 = \frac{7 - 3\sqrt{5}}{8}.$$

By Exercise 11,

$$\frac{DF}{ED} = \tau,$$

so since $\triangle ECD \sim \triangle DGF$,

$$\frac{GF}{CD} = \tau.$$

Therefore, $GF = (\sqrt{5} - 1)/4$, so

$$GF^2 = \frac{3 - \sqrt{5}}{8}.$$

Now, because $\sphericalangle EDI$ is right and $ID = 1/2$, we have

$$ED^2 = \frac{5 - 2\sqrt{5}}{4}$$

by the Pythagorean Theorem, and then

$$EC^2 = \frac{3 - \sqrt{5}}{8}.$$

This means that the first requirement of $EC = GF$ is met. Finally, since

$$\frac{DG}{EC} = \tau,$$

we may conclude that $DG = 1/2$, and this is the second requirement.

Exercises

1. Show that $1/\tau = (\sqrt{5} - 1)/2$ and that $-1/\tau$ is a solution to $x^2 - x - 1 = 0$.
2. Let a, b, and c represent lengths in the regular pentagram below.

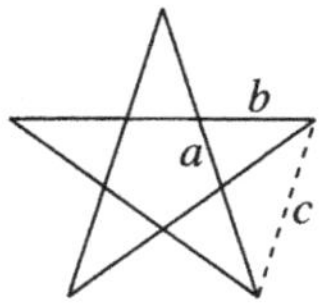

   (a) Prove that each point of the star is an isosceles triangle.
   (b) Show that $b/a = c/b = \tau$.
3. Confirm Equation 4.10.

4. Explain how finding the mean-extreme ratio is equivalent to solving the quadratic equation $x^2 + ax = a^2$.

5. Prove that $x = \tau$, where

$$x = \sqrt{1 + \sqrt{1 + \sqrt{1 + \sqrt{1 + \cdots}}}}.$$

6. Again, prove that $x = \tau$, where x is the **continued fraction**

$$x = \cfrac{1}{1 + \cfrac{1}{1 + \cfrac{1}{1 + \cfrac{1}{\ddots}}}}$$

7. Show that A cuts $\overline{FE}$ in mean and extreme ratio in Figure 4.11.

8. Show that the rectangles $FEDI$, $AEHB$, $FACI$, and $GBCI$ in Figure 4.11 are golden rectangles.

9. Given the logarithmic spiral $r = ae^{b\theta}$ in polar coordinates, let $P(r_0, \theta_0)$ be a point on the spiral and let O be the origin.

(a) Show that the distance from O to P along the curve is finite.

(b) For any other point $P'(r_1, \theta_1)$ on the curve, show that the angle made by the tangent line at P and $\overline{OP}$ is the same as the angle made by the tangent line at P' and $\overline{OP'}$.

10. Show that the product of a length of an edge of a regular pentagon and the golden ratio equals the length of one of its diagonals as noted in Equation 4.12.

11. In regular pentagon $ABCDE$, let G be the midpoint of $\overline{DC}$ and F the intersection of $\overline{AG}$ and $\overline{EB}$. Prove that $FG/AF = \tau$.

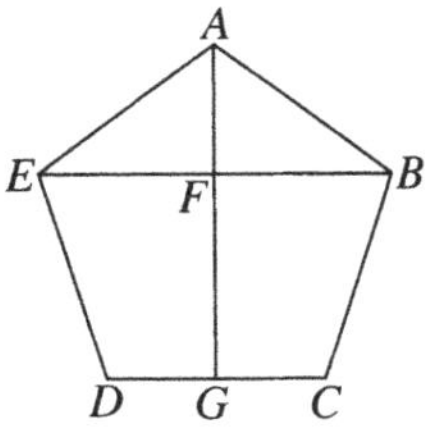

12. A **golden triangle** is an isosceles triangle such that the ratio of the length of a leg to the length of its base is τ.

(a) Construct a golden triangle.

(b) Show that the triangle formed by joining the endpoints of the sides of a regular pentagon with the opposite vertex is a golden triangle.

13. Construct a regular decagon.

14. Use the diagram below to construct a regular tetrahedron.

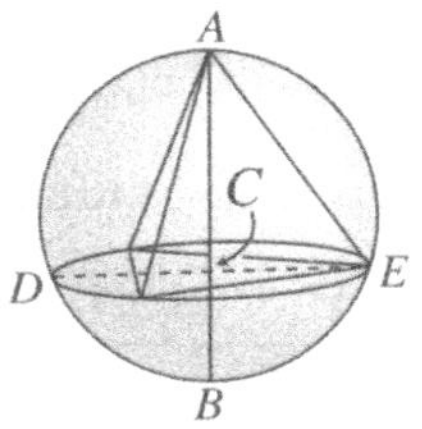

15. Confirm the equations for CD^2, GF^2, ED^2, EC^2, and DG on pages 118 and 120.

16. For a dodecahedron with unit edges, show that its volume is given by

$$V = \frac{15 + 7\sqrt{5}}{4}$$

and its surface area is given by

$$S = 3\sqrt{25 + 10\sqrt{5}}.$$

CHAPTER 5

EUCLID

There were many early Greek mathematicians and philosophers who made contributions to the foundations of geometry: Thales, Mamercus, Hippias of Elis, Pythagoras, Anaxagoras of Clazomenae, Oenopides of Chios, Hippocrates of Chios, Theodorus of Cyrene, Plato, Leodamas of Thasos, Archytas of Tarentum, Theaetetus of Athens, Neoclides, Leon, Eudoxus of Cnidus, Amyclas of Heracleia, Menaechmus, Theudius of Magnesia, Athenaeus of Cyzicus, Hermotimus of Colophon, Phillippus of Mende (Proclus 1970, 65–68). We have already encountered many of them including Hippocrates of Chios. He studied the Delian problem (page 104). He also made significant contributions to the foundations of geometry. It seems that he arrived in Athens around 450 B.C. and lived there for about 20 years. He initially traveled there to settle a lawsuit in which he attempted to recover money lost to some pirates. He was soon attracted to the Athenian philosophical community and began a study of geometry. As Thales is credited with introducing proof to geometry, Hippocrates is credited with introducing both *reductio ad absurdum* to mathematics and the ordering of theorems so that earlier results can be used in the proofs of later ones. He summarized his thought in a book he called the *Elements*. The title refers to the basic results of geometry. As matter consists of atoms, these fundamental propositions are the building blocks for all of geometry (Heath 1921a, 183; Kline 1972, 40).

5.1 THE ELEMENTS

The work of Hippocrates and the other early geometers survive only as fragments or as sources for later works, one of which has survived in its entirety to today. It was written by Euclid and like the work of Hippocrates is called the *Elements*. Euclid lived around 300 B.C. He may have been a Platonist and probably learned geometry from Plato's students in Athens. However, Euclid later took up residence across the Mediterranean in Alexandria, and this is where his influence grew. Many learned geometry from him or his students. Tradition has it that he was a kind teacher and respectful of those mathematicians who came before him. Like Plato, he regarded the study of geometry as worthy for its own sake. This is seen in a story that also shows Euclid's sharp wit. It is recounted by Stobaeus (Heath 1921a, 354, 356–357):

> Someone who had begun to read geometry with Euclid, when he had learnt the first theorem, asked Euclid, "what shall I get by learning these things?" Euclid called his slave and said, "Give him threepence, since he must make gain out of what he learns."

Euclid's *Elements* took many of the earlier works, including theorems of Eudoxus and Theaetetus, and organized them into 13 books (Proclus 1970, 67). Each was carefully structured and focused on a particular topic:

1. Triangles, angles, parallel lines, parallelograms, and the Pythagorean Theorem.
2. Rectangles and a generalization of the Pythagorean Theorem that we know as the Law of Cosines.
3. Circles.
4. Inscribing and circumscribing circles and polygons.
5. Proportions of magnitudes.
6. Similar figures.
7. Arithmetic treated geometrically.
8. More proportions.
9. Even more proportions and some number theory, including even, odd, and triangular numbers.
10. Incommensurable line segments.
11. Fundamentals of solid geometry.
12. Areas and volumes.
13. The Platonic solids.

Euclid began each book by enumerating the key definitions. He then identified the rules of the game by listing any needed postulates and definitions. Then following Hippocrates' deductive structure, Euclid began to prove theorems. From the simple to more complex, we find many familiar propositions. Some have proofs that we have seen before in high school, whereas some are surprising. When Euclid was finished, he had not written the first geometry textbook, but it was probably the best of his time because none of the others survived. Euclid is not perfect, as we shall see, but the *Elements* has served as the basis for geometric work for over 2000 years.

Definitions

The contents of the first book of the *Elements* contains many of the results that comprise an introductory geometry course. Euclid (1925, I Definitions, 153–154) begins by introducing his terminology.

A **point** is that which has no part.

Aristotle used the term *magnitude* to refer to continuous values that measure length, area, or volume (see Section 3.1). Euclid made this idea more precise by writing that a magnitude a is a **part** of b if there exists a number k such that $b = ka$ (Euclid 1925, V Definitions, 113). We may conclude that all magnitudes have parts. To see this, if the magnitude AB is represented by the line segment $\overline{AB}$, the midpoint C of $\overline{AB}$ divides the magnitude into two equal parts. We may then write $AB = 2AC$. We cannot do this with a point because it has no part, so it cannot be measured. If it cannot be measured, it has no magnitude. Therefore, a point has no length, width, weight, time, temperature. Euclid is trying to paint a picture similar to that when we say a that point has neither size nor dimension.

A **line** is a breadthless length.

A line does have a magnitude, its length. The length may be infinite or finite. If it is finite, it has **endpoints**, or **extremities** as Euclid calls them. Euclid makes a distinction between a general line and one that is straight. He says that:

A **straight line** is a line that lies evenly with the points on itself.

We will use the term **curve** for a line that is not necessarily straight and simply write *line* when a straight line is intended.

There have been many attempts to interpret what Euclid meant by his concept of straight. Proclus (1970, 109) gives an explanation consistent with the definition proposed by Archimedes:

> Euclid gives the definition of the straight line that we have set forth above, making clear by it that the straight line alone covers a distance equal to that between the points that lie on it. For the interval between any two points is the length of the line that these points define, and this is what is meant by "lying evenly with the points on itself." If you take two points on a circle or any other kind of line, the length of the line between the two points taken is greater than the

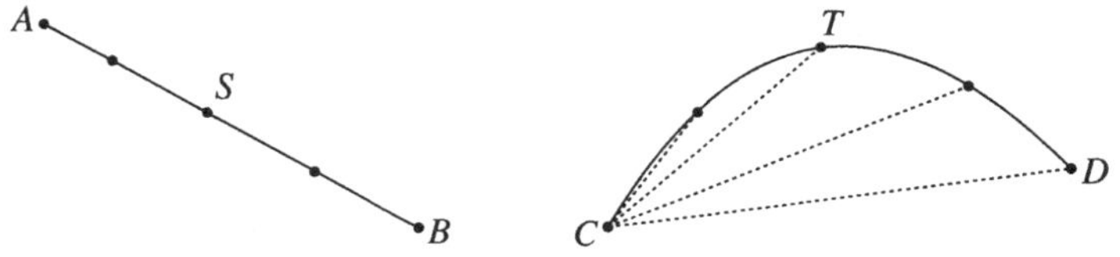

Figure 5.1 Euclid's lines, both straight and curved.

> distance between them. This seems to be a characteristic of every line except the straight. Hence it accords with a common notion that those who go in a straight line travel only the distance they need to cover, as men say, whereas those who do not go in a straight line travel farther than is necessary.

Randomly choose a point on $\overline{AB}$ and call it S (Figure 5.1). The distance between A and S is AS. If a point moves along $\overline{AB}$ from A to S, the distance traveled will be equal to AS. The conclusion is that $\overline{AB}$ is straight. On the other hand, if a point T is chosen on the curve from C to D, the distance along the curve from C to T is greater than the distance CT. This is because this line is not straight.

The key distinction to Euclid between a straight and a curved line is that the true distance between two points is found along a straight line while traveling on a curve yields a longer distance. Stated another way, the length of the straight-line segment between two points is the shortest distance between them. Given two points on any surface, a **geodesic** is the shortest path between the points. On a sphere the geodesic between two points lies along a circle on its surface that shares its center with that of the sphere. Such a circle is known as a **great circle**. On a plane the geodesic between two points is the straight-line segment connecting them.

A **surface** is that that has length and breadth only.

As with lines, a surface may be infinite or finite. If the surface is finite, the extremities of the surface are lines. A surface may also be flat or curved. Euclid then repeats his wording:

A **plane surface** is a surface which lies evenly
with the straight lines on itself.

We will modify Proclus's interpretation of what a curve is to the case of surfaces. Take surface Σ as illustrated in Figure 5.2. The distance from the line segment through A to the line segment through B along the surface is greater than the distance AB. We then conclude that Σ is curved. For the surface to be straight, it is necessary that any two such distances are equal.

Postulates and Common Notions

There is problem with all of this, however. Consider again the term *line*. If we already have an understanding of what a line is, we understand what Euclid means

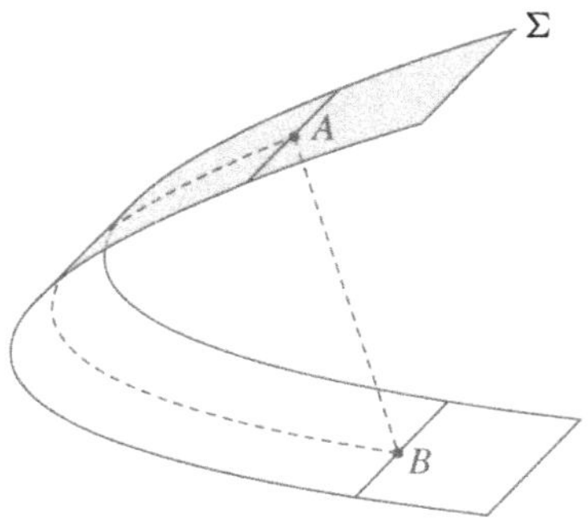

Figure 5.2 A curved surface.

by "breadthless length." This descriptions fits our mental picture of an infinitely thin pencil mark. If, on the other hand, this is our first encounter with the concept of a line, then it is also our first encounter with the combined ideas of *breadthless* and *length*. How should Euclid define these terms so that we understand? He might try *without width* for *breadthless* and tell us that *length* refers to a *straight measurement*, but attempts like these require a mental picture of a line that we do not have. His only recourse is to tell us some properties of a line and hope that we comprehend. He does this with his postulates (Euclid 1925, I Postulates, 154–155). We state them in Euclid's words as translated by Heath.

POSTULATES 5.1.1

1. To draw a straight line from any point to any point.
2. To produce a finite straight line continuously in a straight line.
3. To describe a circle with any center and distance.
4. That all right angles are equal to one another.
5. That if a straight line falling on two straight lines make the interior angles on the same side less than two right angles, the two straight lines, if produced indefinitely, meet on that side on which are the angles less than the two right angles.

Two of the postulates tell us about circles and right angles. For the moment, we will rely on our memory for the definition of these terms. We shall see Euclid's definitions in the next section. The other three postulates deal with straight lines. In Postulate 1 we learn that we can draw them between points, and Postulate 2 tells us that we can construct line segments. The fifth postulate gives the most detail about the behavior of the lines. It is known as the **Parallel Postulate**. This is a bit of a misnomer because it describes when two lines are not parallel.

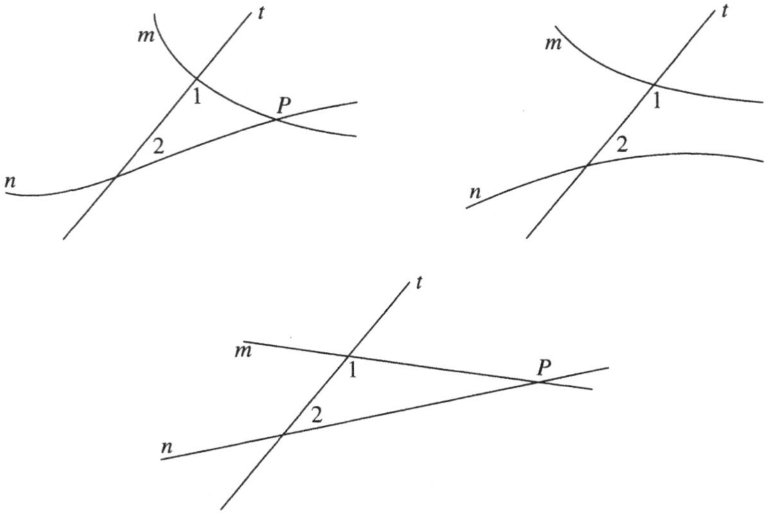

Figure 5.3 Curved and straight lines cut by a transversal.

■ DEFINITION 5.1.2

Two lines in a plane are **parallel** if they do not intersect.

For example, take two lines and label them m and n as in Figure 5.3. Draw a third line t, called a **transversal**, intersecting m and n so that the sum of the measures of $\sphericalangle 1$ and $\sphericalangle 2$ is less than $180°$. If m and n are curved, they may or may not meet each other, but if they are straight, the postulate states that they must intersect. The Parallel Postulate, more than the others, tells us how straight lines behave.

Euclid's definitions of point and line are more description than definition. They aid us in imagining what these objects are like. These sentences do nothing more. Euclid does not rely on these descriptions at any place in the *Elements*. Because of this, we may consider the words *point* and *line* as **undefined terms**. It is not that they do not have meaning. It is that their meaning is not given by definitions. Instead, their meaning is inferred from the postulates and results proven using the postulates.

Undefined terms are a necessity in any logical system. If all of the terms are defined, one of two undesirable events will occur:

- There will be an infinite sequence of unique terms with each one used to define an earlier term. For instance, suppose that we want to define term T_0. We say that T_0 is a T_1. To explain what a T_1 is, we say that it is a T_2. Continuing in this manner we have $T_0, T_1, T_2, T_3, \ldots$, all distinct. This is known as an **infinite regression**. In this case one never arrives at a final meaning of the term.

- There will be a finite sequence of terms in which there are two terms that are used to define each other. For example, suppose again that we want to define

T_0. We say that T_0 is a T_1. To explain what a T_1 is, we say that it is a T_2. Continuing in this manner, we eventually cannot think of new terms, so we say that T_n is a T_0. We have then explained T_0 using T_n and T_n using T_0. This is called a **circular definition.**

In addition to the postulates, Euclid enumerates rules that he calls **common notions**. They are assumptions about magnitudes that resemble the basic facts of algebra. As with the postulates, these are given as translated by Heath (Euclid 1925, I Common Notions, 155).

■ **COMMON NOTIONS 5.1.3**

1. Things that are equal to the same thing are also equal to one another.
2. If equals be added to equals, the wholes are equal.
3. If equals be subtracted from equals, the remainders are equal.
4. Things that coincide with one another are equal to one another.
5. The whole is greater than the part.

Exercises

1. Give examples from the physical world that would be instances of the following terms.

(a) *point*
(b) *straight line*
(c) *curve*
(d) *nonstraight geodesic*
(e) *plane surface*
(f) *nonplane surface*
(g) *parallel lines*
(h) *nonparallel lines*

2. Define the terms *triangle*, *house*, *cousin*, and *happiness*. Write each of the definitions as sentences in the form, *A triangle is. . .* .

3. Write definitions for each of the nouns in the definitions that you wrote for Exercise 2. For these new definitions, avoid including previously used nouns for as long as possible. How many sentences did you write? Were all of the definitions eventually circular? Could any of these have continued without end?

4. Is it possible to develop a sequence of definitions that would serve as an example of an infinite regress? Explain.

5. Give an example of an undefined term that you use in daily conversation.

6. Give algebraic examples of each of the Common Notions.

7. Give geometric examples of each of the Common Notions.

5.2 CONSTRUCTIONS

A **construction** is a method that demonstrates the existence of a geometric object. The postulates allow for only two different tools. A straightedge will produce representations of line segments, and a compass is for circles. Euclid allows line segments to be extended. This is possible because a line segment is a portion of a line, and Euclid assumes that lines have infinite length. He does not allow line segments to be marked off with rulers, however. If measurements must be made on a segment, the divisions should be made by a compass. Remember that in Euclid's day the compass was collapsible. This means that once a circle was drawn the compass would fold up so that another circle of the same size could not be drawn reliably. This may have simply been a matter of technology, but it was also in line with Plato's view. The ideal geometric objects do not move. Since the straightedge and compass are used to represent these objects, the constructions should also not "move."

To be precise the construction does not produce a figure. Instead, it creates the outline of a figure. The definition we give here as well as the other definitions of this section are from Euclid (1925, I Definitions, 153–154). We have previously used many of these terms based on our memory of their meanings. We now set them down as Euclid did.

■ **DEFINITION 5.2.1**

A **boundary** is that which is an extremity of anything, and a **figure** is that which is contained by any boundary or boundaries.

Modern usage views the figure with what Euclid called the boundary; and what Euclid called the figure, we usually call the **interior** of the figure. A figure lying in a plane is called a **plane figure**. The boundary of a plane figure divides the plane into two parts. The part that is not associated with the figure is called its **exterior**. We can now define the circle.

■ **DEFINITION 5.2.2**

A **circle** is a plane figure contained by one line such that all the straight lines falling upon it from one point among those lying within the figure are equal to one another, and the point is called the **center** of the circle.

A **diameter** of the circle is any straight line drawn through the center and terminated in both directions by the circumference of the circle. The boundary of a circle is called its **circumference**. A segment joining the center of the circle to a point on its circumference is called a **radius**.

Unlike circles, there are some figures that have corners. The next definition is an attempt to give a name to these corners. Notice that the definition allows for angles to consist of lines that are not straight, as in Figure 5.3. This is important, for it will allow us to study angles on surfaces that are not flat, such as a sphere.

■ **DEFINITION 5.2.3**

A **plane angle** is the inclination to one another of two lines in a plane that meet one another and do not lie in a straight line. When the lines containing the angles are straight, the angle is called **rectilineal**.

Notice that the definition excludes the idea of a straight angle. In other words, if A, B, and C all lie on the same line, we may say that they are **collinear**, but they do not form an angle.

Let $\overline{AB}$ and $\overline{BC}$ be two line segments that meet at B. They form an angle that we will label $\sphericalangle ABC$. The point B is called the **vertex** of the angle. The angle opens at a particular rate that describes the angle uniquely. This rate is called the **measure** of the angle and is denoted by $\mathrm{m}\sphericalangle ABC$. This measure is a magnitude. Today we measure angles using radians or degrees. Euclid had no such thing. If two angles had the same measure, Euclid would simply say that the were "equal" and not assign a number. He would add and subtract them as if they were numbers, but he would never assign numbers to the results. If two angles have the same measure, they look identical. If they look the same, we say that they are **congruent**. If we need to make calculations based on angles, we will use their measures. Otherwise, we use the concept of congruence. In summary, we may write

$$\sphericalangle ABC \cong \sphericalangle DEF \text{ if and only if } \mathrm{m}\sphericalangle ABC = \mathrm{m}\sphericalangle DEF.$$

We are now ready to define Euclid's fundamental angle. Notice that since Euclid did not have degrees or radians, $\pi/2$ and $180°$ are not part of the definition.

■ **DEFINITION 5.2.4**

When a straight line is set up on a straight line that makes the adjacent angles equal to one another, each of the equal angles is **right**, and the straight line standing on the other is called a **perpendicular** to that on which it stands.

If $\overline{AB}$ and $\overline{CD}$ meet at E to form right angles, we may conclude that

$$\sphericalangle AEC \cong \sphericalangle CEB \cong \sphericalangle BED \cong \sphericalangle DEA.$$

Here are two other types of angles besides the right angle. Although he never assigns a numerical value to these measures, Euclid uses the measures of the angles to compare them to each other.

■ **DEFINITION 5.2.5**

- An $\sphericalangle RST$ whose measure is greater than a right angle is called **obtuse**. This means that there exists $\overline{SU}$ in the interior of $\sphericalangle RST$ such that $\sphericalangle UST$ is right.

- An $\sphericalangle RST$ whose measure is less than a right angle is called **acute**, so there exists $\overline{SU}$ in the exterior of $\sphericalangle RST$ so that $\sphericalangle UST$ is right.

Suppose that $\angle AEC$ is right. Using $m\angle RST$ to represent the measure of $\angle RST$, if $\angle RST$ is obtuse, we write $m\angle RST > m\angle AEC$, and if $\angle RST$ is acute, we write $m\angle RST < m\angle AEC$.

Angles and Segments

We are finally ready for our first construction. These constructions and the remaining results of the chapter are from Euclid (1925).

■ **CONSTRUCTION 5.2.6 [*Elements* I.1]**

Construct an equilateral triangle on a given line segment.

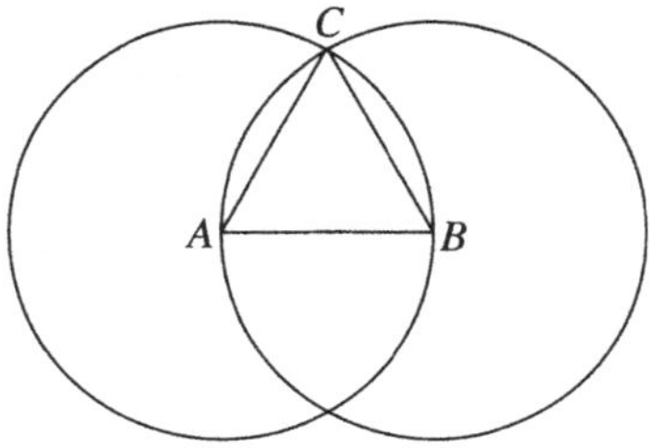

PROOF

We are given $\overline{AB}$. By Postulate 3 construct two circles: one with center A and radius $\overline{AB}$ and the other with center B and the same radius. Let the circles meet at point C. Since $\overline{AC}$ and $\overline{AB}$ are radii of $\odot A$, $\overline{AC} \cong \overline{AB}$, and since $\overline{BC}$ and $\overline{AB}$ are radii of $\odot B$, $\overline{BC} \cong \overline{AB}$. Common Notion 1 guarantees that $\overline{AC} \cong \overline{BC}$. Hence, $\triangle ABC$ is equilateral. ■

Often it is clear that Euclid leaves out details in his proofs. He probably expected his readers, those who were learning geometry for the first time, to prove these missing steps for themselves. What about in the previous proof? Although we have little doubt that working with a compass will yield the equilateral triangle, how do we know that the circles must meet at point C? Our first instinct may be to write the equations for the two circles and then find their points of intersection algebraically. The problem with this solution is that it was not available to Euclid. The solution should be that we introduce a new postulate that allows us to conclude that a point of intersection exists for the two circles. Unfortunately, Euclid has no such postulate, even though he assumes that

lines are continuous

and

if lines look like they intersect, they do.

Not including such unstated assumptions among the postulates is a logical error in Euclid's system.

The next theorem states that it is possible to copy a line segment using only a straightedge and (collapsible) compass. Therefore, in practice we can cheat a little by copying line segments by using our compass as a ruler. Although we will neither be true to Plato nor to Euclid's postulates, we can be confident that our result is provable using only his tools.

■ CONSTRUCTION 5.2.7 [*Elements* I.2]

Using a given point as one of its endpoints, construct a line segment that is congruent to a given line segment.

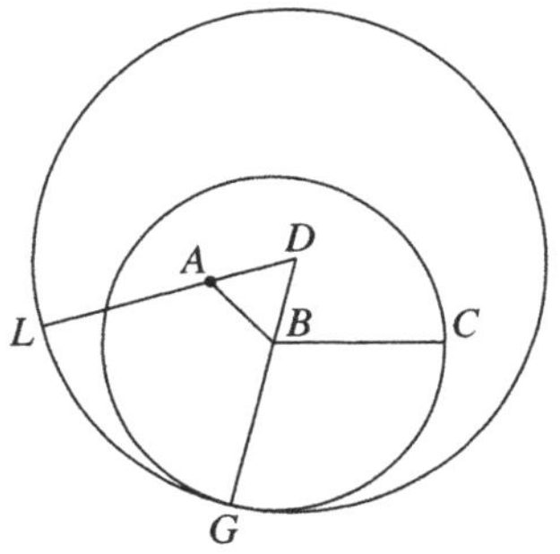

PROOF

Let $\overline{BC}$ be the line segment that we want to copy at A. Construct equilateral $\triangle ABD$ (Construction 5.2.6) and then construct $\odot B$ with radius $\overline{BC}$ by Postulate 3. Extend $\overline{BD}$ to intersect $\odot B$ at G as in the diagram. Next, construct $\odot D$ with radius $\overline{GD}$. Extend $\overline{AD}$ to intersect this circle at L. We claim that $\overline{LA}$ is our desired line segment. To prove this, first note that $\overline{LD} \cong \overline{GD}$ since they are both radii of $\odot D$. Because $LD = LA + AD$ and $GD = GB + BD$ and $AD = BD$, we conclude that $LA = GB$ because of Common Notion 3. However, $GB = BC$ because they are both radii of the same circle, so $LA = BC$ by Common Notion 1. Thus, $\overline{LA} \cong \overline{BC}$. ■

Our next construction requires the use of a definition within its proof. Take two triangles and label their vertices as in Figure 5.4. We will designate these two triangles $\triangle ABC$ and $\triangle DEF$.

■ DEFINITION 5.2.8

We say that $\triangle ABC$ and $\triangle DEF$ are **congruent** and write $\triangle ABC \cong \triangle DEF$ if and only if

- $\sphericalangle ABC \cong \sphericalangle DEF$.
- $\sphericalangle BCA \cong \sphericalangle EFD$.
- $\sphericalangle CAB \cong \sphericalangle FDE$.
- $\overline{AB} \cong \overline{DE}$.
- $\overline{BC} \cong \overline{EF}$.
- $\overline{AC} \cong \overline{DF}$.

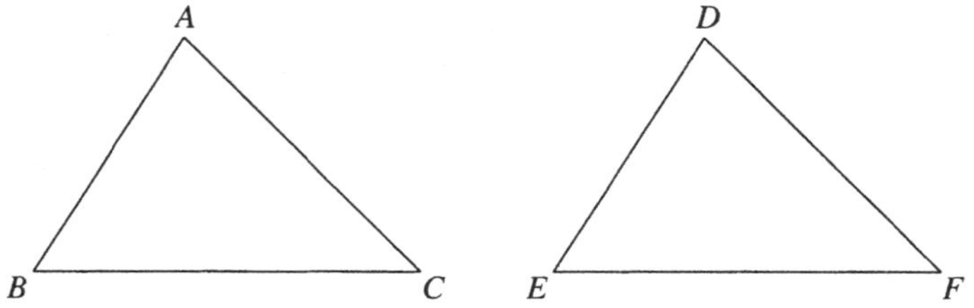

Figure 5.4 Congruent triangles.

The idea behind the definition is that triangles that are congruent look the same. We have four tests to determine whether two triangles are congruent:

SAS, SSS, ASA, AAS.

The statements and proofs for these theorems are Propositions 5.3.2, 5.3.5, and 5.3.6.

The **angle bisector** of $\sphericalangle ABC$ is a line segment $\overline{BF}$ such that $\sphericalangle ABF \cong \sphericalangle FBC$. The next result proves that we can construct these bisectors. The proof requires SSS.

■ **CONSTRUCTION 5.2.9 [*Elements* I.9]**

Construct the bisector of an angle.

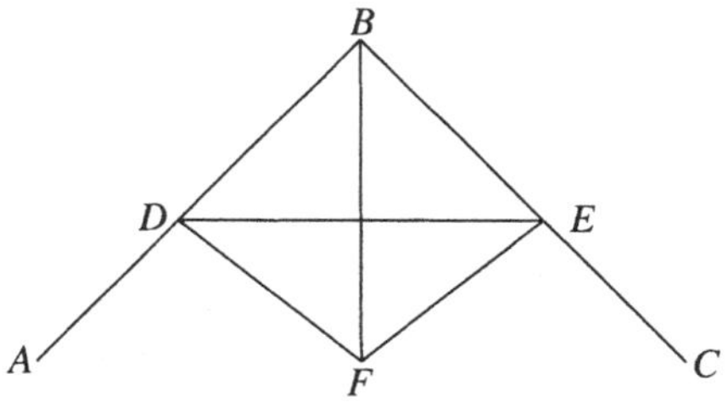

PROOF

We are given $\sphericalangle ABC$. Randomly pick D on $\overline{AB}$ and then take E on $\overline{BC}$ such that $\overline{DB} \cong \overline{BE}$ (Construction 5.2.7). Join $\overline{DE}$ by Postulate 1. By Construction 5.2.6 we may draw equilateral $\triangle DEF$. We assume that F is on the opposite side of $\overline{DE}$ than B. Join $\overline{BF}$. Since $\overline{BF} \cong \overline{BF}$ and $\overline{DF} \cong \overline{EF}$, we have by SSS,

$$\triangle BDF \cong \triangle BEF.$$

Hence, $\sphericalangle ABF \cong \sphericalangle CBF$, and the angle is bisected. ■

The **segment bisector** of $\overline{AB}$ is another $\overline{CD}$ that intersects $\overline{AB}$ at a point M such that $\overline{AM} \cong \overline{MB}$. The point M is called the **midpoint** of $\overline{AB}$. In a circle, the midpoint of a diameter is the center of the circle.

■ **CONSTRUCTION 5.2.10 [*Elements* I.10]**

Construct the bisector of a line segment.

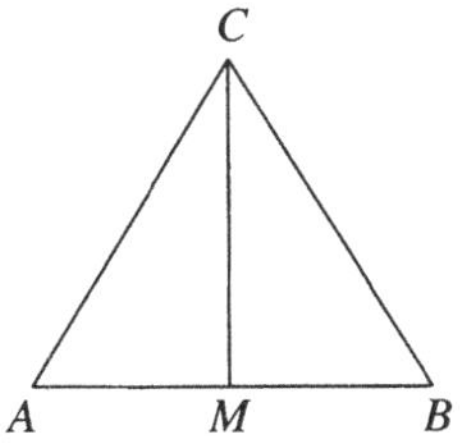

PROOF

Let $\overline{AB}$ be the line segment that we want to bisect. Construct equilateral $\triangle ABC$ (Construction 5.2.6). Bisect $\sphericalangle ACB$ with $\overline{CM}$ such that M is on $\overline{AB}$ (Construction 5.2.9). Now, $\overline{AC}$ is congruent to $\overline{BC}$, $\overline{CM}$ is congruent to $\overline{CM}$, and $\sphericalangle ACM \cong \sphericalangle BCM$, so

$$\triangle ACM \cong \triangle BCM$$

by SAS, and we have $\overline{AM} \cong \overline{BM}$. ■

In *Elements* I.11 and I.12, Euclid proves the basic constructions regarding perpendicular lines. The second result shows how to construct a perpendicular to a line segment through a point not on the line (Exercise 4). The next construction is the first one.

■ **CONSTRUCTION 5.2.11 [*Elements* I.11]**

Construct a perpendicular to a line segment through a point on the line segment.

PROOF

Let C be a point on $\overline{AB}$ and randomly pick D on $\overline{AC}$ different from C. Choose E on $\overline{BC}$ so that $\overline{CD} \cong \overline{CE}$. We may now construct equilateral $\triangle FDE$ and join $\overline{FC}$. Because $\overline{CF}$ is congruent to itself and $\overline{DF} \cong \overline{FE}$, by SSS

$$\triangle DFC \cong \triangle EFC.$$

Therefore, $\sphericalangle DCF \cong \sphericalangle ECF$. Since these two angles are adjacent, $\overline{CF}$ is perpendicular to $\overline{AB}$. ■

CONSTRUCTION 5.2.12 [*Elements* I.12]

Construct a perpendicular to a line segment through a point not on the line segment.

The next construction follows the proof on the Triangle Inequality in the first book of the *Elements*. The Triangle Inequality states that if a, b, and c are sides of a triangle, then $a + b > c$. The Epicureans liked to mock anyone who thought that this theorem required proof because even a donkey will attest to its truth. Place some donkey treats at one end of a corral and a donkey at the other end. Given the option between a straight path to the food and any other, the animal will choose the direct route because it is the shortest. Proclus rebutted the Epicureans by observing that the perceived knowledge of a truth is different from its demonstration (Euclid 1925, vol. 1, 287).

CONSTRUCTION 5.2.13 [*Elements* I.22]

Construct a triangle using line segments that are congruent to three given line segments. The sum of the lengths of any two of the given line segments must be greater than the length of the third.

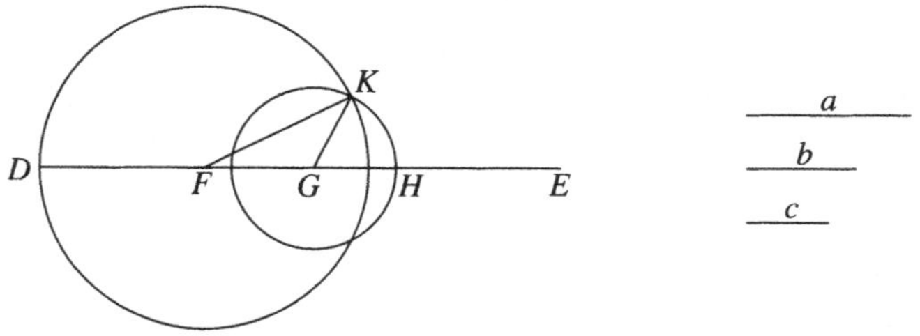

PROOF

Take three line segments of lengths a, b, and c. We will assume that

$$a + b > c, a + c > b, \text{ and } b + c > a.$$

so they can form a triangle. Take $\overline{DE}$ and find F, G, and H on $\overline{DE}$ such that

$$DF = a, FG = b, \text{ and } GH = c.$$

Construct two circles: $\odot F$ with radius $\overline{FD}$ and $\odot G$ with radius $\overline{GH}$. Let K represent the intersection of the circles and join both $\overline{FK}$ and $\overline{GK}$. Since $\overline{FK}$ and $\overline{DF}$ are radii of the same circle, $\overline{FK}$ is congruent to $\overline{DF}$. Similarly, $\overline{GK}$ is congruent to $\overline{GH}$. Therefore, the sides of $\triangle FGK$ are congruent to the three original line segments. ■

We close the section with another familiar construction.

CONSTRUCTION 5.2.14 [*Elements* I.23]

Construct an angle congruent to a given angle on a given point and line segment.

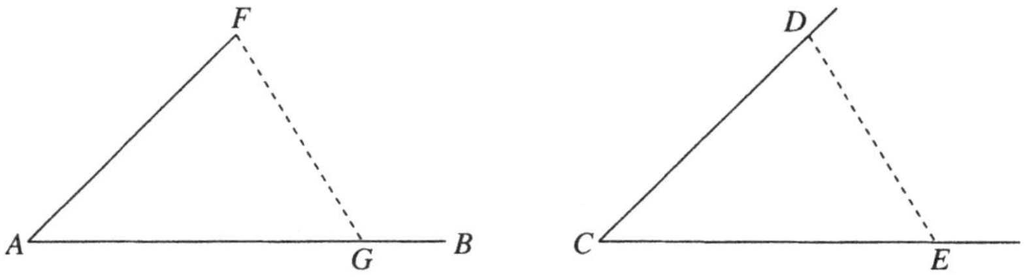

PROOF

We are given $\overline{AB}$ and will copy a given $\measuredangle DCE$ on it at A. Join $\overline{DE}$. Use Construction 5.2.13 to draw $\triangle AFG$ such that $\overline{AG} \cong \overline{CE}$, $\overline{AF} \cong \overline{CD}$, and $\overline{FG} \cong \overline{DE}$. By definition of congruent triangles we have $\measuredangle GAF \cong \measuredangle ECD$. ■

Exercises

1. Show that angle congruence is an equivalence relation. (See Exercise 12 of Section 2.3 for the definition of an equivalence relation.)

2. Given two unequal line segments, construct a copy of the shorter on the longer.

3. Draw three line segments of different length. Let a be the length of the first segment, b the length of the second (with $a > b$), and c the length of the third. Construct segments of lengths $2a$, $b + c$, $a - b$, $c/2$, $c/4$.

4. Prove Construction 5.2.12.

5. Let $\measuredangle A$ be an acute angle. Construct its complement and its supplement. (Recall that two angles are **complementary** if their measures add to a right angle and they are **supplementary** if their measures add to two right angles.)

6. Given $\measuredangle A$ and $\measuredangle B$ such that $m\measuredangle A > m\measuredangle B$, construct an angle with measure:

   (a) $2m\measuredangle A$
   (b) $m\measuredangle A + m\measuredangle B$.
   (c) $m\measuredangle A - m\measuredangle B$.

7. Construct:
   (a) A 45-45-90 triangle
   (b) An isosceles triangle with all of its angles acute
   (c) An isosceles triangle with an obtuse angle
   (d) An angle that measures 22.5°

8. Given $\triangle ABC$, copy its three sides so as to construct $\triangle A'B'C'$ so that $\triangle ABC$ is congruent to $\triangle A'B'C'$. Once completed, copy $\triangle ABC$ three more times by using SAS, ASA, and AAS instead of SSS.

9. Find $\sqrt{2}$ on a number line.

10. What else can be said of the bisector found in Construction 5.2.10.

5.3 TRIANGLES

Although Euclid's *Elements* is a survey of the geometric results of his time, it does contain some original mathematics. This includes the proof of the Pythagorean Theorem that appears at the end of Book I. It is our goal to reproduce that proof. This will require theorems about triangles, angles, and parallelograms, many of which will be familiar. We begin with triangles. We examine how Euclid proved the triangle congruence theorems and then look at the Triangle Inequality.

We start with a definition.

■ **DEFINITION 5.3.1**

Rectilineal figures are plane figures contained by straight lines, **trilateral** figures are those contained by three, and **quadrilateral** figures are contained by four.

A trilateral figure is also known as a **triangle**, and another term for **multilateral** is **polygon**. Euclid distinguishes four types of triangles. An **equilateral triangle** has three congruent sides. An **isosceles triangle** has exactly two congruent sides. We shall call its congruent sides the **legs** of the isosceles triangle, and the third side will be its **base**. A **scalene triangle** has "its three sides unequal." Finally, a **right triangle** is a triangle in which two of its sides form a right angle.

Congruence

Euclid attempts to prove this first proposition about congruent triangles by superimposing one triangle over another. Common Notion 4 does allow us to conclude that two such triangles are congruent if they coincide, but Euclid has no postulate which guarantees that the two seemingly congruent triangles will coincide when one is moved to the same location as the other. For the proof to work it is important that the triangle will not be distorted as it changes position. Euclid used this assumption without first identifying it. Therefore, we shall simply state the theorem without proof, essentially treating it like a postulate. (See Exercise 6 for an outline of Euclid's attempted proof.)

■ **THEOREM 5.3.2 [*Elements* I.4, SAS]**

If two triangles have the two corresponding sides congruent and have the angles contained by those two sides congruent, the triangles are congruent.

Our first triangle theorem that we will prove has been credited to Thales. It uses the concept of **angle addition**. Take two consecutive angles, $\sphericalangle ABD$ and $\sphericalangle DBC$. We have

$$\mathrm{m}\sphericalangle ABC = \mathrm{m}\sphericalangle ABD + \mathrm{m}\sphericalangle DBC.$$

We also have **angle subtraction**,

$$\mathrm{m}\sphericalangle ABD = \mathrm{m}\sphericalangle ABC - \mathrm{m}\sphericalangle DBC.$$

Moreover, the proof requires **segment addition** and **segment subtraction**. Given a segment $\overline{AC}$ containing point B, we have

$$AC = AB + BC$$

and

$$AB = AC - BC.$$

Euclid views angle addition as the joining of two consecutive angles forming a new one, and angle subtraction reverses this process. Similarly, segment addition is viewed as concatenation of line segments, and segment subtraction removes a portion of a line segment from a larger line segment. Euclid does not explicitly state these properties of addition or subtraction, yet he uses them frequently.

THEOREM 5.3.3 [*Elements* I.5]

In an isosceles triangle, the base angles are congruent, and the angles under the base are congruent.

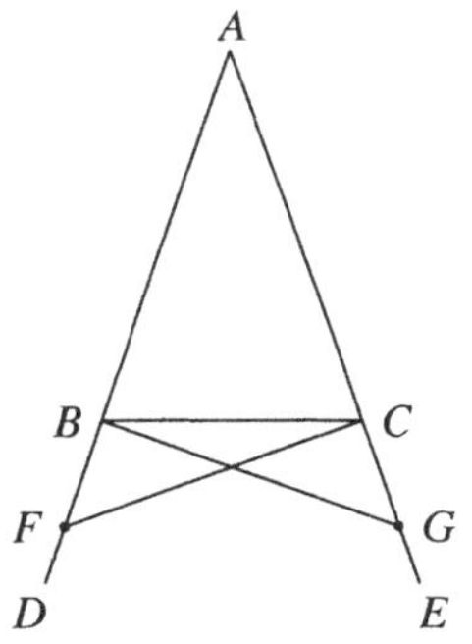

PROOF

Let $\triangle ABC$ be isosceles with $\overline{AB} \cong \overline{AC}$. Extend $\overline{AB}$ to the point D and $\overline{AC}$ to the point E. We must show that

$$\measuredangle ABC \cong \measuredangle ACB$$

and

$$\measuredangle CBD \cong \measuredangle BCE.$$

Randomly choose a point F on $\overline{BD}$. On $\overline{CE}$ find the point G such that $\overline{AG} \cong \overline{AF}$ (Construction 5.2.7). Join F and C to form $\overline{FC}$ and similarly form $\overline{BG}$. Then since $\measuredangle FAC \cong \measuredangle GAB$,

$$\triangle AFC \cong \triangle AGB$$

by SAS. Now, $\overline{FC} \cong \overline{BG}$ and $\measuredangle BFC \cong \measuredangle CGB$. Because $BF = AF - AB$ and $CG = AG - AC$, we conclude that $\overline{BF} \cong \overline{CG}$. Again by SAS we have

$$\triangle BFC \cong \triangle CGB.$$

Therefore, the angles under the base, namely $\sphericalangle FBC$ and $\sphericalangle GCB$, are congruent. Furthermore, since we have

$$\mathrm{m}\sphericalangle ABC = \mathrm{m}\sphericalangle ABG - \mathrm{m}\sphericalangle CBG$$

and

$$\mathrm{m}\sphericalangle ACB = \mathrm{m}\sphericalangle ACF - \mathrm{m}\sphericalangle BCF,$$

we may conclude that $\sphericalangle ABC \cong \sphericalangle ACB$. ■

The proof of the next theorem actually involves a few cases depending on the orientation of C compared to D. As is typical for the proofs found in the *Elements*, Euclid gives only one. It also requires a postulate which guarantees that the angles fall as indicated in the diagram.

■ **THEOREM 5.3.4 [*Elements* I.7]**

Given a line segment $\overline{AB}$, construct a triangle with $\overline{AB}$ as one of its sides. It is impossible to construct a second triangle on the same side of $\overline{AB}$ as the first that is congruent yet different from the first.

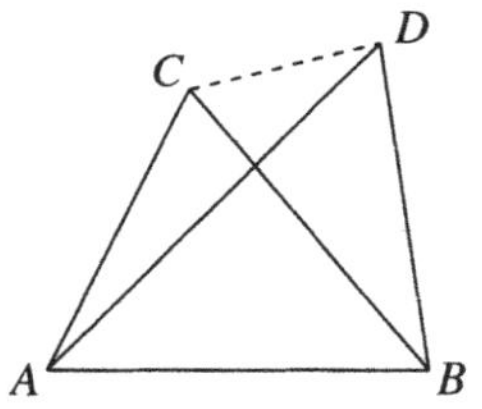

PROOF

Take C and construct $\triangle ABC$. On the same side of $\overline{AB}$ as C, suppose that there exists D such that $C \neq D$, yet $\overline{AC} \cong \overline{AD}$ and $\overline{BC} \cong \overline{BD}$. Join C and D. Since $\triangle ACD$ is isosceles, we have $\sphericalangle ADC \cong \sphericalangle ACD$ by Theorem 5.3.3. Therefore,

$$\mathrm{m}\sphericalangle CDB > \mathrm{m}\sphericalangle ADC = \mathrm{m}\sphericalangle ACD > \mathrm{m}\sphericalangle DCB.$$

However, $\triangle BCD$ is also isosceles, so $\sphericalangle CDB \cong \sphericalangle DCB$ in contradiction to the inequality above. Therefore, C and D are the same point, and the triangles are the same. ■

The next theorem is our second dealing with triangle congruence. Its proof requires SAS.

■ **THEOREM 5.3.5 [*Elements* I.8, SSS]**

If two triangles have corresponding sides congruent, the triangles are congruent.

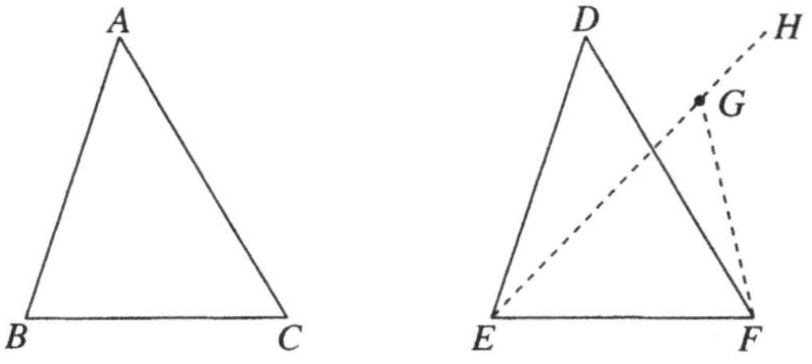

PROOF

Let $\overline{AB} \cong \overline{DE}$, $\overline{BC} \cong \overline{EF}$, and $\overline{AC} \cong \overline{DF}$ in $\triangle ABC$ and $\triangle DEF$. It suffices to show that $\measuredangle B \cong \measuredangle E$, for then by SAS the triangles will be congruent. Euclid proceeds with his argument by moving the first triangle on top of the second. Since we are not willing to assume rigid motion in the plane, we proceed along other lines, yet remain true to the spirit of the proof. Suppose that $\measuredangle B$ is not congruent to $\measuredangle E$. Then, either $m\measuredangle B > m\measuredangle E$ or $m\measuredangle B < m\measuredangle E$. Suppose the latter. This means that there exists H such that $\overline{EH}$ is in the interior of $\measuredangle DEF$ and $\measuredangle HEF$ is congruent to $\measuredangle B$. Find G on $\overline{EH}$ such that $\overline{AB} \cong \overline{EG}$, so $\overline{DE} \cong \overline{EG}$. Join $\overline{GF}$. Then $\triangle ABC \cong \triangle GEF$ by SAS. Thus, $\overline{AC} \cong \overline{GF}$, which means that $\overline{DF} \cong \overline{GF}$. If D and G are different points, we have the impossible construction of Theorem 5.3.4, so we must have $D = G$. Hence, $\measuredangle DEF$ and $\measuredangle GEF$ are the same angle, and we have $\measuredangle B \cong \measuredangle E$. ■

We finish our list of triangle theorems with the two remaining tests for triangle congruence.

■ **THEOREM 5.3.6 [*Elements* I.26, ASA, AAS]**

If two triangles have two pairs of corresponding angles congruent and one pair of corresponding sides congruent, the triangles will be congruent.

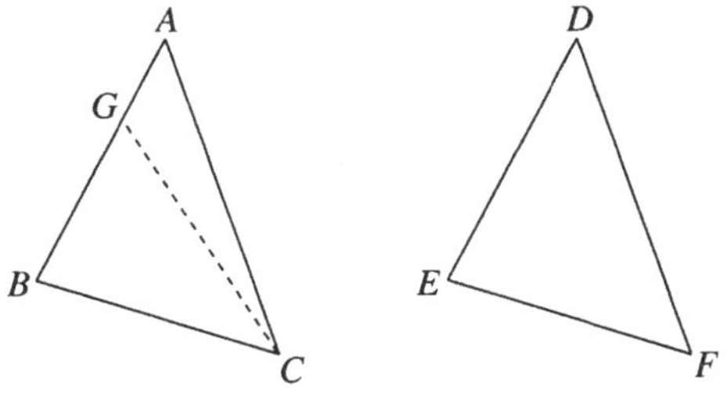

PROOF

We leave the proof of the case when the congruent sides are not between the two angles (AAS) as Exercise 2. To prove the ASA case, suppose that in $\triangle ABC$ and $\triangle DEF$ we have $\measuredangle ABC \cong \measuredangle DEF$, $\measuredangle BCA \cong \measuredangle EFD$, and $\overline{BC} \cong \overline{EF}$. We will show that the triangles are congruent by showing that $\overline{AB}$ is congruent to $\overline{DE}$ and using SAS. To obtain a contradiction, suppose that $AB \neq DE$. Without loss of generality, we may assume that $AB > DE$. Find point G on $\overline{AB}$ such that $\overline{GB} \cong \overline{DE}$. Join $\overline{GC}$ to form $\triangle BCG$. Therefore, $\triangle GBC \cong \triangle DEF$ by SAS. From this we conclude that $\overline{GC} \cong \overline{DF}$ and $\sphericalangle BCG \cong \sphericalangle EFD$. Since $\sphericalangle EFD \cong \sphericalangle BCA$, $\sphericalangle BCG \cong \sphericalangle BCA$, which is impossible. Hence, $\overline{AB} \cong \overline{DE}$ and the triangles are congruent. ■

Inequalities

The *Elements* now turn to the case where the two triangles are not congruent and examine how certain corresponding parts compare to each other. Euclid will prove the famous Triangle Inequality as well as a seemingly innocent but very influential result in Theorem 5.3.12.

■ **THEOREM 5.3.7 [*Elements* I.13]**

If a line segment intersects another line segment forming two angles, the angles are supplementary.

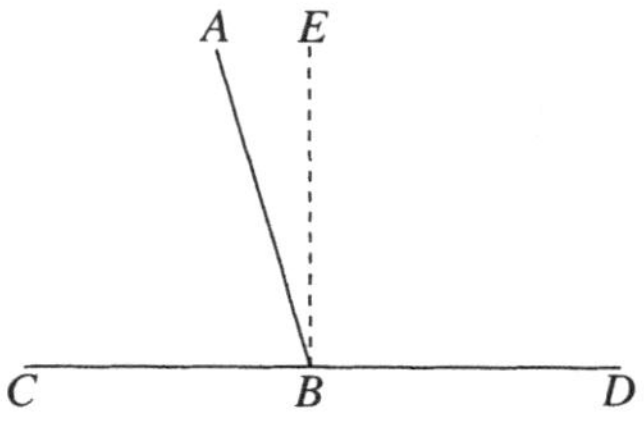

PROOF

Let $\overline{AB}$ and $\overline{CD}$ intersect at B. If $\sphericalangle CBA \cong \sphericalangle ABD$, they are right angles by definition and hence supplementary, so suppose they are not congruent. Construct $\overline{BE}$ perpendicular to $\overline{CD}$. Without loss of generality, $\overline{BE}$ is in the interior of $\sphericalangle ABD$. We see that $\sphericalangle CBE$ and $\sphericalangle EBD$ are right angles. However,

$$\text{m}\sphericalangle CBA + \text{m}\sphericalangle ABE + \text{m}\sphericalangle EBD = \text{m}\sphericalangle CBE + \text{m}\sphericalangle EBD$$

because $\text{m}\sphericalangle CBE = \text{m}\sphericalangle CBA + \text{m}\sphericalangle ABE$. We then have

$$\text{m}\sphericalangle CBA + \text{m}\sphericalangle ABD = \text{m}\sphericalangle CBE + \text{m}\sphericalangle EBD$$

since $\text{m}\sphericalangle ABE + \text{m}\sphericalangle EBD = \text{m}\sphericalangle ABD$. This shows that $\sphericalangle CBA$ and $\sphericalangle ABD$ are supplementary because $\sphericalangle CBE$ and $\sphericalangle EBD$ are right angles. ■

The proof of the next result relies on Theorem 5.3.7. Since its proof is similar, it is left to Exercise 3.

■ **THEOREM 5.3.8 [*Elements* I.14]**

If two line segments intersect a third at the same point so that the two segments lie on opposite sides forming supplementary angles, the two line segments form a line.

The next two theorems are required to prove the Triangle Inequality. As illustrated in the diagram of Theorem 5.3.9, if $\overline{BC}$ is produced to D, then $\measuredangle ACD$ is an **exterior angle** to $\triangle ABC$. On the other hand, $\measuredangle ABC$, $\measuredangle BCA$, and $\measuredangle CAB$ are called the **interior angles** of $\triangle ABC$.

■ **THEOREM 5.3.9 [*Elements* I.16]**

The measure of an exterior angle of a triangle is greater than its opposite interior angles.

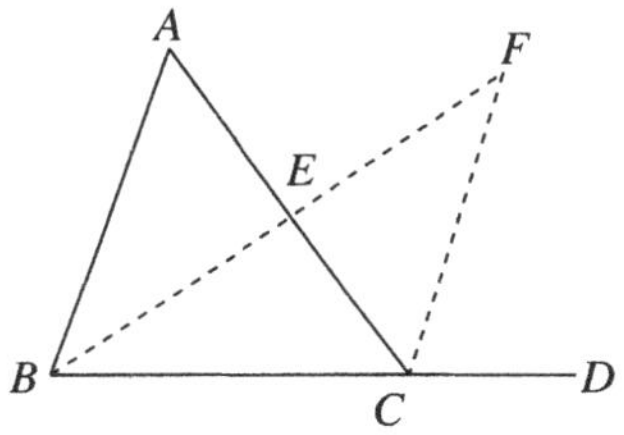

PROOF

We are given $\triangle ABC$ with exterior $\measuredangle ACD$. We must show that the measure of $\measuredangle ACD$ is greater than the measure of $\measuredangle CBA$ and the measure of $\measuredangle BAC$. To do this, let E be the midpoint of $\overline{AC}$. Draw $\overline{BE}$ and extend it to F so that $\overline{BE} \cong \overline{EF}$. Join $\overline{CF}$. Since we also have $\overline{AE} \cong \overline{EC}$ and $\measuredangle AEB \cong \measuredangle FEC$ (Exercise 7), $\triangle ABE \cong \triangle CFE$ by SAS. Hence, $\measuredangle BAE \cong \measuredangle ECF$. But $\text{m}\measuredangle ECD > \text{m}\measuredangle ECF$. Therefore, $\text{m}\measuredangle ACD > \text{m}\measuredangle BAE$. That $\text{m}\measuredangle ACD > \text{m}\measuredangle ABC$ is proven similarly. ■

The second theorem that is needed allows us to compare sides within a triangle relative to their opposite angles.

■ **THEOREM 5.3.10 [*Elements* I.18]**

In any triangle, the greater side is opposite the greater angle.

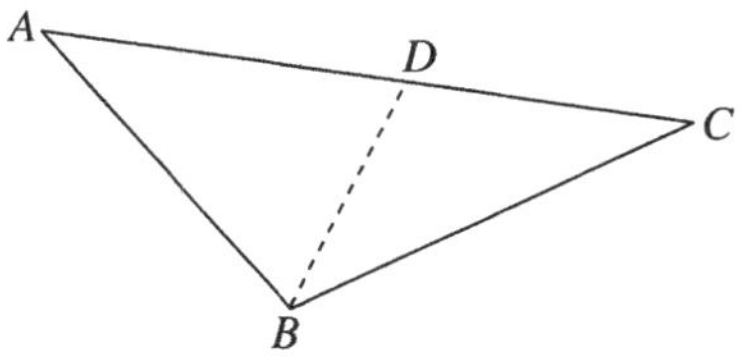

PROOF

Suppose that $AC > AB$ in $\triangle ABC$. Using Euclid's words, "I say that the angle ABC is also greater than the angle BCA." Take D on $\overline{AC}$ such that $\overline{AD}$ is congruent to $\overline{AB}$. First note that $\measuredangle ADB \cong \measuredangle ABD$ since $\triangle ABD$ is isosceles. Also, $\text{m}\measuredangle ADB > \text{m}\measuredangle BCD$ since $\measuredangle ADB$ is exterior to $\triangle BDC$ by

Theorem 5.3.9. This gives m$\sphericalangle ABD >$ m$\sphericalangle ACB$. Therefore, again quoting Euclid, "angle ABC is much greater than angle ACB." ■

■ **THEOREM 5.3.11 [*Elements* I.20, Triangle Inequality]**

In any triangle, the sum of the lengths of any two sides is greater than the length of the remaining one.

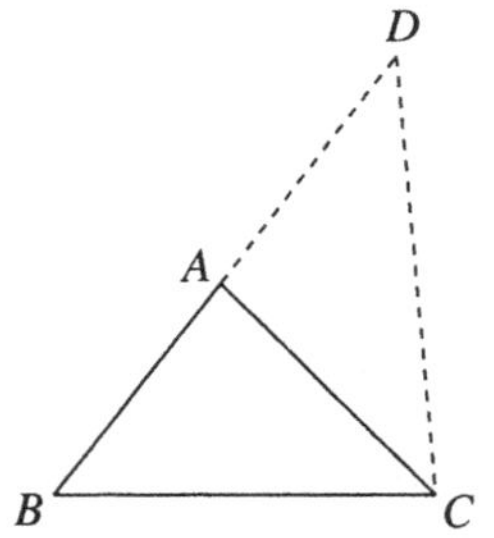

PROOF

From $\triangle ABC$ extend side $\overline{BA}$ to D such that $\overline{AD} \cong \overline{AC}$. Then in isosceles $\triangle ADC$ we have $\sphericalangle ADC \cong \sphericalangle ACD$. Therefore, as in the proof of Theorem 5.3.10, m$\sphericalangle BCD >$ m$\sphericalangle ADC$. Since $\sphericalangle ADC$ and $\sphericalangle BDC$ are the same angle, m$\sphericalangle BCD >$ m$\sphericalangle BDC$. Therefore, $BD > BC$ by Theorem 5.3.10, so $BA + AD > BC$. Since $\overline{AD} \cong \overline{AC}$, we conclude that $BA + AC > BC$. ■

The last inequality of the section gives an upper bound on the sum of any two angles of a triangle.

■ **THEOREM 5.3.12 [*Elements* I.17]**

The sum of any two angles in a triangle is less than that of two right angles.

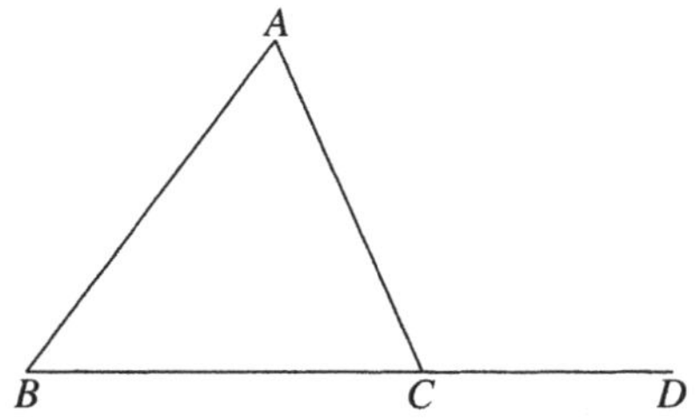

PROOF

Let $\triangle ABC$ be given and produce $\overline{BC}$ to D. Then m$\sphericalangle ABC <$ m$\sphericalangle ACD$ by Theorem 5.3.9. Therefore,

$$\text{m}\sphericalangle ABC + \text{m}\sphericalangle ACB < \text{m}\sphericalangle ACD + \text{m}\sphericalangle ACB,$$

but the right-hand side of the inequality is equal to two right angles. Hence, $\sphericalangle ABC$ and $\sphericalangle ACB$ together are less than two right angles. The rest is proven similarly. ■

Exercises

1. Let $\triangle ABC$ be an isosceles triangle with $\sphericalangle A \cong \sphericalangle B$. If D is on $\overline{AB}$ such that $\overline{CD}$ is perpendicular to $\overline{AB}$, then $\sphericalangle ACD \cong \sphericalangle BCD$.

2. Prove AAS (Theorem 5.3.6).

3. Prove Theorem 5.3.8.

4. Prove that in any triangle the greater angle is opposite the greater side.

5. In $\triangle ABC$ draw $\overline{BE}$ with E on $\overline{AC}$ and $\overline{DC}$ with D on $\overline{BE}$. Prove:

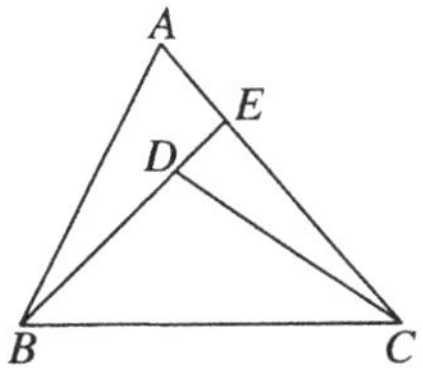

(a) $BD + DC < BA + AC$
(b) $m\sphericalangle BAC < m\sphericalangle BDC$.

6. Euclid's proof of SAS is as follows. Give the reason for each line. The legitimate reasons are "given," the postulates, definitions, basic rules of logic, the Common Notions, Constructions 5.2.6 and 5.2.7, and Exercise 2 from Section 5.2. If there is no legal reason for a line of the proof, explain why.

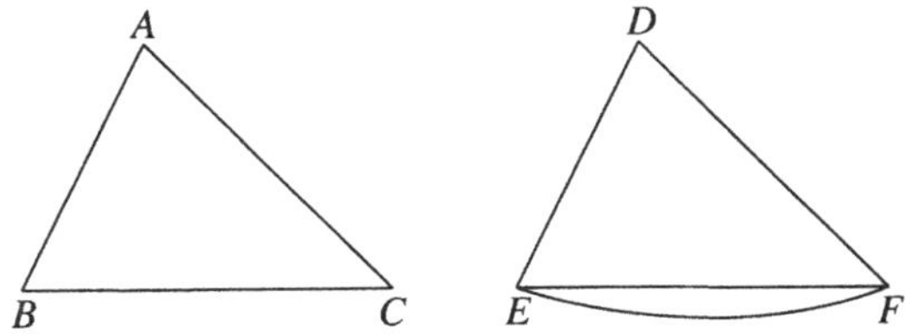

(a) Let $\triangle ABC$ and $\triangle DEF$ be given so that $\overline{AB} \cong \overline{DE}$, $\overline{AC} \cong \overline{DF}$, and $\sphericalangle BAC \cong \sphericalangle EDF$.
(b) Slide $\triangle ABC$ to coincide with $\triangle DEF$ so that A aligns with D and $\overline{AB}$ with $\overline{DE}$.
(c) The point B will then coincide with E.
(d) $\overline{AC}$ will coincide with $\overline{DF}$.
(e) The point C will coincide with point F.
(f) Suppose that $\overline{BC}$ does not coincide with $\overline{EF}$ but with some other segment with E and F as endpoints.
(g) This means that two straight lines enclose a space, which we know is impossible.
(h) Therefore, $\overline{BC}$ coincides with $\overline{EF}$.
(i) Hence, $\triangle ABC \cong \triangle DEF$.

7. When two line segments intersect, the resulting nonadjacent angles are called **vertical angles**. Prove that vertical angles are congruent.

8. Prove that triangle congruence is an equivalence relation (see Exercise 2.3.12).

9. Prove that if a triangle has two congruent angles, the sides that subtend those two angles are congruent.

10. Let $\sphericalangle ADC \cong \sphericalangle BCD$ and $\overline{DA} \cong \overline{BC}$. Prove that $\triangle ADC \cong \triangle BCD$.

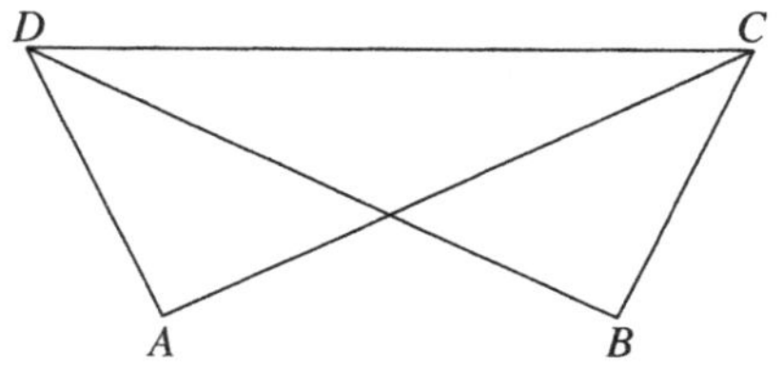

11. We are given $\triangle ABC$, $\triangle A'B'C'$, $\triangle DEG$, and in the last triangle there is drawn a segment from D to F on $\overline{EG}$. Suppose that $\overline{AB} \cong \overline{A'B'} \cong \overline{DE}$, $\overline{AC} \cong \overline{A'C'} \cong \overline{DG}$, and $\sphericalangle ABC \cong \sphericalangle A'B'C' \cong \sphericalangle DEG$. We have set up the conditions for the illegal SSA test for congruence. However, an examination of the diagram will yield an extra condition that can be added to SSA to make it into a legitimate test. What is it? (Do not simply add a hypothesis that will change SSA into SSS, SAS, ASA, or AAS.)

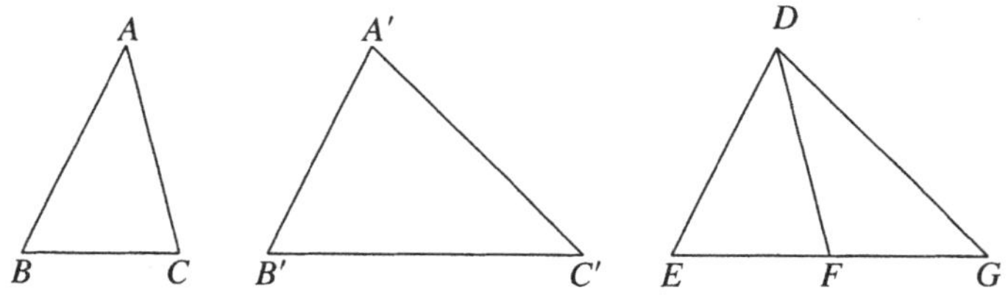

12. Prove that a triangle is an equilateral triangle if and only if it is equiangular.

13. Let $\triangle ABC$ be isosceles with base $\overline{BC}$ and $\overline{AD} \cong \overline{AE}$. Show that we have both $\triangle ABE \cong \triangle ACD$ and $\triangle DBC \cong \triangle ECB$.

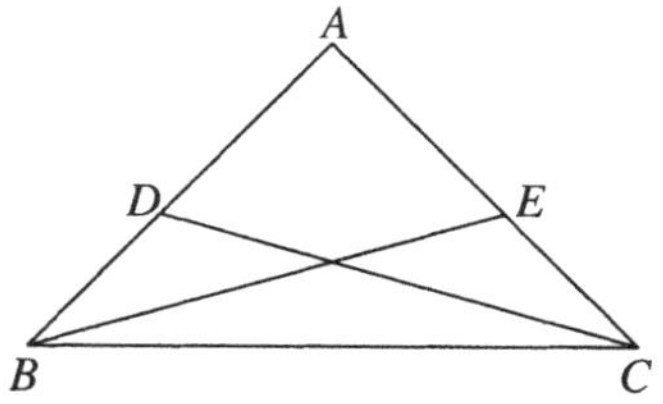

14. For any line l and point P, the **distance** of P to l is the length of the perpendicular segment drawn from P to l. Show that if Q is any point on l, then PQ is greater than or equal to the distance from P to l.

15. Assume that $\overline{DC} \cong \overline{FC}$ and that $\overline{EC}$ bisects $\measuredangle ACB$. Prove that $\triangle AFC$ is congruent to $\triangle BDC$.

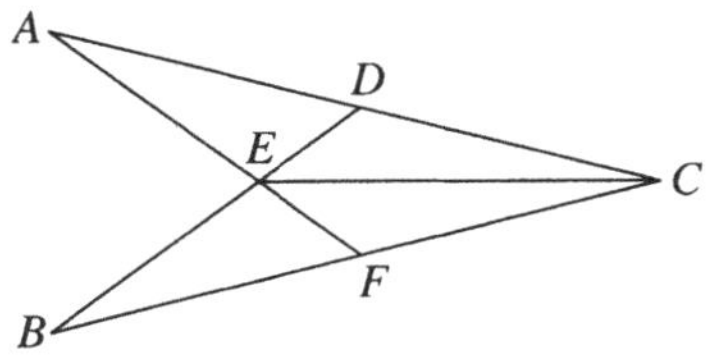

16. We are given that $\measuredangle AEB$ and $\measuredangle ACB$ are right and $\overline{ED} \cong \overline{DC}$. Show that $\triangle AED \cong \triangle BCD$.

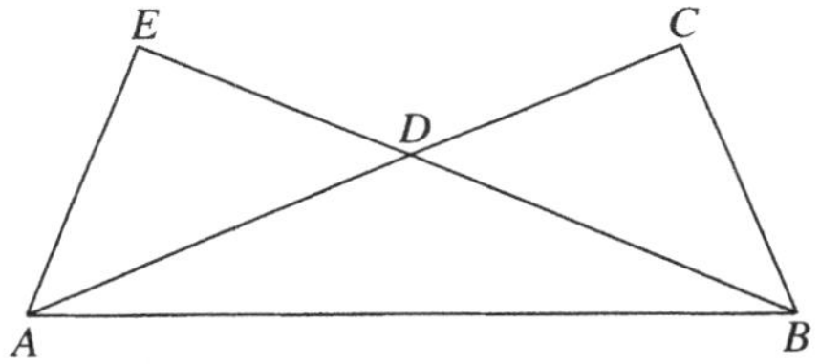

17. Prove the following inequality: If a triangle contains a right or obtuse angle, the measure of the angle is greater than the measures of the other two angles.

18. Use the Triangle Inequality (Theorem 5.3.11) to show that $|a + b| \leqslant |a| + |b|$ for all real numbers a and b.

19. In quadrilateral $ABCD$, $\overline{BD}$ is a diagonal, $\overline{AB}$ is perpendicular to $\overline{BD}$, and $\overline{BC}$ is perpendicular to $\overline{CD}$. Demonstrate that $AD > CD$.

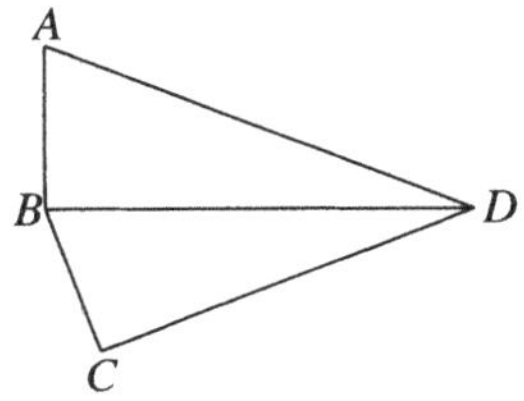

20. Explain what the phrase *without loss of generality* must mean in the proof of Theorem 5.3.6.

5.4 PARALLEL LINES

The first part of Book I of the *Elements* deals with triangles. Euclid next moves to a study of parallel lines. Let l and m be cut by transversal t as in Figure 5.5. We recall the terminology for the angles formed by these lines:

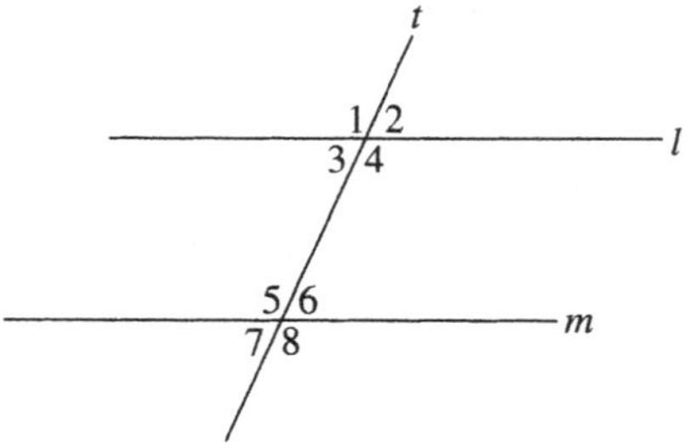

Figure 5.5 Lines l and m cut by transversal t.

angle	name	angle	name
$\sphericalangle 3$ and $\sphericalangle 6$	alternate interior	$\sphericalangle 3$ and $\sphericalangle 7$	corresponding
$\sphericalangle 4$ and $\sphericalangle 5$	alternate interior	$\sphericalangle 4$ and $\sphericalangle 8$	corresponding
$\sphericalangle 1$ and $\sphericalangle 8$	alternate exterior	$\sphericalangle 1$ and $\sphericalangle 7$	exterior, same side
$\sphericalangle 2$ and $\sphericalangle 7$	alternate exterior	$\sphericalangle 2$ and $\sphericalangle 8$	exterior, same side
$\sphericalangle 1$ and $\sphericalangle 5$	corresponding	$\sphericalangle 3$ and $\sphericalangle 5$	interior, same side
$\sphericalangle 2$ and $\sphericalangle 6$	corresponding	$\sphericalangle 4$ and $\sphericalangle 6$	interior, same side

With this in mind we begin with the fundamental theorem.

■ **THEOREM 5.4.1 [*Elements* I.27, 28]**

If two lines are intersected by a transversal such that

- alternate interior angles are congruent,
- alternate exterior angles are congruent,
- corresponding angles are congruent,
- interior angles on the same side are supplementary, or
- exterior angles on the same side are supplementary,

the lines are parallel.

PROOF

We will prove the first and leave the others to Exercise 6. Let $\overline{AB}$ and $\overline{CD}$ be cut by transversal $\overline{EF}$ as in the diagram for Theorem 5.4.2. Suppose that $\sphericalangle AEF \cong \sphericalangle DFE$. We claim that the lines are parallel. To prove this, suppose that, instead, they will meet if extended far enough at some point G as in the diagram. We therefore have $\triangle GEF$, and $\sphericalangle AEF$ is exterior to it. This means that we should have $m\sphericalangle AEF > m\sphericalangle EFG$, since $\sphericalangle EFG$ is an opposite interior angle to $\sphericalangle AEF$ (Theorem 5.3.9). However, they are congruent, so the lines cannot meet. ■

Using *reductio ad absurdum*, we can prove the converse of Proposition 5.4.1. It is Euclid's first application of the Parallel Postulate (5.1.1.5).

THEOREM 5.4.2 [*Elements* I.29]

If two parallel lines are intersected by a transversal, then

- alternate interior angles are congruent,
- alternate exterior angles are congruent,
- corresponding angles are congruent,
- interior angles on the same side are supplementary, and
- exterior angles on the same side are supplementary.

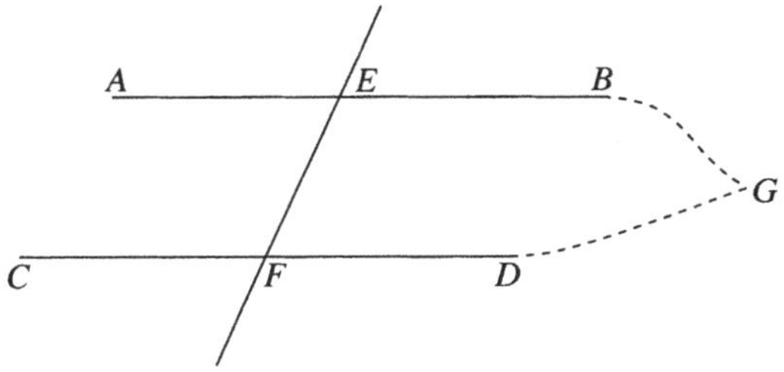

PROOF

Let $\overline{AB}$ and $\overline{CD}$ be parallel, cut by transversal $\overline{EF}$. We will show that alternate interior angles are congruent by showing $\sphericalangle AEF \cong \sphericalangle EFD$. The other parts of the proof will be left to Exercise 6. Suppose they are not congruent. Instead, let $m\sphericalangle AEF > m\sphericalangle EFD$. (The other case is proven similarly.) Therefore,

$$m\sphericalangle AEF + m\sphericalangle BEF > m\sphericalangle EFD + m\sphericalangle BEF.$$

However, $\sphericalangle AEF$ and $\sphericalangle BEF$ are supplementary by Theorem 5.3.7. Thus, $m\sphericalangle EFD + m\sphericalangle BEF$ is less than two right angles, so if produced indefinitely, $\overline{AB}$ and $\overline{CD}$ will meet on that side of $\overline{EF}$ by the Parallel Postulate, a contradiction of the hypothesis. Hence, $\sphericalangle AEF \cong \sphericalangle EFD$. ■

The following takes the result of Theorem 5.3.9 and makes it more precise using the Parallel Postulate. Since Theorem 5.4.3 implies Theorem 5.3.9, many believe that Euclid had his doubts about the Parallel Postulate. It is as if Euclid was delaying the use of it as long as it was reasonable to do so. This need not be the case, though. If the first subject of Book I is triangles and the next is parallel lines, it makes sense to keep the sections as self-contained as possible. Such a careful sequencing of the propositions helps the student learn the material. At any rate, here is the theorem.

THEOREM 5.4.3 [*Elements* I.32]

In any triangle:

- The measure of an exterior angle equals the sum of the two opposite interior angles.
- The sum of the measures of the interior angles equals two right angles.

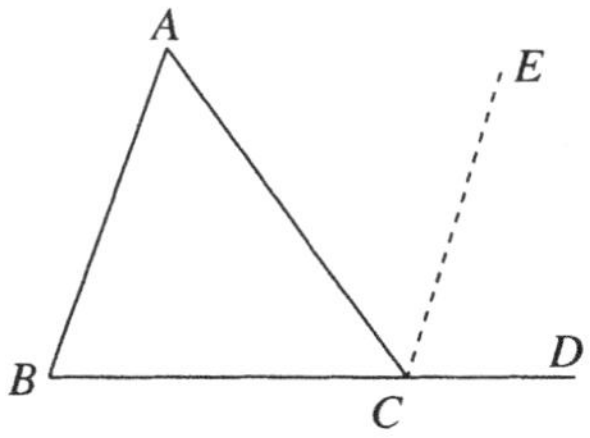

PROOF

Let $\sphericalangle ACD$ be exterior to $\triangle ABC$. We show first that $m\sphericalangle ACD = m\sphericalangle BAC + m\sphericalangle ABC$. Construct $\overline{CE}$ parallel to $\overline{AB}$ [Exercise 3(a)]. Since $\sphericalangle BAC$ and $\sphericalangle ACE$ are alternate interior angles with $\overline{AC}$ as the transversal, these angles are congruent (Theorem 5.4.2). Furthermore, $\sphericalangle ECD \cong \sphericalangle ABC$ by the same theorem since they are corresponding angles with transversal $\overline{BD}$. By angle addition we then have

$$m\sphericalangle ACD = m\sphericalangle ACE + m\sphericalangle ECD,$$

but this is what we want because

$$m\sphericalangle ACE + m\sphericalangle ECD = m\sphericalangle BAC + m\sphericalangle ABC.$$

To see that the sum of the interior angles is the same as two right angles, we continue by noting that

$$m\sphericalangle ACB + m\sphericalangle ACD = m\sphericalangle ACB + m\sphericalangle BAC + m\sphericalangle ABC.$$

The right side is the sum of the interior angles. The left is equal to two right angles because $\sphericalangle ACB$ and $\sphericalangle ACD$ are supplementary. ■

A common alternate proof that the sum of the internal angles of a triangle equals two right angles is attributed to the Pythagoreans (Euclid 1925, Vol. I, 320). Given $\triangle ABC$, construct $\overline{DE}$ through C parallel to $\overline{AB}$ as in Figure 5.6. By Theorem 5.4.2, $\sphericalangle DCA \cong \sphericalangle BAC$ and $\sphericalangle ECB \cong \sphericalangle ABC$. However,

$$m\sphericalangle DCA + m\sphericalangle ACB + m\sphericalangle BCE = \text{two right angles}.$$

Therefore, we may substitute to obtain

$$m\sphericalangle BAC + m\sphericalangle ACB + m\sphericalangle ABC = \text{two right angles}.$$

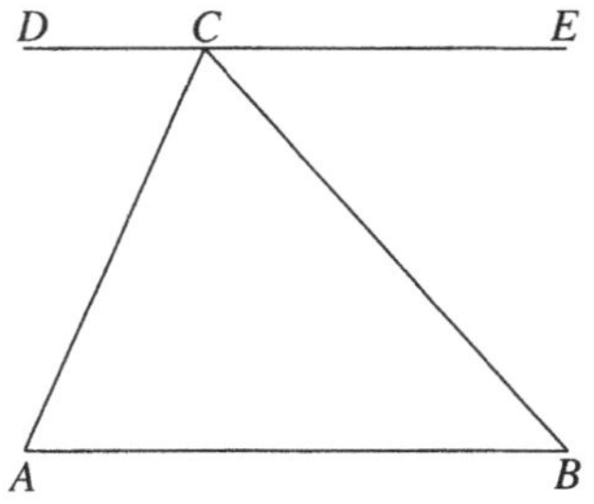

Figure 5.6 The Pythagorean proof regarding the angle sum of a triangle.

Since a diagonal of a quadrilateral divides it into two triangles, we easily obtain the first result of the corollary. The generalization to convex n-gons is the second part. Remember that a polygon is **convex** if given any two points A and B in its interior, $\overline{AB}$ completely lies in the interior of the polygon.

■ **COROLLARY 5.4.4**

- The sum of the measures of the interior angles of a quadrilateral equals four right angles.
- The sum of the measure of the interior angles of a convex n-gon equals $n - 2$ right angles.

Areas and Parallelograms

Once the basic properties of parallel lines are established, Euclid can then define polygons that consist of parallel lines.

■ **DEFINITION 5.4.5**

A **parallelogram** is a quadrilateral in which opposite sides are parallel. Two sides of a quadrilateral are called **opposite** if they do not have a common endpoint, else they are **adjacent**. A segment that joins opposite vertices is called a **diagonal**.

A **rectangle** is a parallelogram with a right angle, and a **square** is a rectangle with two consecutive congruent sides.

We begin by proving two basic properties of parallelograms. It will require the Parallel Postulate. It also involves the concept of area. To this point we have worked with area based on our memory and intuition. At this point let us make the concept more formal. We will say that the **area** of a plane figure F is a number that represents the amount of two-dimensional space that F occupies. We will assume that area has the following properties:

- The area of F is a positive number and will continue to be denoted by area(F).
- If F is divided into nonoverlapping figures $F_1, F_2, \ldots, F_n$ such that the result of combining $F_1, F_2, \ldots, F_n$ is F,

$$\text{area}(F) = \text{area}(F_1) + \text{area}(F_2) + \cdots + \text{area}(F_n).$$

- Congruent figures have equal area.
- The area of a rectangle is the product of the lengths of two consecutive sides.

With this understanding we are now ready to tackle the theorem.

THEOREM 5.4.6 [*Elements* I.34]

In a parallelogram, opposite sides and opposite angles are congruent, and the diameter bisects the area.

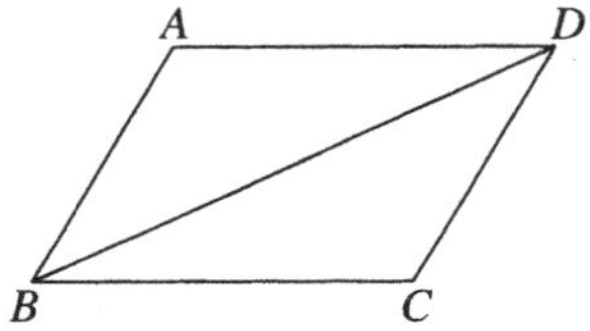

PROOF

Take $\square ABCD$ with diagonal $\overline{BD}$. This means that $\overline{AB}$ is parallel to $\overline{CD}$ and $\overline{AD}$ is parallel to $\overline{BC}$. The alternate interior angles $\angle ADB$ and $\angle DBC$ are congruent by Theorem 5.4.2 with $\overline{BD}$ as the transversal. Similarly, we have $\angle ABD \cong \angle BDC$. Hence, since $\overline{BD} \cong \overline{BD}$ we have

$$\triangle ADB \cong \triangle CBD$$

by ASA. We conclude that

$$\overline{AD} \cong \overline{BC}, \overline{AB} \cong \overline{DC}, \text{ and } \angle DAB \cong \angle BCD$$

by the definition of congruent triangles. Moreover, by Common Notion 2, we see that $\angle ADC \cong \angle ABC$. Finally, since the triangles are congruent and the area of the parallelogram consists entirely of those triangles, the diameter must divide the area of the parallelogram in two. ■

We leave it to Exercise 16 to prove that in a parallelogram consecutive angles are supplementary and the diagonals bisect each other.

Take two parallel lines and call them l and m. A pair of parallelograms that are constructed in such a way that each has a pair of opposite sides with one lying on l and the other lying on m are said to be **in the same parallels**. See the diagram for Theorem 5.4.7 for an illustration. This is Euclid's method to guarantee that the two parallelograms have the same height. When this happens we know something about their areas. The proof of the next theorem requires the use of **trapezoids**, quadrilaterals with exactly one pair of parallel sides (Exercise 18).

■ THEOREM 5.4.7 [*Elements* I.35]

Parallelograms that share a common side and are in the same parallels have the same area.

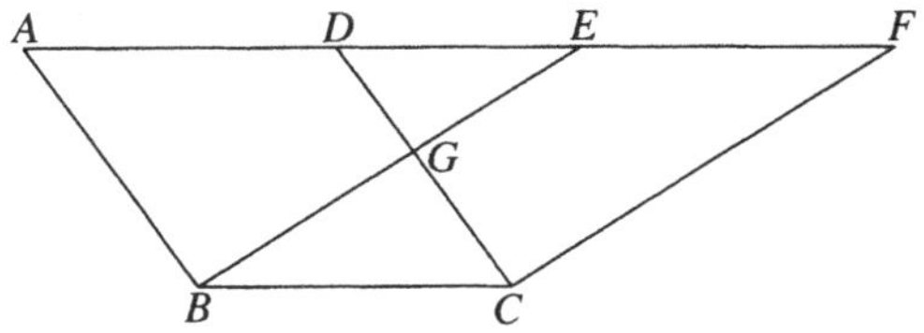

PROOF

Assume that $\overline{AF}$ is parallel to $\overline{BC}$. Let $\square ABCD$ and $\square EBCF$ have common side $\overline{BC}$ and let both $\overline{AD}$ and $\overline{EF}$ be concurrent with $\overline{AF}$. We know that

$$\overline{AD} \cong \overline{BC}$$

and

$$\overline{EF} \cong \overline{BC}$$

by Theorem 5.4.6. Therefore, $AD = EF$, and we have

$$AD + DE = DE + EF.$$

This yields $AE = DF$ by Segment Addition. Now $\sphericalangle FDC \cong \sphericalangle EAB$ since they are corresponding angles. This gives us

$$\triangle EAB \cong \triangle FDC$$

by SAS. If we subtract the area of $\triangle DEG$ from the areas of both of the triangles, we find that trapezoid $ABGD$ has the same area as trapezoid $EGCF$. Adding the area of $\triangle GBC$ to each trapezoid, we find that our original parallelograms have equal area. ■

Triangles are **in the same parallels** if there are two parallel lines such that one side of each triangle is on one of the lines and the opposite vertices are on the other parallel.

■ THEOREM 5.4.8 [*Elements* I.37]

Triangles that share the same base and are in the same parallels have the same area.

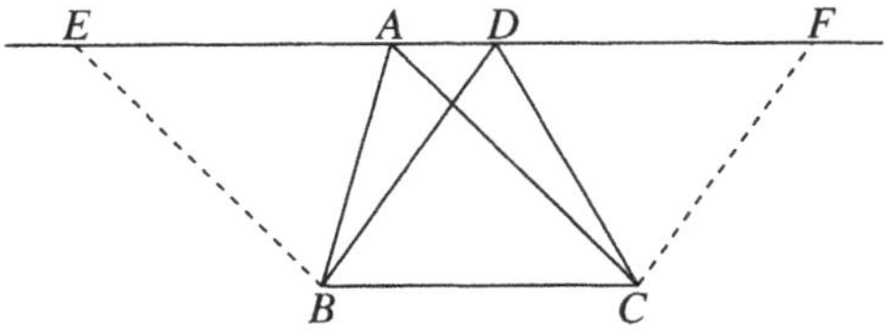

PROOF

Let $\overline{AD}$ be parallel to $\overline{BC}$ and let $\triangle ABC$ and $\triangle DBC$ share base $\overline{BC}$. Extend $\overline{AD}$ to E and F such that $\overline{BE}$ is parallel to $\overline{AC}$ and $\overline{CF}$ is parallel to $\overline{BD}$. Appealing to Theorem 5.4.7, we observe that $\square EBCA$ has the same area as $\square DBCF$. Further, the area of $\triangle ABC$ equals half of the area of $EBCA$ since $\overline{AB}$ is the parallelogram's diagonal (Theorem 5.4.6). Similarly, the area of $\triangle DBC$ is half the area of $\square DBCF$. Therefore, $\triangle ABC$ and $\triangle DBC$ have the same area. ■

The proof of the following is in the spirit of Theorem 5.4.8, so we state it without proof.

■ **THEOREM 5.4.9 [*Elements* I.41]**

If a parallelogram has the same base with a triangle and is in the same parallels, then the area of the parallelogram is double that of the triangle.

We close this discussion by observing that the proof of Theorem 5.4.8 requires that the area of $\triangle ABC$ is equal to half that of $\square EBCA$. If we call $\overline{BC}$ the **base** of both the triangle and the parallelogram, the perpendicular distance between the base and $\overline{EA}$ is the **height** of both the triangle and the parallelogram. Since by Exercise 11 we know that

$$\text{area}(\square EBCA) = \text{base} \times \text{height},$$

we may conclude that

$$\text{area}(\triangle ABC) = \tfrac{1}{2}\text{area}(\square EBCA) = \tfrac{1}{2} \times \text{base} \times \text{height},$$

the familiar formula for the area of a triangle.

Similar Triangles

Since we needed the Parallel Postulate to prove that the angle sum of a triangle is equal to two right angles, it should not be surprising that the Postulate plays a role in the study of similar triangles. Watch carefully for the appearance of the Postulate.

■ **DEFINITION 5.4.10**

We say that $\triangle ABC$ is **similar** to $\triangle DEF$ and write $\triangle ABC \sim \triangle DEF$ if and only if

$$\measuredangle ABC \cong \measuredangle DEF,\ \measuredangle BCA \cong \measuredangle EFD,\ \measuredangle ACB \cong \measuredangle DFE$$

and

$$\frac{AB}{DE} = \frac{BC}{EF} = \frac{AC}{DF}.$$

When these last equations hold, we say that the corresponding sides are **proportional**.

Notice that if the corresponding sides of $\triangle ABC$ and $\triangle DEF$ are proportional, we may also write

$$\frac{AB}{BC} = \frac{DE}{EF}, \frac{AB}{AC} = \frac{DE}{DF}, \text{ and } \frac{BC}{AC} = \frac{EF}{DF}.$$

Our goal will be to demonstrate that for two triangles to be similar it suffices to show that corresponding angles are congruent (and at that, only two such pairs are needed) or show that corresponding sides are proportional. We start by examining what happens when we cut a triangle by a line parallel to its base.

THEOREM 5.4.11 [*Elements* VI.2]

A line that intersects two of the sides of a triangle is parallel to the third side if and only if it cuts the sides of the triangle proportionally.

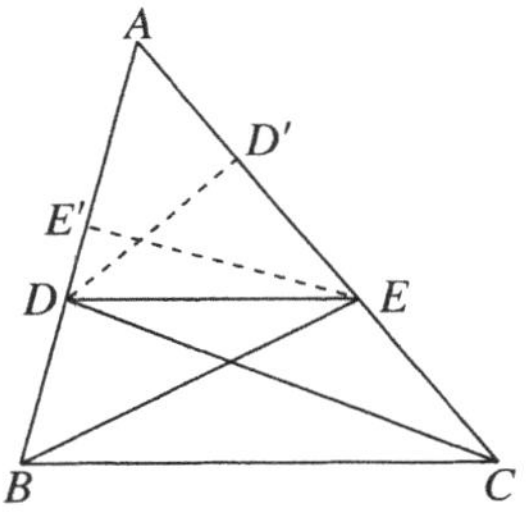

PROOF

We will prove sufficiency. In $\triangle ABC$, let D be on $\overline{AB}$ and E on $\overline{AC}$ such that $\overline{DE}$ is parallel to $\overline{BC}$. Join $\overline{BE}$ and $\overline{DC}$. We see that $\triangle BDE$ and $\triangle CED$ share a common base and are in the same parallels. Thus, they have the same area (Theorem 5.4.8), and we may conclude that

$$\frac{\text{area}(\triangle BDE)}{\text{area}(\triangle ADE)} = \frac{\text{area}(\triangle CED)}{\text{area}(\triangle ADE)}.$$

Now let $\overline{E'E}$ be perpendicular to $\overline{AB}$ and $\overline{DD'}$ perpendicular to $\overline{AC}$. Since

$$\text{area}(\triangle BDE) = \tfrac{1}{2}BD \cdot E'E$$

and

$$\text{area}(\triangle ADE) = \tfrac{1}{2}AD \cdot E'E,$$

we conclude that

$$\frac{\text{area}(\triangle BDE)}{\text{area}(\triangle ADE)} = \frac{BD}{AD}.$$

Similarly using the height DD' yields

$$\frac{\text{area}(\triangle CED)}{\text{area}(\triangle ADE)} = \frac{EC}{AE}.$$

Hence,

$$\frac{BD}{AD} = \frac{EC}{AE}. \blacksquare$$

Multiple applications of Theorem 5.4.11 yield the next result (Exercise 20). This theorem tells us that to show similarity, it is enough to check that corresponding angles are congruent.

■ **THEOREM 5.4.12 [AAA, *Elements* VI.4]**

If the corresponding angles of two triangles are congruent, the corresponding sides are proportional.

Since the sum of the internal angles of a triangle must equal two right angles, AAA implies AA.

■ **COROLLARY 5.4.13 [AA]**

If two pairs of corresponding angles of two given triangles are congruent, the triangles are similar.

We now show that proportional corresponding sides implies similarity.

■ **THEOREM 5.4.14 [*Elements* VI.5]**

If the corresponding sides of two triangles are proportional, the corresponding angles are congruent.

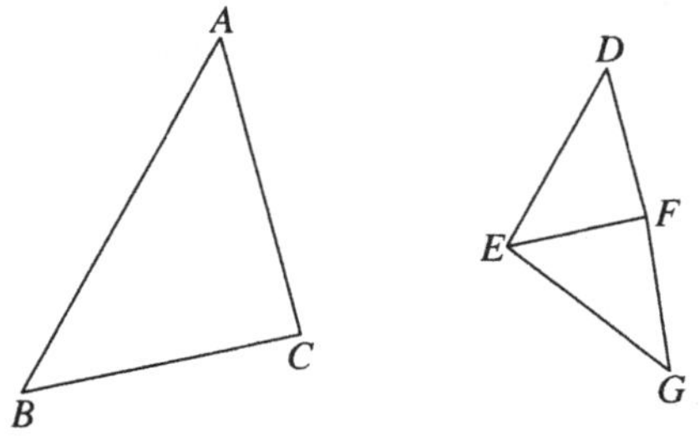

PROOF

Suppose that we are given $\triangle ABC$ and $\triangle DEF$ such that

$$\frac{AB}{BC} = \frac{ED}{EF}, \frac{AB}{AC} = \frac{ED}{DF}, \text{ and } \frac{BC}{AC} = \frac{EF}{DF}.$$

We must show that the corresponding angles are congruent. To do so, find point G so that $\sphericalangle FEG \cong \sphericalangle ABC$ and $\sphericalangle EFG \cong \sphericalangle ACB$. From this we conclude that $\triangle ABC \sim \triangle GEF$ by AA. Therefore, $\sphericalangle BAC \cong \sphericalangle FGE$ and

$$\frac{AB}{BC} = \frac{EG}{EF},$$

so we obtain

$$\frac{ED}{EF} = \frac{EG}{EF}.$$

Hence, $\overline{ED} \cong \overline{EG}$. Similarly, we obtain $\overline{DF} \cong \overline{FG}$. Since $\overline{EF} \cong \overline{EF}$, we know that $\triangle DEF \cong \triangle GEF$ by SSS. By the definition of triangle congruence we then have $\measuredangle DEF \cong \measuredangle GEF$, $\measuredangle EFD \cong \measuredangle EFG$, and $\measuredangle FDE \cong \measuredangle FGE$. From this we conclude that since the corresponding angles of $\triangle ABC$ and $\triangle GEF$ are congruent and the corresponding angles of $\triangle GEF$ and $\triangle DEF$ are also congruent, we know that the corresponding angles of $\triangle ABC$ and $\triangle DEF$ are congruent since congruence of angles is an equivalence relation (Exercise 1 in Section 5.2). ∎

Exercises

1. Suppose that polygons $ABCDE$ and $ABFGH$ are both convex. Must polygon $AEDCBFGH$ be convex? Explain.

2. Prove that the interior angles of a convex regular n-gon each has a measure equal to $180(n-2)/n$ degrees.

3. Construct:
   (a) through a given point a line that is parallel to a given line.
   (b) angles that measure one third and two thirds of a right angle.
   (c) angles that measure $75°$ and $105°$.
   (d) a 30-60-90 triangle.
   (e) a square.
   (f) a parallelogram that is not a rectangle.
   (g) a regular hexagon.
   (h) a regular octagon.
   (i) a regular dodecagon (12-gon).

4. Given points A, B, and E, construct $\square ABCD$ such that $\overline{BC} \cong \overline{AE}$.

5. Prove that if l and m intersect in a point, any perpendicular to l will intersect any perpendicular to m.

6. Prove the remaining parts of Theorems 5.4.1 and Theorem 5.4.2.

7. Prove Corollary 5.4.4.

8. Prove for all distinct lines l, m, and n:
   (a) If l and m are parallel to n, then l is parallel to m.
   (b) If l and m are both perpendicular to n, then l and m are parallel.
   (c) If l and m are both parallel and l is perpendicular to n, then m is perpendicular to n.

9. Explain why the acute angles of a right triangle are complementary.

10. Prove that if three or more parallel lines intercept a transversal in congruent segments, the parallel lines intercept all transversals in congruent segments.

11. Using the definition of area on page 151, explain why the area of a parallelogram is equal to the product of its base and its height.

12. Let $ABCD$ be a quadrilateral such that $\overline{AB}$ and $\overline{DC}$ are both congruent and parallel. Demonstrate that $\overline{AD}$ and $\overline{BC}$ are both congruent and parallel.

13. Prove:

(a) Two parallelograms that are in the same parallels and have congruent bases that lie on one of the parallels have the same area.

(b) Two triangles that are in the same parallels and have congruent bases that lie on one of the parallels have the same area.

(c) Two triangles that have the same area, have a common base, and lie on the same side of the common base are in the same parallels.

14. Let $\overline{AC}$ be a diagonal of $\square ABCD$, $\overline{EG}$ parallel to $\overline{AB}$, $\overline{HF}$ parallel to $\overline{DA}$, and the three segments $\overline{EG}$, $\overline{HF}$, and $\overline{AC}$ intersect at I as in the diagram. Show that the area of $\square EIHD$ equals that of $\square FBGI$.

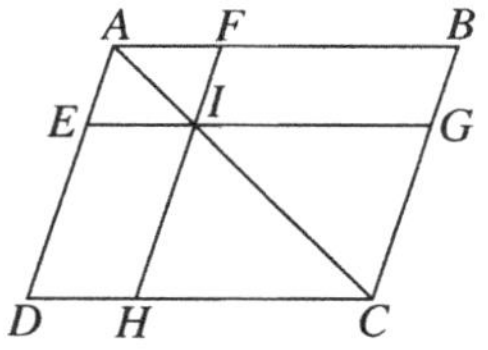

15. Construct a parallelogram that has an angle congruent to a given angle and has the same area as a given triangle.

16. Prove that in a parallelogram consecutive angles are supplementary and the diagonals bisect each other.

17. Recall that the distance from a point A to a line l is the length of the perpendicular segment from A to l. Prove that if l and m are parallel, there exists a unique positive integer d such that the distance from any point on l to a point on m is d. (This means that l and m are **equidistant**.)

18. The parallel sides of a trapezoid are its **bases** and the nonparallel sides are its **legs**. If the legs are congruent, the trapezoid is **isosceles**. Prove the following:

(a) The angles adjacent to a common base are congruent if and only if the trapezoid is an isosceles trapezoid.

(b) The diagonals of trapezoid are congruent if and only if the trapezoid is isosceles.

(c) The line segment joining the midpoints of the legs of a trapezoid (called the **median** of the trapezoid) is parallel to its bases.

(d) The length of the median of a trapezoid equals the arithmetic mean of the lengths of the two bases.

19. Given $\triangle ABC$ and $\overline{DA}$ with D on $\overline{BC}$. Prove that $\overline{DA}$ bisects $\sphericalangle BAC$ if and only if $BD/DC = BA/AC$.

20. Prove AAA (Theorem 5.4.12).

21. Given two segments $\overline{AC}$ and $\overline{DF}$ and a point B on $\overline{AC}$, find E on $\overline{DF}$ such that $AB/BC = DE/EF$.

22. Prove: If two triangles have a pair of corresponding sides proportional and the pair of included angles congruent, the triangles are similar.

23. Two polygons are **similar** if all pairs of corresponding angles are congruent and all pairs of corresponding sides are proportional. Is similarity of polygons an equivalence relation? Explain with a proof or a counterexample.

24. Prove that a pair of similar convex polygons can each be divided into the same number of triangles so that corresponding triangles are similar and their sides are in the same ratio as the sides of the given polygons.

25. Given $\overline{AB}$ and $\overline{CD}$, find $\overline{EF}$ such that

$$\frac{AB}{CD} = \frac{CD}{EF}.$$

(The magnitude EF is called a **third proportional.**)

5.5 CIRCLES

The theorems from Euclid's *Elements* that we have studied thus far have been mainly from the first book. Although the results of this section are not needed for Euclid's proof of the Pythagorean Theorem, these results from the third and fourth books of the *Elements* are surprisingly related to the Parallel Postulate. It is therefore appropriate that we mention these results concerning circles at this point. We will prove the famous theorems regarding tangents and secants, perform some circumscribing and inscribing constructions, and see that we need only three points to identify a circle uniquely.

Tangent and Secant Theorems

A **tangent** line or line segment to a given circle will intersect the circle in exactly one point, while a **secant** line or line segment to a circle will intersect it in exactly two points. A **chord** is a secant segment whose endpoints are on the circle. There is no line that intersects a circle in more than two points. The first two results of this section involve relationships between secants and tangents and the radii of the circle.

■ **THEOREM 5.5.1 [*Elements* III.3]**

If a segment drawn from the center of a circle meets a chord of the circle that does not contain the center in a right angle, the segment bisects the chord.

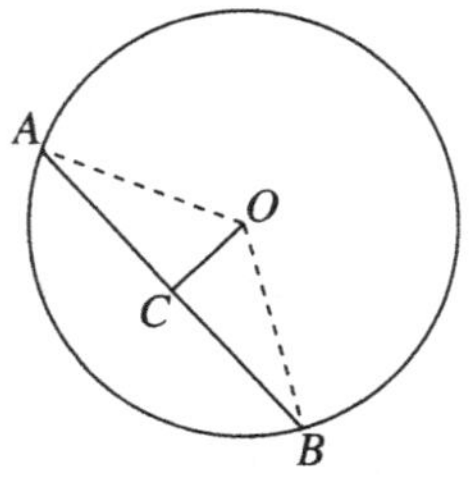

PROOF

Let $\overline{AB}$ be a chord of $\odot O$. Let C be on $\overline{AB}$ such that $\overline{OC}$ is perpendicular to the chord. Join $\overline{OA}$ and $\overline{OB}$. Since $\overline{OA} \cong \overline{OB}$, $\overline{OC} \cong \overline{OC}$, and both $\triangle OAC$ and $\triangle OBC$ are right triangles, these two triangles are congruent (HL, Exercise 5.6.4). Therefore, $\overline{AC} \cong \overline{BC}$. ■

We should note that the previous proof relied on the HL test for congruence. The proof of this test needs the Pythagorean Theorem, and the Pythagorean Theorem needs the Parallel Postulate, as we shall see.

■ **THEOREM 5.5.2 [*Elements* III. 18]**

A tangent to a circle is perpendicular to the radius at the point of tangency.

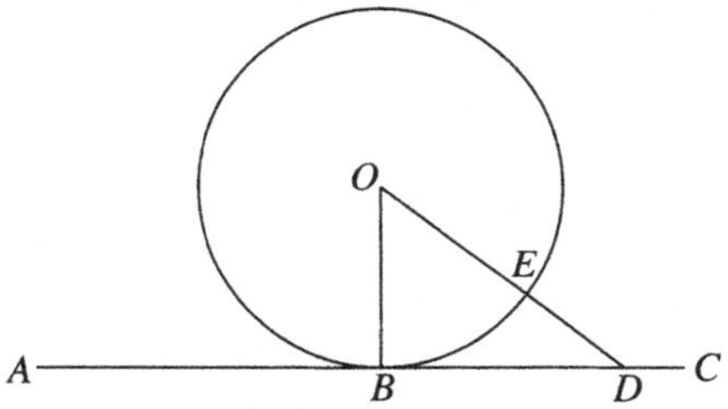

PROOF

Let $\overline{AC}$ be tangent to $\odot O$ at B. Join $\overline{OB}$. Suppose that $\overline{OB}$ is not perpendicular to $\overline{AC}$. Construct $\overline{OD}$ to be perpendicular to $\overline{AC}$ by Construction 5.2.12, where D is the intersection of the two segments. By supposition, $\overline{OD}$ and $\overline{OB}$ are different line segments. Since $\measuredangle ODB$ is right, $\measuredangle OBD$ must be acute (Theorem 5.3.12). Therefore, by Theorem 5.3.10, $OB > OD$. However, $\overline{OD}$ meets the circle in some point E so that

$$OD = OE + ED.$$

Since $\overline{OB}$ and $\overline{OE}$ are both radii,

$$OD = OB + ED.$$

This is impossible. Hence, $\overline{OB}$ is perpendicular to $\overline{AC}$. ■

The next three results will prove very powerful and enable us to write equations based on the lengths of the line segments formed by their intersection with a given circle. The first involves the intersection of two chords, the second of two secants, and the last of a tangent and a secant. We prove the first and third, and leave the proof of the second to Exercise 3.

■ THEOREM 5.5.3 [Chord Theorem, *Elements* III.35]

If $\overline{AB}$ and $\overline{CD}$ are two chords of a circle intersecting at a point E in the interior of the circle,

$$AE \cdot EB = CE \cdot ED.$$

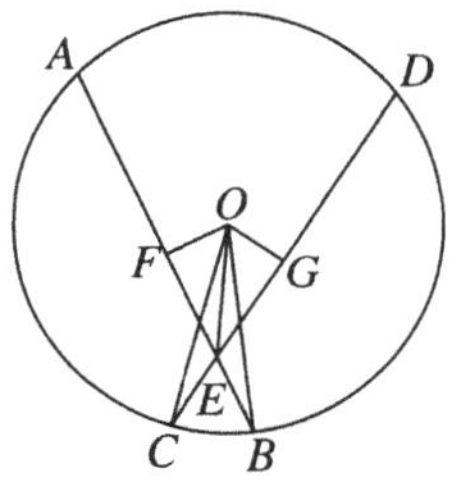

PROOF

If the point E that is the intersection of the two chords is the center of the circle, then $\overline{AE}$, $\overline{EB}$, $\overline{CE}$, and $\overline{ED}$ are all radii, and the equation quickly follows. So, suppose that the center O of the circle does not lie on either chord. (The case when exactly one chord contains the center is proven similarly.) Construct $\overline{OF}$ perpendicular to $\overline{AB}$ and $\overline{OG}$ perpendicular to $\overline{CD}$ with F on $\overline{AB}$ and G on $\overline{CD}$. Join $\overline{OB}$, $\overline{OC}$, and $\overline{OE}$. By Exercise 1(a) we have

$$AE \cdot EB + FE^2 = FB^2. \tag{5.1}$$

Since the rest of the proof is very similar to the proof of the Tangent-Secant Theorem (5.5.5), the remaining details are left to Exercise 2. ■

■ THEOREM 5.5.4 [Secant Theorem]

Let $\overline{AE}$ and $\overline{CE}$ be secants to a circle. If $\overline{AE}$ intersects the circle in A and B and that $\overline{CE}$ intersects it at C and D, then

$$AE \cdot BE = CE \cdot DE.$$

■ THEOREM 5.5.5 [Tangent-Secant Theorem, *Elements* III.36]

If D is a point taken outside a circle such that $\overline{CD}$ is tangent to the circle at C and $\overline{AD}$ is a secant intersecting the circle at A and B, then

$$AD \cdot BD = CD^2.$$

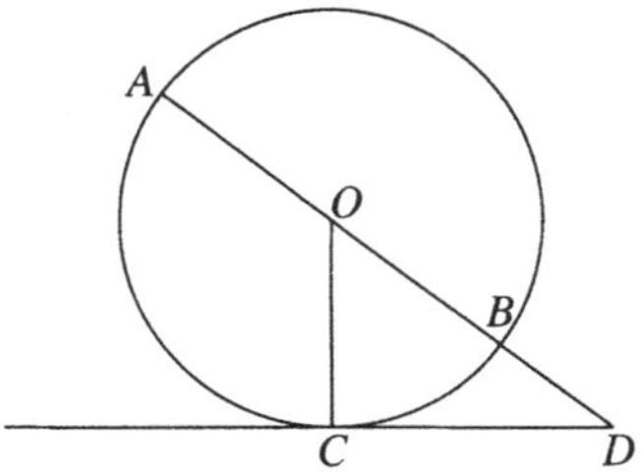

PROOF

Assume the antecedent of the theorem. We have two cases to consider. First let the center O of the circle lie on $\overline{AB}$. Join $\overline{OC}$ to form a right angle with $\overline{CD}$ by Theorem 5.5.2. We then have

$$AD \cdot BD + OB^2 = OD^2 \tag{5.2}$$

by Exercise 1(b). However, the radii $\overline{OC}$ and $\overline{OB}$ are congruent. Thus,

$$AD \cdot BD + OC^2 = OD^2.$$

By the Pythagorean Theorem we know that

$$OD^2 = CD^2 + OC^2,$$

so we have

$$AD \cdot BD + OC^2 = CD^2 + OC^2.$$

When we subtract OC^2 from both sides of the equation, we have the result.

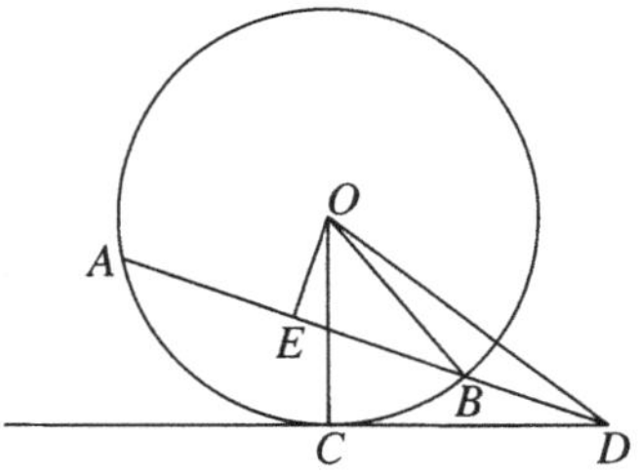

Now assume that O is not on $\overline{AB}$. Drop a perpendicular from O to $\overline{AD}$ at E and join $\overline{OB}$, $\overline{OC}$, and $\overline{OD}$. By Theorem 5.5.1, the point E bisects $\overline{AB}$, and again by Exercise 1(b),

$$AD \cdot BD + EB^2 = ED^2. \tag{5.3}$$

Adding OE^2 to both sides of the equation yields

$$AD \cdot BD + EB^2 + OE^2 = ED^2 + OE^2.$$

Because $EB^2 + OE^2 = OB^2$ and $ED^2 + OE^2 = OD^2$,

$$AD \cdot BD + OB^2 = OD^2.$$

However, $\overline{OB} \cong \overline{OC}$, so

$$AD \cdot BD + OC^2 = OD^2.$$

Furthermore, $OD^2 = CD^2 + OC^2$. Hence,

$$AD \cdot BD + OC^2 = CD^2 + OC^2,$$

and from here we again conclude that

$$AD \cdot BD = CD^2. \blacksquare$$

Circumscribing and Inscribing

Book IV of the *Elements* focuses on inscribing circles within and circumscribing circles about polygons. Euclid includes results about squares and pentagons. He also writes on triangles. Three examples are given here.

■ **CONSTRUCTION 5.5.6 [*Elements* IV.3]**

Circumscribe about a given circle a triangle that is similar to a given triangle.

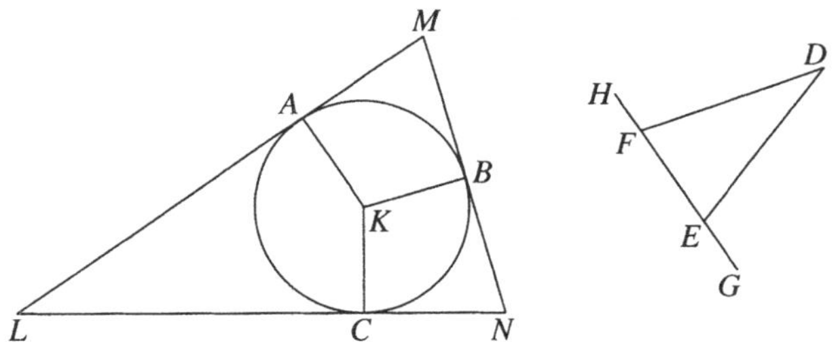

PROOF

Take any circle and $\triangle DEF$. Extend $\overline{FE}$ to $\overline{HG}$ as in the diagram. Let K be the center of the circle (Exercise 4). We will now copy the two external angles from $\triangle DEF$. Choose A and B on the circle such that

$$\sphericalangle AKB \cong \sphericalangle DEG.$$

Next, find C on the circle such that

$$\sphericalangle BKC \cong \sphericalangle DFH.$$

Use Construction 5.2.14 to do this. Let $\overline{LM}$ be tangent to the circle at A, $\overline{MN}$ tangent at B, and $\overline{NL}$ tangent at C. Appealing to Theorem 5.5.2, we see that $\overline{LM}$ is perpendicular to $\overline{AK}$ and $\overline{MN}$ is perpendicular to $\overline{BK}$. Thus, $\sphericalangle KAM$ and $\sphericalangle KBM$ are supplementary, which implies that $\sphericalangle AKB$ and $\sphericalangle AMB$ are supplementary by Corollary 5.4.4. In our given triangle, $\sphericalangle DEG$ and $\sphericalangle DEF$ are also supplementary. Hence

$$m\sphericalangle AKB + m\sphericalangle AMB = m\sphericalangle DEG + m\sphericalangle DEF.$$

Because $\measuredangle AKB \cong \measuredangle DEG$, we have $\measuredangle LMN \cong \measuredangle DEF$. A similar argument will yield $\measuredangle LNM \cong \measuredangle DFE$. Thus, two pairs of corresponding angles are congruent, so ΔLMN is similar to ΔDEF. ■

Heron's proof of his famous equation for the area A of a triangle with sides of length a, b, and c, and semiperimeter $s = (a + b + c)/2$,

$$A = \sqrt{s(s-a)(s-b)(s-c)},$$

involves a circle inscribed within the triangle of interest. Due to space constraints, it was excluded from this text, yet it is a very nice proof and worthy of study. See Heath's version in Euclid (1925, Vol. 2, 87–88). We will, however, give Euclid's construction of a circle inscribed within a triangle.

■ CONSTRUCTION 5.5.7 [*Elements* IV.4]

Inscribe a circle within a given triangle.

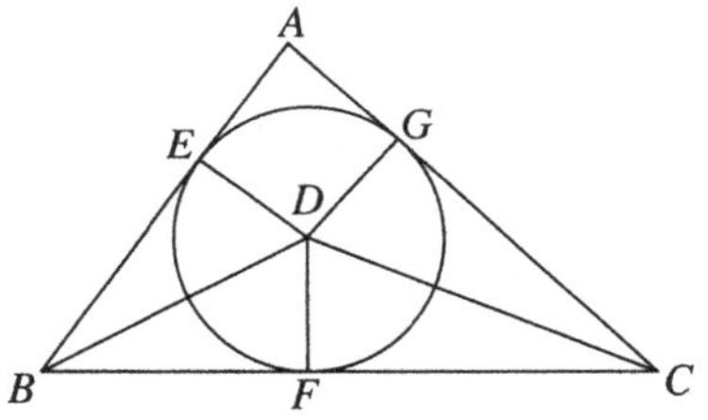

PROOF

We are given ΔABC. Bisect $\measuredangle ABC$ and $\measuredangle ACB$ with $\overline{BD}$ and $\overline{CD}$, respectively. Construct $\overline{DE}$ perpendicular to $\overline{AB}$, $\overline{DF}$ perpendicular to $\overline{BC}$, and $\overline{DG}$ perpendicular to $\overline{AC}$. Since $\measuredangle ABD \cong \measuredangle CBD$, $\measuredangle BED \cong \measuredangle BFD$, and $\overline{BD} \cong \overline{BD}$, we have

$$\Delta EBD \cong \Delta FBD$$

by AAS. Therefore, $\overline{DE} \cong \overline{DF}$. Following a similar argument we find that $\overline{DG} \cong \overline{DF}$. We may now draw $\odot EFG$ with D as its center and $\overline{DE}$ as its radius. Since $\overline{DE}$, $\overline{DF}$, and $\overline{DG}$ are congruent, $\overline{DF}$ and $\overline{DG}$ also serve as radii to the desired circle. Furthermore, since the segments are perpendicular to the sides of the given triangle, these sides are tangent to the circle. Hence, $\odot EFG$ is inscribed within ΔABC. ■

The point D in the proof above is known as the **incenter** of the triangle. The point F found in the next proof is called the **circumcenter** of the triangle, and the circle is called the **circumcircle**.

■ CONSTRUCTION 5.5.8 [*Elements* IV.5]

Circumscribe a circle about a given triangle.

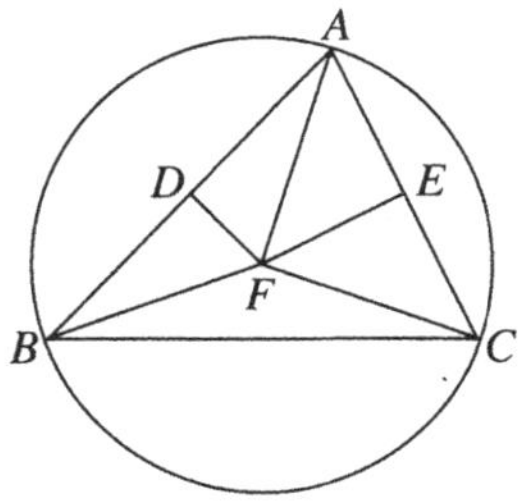

PROOF

Take $\triangle ABC$ and construct the perpendicular bisector of $\overline{AB}$ at midpoint D and the perpendicular bisector of $\overline{AC}$ at midpoint E. These bisectors meet at a point F, by Exercise 5 of Section 5.4. We have three cases to consider: F is in the interior of $\triangle ABC$, F is in the exterior of $\triangle ABC$, and F is on one of the sides of $\triangle ABC$. We prove the first case here and leave the others to Exercise 12. Join $\overline{FA}$, $\overline{FB}$, and $\overline{FC}$. Since $\overline{AD} \cong \overline{DB}$, $\overline{DF} \cong \overline{DF}$, and $\sphericalangle BDF \cong \sphericalangle ADF$, by SAS we may conclude that

$$\triangle FDB \cong \triangle FDA.$$

Thus, $\overline{AF} \cong \overline{BF}$. We can also show that $\overline{AF} \cong \overline{CF}$ and $\overline{BF} \cong \overline{CF}$. We may then construct the circle with center F and radius $\overline{AF}$. This circle will intersect the given triangle only at the points A, B, and C, so the circle has been circumscribed about $\triangle ABC$. ■

The previous constructions all produce a unique circle because

three noncollinear points are enough to determine a circle. (5.4)

The proof of this result requires the next theorem.

■ **THEOREM 5.5.9 [*Elements* III.10]**

Two distinct circles meet in at most two points.

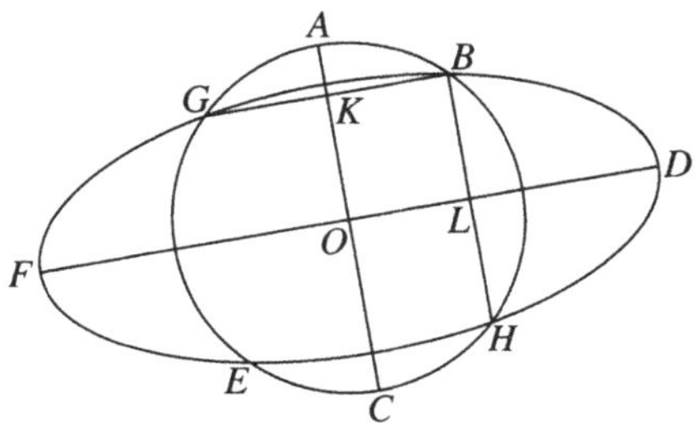

PROOF

Suppose that $\odot ABC$ and $\odot DEF$ meet at G, B, and H. Join $\overline{HB}$ and $\overline{BG}$. Let K be the midpoint of $\overline{GB}$ and L the midpoint of $\overline{BH}$. Assume that

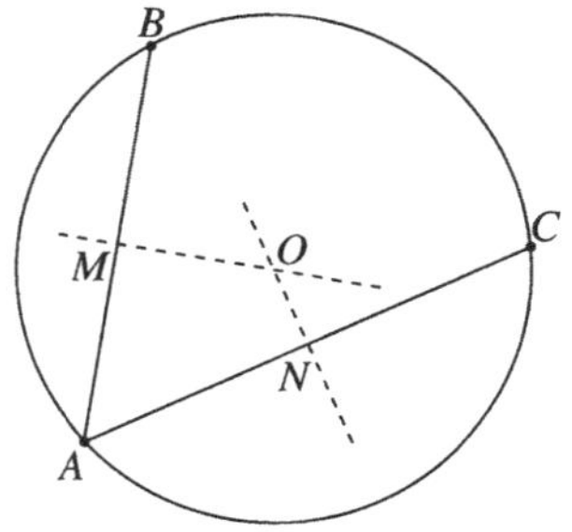

Figure 5.7 Three noncollinear points determine a circle.

$\overline{AKC}$ is perpendicular to $\overline{GB}$ and $\overline{FLD}$ is perpendicular to $\overline{BH}$. Let O be the intersection of $\overline{AC}$ and $\overline{FD}$ (Exercise 5 of Section 5.4). Therefore, by Exercise 5 the center of $\odot ABC$ is on both $\overline{AC}$ and $\overline{FD}$. This means that the center of $\odot ABC$ is O. Similarly, the center of $\odot DEF$ is also O, but this is impossible by Exercise 9. ■

Now take three noncollinear points A, B, and C. Join $\overline{AB}$ and $\overline{AC}$. Let M be the midpoint of $\overline{AB}$ and N the midpoint of $\overline{AC}$ (Figure 5.7). Since the points are not collinear, the perpendicular bisectors of the two segments intersect in a point O, as in the proof of Theorem 5.5.9. Since A and B are equidistant to all points on $\overline{OM}$ and since A and C are equidistant to all points on $\overline{ON}$, the circle with center O and radius equal to $OA = OB = OC$ will pass through the three points. This shows that through three points there exists a circle. That A, B, and C lie on only one circle follows from Theorem 5.5.9. Thus, we have proven Proposition 5.4.

Exercises

1. Confirm algebraically:
   (a) Equation 5.1 in the proof of Theorem 5.5.4
   (b) Equations 5.2 and 5.3 in the proof of Theorem 5.5.5

2. Finish the proof of the Chord Theorem (5.5.3).

3. Prove the Secant Theorem (5.5.4).

4. Construct the center of a given circle.

5. Prove: If a given secant line of a circle is the perpendicular bisector of a chord of the circle, the center of the circle lies on the given secant.

6. Suppose that in a particular circle two secant lines intersect each other in the interior of the circle but neither contains the center of the circle. Prove that the secants do not bisect each other.

7. We say that two circles are **tangent** to each other if they intersect in a single point. Prove that tangent circles do not share a common center.

8. Given $\odot O$ and P on the circle, prove that there exists exactly one tangent line to $\odot O$ at P. Is there exactly one tangent circle to $\odot O$ at P?

9. If two distinct circles meet in two points, the circles do not have a common center.

10. Prove that the segment that joins A and D in Construction 5.5.7 bisects $\sphericalangle EAG$.

11. Prove: The three perpendicular bisectors of the sides of a triangle are concurrent.

12. Prove the other two cases in the proof of Theorem 5.5.8.

13. Let C be external to $\odot O$ and let A and B be two distinct points on the circumference of $\odot O$ such that $\overline{AC}$ and $\overline{BC}$ are tangents to the circle. Show that $\overline{AC} \cong \overline{BC}$.

14. Prove that a circle and a line can intersect in at most two points.

15. The **median** of a triangle is a segment that joins a vertex to the midpoint of the opposite side. Prove that the three medians of a triangle meet in a single point, called the **centroid** of the triangle. (*Hint:* Use Theorem 5.4.11.)

16. Do the following:
    (a) Construct a chord of a circle that is congruent to a given segment. (The given segment cannot be longer that the diameter of the circle.)
    (b) Inscribe within a circle a triangle similar to a given triangle.
    (c) Inscribe a square within a given circle.
    (d) Circumscribe a square about a given circle.
    (e) Inscribe a circle within a given square.
    (f) Circumscribe a circle about a given square.
    (g) Construct an isosceles triangle in which the base angles are twice the measure of the third angle.
    (h) Inscribe a regular pentagon in a given circle.
    (i) Circumscribe a regular pentagon about a given circle.
    (j) Inscribe a circle within a regular pentagon.
    (k) Circumscribe a circle about a regular pentagon.
    (l) Inscribe a regular hexagon within a given circle.
    (m) Inscribe a regular 15-gon within a circle.
    (n) Inscribe a regular 30-gon within a circle.

5.6 THE PYTHAGOREAN THEOREM REVISITED

There is something exceptional about the Pythagorean Theorem. It was known by many different cultures throughout the ancient world. Some knew only special cases, whereas others knew its completely general form. Some of its proofs were very easy to understand, others not. What makes the theorem stand out is its simplicity, but from a theorem that even grade-school students can understand comes a result that was able to overthrow a philosophy. The importance of such a theorem is often validated by naming it after a renowned mathematician, as is the case with the Pythagorean Theorem. This, no doubt, contributes to its fame. Association with the mythical Pythagoras can do nothing but elevate its standing. Adding to the testimony regarding its almost legendary status in mathematics is that through the ages it has received other names. It has been called *The Theorem of the Bride* because some later Pythagoreans gave the numbers 3, 4, and 5 matrimonial interpretations. The first odd number, 3, represents maleness, the square of two, 4, represents the female, and the sum of two and three, 5, represents their offspring. Other names include *The Bride's Chair*, because the diagram in Euclid's proof resembles the chair on which an Eastern bride would be carried to her wedding; the *Dulcarnon*, which is from the Persian meaning "two-horned"; and *Francisci turnica*, since the figure resembles a monk's hood (Euclid 1925, vol. 1, 417–418).

■ THEOREM 5.6.1 [*Elements* I.47, Pythagorean Theorem]

In right triangles the area of the square on the hypotenuse is equal to the sum of the areas of the squares on the legs.

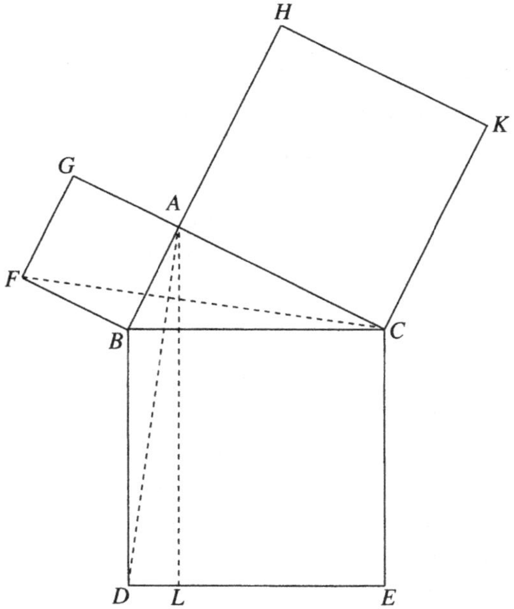

PROOF

Given $\triangle ABC$ with $\measuredangle BAC$ right, construct squares $ABFG$, $BCED$, and $ACKH$ (Exercise 3 in Section 5.4). Draw $\overline{AL}$ parallel to $\overline{BD}$ such that L is on $\overline{DE}$. Finally, join $\overline{AD}$ and $\overline{FC}$. Since $\measuredangle BAC$ and $\measuredangle BAG$ are both right angles, the points G, A, and C are collinear (Theorem 5.3.8). Similarly, B, A, and H are on the same line segment. Furthermore, $\sphericalangle DBC$ and $\sphericalangle FBA$ are both right angles, so $\sphericalangle DBC \cong \sphericalangle FBA$. Hence,

$$m\sphericalangle ABC + m\sphericalangle DBC = m\sphericalangle ABC + m\sphericalangle FBA,$$

so $\sphericalangle DBA \cong \sphericalangle FBC$ by Common Notion 2. Since $\overline{DB} \cong \overline{BC}$ and $\overline{FB} \cong \overline{BA}$, we conclude that $\triangle ABD \cong \triangle FBC$ by SAS. Now, the area of $\square BL$ is double that of $\triangle ABD$ because they share base $\overline{BD}$ and are within the parallels $\overline{BD}$ and $\overline{AL}$ (Theorem 5.4.9). Similarly, the area of $\square BAGF$ is twice the area of $\triangle FBC$. Thus, since $\triangle ABD \cong \triangle FBC$, $\square BL$ has the same area as $\square BAGF$. An argument similar to the previous one will show that the $\square CL$ has the same area as $\square ACKH$. Since $\square BDEC$ has an area equal to the sum of the areas of $\square BL$ and $\square CL$, we have proven the result. ■

In Section 4.1 we have already seen various proofs of Theorem 5.6.1. Euclid's demonstration is purely geometric. It does not rely on ratios but instead depends on an ingenious application of the Parallel Postulate. In the words of Proclus:

> If we listen to those who like to record antiquities, we shall find them attributing this theorem to Pythagoras and saying that he sacrificed an ox on its discovery. For my part, though I marvel at those who first noted the truth of this theorem, I admire more the author of the *Elements*, not only for the very lucid proof by which he made it fast, but also because in the sixth book he laid hold of a theorem even more general than this and secured it by irrefutable scientific arguments. For in that book he proves generally that in right-angled triangles the figure on the side that subtends the right angle is equal to the similar and similarly drawn figures on the sides that contain the right angle. (Proclus 1970, 426)

The proposition to which Proclus refers is Theorem 5.6.4 below. Its proof requires that we first prove two propositions.

■ **THEOREM 5.6.2 [*Elements* VI.19]**

The ratio of the areas of two similar triangles equals the ratio of the squares of two corresponding sides.

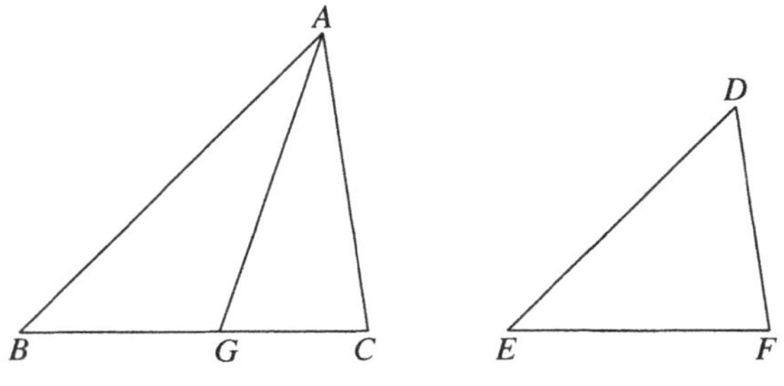

PROOF

Let $\triangle ABC \sim \triangle DEF$. Find G on $\overline{BC}$ such that

$$\frac{AB}{DE} = \frac{BC}{EF} = \frac{EF}{BG} \tag{5.5}$$

by Exercise 25 of Section 5.4. Join $\overline{AG}$. By Exercise 5 we may conclude that

$$\text{area}(\triangle ABG) = \text{area}(\triangle DEF). \tag{5.6}$$

Since $\triangle ABC$ has the same height as $\triangle ABG$,

$$\frac{\text{area}(\triangle ABC)}{\text{area}(\triangle ABG)} = \frac{BC}{BG}.$$

Therefore,

$$\frac{\text{area}(\triangle ABC)}{\text{area}(\triangle DEF)} = \frac{\text{area}(\triangle ABC)}{\text{area}(\triangle ABG)} = \frac{BC}{EF} \cdot \frac{EF}{BG} = \frac{BC^2}{EF^2},$$

where the last equality holds by Equation 5.5. ■

The previous result along with Exercise 24 (Section 5.4) is enough to prove:

■ **COROLLARY 5.6.3**

If CD is the mean proportional between AB and EF, then for any two similar, convex polygons with $\overline{AB}$ and $\overline{CD}$ as corresponding sides,

$$\frac{AB}{EF} = \frac{\text{area}(\text{polygon on } \overline{AB})}{\text{area}(\text{polygon on } \overline{CD})}.$$

We may now prove Euclid's generalization of the Pythagorean Theorem.

■ **THEOREM 5.6.4 [*Elements* VI.31]**

In right triangles the area of the convex polygon with one side on the hypotenuse is equal to the sum of the areas of the convex polygons with one side each on the legs provided that the polygons on the legs are similar to the polygon on the hypotenuse.

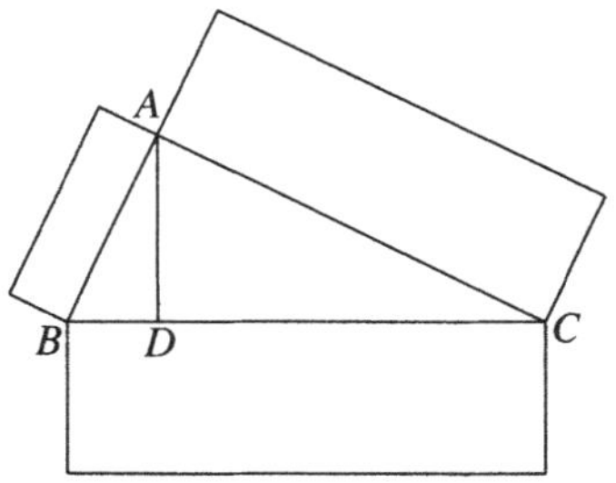

PROOF

We are given $\triangle ABC$ with $\sphericalangle BAC$ right. Draw convex polygons on the legs and hypotenuse so that the polygons on the legs are similar to the polygon on the hypotenuse. (In the diagram the polygons are rectangles, but this will not be used in this proof.) Find D on $\overline{BC}$ such that $\overline{AD}$ is perpendicular to the hypotenuse. As we have seen on page 100,

$$\triangle ABC \sim \triangle DBA \sim \triangle DAC.$$

Hence,

$$\frac{BC}{BA} = \frac{BA}{BD},$$

so by Corollary 5.6.3,

$$\frac{BC}{BD} = \frac{\text{area(polygon on } \overline{BC})}{\text{area(polygon on } \overline{BA})}.$$

Similarly,

$$\frac{BC}{DC} = \frac{\text{area(polygon on } \overline{BC})}{\text{area(polygon on } \overline{AC})}.$$

Therefore,

$$\frac{BC}{BD + DC} = \frac{\text{area(polygon on } \overline{BC})}{\text{area(polygon on } \overline{BA}) + \text{area(polygon on } \overline{AC})}.$$

However, $BC = BD + DC$, which allows us to conclude that

$$\text{area(polygon on } \overline{BC}) = \text{area(polygon on } \overline{BA}) + \text{area(polygon on } \overline{AC}). \blacksquare$$

Exercises

1. Prove the converse to the Pythagorean Theorem (Euclid 1925, I.48): If in $\triangle ABC$, $AC^2 = AB^2 + BC^2$, then $\sphericalangle ABC$ is right.

2. Use the Pythagorean Theorem to confirm these two results:
   (a) In an isosceles right triangle, the length of the hypotenuse is $\sqrt{2}$ times the length of either leg.
   (b) In a right triangle where one acute angle is twice the measure of the other acute angle, the length of the hypotenuse is twice that of the shorter leg, and the length of the longer leg is $\sqrt{3}$ times the length of the shorter leg.

3. Construct a square that has area equal to that of a given rectangle.

4. [HL] Suppose that the hypotenuse and leg of one triangle is congruent to the hypotenuse and leg of another triangle. Prove that these triangles are congruent.

5. Demonstrate Equation 5.6 in the proof of Theorem 5.6.2.

6. Prove Corollary 5.6.3.

7. Prove that Theorem 5.6.4 implies Theorem 5.6.1.

8. Euclid (1925, II.12, 13) proves what we call the **Law of Cosines**. Using modern terminology: Given $\triangle ABC$ with $a = BC$, $b = AC$, and $c = AB$,

$$c^2 = a^2 + b^2 - 2ab\cos C.$$

He proves the theorem by examining triangles with an obtuse angle and those with only acute angles separately. Prove the following steps in his proof.

(a) If C lies on $\overline{AB}$, then $AB^2 = AC^2 + CB^2 + 2AC \cdot CB$.

(b) Let $\triangle ABC$ have $\sphericalangle BAC$ obtuse. Let D be on $\overline{AC}$ produced beyond A such that $\overline{BD}$ and $\overline{AC}$ produced are perpendicular. From this we conclude that $BC^2 = BA^2 + AC^2 + 2AC \cdot DA$.

(c) If C lies on $\overline{AB}$, then $AB^2 + CB^2 = 2AB \cdot CB + AC^2$.

(d) Let $\triangle ABC$ have only acute angles. If D is on $\overline{BC}$ so that $\overline{AD}$ is perpendicular to $\overline{BC}$, then $AC^2 = BC^2 + BA^2 - 2BC \cdot BD$.

9. Confirm that Exercises 8(b) and 8(d) are equivalent to our version of the Law of Cosines.

10. **Ptolemy's Theorem** states that if quadrilateral $ABCD$ is inscribed within a circle, then $AC \cdot BD = AB \cdot CD + BC \cdot DA$ (Maor 2007, 102–104). Prove:

(a) Ptolemy's Theorem.

(b) that Ptolemy's Theorem implies the Pythagorean Theorem.

11. The following is attributed to Leonardo Da Vinci (Maor 2007, 104–106). Let $\triangle ABC$ have a right angle at A. On the sides of the triangle construct squares $KCAG$, $CEDB$, and $ABIH$. Join $\overline{GH}$ and construct $\triangle FDE$ to be congruent to $\triangle ABC$. Prove the following steps that lead to a proof of the Pythagorean Theorem.

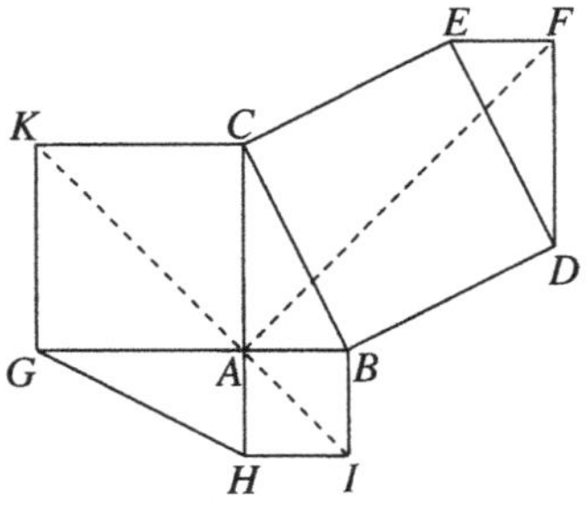

(a) $KGHI \cong KCBI$ and $ACEF \cong FDBA$.

(b) area$(ACEFDB)$ = area$(KGHIBC)$.

(c) area$(CEDB)$ = area$(KCAG)$ + area$(ABIH)$.

12. Prove this theorem due to Pappus (Maor 2007, 58–59): On $\triangle ABC$ construct $\square ABDE$ and $\square ACFG$. Extend $\overline{DE}$ and $\overline{FG}$ to meet at point H. Draw $\overline{BM} \cong \overline{CN}$, where each is parallel and congruent to $\overline{HA}$. This produces $\square BMNC$. We conclude that area$(\square BMNC)$ = area$(\square ABDE)$ + area$(\square ACFG)$.

13. Go back through the proofs beginning with Section 5.4 and identify each step that required the Parallel Postulate.

CHAPTER 6

ARCHIMEDES

We know little about the life of Archimedes. He was evidently born in 287 B.C. in Syracuse on the island of Sicily. His mother is unknown. His father, Phidias, was an astronomer and either a relative or a friend of Hieron II, the king of Syracuse, and Hieron's son, Gelon. We know that Archimedes visited Egypt to study at Alexandria. It was on this trip when he probably invented what we call the Archimedean screw. It is a device designed to raise water from a lower elevation to a higher one. No doubt more important to Archimedes, it was at Alexandria where he met the astronomer Conon of Samos; Dositheus, his student; and Eratosthenes of Cyrene. All three would become lifelong friends of Archimedes, even though most of his life Archimedes lived in Syracuse (Heath 1921b, 16).

Outside of his writings, much of what we know about Archimedes comes from *Marcellus* by the ancient historian Plutarch (1932, 368–388). He wrote this work in A.D. 75 about the life of Marcus Claudius Marcellus, a commander in the Roman army. Plutarch noted that he had the reputation as a fierce and skillful warrior. However, aside from the battlefield, Marcellus was modest and respectable. He was also a learned man who had studied the Greek teachings. In 214 B.C., a year after the death of Hieron, Marcellus laid siege to Syracuse. Plutarch details the situation. The Carthaginians had taken control of Sicily, and the Syracusans had aligned themselves with the invaders. In response, the Roman senate sent Marcellus and his army to Sicily. They first took the city of Leontini. Despite the fact that Marcellus took no prisoners except those who had deserted the city, the people of Syracuse received word that the Romans had killed every last adult. They therefore readied themselves to resist the Romans (Plutarch 1932, *Marcellus*, 376–377).

And resist them they did. Marcellus brought forces to bear upon Syracuse by both land and sea. Each attack was repelled. His land forces were met with huge rocks flung down upon them by catapults, and through holes in the walls of the city rained arrows and missiles. Some were designed to hit near targets and others far ones. Marcellus' warships received a similar fate. Machines dropped massive weights on the ships, and giant levers yanked them from the water. Each of these war devices was designed at an earlier date by Archimedes at the command of King Hieron. Together, the machines dealt a terrible blow to the Romans. Plutarch (1932, *Marcellus*, 378) states that things went so badly during their attacks on the city that

> such terror had seized upon the Romans, that, if they did but see a little rope or a piece of wood from the wall, instantly crying out, that there it was again, Archimedes was about to let fly some engine at them, they turned their backs and fled.

Marcellus ordered his ships to pull back and his troops to lay siege.

After two years, the Syracusans finally made a mistake. During the siege the leaders of Syracuse negotiated from time to time with Marcellus in an attempt to end the hostility. During these meetings Marcellus noticed a tower that was perfectly situated for a secret assault. He just needed the right time. The people of Syracuse gave it to him. During the feast of Diana, there was a festival in the city. While they were busy playing and drinking, the Romans attacked, and by daybreak the city was theirs. Syracuse would be plundered but not razed to the ground. Those were the orders of Marcellus. He was willing to let his men take bounty but not to hurt the citizens of Syracuse. Unfortunately, in the confusion of the day, Archimedes was killed. The stories vary. Some say that he was studying some figures that he had drawn on the ground and was simply killed by a soldier who did not know who he was. Others have Archimedes working on a problem, and when the soldier approached him, Archimedes insisted that the soldier wait until he had completed the problem. At that refusal to cooperate, the soldier killed Archimedes (Archimedes 1897, xv, xvii). However it happened, the event had a profound effect on Marcellus:

> Certain it is that his death was very afflicting to Marcellus; and that Marcellus ever after regarded him that killed him as a murderer; and that he sought for his kindred and honoured them with signal favours. (Plutarch 1932, *Marcellus*, 380)

6.1 THE ARCHIMEDEAN LIBRARY

Unlike his life, we do know much about the work of Archimedes because most of it has survived. In this section we survey these works. The order in which they were written has been reconstructed in various sources. We will rely on Heath (1921b, 22). The first of Archimedes' works is a two-volume piece entitled *On Plane Equilibriums*

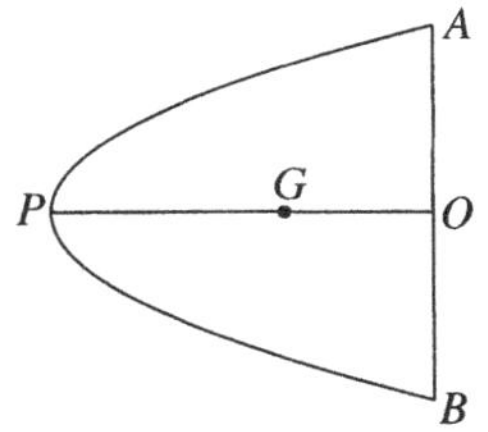

Figure 6.1 G is the center of gravity of a parabolic section.

(Heath 1921b, 75–81). The first book gives a set of postulates from which the famous Law of the Lever is derived. Archimedes referred to this principle when he wrote to Hieron that given "a place to stand on! and I will move the earth" (Dijksterhuis 1987, 14). Results concerning centers of gravity of some simple geometric shapes then follow the law. The second book is dedicated to finding the center of gravity of a **parabolic section** (Figure 6.1). He proved that if G is the center of gravity of a parabolic section and $\overline{PO}$ is its axis of symmetry, then G is on $\overline{PO}$ and

$$\frac{PG}{GO} = \frac{3}{2}. \tag{6.1}$$

Archimedes next determined the area of a parabolic section in his *Quadrature of the Parabola* (Heath 1921b, 85–91). He wrote the work as a letter to Dositheus upon the death of Conon. He opens:

> When I heard that Conon, who was my friend in this lifetime, was dead, but that you were acquainted with Conon and withal versed in geometry, while I grieved for the loss not only of a friend but an admirable mathematician, I set myself the task of communicating to you, as I had intended to send to Conon, a certain geometrical theorem which had not been investigated before but has now been investigated by me, and which I first discovered by means of mechanics and then exhibited by means of geometry. (Archimedes 1897, *Quadrature of the Parabola*, Introduction, 233)

To find the area, Archimedes began by inscribing a triangle within the parabolic section so that one vertex of the triangle is at the vertex P of the parabola, and the base of the triangle is on $\overline{AB}$ (Figure 6.2). As before, $\overline{PO}$ is the axis of symmetry. We focus our work on the top half of the section. Find the midpoint M of $\overline{AO}$ and draw $\overline{RM}$ parallel to $\overline{PO}$ with R on the parabola. We will show that

$$4\text{area}(\triangle PRA) = 2\text{area}(\triangle PAM) = \text{area}(\triangle PAO). \tag{6.2}$$

The second equality follows because

$$2\text{area}(\triangle PAM) = AM \cdot PO = \frac{AO \cdot PO}{2} = \text{area}(\triangle PAO).$$

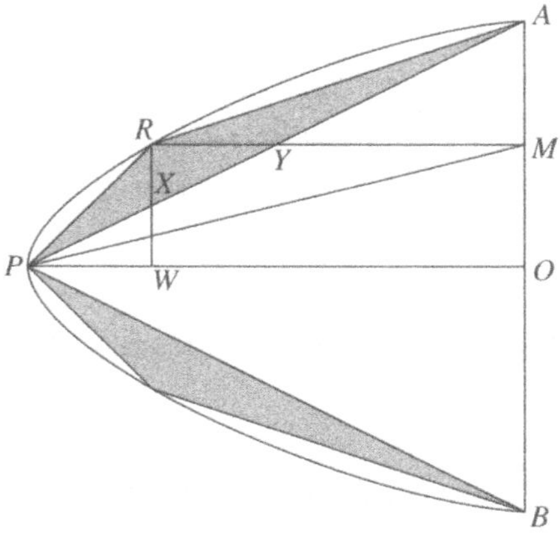

Figure 6.2 The quadrature of the parabola.

To see the first equality, join $\overline{PM}$ and then draw $\overline{RW}$ perpendicular to $\overline{PO}$ at W. Denote the intersection of $\overline{RW}$ and $\overline{PA}$ by X. Since by Exercise 2 we have

$$RX = \frac{AM}{2}, \tag{6.3}$$

we may conclude that

$$\begin{aligned} 2\text{area}(\triangle PRA) &= 2\text{area}(\triangle PRX) + 2\text{area}(\triangle ARX) \\ &= RX \cdot PW + RX \cdot WO \\ &= RX \cdot PO \\ &= \frac{AM \cdot PO}{2} \\ &= \text{area}(\triangle PAM). \end{aligned}$$

Repeat this construction on the bottom half of the parabola. Let A_0 represent the area of $\triangle PAB$ and A_1 denote the combined areas of the two shaded triangles (the triangles **constructed on** $\triangle PAB$); then, by Equation 6.2,

$$A_1 = \frac{1}{4}A_0.$$

When we repeat the construction above on the two shaded triangles, we find that the total area of the four triangles constructed on the shaded triangles is a fourth of A_1. Continue constructing triangles so that if A_i represents the area of triangles constructed on given triangles as above,

$$A_i = \frac{1}{4}A_{i-1}.$$

We can now take Archimedes result and apply methods of calculus to find that the area of the parabola is given by the infinite geometric series,

$$\sum_{i=0}^{\infty} A_i = \sum_{i=0}^{\infty} \left(\frac{1}{4}\right)^i A_0 = \frac{4}{3} A_0.$$

Next in the Archimedean library comes *The Method* and the two books of *On the Sphere and Cylinder* (Heath 1921b, 27–50). *The Method* is a letter written to Eratosthenes. Both books of *On the Sphere and Cylinder* are addressed to Dositheus. The first proves the formulas for the volume and surface area of a sphere. The second deals with a few problems and theorems concerning spheres cut by planes.

In *On Spirals*, Archimedes studies a mechanical curve that is believed to have been discovered by Conon, but we know it as the **spiral of Archimedes** (Heath 1921b, 64–75). He defines the spiral as follows:

> If a straight line drawn in a plane revolves at a uniform rate about one extremity which remains fixed and returns to the position from which it started, and if, at the same time as the line revolves, a point moves at a uniform rate along the straight line beginning from the extremity which remains fixed, the point will describe a spiral in the plane. (Archimedes 1897, *On Spirals*, Definition 1, 165)

The equation $r = a\theta$ in polar coordinates describes such a curve. It is illustrated in Figure 6.3. Call $\overline{OA}$ the **initial line**. The **first circle** is the circle with center O and with the initial line as its radius. Archimedes in *On Spirals* finds the area between one turn of the spiral and the initial line. He divides the first circle into n sectors of equal area. Within each of these sectors he constructs two more sectors: one that is inscribed within the spiral and one that is circumscribed about it (Figure 6.4). Define

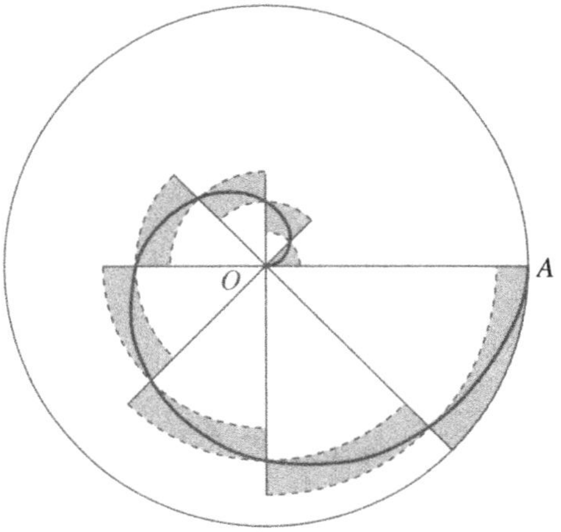

Figure 6.3 The spiral of Archimedes.

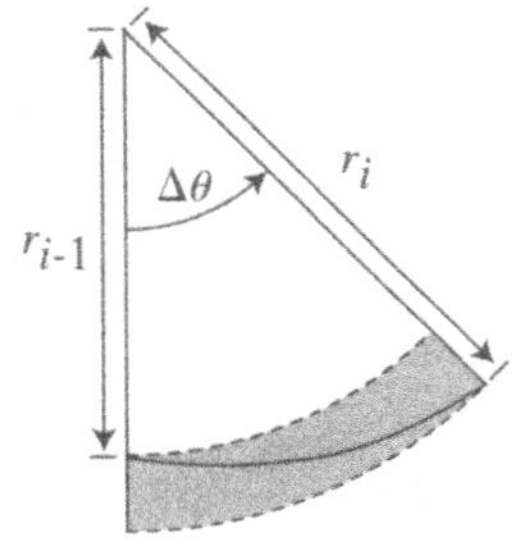

Figure 6.4 Inscribed and circumscribed sectors.

$$\Delta\theta = \frac{2\pi}{n},$$

and for $i = 0, 1, \ldots, n$ let

$$r_i = ai\Delta\theta.$$

We then have

$$\text{area(inscribed sector)} = \frac{1}{2}r_{i-1}^2\Delta\theta \tag{6.4}$$

and

$$\text{area(circumscribed sector)} = \frac{1}{2}r_i^2\Delta\theta \tag{6.5}$$

for $i = 1, 2, \ldots, n$.

Now let A represent the area enclosed by one turn of the spiral and A_i represent the ith slice of that area. By Equations 6.4 and 6.5, we see that

$$\frac{1}{2}r_{i-1}^2\Delta\theta \leqslant A_i \leqslant \frac{1}{2}r_i^2\Delta\theta,$$

and then

$$\sum_{i=1}^{n}\frac{1}{2}r_{i-1}^2\Delta\theta \leqslant \sum_{i=1}^{n}A_i \leqslant \sum_{i=1}^{n}\frac{1}{2}r_i^2\Delta\theta.$$

When we let $n \to \infty$, we obtain

$$\lim_{n\to\infty}\sum_{i=1}^{n}\frac{1}{2}r_{i-1}^2\Delta\theta = \lim_{n\to\infty}\sum_{i=1}^{n}\frac{1}{2}r_i^2\Delta\theta = \int_0^{2\pi}\frac{1}{2}a^2\theta^2\,d\theta.$$

Since

$$\int_0^{2\pi}\frac{1}{2}a^2\theta^2\,d\theta = \frac{4}{3}\pi^3a^2, \tag{6.6}$$

we conclude by the Sandwich Theorem that

$$A = \frac{4}{3}\pi^3a^2 = \frac{1}{3}\pi(2\pi a)^2 = \frac{1}{3}\pi OA^2.$$

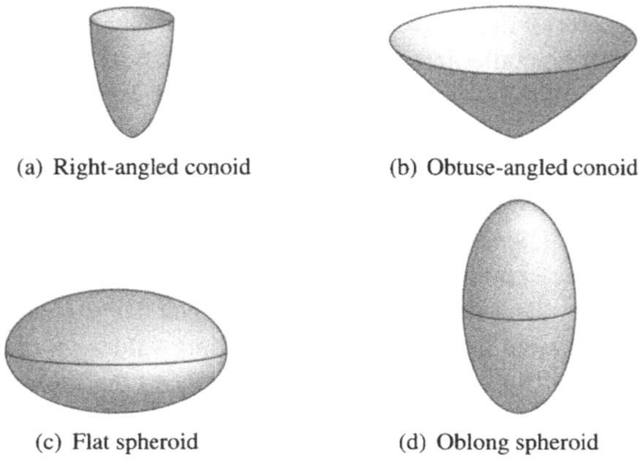

Figure 6.5 Conoids and spheroids with a vertical axis of revolution.

Archimedes next studies volumes of revolution. In particular he will examine the **right-angled conoid** (which is also known as the **paraboloid of revolution**), the **obtuse-angled conoid** (or the **hyperboloid of revolution**), the **flat spheroid** (an **ellipsoid of revolution** rotated about the minor axis), and the **oblong spheroid** (an ellipsoid of revolution rotated about the major axis). These loci are illustrated in Figure 6.5. He does this in his work named *On Conoids and Spheroids* (Heath 1921b, 56–64). His theorems include results on what locus results when a conoid or spheroid is cut by a plane and also results about their volumes. For example:

> Any segment of a paraboloid of revolution is half as large again as the cone or segment of a cone which has the same base and the same axis. (Archimedes 1897, *On Conoids and Spheroids* 21–22, 131)

We can view the right-angled conoid with height h and base radius r as the solid of revolution of the parabola given by

$$x^2 = \frac{r^2}{h}y$$

in the first quadrant. Using a little calculus, the volume obtained by revolving this curve about the y-axis is

$$\int_0^h \frac{\pi r^2}{h} y\, dy = \frac{1}{2}\pi r^2 h. \tag{6.7}$$

Because the volume of the cone with the same height and base is $\pi r^2 h/3$, we see that the volume of the paraboloid of revolution is 3/2 times the volume of the corresponding cone.

In *On Floating Bodies* Archimedes introduces the basics of the science that we now call **hydrostatics** (Heath 1921b, 91–97). If the legend is to be believed, King Hieron commissioned a solid gold wreath to be made. The craftsman instead used some silver in the interior of the wreath. Suspecting that he had been cheated, Hieron asked Archimedes if he could determine whether it was pure gold without melting it down. Shortly thereafter, as Archimedes entered a full tub to bathe, he was inspired by watching the water overflow as he entered. Realizing that he simply needed to obtain some solid gold that weighed as much as the wreath and then compare how much water was displaced by each to answer the question, he leapt from the tub and ran through the streets naked crying, "*Eureka!*" ("I have found it!")

There are two more extant works by Archimedes. The *Measurement of a Circle* gives an estimate for the value of π, and in the *The Sand-reckoner* Archimedes describes a number system that is capable of handling values up to 8×10^{16} (Heath 1921b, 50–56, 81–85). Based on his model, he deduced that his system could give the number of grains of sand needed to fill the universe. Interestingly enough, his model was based on the work of Aristarchus of Samos, who believed that the sun was the center of the universe. Although he used it for his book, Archimedes rejected Aristarchus's system because he observed no **parallax**, the apparent shift in the position of a star due to the earth revolving around the sun. Aristarchus rebutted that the reason that there is no observed parallax was because the distance to the fixed stars was too great.

Concerning Archimedes' lost books, there is a fragment that goes by the name of *Stomachion*; a book on polyhedra; one on arithmetic; *On Levers*; *On Centers of Gravity*; a book on optics; one on building a mechanical sphere to represent the solar system, called *On Sphere-making*; another *Method*; and a writing on the calendar. These have all been mentioned by various writers, but their exact content would only be a guess (Heath 1921b, 23–25).

Exercises

1. Prove that the center of gravity of a parabolic section is found on its axis of symmetry at a distance from the vertex that is described in Equation 6.1.

2. Confirm Equation 6.3.

3. Find the area of each region:
   (a) Enclosed by the ellipse
   $$\frac{x^2}{a^2} + \frac{y^2}{b^2} = 1.$$
   (b) Bounded by the hyperbola
   $$\frac{x^2}{a^2} - \frac{y^2}{b^2} = 1$$
   $(x > 0)$ and the line $x = c$ $(c > a)$.

4. Estimate the area enclosed by one turn of the Archimedean spiral using:
   (a) six inscribed sectors (Equation 6.4).
   (b) eight circumscribed sectors (Equation 6.5).

5. Prove Equation 6.6 using:
   (a) the First Fundamental Theorem of Calculus.
   (b) the limit definition of the definite integral.

6. Answer the following about the Archimedean spiral.
   (a) Find the area enclosed by two turns of the spiral.
   (b) Find a formula in Cartesian coordinates that describes the spiral.
   (c) Locate the extrema of one turn of the spiral.
   (d) Calculate the length of one turn of the spiral.

7. Confirm Equation 6.7.

8. Find the volume of each conoid or spheroid obtained by revolving the region bounded by the given curves about the x-axis.
   (a) $y = 1/x, x = 0, x = 4, y = 0$ (obtuse-angled conoid)
   (b) $9x^2 + 16y^2 = 144$ (flat spheroid)
   (c) $36x^2 + 25y^2 = 900$ (oblong spheroid)

9. The *Book of Lemmas* is a collection of elementary theorems that survives only as a Latin translation of an Arabic edition. Since Archimedes' name appears in the text, it cannot be directly attributed to him. However, it is likely that the source of the problems date to the great mathematician (Boyer and Merzbach 1991, 133-134; Heath 1921b, 101–103). Solve the following from the *Book of Lemmas*.
   (a) Let semicircle ADC have diameter $\overline{AC}$. Let B be on $\overline{AC}$ so that $\overline{DB}$ is perpendicular to $\overline{AC}$. Draw two semicircles, one on $\overline{AB}$ and the other on $\overline{BC}$. Show that the area in gray bounded by the three semicircles (called the "shoemaker's knife") equals the area of the circle with diameter $\overline{DB}$.

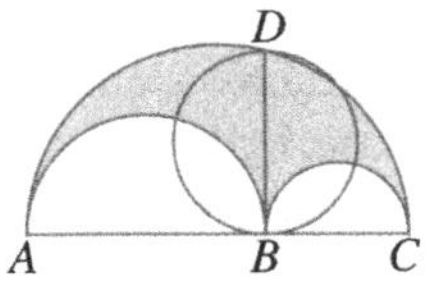

   (b) Start with semicircle AGD on diameter $\overline{AD}$. Let F be the midpoint of the diameter. Find B and C on $\overline{AD}$ so that $\overline{AB} \cong \overline{CD}$. Assume that B and C are on opposite sides of F. Draw semicircles on $\overline{AB}$ and $\overline{CD}$ above $\overline{AD}$ and a semicircle on $\overline{BC}$ below. Let E be on this last semicircle so that $\overline{GE}$ is perpendicular to $\overline{AD}$ through F. Show that the area in gray bounded by the semicircles (called the "salt-cellar") equals the area of the circle with diameter $\overline{GE}$.

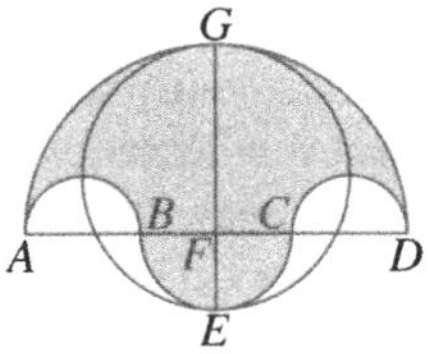

10. This is the theorem of the "broken chord" of Archimedes: Let $\overline{AB}$ and $\overline{BC}$ be chords of a circle. Assume that $BC > AB$. Let D be the midpoint of $\widehat{ABC}$ and drop a perpendicular from D to $\overline{BC}$ at F. Show that $AB + BF = FC$ (that F is the midpoint of the broken chord).

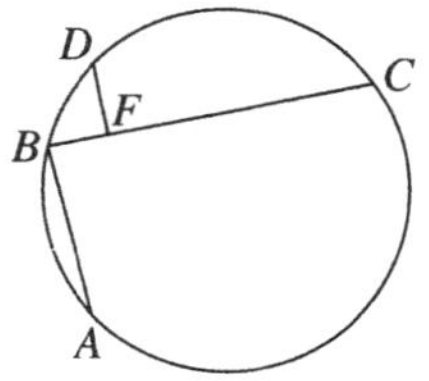

6.2 THE METHOD OF EXHAUSTION

In a city in southwest Asia Minor called Cnidus circa 408 B.C. was born the second greatest geometer of ancient Greece, Eudoxus. When he was 23 years old, Eudoxus traveled to Italy to learn geometry from Archytas. After only a few months he decided to move on to Plato's Academy in Athens. The story goes that he was too poor to live in the city, so Eudoxus lived two miles away in the port city of Piraeus. After Athens, his travels took him to Egypt, where he learned astronomy from the priests of Heliopolis and then to Asia Minor. At the age of 40, Eudoxus returned to Athens and set up a school that for a short time rivaled that of Plato. Eudoxus died in 355 B.C. Near the end of his life he returned home to Cnidus, where he was elected to public office (Heath 1921a, 322).

During his lifetime, Eudoxus's research went beyond the study of pure mathematics. Although none of his works survive, we know that he was the first to bring first mathematical approaches to astronomy, and as such he made two fundamental contributions. First, Eudoxus took the Pythagorean belief in a spherical universe and made it into a system. This led to his studies of the constellations in *Phenomena* and *Mirror*, and it helped establish some basic tenets of Greek astronomy. He assumed, for example, that the earth was a sphere placed at the center of an immense spherical universe. The earth did not move, but the celestial sphere revolved once daily around it. His second contribution added to the first in that he sought to explain the motions of the planets. He did it by placing the main orbit of a planet on a sphere, and on that sphere he placed others that helped to explain their retrograde motion. He wrote on this in his work titled *On Speeds*. His model had its limitations and critics, but what was significant was that his model no longer was based on mythology like that of the Babylonians or Egyptians. It was now the domain of geometry (Evans 1998, 75–76, 306, 309–310).

Eudoxus also made two fundamental contributions to geometry. The first involved his research involving Pythagorean mathematics, which included their theory of proportions. The Pythagorean theory was intertwined with their initial belief that all lengths were rational. The discovery that the hypotenuse of an isosceles right triangle is incommensurable with the length of one of its legs brought their entire system into doubt. How could any of the results of Pythagorean mathematics be trusted when the theory assumed that all numbers were rational? To make matters worse, the collection of the irrational grew. Theodorus, who lived around 450 B.C. and was a tutor of Plato, proved that in addition to $\sqrt{2}$ the numbers

$$\sqrt{3},\ \sqrt{5},\ \sqrt{6},\ \sqrt{7},\ \sqrt{8},\ \sqrt{10},\ \sqrt{11},\ \sqrt{12},\ \sqrt{13},\ \sqrt{14},\ \sqrt{15},\ \sqrt{17}$$

were each irrational. Theodorus's student, Theaetetus of Athens, generalized the result and showed that $\sqrt{n}$ is irrational for each number n that is not a perfect square. Even Plato added fuel to the fire when he noted that it is possible to add two irrational values and have a rational sum. Not only did rational values lead to the construction of the irrational using the Pythagorean Theorem, but calculation with irrational values could yield rational results. There appeared to be a close relation between these two types of numbers. This made ignoring the irrationals and only working with rational numbers difficult, but the early Greeks did not know how to handle both types of values within the same system. They needed someone to put their mathematics on a sure logical footing (Heath 1921a, 156, 203, 209–210).

For this Eudoxus is responsible. He did it by shifting the focus away from numbers to what he called **magnitudes**. A magnitude is essentially a measurement that we often associate with a positive real number. However, magnitudes were not numbers because to the Greeks numbers were discrete. Instead, they were characteristics of geometric objects that could vary continuously. Examples of magnitudes include length, area, volume, and weight. This new system solved the problem of irrational values by ignoring them, and it did this by ignoring numbers altogether.

In shifting mathematics away from numbers to magnitudes, Eudoxus made the study of mathematics the study of geometry and made geometry the standard for rigor for centuries to come. Even Isaac Newton in 1687 felt required to meet these standards in his *Principia Mathematica* by giving geometric demonstrations for results that he had first proven algebraically. We are led to wonder what would have happened if the Pythagoreans would have been able to incorporate the irrational numbers into their mathematics. Their number theory may have developed into a full algebraic system that, when combined with their geometry, may have led them even to the calculus. Did Eudoxus' solution prevent this? His theory of magnitudes did give geometry a solid foundation and enable the Greeks to push the subject to its limits, but it was certainly at a cost (Kline 1972, 48–49).

The second significant contribution of Eudoxus also involved magnitudes. Around 450 B.C. a contemporary of Socrates by the name of Antiphon used inscribed polygons to describe the area of a circle. Take a circle and inscribe within it an equilateral triangle (see Figure 6.6). On each side of the triangle construct an isosceles triangle with the vertex on the included arc. The result is a regular hexagon. On each of the sides of the hexagon, again draw an isosceles triangle with its vertex on the included arc. Now we have a regular dodecagon. If this process is continued indefinitely, Antiphon believed that at some point the entire area would be exhausted and we would have a polygon that has area equal to that of the given circle (Heath 1921a, 219).

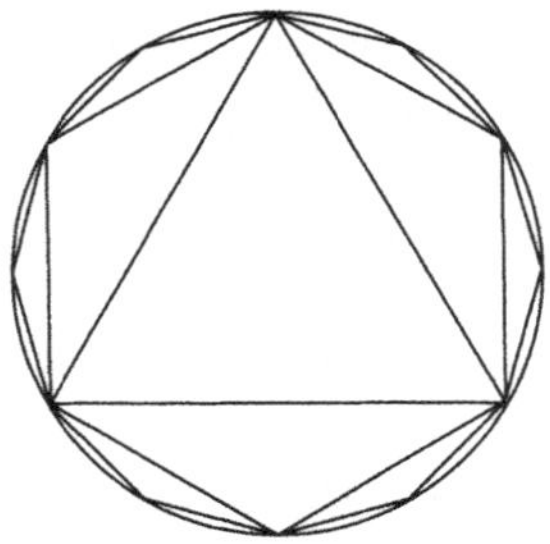

Figure 6.6 Antiphon's method to find the area of a circle.

We know that this process leads to no such polygon, but we do know that we can find an inscribed polygon with area that is as close to the area of the circle as we want. Eudoxus also knew this. At the beginning of his proof he draws a crucial conclusion based on a statement that has become known as the **Postulate of Archimedes**.

■ **POSTULATE 6.2.1 [Archimedes]**

Given magnitudes a and b such that if $a > b$, there exists a positive integer k so that $kb > a$.

The notation kb in the definition refers to the magnitude b being added to itself k times. That is, the segment that represents b has $k - 1$ copies concatenated to itself.

To prove his theorem regarding the area of a circle, Eudoxus needed a result whose demonstration is preserved by Euclid. We should note that "removing a magnitude from a magnitude" means that a portion of a given magnitude is taken away so that what remains is still a magnitude.

■ **LEMMA 6.2.2 [*Elements* X.1]**

Given $AB > CC'$, if from AB is removed a magnitude greater than its half, from the remainder is removed a magnitude greater than its half, and if this process is continued, there will be left some magnitude which is less than CC'.

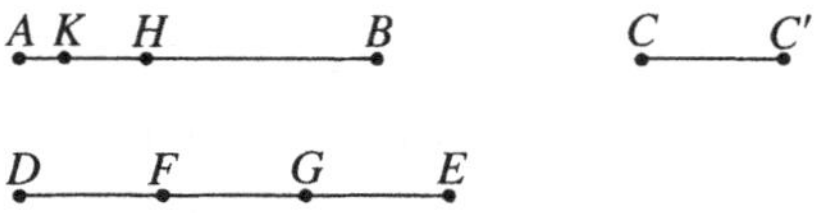

PROOF

We are given $\overline{AB}$ and $\overline{CC'}$ with $CC' < AB$. There is a multiple of CC' that is greater than AB by Archimedes' Postulate. Let DE be that multiple. Divide $\overline{DE}$ into parts congruent to $\overline{CC'}$, say $\overline{DF}$, $\overline{FG}$, and $\overline{GE}$. Begin by finding H on $\overline{AB}$ so that

$$HB > \frac{1}{2}AB.$$

It is possible to find H because we simply take it on the segment which is to the left of the midpoint of $\overline{AB}$. We then have $AH < DG$ because

$$AH = AB - HB < \frac{1}{2}AB < \frac{1}{2}DE < DE - GE = DG.$$

Next find K on $\overline{AH}$ such that

$$KH > \frac{1}{2}AH,$$

and we then have $AK < DF$ since

$$AK = AH - KH < \frac{1}{2}AH < \frac{1}{2}DG = DG - FG = DF.$$

Therefore, we have found a magnitude within AB which is less than the desired magnitude because $DF = CC'$. If we had needed to divide $\overline{DE}$ into more than three subdivisions, we would continue this process until the number of divisions of $\overline{AB}$ equals the number of divisions of $\overline{DE}$. ■

The lemma works in this way. Let a and b be magnitudes. The lemma allows us by successive subtractions off a to find another magnitude that is less than b. This method constructs the following sequence:

$$\begin{aligned}
a_1 &= a \\
a_2 &< \frac{a_1}{2} \\
a_3 &< \frac{a_2}{2} < \frac{a_1}{4} \\
a_4 &< \frac{a_1}{8} \\
&\vdots \\
a_n &< \frac{a_1}{2^{n-1}} \\
&\vdots
\end{aligned}$$

Continuing, we find that there exists a positive integer N such that

$$a_N < \frac{a}{2^N} < b.$$

The acceptance of the Postulate of Archimedes is a denial of the existence of a magnitude, known as an **infinitesimal**, that is less than every other magnitude. Both the affirmation and the denial of the existence of infinitesimals have appeared at various times in the history of mathematics, and they played a particularly important role in the development of the calculus.

Area and Circumference

Any proof that relies on Lemma 6.2.2 is said to use the **method of exhaustion**. Eudoxus used this technique to prove many theorems. These include that the volumes of two spheres are to each other as the cubes of their radii, the volume of a pyramid equals one third the volume of a prism of the same base and the same height, and the volume of a cone is one third the volume of a cylinder with the same base and of the same height (Kline 1972, 50). He also used it to prove the next theorem.

■ **THEOREM 6.2.3 [*Elements* XII.2]**

The areas of two circles are to one another as the squares of their diameters.

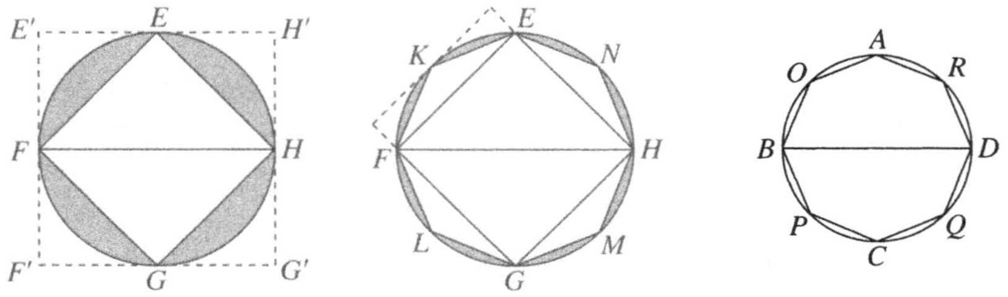

PROOF

Take two circles and call them $\odot ABCD$ and $\odot EFGH$, where $\overline{BD}$ is a diameter of the first circle and $\overline{FH}$ is a diameter of the second. We need to show

$$\frac{\text{area}(\odot ABCD)}{\text{area}(\odot EFGH)} = \frac{BD^2}{FH^2}.$$

Suppose not, and let s be an area that is either less than or greater than the area of $\odot EFGH$, so that

$$\frac{\text{area}(\odot ABCD)}{s} = \frac{BD^2}{FH^2}. \tag{6.8}$$

We will show that the first case leads to a contradiction and leave the second to Exercise 6. So assume that

$$s < \text{area}(\odot EFGH).$$

We will apply the procedure of Lemma 6.2.2 to the magnitudes

$$\text{area}(\odot EFGH)$$

and

$$\text{area}(\odot EFGH) - s.$$

To do this, first construct $\square EFGH$ inscribed within $\odot EFGH$ and $\square E'F'G'H'$ circumscribed about it. Then appealing to Exercise 5, we have

$$2\text{area}(\square EFGH) = \text{area}(\square E'F'G'H') > \text{area}(\odot EFGH), \tag{6.9}$$

so

$$\text{area}(\square EFGH) > \frac{1}{2}\text{area}(\odot EFGH).$$

Take $\square EFGH$ and remove it from $\odot EFGH$, resulting in the shaded area in the first figure. We next construct isosceles ΔFKE with K on the circle. We observe that

$$\text{area}(\Delta FKE) = \frac{1}{2}\text{area}(\sqsubset FKE) > \frac{1}{2} \text{ the area enclosed by } \widehat{FKE}.$$

This is because if a tangent is drawn at K and a rectangle is constructed on the tangent with base $\overline{FE}$, the area of ΔEKF is half that of the rectangle. Repeat this with the isosceles ΔENH, ΔHMG, and ΔGLF. Letting a represent the sum of the areas of these four triangles, we conclude that

$$a > \frac{1}{2}\text{area(shading on the right)},$$

and we remove a from the shading in the first figure resulting in the shading of the second. We continue this process of constructing isosceles triangles so that at each step an area is removed that is more than half of the area that remains from the previous step. Lemma 6.2.2 tells us that eventually the remaining area will be less than the area of $\odot EFGH$ minus s. Suppose that the process stops as with the second diagram on page 186 so that

$$\text{area(shading on the right)} < \text{area}(\odot EFGH) - s.$$

Since the area of polygon $EKFLGMHNE$ is equal to the area of $\odot EFGH$ less the shading, we conclude that

$$\text{area}(EKFLGMHN) > s. \tag{6.10}$$

We now need to include $\odot ABCD$ so that we can set up the ratio. Construct the polygon $AOBPCQDR$ so that it is similar to the polygon $EKFLGMHN$ and inscribed within $\odot ABCD$. This yields

$$\frac{BD^2}{FH^2} = \frac{\text{area}(AOBPCQDR)}{\text{area}(EKFLGMHN)} \tag{6.11}$$

by Exercise 5, and then by Equation 6.8 we have

$$\frac{\text{area}(\odot ABCD)}{s} = \frac{\text{area}(AOBPCQDR)}{\text{area}(EKFLGMHN)}.$$

Since the area of $\odot ABCD$ is greater than the area of $AOBPCQDR$, s must be greater than the area of $EKFLGMHN$, a contradiction of Inequality 6.10. Therefore, s cannot be less than the area of $\odot EFGH$. This combined with Exercise 6 proves that $s = \text{area}(\odot EFGH)$. ■

The theorem can be used to derive the formula for the area of a circle. Let A be the area of a randomly selected circle and r its radius. The theorem shows that the ratio $A/4r^2$ is the same number no matter which circle is chosen. Call this number $\pi/4$ and write

$$\frac{A}{4r^2} = \frac{\pi}{4},$$

from which we quickly derive

$$A = \pi r^2. \tag{6.12}$$

Now to find the circumference. Archimedes provided the calculation in his work *Measurement of a Circle* (Archimedes 1897, 91–98).

THEOREM 6.2.4 [*Measurement of a Circle* 1]

The area of a circle is equal to the area of a right triangle in which the length of one leg equals the radius of the circle and the other leg equals the circumference of the circle.

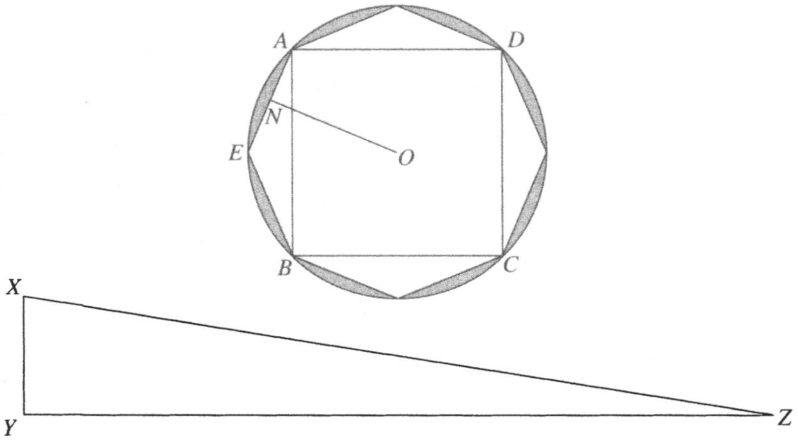

PROOF

The proof uses the method of exhaustion. Let $\odot ABCD$ be the given circle and let $\triangle XYZ$ be such that XY equals the radius of the circle and YZ equals its circumference. To obtain a contradiction, first suppose that the area of the circle is greater than the area of the triangle. Inscribe $\square ABCD$ within $\odot ABCD$. Bisect the arcs cut by $\overline{AB}$, $\overline{BC}$, $\overline{CD}$, and $\overline{DA}$. Join the resulting midpoints to form a regular octagon. Continue this process until we find a regular polygon inscribed within the circle such that its area is a and

$$\text{area}(\odot ABCD) - a < \text{area}(\odot ABCD) - \text{area}(\triangle XYZ)$$

by Lemma 6.2.2. Thus, we have

$$a > \text{area}(\triangle XYZ).$$

Suppose that the polygon with area a is the polygon in the diagram. Let $\overline{AE}$ be a side of the polygon. Find N on $\overline{AE}$ such that $\overline{ON}$ is perpendicular to $\overline{AE}$. Then ON is less than the radius of $\odot ABCD$ and thus less than XY. Furthermore, the perimeter of the polygon is less than the circumference, so it is also less than YZ. Therefore, since the area of a regular polygon equals half the product of its apothem and its perimeter,

$$a < \text{area}(\triangle XYZ),$$

a contradiction. The case when the area of the circle is less than the area of the triangle is similarly brought to a contradiction by using circumscribed polygons (Exercise 7). ■

In modern notation the theorem states that if A is the area of a circle, C its circumference, and r is its radius,

$$A = \frac{Cr}{2}.$$

By Equation 6.12 we conclude that $\pi r^2 = Cr/2$. Hence,

$$C = 2\pi r,$$

and we see further that π represents the ratio of a circle's circumference to its diameter.

Estimating Pi

Now that we have formulas for the area and circumference of a circle, it is natural to try to determine the value of π. Archimedes succeeded in finding an estimate in his *Measurement of a Circle*. His used the strategy found in the method of exhaustion.

■ **THEOREM 6.2.5 [*Measurement of a Circle* 3]**

The circumference of a circle is greater than $3\frac{10}{71}$ times its diameter but less than $3\frac{1}{7}$ times its diameter.

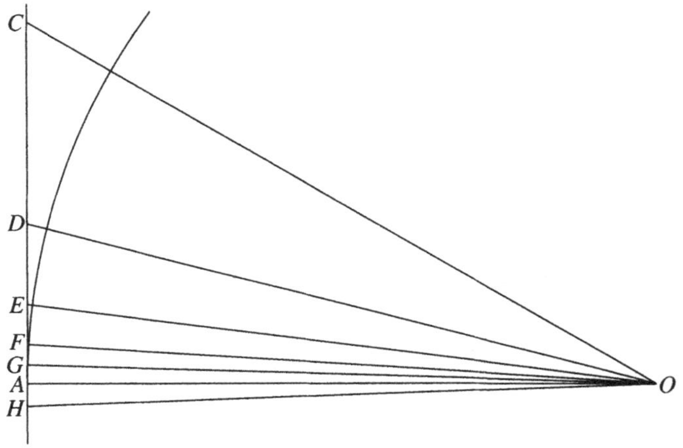

PROOF

We are given $\odot O$ with radius $\overline{OA}$. The goal is to both inscribe and circumscribe a regular 96-gon about the circle and to use those polygons to estimate π. Find C so that $\overline{AC}$ is tangent to $\odot O$ and $\measuredangle COA$ is a third of a right angle. We may view $\overline{AC}$ as half of one side of a regular dodecagon circumscribed about the circle. Now construct a sequence of angle bisectors:

- Bisect $\measuredangle COA$ with $\overline{OD}$.
- Bisect $\measuredangle DOA$ with $\overline{OE}$.
- Bisect $\measuredangle EOA$ with $\overline{OF}$.
- Bisect $\measuredangle FOA$ with $\overline{OG}$.

Assume that D, E, F, and G are on $\overline{AC}$. Therefore, $m\measuredangle GOA$ is $1/48$ that of a right angle. Extend $\overline{GA}$ to H so that $\overline{GA} \cong \overline{AH}$ and $\measuredangle GOA \cong \measuredangle HOA$. We see that $m\measuredangle GOH$ is $1/24$ that of a right angle, so $\overline{GH}$ can be considered one side of a regular 96-gon circumscribed about $\odot O$. Its perimeter is $96GH$. We therefore conclude that

$$\pi = \frac{\text{circumference of } \odot O}{\text{diameter of } \odot O} < \frac{96GH}{2OA} = \frac{96GA}{OA}. \tag{6.13}$$

We need to find an estimate for GA/OA. In order to accomplish this, begin by noticing that the angles constructed above yield a sequence of ratios,

$$\frac{OA}{CA}, \frac{OA}{DA}, \frac{OA}{EA}, \frac{OA}{FA}, \frac{OA}{GA},$$

where each numerator is the radius of the circle and each denominator is half the length of a side of a regular polygon that has been circumscribed about the circle. Note that since $\triangle COA$ is a 30-60-90 triangle,

$$\frac{OC}{CA} = 2 \text{ and } \frac{OA}{CA} = \sqrt{3}.$$

Hence,

$$\frac{OC}{CA} = \frac{306}{153} \tag{6.14}$$

and

$$\frac{OA}{CA} > \frac{265}{153}. \tag{6.15}$$

The fraction $265/153$ is chosen to approximate $\sqrt{3}$, and $306/153$ is chosen to have the same denominator as the first fraction, yet equal 2.

For the second ratio of our sequence, because $\overline{OD}$ bisects $\measuredangle COA$, we may appeal to Exercise 19 of Section 5.4 and conclude that

$$\frac{OC}{OA} = \frac{CD}{DA}.$$

Therefore,

$$\frac{OC + OA}{OA} = \frac{CD + DA}{DA} = \frac{CA}{DA},$$

and we have *permutando*,

$$\frac{OA}{DA} = \frac{OC + OA}{CA} = \frac{OC}{CA} + \frac{OA}{CA} > \frac{306}{153} + \frac{265}{153}.$$

The last inequality follows from Equation 6.14 and Inequality 6.15. Therefore, we have

$$\frac{OA}{DA} > \frac{571}{153}$$

as a lower bound for the second term of the sequence. By the Pythagorean Theorem we know that $OD^2 = OA^2 + DA^2$. Hence,

$$\frac{OD^2}{DA^2} = \frac{OA^2 + DA^2}{DA^2} > \frac{571^2 + 153^2}{153^2} = \frac{349,450}{23,409},$$

so we have

$$\frac{OD}{DA} > \frac{591\frac{1}{8}}{153}.$$

This allows us to find a lower bound for the third term of the sequence because we may repeat the argument above to find that

$$\frac{OA}{EA} = \frac{OD}{DA} + \frac{OA}{DA},$$

and then

$$\frac{OA}{EA} > \frac{1162\frac{1}{8}}{153} \text{ and } \frac{OE}{EA} > \frac{1172\frac{1}{8}}{153}. \tag{6.16}$$

Similarly,

$$\frac{OA}{FA} > \frac{2334\frac{1}{4}}{153} \text{ and } \frac{OF}{FA} > \frac{2339\frac{1}{4}}{153}, \tag{6.17}$$

and finally,

$$\frac{OA}{GA} > \frac{4673\frac{1}{2}}{153}. \tag{6.18}$$

From Inequalities 6.13 and 6.18, we conclude that

$$\pi < \frac{96GA}{OA} < \frac{96 \cdot 153}{4673\frac{1}{2}} = 3 + \frac{667\frac{1}{2}}{4673\frac{1}{2}} < 3 + \frac{667\frac{1}{2}}{4672\frac{1}{2}} = 3\frac{1}{7} \tag{6.19}$$

A similar argument using inscribed polygons will yield the lower limit. ■

Without decimal notation, calculators, or computers, Archimedes found upper and lower bounds on π equivalent to

$$3.140845070 < \pi < 3.142857143.$$

Furthermore, the upper bound of π is the famous Archimedean approximation:

$$\pi \approx \frac{22}{7}.$$

Exercises

1. Theodorus showed that $\sqrt{n}$ is irrational when the radicand n is not a perfect square and $2 \leqslant n \leqslant 17$. Use the diagram below to make a conjecture on why Theodorus stopped at 17.

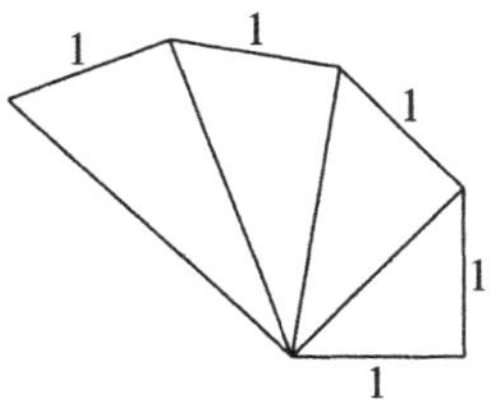

2. Use Antiphon's method of finding the area of a circle to estimate π accurate to five decimal places. What is the minimum number of sides of the inscribed polygon that is necessary to accomplish this?

3. Show that the proof of Lemma 6.2.2 still works if $\overline{DE}$ is divided into an even number of segments.

4. Identify the unstated assumptions in Lemma 6.2.2.

5. Prove Equations 6.9 and 6.11 of the proof of Theorem 6.2.3.

6. Prove that the assumption $s > \text{area}(\odot EFGH)$ also leads to a contradiction in the proof of Theorem 6.2.3.

7. In Archimedes' proof of Theorem 6.2.4, prove that the case when the area of the circle is less than the area of the triangle yields a contradiction using circumscribed polygons.

8. Burton (1985, 126) observes that Archimedes used but never proved that

$$\frac{1351}{780} > \sqrt{3} > \frac{265}{153}, \tag{6.20}$$

and he conjectured that Archimedes would have demonstrated the left-hand inequality by relying on the following two inequalities. Show both.

(a) $26 - \frac{1}{52} = \sqrt{26^2 - 1 + \left(\frac{1}{52}\right)^2} > \sqrt{26^2 - 1}.$

(b) $\frac{1351}{780} = \frac{1}{15}\left(26 - \frac{1}{52}\right) > \frac{\sqrt{26^2 - 1}}{15} = \sqrt{3}.$

9. Prove the right-hand side of Inequality 6.20 by replacing $1/52$ with $1/51$ in the inequalities of Exercise 8.

10. Given a circle with unit radius:
   (a) Construct an octagon inscribed within the circle.
   (b) Construct an octagon circumscribed about the circle.
   (c) Use the two octagons to approximate the area of the circle and then use this result to approximate the value of π.
   (d) What size n-gon is required so that the difference between the estimate for π and its true value is less than 0.001?

11. Use calculus to prove that $A = \pi r^2$ and $C = 2\pi r$.

12. Check that Inequalities 6.16, 6.17, 6.18, and 6.19 on page 191 hold.

13. Show that the ratio of the area of a circle to the square of its diameter is approximately $11/14$ without a calculator.

14. The volume of a cone is one third the volume of a cylinder with the same base and of the same height.
   (a) Write an algebraic equation for this geometric expression.
   (b) Use calculus to prove it.

15. The volumes of two spheres are to one another as the cubes of their radii.
   (a) Write an algebraic equation for this geometric expression.
   (b) Use calculus to prove it.

6.3 THE METHOD

Take a two-by-four and place it across a sawhorse. What we have is a crude **lever** with the sawhorse as the **fulcrum**. Put a brick on one end of the two-by-four and three bricks at the other end. We know that for this to balance, we must move the lumber so that the distance from the fulcrum to the lone brick is longer than that to the three bricks. More precisely, if m_1 is the mass of the single brick and d_1 is its distance to the fulcrum, and if m_2 is the mass of the three bricks with d_2 the distance from the bricks to the fulcrum, then for the lever to balance we need

$$\frac{m_1}{m_2} = \frac{d_2}{d_1}. \tag{6.21}$$

Now make the situation abstract. Represent the two-by-four as a line segment and the sawhorse as a point on the segment. Treat the masses as if they are concentrated in a point. We say that this **system** is in **equilibrium** if the lever balances, and it will balance if and only if Equation 6.21 holds. This proposition is known as the **Law of the Lever**, and Archimedes gave a nearly complete proof of it in his *On the Equilibrium of Planes* (Archimedes 1897, 189–220).

The Law of the Lever was known before the time of Archimedes. The significance of *On the Equilibrium of Planes* was not the novelty of the concept but how it was presented. Archimedes was the first to derive the Law of the Lever from a set of postulates. Each of these assumptions was chosen by Archimedes because he considered each to be clearly true based on his own experience. In so doing, Archimedes became the first Greek to treat what happened on the earth mathematically. Plato viewed the universe as an imperfect representation of the realm of forms, and mathematics was the way to escape this world and learn about the true reality. Aristotle believed that one learns truth by observation, but although his *Physics* dealt with natural phenomena, it was based on nonmathematical intuition. Archimedes followed neither one. Instead, he believed that natural phenomena can be studied using the method of the *Elements*. From agreed upon and simple first principles that reflect nature, laws of the universe can be deduced mathematically because mathematics perfectly represents the world in which we live (Boyer and Merzbach 1991, 121–122).

Center of Gravity

The point at which we find the fulcrum of a system of weights that is in equilibrium is called the **center of gravity** of the system. We assume that the total weight of the system is concentrated at this point. We now generalize this concept of balance to other situations. We first deal with a single object. Take either a plane or solid figure. If it is only two-dimensional, view it as having a nearly infinitesimal thickness and uniform density. Such an object is called a **lamina**. The ancient Greeks knew of three possible interpretations for what it meant to balance such an object (Figure 6.7). The first two have the object being suspended in some way. The **horizontal suspension** point is the location where an object must be attached to a wire so that it is perfectly balanced in the horizontal direction. Now take the object and hang it vertically by

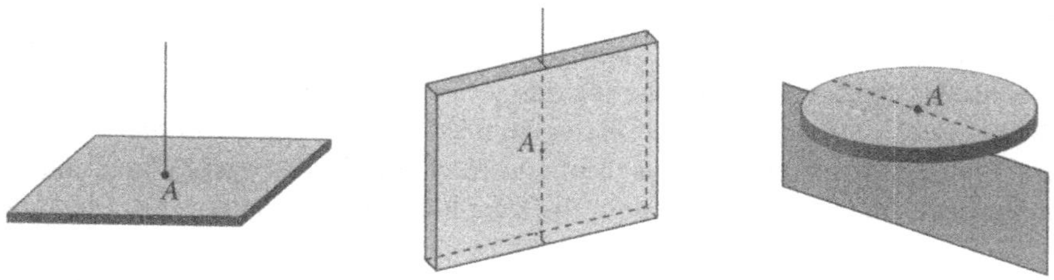

(a) Horizontal suspension. (b) Vertical suspension. (c) Balance.

Figure 6.7 Centers of gravity.

the wire. From the point connected to the wire draw a vertical line. Remove the object, rotate it a few degrees, and then hang again vertically so that it balances. When the second vertical line is drawn it will intersect the first in the **vertical suspension** point. Finally, imagine an extremely thin metal sheet standing upright. The intersection of all lines on which the object will balance on that sheet we call the **balance** point. Although experiment would confirm our guess, we assume that these three interpretations result in the same point that we define to be the center of gravity of the object (Stein 1999, 15–17). We also make the following two assumptions:

■ **POSTULATES 6.3.1 [*On the Equilibrium of Planes* I, Postulates]**

1. The centers of gravity of similar figures are in the same relative positions.
2. The center of gravity of a convex polygon must lie within the figure.

We now follow Archimedes to generalize the center of gravity to any number of weights. He began by proving the following special case. See Exercise 2.

■ **THEOREM 6.3.2 [*On the Equilibrium of Planes* I.4, 5]**

- If the centers of gravity of two equal weights are not at the same point, the weights balance at the midpoint of the segment joining their centers of gravity.
- If three equal weights have their centers of gravity on a straight line at equal distances, the center of gravity of the system will coincide with the center of gravity of the middle weight.

Theorem 6.3.2 can be generalized to any number of weights. We deal with the even number case here and leave the odd case to Exercise 4. Since the two-weight result is the first part of Theorem 6.3.2, to prove the result for an even number of weights, we start with four.

■ **THEOREM 6.3.3 [*On the Equilibrium of Planes* I.5 Corollary]**

If an even number of equal weights have their centers of gravity on a straight line such that the distances between any pair of consecutive centers of gravity are the same, the midpoint of the segment joining the middle two centers of gravity is the center of gravity of the entire system.

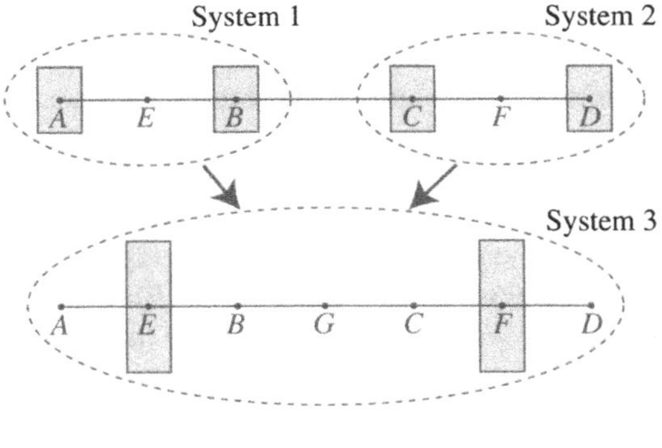

PROOF

Let weights of equal magnitude be placed on a line segment so that their centers of gravity are at points A, B, C, and D. Let E be the midpoint of $\overline{AB}$, and let F be the midpoint of $\overline{CD}$. By Theorem 6.3.2, the total mass of System 1 is concentrated at E, and the total mass of System 2 is at F. Now the lever can be viewed as in System 3, with G being the midpoint of $\overline{EF}$. Since the mass at E equals the mass at F, the system is in equilibrium with center of gravity G. Since $\overline{EB} \cong \overline{CF}$ and $\overline{EG} \cong \overline{GF}$, we have $\overline{BG} = \overline{GC}$. Hence, G is the midpoint of $\overline{BC}$. Generalize this procedure to complete the proof. ■

As an example of how Archimedes worked with centers of gravity, we shall find the center of gravity of a parallelogram.

■ **THEOREM 6.3.4 [*On the Equilibrium of Planes* I.9]**

The center of gravity of a parallelogram lies on the segment joining the midpoints of opposite sides.

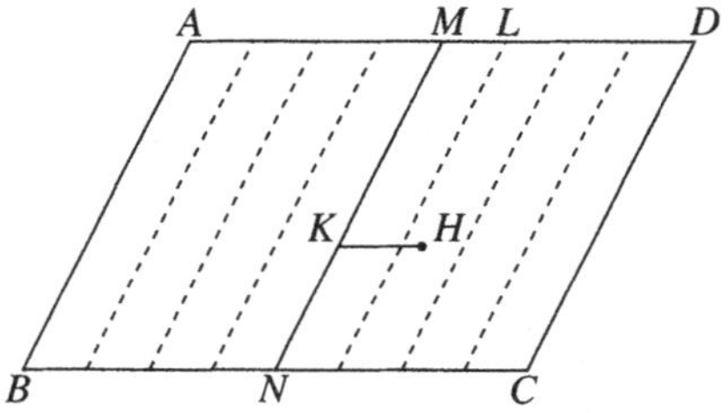

PROOF

Let $ABCD$ be a parallelogram. Assume that M is the midpoint of $\overline{AD}$ and N is the midpoint of $\overline{BC}$. Join $\overline{MN}$. Designate the center of gravity of the parallelogram by H. We must prove that H is on $\overline{MN}$, so suppose it lies elsewhere inside the parallelogram (Postulate 6.3.1.2). Find K on $\overline{MN}$ so that $\overline{KH}$ is parallel to $\overline{AD}$. Find the midpoint of $\overline{MD}$. This divides the segment into two segments. Find the midpoint of one of these resulting segments and divide again. By continually finding midpoints in this fashion, we can locate L on $\overline{MD}$ such that $ML < KH$ (Lemma 6.2.2). Using this length, divide both $\overline{AM}$ and $\overline{MD}$ into consecutive congruent line segments of length ML. At the endpoints of each of these congruent segments, draw lines parallel to $\overline{MN}$. The resulting parallelograms are congruent. Therefore, their centers of gravity are in the same relative position (Postulate 6.3.1.1). This means that we have an even number of magnitudes of equal weight with collinear centers of gravity. Thus, by Theorem 6.3.3 the center of gravity of the system of parallelograms is on the midpoint of the line segment joining the centers of gravity of the middle two parallelograms, but this is impossible because H does not lie within either of these parallelograms. We are forced to conclude that H lies on $\overline{MN}$. ■

The last proposition quickly yields the next, so its proof is left to Exercise 5.

■ **THEOREM 6.3.5**

The center of gravity of a parallelogram lies at the intersection of its diagonals.

The Letter to Eratosthenes

Nearly all of the extant works of the ancient Greek mathematicians came down to us in their finished, polished form. By our own experience, work in mathematics does not flow directly from theorem to finished proof in one sitting. It may take hours, days, weeks, and in some cases years to write a proof. There are times of progress and times of stagnation. Often, a promising argument leads to a dead end, while an idea out of nowhere may be exactly what is needed. It would be fascinating and highly instructive if we could find a book of notes written by someone like Eudoxus that would illuminate the process by which he came by his results. In the case of Archimedes, *The Method* is that example. As with many of the works of Archimedes, *The Method* is actually a letter. It was written to his good friend Eratosthenes. More important, Archimedes knew that Eratosthenes was the perfect person to which to send this.

> Seeing moreover in you, as I say, an earnest student, a man of considerable eminence in philosophy, and an admirer [of mathematical inquiry], I thought fit to write out for you and explain in detail in the same book the peculiarity of a certain method, by which it will be possible for you to get a start to enable you to investigate some of the problems in mathematics by means of mechanics. (Archimedes 1897, *The Method*, Introduction, 13).

Eratosthenes was born in Cyrene around 284 B.C. Beside mathematics he studied philosophy and poetry. He combined his interests and wrote a summary of the mathematics needed for the study of Platonic philosophy. It included the tale of the Delians with their problem of doubling the volume of their altar (Section 4.3). His *On Means* was included in the *Treasury of Analysis* of Pappus. He studied geography and wrote *On the Measurement of the Earth*. Because of his expertise, he was invited to Egypt by Ptolemy Evergetes around the year 244 B.C. to tutor his son. He later became the librarian at Alexandria. Although he was expert in many fields, he was never viewed as the leading scholar in any of them. Mathematics is a case in point, where he was second to Archimedes. Because he always seemed to come in second, he gained the nickname *Beta* (Heath 1921b, 104–107).

Eratosthenes is famous for two main contributions that are still taught today. The first is in the field of number theory. The ancient Greeks did study numbers, including the prime numbers (page 47). Eratosthenes' contribution is an algorithm designed to identify the primes in a list of given numbers. It is known as the **sieve of Eratosthenes** (Figure 6.8). The sieve works like this. Make a list of consecutive numbers starting with 2 and ending with n. Circle 2 but cross out all other multiples of two. Next, circle 3 (the next number that is neither circled nor crossed-out) and cross out all other multiples of three. Repeat this process until the next available number is greater than $\sqrt{n}$. At his point, as seen in Figure 6.8, all circled and nonmarked numbers are prime.

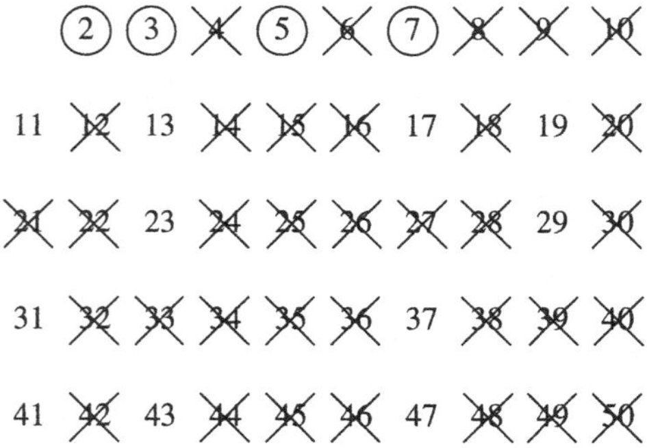

Figure 6.8 The sieve of Eratosthenes.

The second of Eratosthenes' contributions is usually taught in trigonometry or high school geometry courses. It was known that on the summer solstice the sun would be directly above Syene (now Aswan) in Egypt. This was confirmed by noting that the sun could be seen at the bottom of a well. Also on the summer solstice, at Alexandria the sun formed an angle equivalent to one fiftieth of a circle. This is the angle α in Figure 6.9. If we suppose that the rays of the sun form parallel lines, we may conclude that the measure of the arc between the two cities is also α. Estimate the distance between Syene and Alexandria to be 5000 stadia. This is based on how long it took a caravan with camels to travel between those two cities (Kline 1972, 160–161). Since the cities were approximately on the same longitudinal line and α represents one fiftieth of a circle, the circumference of the globe can be calculated to be approximately 250,000 stadia. Assuming that a stadia is equivalent to 517 feet, Eratosthenes' measurement equals 24,479 miles. The earth's true circumference is approximately 24,900 miles (24,902 miles at the equator and 24,860 miles pole to pole), so his measurement is off by 2%, the best among the ancient Greeks.

We see then that Eratosthenes had the intellectual résumé to appreciate and use the method that Archimedes was about to show him. Archimedes continues:

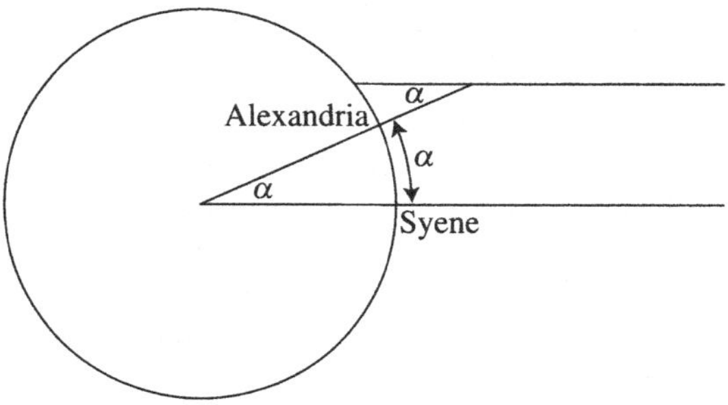

Figure 6.9 Eratosthenes and the circumference of the earth.

> This procedure, is I am persuaded, no less useful even for the proof of the theorems themselves; for certain things first became clear to me by a mechanical method, although they had to be demonstrated by geometry afterwards because their investigation by the said method did not furnish an actual demonstration. But it is of course easier, when we have previously acquired, by the method, some knowledge of the questions, to supply the proof than it is to find it without any previous knowledge. This is a reason why, in the case of the theorems the proof of which Eudoxus was the first to discover, namely that the cone is a third part of the cylinder, and the pyramid of the prism, having the same base and equal height, we should give no small share of the credit to Democritus, who was the first to make the assertion with regard to the said figure though he did not prove it. (Archimedes 1897, *The Method*, Introduction, 13)

Archimedes told Eratosthenes that the first theorem that he discovered using the method is the formula on page 177 for the area of a parabola. By repeating the method for different questions he convinced himself of the truth of many other results that he outlined for Eratosthenes. We will examine one of these by following Archimedes' line of thought as he uses his method to convince himself of the truth of his formula regarding the volume of a sphere, namely that

> the volume of a sphere is four times the volume of a cone with base area equal to that of a great circle of the sphere and height equal to its radius.

We start with the preliminary constructions. Let $\odot ABCD$ be a great circle of a sphere with perpendicular diameters $\overline{AC}$ and $\overline{BD}$ (Figure 6.10). Construct a circle with diameter $\overline{BD}$ and perpendicular to $\overline{AC}$. With this circle as its base, draw a cone with vertex A. Extend this cone to a new base that is a circle perpendicular to $\overline{AC}$ and with diameter $\overline{ECF}$. On this circle, construct a cylinder with height equal to AC. Finally, extend $\overline{AC}$ beyond A to H such that $\overline{HA} \cong \overline{AC}$. We shall view $\overline{HC}$ as a lever with fulcrum A.

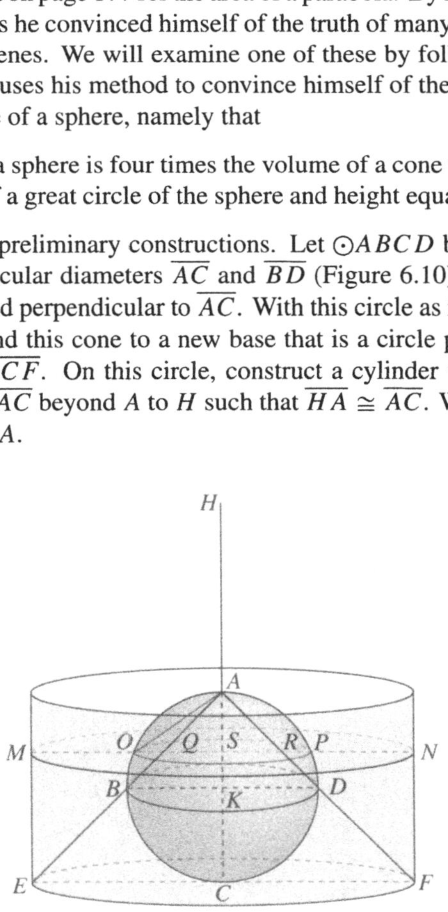

Figure 6.10 The sphere and its cone in a cylinder.

Next take a plane parallel to $\odot BD$ and intersect the cylinder. This forms a circle with diameter $\overline{MN}$. Call the points where $\overline{MN}$ intersects the sphere O and P. Further, $\overline{MN}$ intersects $\overline{AC}$ at S, $\overline{AE}$ at Q, and $\overline{AF}$ at R. Join $\overline{AO}$. This plane also intersects the sphere, forming a circle with diameter $\overline{OP}$, and it intersects the cone, forming a circle with diameter $\overline{QR}$.

Now that the setup is complete,

$$\overline{MS} \cong \overline{AC} \text{ and } \overline{QS} \cong \overline{AS}. \tag{6.22}$$

Therefore,

$$MS \cdot QS = AC \cdot AS = AO^2 = OS^2 + QS^2.$$

The last equality follows by the Pythagorean Theorem, and the second-to-last equality holds because $\triangle OAC \sim \triangle SAO$. Then, since $\overline{HA} \cong \overline{AC}$,

$$\frac{HA}{AS} = \frac{AC}{AS} = \frac{MS}{QS} = \frac{MS \cdot MS}{MS \cdot QS} = \frac{MS^2}{OS^2 + QS^2} = \frac{MN^2}{OP^2 + QR^2}.$$

Therefore,

$$\frac{HA}{AS} = \frac{\text{area(circle with diameter } \overline{MN})}{\text{area(circle with diameter } \overline{OP}) + \text{area(circle with diamter } \overline{QR})},$$

This means that if we take lever $\overline{HS}$ and place $\odot MN$ at S and both $\odot OP$ and $\odot QR$ at H, it balances at fulcrum A (Figure 6.11). The above can be repeated for any plane that cuts the cylinder parallel to its base. Archimedes now argues that since all of the slices of the cylinder at each S balance with the corresponding slices of the sphere and cone when placed at H with fulcrum A, the same lever will balance when the

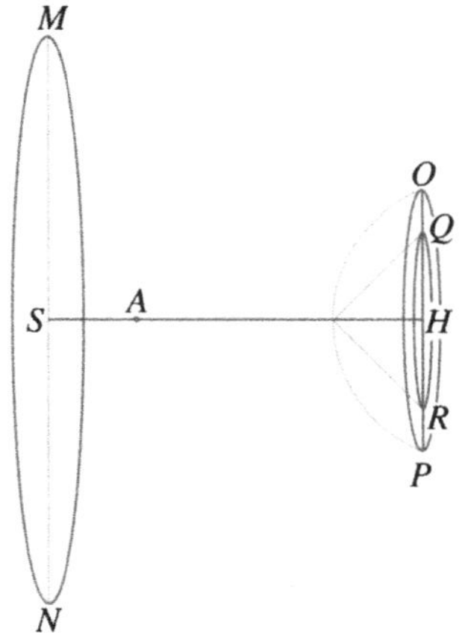

Figure 6.11 Circles formed by the plane in balance.

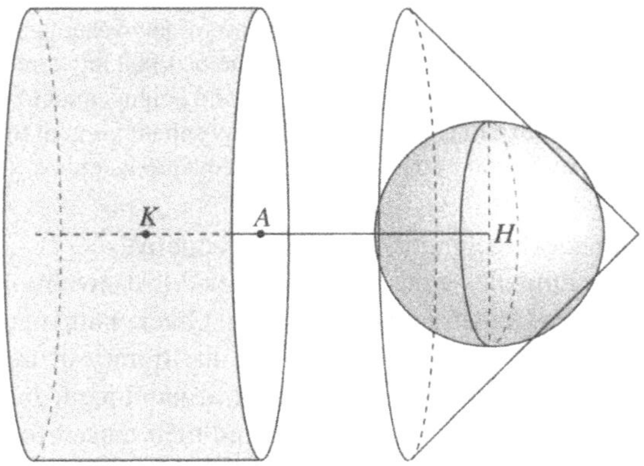

Figure 6.12 The cylinder balancing the sphere and cone.

entire cylinder is placed so that its center of gravity is at K while the sphere and cone are placed with their centers of gravity at H (Figure 6.12). By the Law of the Lever,

$$\frac{HA}{AK} = \frac{\text{volume(cylinder)}}{\text{volume(sphere)} + \text{volume(cone } AEF)}.$$

Since $HA = 2AK$,

$$\text{volume(cylinder)} = 2\text{volume(sphere)} + 2\text{volume(cone } AEF).$$

By Exercise 14 of Section 6.2,

$$\text{volume(cylinder)} = 3\text{volume(cone } AEF).$$

Thus,

$$\text{volume(cone } AEF) = 2\text{volume(sphere)},$$

but since $EF = 2BD$,

$$\text{volume(cone } AEF) = 8\text{volume(cone } ABD).$$

We may now conclude that

$$\text{volume(sphere)} = 4\text{volume(cone } ABD),$$

which is the formula.

At this point Archimedes observed that this method, which brought him to the formula for the volume of a sphere, also allowed him to derive the formula for the surface area of a sphere. He wrote:

> [f]rom this theorem, to the effect that a sphere is four times as great as the cone with a great circle of the sphere as base and with height equal to the radius of

> the sphere, I conceived the notion that the surface of any sphere is four times as great as a great circle in it; for, judging from the fact that any circle is equal to a triangle with base equal to the circumference and height equal to the radius of the circle, I apprehended that, in like manner, any sphere is equal to a cone with base equal to the surface of a sphere and height equal to its radius. (Archimedes 1897, *The Method*, 2, 20–21)

It seems that Archimedes' technique was quite productive.

The Method of Archimedes would have remained lost save by happy chance. In 1899, John L. Heiberg, a Danish scholar in ancient Greek mathematics and specialist on Archimedes, learned of a manuscript held at the library of the Holy Sepulchre at Jerusalem. The document was a **palimpsest**, a parchment manuscript that has had its original writing washed or scraped off and then reused for another purpose. In this case, the work was an Eastern Orthodox prayer book dating to about the thirteenth century. Fortunately, the original writing, dating to the tenth century, was not perfectly removed. Because of this, some of the original was translated and was seen to be a copy of a work of Archimedes. Hearing this, Heiberg had to see it, and he did in 1906 and 1908 at Constantinople. Heiberg translated what he could of the old text and discovered that the palimpsest contained copies of many Archimedean works. It includes *On the Sphere and Cylinder*, most of *On Spirals*, and parts of both *Measurement of a Circle* and *On the Equilibrium of Planes*, not to mention the only known Greek copy of *On Floating Bodies*. More important, it contains nearly all of *The Method*. This work of Archimedes had been cited by various writers, including some quotations, but the exact content had been unknown. The good fortune of the Heiberg discovery enables us to see some of the ideas behind Archimedes' great work. For further details see Netz and Noel (2007), Dijksterhuis (1987), and Archimedes (1897, The *Method* of Archimedes).

Exercises

1. Masses of a and b are placed at the ends of lever $\overline{AB}$ with fulcrum F.

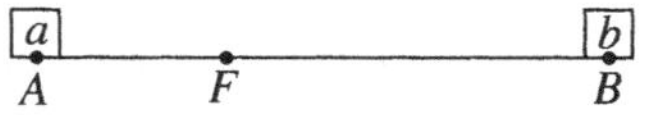

   (a) If $a = 3$, $b = 7$, and $AF = 10$, what is FB?
   (b) If a is three times as massive as b when $AF = 20$, what is FB?
   (c) If the lever balances in its current configuration and then a mass b is added to the left-hand side, how far to the left must F be moved to make the lever balance?

2. Prove Theorem 6.3.2.

3. Rewrite the proof of Theorem 6.3.3 for the case of six equally spaced weights on a lever.

4. Prove that if five weights have their centers of gravity on a straight line such that the distances between any pair of consecutive centers of gravity are the same, the center of gravity of the middle segment is the center of gravity of the entire system.

5. Prove Theorem 6.3.5.

6. Show using any method that the center of gravity of:
   (a) a circle is its center.
   (b) a triangle is the intersection of two segments each joining a vertex with the midpoint of the opposite side.

7. Construct the center of gravity of a rectangle, a circle, and a triangle.

8. Use the sieve of Eratosthenes to find all prime numbers between 51 and 150.

9. The circumference of the earth's equator is 24,902 miles and the circumference of the great circle that has its diameter as the segment with its endpoints at the poles is 24,860 miles. Find the approximate volume of the earth in cubic miles.

10. Put a coordinate system on Figure 6.10 so that the center of the sphere K is the origin, D is $(0, a, 0)$, and A is $(0, 0, a)$. Find equations for:
   (a) $\overline{AF}$
   (b) $\odot OP$ where S is $(0, 0, s)$
   (c) $\odot QR$
   (d) $\odot MN$

11. Continue with Exercise 10.
   (a) Find the areas of $\odot OP$, $\odot QR$, and $\odot MN$
   (b) Show that HA/AS equals

$$\frac{\text{area(circle with diameter } \overline{MN})}{\text{area(circle with diameter } \overline{OP}) + \text{area(circle with diamter } \overline{QR})},$$

   (c) Compute the volume of the sphere $ABCD$, the cone AEF, and the cylinder with height AC and base $\odot EF$ by integration.
   (d) Use the volume just found to show that

$$\frac{HA}{AK} = \frac{\text{volume(cylinder)}}{\text{volume(sphere)} + \text{volume(cone } AEF)}.$$

12. Use Archimedes' method to show the following:
   (a) The volume of a cylinder with base equal to a great circle of a sphere and height equal to its diameter is $3/2$ times the volume of the sphere.
   (b) The volume of a right-angled conoid between its vertex and a plane perpendicular to its axis is $3/2$ times the volume of a cone that has the same base and the same height as the segment of the conoid.

13. Find the centers of gravity of the given solids using any method.
   (a) A right-angled conoid.
   (b) A hemisphere.
   (c) An obtuse-angled conoid.
   (d) A spheroid (oblong or flat).

6.4 PRELIMINARIES TO THE PROOF

Although we use similar methods in integral calculus today, Archimedes did not regard *The Method* as a legitimate means of proof. It was simply a tool for finding formulas. Once he had one in hand, Archimedes would then resort to Euclid's geometry for the final demonstration. In the first book of *On the Sphere and Cylinder*, Archimedes gave his geometric proofs for the formulas for the volume and the surface area of a sphere (Archimedes 1897, 1–55). As we shall see, their justifications are quite involved and will cover two sections.

Inscribing and Circumscribing Regular Polygons

We begin with two technical lemmas.

■ **LEMMA 6.4.1 [*On the Sphere and Cylinder* I.2]**

If $AB > d$, there exists $\overline{EG}$ and $\overline{HG}$ so that EG is greater than HG and $EG/HG < AB/d$.

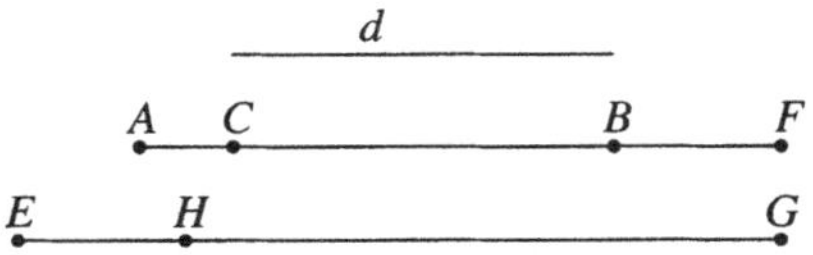

PROOF

Let AB and d represent two magnitudes such that $AB > d$. Find C on $\overline{AB}$ such that $CB = d$. Let F extend $\overline{AB}$ so that $AF = nAC$, where n is the smallest number required for nAC to be greater than d. Now, take any other line segment. Call it $\overline{HG}$. Find E, extending $\overline{HG}$ such that $HG = nEH$. Since $EH/HG = 1/n$ and $AC/AF = 1/n$, we have

$$\frac{EH}{HG} = \frac{AC}{AF}.$$

Then these magnitudes are proportional *componendo*, namely

$$\frac{EH + HG}{HG} = \frac{AC + AF}{AF}.$$

Because $AF > CB$, we also have

$$\frac{AC}{AF} < \frac{AC}{CB}.$$

Finally, since $EH + HG = EG$ and $AC + CB = AB$, we may conclude that

$$\frac{EG}{HG} = \frac{AC + AF}{AF} = \frac{AC}{AF} + \frac{AF}{AF} < \frac{AC}{CB} + \frac{CB}{CB} = \frac{AB}{CB}.$$

Since $CB = d$, we see that $\overline{EG}$ and $\overline{HG}$ are the line segments desired. ■

■ **LEMMA 6.4.2 [*On the Sphere and Cylinder* I.3]**

Given $a > b$ and a circle, it is possible to inscribe a regular polygon P_I within the circle and to circumscribe another regular polygon P_C about it so that

$$\frac{\text{side of } P_C}{\text{side of } P_I} < \frac{a}{b}.$$

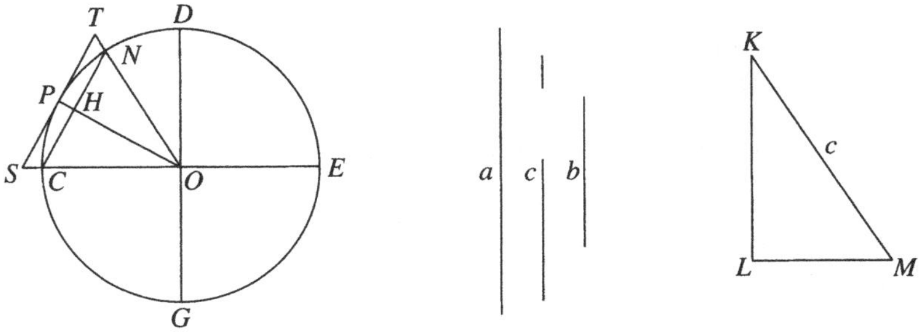

PROOF

Take $a > b$ and let $\overline{DG}$ and $\overline{CE}$ be perpendicular diameters of $\odot O$. By Lemma 6.4.1, find line segments with lengths c and KL such that $c > KL$ and

$$\frac{c}{KL} < \frac{a}{b}.$$

Construct $\overline{LM}$ so that $\overline{LM}$ is perpendicular to $\overline{KL}$ and $KM = c$ (see Exercise 3). Bisect $\measuredangle DOC$ continually until we have $\measuredangle NOC$, so that N is on the circle and

$$\text{m}\measuredangle NOC = \frac{\text{m}\measuredangle DOC}{2^n} < 2\text{m}\measuredangle LKM$$

for some n. Join $\overline{CN}$. This will be a side of a regular 2^{n+2}-gon inscribed within the circle. Next let radius $\overline{OP}$ bisect $\measuredangle NOC$. Notice that $\overline{OP}$ also bisects $\overline{NC}$ at H in right angles. Construct tangent $\overline{ST}$ at P where S is on an extension of $\overline{OC}$ and T on an extension of $\overline{ON}$. This makes $\overline{ST}$ into a side of a circumscribed regular 2^{n+2}-gon. Since $\text{m}\measuredangle NOC < 2\text{m}\measuredangle LKM$, we conclude that

$$\text{m}\measuredangle HOC < \text{m}\measuredangle LKM.$$

Therefore, by Exercise 4,

$$\frac{OC}{OH} < \frac{KM}{KL}. \tag{6.23}$$

Since $\triangle OHC \sim \triangle OPS$, we may conclude that

$$\frac{SP}{CH} = \frac{OP}{OH}.$$

Since $ST = 2SP$, $CN = 2CH$, and $OP = OC$, we have

$$\frac{ST}{CN} = \frac{SP}{CH} = \frac{OP}{OH} = \frac{OC}{OH},$$

and since $KM = c$,

$$\frac{ST}{CN} < \frac{c}{KL} < \frac{a}{b}. \blacksquare$$

We now use the two lemmas to prove that we can find two regular polygons, one inscribed within a circle and the other circumscribed about it, such that ratio between their areas is as small as we want.

■ **THEOREM 6.4.3 [*On the Sphere and Cylinder* I.5]**

Given $a > b$ and a circle, it is possible to inscribe a regular polygon P_I within the circle and to circumscribe another P_C about it so that

$$\frac{\text{area}(P_C)}{\text{area}(P_I)} < \frac{a}{b}.$$

PROOF

Let $\odot A$ be the circle and suppose that a and b are magnitudes such that $a > b$. By Lemma 6.4.1 let c and d be lengths of segments such that $c > d$ and

$$\frac{c}{d} < \frac{a}{b},$$

Let m be the mean proportional between c and d. That is,

$$\frac{c}{m} = \frac{m}{d}.$$

By Lemma 6.4.2 we can inscribe a regular polygon P_I within $\odot A$ and circumscribe a regular polygon P_C about the circle so that

$$\frac{\text{side of } P_C}{\text{side of } P_I} < \frac{c}{m}.$$

This leads to

$$\frac{(\text{side of } P_C)^2}{(\text{side of } P_I)^2} < \frac{c^2}{m^2}.$$

Because P_C and P_I are similar, by Exercise 5,

$$\frac{(\text{side of } P_C)^2}{(\text{side of } P_I)^2} = \frac{\text{area}(P_C)}{\text{area}(P_I)}. \tag{6.24}$$

Hence, since $m^2 = cd$,

$$\frac{\text{area}(P_C)}{\text{area}(P_I)} < \frac{c^2}{m^2} = \frac{c}{d} < \frac{a}{b}. \blacksquare$$

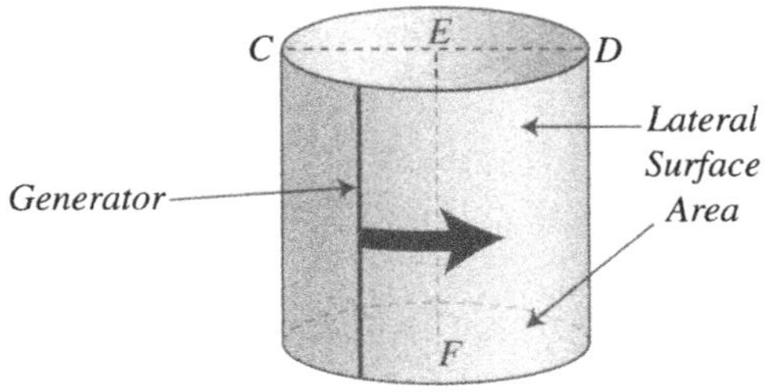

Figure 6.13 The lateral surface area and a generator of a right circular cylinder.

Cylinders and Cones

Recall that that the lateral surface area of a solid with at least one base is the surface area of the solid excluding the area of its base(s) (Exercise 4 in Section 1.1). The **generator** of a right circular cylinder is a line segment on its lateral surface that is perpendicular to its base and equal in length to the height of the cylinder (Figure 6.13). The generator received its name because the cylinder can be viewed as the result of rotating the generator through a complete circle about the center of the base. Finally, the **axis** of a cylinder is the line segment joining the centers of its two bases.

■ **THEOREM 6.4.4 [*On the Sphere and Cylinder* I.13]**

The lateral surface area of any right circular cylinder is equal to the area of a circle whose radius is the mean proportional between the height of the cylinder and the diameter of its base.

PROOF

Take a right circular cylinder with lateral surface area s, base $\odot E$, and axis $\overline{EF}$ as in Figure 6.13. Let m be the mean proportional between EF and CD. In other words,

$$\frac{EF}{m} = \frac{m}{CD}.$$

Then let $\odot B$ denote a circle with radius m. We will show that $\text{area}(\odot B) = s$ by contradiction. We have two cases.

First assume that $\text{area}(\odot B) < s$. Let P_C represent a regular polygon circumscribed about $\odot B$ and P_I represent a regular polygon inscribed within $\odot B$ such that

$$\frac{\text{area}(P_C)}{\text{area}(P_I)} < \frac{s}{\text{area}(\odot B)}. \tag{6.25}$$

This follows from Theorem 6.4.3. Circumscribe a regular polygon about $\odot E$ that is similar to P_C. Call this polygon P_E. Construct a prism on P_E with the same height as the cylinder. This prism circumscribes the cylinder.

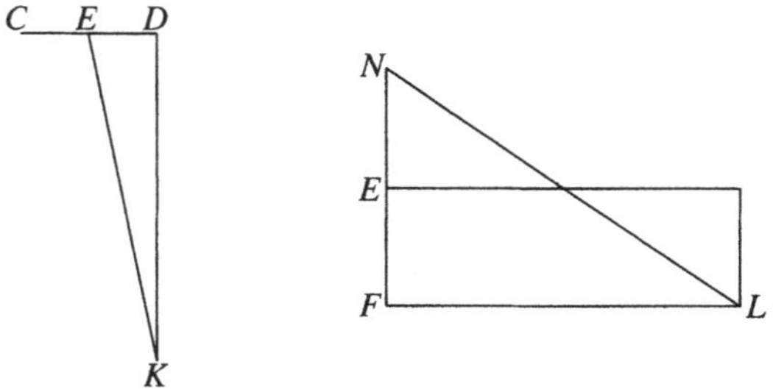

We now need some line segments. At $\overline{CD}$, construct $\overline{KD}$ perpendicular to $\overline{CD}$, and at $\overline{EF}$, construct $\overline{FL}$ perpendicular to $\overline{EF}$ such that both DK and FL equal the perimeter of P_E. Join $\overline{EK}$ remembering that E is the midpoint of $\overline{CD}$. We then have:

- The area of $\triangle KDE$ equals the area P_E (Exercise 1).
- The area of parallelogram EL equals the lateral surface of the prism.

Next, extend $\overline{EF}$ to N such that $\overline{EF} \cong \overline{EN}$ and join $\overline{NL}$. Since P_E is similar to P_C, we have

$$\frac{\text{area}(\triangle KDE)}{\text{area}(P_C)} = \frac{\text{area}(P_E)}{\text{area}(P_C)} = \frac{ED^2}{m^2} = \frac{ED^2}{EF \cdot CD}.$$

However,

$$\frac{\text{area}(\triangle KDE)}{\text{area}(\triangle LFN)} = \frac{ED \cdot DK}{FN \cdot FL} = \frac{ED}{FN} = \frac{ED^2}{2EF \cdot ED}.$$

Since $CD = 2ED$,

$$\text{area}(P_C) = \text{area}(\triangle LFN) = \text{area(parallelogram } EL),$$

but the area of parallelogram EL equals the lateral surface area of the prism, so we have

$$\text{area}(P_C) = \text{lateral surface area of the prism.}$$

However, by Inequality 6.25 we have

$$\frac{\text{lateral surface area of the prism}}{\text{area}(P_I)} < \frac{s}{\text{area}(\odot B)},$$

which yields

$$\frac{\text{lateral surface area of the prism}}{s} < \frac{\text{area}(P_I)}{\text{area}(\odot B)}.$$

This is impossible, since the lateral surface area of the prism is greater than s, yet the polygon inscribed within $\odot B$ has area less than the area of $\odot B$.

Therefore, area($\odot B$) $\geqslant s$. To prove that area($\odot B$) $\leqslant s$, construct polygons so that

$$\frac{\text{area}(P_C)}{\text{area}(P_I)} < \frac{\text{area}(\odot B)}{s}$$

and then proceed in a manner similar to the above. ■

To see what Theorem 6.4.4 means algebraically, let h be the height of a given cylinder and r the radius of its base. Let m be the mean proportional between h and $2r$. Therefore, $m^2 = 2rh$, and the area of the lateral surface area of the cylinder is given by the formula

$$\text{lateral surface area of a cylinder} = \pi m^2 = 2\pi rh.$$

We call the length of a generator of a right circular cone its **slant height**. If r represents the radius of the cone's base and s is its slant height, then

$$\text{lateral surface area of a cone} = \pi rs.$$

The next theorem states this equation in the language of geometry. Its proof is similar to the previous result.

■ **THEOREM 6.4.5 [*On the Sphere and Cylinder* I.14]**

The lateral surface area of a right circular cone is equal to the area of a circle whose radius is the mean proportional between its slant height and the radius of its base.

Solid Rhombi

In Figure 6.14 we see an object with a name that is unique to Archimedes. A **solid rhombus** (or simply a **rhombus**) is a solid consisting of two cones on opposite sides of the same base. Although not necessary, we shall assume that a solid rhombus is constructed from right cones. This figure will be crucial to Archimedes' justification of his spherical volume formula.

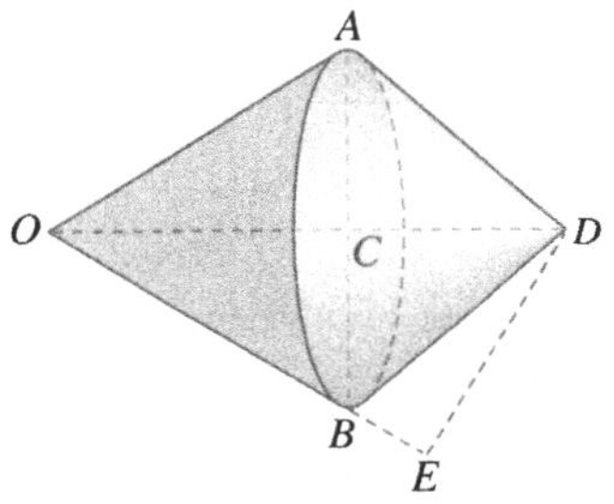

Figure 6.14 $ADBO$ is a solid rhombus of Archimedes.

THEOREM 6.4.6 [*On the Sphere and Cylinder* I.18]

The volume of a solid rhombus is equal to the volume of a cone with the following properties:

- Its base has area equal to the lateral surface area of one of the cones comprising the rhombus.
- Its height is equal to the length of the perpendicular drawn from the vertex of the other cone to one side of the first.

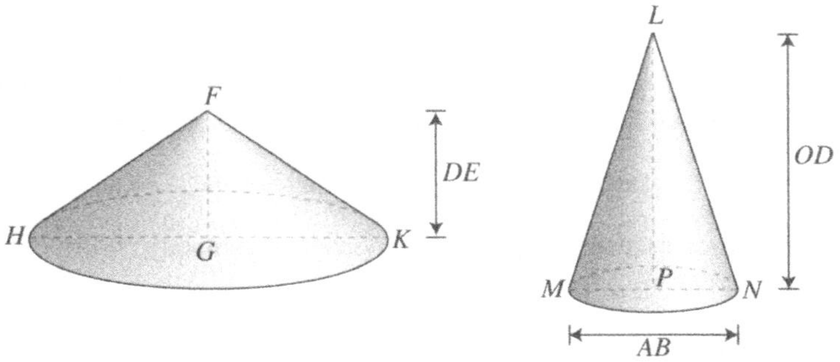

PROOF

Let O and D be the vertices of two cones sharing a base with diameter $\overline{AB}$ and center C (Figure 6.14). Let E be chosen to extend $\overline{OB}$ so that $\overline{DE}$ is perpendicular to $\overline{OE}$. Take another cone FHK with:

- Base equal to the lateral surface area of cone OAB
- Height FG equal to DE

We claim that the volume of cone FHK equals the volume of the rhombus. To see this, construct cone LMN such that:

- The area of its base is equal to the area of $\odot AB$.
- Its height LP equals OD.

Therefore,

$$\frac{LP}{CD} = \frac{OD}{CD}.$$

Because the volumes of two cones with bases of equal area are to one another as their heights and Exercise 9,

$$\frac{LP}{CD} = \frac{\text{volume(cone } LMN)}{\text{volume(cone } DAB)}$$

and

$$\frac{OD}{CD} = \frac{\text{volume(rhombus } OADB)}{\text{volume(cone } DAB)}, \tag{6.26}$$

so

$$\text{volume(rhombus } OADB) = \text{volume(cone } LMN). \tag{6.27}$$

Since $AB = MN$ and the lateral surface area of cone OAB equals the area of the base of cone FHK,

$$\frac{\text{area(base of } FHK)}{\text{area(base of } LMN)} = \frac{\text{lateral surface area of } OAB}{\text{area(base of } OAB)} = \frac{OB}{BC} = \frac{OD}{DE} = \frac{LP}{FG}.$$

The second equality follows by Exercise 10, the third by similar triangles, and the last by hypothesis. Hence, cones FHK and LMN have the same volume, so appealing to Equation 6.27, the volume of cone FHK is the same as the volume of the given rhombus. ■

Also important to Archimedes' argument is the last solid of this section. Take solid rhombus $OACB$ (see the diagram in the next theorem). Cut cone OAB by plane Π that is parallel to $\odot AB$ and then remove the newly formed solid rhombus $ODCE$ from $OACB$. We shall invent some terminology and call the resulting solid a **cored rhombus**. We need to know its volume.

■ **THEOREM 6.4.7 [*On the Sphere and Cylinder* I.20]**

The volume of a cored rhombus is equal to the volume of a cone with the following properties:

- Its base has area equal to the lateral surface area of the frustum.
- Its height is equal to the length of the perpendicular drawn from the vertex of the other cone to one side of the frustum.

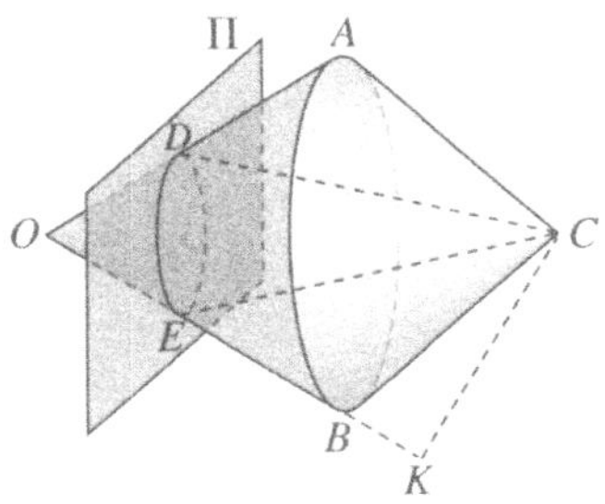

PROOF

Let cone ACB and frustum $DABE$ have $\odot AB$ as a common base. Construct $\overline{CK}$ perpendicular to $\overline{OB}$ produced. We will show that the volume of the cored rhombus equals the volume of a cone FGH with base $\odot GH$ and with the following properties:

- Base area equal to the lateral surface area of the frustum $DABE$
- Height equal to CK

To prove this, construct two cones.

- The first we will call LMN. The area of its base $\odot MN$ will equal the lateral surface area of cone OAB and have height CK.
- The second we will call PQR. The area of its base $\odot QR$ will be equal to the lateral surface area of ODE and have height CK.

Notice that the lateral surface area of cone OAB is the sum of the lateral surface areas of cone ODE and frustum $DABE$. Therefore,

$$\text{area}(\odot MN) = \text{area}(\odot QR) + \text{area}(\odot GH).$$

Since these cones have the same height, we may conclude that

$$\text{volume}(\text{cone } LMN) = \text{volume}(\text{cone } PQR) + \text{volume}(\text{cone } FGH).$$

However, the volume of cone LMN equals the volume of solid rhombus $OACB$ and cone PQR has volume equal to solid rhombus $ODCE$ (Theorem 6.4.6), so the volume of cone FGH equals the volume of the cored rhombus. ■

But what is the lateral surface area of the frustum?

■ **THEOREM 6.4.8 [*On the Sphere and Cylinder* I.16]**

The lateral surface area of the frustum of a cone equals the area of a circle with radius equal to the mean proportional between the slant height of the frustum and the sum of the radii of its bases.

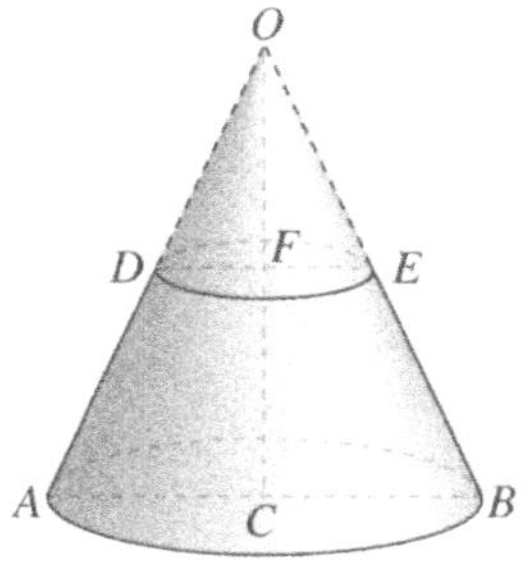

PROOF

Let F be the center of $\odot DE$ and C the center of $\odot AB$, so that O, F, and C are all on the axis of cone OAB. By Theorem 6.4.5, the lateral surface area of cone OAB is equal to the area of a circle with the square of its radius being $OA \cdot AC$. Similarly, the lateral surface area of cone ODE is equal to the area of a circle with the square of its radius equal to $OD \cdot DF$. The lateral surface area of the frustum is the difference between these two areas. Now calculate the difference between the squares of the radii:

$$\begin{aligned} OA \cdot AC - OD \cdot DF &= (OD + AD) \cdot AC - OD \cdot DF \\ &= OD \cdot AC + AD \cdot AC - OD \cdot DF \\ &= OA \cdot DF + AD \cdot AC - OD \cdot DF \\ &= AD \cdot DF + AD \cdot AC \\ &= AD \cdot (DF + AC). \end{aligned}$$

The third equality follows because $\triangle OAC \sim \triangle ODF$, which implies that $OD \cdot AC = OA \cdot DF$.

Now if we subtract the areas of the two circles, the result is the area of a circle with the square of its radius equal to $AD \cdot (DF + AC)$. Therefore, this radius is the mean proportional between AD and $DF + AC$. Since AD is the slant height of the frustum and $DF + AC$ is the sum of the radii of the bases of the two cones, we have proven the theorem. ■

Exercises

1. Demonstrate that the area of a regular convex polygon equals half the product of its perimeter and its apothem.

2. Confirm Lemma 6.4.2 and Theorem 6.4.3 by finding the number of sides of two similar regular polygons P_I and P_C such that P_I is inscribed within a circle and P_C is circumscribed about the same circle and
   (a) side of P_C/side of $P_I < 1.00005$.
   (b) side of P_C/side of $P_I < 1.000001$.
   (c) $\text{area}(P_C)/\text{area}(P_I) < 1.000001$.
   (d) $\text{area}(P_C)/\text{area}(P_I) < 1.0000005$.

3. Let $c > a$. Construct a right triangle with a leg of length a and a hypotenuse of length c.

4. Prove Inequality 6.23.

5. Prove Equation 6.24.

6. A cylinder has a height of 10 and a base of radius 3. Find the cylinder's
   (a) lateral surface area.
   (b) surface area.
   (c) volume.

7. A cone has a height of 7 and a base of radius 4. Find the cone's
   (a) slant height.
   (b) lateral surface area.
   (c) surface area.
   (d) volume.

8. Let a right circular cylinder and a right cone have bases of radius r. Let h be the height of the cylinder and s the slant height of the cone. Use calculus to show that the lateral surface area of the cylinder is $2\pi rh$ and that of the cone is πrs.

9. Prove:
   (a) The volumes of two cones with bases of equal area are to one another as their heights.
   (b) Equation 6.26

10. Show that the ratio of the surface of a right cone to the area of its base is equal to the ratio of the slant height of the cone to the radius of its base.

11. A solid rhombus consists of two right cones, one with a slant height of 6 and the other with a height of 8. The common base of the cones has a diameter of 10. Find the rhombus's
   (a) surface area.
   (b) volume.

12. Using the theorems of the section, write algebraic formulas for
   (a) the volume of a solid rhombus.
   (b) the volume of a cored rhombus.
   (c) the surface area of the frustum of a cone.

6.5 THE VOLUME OF A SPHERE

Archimedes' proof of the formula for the volume of a sphere is an elegant example of the method of exhaustion. In *On the Sphere and Cylinder*, he constructs two regular polygons, one that is inscribed within and the other that is circumscribed about a great circle of a given sphere. When the polygons are revolved about a diameter of the circle, the sphere fits between the two volumes of revolution, which enables Archimedes to find his prized result (Archimedes 1897, 32–44).

Revolve a Polygon

Begin by taking a regular n-gon where n is a multiple of 4. We shall call such a polygon **standard**. Assume that A and A' are opposite vertices. Let $B, C, M, N, Y, \ldots$ be vertices of the n-gon on one side of $\overline{AA'}$ and $B', C', M', N', Y', \ldots$ be vertices on the other side. If we revolve the polygon about $\overline{AA'}$, the result is as displayed in the diagram for the next theorem with $n = 12$. Archimedes began by finding the surface area for this solid.

■ **THEOREM 6.5.1 [*On the Sphere and Cylinder* I.24]**

The surface area of the solid of revolution of standard polygon $ABC \ldots A' \ldots C'B'$ about $\overline{AA'}$ is equal to a circle with a radius whose square is

$$AB \cdot (BB' + CC' + MM' + NN' + \cdots + YY').$$

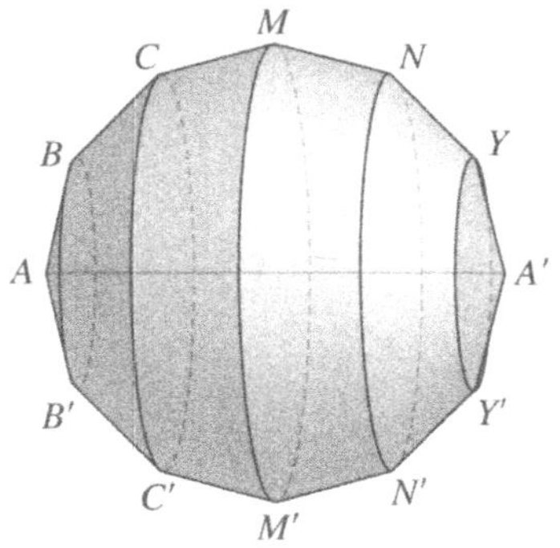

PROOF

The surface of the solid is comprised of parts of the surfaces of various cones. Appealing to Theorem 6.4.5, the lateral surface area of cone ABB' is equal to

$$\pi \cdot AB \cdot \frac{1}{2}BB',$$

and the lateral surface area of the frustum $BB'C'C$ equals

$$\pi \cdot BC \cdot \frac{1}{2}(BB' + CC'),$$

by Theorem 6.4.8. Continuing in this manner, we find the surface areas of the remaining portions of the revolved polygon. Since the slant heights of the cones and frustums are of the same length,

$$AB \cdot \frac{1}{2}BB' + BC \cdot \frac{1}{2}(BB' + CC') + \cdots$$
$$= AB \cdot \frac{1}{2}BB' + AB \cdot \frac{1}{2}BB' + AB \cdot \frac{1}{2}CC' + \cdots .$$

Thus, the surface area of the solid is equal to

$$\pi \cdot AB \cdot (BB' + CC' + MM' + NN' + \cdots + YY'). \blacksquare$$

Now for the volume generated by the polygon.

■ **THEOREM 6.5.2 [*On the Sphere and Cylinder* I.26]**

The volume of revolution of a standard polygon is equal to the volume of a cone with these properties:

- Its base has area equal to the surface area of revolution.
- Its height is equal to the length of the apothem.

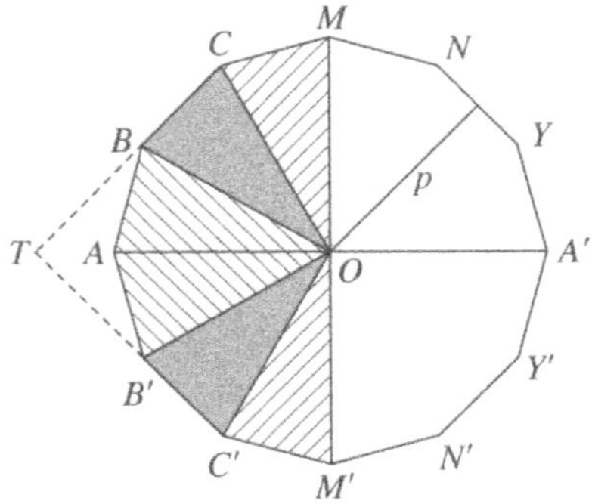

PROOF

Let the standard polygon $ABC \ldots A' \ldots C'B'$ be revolved about $\overline{AA'}$. Let O be the midpoint of $\overline{AA'}$ and let p be the length of the apothem of the polygon. It is enough to examine what happens on the left-hand side, for the two halves are symmetrical. We have two cases to consider.

- $OBAB'$ generates a solid rhombus. By Theorem 6.4.6, its volume is equal to the volume of a cone with base area equal to the lateral surface area of cone generated by ABB', and its height is p since it is the length of a perpendicular to $\overline{AB}$ through O.

- Extend both $\overline{BC}$ and $\overline{B'C'}$ to meet at T. Then $OCTC'$ generates a solid rhombus. Remove the solid rhombus generated by $OBTB'$. Theorem 6.4.7 states that the resulting cored rhombus has volume equal to that of a cone with base area equal to the lateral surface area of frustum $BB'C'C$ and with height p. Continue forming cored rhombi until the vertical diameter $\overline{MM'}$ of the sphere is reached.

Sum up the volumes on both left and right sides of $\overline{MM'}$. Each of these volumes equals the volume of a cone, and each of these cones has height p. Adding these volumes we find a cone with height p and base area equal to the sum of the base areas of the cones, and this sum equals the surface area generated by revolving the polygon. Therefore, the volume of this new cone equals the volume of revolution of the polygon. ∎

Inscribe the Polygon

We start by proving an identity from Archimedes (1897, *On the Sphere and Cylinder* I.21). Let $ABC \ldots A' \ldots C'B'$ be a standard regular polygon (Figure 6.15). Join $\overline{BB'}$, $\overline{CC'}$, ... Notice that these line segments are perpendicular to $\overline{AA'}$, and we find that $\overline{AB}$ is parallel to $\overline{B'C}$ (Exercise 4). Therefore, $\sphericalangle ABB' \cong \sphericalangle CB'B$. Since $\sphericalangle BFA$ and $\sphericalangle B'FK$ are both right angles,

$$\triangle ABF \sim \triangle KB'F.$$

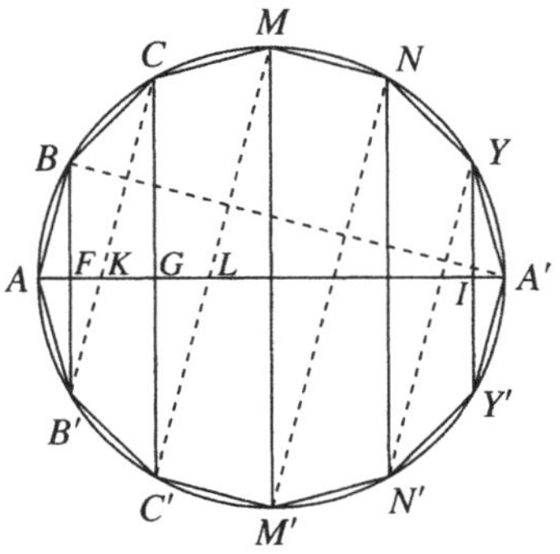

Figure 6.15 The inscribed dodecagon.

Similarly,

$$\triangle KB'F \sim \triangle KCG,$$

$$\triangle KCG \sim \triangle LC'G,$$

and so on. Therefore,

$$\frac{BB' + CC' + \cdots}{AA'} = \frac{BF + B'F + CG + C'G + \cdots}{AF + KF + KG + GL + \cdots} = \frac{BF}{AF} = \frac{A'B}{AB},$$

where the second equality holds since

$$\frac{BF}{AF} = \frac{B'F}{KF} = \frac{CG}{KG} = \frac{C'G}{LG} = \cdots$$

and the last holds because $\triangle BFA \sim \triangle A'BA$. Hence, we have the following identity:

$$AB \cdot (BB' + CC' + \cdots + YY') = AA' \cdot A'B. \tag{6.28}$$

Following Archimedes, we will sandwich the sphere between two solids. By the method of exhaustion, we will assume that the volume of the sphere is not equal to the inscribed and circumscribed volumes of revolution. This assumption will lead to a contradiction as we check the cases when the volume of the sphere is less than or greater than the desired volumes. To accomplish this we need two theorems that give bounds on the volumes of revolution of the polygons.

■ **LEMMA 6.5.3 [*On the Sphere and Cylinder* I.25]**

The surface area of a standard polygon revolved about its axis and inscribed within a sphere is less than four times the area of a great circle of the sphere.

PROOF

Inscribe a standard polygon within a great circle of a sphere and rotate about the diameter of the circle as in Figure 6.15. Take $\odot R$ with radius of length r such that

$$r^2 = AB \cdot (BB' + CC' + \cdots + YY').$$

By Theorem 6.5.1, the surface area generated by rotating the polygon equals the area of $\odot R$. We then have by Equation 6.28 that

$$r^2 = AA' \cdot A'B,$$

and then, as seen in Figure 6.15,

$$r^2 < (AA')^2.$$

Therefore,

$$\pi r^2 < \pi (AA')^2 = 4 \left[\pi \left(\frac{AA'}{2} \right)^2 \right].$$

This is what we needed to prove because $\pi (AA'/2)^2$ is the area of a great circle of the sphere. ■

Archimedes used this proposition to draw a similar conclusion about its volume.

■ **THEOREM 6.5.4 [*On the Sphere and Cylinder* I.27]**

The volume of a standard polygon revolved about its axis and inscribed within a sphere is less than four times the volume of a cone with

- base area equal to a great circle of the sphere, and
- height equal to a radius of the sphere.

PROOF

Consider two cones:

- By Theorem 6.5.2, the volume of the rotated polygon equals the volume of a cone with base area equal to the surface area of the rotated polygon and with height equal to the apothem of the polygon. Call this cone R.
- Let S represent a cone whose base has area equal to a great circle of the sphere and whose height equals the radius of the sphere.

Appealing to Theorem 6.5.3, we have

$$\text{area(base of } R) < 4 \cdot \text{area(base of } S).$$

Since the length of the apothem of the polygon is less than the radius of a great circle of the sphere, we conclude that

$$\text{volume of revolution of the polygon} < 4 \cdot \text{volume}(S).$$ ■

Circumscribe the Polygon

For the second half of the method of exhaustion, Archimedes circumscribed a volume of revolution about the sphere. Before we can find a bound on its volume, we first need a lemma that Archimedes assumed without proof. The proof that we will use here will follow the argument of Eutocius, a commentator on the works of Archimedes who lived around A.D. 500 (Netz 2004, 263; Burton 1985, 193).

■ **LEMMA 6.5.5 [Eutocius]**

If a standard polygon with side $\overline{AB}$ is circumscribed about a circle with center O and diameter $\overline{AA'}$, and if P is the point of intersection of $\overline{AB}$ with the circle, then $A'B = 2OP$.

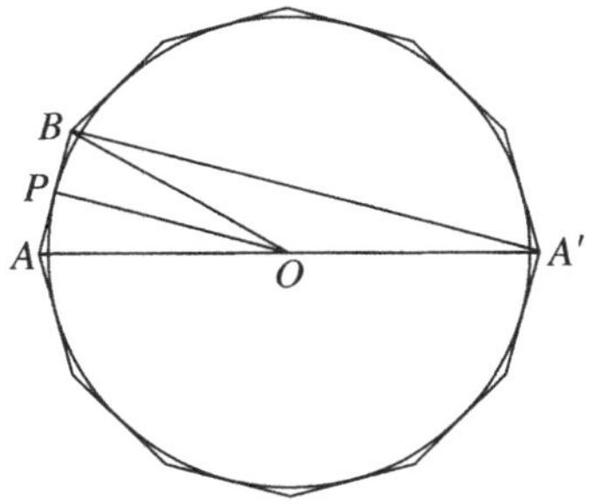

PROOF

Join $\overline{OP}$ and $\overline{OB}$. Since $\overline{AB}$ is tangent at P, $\overline{OP}$ is perpendicular to $\overline{AB}$. Observe that P is the midpoint of $\overline{AB}$. This is because in the regular polygon we have $\overline{OA} \cong \overline{OB}$, and this yields $\overline{AP} \cong \overline{PB}$ by the Pythagorean Theorem. Notice that $m\sphericalangle BOA = 2m\sphericalangle POA$ and $m\sphericalangle BOA = 2m\sphericalangle BA'A$, so $\sphericalangle POA \cong \sphericalangle BA'A$. From this we conclude that $\overline{OP}$ is parallel to $\overline{A'B}$. Therefore, because $\triangle ABA' \sim \triangle APO$, we have

$$\frac{AA'}{OA} = \frac{A'B}{OP}.$$

However, $AA' = 2OA$, which yields $A'B = 2OP$. ■

■ **THEOREM 6.5.6 [*On the Sphere and Cylinder* I.30]**

The volume of a standard polygon revolved about its axis and circumscribed about a sphere is greater than four times the area of a great circle of the sphere.

PROOF

Let $ABC \ldots A' \ldots C'B'$ be a standard polygon that is circumscribed about $\odot O$. Appealing to Theorem 6.5.1 we see that the surface of revolution of the polygon is equal to the area of a circle with radius r such that

$$r^2 = AB \cdot (BB' + CC' + \cdots).$$

Therefore, by Equation 6.28,

$$r^2 = AA' \cdot A'B > A'B \cdot A'B.$$

If we let P be the point on $\overline{AB}$ that intersects $\odot O$, by Eutocius's lemma (6.5.5),

$$r^2 > 4OP^2,$$

which implies that

$$\pi r^2 > 4\pi \cdot OP^2.$$

Hence, the area of the surface of revolution must then be greater than four times the area of a great circle of the sphere. ■

Finish the Proof

We start with a lemma.

■ **LEMMA 6.5.7 [*On the Sphere and Cylinder* I.32]**

Inscribe a standard polygon within a great circle of a sphere and circumscribe a polygon similar to the first about the circle. Let ZY represent the length of a side of the inscribed polygon and AB the length of a side of the circumscribed polygon. If both polygons are rotated together to inscribe and circumscribe the sphere, then

$$\frac{\text{circumscribed volume}}{\text{inscribed volume}} = \frac{AB^3}{ZY^3}.$$

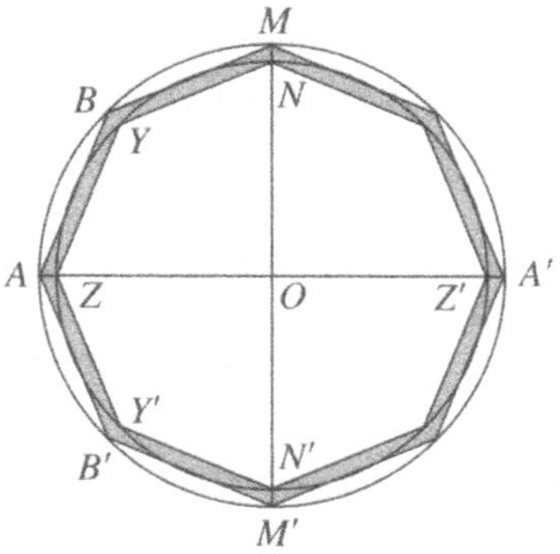

PROOF

First construct $\odot R$ and $\odot S$ such that

- The area of $\odot R$ equals the surface area of the circumscribed solid.
- The area of $\odot S$ equals the surface area of the inscribed solid.

Next we define two cones.

- Let cone C_R have base $\odot R$ and height OZ.
- Let cone C_S have base $\odot S$ and height p, where p is the length of the perpendicular from O to $\overline{ZY}$.

Then, the volume of cone C_S is the same as the volume of the inscribed solid (Theorem 6.5.2), and since the circumscribed solid can be inscribed within a larger circle and since OZ is the length of the circumscribed polygon's apothem, we may also conclude that the volume of C_R equals the volume of the circumscribed solid. Now by Theorem 6.5.1, the square of the radius of $\odot R$ is

$$AB \cdot (BB' + MM' + \cdots),$$

and the square of the radius of $\odot S$ is

$$ZY \cdot (YY' + NN' + \cdots).$$

Since the polygons are similar,

$$\frac{AB}{ZY} = \frac{BB' + MM' + \cdots}{YY' + NN' + \cdots}.$$

Therefore,

$$\frac{\text{area}(\odot R)}{\text{area}(\odot S)} = \frac{AB^2}{ZY^2}.$$

Hence, by Exercise 6 we may conclude that

$$\frac{\text{volume}(C_R)}{\text{volume}(C_S)} = \frac{AB^3}{ZY^3}, \tag{6.29}$$

and this is the equation that we sought. ■

Archimedes now has all of the tools necessary to prove his formula.

■ **THEOREM 6.5.8 [*On the Sphere and Cylinder* I.34]**

The volume of a sphere is equal to four times the volume of a cone with the properties that:

- Its base is equal to the area of a great circle of the sphere.
- Its height is equal to the radius of the sphere.

PROOF

We use the diagram for Lemma 6.5.7. Let the sphere have great circle $ZNZ'M'$ and let V be the cone such that:

- Its base area is four times the area of $\odot ZNZ'M'$.
- Its height equals the radius of the sphere.

Suppose that the result does not hold. We then have two cases.

First suppose that the volume of the sphere is greater than the volume of cone V. By Lemma 6.4.1, find magnitudes a and b such that $a > b$ and

$$\frac{a}{b} < \frac{\text{volume(sphere)}}{\text{volume}(V)}.$$

Let d and e be two arithmetic means (page 94) placed in progression between a and b. As noted in Exercise 7, this means that $a > d > e > b$ such that

$$a - d = d - e. \tag{6.30}$$

Circumscribe a regular polygon about a great circle and inscribe a similar regular polygon within the circle such that

$$\frac{AB}{ZY} < \frac{a}{d}$$

by Lemma 6.4.2. Rotate both polygons about the diameter $\overline{AA'}$. Appealing to Lemma 6.5.7, we now find that

$$\frac{\text{circumscribed volume}}{\text{inscribed volume}} = \frac{AB^3}{ZY^3}.$$

Eutocius showed using Equation 6.30 that $a^3/d^3 < a/b$ (Exercise 7). Thus,

$$\frac{\text{circumscribed volume}}{\text{inscribed volume}} < \frac{a^3}{d^3} < \frac{a}{b} < \frac{\text{volume(sphere)}}{\text{volume}(V)},$$

but

$$\text{the circumscribed volume} > \text{volume(sphere)}$$

and

$$\text{the inscribed volume} < \text{volume}(V)$$

by Theorem 6.5.4. This yields

$$\frac{\text{circumscribed volume}}{\text{inscribed volume}} > \frac{\text{volume(sphere)}}{\text{volume}(V)},$$

a contradiction. The case when the volume of the sphere is less than four times the volume of the described cone also leads to a contradiction. This is left to Exercise 8. ■

Since the volume of a cone is one third that of a cylinder, Archimedes finishes.

■ **COROLLARY 6.5.9**

The volume of a sphere is two thirds the volume of a cylinder with base area equal to that of a great circle of the sphere and with height equal to the sphere's diameter.

In other words, letting r represent the radius of a given sphere,

$$\text{volume(sphere)} = \frac{2}{3}[\pi r^2(2r)] = \frac{4}{3}\pi r^3.$$

Along the way, Archimedes also proved that the surface area of a sphere is four times the area of a great circle of the sphere: namely,

$$\text{surface area(sphere)} = 4\pi r^2.$$

Netz (2004, 146–147) notes that as one reads the original Greek there is a sense of anticipation as the main result nears and then one of accomplishment when it arrives. This is seen in the words that Archimedes chose (Netz calls them "crisp"), and it is seen in the manner in which he pulled all of the details together at the end in a elegant little package. In the proof of Theorem 6.5.8 he even made use of Lemmas 6.4.1 and 6.4.2, technical results that we met at the very beginning of this journey. It is as if Archimedes cannot wait to write it down and show off what he has discovered. Indeed, the formula for the volume of a sphere was Archimedes' proudest accomplishment as a mathematician, so much so that he told his friends that he wanted the sphere with the cylinder on his tomb. Approximately 130 years after the death of Archimedes, the great Roman political philosopher Cicero recounted this story:

> I will present you with a humble and obscure mathematician of the same city, called Archimedes, who lived many years after; whose tomb, overgrown with shrubs and briers, I in my quæstorship discovered, when the Syracusans knew nothing of it, and even denied that there was any such thing remaining; for I remembered some verses, which I had been informed were engraved on his monument, and these set forth that on the top of the tomb there was placed a sphere with a cylinder. When I had carefully examined all the monuments (for there are a great many tombs at the gate Achradinæ), I observed a small column standing out a little above the briers, with the figure of a sphere and a cylinder upon it; whereupon I immediately said to the Syracusans—for there were some of their principal men with me there—that I imagined that was what I was inquiring for. Several men, being sent in with scythes, cleared the way, and made an opening for us. When we could get at it, and were come near to the front of the pedestal, I found the inscription, though the latter parts of all the verses were effaced almost half away. Thus one of the noblest cities of Greece, and one which at one time likewise had been very celebrated for learning, had known nothing of the monument of its greatest genius, if it had not been discovered to them by a native of Arpinum. (Cicero 1951, *Tusculan Disputations* V.23)

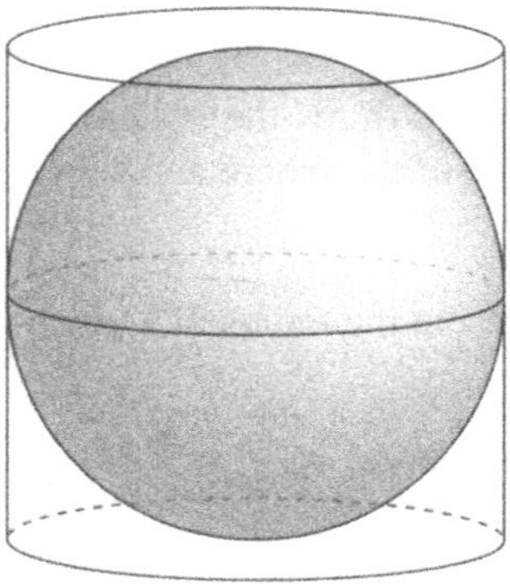

Exercises

1. Use Theorem 6.5.2 to write a formula for the volume of revolution of a standard polygon.

2. Check the formula found in Exercise 1 for the volume of revolution of a standard polygon with 12 sides and radius of revolution equal to 1. Compare this volume with the volume of the unit sphere.

3. Show that the sum of the volumes of cones with height h is equal to a cone of height h having a base with area equal to the sum of the areas of the bases of the original cones.

4. Prove that $\overline{AB}$ is parallel to $\overline{B'C}$ in Figure 6.15.

5. Prove that Theorem 6.5.8 implies Corollary 6.5.9.

6. Confirm Equation 6.29.

7. Let $a > d > e > b$ such that $a - d = d - e$.
   (a) Explain how to find d and e given a and b.
   (b) Show that $a^3/d^3 < a/b$.

8. Let V be the cone in the proof of Theorem 6.5.8. Assume that the volume of the sphere is less than four times the volume of the cone V and choose a and b such that a is greater than b and

$$\frac{a}{b} < \frac{\text{volume}(V)}{\text{volume}(\text{sphere})}.$$

Modify the proof of Theorem 6.5.8 as needed to derive a contradiction.

PART III

ENLIGHTENMENT

CHAPTER 7

FRANÇOIS VIÈTE

The Middle Ages did not produce many mathematicians of note. One famous exception is Leonardo of Pisa (Section 4.4). Another worthy of mention is Nicole Oresme (*c.* 1323–1382). Oresme was a French scholar and bishop of the Catholic church in Lisieux whose interest in mathematics led to early developments in ideas that we now take for granted. He established rules for exponents, including fractional powers for the first time. Later he attempted a generalization to irrational powers but was limited by his lack of an adequate notation. Another of his contributions was possibly inspired by the work of the ancient Greeks. Dating back to Aristotle people believed that when a substance lost some or all of a quality, another quality would rush to take its place. For instance, heat was viewed as a quality. When a hot piece of metal cooled, it lost the quality that made it hot and gained the quality that made it not as hot. Oresme and many of his scholastic colleagues rejected this concept. Instead of viewing heat as a quality, they considered it a quantity. Since quantity is a magnitude and since magnitudes have a geometric representation, Oresme concluded that quantity should be represented graphically. He drew a horizontal line and called it the **longitude**, and perpendicular to this he drew vertical lines called **latitudes** that represented the quantities of interest. Oresme then used his new system to study various types of motion. He let distances along the longitude represent time and

Revolutions of Geometry. By Michael O'Leary

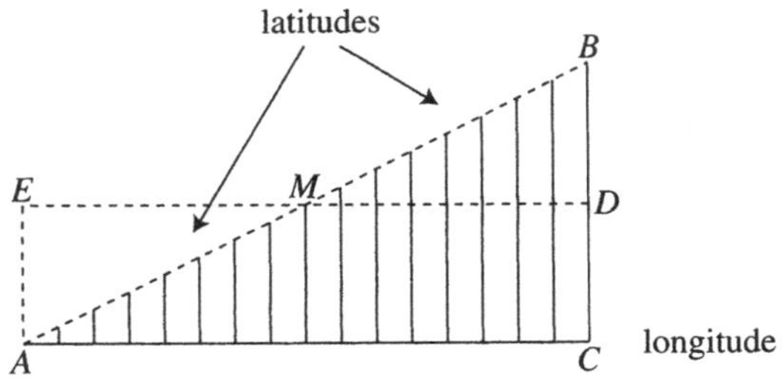

Figure 7.1 The longitude and latitudes of Oresme.

the heights of the latitudes represent velocities. When the velocities had a constant acceleration, he saw that his representation formed a right triangle (Figure 7.1). If the motion was uniform, a rectangle was the result. Let the right triangle be $\triangle ABC$ with $\overline{AB}$ as the hypotenuse. Suppose that M is the midpoint of $\overline{AB}$ and through M draw $\overline{ED}$ so that $ACDE$ is a rectangle. Oresme saw that the area of $\square ACDE$ equals the area of $\triangle ABC$ and $BC = 2DC$. Since the areas of these geometric shapes represent the distance traveled, a basic fact of the integral calculus later developed by Newton and Leibnitz some 330 years later, Oresme concluded that the straight-line distance covered by an object traveling at a uniformly accelerating rate is equal to the distance traveled by another object with a uniform velocity equal to half the final velocity of the first object (Boyer and Merzbach 1991, 262–263; Kline 1972, 210–211).

Although Oresme's longitudes and latitudes did have some influence on the work of Galileo, it otherwise had little impact in scientific circles. This is typical of the Middle Ages. It was not that there were no mathematical pursuits, it was that they did not go far. This continued into the Renaissance. The rediscovery of the great ancient teachings of the Greeks led to an explosion of learning throughout Europe. Many of the names are familiar. This was the time of Copernicus, Galileo, da Vinci, Michelangelo, Machiavelli, Luther, Erasmus, More, Raphael, Shakespeare, and Dürer, among others. The Renaissance saw the rise of the scientific method, that conclusions about the world should be based on experiment and observation and not on mere speculation. It saw great developments in art and reformation of religion. Despite all of this, there was little mathematical progress, especially in geometry. Mathematics needed a new discovery to awaken it from its doldrums. It was found at the end of the Renaissance by François Viète.

Viète was born in Fontenay-le-Comte, Poitou, France in 1540. He received his law degree in 1560 from the University of Poitiers. Evidently a successful lawyer counting the queen of Austria among his clients, in 1564 Viète became the advisor of Antoinette d'Aubeterre and a tutor for her daughter Catherine. This was the beginning of a long career serving in the highest levels of French government. His association with Catherine eventually brought him to Paris, and he would live there for the next four years. Although he was a lawyer by profession, Viète was very interested in mathematics. During his years in Paris he had enough free time to study and do research. He had a particular interest in trigonometry and astronomy. He used trigonometry to solve algebraic equations, and his terminology would be familiar to us except that he used the word *Prosinus* to name tangent and *Transsinuosa* for secant. The abbreviations for the trigonometric functions were a later innovation due to the physician and mathematician Thomas Finck, who is also credited with the first use of the terms *tangent* and *secant* in 1583. Viète would reject Finck's choice because these terms already had meaning from geometry (Viète 1983, 1–3; Cajori 1928, vol. II, 150).

His mathematical diversions ended when he became a member of the Bretagne *parlement* under the appointment of Charles IX in 1573. He would later serve under both Henri III and Henri IV. Under the former he would make his name as a code breaker during the war with the Spanish. By our standards the codes were not very sophisticated, but in those days the Spanish were sure that Viète was in league with the devil. While counselor to Henry IV, the Dutch ambassador to France insulted French mathematicians by claiming that none could find a solution to a 45-degree polynomial equation posed by the Belgium mathematician Adriaan van Roomen. When Viète heard the problem, within minutes he had one solution, and by the next day he had found 22. His answers came when he realized that there was a trigonometric relationship among the solutions. When van Roomen heard the result of the challenge, the two began corresponding and would eventually become good friends (Boyer and Merzbach 1991, 310–311).

7.1 THE ANALYTIC ART

Other than two periods covering approximately 12 years, Viète spent little time researching mathematics. His became the life of the professional politician. Nonetheless, his mathematical contributions were significant. Not only did he advance trigonometry, he also made a subtle but important change to algebra. He was the first to use letters consistently to represent magnitudes in a way that identified their different roles in an equation. We have already seen letters used to represent values. This dates back to the days of Euclid. However, there was no distinction between values that were known and those that were unknown. No distinction was made between variables and coefficients. This led to a clumsy situation in algebra. To find solutions to an equation like a cubic, early algebraists would first write (using modern notation) $x^3 + 2x = 1$ and then apply a method to find any solutions. After another example or two, they expected the reader to understand the strategy used. No generalized formulas or methods were developed to help find solutions (Mahoney 1994, 35). Viète remedied this.

> In order to assist this work by another device, given terms are distinguished from unknown [terms] by constant, general and easily recognized symbols, as (say) by designating unknown magnitudes by the letter A and the other vowels E, I, O, U and Y and given terms by the letters B, G, D, and the other consonants. (Viète 1983, 24)

His equations, though, would have only one vowel because he only studied equations with one unknown (Boyer 1956, 59–60). One such example is $ba^2 + d^2a = z^3$.

The a represents the unknown quantity, and the others are **parameters**, arbitrary constants that allow the equation to be viewed as a family of equations as opposed to a single instance. Viète wrote this equation as

B in A quadratum plus D plano in A aequari Z solido.

In another work we find

A cubus, +A quadrato in B ter, +A in B quadratum ter, +B cubo.

This is $a^3 + 3a^2b + 3ab^2 + b^3$ using modern notation (Cajori 1928, vol. I, 184). Notice that his innovation only went so far. Writing the words *cubus* and *quadrato* instead of using notation was a remnant of the past. As with most transitional figures, his work had elements of both old and new.

We have to be careful when interpreting Viète's expressions and equations. His letters represent magnitudes, and the operations have a geometric interpretation. Consider again the expression $a^3 + 3a^2b + 3ab^2 + b^3$. Each of the letters represents a length, so the terms represent volumes. Since the terms are all of the same type, they can be added together. Viète called such terms **homogeneous**. An expression such as $a^2 + ab^2$ is not homogenous because the first term represents an area and the second a volume. Such an expression is banned from his algebra. Viète (1983, 15) codified this restriction in the **Law of Homogeneous Terms**:

The terms of an expression must be homogeneous.

Pappus

The Law of Homogeneous Terms played an important role in Viète's *Introduction to the Analytic Art*, published in 1591. In it he introduced his new algebraic system. He did not call his work *algebra* because it resembled only slightly any algebraic methods then taught. Instead, he called it the **analytic art** or simply **analysis** (Mahoney 1994, 34). Viète's discovery came from a study of the work of Pappus, a Greek philosopher and mathematician who lived near the end of the third century. At the time, interest in Greek geometry had waned, and it was Pappus who led a small revival in the subject. Unfortunately, most of his writings are lost except for some fragments. It is believed that Pappus wrote a number of commentaries, including an extensive one on the *Elements*. The most important of his works that has survived is called the *Mathematical Collection*. Probably assembled by an editor shortly after his death, the eight books of the *Mathematical Collection* have no central theme or flow. Its probable purpose was to faithfully preserve the writings of Pappus (Heath 1921b, 355–357; Boyer and Merzbach 1991, 164, 172–173, 184; Pappus 1986, 1–3).

The contents of the seventh book of the *Mathematical Collection*, known as the *Treasury of Analysis*, played a central role in Viète's studies. The purpose of the *Treasury* was to enumerate results that would help further geometric research. These were lemmas considered by Pappus to be missing steps in the proofs of his predecessors, such as Euclid, Eratosthenes, and Apollonius. Since many of the works to which Pappus referred were lost, the *Treasury* provided crucial information regarding their content. One of the extant works found referenced in the *Treasury* is Euclid's *Data*. It provided the core of the content of the *Treasury*. In the *Data* Euclid provided details concerning the constructibility of certain geometric objects. If a problem of interest could be reduced to one in his list, either the answer to the question would be known or, at minimum, a reasonable strategy would be suggested (Pappus 1986, 68; Mahoney 1994, 30; Heath 1921b, 401).

The mathematicians of Viète's day read the ancient Greeks and found themselves frustrated. They did not understand how it was possible for the Greeks to have arrived at the results that they did. When Viète began to study the *Mathematical Collection*, he believed that the answer was to be found in the "Field of Analysis." In the introduction to the *Mathematical Collection* addressed to his son Hermodorus, Pappus wrote of two important geometric proof techniques. Pappus explains that

> analysis is the path from what one is seeking, as if it were established, by way of its consequences, to something that is established by synthesis. That is to say, in analysis we assume what is sought as if it has been achieved, and look for the thing from which it follows, and again what comes before that, until by regressing in this way we come upon some of the things that are already known, or that occupy the rank of a first principle. We call this kind of method "analysis," as if to say *anapalin lysis* (reduction backward). In synthesis, by reversal, we assume what was obtained last in the analysis to have been achieved already, and setting now in natural order, as precedents, what before were following, and fitting them to each other, we attain the end of the construction of what was sought. This is what we call "synthesis." (Pappus 1986, *Mathematical Collection*, Introduction, 82)

Synthesis is by and far the most common method we find when reading the Greek mathematicians. The proofs begin with given hypotheses and argue to a conclusion. Analysis works in the reverse. From an assumed conclusion, the geometer works to find hypotheses that will yield the conclusion. The expectation is that the steps will reverse, yielding a synthetic proof. Viète interpreted Pappus's words to mean that the mystery to understanding how the Greeks arrived at their results was that they used algebra, and this belief served as the foundation for Viète's analytic art (Mahoney 1994, 31–32; Viète 1983, 11 n. 3, 29–30).

Analysis

Pappus divided analysis into two different types. The first he called **theorematic** (***zetetics***). The goal of this kind of analysis is to prove propositions. Suppose that *if* p *then* q is to be proven. The method of theorematic analysis states that q is first assumed. If p is then deduced from that supposition, reverse the steps to obtain a synthetic proof of *if* p *then* q. Although, it is often the case that a minor adjustment will save the proof, Pappus fails to mention that the steps are not necessarily reversible! However, if a sequence of deductions leads to a false proposition, q must be false, and the theorem is doomed.

The second type of analysis Pappus named **problematic** (***poristics***). Here the goal is to perform a requested task or solve a problem. Again, begin by supposing that the solution has been achieved and work backward. If this assumption leads to an impossibility, the task is impossible to complete. If the assumption leads to a situation that is possible, the job at hand is possible and can (hopefully) be accomplished by reversing the steps (Pappus 1986, *Mathematical Collection* VII.2, 82–83). Theorematic analysis was not very common in Greek mathematics, but problematic analysis was, particularly regarding locus problems (Pappus 1986, 67).

Inspired by what he had found, Viète (1983, 12) took the types of analysis and reinterpreted them for his analytic art. He started with the theorematic.

DEFINITION 7.1.1

Zetetics is when an equation is set up to solve a problem.

The rules of zetetics govern the manner by which equations can be written. The Law of Homogenous Terms and the convention of using vowels to represent unknown values are included. Since letters represent magnitudes, Viète insists that

a represents length,
a^2 represents area, and
a^3 represents volume.

Viète enumerates a total of 13 rules of zetetics that allow one to replace one equation with another. Once the equation is set up properly, it is assumed to be true, and work may proceed to find the unknown (Viète 1983, 23–27; Boyer and Merzbach 1991, 305). What follows are three examples of his basic rules.

THEOREM 7.1.2 [Transposition]

If $a^2 - d^p = g^2 - ba$, then $a^2 + ba = g^2 + d^p$.

THEOREM 7.1.3 [Depression]

If $a^3 + ba^2 = z^p a$, then $a^2 + ba = z^p$.

THEOREM 7.1.4 [Reduction]

If $ba^2 + d^p a = z^s$, then $a^2 + \frac{d^p a}{b} = \frac{z^s}{b}$.

The p exponent translates the word *planum*, meaning that d^p is a two-dimensional value, and the s on z^s translates the Latin *solidum*, meaning that this is a three-dimensional value. For us these rules are commonplace, but in Viète's time they were revolutionary. We should note, though, that he proved his rules as we would. For example, to prove transposition, Viète took $d^p + ba$ and added it to both sides of

$$a^2 - d^p = g^2 - ba.$$

Since $d^p + ba$ is a plane value and both sides of the equation are plane values, this addition is legal by the Law of Homogeneous Terms.

Viète next redefines problematical analysis.

■ **DEFINITION 7.1.5**

Poristics is when the truth of a theorem is tested using an equation.

Viète explains what he means:

> Such [theorems], although they are demonstrated by and grounded in zetetics, are still subject to the rules of synthesis, which is rated the most rational method of demonstration. If necessary, they are confirmed by it, this being a great miracle of the inventive art. So the footsteps of analysis are retraced. This is itself [a form of] analysis and, thanks to the introduction of symbolic logistic, is no longer difficult. But if some unfamiliar discovery is presented or has been stumbled on fortuitously and its truth must be weighed and inquired into, the poristic way should first be tried. It will be easy to return to synthesis later on. (Viète 1983, 28)

Manipulating algebraically the equation found by zetetics often leads to a solution, but sometimes answers present themselves without much effort. In these cases we can save time by checking whether our perceived solution does in fact work. In other instances, we find that we simply must check our answers. An example of poristics with which we are familiar is when we have extraneous solutions to an equation. If we needed to solve an equation such as $x = \sqrt{15 - 2x}$, by substituting back into the equation, we would find that $x = 3$ is a solution but $x = -5$ is not.

The last type of analysis is Viète's own creation.

■ **DEFINITION 7.1.6**

Exegetics is when the unknown value is actually found.

Once the equation is changed into an appropriate form, the mathematician can determine the solution and make sense of it. If the magnitudes were geometric, the final answer should be interpreted as a segment, plane or solid, possibly involving a construction. On the other hand, if the magnitudes were simply numeric, the solution should be given in terms appropriate to the original question, possibly involving a clarifying example. A contemporary example of exegetics is when we include appropriate units, such as "miles," with the answer to a problem (Viète 1983, 28–29).

Once Viète established the foundations of the analytic art, he complemented them with a full system of methods needed to solve equations. He called this the **symbolic logistic**. It is a system of rules that enable one to add, subtract, multiply, and divide symbolically. The rules include familiar ones, such as when a positive magnitude is multiplied by a negative magnitude, the result is negative. Rules for exponents are also given even though Viète does not use the exponential notation with which we are familiar. The symbolic logistic was the first modern theory of equations (Mahoney 1994, 36).

To illustrate his system, take the technique that Viète called **syncrisis**. Suppose that we are given

$$b^{n-m}a^m - a^n = z^n$$

with $n > m > 0$. The goal of syncrisis is to express z^n in terms of unknown values only. To do this, assume that e is another solution. This is reasonable because a polynomial equation of degree n has up to n solutions. If we are unlucky and there are not at least two distinct solutions, other methods will have to be applied. Now write

$$b^{n-m}e^m - e^n = z^n,$$

and from here we may conclude that

$$b^{n-m}a^m - a^n = b^{n-m}e^m - e^n$$

since "things equal to the same thing are equal to each other." Since the values for a and e are arbitrary, we may assume that $a > e$ so that we may subtract e from a. Remember that these unknowns still represent magnitudes as the Greeks understood them, so Viète excludes negative values from his algebra. We continue and write

$$b^{n-m}a^m - b^{n-m}e^m = a^n - e^n.$$

Dividing through by $a^m - e^m$ yields

$$b^{n-m} = \frac{a^n - e^n}{a^m - e^m}.$$

We may now substitute into the original equation and find that

$$z^n = \frac{a^n e^m - a^m e^n}{a^m - e^m}.$$

To illustrate the method, Viète takes $n = 2$ and $m = 1$, giving the equation

$$ba - a^2 = z^2.$$

Proceeding by syncrisis, we see that

$$z^2 = \frac{a^2 e - ae^2}{a - e} = ae.$$

Therefore, we conclude that z^2 is "the product of the two roots being sought and arises from dividing the difference between the product of the square of one and the root of the other by the difference between the roots, as was also proposed in general terms" (Viète 1983, 208–210).

Exercises

1. Explain how solving a linear equation in x by working to get x by itself is an example of analysis.

2. Use syncrisis to find the solutions to $12a - a^2 = 20$.

3. Given $a^n - b^{n-m}a^m = z^n$ and $b^{n-m}e^m - e^n = z^n$, show by syncrisis that

$$a^n + \frac{a^m(e^n + a^n)}{e^m - a^m} = z^n.$$

4. From *Five Books of Zetetica* (Viète 1983, 101, 119, 142):
   (a) Let b^p be the product of two numbers such that the ratio of the greater to the lesser is s/r. Find the numbers.
   (b) Find the dimensions of two right triangles such that the product of their heights plus the product of their bases is a square.
   (c) Given $d - a$ and $c - d$, find a, b, c, and d so that b and c are continued mean proportionals between a and d (see page 104).

5. From *Understanding and Amendment of Equations* (Viète 1983, 21, 184, 199, 202), prove:
   (a) If $a^3 - 3b^2a = z^s$ and $a + b = e$, then $e^3 - 3be^2 = z^s - 2b^3$.
   (b) If $a^2 + ba = z^p$, then $a(b^2 + z^p) - a^3 = bz^p$.
   (c) If $a^2 - ba = s^p - d^p$, then $a^4 + a^2(2d^p - b^2) - 2bs^pa = s^{pp} - d^{pp}$ (where pp refers to *plano-plane*, that is, the product of two plane magnitudes).

6. Prove the following from Viète (1983, 50).
   (a) $(a+b)(a^n - a^{n-1}b + a^{n-2}b^2 - \cdots \pm b^n) = \begin{cases} a^{n+1} + b^{n+1}, & n \text{ is even} \\ a^{n+1} - b^{n+1}, & n \text{ is odd.} \end{cases}$
   (b) If $(a+b)^n$ and $(a-b)^n$ have an even number of terms in their expansion,

$$\frac{a-b}{a+b} = \frac{a^n - a^{n-1}b + a^{n-2}b^2 - \cdots - b^n}{a^n + a^{n-1}b + a^{n-2}b^2 + \cdots + b^n}.$$

7. Suppose that $a^3 + z^s = 3b^2a$ has two positive roots, c and d. Show that both $3b^2 = a_1^2 + a_1a_2 + a_2^2$ and $z^s = a_1a_2^2 + a_2a_1^2$.

8. Viète used the diagram to show that

$$\sin x + \sin y = 2\sin\frac{x+y}{2}\cos\frac{x-y}{2}$$

by letting $AB = \sin x$ and $CD = \sin y$ (Boyer and Merzbach 1991, 308–309). Explain why this works.

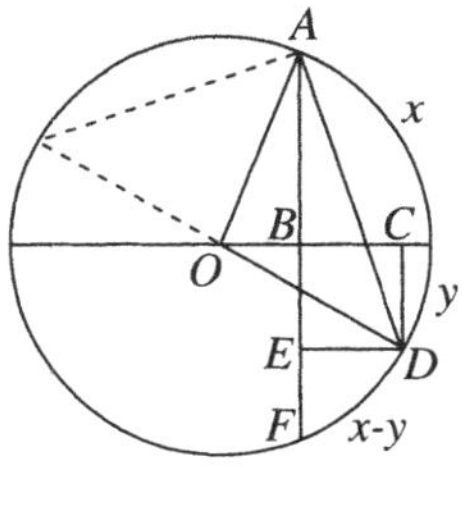

9. Use the method of Exercise 8 to show the following trigonometric identities.
   (a) $\sin x - \sin y = 2\cos\frac{x+y}{2}\sin\frac{x-y}{2}$.
   (b) $\cos x + \cos y = 2\cos\frac{x+y}{2}\cos\frac{x-y}{2}$.

7.2 THREE PROBLEMS

Viète made other contributions to mathematics besides his analytic art. In the field of geometry he worked issues related to the three famous Greek construction problems. The first is the Delian problem. The other two are known as **squaring the circle** and **trisecting an angle**. To square a circle means to find a square with the area of a given circle, while to trisect an angle means to divide an arbitrary angle into three congruent angles. These three problems cannot be constructed with a compass and straightedge alone, so they require other methods.

Square the Circle

Viète used his analysis to give the first equation for π involving an infinite sequence of algebraic operations. If he could find its value, he could square the circle by computing $\sqrt{\pi}$. Up to his day there were many estimates for π. Simply use Archimedes' method with regular polygons with more and more sides inscribed in and circumscribed about a circle. Viète's approach found this equation:

$$\frac{2}{\pi} = \sqrt{\frac{1}{2}}\sqrt{\frac{1}{2}+\frac{1}{2}\sqrt{\frac{1}{2}}}\sqrt{\frac{1}{2}+\frac{1}{2}\sqrt{\frac{1}{2}+\frac{1}{2}\sqrt{\frac{1}{2}}}}\cdots.$$

To represent this infinite product, Viète wrote the first four terms (without the square root sign but with an l for *latus*) and then noted, "*in infinitum observata uniformi methodo*" (Cajori 1928, vol. II, 78, 185). Boyer and Merzbach (1991, 321) note that this equation is closely linked to a curve called the **quadratrix**. The curve was discovered by Hippias around 475 B.C. Hippias was born in Elis in southern Greece, and he was a Sophist. Plato included him in two of his dialogues, but his portrayal was not very flattering (Plato 1961, *Hippias Major* and *Hippias Minor*).

The quadratrix was the first locus to be described by a Greek mathematician beyond the simple line and circle. It works like this: Take $\square ABCD$ and let $\widehat{BD}$ be the first quadrant of $\odot A$ with radius $\overline{AB}$. If $\overline{BC}$ moves uniformly to $\overline{AD}$ in the same amount of time as radius $\overline{AB}$ turns to $\overline{AD}$, the result is curve BHG in Figure 7.2. Letting HF represent the height of $\overline{BC}$ at a given point in time, due to the uniform motion, we see that

$$\frac{HF}{AB} = \frac{\theta}{\pi/2}.$$

Since $HF = AH \sin\theta$,

$$\frac{AH\sin\theta}{\theta} = \frac{2AB}{\pi}.$$

Letting $a = AB$ and $r = AH$, this becomes

$$r = \frac{2a\theta}{\pi\sin\theta} \qquad (7.1)$$

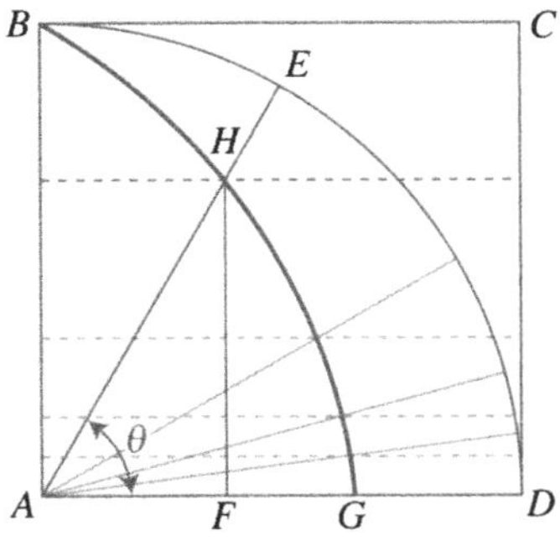

Figure 7.2 The quadratrix of Hippias.

in polar form. Therefore, because

$$\lim_{\theta \to 0+} \frac{\theta}{\sin \theta} = 1,$$

there exists a point G on $\overline{AD}$ such that H lands on G when $\theta = 0$ and

$$AG = \frac{2a}{\pi}. \tag{7.2}$$

This equation means that the quadratrix has given us a segment that we can now use to construct a segment of length π with only a straightedge and compass. Namely,

$$\pi = \frac{2a}{AG},$$

and from this we can construct a segment of length $\sqrt{\pi}$. This is what we need to square the circle (Heath 1921a, 226–227).

Back to Viète's formula. Define the sequence r_n using Equation 7.1 with $a = 1$ and $\theta = \pi/2^n$,

$$r_n = \left(2^{n-1} \sin \frac{\pi}{2^n}\right)^{-1}$$

This equation amounts to taking a right angle, continually bisecting it, and finding the corresponding value on the quadratrix. Furthermore, observe that $r_1 = 1$ and by a double-angle formula,

$$\sin \frac{\pi}{2^{n-1}} = 2 \sin \frac{\pi}{2^n} \cos \frac{\pi}{2^n}.$$

This gives

$$\left(2 \sin \frac{\pi}{2^n}\right)^{-1} \left(\sin \frac{\pi}{2^{n-1}}\right) = \cos \frac{\pi}{2^n}.$$

Therefore,

$$\frac{r_n}{r_{n-1}} = \left(2^{n-1} \sin \frac{\pi}{2^n}\right)^{-1} \left(2^{n-2} \sin \frac{\pi}{2^{n-1}}\right) = \cos \frac{\pi}{2^n}. \tag{7.3}$$

We now proceed by examining the infinite product:

$$\frac{r_2}{r_1} \cdot \frac{r_3}{r_2} \cdot \frac{r_4}{r_3} \cdots = \lim_{n\to\infty} \left(r_2 \cdot \frac{r_3}{r_2} \cdot \frac{r_4}{r_3} \cdots \frac{r_n}{r_{n-1}}\right) = \lim_{n\to\infty} r_n = \frac{2}{\pi}.$$

The last equality holds due to the limit implied by Equation 7.2. Substituting with Equation 7.3 gives

$$\cos \frac{\pi}{4} \cdot \cos \frac{\pi}{8} \cdot \cos \frac{\pi}{16} \cdots = \frac{2}{\pi},$$

and we then arrive at Viète's formula because

$$\cos \frac{\pi}{4} = \sqrt{\frac{1 + \cos \pi/2}{2}} = \sqrt{\frac{1}{2}}$$

$$\cos \frac{\pi}{8} = \sqrt{\frac{1 + \cos \pi/4}{2}} = \sqrt{\frac{1}{2} + \frac{1}{2}\sqrt{\frac{1}{2}}}$$

$$\cos \frac{\pi}{16} = \sqrt{\frac{1 + \cos \pi/8}{2}} = \sqrt{\frac{1}{2} + \frac{1}{2}\sqrt{\frac{1}{2} + \frac{1}{2}\sqrt{\frac{1}{2}}}}$$

$$\vdots$$

Duplicate the Cube

Viète solved the problem of duplicating the cube by again finding inspiration from the seventh book of the *Mathematical Collection* of Pappus. There he found problems that dealt with ***neusis***, a term that means "verging" or "inclination." According to Pappus, if a line passes through a point when produced, the line makes a neusis on that point. The general problem is that given two lines, place between them a segment of a given magnitude so that the segment intersects a particular point. The *neusis* often assumed that the segment had to be shifted from some other position so that it would include the desired point. Many Greek geometers used a *neusis* without explicitly stating it as a postulate or proving it as a theorem. For example, Archimedes did this in his work *On Spirals* (Pappus 1986, *Mathematical Collection*, Introduction, 112; Boyer 1956, 143; Archimedes 1897, c-ci).

To solve a *neusis* problem, one needs a tool besides a straightedge and compass. The ancient Greek geometer Nicomedes, who lived around 200 B.C., provided such a tool. Start with a horizontal **ruler** labeled $\overline{AB}$ (Figure 7.3). At its midpoint F attach another ruler perpendicular to the first. Call it $\overline{FE}$. On this second ruler fix a small peg at point C called the **pole**, and on C attach a third ruler that itself has a peg at D and at the other end a slot in which the peg at C can run. The peg of the third

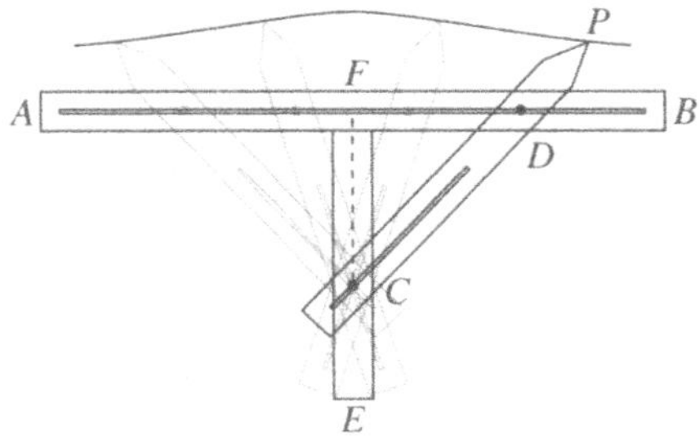

Figure 7.3 The first conchoid of Nicomedes.

ruler slides in a groove in ruler $\overline{AB}$. As the peg D slides along $\overline{AB}$, the endpoint P of the third ruler describes a locus of points that is known as the **first conchoid** of Nicomedes. If we let $DP = a$ (the **distance**) and $FC = b$ (Figure 7.4), then using polar coordinates the curve is given by the equation

$$r = a + b \csc \theta. \tag{7.4}$$

The derivation is left for Exercise 7. Proclus noted that there were other conchoids, but the details of the work of Nicomedes are no longer extant (Heath 1921a, 238–240).

Related to the conchoid is a curve known as a **cissoid**. The credit of the discovery of this curve is due to Diocles, who around 100 B.C. wrote *On Burning-glasses* in which he gave a solution to the duplication of the cube problem using this curve. The details of his solution are left to Exercise 11. To define the curve, take a circle with radius $\overline{AB}$ and construct $\overline{CD}$ to be the perpendicular bisector of the diameter (Figure 7.5). Find G and H on $\overline{AB}$ so that $\overline{GC} \cong \overline{CH}$ and erect perpendiculars at G and H that meet the circumference of the circle at E and F, respectively. Join $\overline{EB}$ and let P be the intersection of $\overline{EB}$ with $\overline{HF}$. The locus of all points P formed in this manner as different points G and H are found on $\overline{AB}$ with $\overline{GC} \cong \overline{CH}$ is the cissoid (Heath 1921a, 264–265).

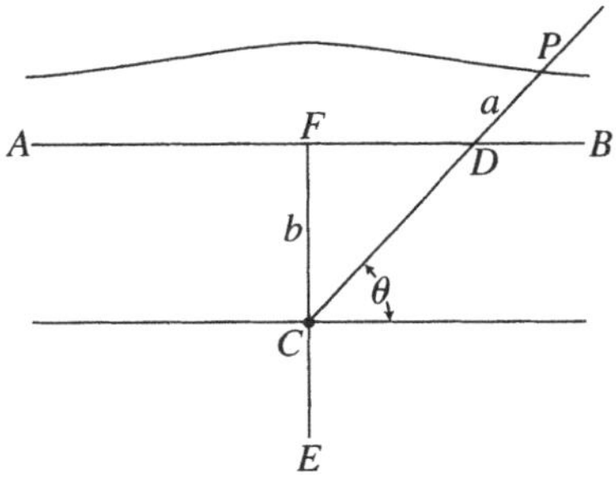

Figure 7.4 Deriving the polar equation for the first conchoid.

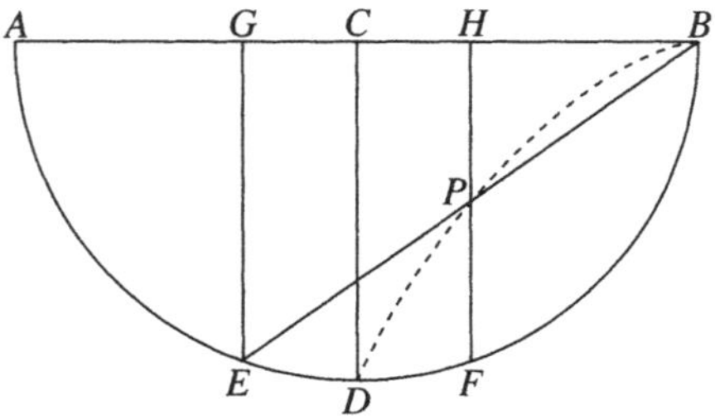

Figure 7.5 The cissoid of Diocles.

When the cissoid of Diocles is generalized, we see the the conchoid is a type of cissoid in this new sense. Let c and c' be two curves and fix a point B that we will call the **pole**. Take any point E on c and find the line l that joins B and E. Let E' be the intersection of l and c' and take P on l such that $BP = EE'$. The locus of all points P is the **generalized cissoid** (Figure 7.6). If c is a circle, B is a point on c, and c' is the tangent line to c that is opposite B, the resulting curve is the cissoid of Diocles. The conchoid of Nicomedes is a generalized cissoid when c is a line, c' a circle, and B the center of the circle. See Exercise 12.

Inspired by the *neusis* problems, Viète took what he learned from Pappus and expanded Euclid's postulate set to include the following three propositions. These and the proofs that follow are from Viète's *A Supplement to Geometry* (Viète 1983, 388–398).

■ POSTULATE 7.2.1

Through a point not on two given lines, there exists a third line that intersects the first two lines, forming a segment of any possible length.

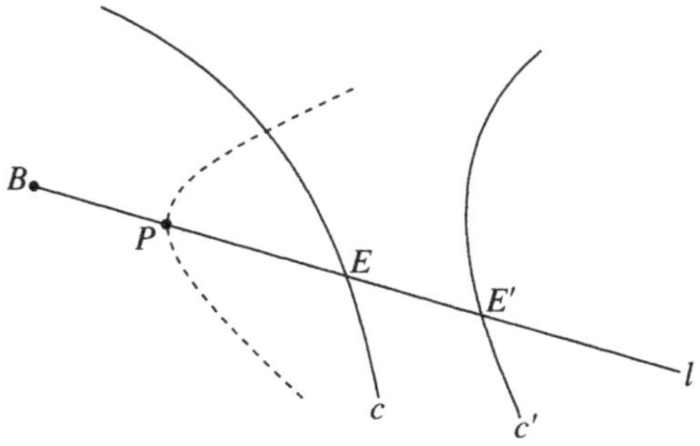

Figure 7.6 The generalized cissoid.

POSTULATE 7.2.2

Given a point on the exterior of an angle such that there exists lines through the point intersecting both sides of the angle, there exists a line through the point that intersects both sides of the angle, forming a segment of any possible length.

POSTULATE 7.2.3

There exists a line from any given point within or on a circle to any line that intersects the circle, forming a segment of any possible length.

The conchoid provided justification for Viète that these postulates were reasonable. For example, we can use it to confirm Postulate 7.2.1. Let the pole C be the given point and the distance a the given length. Suppose that one of the lines is $\overline{AB}$ and the other line l intersects the generated conchoid. If we move the ruler so that P is at the intersection of the conchoid and l, the line formed by the ruler is the desired line that passes through C and intersects the two lines forming a segment of length a.

Nicomedes used his curve to duplicate the cube and trisect an angle (Exercises 9 and 10). Viète used his *neusis* postulates to do the same. First, he duplicated the cube by finding two mean proportionals. We start with a lemma.

LEMMA 7.2.4 [*A Supplement to Geometry*, IV]

Let B be on the exterior of $\odot O$. If $\overline{BD}$ intersects $\odot O$ at C and D and $\overline{BF}$ intersects $\odot O$ at E and F such that

$$BC \cdot BE = CD \cdot EF,$$

$\overline{BC}$ and $\overline{BE}$ will be two continued mean proportionals between $\overline{CD}$ and $\overline{EF}$.

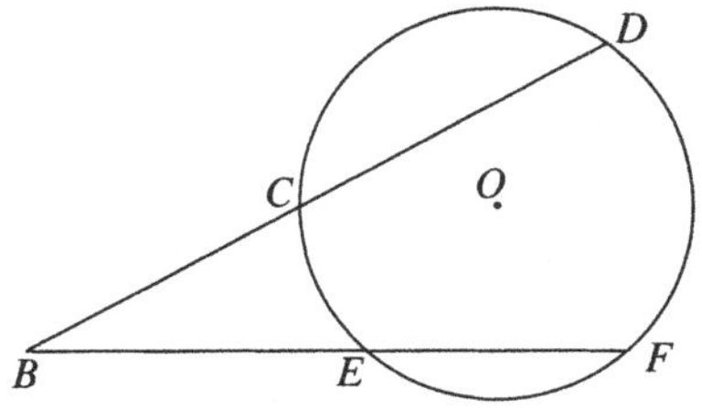

PROOF

Let $\overline{BD}$ intersect $\odot O$ at C and D and $\overline{BF}$ intersect the circle at E and F. By hypothesis we have

$$\frac{CD}{BE} = \frac{BC}{EF}.$$

Since $BC = BD - CD$ and $EF = BF - BE$, we have

$$\frac{CD}{BE} = \frac{BD}{BF}.$$

By the Secant Theorem (5.5.4),

$$\frac{BD}{BF} = \frac{BE}{BC}.$$

Hence,

$$\frac{CD}{BE} = \frac{BE}{BC} = \frac{BC}{EF}. \blacksquare$$

Viète uses the lemma to prove that one can construct (using his additional postulates) two mean proportionals between two given magnitudes, and thus one can duplicate the cube.

■ **CONSTRUCTION 7.2.5 [*A Supplement to Geometry*, V]**

Find the two mean continued proportionals between two given magnitudes.

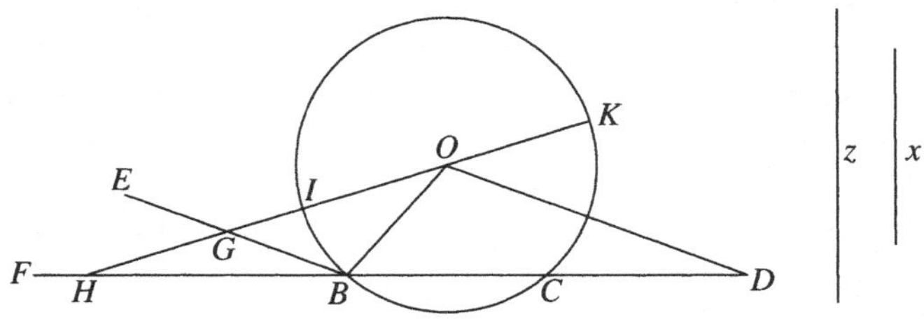

PROOF

We are given line segments of length x and z. Draw $\odot O$ with radius $\overline{OB}$ such that $OB = z/2$. Next draw chord $\overline{BC}$ such that $BC = x$ by Postulate 7.2.3. Extend $\overline{BC}$ to D so that C is the midpoint of $\overline{BD}$. Join $\overline{OD}$. Let $\overline{BE}$ be parallel to $\overline{OD}$ and extend $\overline{BD}$ indefinitely to F beyond B. By Postulate 7.2.2, find H on $\overline{BF}$ such that $\overline{OH}$ meets $\overline{BE}$ at G with $\overline{HG} \cong \overline{OB}$. Finally, let $\overline{OH}$ be extended to meet $\odot O$ at K and let I be the intersection of $\overline{OH}$ with $\odot O$. Notice that we have

$$\frac{HG}{HB} = \frac{OG}{BD} \tag{7.5}$$

since $\overline{OD}$ is parallel to $\overline{BG}$ (Exercise 6), and because $HG = OB$ we have

$$\frac{HG}{IK} = \frac{BC}{BD}. \tag{7.6}$$

Because of Equations 7.5 and 7.6, we see that

$$HB \cdot OG = IK \cdot BC.$$

Because $HG = OI$, we have $OG = HI$. Hence,

$$HB \cdot HI = IK \cdot BC.$$

By the last equation we see that the conditions to Lemma 7.2.4 are satisfied. Therefore, HB and HI are the two mean continued proportionals desired because $IK = z$ and $BC = x$. ■

Trisect an Angle

The third postulate allows Viète to trisect an arbitrary angle. The proof of this theorem is similar to that of Archimedes (1897, *Book of Lemmas*, Proposition 8, 309).

■ **CONSTRUCTION 7.2.6 [*A Supplement to Geometry*, IX]**

Trisect a given angle.

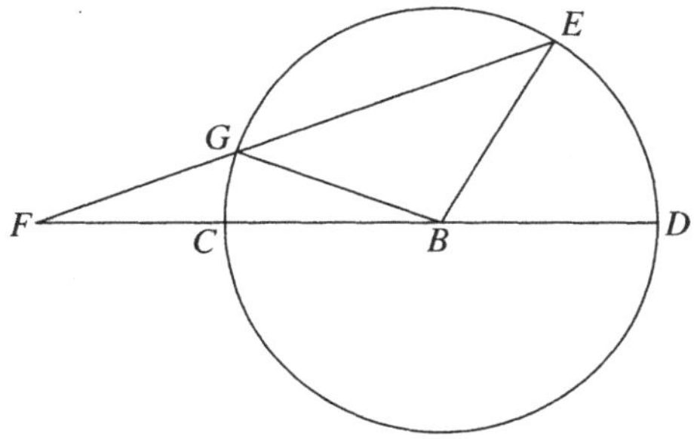

PROOF

Let $\measuredangle A$ be the given angle. If $\measuredangle A$ is a right angle, it is a theorem of Euclid's geometry that $\measuredangle A$ can be trisected. Suppose, then, that $\measuredangle A$ is acute. The obtuse case will be similar to what follows.

Construct $\odot B$ with diameter $\overline{CD}$. Find E on the circumference such that $\measuredangle EBD \cong \measuredangle A$. Find F on an extension of $\overline{CD}$ beyond C so that $\overline{EF}$ intersects the circle at G and $\overline{FG} \cong \overline{BC}$ (Postulate 7.2.3). To trisect $\measuredangle EBD$, first join $\overline{BG}$. This yields the isosceles $\triangle FGB$. The details that follow are left unstated by Viète. They are due to Witmer (Viète 1983, 398). Now $\measuredangle BGE$ is an exterior angle of $\triangle BGF$. Hence,

$$\mathrm{m}\measuredangle BGE = \mathrm{m}\measuredangle BFG + \mathrm{m}\measuredangle GBF = 2\mathrm{m}\measuredangle GBF$$

since the base angles are congruent. However, $\measuredangle BGE \cong \measuredangle BEG$ because $\triangle BEG$ is isosceles. This gives

$$\begin{aligned}\mathrm{m}\measuredangle GBE &= \text{two right angles} - \mathrm{m}\measuredangle BGE - \mathrm{m}\measuredangle BEG \\ &= \text{two right angles} - 4\mathrm{m}\measuredangle GBF.\end{aligned}$$

Furthermore,

$$\mathrm{m}\measuredangle GBD = \text{two right angles} - \mathrm{m}\measuredangle GBF.$$

Thus,

$$\begin{aligned}\mathrm{m}\measuredangle EBD &= \text{two right angles} - \mathrm{m}\measuredangle GBE - \mathrm{m}\measuredangle GBF \\ &= \text{two right angles} - (\text{two right angles} - 4\mathrm{m}\measuredangle GBF) - \mathrm{m}\measuredangle GBF \\ &= 3\mathrm{m}\measuredangle GBF.\end{aligned}$$

Hence, $\measuredangle A$ is trisected. ■

Exercises

1. Suppose that for all circles, there exists a rectangle with area equal to the circle. From this prove that every circle can be squared.

2. Use the spiral of Archimedes to square the circle and trisect an angle.

3. Find the equation in rectangular coordinates for the quadratrix and graph it.

4. The quadratrix can be used to trisect an angle. Take $\measuredangle GAH$ and let H lie on the quadratrix. Find M on $\overline{BA}$ and N on $\overline{CD}$ such that $\overline{MN}$ is parallel to $\overline{AD}$. Trisect both $\overline{MA}$ at points M' and M''. Do the same for $\overline{ND}$ at N' and N''. Let H' be the intersection of the quadratrix with $\overline{M'N'}$ and H'' be the intersection between the quadratrix and $\overline{M''N''}$.

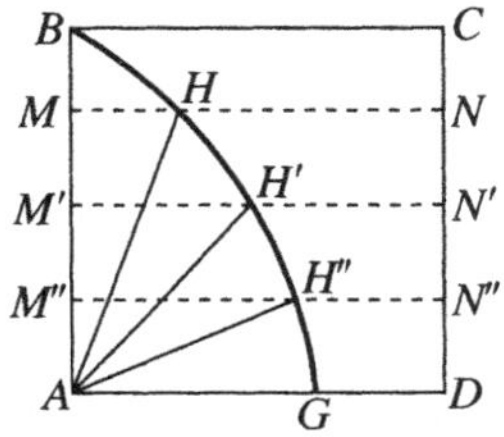

(a) Prove that $\measuredangle GAH'' \cong \measuredangle H''AH' \cong \measuredangle H'AH$.
(b) Perform the construction to trisect a randomly selected acute angle.

5. Use the quadratrix to divide an angle that measures half a right angle into five congruent angles.

6. Confirm Equation 7.5 in the proof for Construction 7.2.5.

7. Derive Equation 7.4. This is illustrated in Figure 7.4.

8. The conchoid of Nicomedes can be described by the equation

$$(x^2 + y^2)(x - b)^2 - a^2x^2 = 0.$$

(a) Show how this equation is derived from Equation 7.4.
(b) Graph this equation.
(c) Find the area of the loop when $b/a = 1/2$.
(d) Generalize the argument on page 239 to generate the loop.

9. Nicomedes used the conchoid to duplicate the cube (Heath 1921a, 260–262). To do this it suffices to find two continued mean proportional between two segments $\overline{AB}$ and $\overline{BC}$. Draw them at right angles. Find J so that $AJCB$ is a rectangle. Let D be the midpoint of $\overline{AB}$ and E the midpoint of $\overline{BC}$. Join $\overline{DJ}$ and produce it beyond D to meet $\overline{BC}$ produced at G. Construct $\overline{EF}$ such that $\overline{FC} \cong \overline{AD}$. Join $\overline{GF}$. Construct $\overline{CH}$ parallel to $\overline{GF}$.

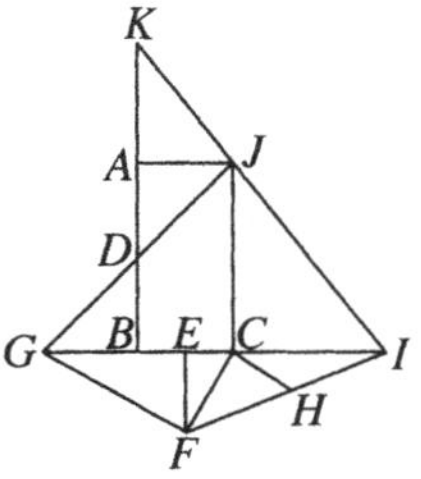

(a) Draw $\overline{FI}$ and assume that H is on it so that I is on $\overline{BC}$ produced and $\overline{HI} \cong \overline{FC}$. An appropriate conchoid will enable us to find I. Give the pole, ruler, and distance of this conchoid.

(b) Join $\overline{JI}$ and extend it to meet $\overline{AB}$ produced at K. Prove that CI and AK are two continued mean proportionals between AB and BC.

10. The conchoid can be used to trisect an angle (Heath 1921a, 236, 238). Given $\measuredangle ABC$, draw $\overline{AC}$ perpendicular to $\overline{BC}$. Find D so that $DACB$ is a rectangle. Let E be on $\overline{DA}$ produced. To find the exact location of E, draw the conchoid with pole B, ruler $\overline{AC}$, and distance $2AB$. Join $\overline{BE}$ and let G be the intersection of $\overline{BE}$ with $\overline{AC}$. Let F be the midpoint of $\overline{DE}$. Prove that $\text{m}\measuredangle ABC = 3\text{m}\measuredangle GBC$.

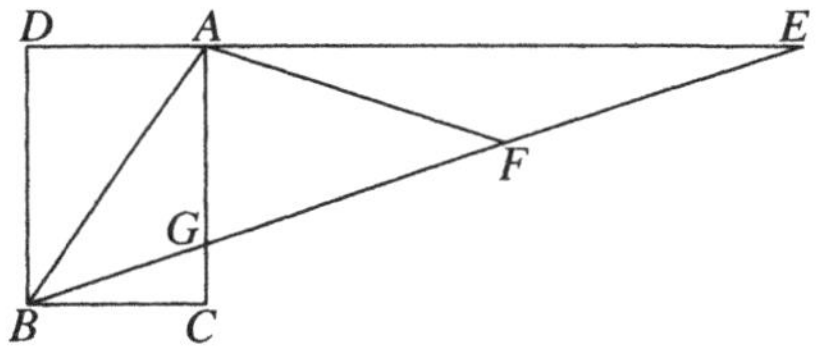

11. The cissoid of Diocles is illustrated in Figure 7.5. The point P is a point on the curve. Let $r = PB, \theta = \text{m}\measuredangle ABP$, and $AB = 1$.

(a) Show that the equation in polar coordinates for the cissoid of Diocles is $r = \sec\theta - \cos\theta$.

(b) Prove that the equation in part (a) is equivalent to $r = \tan\theta \sec\theta$.

(c) Show that the equation for this cissoid is $x(x^2 + y^2) - y^2 = 0$.

(d) Solve the Delian problem using the cissoid by showing that HF and HB are two continued mean proportionals between AH and HP.

12. Show that both the conchoid of Nicomedes and the cissoid of Diocles are generalized cissoids.

13. Given a right triangle, find another right triangle with the same height such that the product of the difference between their bases and the difference between their hypotenuses is equal to the area of a given rectangle.

14. If there exists an isosceles triangle and a line from an endpoint of the base to the opposite leg of length equal to that leg, prove that the exterior angle formed by the base and the line has measure equal to three times one of the base angles.

7.3 CONIC SECTIONS

Viète's analytic art was initially designed to study equations with one indeterminant. The natural next step is to generalize this to two indeterminants. This new system would be able to handle lines as well as higher curves that include the parabola, hyperbola, and ellipse. These familiar curves are called **conic sections** because they are the loci formed when a circular cone is cut by a plane (Figure 7.7).

It is believed that the insight that led to the conic sections was due to a student of Eudoxus and tutor of Alexander the Great by the name of Menaechmus. He lived about 350 B.C. and he was interested in the Delian problem. He did not limit himself to the straightedge and compass, but instead, used other curves. A letter to Ptolemy III Euergetes of Egypt purportedly written by Eratosthenes enumerates various solutions to the doubling-the-cube problem. The writer includes those by Archytas and Eudoxus using "so-called curved lines" that cannot be constructed with straightedge and compass alone. He then states that

> as it turned out, all their solutions were theoretical, and no one of them was able to give a practical construction for ordinary use, save to a certain small extent Menaechmus, and that with difficulty. (Heath 1921a, 244–246)

There were other notable Greeks who studied the conic sections. Aristaeus wrote *Solid Loci* in five books, and Euclid wrote *Conics* in four, but the culmination of the Greek theory of these loci was written by Apollonius, who lived from about 262 to 190 B.C. Not much is known about his life. He was born in Perga in northwest Asia Minor. At an early age he traveled to Alexandria and learned geometry from the students of Euclid. The result of his studies was his masterpiece entitled *Conic Sections*. Not only did it summarize the work in the field up to his time, but his systematic approach and new contributions earned him the title "Great Geometer." His expertise in geometry was nearly matched by his skill as an astronomer. It is likely that Apollonius was the first to develop a theory of epicycles to explain the retrograde motion of the planets (Kline 1972, 89–90; Heath 1921b, 116–121; Pannekoek 1961, 134).

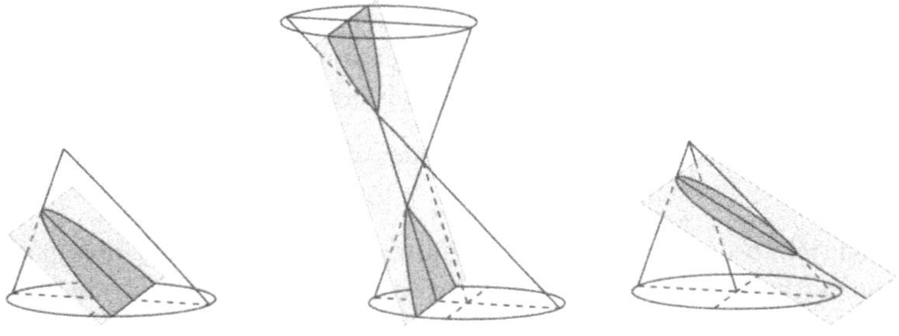

(a) Parabola (b) Hyperbola (c) Ellipse

Figure 7.7 The conic sections.

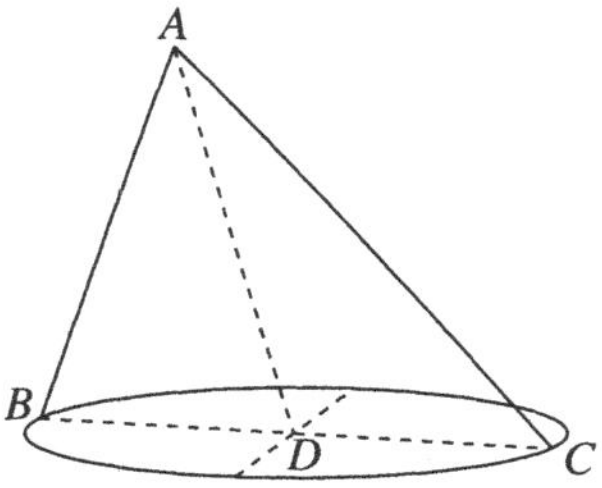

Figure 7.8 An oblique cone with axial $\triangle ABC$ and axis $\overline{AD}$.

The earlier Greeks used right circular cones to define the conic sections. Apollonius was the first to generalize the definition to allow nonright (**oblique**) cones, as in Figure 7.8. Take a circular cone and find a diameter of its base. Join the vertex to the endpoints of the diameter. Call the resulting triangle an **axial triangle** of the cone. The line segment joining the vertex to the center of the cone's base is called the **axis** of the cone. We may now define the conic sections (Figure 7.7).

■ **DEFINITION 7.3.1**

The locus of the points of intersection of a cone with a plane that is parallel to one of the sides of an axial triangle is called a **parabola**.

■ **DEFINITION 7.3.2**

The locus of the points of intersection of a cone with a plane that intersects both legs of an axial triangle is called an **ellipse**.

Finally, place a copy of the cone so that the two cones share a common vertex and their axes form a straight line. This is called a **double-napped cone**. The axes of the cones form a straight line that serves as the axis of the double cone.

■ **DEFINITION 7.3.3**

The locus of the points of intersection of a double cone with a plane that is parallel to its axis is called a **hyperbola**.

We call the plane that intersects the cone and determines the conic section a **cutting plane**. This plane can also intersect the cone so that the locus of common points is one of the sides of an axial triangle, is a circle, or contains only the cone's vertex. For this reason, lines, circles, and points can be considered conic sections.

The Parabola

Apollonius (1889, I.1–14) proceeds to prove the basic results for the three conic sections. To identify the main components of these loci, he introduced some terminology. Let $\triangle ABC$ be the axial triangle of the cone that is perpendicular to the cutting plane, as in the diagram for Theorem 7.3.4.

- The cutting plane intersects the base of the cone in $\overline{ED}$. The intersection of the plane of the axial triangle and the cutting plane is called the **diameter** (or **axis**) of the conic section. The segments that are parallel to $\overline{ED}$ and have their endpoints on the locus and are bisected by the diameter are called **chords**.
- A point P that is located where the diameter intersects either $\overline{AB}$ or $\overline{AC}$ is called a **vertex**.
- The **ordinate** $\overline{QV}$ is half of a chord and has one endpoint on the diameter. The ordinate is used to measure the distance from the diameter to the locus.
- The segment $\overline{PV}$ that is cut off by the ordinate is called the **abscissa**.
- $\overline{PL}$ plays a part similar to what we saw with Archimedes work (for example, on page 210). This segment is called the **parameter of the ordinates** or the ***latus rectum***. The parameter controls the shape of the conic section.

With this terminology, we are ready to give Apollonius's description of the parabola.

■ THEOREM 7.3.4 [*Conics* I.11]

Cut cone ABC with a plane so that the resulting section is parallel to $\overline{AC}$. Let $\overline{PM}$ be the diameter of the section, and let $\overline{QV}$ be any ordinate to the diameter with V on $\overline{PM}$. If $\overline{PL}$ is chosen such that it is perpendicular to $\overline{PM}$ and $BC^2/(AC \cdot BA) = PL/PA$, then

$$QV^2 = PL \cdot PV.$$

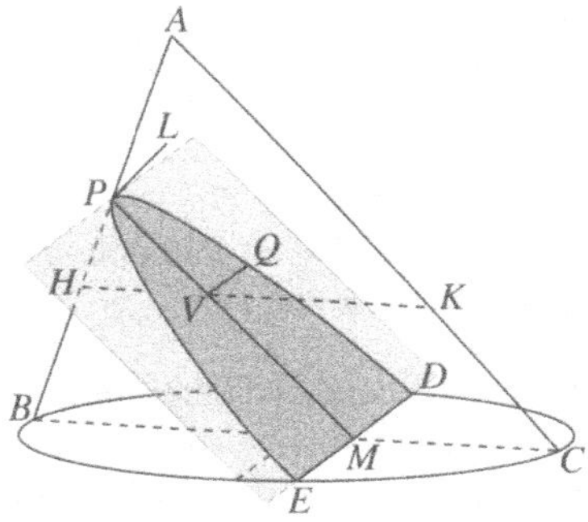

PROOF

In the parabola $\overline{PM}$ is parallel to $\overline{AC}$ and $\overline{HK}$ is parallel to $\overline{BC}$. To find the relationship between PV and QV, note that

$$\frac{HV}{PV} = \frac{BC}{AC} \tag{7.7}$$

because $\triangle PHV \sim \triangle ABC$. Furthermore, by Theorem 5.4.11,

$$\frac{VK}{PA} = \frac{MC}{PA} = \frac{BC}{BA}. \tag{7.8}$$

Therefore, by the Chord Theorem (5.5.3)

$$QV^2 = HV \cdot VK,$$

so Equations 7.7 and 7.8 yield

$$\frac{QV^2}{PV \cdot PA} = \frac{HV \cdot VK}{PV \cdot PA} = \frac{BC^2}{AC \cdot BA} = \frac{PL}{PA} = \frac{PL \cdot PV}{PA \cdot PV}.$$

This allows us to conclude that $QV^2 = PL \cdot PV$. ■

Apollonius is responsible for naming this section. He called it a parabola because QV^2 equals the area of the rectangle *applied* ($\pi\alpha\rho\alpha\beta\acute{\alpha}\lambda\lambda\epsilon\iota\nu$) to PL and PV (Heath 1921b, 138). Furthermore, if we view the point V as moving along $\overline{PM}$ away from vertex, we may view the diameter as the x-axis. With this interpretation, if we let $x = PV$, $y = QV$ and $p = PL/4$, we have

$$y^2 = 4px, \tag{7.9}$$

the familiar equation for a parabola.

The Hyperbola

We now turn to the hyperbola. We will show its general form and describe its asymptotes. The equation that we will find is for one **branch** of the hyperbola, but it does hold for both branches when taken separately. The diameter of the hyperbola is illustrated in the diagram of Theorem 7.3.5 as containing $\overline{PP'}$, with **center** O being the midpoint of $\overline{PP'}$.

■ **THEOREM 7.3.5 [*Conics* I.12]**

Cut cone ABC with a plane not parallel to $\overline{AC}$ but intersecting it at a point P' on $\overline{AC}$ produced beyond A. Let P be the intersection of the cutting plane and $\overline{AB}$. Let $\overline{PM}$ be the diameter of one branch of the section and let $\overline{QV}$ be any ordinate with V on $\overline{PM}$. Let $\overline{PL}$ be perpendicular to the diameter such that

$$\frac{BF \cdot FC}{AF^2} = \frac{PL}{PP'},$$

where $\overline{AF}$ is parallel to $\overline{PM}$ and F is the midpoint of $\overline{BC}$. If $\overline{RV}$ is parallel to $\overline{PL}$ and $\overline{P'L}$ is joined and extended to intersect $\overline{RV}$ at R, then

$$QV^2 = RV \cdot PV.$$

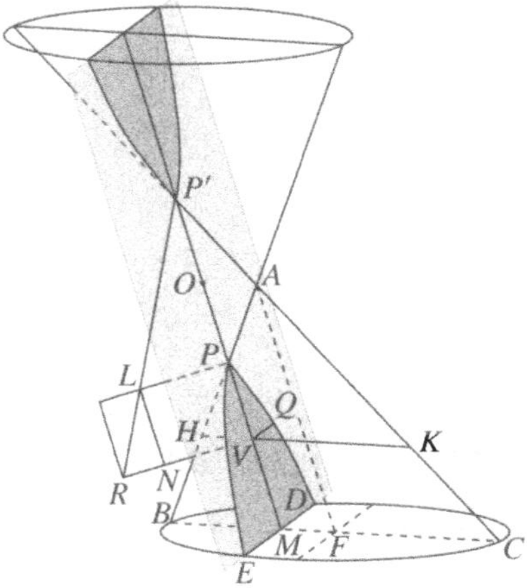

PROOF

By similar triangles we see that

$$\frac{HV}{PV} = \frac{BF}{AF}$$

and

$$\frac{VK}{P'V} = \frac{FC}{AF}.$$

Therefore, by the Chord Theorem (5.5.3),

$$\frac{QV^2}{PV \cdot P'V} = \frac{HV \cdot VK}{PV \cdot P'V} = \frac{BF \cdot FC}{AF^2}.$$

Since

$$\frac{PL}{PP'} = \frac{RV}{P'V} = \frac{PV \cdot RV}{PV \cdot P'V}, \tag{7.10}$$

we may conclude that

$$\frac{QV^2}{PV \cdot P'V} = \frac{PV \cdot RV}{PV \cdot P'V}. \tag{7.11}$$

Hence,

$$QV^2 = RV \cdot PV. \tag{7.12}$$

■

Apollonius called this section the hyperbola because the square of the ordinate equals the area of a rectangle with base equal to the abscissa and height that *overlaps* (ὑπερβάλλει) the parameter.

To see that this is equivalent to the formula that we know from analytic geometry, use the diagram from the last proof. Let $h = OP$ and $p = PL$, the parameter of the ordinates. For any V on $\overline{PM}$, let $x = PV$ (the abscissa) and $y = QV$ (the ordinate). To write an expression for RV, notice that the ratio RN/x is $p/2h$, so

$$RV = \frac{p}{2h}x + p.$$

Substituting into (7.12), we obtain

$$y^2 = x\left(\frac{p}{2h}x + p\right). \tag{7.13}$$

Hence,

$$\frac{(x+h)^2}{h^2} - \frac{2y^2}{ph} = 1. \tag{7.14}$$

We should also note at this time that for any constant k, an equation of the form

$$xy = k$$

is also a hyperbola (Exercise 10).

Not only did the ancient Greeks have an equation for the hyperbola, but they could identify its asymptotes. Their equations are

$$y = \pm\sqrt{\frac{p}{2h}}(x+h) \tag{7.15}$$

when we use Equation 7.14 (Exercise 6). Therefore, to construct the asymptote we must find a point S in the cutting plane such that $\overline{PS}$ is perpendicular to $\overline{OP}$ and

$$PS^2 = \frac{ph}{2} = \frac{PL \cdot PP'}{4}.$$

This explains the condition found by Apollonius in the next proposition.

THEOREM 7.3.6 [*Conics* II.1, 2, 15, 17, 21]

Take diameter $\overline{PP'}$ of a hyperbola with parameter p. If on the tangent at P there are points S and T on either side of P such that

$$PS^2 = PT^2 = \frac{p \cdot PP'}{4},$$

the $\overline{OS}$ and $\overline{OT}$ produced will not intersect the hyperbola. Moreover, the opposite branch will have the same asymptotes.

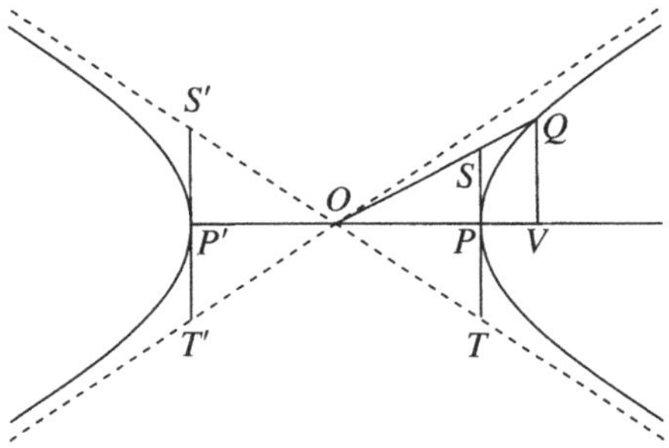

PROOF

Let points S and T be as described. We show first that the line through O and S does not intersect the hyperbola. To show this, suppose that the $\overline{OS}$ produced intersects the hyperbola at Q. Draw the ordinate $\overline{QV}$ and notice that it is parallel to $\overline{ST}$. We now have

$$\frac{p}{PP'} = \frac{p \cdot PP'}{(PP')^2} = \frac{4PS^2}{(2OP)^2} = \frac{PS^2}{OP^2} = \frac{QV^2}{OV^2}.$$

The last equality holds because $\triangle OSP \sim \triangle OQV$. However, by Equations 7.10 and 7.11,

$$\frac{QV^2}{PV \cdot P'V} = \frac{p}{PP'}. \tag{7.16}$$

Therefore, we may conclude that

$$OV^2 = PV \cdot P'V.$$

Now $PV = OV - OP$ and $P'V = OV + OP$, which gives

$$PV \cdot P'V = OV^2 - OP^2. \tag{7.17}$$

So we may write

$$OV^2 = OV^2 - OP^2.$$

This is impossible, so the $\overline{OS}$ produced cannot intersect the hyperbola.

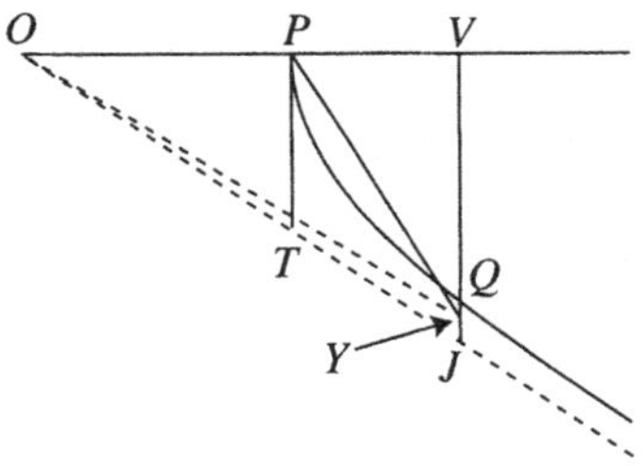

To prove that the $\overline{OT}$ (or $\overline{OS}$) produced approaches the hyperbola asymptotically, we will show that every line in the interior of $\sphericalangle TOP$ will intersect the hyperbola. Instead, suppose that there exists Y on the interior of $\sphericalangle TOP$ such that $\overline{OY}$ does not intersect the hyperbola. By the Parallel Postulate, there is a line through P that is parallel to $\overline{OT}$ but intersects $\overline{OY}$. Let the point of intersection be Y. Find V on the diameter and Q on the hyperbola such that $\overline{QV}$ is parallel to $\overline{PT}$ and the $\overline{QV}$ produced is incident with Y and intersects the $\overline{OT}$ produced at J. This means that $PYJT$ is a parallelogram, so $\overline{PY} \cong \overline{TJ}$ and $\overline{PT} \cong \overline{YJ}$. Therefore,

$$QJ > PT \text{ and } VJ + QV > PT$$

from which we conclude that

$$QJ \cdot (VJ + QV) > PT^2. \tag{7.18}$$

However,

$$\frac{VJ^2}{OV^2} = \frac{PT^2}{OP^2} = \frac{p}{PP'}, \tag{7.19}$$

where the last equality follows by the hypothesis on PT^2 and the fact that $2OP = PP'$. Equations 7.16, 7.17 and 7.19 yield

$$\frac{VJ^2}{OV^2} = \frac{QV^2}{OV^2 - OP^2} = \frac{VJ^2 - QV^2}{OP^2}.$$

Hence,

$$\frac{PT^2}{OP^2} = \frac{VJ^2 - QV^2}{OP^2},$$

which gives

$$PT^2 = VJ^2 - QV^2 = QJ \cdot (VJ + QV),$$

and this is a contradiction of Equation 7.18.

That $\overline{OS'}$ and $\overline{OT'}$ are asymptotes of the other branch of the hyperbola is readily seen by repeating the arguments above. Since $\overline{ST}$ and $\overline{S'T'}$ are parallel and $SP = P'T'$, the points T', O, and S must be collinear. Similarly, S', O, and T are collinear. Thus, the two branches share the same asymptotes. ■

We note that the asymptotes for the hyperbola $xy = k$ are the x- and y-axes.

The Ellipse

Apollonius has one more conic section to consider. Its equation and derivation are similar to those of the hyperbola. Its diameter is $\overline{PP'}$, and its center is the midpoint of this segment.

■ **THEOREM 7.3.7 [*Conics* I.13]**

Cut cone ABC with a plane so that the resulting section is not parallel to $\overline{AC}$. Let the diameter $\overline{PM}$ intersect $\overline{AB}$ at P and $\overline{AC}$ at a point P'. Construct $\overline{AF}$ parallel to $\overline{PM}$, meeting the $\overline{BC}$ produced at F. Draw $\overline{PL}$ perpendicular to the diameter such that

$$\frac{PL}{PP'} = \frac{BF \cdot FC}{AF^2}.$$

Join $P'L$ and let $\overline{VR}$ be parallel to $\overline{PL}$ and intersect $\overline{P'L}$ at R. We conclude that

$$QV^2 = RV \cdot PV.$$

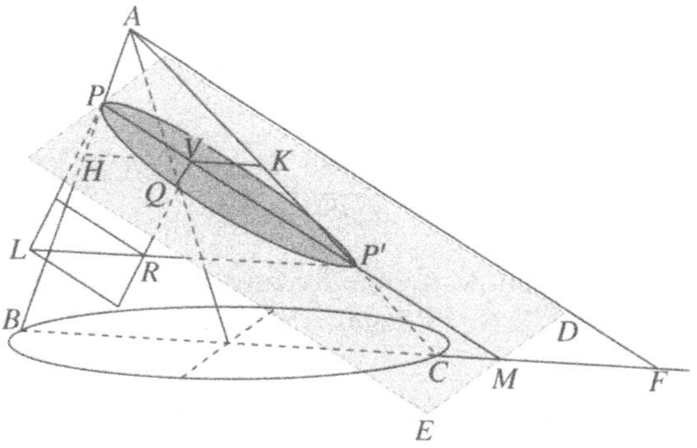

PROOF

Draw $\overline{HK}$ through V and parallel to $\overline{BC}$. Again using the Chord Theorem (5.5.3), we have

$$QV^2 = HV \cdot VK.$$

Because $\triangle HPV \sim \triangle BAF$,

$$\frac{HV}{PV} = \frac{BF}{AF}.$$

Since $\sphericalangle KVP' \cong \sphericalangle AFC$ and $\sphericalangle VKP' \cong \sphericalangle ACF$, we conclude that $\triangle VKP'$ is similar to $\triangle FCA$. Hence,

$$\frac{VK}{P'V} = \frac{FC}{AF}.$$

The latter two equations yield

$$\frac{HV \cdot VK}{PV \cdot P'V} = \frac{BF \cdot FC}{AF^2}.$$

Appealing to the fact that $\triangle LPP' \sim \triangle RVP'$ and the hypothesis of the theorem,

$$\frac{QV^2}{PV \cdot P'V} = \frac{HV \cdot VK}{PV \cdot P'V} = \frac{PL}{PP'} = \frac{RV}{P'V} = \frac{PV \cdot RV}{PV \cdot P'V}.$$

We conclude that $QV^2 = PV \cdot RV$. ■

Apollonius calls this conic an ellipse because the equation shows that the square of the ordinate is equal to the area of the rectangle with base length equal to the abscissa and length that lies along the parameter but *falls short of it* (ἐλλείπει). This is evident in the equation when it is noted that

$$QV^2 = PV \cdot RV = PV \cdot PL - PV(PL - RV).$$

Moreover, if we set

$$y = QV, x = PV, p = PL, \text{ and } d = PP',$$

we may describe the ellipse only in terms of its parameter and diameter with the equation

$$y^2 = px - \frac{p}{d}x^2. \tag{7.20}$$

Exercises

1. It is unknown exactly how Menaechmus identified the conic sections. Heath (1921b, 111–112) conjectured that Menaechmus described the parabola like this: Start with a right circular cone OQR with $\sphericalangle QOR$ right. Cut the cone with a plane perpendicular to $\overline{OQ}$ and $\triangle OQR$. Choose B on $\overline{OQ}$ and C on $\overline{OR}$ so that $\overline{BC}$ is parallel to $\overline{QR}$. Let $\overline{AG}$ be in that plane with the properties that A lies on $\overline{OQ}$ and $\overline{CG}$ is perpendicular to $\overline{BC}$. Drop a perpendicular from O to $\overline{QR}$ intersecting $\overline{AG}$ at L. Let P be on the intersection of the cone with the plane. This means that P is on the circle with diameter $\overline{BC}$. Take N on $\overline{BC}$ so that $\overline{PN}$ is perpendicular to $\overline{BC}$. Construct $\overline{AD}$ parallel to $\overline{BC}$ and $\overline{DF}$ parallel to $\overline{OL}$ with D on $\overline{OR}$ and F on $\overline{AG}$.

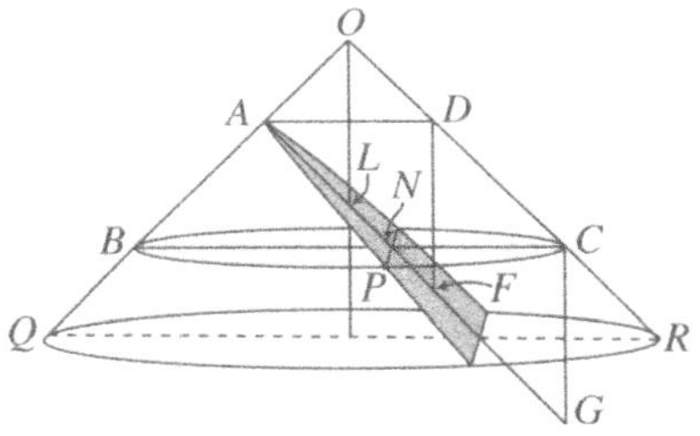

(a) Why does $PN^2 = BN \cdot NC$?
(b) Show that $BN/NA = NG/NC$.
(c) Prove that $PN^2 = 2AL \cdot NA$.

2. Menaechmus used conic sections to construct two continued mean proportionals between two given segments. Thus, he was able to duplicate the cube (Heath 1921a, 253–255). Take $\overline{OA}$ and $\overline{OB}$ so that they meet at O in right angles and $OA = 2OB$.

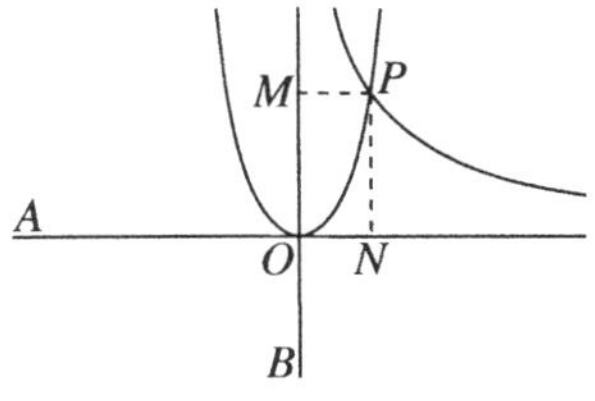

(a) Proceed by analysis. Suppose that we are able to find the two mean proportionals x and y between these two magnitudes. Find M extending $\overline{OB}$ and N extending $\overline{OA}$ such that $x = OM$ and $y = ON$. Construct $\square OMPN$. Then by assumption, $OB/ON = ON/OM = OM/OA$.

Show that P lies on the parabola with vertex O and axis $\overline{OM}$ and on the hyperbola with $\overline{OM}$ and $\overline{ON}$ as asymptotes.

(b) Write a synthetic proof that OM and ON are the sought-after mean proportionals.

3. Describe a cone that when appropriately cut by a plane yields the parabola given by $y^2 = x$. Repeat the problem for $y^2 = 2x$.

4. Find possible lengths for all of the segments in the figure for Theorem 7.3.4 if the parameter $PL = 4$.

5. Show that Equation 7.14 follows from Equation 7.13.

6. Confirm that the lines identified in Equation 7.15 are the asymptotes of the hyperbola of Equation 7.14.

7. Use Equation 7.13 to explain Apollonius's choice of the name *hyperbola* and Equation 7.20 to explain his choice of the word *ellipse*.

8. Prove that if a line through the center of a hyperbola meets one branch, it will meet the other branch.

9. Using the notation of Theorems 7.3.5 and 7.3.7, prove that the following hold true in any hyperbola or ellipse.

(a) QV^2 varies directly as $PV \cdot P'V$.

(b) A chord that contains the conic's center O is bisected by O.

10. Explain why the locus described by the equation $xy = k$ is a hyperbola.

11. For each of the following common equations for the conic sections, find the expression that represents the value of the parameter p.

(a) $x = a(y-k)^2 + h$.

(b) $\dfrac{(x-h)^2}{a^2} - \dfrac{(y-k)^2}{b^2} = 1$.

(c) $\dfrac{(x-h)^2}{a^2} + \dfrac{(y-k)^2}{b^2} = 1$.

12. Explain how it would be possible for the ordinate not to be perpendicular to the diameter of a conic section.

13. Let $\overline{PP'}$ be a diameter of an ellipse and O its center. Show that each chord that contains O is a diameter of the ellipse. (A pair of diameters with the property that each bisects all chords parallel to the other are called **conjugate diameters.**)

14. Construct the conjugate to the diameter of a hyperbola.

15. Given that a diameter of the ellipse $x^2/a^2 + y^2/b^2 = 1$ has a slope of 5, find the equation of the conjugate diameter.

16. Two conic sections are **similar** if all corresponding ordinates that are placed at proportional distances on the diameter from the vertex are proportional. Prove that:

(a) All parabolas are similar.
(b) A parabola cannot be similar to a hyperbola.
(c) An ellipse cannot be similar to a hyperbola.

7.4 THE ANALYTIC ART IN TWO VARIABLES

Viète died on February 23, 1603 in Paris. Although he made contributions to synthetic geometry, it is his analytic art that would set the stage for an explosion of mathematical research, but first his symbolic technique must be pushed to include more than one variable. It would take the work of two great thinkers, one a philosopher and the other a lawyer but both excellent mathematicians, to make the required leap. We first meet the lawyer.

Many consider Pierre de Fermat the "prince" of the amateur mathematicians, but the turn of the seventeenth century saw mathematics just beginning to be a unified field. Since the time of the decline of Greek mathematics, those who pursued the subject would write an isolated text or compose some letters, but organized research did not exist. It was through the efforts of people like Fermat and some of his associates that the field began to take the form that we know today. Should, then, Fermat be considered the "prince of the amateurs"? If independently discovering analytic geometry, founding modern number theory, contributing to the early development of probability theory, and being one or two steps away from the calculus is enough to warrant the title, then he is deserving. Since in those days *amateur mathematician* was redundant, it might simply be best to say that Fermat should rank as one of the greatest mathematicians of all time.

Fermat was born in Toulouse, France. The exact date is not know with certainty, but he was baptized on August 20, 1601. His parents were Dominique and Claire; he was a wealthy leather merchant and second consul of Beaumont-de-Lomagne, and she was the daughter of prominent parents. Fermat had a brother and two sisters. Little is known of his private life, especially his childhood. Fermat probably grew up in Beaumont. If this was so, he would have attended the Franciscan school at the monastery of Grandselve. At the University of Orléans he studied law, and by January 1631 he had purchased the offices of *conseiller* and *commissaire* from the widow of the previous holder. In April of that year Fermat married Louise de Long, and they would have five children: Clément-Samuel, the eldest, Jean, Claire, Cathérine, and Louise (Mahoney 1994, 15–16).

In his spare time Fermat studied science. He was particularly interested in optics. The **principle of least time**, that a beam of light will always travel the quickest route between two points, is due to him. However, before arriving in Bordeaux in the late 1620s, Fermat had never studied mathematics proper. Here he encountered the work of François Viète. He read the *Introduction to the Analytic Art*, and although he never met Viète, he was taught by his students. This encounter impressed upon the new mathematician a particular approach to the field. Like his master, Fermat gained an appreciation for the writings of the ancient Greeks. He was particularly interested in the work of Apollonius, Pappus, Diophantus, and Archimedes. For example, both direct and indirect references to the *Quadrature of the Parabola*, *On Spirals*, and the *Sphere and Cylinder* are found in Fermat's works. His dedication was so sincere that throughout his career, Fermat even used Viète's cumbersome notation (Mahoney 1994, 14, 26–27; Kline 1972, 274).

His interest in the ancients led Fermat to become involved in reconstructing Greek texts. The study of the classical Greeks was a bit of an academic craze, and it was common for mathematicians to attempt to restore lost works. Fermat's contributions included the *Solid Loci* of Aristaeus, the *Porisms* and *Surface Loci* of Euclid, and the *Plane Loci* of Apollonius. The content of these works were various categories of curves, which the classical Greek geometers characterized depending on how they were constructed. The basic class, known as **plane loci**, were those that could be constructed with a straightedge and compass. These included lines and circles. The next category was the **solid loci**. Their construction required the use of one or more conic sections. The last class was the **linear loci**, which included the more complicated curves, such as spirals, conchoids, and cissoids (Heath 1921a, 218).

Producing the reconstruction of *Plane Loci* was difficult for Fermat, but its completion was rewarding. Along with the *Conics*, it formed the basis for Fermat's *Introduction to Plane and Solid Loci*, which he finished in 1629. He believed that although the mathematics of Apollonius were impressive, the results were not general enough to be useful. This opinion motivated Fermat to use algebraic equations to describe the loci studied by Apollonius. Fermat used Viète's methods as a foundation to his work, but since Viète's equations involved only a single variable, it was necessary for Fermat to extend the analytic art to the two-variable case (Mahoney 1994, 40, 51, 77–78).

The Analysis of Conics

The following results are adapted from Fermat's *Introduction to Plane and Solid Loci* as translated by Joseph Seidlin in Smith (1959, 389–396). Fermat begins:

> None can doubt that the ancients wrote on loci. We know this from Pappus, who, at the beginning of Book VII, affirms that Apollonius had written on plane loci and Aristæus on solid loci. But, if we do not deceive ourselves, the treatment of loci was not an easy matter for them. We can conclude this from the fact that, despite the great number of loci, they hardly formulated a single generalization, as will be seen later on. We therefore submit this theory to an apt and particular analysis which opens the general field for the study of loci.
>
> Whenever two unknown magnitudes appear in a final equation, we have a locus, the extremity of one of the unknown magnitudes describing a straight line or curve. The straight line is simple and unique; the classes of curves are indefinitely many—circle, parabola, hyperbola, ellipse, etc.

Algebraists had studied such equations prior to this, but here Fermat became one of first to give geometric meaning to the equations. He is replacing the Greek description of loci with one based on algebra. Fermat continues:

> It is desirable, in order to aid the concept of equation, to let the two unknown magnitudes form an angle, which usually we would suppose to be a right angle, with the position and the extreme point of one of the unknown magnitudes

> established. If neither of the two unknowns is greater than a quadratic, the locus will be plane or solid. . . .

Fermat is setting up a coordinate system, but it has only one horizontal axis, which Fermat chooses before he draws the locus. The ordinates are usually placed at right angles to the abscissa, but Fermat's description does not require it. This is because Apollonius did not require it. All of the results of the *Conic Sections* apply equally when the ordinates are **oblique**.

To set this up, fix $\overline{NM}$ and view N as the origin. Let a and e represent unknowns. Find X on $\overline{NM}$ and Y not on $\overline{NM}$ such that $NX = a$ and $XY = e$, as in Figure 7.9. As the value of a changes, the abscissa NX changes. Since e depends on a by way of an equation, the ordinate XY also changes. The point Y then traces out a locus. In the *Introduction*, Fermat used equations in two variables of first or second degree to describe the conic sections. Mahoney (1994, 83) gives the list:

$ad = be$	line
$ae = z^2$	hyperbola
$a^2 \pm ae = re^2$	line
$a^2 = re$	parabola
$z^2 - a^2 = e^2$	circle
$z^2 - a^2 = re^2$	ellipse
$z^2 + a^2 = re^2$	hyperbola

Notice the influence of Viète in the notation. For example, Fermat denotes unknown values with vowels. Furthermore, in the equation $ae = z^2$ Fermat writes z^2 because of the Law of Homogeneity. Since ae represents a two-dimensional magnitude on the left, there must be a two-dimensional magnitude on the right. The exponential notation is due to Descartes, and is used here only to aid reading. Fermat wrote z *quad* where we write z^2 (Mahoney 1994, 35).

As a quick example, let d be a constant. Take $\overline{PV}$ perpendicular to $\overline{PZ}$ and construct a parabola with diameter $\overline{PV}$ and parameter d (Figure 7.10). Let Q be a point on the parabola and assume that $\overline{QV}$ is parallel to $\overline{ZP}$, so

$$QV^2 = d \cdot PV$$

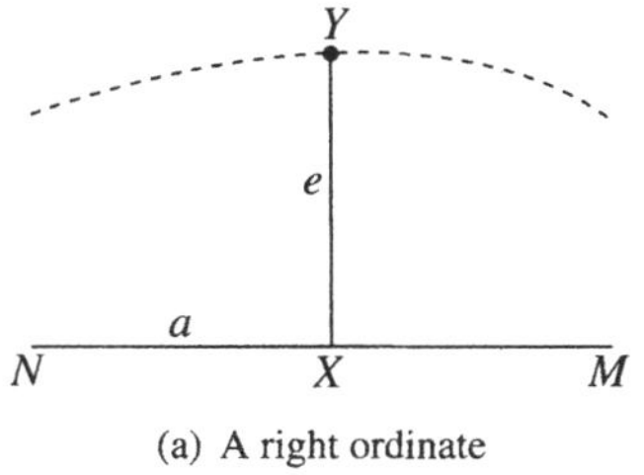

(a) A right ordinate

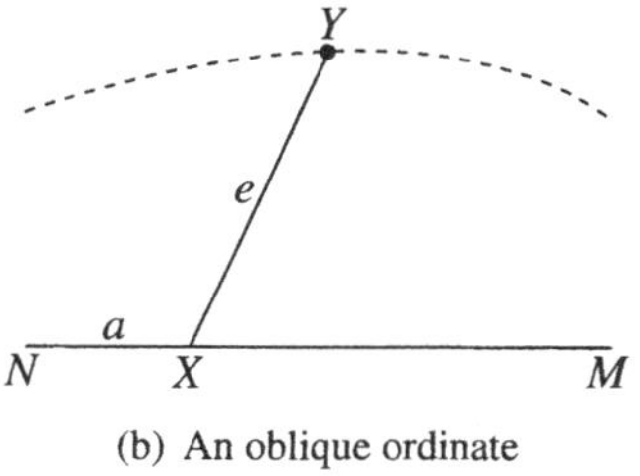

(b) An oblique ordinate

Figure 7.9 Fermat's uniaxial coordinate geometry.

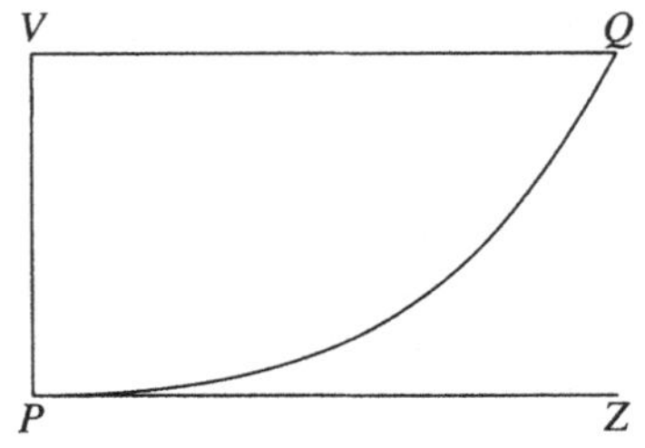

Figure 7.10 The locus of points described by $a^2 = de$ is a parabola.

by Theorem 7.3.4. Therefore, if we let $a = QV$ and $e = PV$, we see that the constructed parabola is the locus described by

$$a^2 = de.$$

For a more involved example, Fermat showed that

$$ae = z^2 \tag{7.21}$$

describes a hyperbola. For Fermat to prove his claim he must construct a locus of points in which Equation 7.21 holds and then demonstrate that the locus is a hyperbola. Modifying the traditional use of the term further even than Viète does, Fermat calls the procedure that leads from the equation to the locus **analysis**. The **synthesis** is the demonstration that a conic section yields a degree one or degree two indeterminant equation (Mahoney 1994, 85).

For the analysis, start with $\sphericalangle SPM$ and view P as the origin. We are given $ae = z^2$ and will interpret a as the distance PV, with V on $\overline{PM}$, and e as the distance QV, with $\overline{QV}$ parallel to $\overline{PS}$ (Figure 7.11). Now find points N and O such that $\overline{MN}$ is parallel to $\overline{PS}$, O is on $\overline{PS}$, $\overline{PO} \cong \overline{MN}$ and

$$MN \cdot ON = z^2.$$

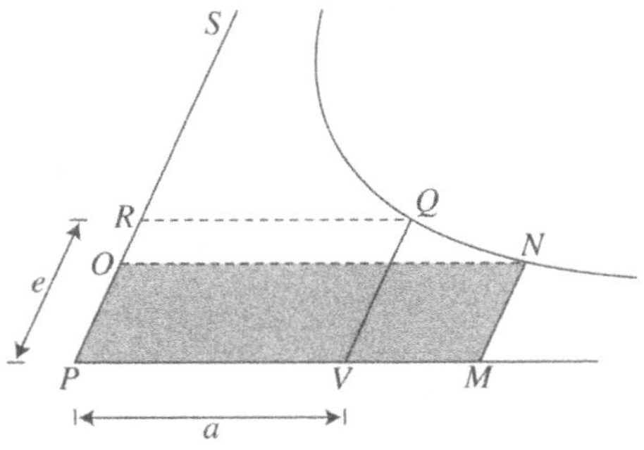

Figure 7.11 The locus of points described by $ae = z^2$ is a hyperbola.

The locus of points Q such that there exists V on $\overline{PM}$ and R on $\overline{PS}$ such that $VQRP$ is a parallelogram and

$$QV \cdot QR = MN \cdot NO$$

is seen to be the curve described by $ae = z^2$. Hence, the equation gave rise to a locus of points that is a hyperbola (Theorem 7.3.5). Its asymptotes are $\overline{PM}$ and $\overline{PS}$.

The synthesis for the basic hyperbola is straightforward. Take the equation $QV \cdot PV = z^2$ and view PV as the abscissa and QV as the ordinate. Letting $a = PV$ and $e = QV$, we find that $ae = z^2$. This means that the locus can be described by the equation.

For one last example, Fermat next proves that the equation

$$d^2 + ae = ra + se \tag{7.22}$$

also describes a hyperbola. Begin by solving for d^2 to obtain

$$d^2 = ra + se - ae.$$

Next find a rectangle such that its area includes the terms of d^2. A rectangle with dimensions of $a - s$ by $r - e$ meets our requirement because

$$(a - s)(r - e) = ra + se - ae - rs = d^2 - rs.$$

To begin the construction, let $\sphericalangle SPT$ be given. Take $\overline{QV}$ to be parallel to $\overline{PS}$, and let $PV = a$ and $QV = e$. Draw $\overline{SR}$ and $\overline{PT}$ so that they are parallel to each other. Let $PS = r$ and take O on $\overline{PT}$ so that $PO = s$ (Figure 7.12). At O draw $\overline{OM}$ parallel to $\overline{PS}$ with M on $\overline{SR}$, and produce $\overline{QV}$ to R beyond Q. Because $PO = s$ and $PV = a$,

$$a - s = OV = MR.$$

Similarly, since $PS = VR = r$ and $QV = e$, we have

$$r - e = RQ.$$

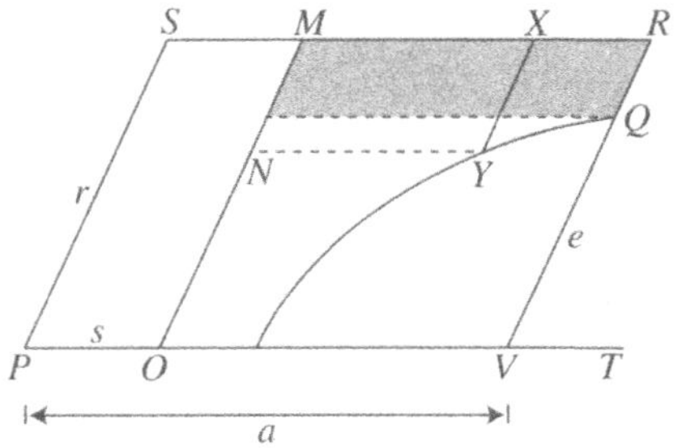

Figure 7.12 Fermat's analysis of $d^2 + ae = ra + se$.

Therefore, the product $MR \cdot RQ$ equals constant $d^2 - rs$. Finally, take points X and Y so that X is on $\overline{MR}$ and $\overline{XY}$ is parallel to $\overline{PS}$. View MX as the abscissa and XY as the ordinate. Then

$$MX \cdot XY = MR \cdot RQ$$

describes a hyperbola containing Q. Its asymptotes are $\overline{MR}$ and $\overline{OM}$.

Focus and Directrix

Near the end of book VII of the *Mathematical Collection* we find Pappus's contribution to conic sections. Although it is common for us to refer to the focus and the directrix of a conic, particularly with respect to the parabola, there are only three extant references to the directrix property in the ancient Greek literature. One is from a fragment of Anthemius on burning mirrors and two by Pappus (Pappus 1986, *Mathematical Collection*, VII.235–238; Heath 1921a, 243; Heath 1921b, 119–121). There he gives a complete proof of the following proposition:

> If the distance of a point from a fixed point is in a given ratio to its distance from a fixed line, the locus of the point is a conic section which is an ellipse, a parabola, or a hyperbola according as the given ratio is less than, equal to, or greater than, unity.

This result is given as a lemma to Euclid's lost work *Surface Loci*, so it is probable that Euclid and thus Apollonius knew this proposition. The fixed point is called the **focus**, and the fixed line is the **directrix**. The ratio is called the **eccentricity** of the conic section. Denote this ratio by e. It uniquely identifies the three different types of conics:

ellipse	$e < 1$,
parabola	$e = 1$,
hyperbola	$e > 1$.

Let us proceed by analysis and show that the focus-directrix definition of the conics yield Apollonius's definition. However, we will return to letting x and y represent indeterminants.

Since the derivation for $e = 1$ is commonly found in precalculus texts, we will show that an ellipse is the result when $e < 1$. Let C be the focus and let the directrix be a line through D. The locus of points P such that

$$\frac{PC}{PQ} = e < 1 \tag{7.23}$$

is illustrated in Figure 7.13, where Q is the point on the directrix so that $\overline{PQ}$ is perpendicular to $\overline{QD}$. Now find A and A' so that they are both collinear with C and D and

$$\frac{AC}{AD} = \frac{A'C}{A'D} = e.$$

Notice that if $e \geqslant 1$, the point A' does not exist. Let O be the midpoint of $\overline{A'A}$ and make the following observation:

$$\frac{OC}{OA} = \frac{2OC}{A'A} = \frac{A'C - AC}{A'D - AD} = \frac{eA'D - eAD}{A'D - AD} = e.$$

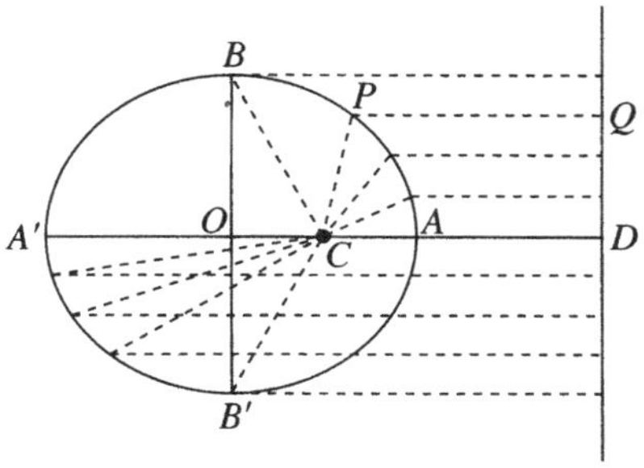

Figure 7.13 An ellipse described by a focus and a directrix.

We now find an equation to describe the curve generated by Equation 7.23. View O as the origin, $\overline{A'D}$ as the x-axis, and $\overline{BB'}$ as the y-axis. Let $a = OA = OA'$ so that A is the point $(a, 0)$ and A' is the point $(-a, 0)$. Since $OC = eOA$, the focus C is represented by $(ae, 0)$, and because

$$OA + AD = \frac{OC}{e} + \frac{AC}{e} = \frac{a}{e},$$

the equation for the directrix is

$$x = \frac{a}{e}.$$

Letting a random point P on the ellipse be represented by (x, y) yields

$$PC^2 = (x - ae)^2 + y^2,$$

and

$$PQ^2 = (x - a/e)^2.$$

Because $PC^2 = e^2 PQ^2$, the equation for the locus is

$$(x - ae)^2 + y^2 = e^2(x - a/e)^2.$$

From this we obtain

$$x^2(1 - e^2) + y^2 = a^2(1 - e^2),$$

from which we conclude

$$\frac{x^2}{a^2} + \frac{y^2}{a^2(1 - e^2)} = 1.$$

Let $b^2 = a^2(1 - e^2)$, and we have

$$\frac{x^2}{a^2} + \frac{y^2}{b^2} = 1, \tag{7.24}$$

Therefore,

$$y^2 = b^2 - \frac{b^2 x^2}{a^2}.$$

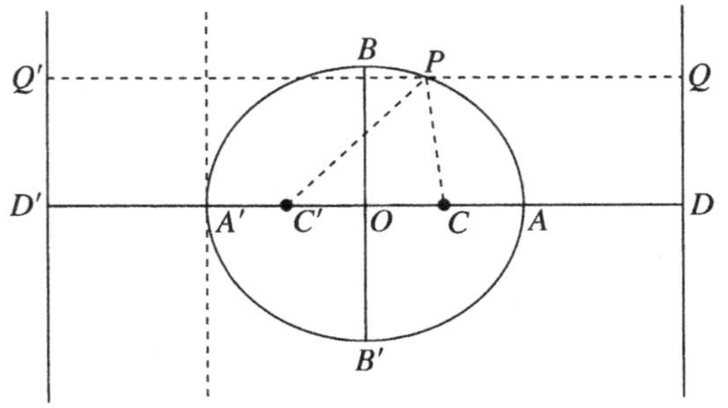

Figure 7.14 An ellipse described by two foci and two directrices.

To see that this is an ellipse, we compare it with Equation 7.20. However, the abscissa of Apollonius began at A', not O, so replace x with $x - a$ and calculate

$$y^2 = b^2\left(1 - \frac{(x-a)^2}{a^2}\right) = b^2\left(1 - \frac{x^2 - 2ax + a^2}{a^2}\right) = b^2\left(\frac{x^2}{a^2} - \frac{2x}{a}\right).$$

Hence, with A' as the origin the equation becomes

$$y^2 = \frac{2b^2}{a}x - \frac{b^2}{a^2}x^2.$$

This is an ellipse with a parameter of the ordinates equal to $2b^2/a$.

Because of the symmetry of Equation 7.24, we can find points C' on $\overline{A'A}$ such that $\overline{A'C'} \cong \overline{CA}$ and D' on $\overline{A'A}$ extended to the left so that $\overline{D'A'} \cong \overline{AD}$ (Figure 7.14). Then C' is another focus that will also generate the ellipse with the line through D' perpendicular to $\overline{A'A}$ serving as the directrix. Hence, any point P on the ellipse will also satisfy

$$\frac{PC'}{PQ'} = e,$$

where Q' is on the left directrix so that $\overline{PQ'}$ is perpendicular to $\overline{D'Q'}$. From this we see that

$$PC' + PC = e(PQ' + PQ) = eQ'Q.$$

That is, the sum of the distances from a point on the locus to the two foci is constant.

Not only does Fermat's analytic geometry simplify study of the conic sections, but the methods of algebra enable the mathematician to write down the thoughts that lead to the solution. The Greek geometers' published proofs left the method that led them to their discoveries hidden from view. Algebra, on the other hand, treats the reader as a student by displaying the details for all to see. Fermat, though, was inspired by their work (Mahoney 1994, 93, 115). He translated the abscissa and ordinate of Apollonius into independent and dependent variables, so the algebra seems to have been there all along.

After 1631 Fermat lived in Toulouse for the most part. During this time he often traveled to Castres to plead cases before the *Commission de la Chambre de l'Édit* of Nantes. In 1632 he joined the commission, a position that he held for the rest of his life. Fermat had been in good health until around 1653, when he caught the plague and nearly died. He recovered but fell ill again in 1660, and his health never fully returned. He died in Castres on January 12, 1665 (Mahoney 1994, 17).

Exercises

1. Use Fermat's technique to show that:
   (a) $ad = be$ is a line.
   (b) $z^2 - a^2 = re^2$ is an ellipse.
   (c) $z^2 - a^2 = e^2$ is a circle.

2. Graph the following conic sections using oblique coordinates. Assume that the angle that the ordinate makes with the abscissa is 45°.
   (a) $y^2 = 20x$
   (b) $y^2 = 4x + x^2$
   (c) $y^2 = 8x - 3x^2$

3. Find the length of the parameter of the ordinates and the distance of the vertex to the focus for each of the conic sections in Exercise 2.

4. Graph each conic based on the given distance d between the focus and the directrix and the given eccentricity e.
   (a) $d = 2, e = 1/3$
   (b) $d = 4, e = 1$
   (c) $d = 1, e = 4$

5. Find conditions on the coefficients of $ax^2 + by^2 + cxy + dx + ey + f = 0$ so that the resulting conic section is a
   (a) point
   (b) circle
   (c) ellipse
   (d) hyperbola
   (e) parabola

6. Prove that the length of the parameter of the ordinates of each of the three conic sections equals the length of the segment that joins two points of the section, contains the focus, and is parallel to the directrix.

7. Use the focus-directrix definition to derive the equation $y^2 = 4px$ for a parabola.

8. Find the points of intersection of $y^2 = 4px$ and $x^2/a^2 + y^2/b^2 = 1$.

9. Let $2a$ be the length of the major axis, $2b$ the length of the minor axis, and $2c$ the distance between the foci. Using the definition of an ellipse as being the locus of all points P such that the sum of the distances of P to the two foci is constant, derive

the equation for the ellipse:

$$\frac{x^2}{a^2} + \frac{y^2}{b^2} = 1.$$

10. The equation

$$\frac{x^2}{a^2} - \frac{y^2}{b^2} = 1$$

describes a hyperbola.

(a) Explain how this equation can be derived in a manner similar to the ellipse, namely that an ellipse is the locus of all points P such that the difference in the distances between P and two foci is constant.

(b) Find the length of its parameter.

(c) Show that $b^2 = c^2 - a^2$.

(d) Show that its asymptotes are $y = \pm(b/a)x$.

11. Describing the parameter of the ordinates as in Exercise 6, let P and P' be the intersections of the parabola $(y - 3)^2 = 8x$ and its parameter. Find the point on the line $y = 3$ that is the intersection of the two tangents of the parabola at P and P'.

12. Show that the angle formed by the tangent to an ellipse at P and the segment joining P to the focus is constant.

13. Let P_1 and P_2 be two points on a parabola with focus F. Suppose that $\overline{Q_1P_1}$ and $\overline{Q_2P_2}$ are lines that are parallel to the parabola's axis of symmetry and let $\overline{R_1S_1}$ and $\overline{S_2R_2}$ be tangent to the parabola at P_1 and P_2, respectively. Show that if $\measuredangle R_1P_1Q_1 \cong \measuredangle E_1P_1S_1$ and $\measuredangle R_2P_2Q_2 \cong \measuredangle E_2P_2S_2$ as in the diagram, F is on both $\overline{P_1E_1}$ and $\overline{P_2E_2}$ (produced). This demonstrates that any beam of light that is parallel to the axis of the parabola and in the same plane will reflect off the parabola making congruent angles with the tangent and move toward the focus.

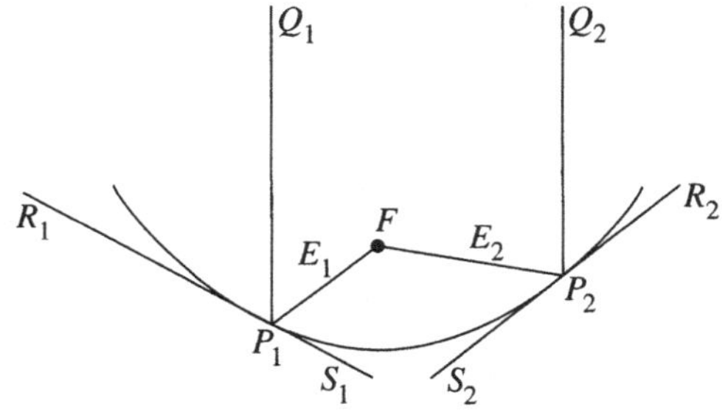

CHAPTER 8

RENÉ DESCARTES

Anselm of Canterbury lived from 1033 to 1109, and he was the greatest theologian of his time. He admired the work of Augustine, and like his hero, he viewed himself as a defender of Christianity. To this end, we wrote many works, the two most significant titled *On the Being of God* and *Discourse on the Existence of God.* Combining philosophical argument with his religion, he sought to prove the existence of God by various arguments. This way of thinking would eventually dominate the theology of the Middle Ages as a system known as **scholasticism**. The leaders of the movement did not seek to discover new religious truth but desired to harmonize the teachings of the Roman Catholic Church with reason. In so doing, the scholastics created an elaborate theological system that attempted to answer all questions asked of them by combining the philosophy of Aristotle with the theology of Rome. In line with the teachings of Anselm, they believed that revelation and science form a consistent system. Since God was the author of both Scripture and nature, there should be no contradiction. Notable scholastic thinkers who taught this viewpoint include Thomas Aquinas and Peter Lombard (Schaff 1907, 587–594).

It is into this world that René Descartes was born. The date was March 31, 1596, and the town was La Haye in Touraine, France. His mother, Jeanne, died in childbirth 14 months later. His father, Joachim, was the councilor at the *parlement* of Rennes in Brittany. René had two older siblings, a brother named Pierre and a sister, Jeanne. Descartes was a sickly child, so his father delayed sending him to school until he would be strong enough to manage the stresses of school life. During these early years he was nurtured primarily by his maternal grandmother and a nurse, both of whom he would cherish his entire life (Vrooman 1970, 20–23).

Joachim recognized in the young René an intellectual ability that he believed should be cultivated, so he decided to send the "little philosopher," as he called him, to the best school in France. The year is 1606, and the school was La Flèche in Anjou. It was established by Henri IV, who desired it to be the best school of its kind, and it was run by the Society of Jesus. The Jesuits were Rome's answer to the Reformation. Their mission was to strengthen the faith of Catholics and to fight heresy. They did this primarily through education. At La Flèche this began with a year of introductory classes, followed by three years of grammar, closing with a year of rhetoric. Along the way Latin and Greek were studied. Most students left at this point to attend other universities, but Descartes stayed. His studies now turned to philosophy. The curriculum included study of the traditional liberal arts with some modification. During the first five years, Descartes studied grammar and rhetoric. During his last three years the La Flèche curriculum included geometry, arithmetic, music, astronomy, and dialectic (Gaukroger 1995, 38-39, 41-42, 48, 58; Boyer and Merzbach 1991, 335).

Descartes finished his career at La Flèche in 1614 and then studied law at Poitiers, but law was not for Descartes. It was his Jesuit education that had the greatest impact on him but not as one might think. There was order, method, and certainty within scholasticism, and Descartes was a devout Catholic. The method they taught him, though, did not satisfy him. Instead of embracing Aristotle and Aquinas, he became skeptical of the current philosophies. This was the time when the dispute between the Protestants and Catholics was at its highest (Kline 1972, 305). Luther had died in 1546 and Calvin in 1564. The Council of Trent had laid down the law in 1563, and Martin Chemnitz responded in 1573 with his four-volume *Examination of the Council of Trent*. War in Europe would ebb and flow, with countries aligning along religious lines. In the midst of all of this, instead of feeling learned, Descartes felt ignorant. There was no certainty to be found, neither in the writings of the great philosophers nor in that of the scholastics. He concluded that there must be more than what the Jesuits and their curriculum had to offer (Vrooman 1970, 40, 42-43; Descartes 1985, *Discourse* I, 266)

Descartes therefore decided to move to Paris. There he experienced life by making some questionable friends and putting his mathematical skills to good use by becoming expert at gambling. However, his Catholic upbringing tempered his time in Paris. It was also during this time that Descartes met two men who would remain his friends throughout his life: Claude Mydorge and Father Marin Mersenne. Both attended La Flèche, but it is doubtful that Descartes knew them there. Mersenne was a friar in the Minim order, a branch of the Franciscans founded in 1435 and confirmed by the Pope in 1474 (Jackson 1953, "Minim," vol. VII, 383). He made minor contributions to mathematics, particularly in number theory. However, his main contribution to the field was that he was very skilled at understanding the arguments of other mathematicians. By way of an extensive correspondence with the great thinkers of his day, Mersenne was the main means of spreading information regarding new mathematical results. Starting in 1588 and lasting almost to his death in 1648, he held regular meetings of mathematicians at the Minim monastery in Paris. Fermat was a part of this group. Descartes' other friend, Mydorge, was born into one of France's wealthiest families. His position gave him the leisure time to focus on mathematics. He specialized in the conic sections, primarily because of his interest in optics. Later, Descartes and Mydorge would collaborate in this field (Gaukroger 1995, 59, 132, 138–141; Burton 1985, 470–471, 479; Kline 1972, 295, 304).

Because of his poor health as a child, Descartes habitually slept late every morning, and upon awaking he would stay in his bed reading and studying. City life did not afford him the solitude required for this daily meditation. Now dissatisfied with both his education and Paris, he decided to see the world by volunteering his services to the army of Prince Maurice of Holland. Religious lines in Europe were not as tightly drawn as one might expect. The Dutch war against the Spanish was a Protestant struggle, but because Spain was also the enemy of France at the time, it was not uncommon for Catholic French men to fight alongside the Dutch. When Descartes arrived in Breda in the Netherlands, it was his fortune that the fighting had ceased, so he was alone, away from home, and bored. It was here on November 10, 1618 where he met Isaac Beeckman (Vrooman 1970, 47).

This meeting took place as the two men stood examining a mathematical problem posted on a bulletin board. It was written in Flemish, so Descartes could not read it. He asked Beeckman, who he did not know at the time, for the problem's translation. This chance encounter led to conversation, and although Beeckman was eight years older than Descartes, they soon discovered that they had much in common. Beeckman had once studied theology and medicine, but his current interests were mathematics and physics, particularly mechanics. Each was happy to have found someone with an interest in the sciences, and they soon became friends. Their relationship, though, revolved around mathematics. Beeckman would pose problems to the younger Descartes, and then Descartes would solve them. Beeckman sometimes knew the answer ahead of time, and at other times he was simply eager for a solution that he knew Descartes would be able to find. Besides these problems, Descartes worked with Beeckman on three major projects: a mathematical study of musical intervals, a study of falling objects, and problems in hydrostatics. Descartes had always shown mathematical ability. These days of problem solving would serve to hone his skills (Gaukroger 1995, 68-70, 74-89).

8.1 COMPASSES

Beeckman would soon leave Breda. This led to a correspondence between Beeckman and Descartes that lasted into the 1620s. In one of his letters, Descartes mentioned that he had plans to write a book on mechanics and one on geometry. Of course, Beeckman thought that the mechanics text was a good idea, but Descartes focused his study on mathematics instead of physics. These studies led Descartes to invent

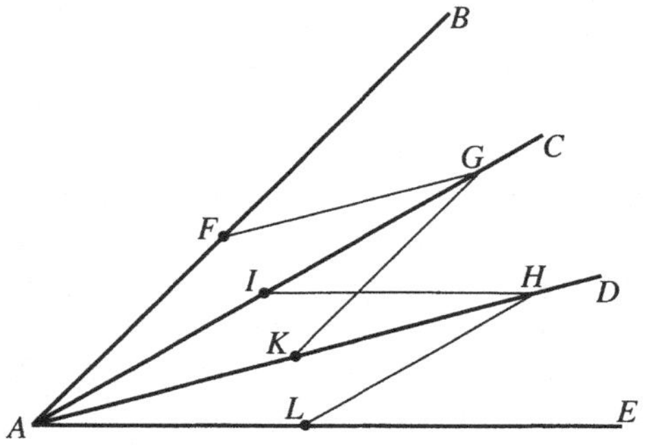

Figure 8.1 Descartes' trisection compass.

various types of compasses that he would then use to solve problems, both geometrical and algebraic. His **trisection compass** (Figure 8.1) enabled him to trisect any angle. He constructed it in such a way that when it opened, the three angles remained congruent to each other. There were pivots at F, I, K, and L, and there were slides at G and H. Trisection was easy because all one needed to do was to lay $\overline{AE}$ along one side of the angle and then open it so that $\overline{AB}$ coincided with the other side of the angle. Place marks at C and D, and you have your trisection.

Descartes constructed other compasses, but it was his modification of the **mesolabe** of Eratosthenes that gave him the most satisfaction. The mesolabe was a device used to find the two mean proportionals between given magnitudes. As we have seen on page 104, this means that he was able to duplicate the cube. To build this device, start with rectangle $AXYE$ as in Figure 8.2. Within the rectangle construct three congruent triangles, $\triangle AMF$, $\triangle MNG$, and $\triangle NQH$, with $\overline{MF}$, $\overline{NG}$, and $\overline{QH}$ perpendicular to $\overline{AX}$. The leftmost $\triangle AMF$ is fixed while the other two slide and are able to overlap each other. A diagonal beam $\overline{AY}$ is able to slide along the base with a pivot at A. Given magnitudes AE and a, slide the beam and the rightmost triangle, now described by $\triangle N'QH$, so that $DH = a$. The beam now intersects the base at point K. Take C to be the intersection of $\overline{AK}$ with $\overline{N'H}$. Finally, slide the middle triangle, which is now $\triangle M'NG$, so that C is on $\overline{NG}$. Let B denote the intersection of $\overline{AK}$, $\overline{MF}$, and $\overline{M'G}$. That these three segments intersect in a point is Exercise 2.

We claim that BF and CG are the desired mean proportionals. First note that by similar triangles,

$$\frac{EK}{FK} = \frac{AK}{BK} = \frac{FK}{GK}.$$

We also have

$$\frac{EK}{FK} = \frac{AE}{BF}$$

and

$$\frac{FK}{GK} = \frac{BF}{CG}.$$

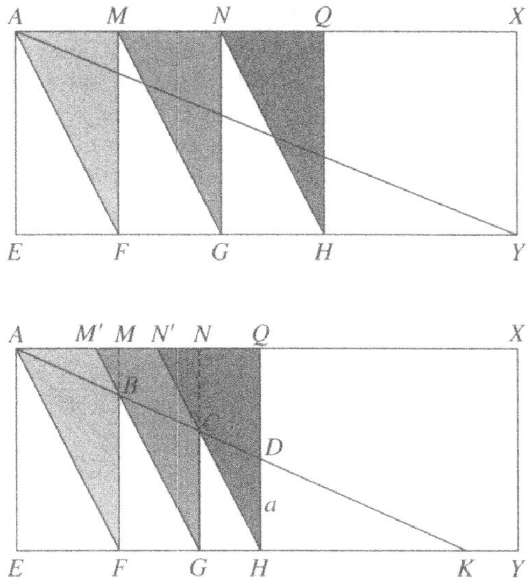

Figure 8.2 Eratosthenes' mesolabe.

Therefore,

$$\frac{AE}{BF} = \frac{BF}{CG}.$$

Similarly, we may conclude that

$$\frac{BF}{CG} = \frac{CG}{DH},$$

which yields the continued proportion,

$$\frac{AE}{BF} = \frac{BF}{CG} = \frac{CG}{DH}.$$

Hence, BF and CG are the two mean proportionals between AE and $DH = a$ (Heath 1921a, 258–260; Gaukroger 1995, 95).

Eratosthenes' original mesolabe can be used to solve certain quadratic equations, but Descartes used his version to solve cubics. In fact, Descartes believed that his adaptation could be generalized to solve an equation in one variable of any degree. Even though Descartes wrote to Beeckman about his mesolabe (Figure 8.3) shortly after its invention, its best description is found in Descartes' *Geometry*:

> Consider the lines AB, AD, AF, and so forth, which we may suppose to be described by means of the instrument YZ. This instrument consists of several rulers hinged together in such a way that YZ being placed along the line AN the angle XYZ can be increased or decreased in size, and when its sides are together the points B, C, D, E, F, G, H, all coincide with A; but as the size of

the angle is increased, the ruler BC, fastened at right angles to XY at the point B, pushes toward Z the ruler CD, which slides along YZ always at right angles. In like manner, CD pushes DE, which slides along YX always parallel to BC; DE pushes EF; EF pushes FG; FG pushes GH; and so on. Thus we may imagine an infinity of rulers, each pushing another, half of them making equal angles with YX and the rest with YZ. (Descartes 1925, 44–47)

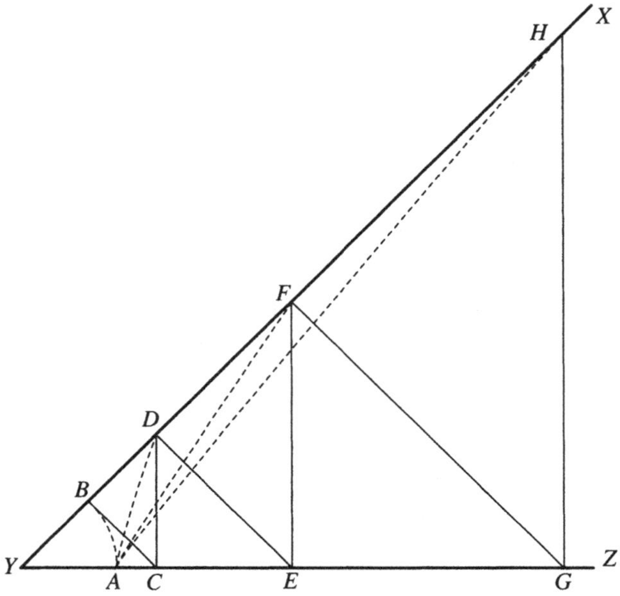

(a) Open 45°

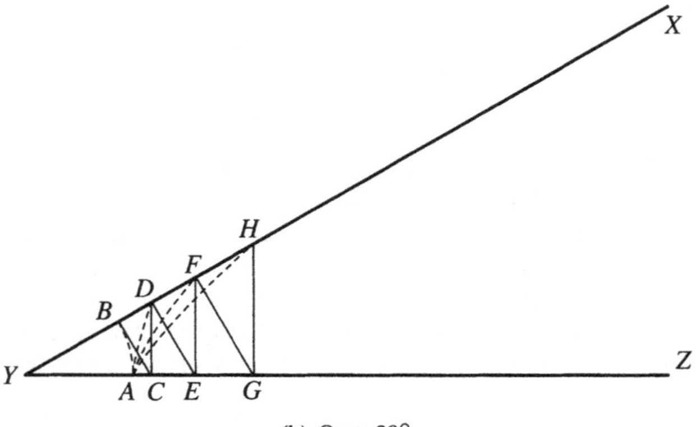

(b) Open 30°

Figure 8.3 Descartes' mesolabe.

No matter the angle at which the mesolabe is opened, we have a sequence of similar triangles. For example, in Figure 8.3, $\triangle YBC \sim \triangle YCD$. Therefore, since this instrument is unbounded in nature, we have an infinite continuous proportion:

$$\frac{YB}{YC} = \frac{YC}{YD} = \frac{YD}{YE} = \frac{YE}{YF} = \frac{YF}{YG} = \frac{YG}{YH} = \cdots,$$

with the first two equalities being the proportions from Eratosthenes' mesolabe.

As an example Descartes used his mesolabe to find a root to the equation $x^3 = x+2$ (Gaukroger 1995, 97-98). Let $YA = 1$ (and then $YB = 1$) and $YC = x$. By the continued proportion, we have

$$YE = \frac{YD^2}{YC}$$

and

$$YD = \frac{YC^2}{YB}.$$

Hence,

$$YE = \frac{YC^3}{YB^2}.$$

Since $YE = YC + CE$ and $YB = 1$,

$$YC^3 = YC + CE.$$

In other words,

$$x^3 = x + CE.$$

If we open the mesolabe so that the length of CE is 2, the magnitude of YC represents a positive, real root of our equation.

Exercises

1. Solve $x^2 = 2$ and $x^3 = 18$ using Eratosthenes' mesolabe.

2. Retrace the steps of the instructions on how to use Eratosthenes' mesolabe on page 270 to show that $\overline{AK}$, $\overline{MF}$, and $\overline{M'G}$ of Figure 8.2 intersect in a single point.

3. After Descartes' discussion of the curves generated by the mesolabe in the third book of the *Geometry*, Descartes solves the equation

$$x^4 - 4x^3 - 19x^2 + 106x - 120 = 0.$$

He notes: "An equation can have as many true roots as it contains changes of sign, from + to − or from − to +; and as many false roots as the number of times two + signs or two − signs are found in succession" (Descartes 1925, 160). This rule is **Descartes' Rule of Signs**.

(a) What did Descartes mean by "false root"?

(b) Find the roots of $x^4 - 4x^3 - 19x^2 + 106x - 120 = 0$.

4. Use Descartes' mesolabe to construct solutions to these equations.
   (a) $x^3 = x - 4$.
   (b) $x^2 = x + 1$.
   (c) $x^4 = x^2 + 5$.

5. As Descartes' mesolabe is opened, the points B, D, F, and H describe curves originating at A (Figure 8.3). Each of these curves can be used to solve geometric problems (Descartes 1925, 47). Confirm algebraically that the curves have the given equations. Assume that the origin is at Y.
   (a) The curve through A and B is a circle with equation $x^2 + y^2 = r^2$, where $r = YA$.
   (b) The curve through A and D is $x^4 = r^2(x^2 + y^2)$, where $YA = YB = r$, $YC = x$, $CD = y$. (*Hint:* YD is needed.)
   (c) If we let $YA = YB = r$, $YE = x$, and $EF = y$, the equation for the curve through A and F is $x^8 = r^2(x^2 + y^2)^3$.
   (d) The equation for the curve through A and H is $x^{12} = r^2(x^2 + y^2)^5$.

8.2 METHOD

Descartes realized that his compass could solve any question that could be restated in terms of mean proportionals. This included polynomial equations of any degree. In his mind he was charting a new course for mathematics. Descartes rejected the arguments of the ancient Greeks. He believed that they must have possessed a method that simplified their work, but instead they published complicated proofs that only served to display their ingenuity. Neither did Descartes accept the methods of the algebraists. He believed they lacked generality and considered their subject to be nothing more than a bag of tricks that are deployed as needed. Descartes wrote:

> [A]s to the Analysis [geometry] of the ancients and the Algebra of the moderns, besides the fact that they embrace only matter the most abstract, such as appear to have no actual use, the former is always so restricted to the consideration of symbols that it cannot exercise the Understanding without greatly fatiguing the Imagination; and in the latter one is so subjected to certain rules and formulas that the result is the construction of an art which is confused and obscure, and which embarrasses the mind, instead of a science which contributes to its cultivation. (Descartes 1985, *Discourse on Method* II, 270)

Descartes believed that a new mathematics was needed that would reveal the hidden method that underlies geometry and could generalize the methods of algebra. He called this system ***mathesis universalis***. As his compass was both a geometric tool that generated loci and an algebraic tool for solving equations, the *mathesis universalis* would blur the distinction between these two subjects. It would be a subject that would encompass all of the mathematical disciplines and provide firm principles that would guide one's work (Gaukroger 1995, 93, 99-103; Descartes 1985, *Rules of the Direction of the Mind* 4.).

On March 26, 1619, Descartes wrote to Beeckman telling him that he was close to a major breakthrough. Later that year Descartes left Breda and found himself with another army, this time in Bavaria. Again, there was a period of inactivity. It appeared that Descartes was destined never to see much action on his campaigns. It was winter, and he was alone with his thoughts, cooped up in a small, one-room shack that was kept quite warm by an iron stove. He spent long stretches of time contemplating. His *mathesis universalis* was a good start, but he now had grander plans. His goal was to develop an overarching science of the sciences. It would be a system that will enable him to know the truth and know it with certainty. Then on November 10 the system gelled. Descartes took the best qualities of mathematics and logic and refined them down to four rules. If he would follow these faithfully, certainty would follow. Descartes' rules are these:

Rule 1. One must not accept anything as true that has not been clearly recognized to be so. That is, one must carefully avoid prejudice in judgment and accept nothing unless it presents itself to one's mind so clearly and distinctly that there could be no doubt about its truth.

Rule 2. One must divide each problem or question into as many parts as possible and in an appropriate order so that the answer may be discovered in the best manner possible.

Rule 3. One must order one's reflections by beginning with the most simple ideas and then proceed by degrees to more and more complex ones. The order in which the ideas are ranked may be natural to their content or purely artificial.

Rule 4. One must exhaust all possibilities and make one's arguments so general that it becomes certain that nothing has been omitted.

The resulting system would govern his thought from that day forward. Descartes called his system the **Method** (Descartes 1985, *Discourse on Method* II, 271).

Both exhausted and exhilarated by his discovery, Descartes fell asleep and dreamt. In his dream he was visited by ghosts. He tried to drive them away but he had a weakness in his right side that hindered him. As he struggled a great wind suddenly spun him around on his left foot several times. Because of the gale, the phantoms no longer seemed a problem, and Descartes found himself walking with great difficulty and leaning to the left because of the weakness. Feeling ashamed at his struggles, he saw that he was at a college and across the way there was a chapel. Continuing to fight the wind, he managed to reach the building. He wanted to go inside to pray but realized that he had passed someone on the way that he recognized but failed to greet. When he tried to go back to find the man, the wind blew him against the wall of the chapel. Then someone else called out, "René Descartes!" This new stranger told him to look for Monsieur N., but then he noticed that the man was surrounded by people who were not affected by the wind. They were able to stand straight, while he kept leaning to the left. At this point the wind finally subsided, and he awoke to find a pain in his right side.

Descartes turned over to his left side and prayed that the dream would have no evil consequences. He worried that it was a sign from God that he had sinned.

He continued to pray and meditate on the nature of good and evil for quite some time before falling back to sleep. Then he dreamt again, but this time his dream consisted merely of a sharp clap of thunder. Descartes awoke immediately and saw sparks before his eyes that made it difficult to see. He blinked a few times and they disappeared. Happy that all now seemed quiet, he went back to sleep.

Descartes dreamt one last time. The third dream began with a book on a table. He was pleased to see that it was a reference book because it might prove to be useful. Descartes noticed that there was another book. It was was titled *Corpus Poetarum.* Randomly opening this book, he read the sentence, "What road in life shall I follow?" At this a stranger appeared and spoke to him a poem, each line of which began with the words "it is and is not." The stranger explained to Descartes that the poem was very famous. Descartes agreed and said that he knew it well. It was from the *Idylls* of Ausonius and can be found in the *Corpus Poetarum.* Descartes went to the anthology and began to search for the poem. The man seemed impressed with the book and wanted to know how Descartes had obtained it. He replied that he did not know, and then the book disappeared and instantly reappeared at the other end of the table. The reference book changed, too. It turned into some other kind of dictionary. Descartes went over to the *Corpus Poetarum* and found the poems by Ausonius, but he could not find the one quoted by the man. Descartes told the stranger that he knew of a better poem. It started with the words, "What road in life shall I follow?" Descartes tried to find this poem in the anthology, but instead he found pages made of copper with portraits in relief upon them. Some of the individuals looked familiar. At that realization, the book and the stranger disappeared.

The dream continued. Descartes doubted that what he saw was real, so he began to interpret the events. He viewed the dictionary as representing the sciences and the *Corpus Poetarum* as the union of philosophy with wisdom. The question "What road in life shall I follow?" he viewed as wise advice. Descartes then awoke, and he continued his interpretation. He believed that the *Corpus Poetarum* actually referred to revelation and inspiration, and the "it is and is not" represents truth and error. Descartes viewed this third dream as guidance for his future, while the first two dreams were a rebuke for his earlier life. The pain in his side was from the devil, who was striving to prevent him from his pursuits, and Descartes was unable to enter the chapel because it was the devil who wanted him to go inside, so God prevented him. The thunder in the second dream was both a punishment for sin and a sign that the spirit of truth would soon descend upon him (Gaukroger 1995, 106–108).

Confirmed by his dreams, Descartes realized that he had reached a pivotal moment in his life and that it was time to leave the hot room.

> Having thus assured myself of these maxims, and having set them on one side along with the truths of religion which have always taken the first place in my creed, I judged that as far as the rest of my opinions were concerned, I could safely undertake to rid myself of them. And inasmuch as I hoped to be able to reach my end more successfully in converse with man than in living longer shut up in the warm room where these reflections had come to me, I hardly awaited the end of winter before I once more set myself to travel. (Descartes 1985, *Discourse on Method* III, 274)

And he traveled for nine years. During those journeys he observed his surroundings, and as each impression and supposed fact impressed upon him, he tested them. Is this clearly the case? Is there any doubt to its truth? If there was any doubt, then he would set that notion to the side. He was careful to note that his employment of doubt is not like that of the skeptics. They doubt for doubt's sake. His doubt is intended to remove all error and eventually, little by little, lead to certainty (Descartes 1985, *Discourse on Method*).

At the end of his travels, he returned to Paris and renewed some old acquaintances. He saw Mersenne and Mydorge as well as the mathematician Gérard Desargues. Unlike his early years in Paris, Descartes now had some renown. Word had spread during his travels about his research, and his friends wanted him to publish. This, combined with the bustle of city life, left little time for solitude and quiet. To find some tranquility, Descartes decided to move to Holland in 1629. He would eventually live in various residences near Amsterdam. He was able to go about his meditations in peace, yet enjoy all of the benefits of a busy city in anonymity. He was also able to visit Beeckman (Vrooman 1970, 71, 75, 81).

Relying on the experiences of his travels, Descartes (1985, *Discourse on Method* IV, 276–277) now decided to put his Method to use. Following the plan of his four rules, he first rids his mind of all of which he has the smallest doubt. Since our senses can deceive us, he rejected all thoughts that were based on what he can see, feel, smell, hear, and taste. Since one can make an error in logic, Descartes rejected all the statements that he had previously accepted by demonstration. Finally, since he can imagine that there are thoughts that can come to us in dreams that are similar to thoughts that come to us while awake, Descartes rejected all images that were in his memory and decided to treat them as illusions.

> But immediately afterwards I noticed that whilst I thus wished to think all things false, it was absolutely essential that the "I" who thought this should be somewhat, and remarking that this truth "I think, therefore I am" was so certain and so assured that all the most extravagant suppositions brought forward by the sceptics were incapable of shaking it, I came to the conclusion that I could receive it without scruple as the first principle of the Philosophy for which I was seeking. (Descartes 1985, *Discourse on Method* IV, 275)

Now that Descartes had identified a proposition that was true, he wanted to know what was it about the sentence that caused it to be true. It seemed to him that there was nothing particularly special about it. There was no distinctive characteristic that could be found simply by examining it, except, of course, for the fact that he was thinking it. For Descartes this was the key. He concluded that

the things that we conceive very clearly and distinctly are all true.

However, as he noted, it is difficult to determine if one is conceiving clearly and distinctly. Simply consider the work and the years that led to this one realization.

He continued to ponder and quickly realized that there was another truth of which he was certain. It was that he did not possess perfect knowledge. This was very obvious! He had just finished rejecting almost everything that he thought he knew. Because of his doubts, he now accepted only two propositions to be true, but upon reflection he found another. We are certainly in a better position when we know that a given proposition is true as opposed to when we are ignorant regarding its truth value. Descartes put it this way:

it is a greater perfection to know than to doubt.

Here Descartes uses the term **perfection** to refer to a characteristic that has the capability of being improved or brought closer to a highest level. If the characteristic reaches that highest level, it is said to have been made **perfect**. For example, take the set of propositions known by an individual. As the person learns, the set expands. Over a lifetime we should expect the set to continue to grow, but we also know that at no point will the set ever encompass all true propositions.

Descartes did not believe that perfect knowledge existed in the world that he observed. When he examined all of the beings with which he had come into contact, none of them seemed to have perfect knowledge. It is certainly the case that the animals he knew did not possess such knowledge. As for the people, they appeared to have varying degrees of knowledge, but since he had no reason to believe that they were any different from him, Descartes concluded that their knowledge must also be incomplete. Based on what little he knew and based on what he perceived to be the case, he had no reason to believe that perfect knowledge existed.

The problem is that he did have such knowledge. Descartes believed that he had a clear and distinct idea of perfect knowledge. If this did not originate in his observations, from whence did it come? Descartes had no reason to believe that it was simply a creation of his own mind. It was not the result of a deduction. Remember that Descartes is applying his Method, and as a result of this he currently has only a few sentences on which to rely. Together they do not imply the existence of perfect knowledge. His belief cannot be the result of his imagination because how could what is perfect originate from what is not? If the source is neither the world nor his intellect, Descartes concluded that it must rest in a nature that is beyond his direct experience. Because his belief is with regard to perfect knowledge, Descartes thought that this nature must epitomize the quality that he is considering. Stated another way,

the concept of a perfection must proceed from some nature that is truly perfect.

This nature he called God, the source of Descartes' conception of perfect knowledge and of all that is perfect.

Since his belief that God exists is grounded differently from beliefs based on direct experience, Descartes searched for an analogy, and he found it in geometry. The subject consists of some fundamental properties, such as continuity, infinite space, divisibility into parts, and movement of figures. These are all assumed to be true by the geometers. However, when one studies their proofs, one is drawn to the certainty of their validity and that the theorems must be true, but nowhere does one syllogistically prove the existence of any of the geometric objects discussed. The triangle is an example. Descartes clearly and distinctly understood that the sum of its internal angles adds to two right angles. However, when he examines the world, he sees no such perfect triangles, yet he believes they exist. The existence of God is like this. It comes from an intellectual pursuit, but it is not the result of observation, and it is not the result of a string of deductions. As the concept of a triangle includes the property concerning its angle sum, so does the concept of a perfect being include the property of existence. Descartes believed that we should be at least as confident of God's being as we are of any geometric proof.

Descartes wrote that those who say that there is no God never raise their minds above their senses. They say that only what that they can see is possible. Since they cannot see God, He cannot exist. Even the scholastics unknowingly think this way when they teach that there is nothing in the intellect that does not come from the senses. Furthermore, if anyone has doubts concerning God, Descartes argued that everything else, including such objects as their body, the stars, and the earth, must also be doubted. It is true that for all practical purposes we may trust that these things exist, but when one analyzes them logically, there is no reason to believe that what appears to be reality is nothing more than a dream. The doubt that Descartes has wanted to remove cannot be eliminated without the acknowledgment of God's existence. At this point Descartes believed that he had found a method to certainty. He can now trust his experience and rebuild the knowledge that he had initially jettisoned at the start of this endeavor.

8.3 ANALYTIC GEOMETRY

In June 1630, Descartes enrolled as a student of mathematics at the University of Leiden. The then chair of the mathematics department was a man named Jacobus Golius. He was a professor of oriental languages, and this no doubt influenced his mathematical interests. He was particularly fond of Apollonius and his work on conic sections. In 1631 he had been drawn to a locus problem that had puzzled the ancient Greeks. Pappus (1986, *Mathematical Collection* VII Introduction, 120) described it as a problem known back to the time of Aristaeus and Euclid. The problem requires certain results on conic sections that Euclid did not have, so the complete solution eluded him. Pappus further claimed that Euclid limited what he wrote on the matter out of respect for the work on conics by Aristaeus. He did not want to upstage such a fine mathematician. Apollonius was another matter. Although he came close to giving a complete description of the locus, Pappus does not have kind words for him because he did not show respect for those who came before him. He wrote that Apollonius "boasts of having augmented instead of acknowledging his indebtedness to the first to have written on it." Pappus then gives his solution to this problem. There are two versions of it. In one the locus is described by three lines. The other has four. The three-line version reads like this:

> If three lines are fixed in position, and from a point is drawn three lines intersecting the fixed lines in given angles, and the ratio of the rectangle contained by two of the lines equals the square of the third, the point will lie on a conic section.

The problem is known as the **Three and Four Line Locus Problem** or simply the **Pappus Problem**. The three-line version is illustrated in Figure 8.4. There are three fixed lines, $\overline{AB}$, $\overline{AD}$, and $\overline{EF}$. The point C describes a locus that is a conic section

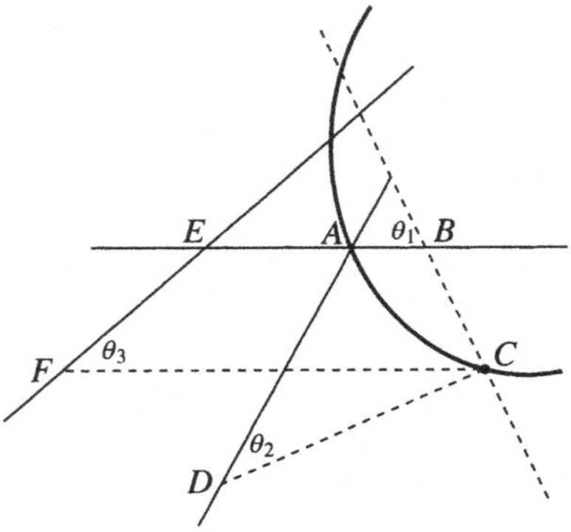

Figure 8.4 The three-line Pappus Problem

as $\overline{CB}$, $\overline{CD}$, and $\overline{CF}$ are drawn in different positions but at constant angles θ_1, θ_2, and θ_3 with the three given lines and with the property that

$$CD \cdot CF = CB^2.$$

In late 1631 Golius sent a copy of the problems to Descartes. It took him between five and six weeks to solve it (Watson 2002, 196; Gaukroger 1995, 210–211; Boyer and Merzbach 1991, 188).

At this point in his life Descartes decided that it was time to publish his research. His book would combine all of his work on optics, weather, and anatomy along with his metaphysical speculations. Its title would be *The World, or a Treatise on Light*. It would be revolutionary. His proof of the existence of God would be the foundation and give his physics certainty. He would prove that the universe was composed simply of matter in motion governed by the laws of physics established by God. In his theory all of the stars and the planets originated as a mechanical product of these laws. God set them in motion and continues to preserve the system, but He is not needed to explain its day-to-day activity. The climax of *The World* would be a description of the human body and how it, too, is the result of the laws of nature. Physically people are no different from animals. What distinguishes a person from an animal is that a person has a soul, but an animal is nothing more than an automaton (Watson 2002, 166–167).

On July 22, 1633, Descartes wrote Mersenne that *The World* was nearly finished and that he would soon send him a copy. His ultimate intention for his work was that it would replace Aristotle's system in the schools of the Church. Descartes had some concerns about this due to the Inquisition. He thought that he had solved his problem by writing *The World* as a "fable." It would be helpful to science because of its explanations of the workings of the universe, but it was still just a fable. On July 23, 1633 everything changed. Descartes learned that the *Dialogue on the Two Chief World Systems* written by Galileo was condemned by the Pope because it taught that the earth revolves around the sun. Galileo had recanted and was sentenced to spend the rest of his life under house arrest. Descartes would later write to Mersenne:

> This has so astonished me that I am quasi-resolved to burn all my papers, or at least to show them to no one. Because I can't imagine that Galileo who is Italian and very well received by the Pope, or so I hear, could have been taken before a criminal court for anything other than that he had without doubt wanted to establish the movement of the earth, which I know was awhile back censored by some Cardinals. But I thought it was accepted even in Rome if one didn't teach it publicly. And I confess that if it is false, so are all the foundations of my philosophy, because it obviously follows from them.

Descartes was devoted to both his Church and his philosophy. He thought that there must be some way out of this predicament (Watson 2002, 166, 168, 170).

The Geometry

His solution was to publish an extremely edited version of *The World.* He would put his metaphysical thoughts, including some autobiographical information, in the *Discourse on Method for Reasoning Well and for Seeking Truth in the Sciences*, and his scientific work would be released as a series of three appendices. The *Meteorology* is a study on atmospheric phenomena, the *Dioptrics* is a work on lenses, including the Law of Refraction, and the *Geometry* is Descartes' new contribution to mathematics and includes his solution to the Pappus Problem (Watson 2002, 182–183, 186).

The *Geometry* of Descartes (1925) is divided into three small books:

- "Problems the Construction of Which Requires Only Straight Lines and Circles"
- "On the Nature of Curved Lines"
- "On the Construction of Solid or Supersolid Problems"

The titles of the first two are essentially self-descriptive of their contents. The third book can best be described as Descartes' theory of equations. In it he constructs solutions to cubics and quartics. Although modern mathematics can be said to be born with Viète, much of the notation and techniques with which we are familiar began with Descartes. His use of x and y to represent indeterminants, the use of exponents, and the habit of moving terms of an equation to one side so that it is set to zero are all due to him (Boyer 1956, 95).

The *Geometry* begins with Descartes plainly stating its fundamental principle:

> Any problem in geometry can easily be reduced to such terms that a knowledge of the lengths of certain straight lines is sufficient for its construction. (Descartes 1925, I, 2)

The term "problem in geometry" refers to the question of the constructibility of a given locus. At first Descartes limited himself to the straightedge and compass, but then he expanded the meaning of construction to include other mechanical devices. To accomplish this, Descartes gave the algebraic operations of addition, subtraction, multiplication, division, and root extraction geometric meaning. We take two numbers that we wish to operate together, and associate them with two line segments

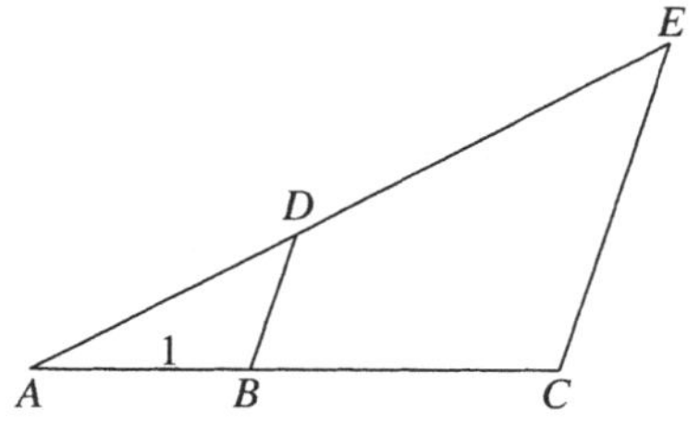

Figure 8.5 $AE = AC \cdot AD$.

with lengths equal to these numbers. We let AC and AD represent these lengths and choose B on $\overline{AC}$ such that $AB = 1$. The constructions are as follows:

- The sum and difference of AC and AD are formed using the standard Greek techniques of concatenation and removal of segments.
- To multiply AC by AD, join $\overline{BD}$ as in Figure 8.5. Construct $\overline{CE}$ parallel to $\overline{BD}$ so that $\overline{AE}$ is an extension of $\overline{AD}$. The length AE is then the desired product. To see this, notice that $\triangle ABD \sim \triangle ACE$. Therefore,

$$\frac{AB}{AC} = \frac{AD}{AE}.$$

  Cross-multiplication will give the result.
- We proceed in a similar manner to divide AC by AD.
- Find the mean proportional between 1 and AD to find the square root of AD. Iterate this method to find higher roots (Exercise 2).

Now that each algebraic operation has a geometric interpretation, we let $a = AC$ and $b = AD$. Descartes then uses some very familiar notation:

$a + b$	a added to b
$a - b$	b subtracted from a
ab	a multiplied by b
a/b	a divided by b
$\sqrt{a}$	extract the square root of a

He also applied higher roots such as $\sqrt[3]{a}$, but his notation was instead $\sqrt{\text{C.}\ a}$ for cube root. Descartes also used the notation a^n to represent a multiplied by itself n times. This enabled Descartes to work with such expressions as a^4, which had no meaning to the ancient Greeks. The phrases *a-squared* and *a-cubed* that label a^2 and a^3, respectively, can still be used, but they have lost the interpretations that the Greeks had placed on them (Kline 1972, 260, 262).

Solving a Quadratic

With the initialization of his method done, Descartes wrote that it is often not necessary to draw lines to find the results of calculations or to solve equations. Instead, use letters as described above and then the method of analysis.

> If, then, we wish to solve any problem, we first suppose the solution already effected, and give names to all the lines that seem needful for its construction,—to those that are unknown as well as to those that are known. Then, making no distinction between known and unknown lines, we must unravel the difficulty in any way that shows most naturally the relations between these lines, until we find it possible to express a single quantity in two ways. This will constitute an equation, since the terms of one of these two expressions are together equal to the terms of the other. (Descartes 1925, I, 6)

Frans van Schooten was a Dutch mathematician and student of Descartes. In 1649 he published an edition of the *Geometry* entitled the *Geometry by René Descartes*. It included supplementary material that provided many of the details that Descartes had left out of the *Geometry*. One such example is the following clarification of Descartes' remarks (Boyer and Merzbach 1991, 371; Descartes 1925, 9, n. 12). Take $\overline{AB}$ containing a point C (Figure 8.6). We want to extend $\overline{AB}$ to D so that the rectangle with sides $\overline{AD}$ and $\overline{BD}$ is equal to the square on $\overline{CD}$. This means that D is collinear with the other points so that

$$AD \cdot BD = CD^2. \tag{8.1}$$

Now follow Descartes' advice. Since both AC and CB are given, let $a = AC$ and $b = CB$ and let the unknown x represent BD. Then

$$AD = a + b + x$$

and

$$CD = b + x.$$

Therefore, by Equation 8.1,

$$ax + bx + x^2 = b^2 + 2bx + x^2,$$

from which follows

$$x = \frac{b^2}{a - b}.$$

Descartes used a similar method to solve the quadratic equation

$$x^2 = ax + b^2. \tag{8.2}$$

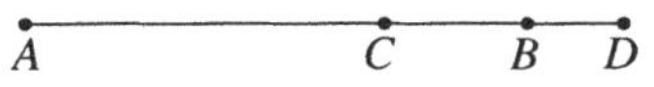

Figure 8.6 Van Schooten's example.

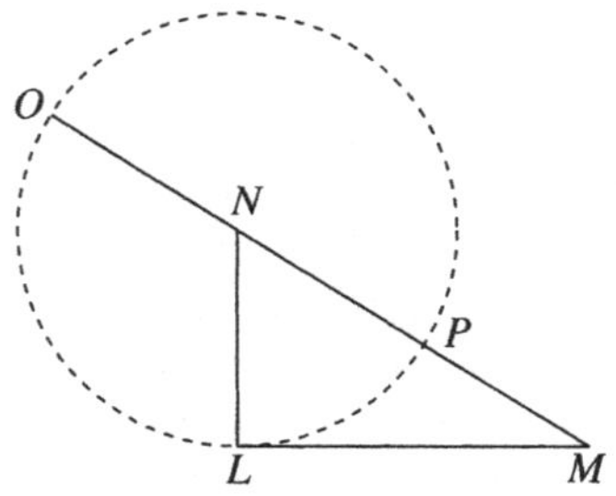

Figure 8.7 Descartes solving a quadratic equation.

To find x, construct right $\triangle LMN$ with a right angle at L, $LM = b$, and $NL = a/2$ (Figure 8.7). Since both a and b are given from the equation, LM and NL can be drawn. Construct $\odot N$ with radius $\overline{LN}$ and extend $\overline{MN}$ to $\overline{MO}$ with O on the circle. By the Tangent-Secant Theorem (5.5.5),

$$LM^2 = OM \cdot PM.$$

So we now identify each of the segments. Let $x = OM$ and notice that $a = OP$, which means that $PM = x - a$. We then see that

$$b^2 = x(x - a) = x^2 - ax,$$

which is equivalent to Equation 8.2, and we have

$$x = ON + MN = \frac{a}{2} + \sqrt{\frac{a^2}{4} + b^2} \tag{8.3}$$

as the positive solution to the equation.

The Four-Line Pappus Problem

Descartes now had the needed machinery to attack the Pappus Problem. We will give his solution to the four-line version (Pappus 1986, *Mathematical Collection* VII Introduction, 120).

> If four lines are fixed in position, and from a point is drawn four lines intersecting the fixed lines in given angles, and the ratio of the rectangle contained by two of the lines equals the rectangle contained by the other two, the point will lie on a conic section.

Let $\overline{AB}$, $\overline{AD}$, $\overline{EF}$, and $\overline{GH}$ be given line segments fixed in position in a plane (Figure 8.8). Let θ_1, θ_2, θ_3, and θ_4 be four chosen angle measures. Extend the segments so that they all intersect the $\overline{AB}$ produced in the points labeled in the diagram. Take C to be the intersection of $\overline{CB}$, $\overline{CD}$, $\overline{CF}$, and $\overline{CH}$ such that

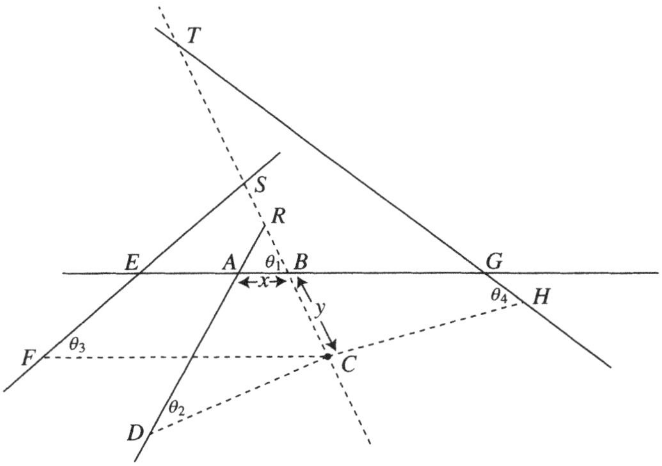

Figure 8.8 The Pappus Problem for four lines.

$$\text{m}\measuredangle ABR = \theta_1, \text{m}\measuredangle CDA = \theta_2, \text{m}\measuredangle EFC = \theta_3, \text{m}\measuredangle CHG = \theta_4,$$

and

$$CB \cdot CF = CD \cdot CH. \tag{8.4}$$

To solve the Pappus Problem, Descartes must show that C traces out a conic section.

Descartes began an analysis of the problem and explained the strategy to find necessary conditions on C:

> First, I suppose the thing done, and since so many lines are confusing, I may simplify matters by considering one of the given lines and one of those to be drawn (as, for example, AB and BC) as the principal lines, to which I shall try to refer all the others. Call the segment of the line AB between A and B, x, and call BC, y. (Descartes 1925, I, 29)

Descartes has set up a pair of axes, although they are not necessarily perpendicular. The point C is located by moving a distance of x from the origin A along $\overline{AB}$ and then y along the line that makes an angle θ_1 with $\overline{AB}$. Essentially, the x is the abscissa and the y is the ordinate of Apollonius (page 248),

Now to find the needed values. We will assume that our situation is as in Figure 8.8. There are other cases that will need to be handled separately (see Exercise 9). Since $\overline{DA}$ meets $\overline{AB}$ and $\overline{BC}$ meets $\overline{AB}$ in given angles, the angles of $\triangle ARB$ are constant regardless of the position of the four lines. Therefore, the ratio RB/AB is constant. Rewrite this using the equation $RB/AB = b/z$. Hence,

$$RB = \frac{b \cdot AB}{z} = \frac{bx}{z}.$$

Since B lies between C and R in the figure,

$$CR = BC + RB = y + \frac{bx}{z}.$$

We also know the angles of $\triangle DRC$, so that CD/CR is determined. Write $CD/CR = c/z$ for some value c but with the same denominator as in RB/AB. We then have

$$CD = \frac{c \cdot CR}{z} = \frac{1}{z}\left(cy + \frac{bcx}{z}\right) = \frac{czy + bcx}{z^2}. \tag{8.5}$$

Since the location of $\overline{AB}$ and $\overline{EF}$ is known at the outset, the value $k = EA$ is known. In our diagram we see that

$$EB = EA + AB = k + x.$$

Since the angles of $\triangle ESB$ are determined, SB/EB is also determined. Find d such that $SB/EB = d/z$. Therefore,

$$SB = \frac{d \cdot EB}{z} = \frac{dk + dx}{z}.$$

Because $CS = CB + SB$,

$$CS = \frac{zy + dk + dx}{z}.$$

We also know CF/CS because the angles of $\triangle FSC$ are determined, so write $CF/CS = e/z$, for some e. Hence,

$$CF = \frac{e \cdot CS}{z} = \frac{ezy + dek + dex}{z^2}. \tag{8.6}$$

Since AG is given, set $l = AG$. Then $BG = l - x$. Because we know that the angles of $\triangle BGT$ will not change, find f so that $BT/BG = f/z$. We may then conclude that

$$BT = \frac{f \cdot BG}{z} = \frac{fl - fx}{z},$$

and then

$$CT = BC + BT = \frac{zy + fl - fx}{z}.$$

Choose g so that we can write $CH/CT = g/z$, and we finally arrive at

$$CH = \frac{gzy + fgl - fgx}{z^2}. \tag{8.7}$$

We now substitute y and Equations 8.6, 8.5, and 8.7 into Equation 8.4:

$$y\left(\frac{ezy + dek + dex}{z^2}\right) = \left(\frac{czy + bcx}{z^2}\right)\left(\frac{gzy + fgl - fgx}{z^2}\right).$$

This yields

$$z^2y(ezy + dek + dex) = cgz^2y^2 + cfglzy - cfgzxy + bcgzxy + bcfglx - bcfgx^2,$$

which leads to the equation for y^2 that is displayed by Descartes,

$$y^2 = \frac{(cfglz - dekz^2)y - (dez^2 + cfgz - bcgz)xy + bcfglx - bcfgx^2}{ez^3 - cgz^2}.$$

However, we want an equation for y. To do this, we will appeal to Descartes' method for solving certain quadratics (page 283). First, simplify the last equation by setting

$$2m = \frac{cfglz - dekz^2}{ez^3 - cgz^2}$$

and

$$\frac{2n}{z} = \frac{dez^2 + cfgz - bcgz}{ez^3 - cgz^2}.$$

From this we conclude that

$$y^2 = 2my - \frac{2n}{z}xy + \frac{bcfglx - bcfgx^2}{ez^3 - cgz^2}.$$

This quadratic in y has

$$y = m - \frac{nx}{z} + \sqrt{\frac{1}{4}\left(2m - \frac{2n}{z}x\right)^2 + \frac{bcfglx - bcfgx^2}{ez^3 - cgz^2}}$$

as a solution by Equation 8.3. This simplifies to

$$y = m - \frac{nx}{z} + \sqrt{m^2 - \frac{2mnx}{z} + \frac{n^2x^2}{z^2} + \frac{bcfglx - bcfgx^2}{ez^3 - cgz^2}}.$$

So that this is easier to analyze, we combine like terms under the radical, remembering that all letters except x represent given constants. Let

$$o = -\frac{2mn}{z} + \frac{bcfgl}{ez^3 - cgz^2}$$

and for some p let

$$\frac{p}{m} = \frac{n^2}{z^2} - \frac{bcfg}{ez^3 - cgz^2}.$$

Therefore, Descartes wrote

$$y = m - \frac{nx}{z} + \sqrt{m^2 + ox + \frac{p}{m}x^2}. \tag{8.8}$$

The Loci Are Conic Sections

Now that Descartes had an expression for the value of y, he laid out the sequence of steps needed to construct it. In so doing he was able to explain why the resulting locus was a conic section. Begin by finding I such that $\overline{AI}$ is parallel to $\overline{BC}$ and $AI = m$. Next choose K on $\overline{BC}$ so that $\overline{IK}$ is both parallel and congruent to $\overline{AB}$ (Figure 8.9). This means that $IK = x$ and

$$BK = m.$$

Take L on $\overline{BC}$ such that

$$LK = nx/z.$$

The exact placement of L and K in Figure 8.9 depends upon the signs of m and nx/z. We also see that $\overline{IL}$ is a fixed line because m is a constant and LK is directly proportional to x. Now extend $\overline{BK}$ to C so that

$$LC = \sqrt{m^2 + ox + \frac{p}{m}x^2}. \tag{8.9}$$

As $\overline{BC}$ moves along at a fixed angle with respect to $\overline{AB}$, we see that $\overline{LC}$ moves along at a fixed angle with respect to $\overline{IL}$. Because $BC = y$, the point C describes the curve that is the solution to the Pappus Problem. Descartes recognized it and drew the following conclusions.

- If $LC = 0$ or the radicand of LC is a perfect square, the locus is a line.
- If $p/m = 0$, the locus is a parabola. This is because

  $$LC^2 = m^2 + ox$$

  is a shifted version of Equation 7.9. To see how this works, consider the locus of points described by

  $$y = 2 - x + \sqrt{4 + x}.$$

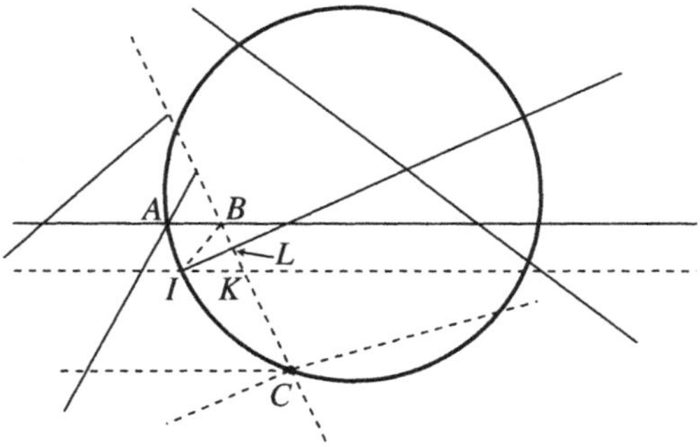

Figure 8.9 The construction of the solution to the Pappus Problem.

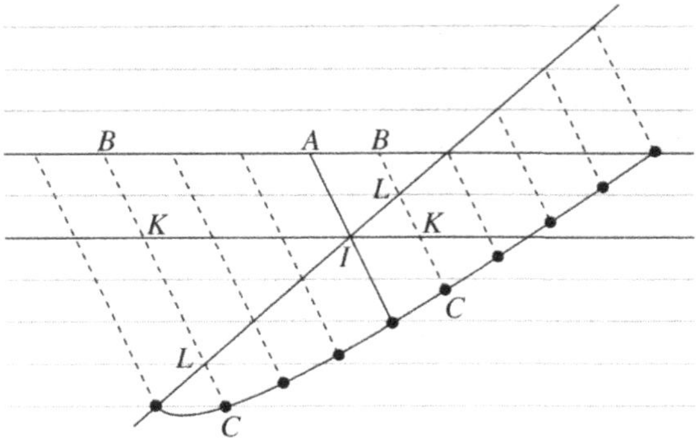

Figure 8.10 The graph of $y = 2 - x + \sqrt{4 + x}$ using oblique coordinates.

That is, $m = 2$, $n = z = o = 1$, and $p = 0$ in Equation 8.8. This means that $BK = 2$ and $LK = x$ (Figure 8.10). Notice that in this oblique coordinate system, $\overline{AB}$ serves as the x-axis and $\overline{AI}$ as the y-axis. Let us now check some points. When $AB = x = 1$, B is one unit to the right of A and we see that $BC = 1 + \sqrt{5}$ and $LC = \sqrt{5}$. When B is three units to the left of A, which means that $x = -3$, we have $BC = 6$ and $LC = 1$. Lastly, we should note that $\overline{IL}$ is the diameter of the parabola.

- If $p/m > 0$, the locus is a hyperbola because Equation 8.9 can be put into the form of Equation 7.13.

- If $p/m < 0$, the locus is an ellipse or a circle because Equation 8.9 can be put into the form of Equation 7.20.

Together with Fermat's work in his *Introduction to Plane and Solid Loci*, the method of Descartes' proof of the Pappus Problem is arguably the most significant revolution in the history of geometry since the Greeks' blending of logic and earth measure. In so doing, Descartes and Fermat made two fundamental contributions. First, they used algebra to represent geometry, following the manner of Viète. In fact, since Descartes claimed to have read Viète only after the publication of his *Geometry*, many accused him of plagiarism, while others viewed the connection as a mere accident—that the time was simply right for the discovery (Watson 2002, 196–197; Mahoney 1994, 32). However, their representations went a step further. While Viète limited himself to the study of equations with one indeterminant, Descartes and Fermat used two. The two indeterminants could then be used to describe an entire curve. Thus, loci no longer need geometric descriptions. They could now be identified with an equation.

Their second contribution is their interpretation of the x and y variables. Descartes and Fermat viewed them as describing distances from an origin. The x played the role of the abscissa, and the y was the ordinate. In Fermat's case, a horizontal axis was set up first, and then the locus was drawn. Descartes, on the other hand, created the locus first, and then attached the axis where it was needed. Nonetheless, in both cases the equation can be used to trace out the locus on paper so that it can be clearly seen. Although it has its roots in the work of the ancients, we can consider the work of Descartes and Fermat as the beginning of **analytic geometry**.

Exercises

1. Let $a = 5$ and $b = 3$ be lengths of two line segments. Construct segments of length $a + b$, $a - b$, ab, and a/b.

2. To construct the square root of a, start with B on $\overline{AC}$ such that $AB = a$ and $BC = 1$. Draw a semicircle with diameter $\overline{AC}$ and erect a perpendicular at B intersecting the semicircle at D.
   (a) Show that $BD = \sqrt{a}$.
   (b) Construct $\sqrt{1 + \sqrt{2}}$ and $\sqrt[4]{2}$.

3. Use Descartes' method to construct solutions to:
   (a) $x^2 = ax - b^2$.
   (b) $x^2 = b^2 - ax$.
   (c) $ax^2 + bx + c = 0$.

4. Show that the locus described by $ax + by + c = 0$ is a line and the locus described by $x^2 + y^2 + dx + ey + f = 0$ is a circle.

5. Use the equations of Exercise 4 to find the points of intersection of:
   (a) two lines.
   (b) a line and a circle.
   (c) two circles.

6. Show that the locus with ordinate as described in Equation 8.9
   (a) is a line if $LC = 0$ or the radicand of LC is a perfect square.
   (b) is a hyperbola if $p/m > 0$.
   (c) is an ellipse or a circle if $p/m < 0$.

7. In Figure 8.8, $\pi - \theta_1$ represents the angle that $\overline{AB}$ makes with $\overline{BC}$. Assume that $\theta_1 = \pi/2$ and graph the following as in Figure 8.10.
   (a) $y = 4 - 3x$.
   (b) $y = 3 + \sqrt{9 + x + 2x^2}$.
   (c) $y = 1 + x + \sqrt{1 - x^2}$.

8. Repeat Exercise 7 for $\theta_1 = \pi/4$, $\theta_1 = \pi/3$, and $\theta_1 = 2\pi/3$.

9. Solve the four-line Pappus Problem (page 284) for these special cases.
   (a) The given lines are parallel to $\overline{AB}$.
   (b) The given lines are parallel to $\overline{BC}$.
   (c) $\overline{AB}$ is perpendicular to $\overline{AC}$.

10. Use Descartes' method to solve the three-line Pappus Problem (page 279).

11. This is a special case of the five-line Pappus Problem. The locus is known as the **trident of Descartes**. Let $\overline{AA'}$, $\overline{BB'}$, $\overline{CC'}$, and $\overline{DD'}$ be perpendicular to $\overline{AD}$. Assume that these four lines are equally spaced at k units apart. For any point P, let x represent the distance from P to $\overline{AA'}$ and let y be the distance from P to $\overline{BD}$, making these two lines the coordinate axes. Let b be the distance from P to $\overline{BB'}$, c the distance from P to $\overline{CC'}$, and d the distance from P to $\overline{DD'}$. Show that the locus of all points P satisfying $bcd = kxy$ is

$$(k + x)(k - x)(2k - x) = kxy$$

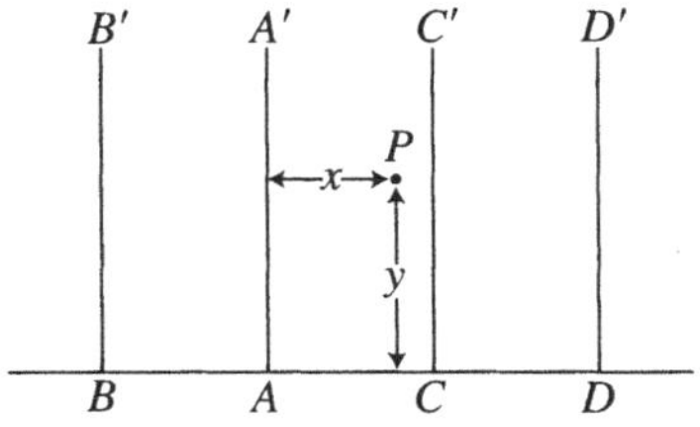

CHAPTER 9

GÉRARD DESARGUES

The group of mathematicians who met regularly under the leadership of Marin Mersenne must be considered one of the most significant gatherings of intellectuals in history. Their work revolutionized mathematics by combining algebra with geometry to found the new field of analytic geometry. However, there was one member of Mersenne's group who was not as interested as his colleagues in these new methods. He still preferred the old synthetic approaches, yet the way in which he would use them would be so out of the mainstream that it would take over 150 years before his methods would be accepted. This nonconformist's name was Gérard Desargues.

Desargues was born in Lyon, France on February 21, 1591. He was an architect, military engineer, and self-taught applied mathematician, who was not concerned with theory for theory's sake. He had studied Apollonius, though. He understood very well the workings of conic sections and wanted to use his knowledge to help those who shared his interests by giving them good mathematical tools with which to work. His research led him to many interesting theorems that he would distribute to his friends or publicize at free lectures that he would lead in Paris (Kline 1972, 288; Boyer and Merzbach 1991, 359).

His landmark work that resulted from his research he titled *Brouillon project d'une atteinte aux événemens des rencontres du cône avec un plan*, which translates as

Proposed Draft of an Attempt to Treat the Results of a Cone Intersecting a Plane. He completed it in 1639. Seeking to present his results to nonmathematicians, Desargues invented his own terminology for standard geometric ideas. He intended to make his results more readable by removing the technical jargon that he believed was a hinderance to the layman. He borrowed expressions from botany and used words such as *palm*, *trunk*, and *tree* to name different types of lines. Unfortunately for Desargues, the new terminology simply confused matters. The craftsmen for whom he intended the work could not follow it in the least, and except for his good friends Mersenne, Descartes, Fermat, and Blaise Pascal, the mathematical community thought that he was crazy. Desargues printed only 50 copies of the *Brouillon Project*, all of which would be quickly lost. Angered at the reception of his work, Desargues left Paris and would live for the rest of his life in Lyon, where he died in the September of 1661 (Kline 1972, 288–289).

Desargues was not without some influence in his day. The French mathematician Philippe de Lahire studied his work and eventually combined Desargues' techniques with those of Descartes. As with Desargues, Lahire's interests included conic sections. He wrote a book titled *Sectiones Conicae* in which he took many of the results of Apollonius and added to them results based on the *Brouillon Project*. He was also one of the first to study surfaces analytically by examining equations in three unknowns. When most wanted no association with the work of Desargues, Lahire believed it to be important. His commitment led him to create his own copy of the *Brouillon Project* by hand (Boyer and Merzbach 1991, 361, 368–369).

9.1 PROJECTIONS

If we were asked to identify the major scholarly activity of the Renaissance, we would probably first choose the development of art. In the Middle Ages artists focused on depicting religious subjects and were not concerned with realism. Instead, their art represented abstract, spiritual ideas and contained much symbolism. Influenced by the newly rediscovered thinking of the ancient Greeks, the artists of the Renaissance desired to study nature and represent it accurately. In particular, they wanted their paintings to project the viewpoint of a person looking out a window onto a scene, and they wanted their representation to resemble that scene as close as possible. This meant that they needed to study geometry so that they could discover the laws of space. The artists of the Renaissance were the applied mathematicians of the time (Kline 1953, 126–127).

This study began in the fifteenth century with Leone Battista Alberti of Genoa. He was a scholar, writer, architect, sculptor, philosopher, cryptographer, and artist who studied the classics of Rome and Greece at the reputable Gasparino Barzizza in Padua and law at the University of Bologna. Sometime after 1430 he moved to Florence. The city had been the home of his family before political defeats forced them into exile. While in Florence, Alberti became an expert on ancient art and architecture. He was instrumental in the designs of many of the great buildings of his time, including Santa Maria Novella and the Rucellai Palace in Florence, the

Malatesta Temple in Rimini, and Sant'Andrea in Mantua. His interests in ancient architecture coincided perfectly with those of the Medici family who were the new ruling class of Florence. Previously, the wealthy inhabitants of the city shied away from public displays of wealth. Led by the Medicis, the upper class now flaunted their wealth and power. This benefited Alberti and gained him fame and influence because his interests in the glories of ancient times fit perfectly with the new attitudes (Grafton 2000, 6–9).

Alberti's contribution to art is what interests us here. Linked to his interest in architecture, Alberti was the first to write a formal introduction on perspective in his 1435 treatise entitled *On Painting*. The term **perspective** refers to the formal science of sight. It was primarily the invention of 1420's Florentine artists, and its goal was to represent a three-dimensional scene on a two-dimensional canvass. This meant that at the core of Alberti's work was geometry, but this was not the abstract geometry of a mathematician but the applied science of a painter. To him, points and lines did not exist in some ideal world but were found on the canvass (Grafton 2000, 120–124; Boyer and Merzbach 1991, 296).

The system of perspective championed by Alberti is known as a **focused system**. He started by choosing the location of the eye of a person observing a scene. Alberti then imagined rays of light streaming from the eye to the scene, forming a **visual pyramid**, its base being some geometric shape that surrounded the scene. We have a geometric term for this.

■ **DEFINITION 9.1.1**

The set of lines joining a single fixed point with each of the points on a given plane figure is called a **projection**.

Alberti then wanted to paint the scene found in the base of the pyramid. To do this he imagined that a plane sliced the projection at some angle. The intersection of the projection with the plane is an image that represented the scene viewed (Grafton 2000, 123; Kline 1953, 132–133). We also have a geometric term for this.

■ **DEFINITION 9.1.2**

A **section** is the locus of points determined by the intersection of a projection and a plane.

All of this is represented in Figure 9.1. The projection is the visual pyramid, which has its vertex at the eye E. The base of the pyramid is $\mathscr{A}$ lying in the plane Λ. The plane K cuts the projection, and the result is section $\mathscr{B}$.

The image found in the section is not a perfect representation of the scene. Not only is it two-dimensional, but the images of the objects in it do not possess their natural shapes. To illustrate this, turn on a flashlight and point it toward the floor. Under normal conditions a cone of light forms with its vertex at the bulb and continues until it hits the floor where it stops abruptly. We can imagine that the floor is a plane that cuts the cone of light. If the light is held at an angle as in Figure 9.2(a), the spot is an ellipse because the plane of the floor intersects the cone at an angle that is not

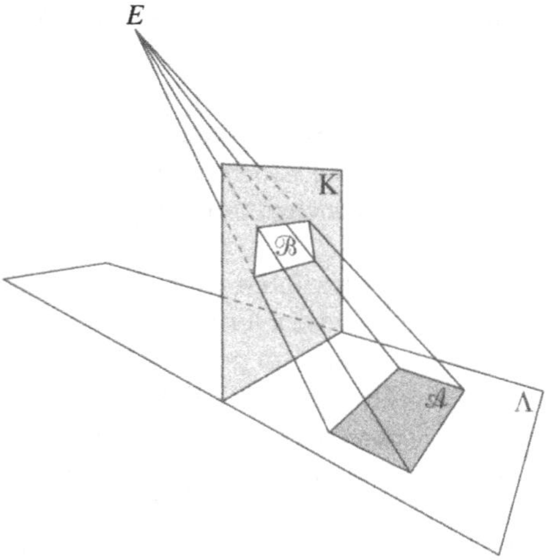

Figure 9.1 The polygon $\mathcal{A}$ and its section $\mathcal{B}$ cutting a visual pyramid.

right. If the flashlight is held so that the axis of the cone of light is perpendicular to the floor, the result is a circle, as illustrated in Figure 9.2(b). We can obtain the other conic sections if we simply hold the flashlight in the correct manner. Think of the cone of light as Alberti's visual pyramid, the spot on the floor as the section, and the flashlight bulb as the eye. The cone of light remains circular, but its image on the floor changes depending on how the light is pointed. Similarly, the image of a scene changes depending on from where we view it. The image in the section represents not the objects but how we view the objects.

There are other circumstances that warrant study. For instance, take the projection and slice it with another plane to form a second section (Figure 9.3). What properties does it share with the first section? Since both projections share the same figure, we

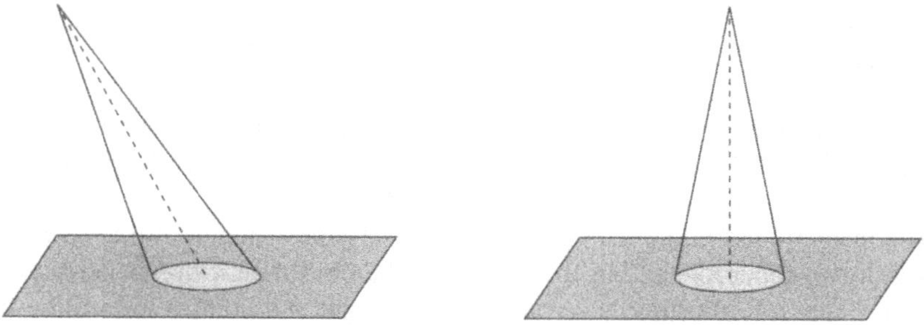

(a) The spot is an ellipse (b) The spot is a circle

Figure 9.2 Shining a flashlight on the floor, or looking at a circle.

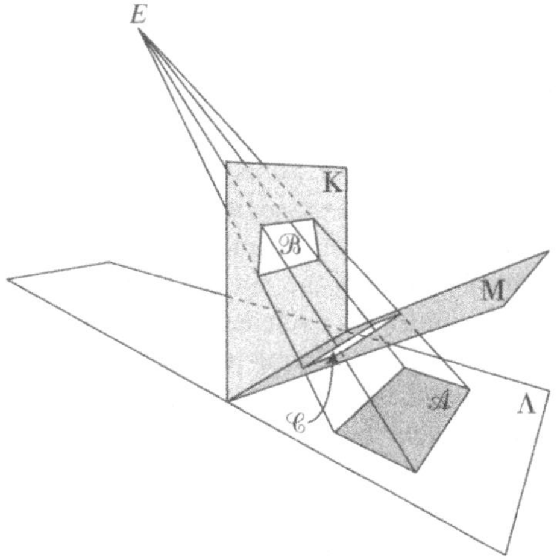

Figure 9.3 The polygon $\mathscr{A}$ and its sections $\mathscr{B}$ and $\mathscr{C}$ with one projection.

would expect the sections to have something in common. Now take the same figure and look at it from two vantage points (Figure 9.4). The result is a second projection. Cut the new projection with a plane and look at the new section, and ask the same questions as before. The field of geometry that seeks the answers to these questions we call **projective geometry**. Although traces can be found in the writings of the Greeks, the subject can be said to have originated with Desargues.

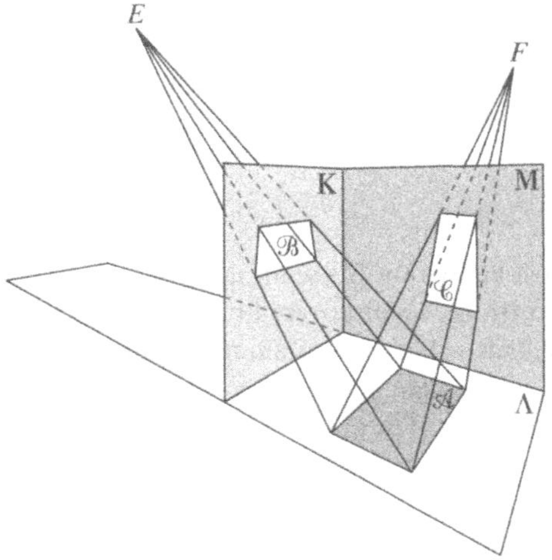

Figure 9.4 The polygon $\mathscr{A}$ and its sections $\mathscr{B}$ and $\mathscr{C}$ with two projections.

Alberti's work on perspective was one of the highest achievements of the Renaissance, but he was not alone. Besides being one of the top architects of the Italian Renaissance, Filippo Brunelleschi is considered to have founded the science of painting. By 1425 he had developed a system of perspective that he taught to Donatello and Masaccio, among others. In his treatise *De Prospettiva Pingendi*, the painter Piero della Francesca took Alberti's work on perspective and modified it into a fuller system. He taught that the physical world could be brought into order by mathematics and represented faithfully through the laws of perspective and solid geometry. Leonardo da Vinci viewed painting as a science that had as its goal a realistic representation of its subject matter. Every painting required a careful preparation in which the geometry and perspective of the scene are carefully mapped. Albrecht Dürer, a German draftsman and painter, shared da Vinci's views on the technical aspects of his art. Dürer firmly believed that it was not enough to make a rough sketch of the perspective of a picture, but instead the artist must have a sound grounding in the workings of geometry. Dürer was so convinced of this that he studied geometry himself and wrote a text on the basics of Euclid's system. Of all the Renaissance artists who studied perspective, Dürer was the most influential (Kline 1953, 132–133; Boyer and Merzbach 1991, 294; Wölfflin 1971, 285).

9.2 POINTS AT INFINITY

Artists use perspective to represent three dimensions on a two-dimensional surface. To do this they draw a line that represents the horizon, and on that line they put a point known as a **vanishing point**. All lines that are to be considered perpendicular to the plane of the canvass intersect at such a point. Since the vanishing point is a key element of representing three dimensions, any geometry that is to model perspective drawing must take the vanishing point into consideration. Desargues knew this and incorporated it into his projective geometry. The method that he chose to represent vanishing points was not only anticipated by the great artists of the Renaissance but also by a particular German nonconformist.

Kepler

Johannes Kepler is best known for his work in astronomy. Using the precise astronomical data of Tycho Brahe, he discovered his three laws of planetary motion, the first being that the planets orbit the sun along elliptical paths with the sun at one focus. Observe that Kepler's theory presupposes a Copernican solar system with the sun, not the earth, at its center. The authorities at the time also noted this assumption of his theory, but fortunately for Kepler, his situation did not become as serious as that of Galileo.

If Kepler had not discovered his three laws, he would still have become a significant figure because of his work in mathematics. An example of Kepler's contributions is his *Introduction to Vitello's Optics* of 1604. Not only a study of optics and parabolic mirrors, the *Introduction* examined all of the conic sections. Ignoring the traditional

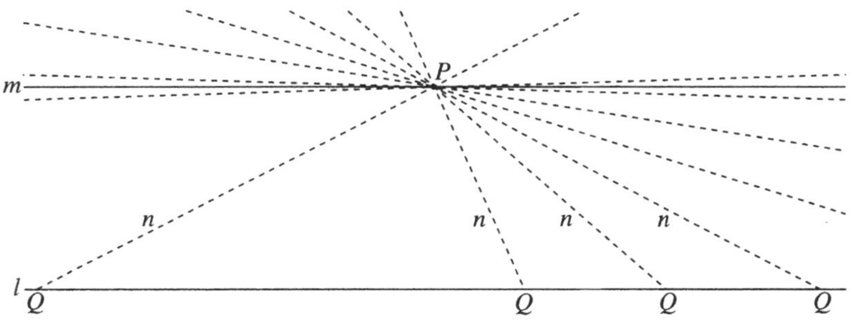

Figure 9.5 Kepler motivates the point at infinity.

classification of parabola, hyperbola, and ellipse, he instead viewed the conic sections as different manifestations of a single curve. What the curve looked like all depended on the location of two points. Each of these Kepler called a **focus**, which is Latin for *hearth*—an appropriate term when one considers what happens at the focus of a parabolic mirror when light from the sun hits it.

Crucial to his description of the conic sections is his **point at infinity**. Take a line l and a point P not on l (Figure 9.5). Let m represent the unique line through P that is parallel to l. Then any other line n through P will intersect l at some Q. As n moves out to the right, its intersection Q with l moves with it. The farther n moves, the closer it comes to m, and the more distant Q is from its original location. Kepler imagined that as n turns so that it coincides with m, the point Q has traveled the extent of l and is now at infinity. However, the line n continues to rotate counterclockwise. As it does, the point of intersection that n makes with l now appears on the left side of l. The motion continues until n and Q return to their original positions. Of course, the point at infinity can be reached by clockwise rotation of n as well. This gives the line the behavior of a circle, with the point at infinity serving as the transition between opposite sides. Indeed, Kepler regarded a line as nothing more than a circle with center at infinity (Kline 1972, 290).

The concept of the point at infinity is crucial to Kepler's understanding of the conic sections. His method starts by fixing one of the foci. In Figure 9.6 a focus is located at the origin in each of the nine graphs. The other focus will be labeled F. At the start F is also at the origin, and the graph is a pair of lines intersecting at the two concurrent foci. Now send the focus F to infinity in the positive direction. As F moves out to the right, the two lines pull apart, creating a hyperbola. F continues to move, and the shape of the hyperbola continues to change. Next imagine that F has reached infinity. The resulting curve is a parabola, a conic section with one of its foci at infinity. Now that the point F is at infinity, it will return to the plane, but this time it comes from the left and forms an ellipse. As F approaches the origin, the ellipse becomes more circular, until eventually it is a circle when F is back at its original position. This process of describing the conic sections is known as Kepler's **principle of continuity** (Boyer and Merzbach 1991, 323).

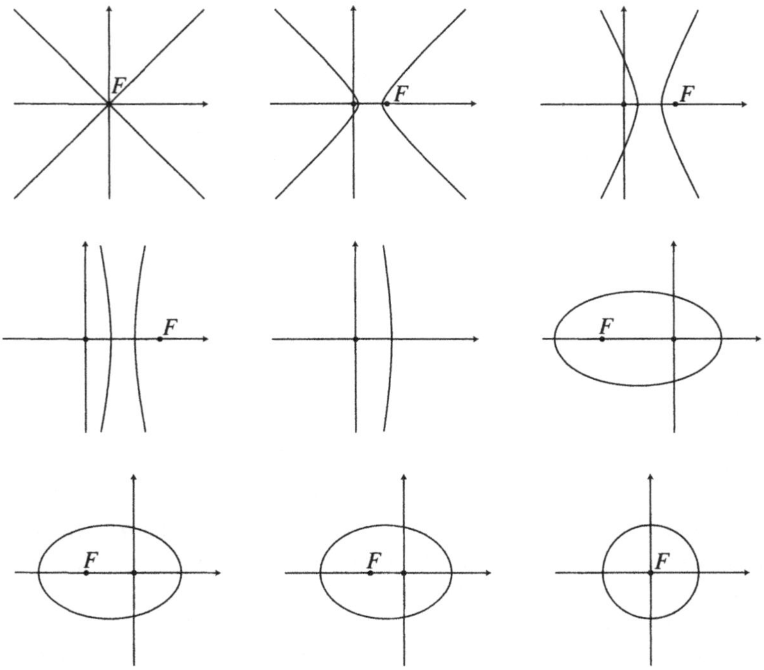

Figure 9.6 Kepler's principle of continuity.

Kepler was one of those rare people who was creative enough to look past the beliefs of his contemporaries and strike out in new paths. His early years certainly showed no signs of this. He was born in Weil in southern Germany in 1571. His parents were poor and not the best role models. His father was an alcoholic who earned a living as a mercenary and then as a barkeep. His mother was unlearned, violent, and so interested in the black arts that she was jailed for being a witch. While he was a young boy, his father required Kepler to help at the bar, and he was eventually taken out of school to work in the fields. To make matters worse, Kepler caught smallpox at the age of 4. The disease damaged his eyesight and crippled his hands (Kline 1972, 242; Burton 1985, 338).

Despite all of this, he matured into an intelligent young man. In 1588 he earned his bachelor's degree at the College of Maulbronn. He then attended the University of Tübingen. He intended to study to become a Lutheran minister, but one day he learned of a new theory of the world that excited him. It was the heliocentric theory of Copernicus. He wanted to prepare a paper and give a presentation on it, but the professor in charge of student presentations refused to give him permission. Disillusioned, Kepler left Tübingen and dropped his plans for the ministry. Instead, he took a position teaching mathematics in Austria. A scientific career became his choice for his life's work (Kline 1972, 243; Burton 1985, 338).

He taught in the city of Grätz. It was under Roman Catholic control, and he was Protestant. For many years life was good, and Kepler had a steady income. He did not have any students because he was not a very good mathematics teacher, but he was well respected. To keep him employed, he was given the task of of publishing the annual almanac. As a mathematician he was expected to know astrology and include his predictions for the upcoming year. Kepler often accurately predicted future events, so he gained a reputation as a fine astrologer. Although Kepler would have preferred not to have been involved with it, astrology did provide a steady income when mathematics and science were not always reliable (Kline 1972, 243; Burton 1985, 338).

His move to Austria would prove to be fateful. In 1598 all of the Protestant clergy and teachers were expelled from Grätz, which came without warning. Because of Kepler's reputation, a month later the church leadership of the city said they would allow him to return, but Kepler was not interested. Another opportunity had arisen. Tycho Brahe had recently learned of Kepler from a work that he had published. He needed an assistant, and Kepler was available. The two worked together in Denmark and then later in Prague. After Brahe died in 1601, Kepler succeeded him as the official mathematician and astrologer for Emperor Rudolph II (Boyer and Merzbach 1991, 243; Kline 1972, 339).

The work that had interested Brahe was titled *Mysterium Cosmographicum*. In it Kepler expresses a belief that the universe was designed based on a preexisting mathematical pattern. Kepler expected this design to be simple and rational. He believed that this gave another reason to accept the planetary system of Copernicus over the complexity of the system of Ptolemy. Seeking to incorporate mathematics further into the divine structure of the solar system, Kepler devised a model based on the five Platonic solids. He imagined that each of the then-known six planets orbited the sun on spheres that were inscribed and circumscribed about a cube, a tetrahedron, a dodecahedron, an icosahedron, and an octahedron. The cube was the outermost solid of the model, and the others were nested one inside the other until the innermost octahedron was reached. Despite being overjoyed with his system, Kepler eventually realized that it did not match his observations, so he abandoned it. He had a sharp imagination which worked well for him, but when data suggested that he had an incorrect theory, Kepler would reject his previous thinking and move forward in new directions (Kline 1972, 243–244; Burton 1985, 338–339).

Projective Planes

Although projective geometry will not have this limitation, we concentrate on the two-dimensional case. This means that we need to find the analog to a projection and section when we limit ourselves to the plane, and we need to determine how to represent points at infinity. We start with the projection.

■ **DEFINITION 9.2.1**

The set of all lines in a plane that pass through a fixed point is called a **pencil**.

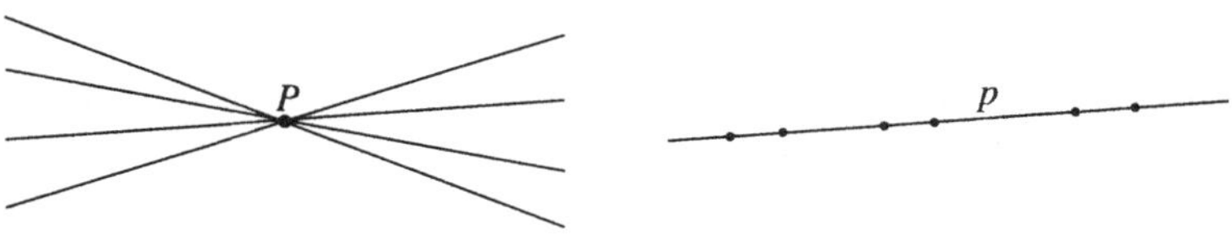

(a) The pencil of lines through P (b) The range of points on p

Figure 9.7 The fundamental objects of projective geometry.

At first encounter this term seems odd because we are used to the word *pencil* as a name for a writing implement. When we check a dictionary, we find that the word is from the Latin meaning "brush," so the term fits [Figure 9.7(a)]. The next definition is the one-dimensional version of a section [Figure 9.7(b)].

■ **DEFINITION 9.2.2**

The set of all points on a line is called a **range**.

Therefore, in the plane a pencil of lines through a point is a projection, and a range is the section formed when a line intersects the lines of a pencil.

Both the pencil and the range are objects in a Euclidean plane, but the plane has no point at infinity. To include this new point in the geometry, we need to describe carefully what objects are to be added and the rules that these objects obey. Let Λ denote a plane from Euclid's geometry. The plane contains certain types of geometric **objects** or **elements**. These are the points and lines that lie in Λ. The postulates and theorems listed in the *Elements* describe what we know about these objects. To distinguish them from the new objects that will be added to the plane, we call any element that lies in Λ an **ordinary** element. If P is a point in Λ or l is a line in this plane, then P is an ordinary point and l is an ordinary line.

Now choose a point O in Λ. For each line l in the pencil through O, add a point Ω_l to the plane. Each of these points will be a "point at infinity." Such a point is called an **ideal point** to distinguish it from an ordinary point of Λ. Since between any two points in Λ there exists a unique line, we want our new points to have the same property. Therefore, we add a "line at infinity" to Λ. It is labeled as ω and called the **ideal line**. All of the ideal points lie on this line.

Our plane is now complete, and we are ready to name it.

■ **DEFINITION 9.2.3**

The plane obtained when the ideal line and its ideal points are joined to a Euclidean plane is called a **projective plane**.

When we imagine the projective plane, we want the ideal points to be the vanishing points. That is, all lines that are parallel in the plane and vanish in the distance meet at the same point on the horizon. This is illustrated by railroad tracks. When we stand looking down at the tracks, we see equidistant lines. When we look toward the horizon, we see that the lines that originally looked parallel now head off and meet in a point. To guarantee that our ideal points have the same property, we set down the following assumption.

■ POSTULATE 9.2.4

For every ordinary line m, the ideal point Ω_l is on m if and only if either m is parallel to l or m and l are the same line.

We have now described a new geometry. We assume that it satisfies the first four of Euclid's postulates (5.1.1). The Parallel Postulate does not hold because in this geometry all of the lines intersect in either an ordinary point or an ideal point. However, we still need to deal with the situation when two ordinary lines are parallel and thus intersect in an ideal point. Since we will not be dealing with angle measure in this new geometry, we replace the Parallel Postulate with one that is a modified version of a statement by John Playfair.

■ POSTULATE 9.2.5

For every ordinary line l and every ordinary point P not on l, there exists a unique ordinary line m on P such that l and m intersect in an ideal point.

Because of the postulates, we use the phrases *l is parallel to m* and *l intersects m at an ideal point* interchangeably.

We will now prove two basic results.

■ THEOREM 9.2.6

An ordinary line contains exactly one ideal point.

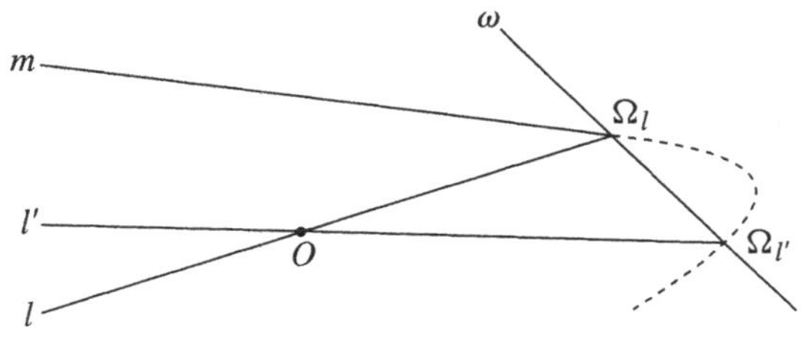

PROOF

Let m be an ordinary line. If m is on O, there exists only one ideal point on m by construction of the projective plane. Suppose then that m is not on O. By Postulate 9.2.5, there exists a line l on O such that m and l are parallel. We conclude by Postulate 9.2.4 that Ω_l is on m. To prove that this is the only ideal point on m, suppose that there exists another line l' through O so that $l \neq l'$ and $\Omega_{l'}$ is on m. This means that m is also parallel to l'. This contradicts Postulate 9.2.5 because l and l' intersect at O. ■

■ **THEOREM 9.2.7**

There is exactly one line common to any two distinct points.

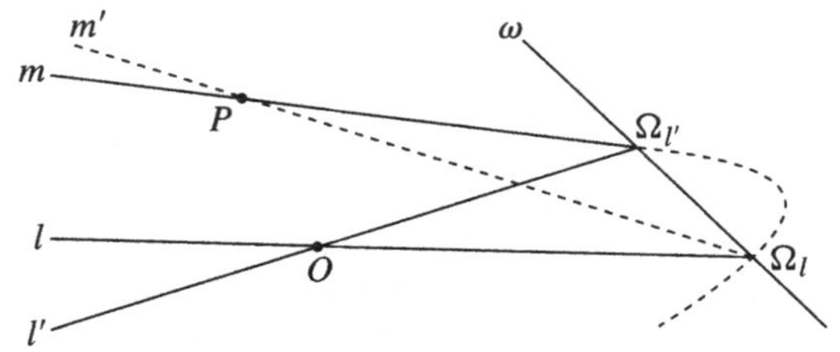

PROOF

We already know that this holds for any two ordinary points, so we have two cases to check. First, let Ω_l and $\Omega_{l'}$ be two distinct ideal points, with l and l' being different lines on O. By construction ω contains these two ideal points. If there exists an ordinary line m such that Ω_l and $\Omega_{l'}$ are on m, this would contradict Theorem 9.2.6.

Second, continue to let Ω_l represent an ideal point and let P be an ordinary point. Postulate 9.2.5 guarantees a line m through P that is parallel to l. Therefore, Ω_l is on m. If there is a second line m' incident with both P and Ω_l, then we have a contradiction of Postulate 9.2.5. ■

We leave the proof of the next result to Exercise 5.

■ **THEOREM 9.2.8**

Two distinct lines intersect in exactly one point, either ordinary or ideal.

We now have a complete picture of a projective plane. It is illustrated in Figure 9.8. Since the lines l, m, n, and p all intersect at the point O, they intersect the ideal line ω at different ideal points. Since l and r intersect at the ideal point Ω_l, these two ordinary lines are parallel despite how they look in the picture.

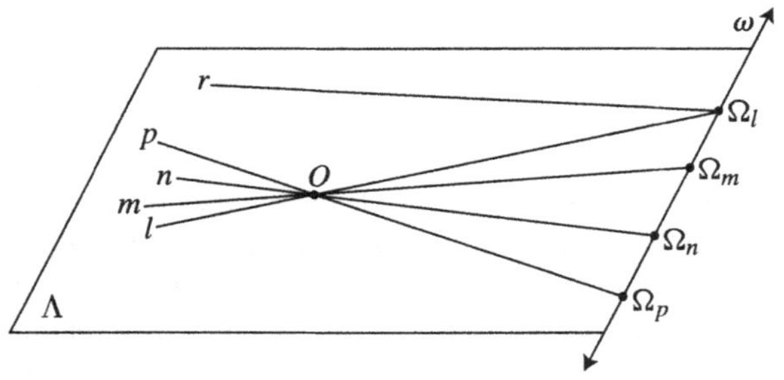

Figure 9.8 A projective plane.

If we need to work in three dimensions, we simply generalize the routine that defined a projective plane. Fix a point O in three-dimensional Euclidean space. For each line l through O, add an ideal point Ω_l, and for each plane Λ through O, add an ideal line ω_Λ. This means that Λ is a projective plane. Finally, we introduce one new object Λ_∞. It is called the **ideal plane**, and it contains all of the ideal lines.

DEFINITION 9.2.9

The space obtained when the ideal points, lines, and plane are joined to Euclidean space is called **projective space**.

All results that are limited to a plane in projective space still can be proven using the postulates for a projective plane. However, our set of assumptions must be expanded with the following postulates when working in three dimensions.

POSTULATES 9.2.10

1. If Λ is an ordinary plane, Ω_l is on Λ if and only if Λ is parallel to l or Λ contains l.

2. Ω_l is on ω_Λ if and only if Λ contains l.

3. If K is an ordinary plane, ω_Λ is on K if and only if K is parallel to Λ or K $= \Lambda$.

4. A point is on Λ_∞ if and only if the point is ideal.

5. A line is on Λ_∞ if and only if the line is ideal.

We also assume that the basic results of Euclidean space hold in projective space, namely that three ordinary, noncollinear points or one ordinary point and one ordinary line not containing the point uniquely determine a plane.

Exercises

1. Draw a parallelogram in a projective plane and a cylinder in projective space.

2. In a plane, let Ω_1 and Ω_2 be distinct ideal points. If Ω_1 is on l_1 and Ω_2 on l_2, can the lines l_1 and l_2 be parallel? Explain.

3. Let Λ be a plane. Let the point O on Λ be used to describe a projective plane as on page 302. Let O' be another point on Λ and use it to describe a projective plane. Show that O and O' generate the same ideal points and the same ideal line.

4. Prove using ideal points: If two distinct lines are parallel to a third line, the lines are parallel. Assume that all three lines lie in the same plane.

5. Prove Theorem 9.2.8.

6. Prove the following about projective space. Use Postulates 9.2.10.
   (a) Two distinct points lie on exactly one line.
   (b) Two distinct lines intersect in exactly one point.

(c) Two distinct planes intersect in exactly one line.
(d) Three noncollinear points lie on exactly one plane.
(e) Two distinct, coplanar lines intersect in exactly one point.
(f) Three distinct planes that do not intersect in a line meet in exactly one point.
(g) Any two distinct lines lie on exactly one plane.
(h) A plane and a line not in the plane intersect in exactly one point.
(i) A line and a point not on the line lie in exactly one plane.

7. Prove that if the ideal point Ω is on the plane Λ and on the line l, either l is on Λ or they have no ordinary points in common.

8. Let A and B lie in the plane Λ. Prove that the line joining A and B lies in Λ.

9.3 THEOREMS OF DESARGUES AND MENELAUS

Probably the best known theorem due to Desargues does not appear in the *Brouillon Project*. The result was instead originally published by an engraver and friend of Desargues by the name of Abraham Bosse in 1648. It is known as Desargues' Theorem and can be considered a result in Euclid's geometry. However, because Theorem 9.2.8 removes any concerns about parallel lines, it is easier to prove it in projective geometry. As we shall see, the theorem holds true in both two and three dimensions, yet it is known that the only way to prove Desargues' Theorem is first to show it for two triangles in different planes and then prove it for two coplanar triangles. For this reason, the proof assumes the properties of projective space found in Exercise 6 of Section 9.2 (Fishback 1962, 47; Boyer and Merzbach 1991, 361).

■ THEOREM 9.3.1 [Desargues]

In a pair of triangles, if the lines joining corresponding vertices are concurrent, then the points of intersection of corresponding sides are collinear.

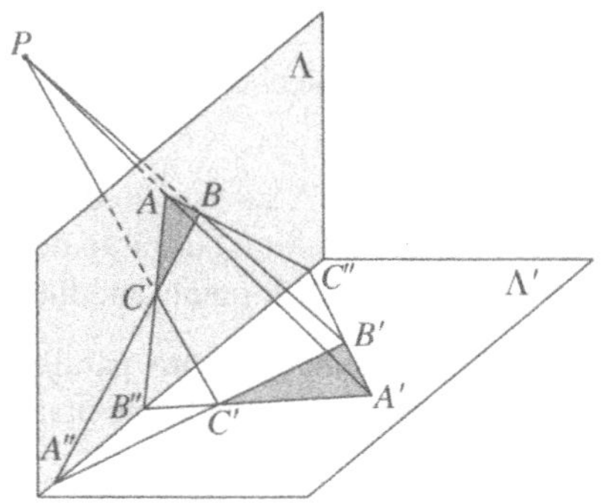

PROOF

We have two cases to consider. First take Λ and Λ' to be two different planes in projective space. Let $\triangle ABC$ lie in plane Λ and $\triangle A'B'C'$ in Λ'. Assume that the $\overline{AA'}$, $\overline{BB'}$, and $\overline{CC'}$ produced intersect at P. Any pair of these three lines determine a plane. Take, for instance, $\overline{AA'}$ and $\overline{CC'}$. They form projective plane $PAA'C'C$. Both $\overline{AC}$ and $\overline{A'C'}$ lie in that plane, so they intersect at a point B'' (Theorem 9.2.8). If the two lines are parallel, B'' is ideal, else it is ordinary. Because $\overline{AC}$ lies in Λ and $\overline{A'C'}$ lies in Λ', their point of intersection B'' lies on the line of intersection of the two planes. Similarly, the other two pairs of sides of the triangles meet at points A'' and C'' on the line of intersection between Λ and Λ'. Thus, the points of intersection of corresponding sides are collinear.

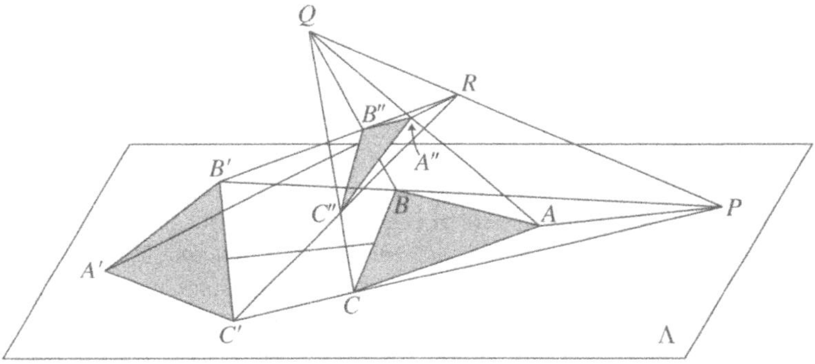

Now let the two triangles lie in the same plane Λ. Assume that the lines joining corresponding vertices meet at a point P in Λ. We will use the two-plane case to prove the result. Pick any point Q not in Λ. Join $\overline{QP}$ and choose R on $\overline{QP}$. Lines $\overline{QA}$ and $\overline{RA'}$ lie in a plane. Therefore, they intersect at a point A'' (Theorem 9.2.8). Similarly, $\overline{QB}$ and $\overline{RB'}$ meet at B'', and $\overline{QC}$ and $\overline{RC'}$ meet at C''. The three points A'', B'', C'' are not collinear. If they were, these three points and Q would all line in a plane that would intersect Λ in a line. This is impossible, because it would mean that the vertices of $\triangle ABC$ would be collinear. Now, $\triangle A''B''C''$ has the property that the lines joining its vertices with the corresponding vertices of $\triangle ABC$ meet at Q and those that meet the corresponding vertices of $\triangle A'B'C'$ intersect at R. Thus, by the two-plane case of the theorem, there are lines l and l' (not pictured) so that the corresponding sides of $\triangle A''B''C''$ and $\triangle ABC$ produced will meet on l, in addition to the corresponding sides of $\triangle A''B''C''$ and $\triangle A'B'C'$ intersecting on l'. As we saw above, it must be the case that these lines are both equal to the line of intersection of the plane $A''B''C''$ and Λ. Since two lines intersect in exactly one point, the point where the $\overline{A''B''}$ and $\overline{AB}$ produced intersect and the point where the $\overline{A''B''}$ and $\overline{A'B'}$ produced intersect must be the same. Thus, $\overline{AB}$ and $\overline{A'B'}$ intersect on l. Similarly, the other pairs of corresponding sides of $\triangle ABC$ and $\triangle A'B'C'$ will meet on l when extended. ■

The converse of Theorem 9.3.1 is also true and due to Desargues. Like the previous theorem, we will also refer to it as Desargues' Theorem. Its proof is left to Exercise 1.

■ **THEOREM 9.3.2 [Desargues]**

In a pair of triangles, if the points of intersection of corresponding sides are collinear, the lines joining corresponding vertices are concurrent.

Menelaus

The Greek astronomer and mathematician named Menelaus was from Alexandria and lived around A.D. 98. He is known for his work in trigonometry. Although many of his works are no longer extant, one that has survived as an Arabic translation is called the *Sphaerica*. It is a treatise in three books. The first book involved studying angles and triangles on the surface of a sphere. Menelaus's goal was to prove results about spherical triangles in the manner of Euclid's study of triangles on a plane. His results included the proof that the sum of the internal angles of a spherical triangle is greater than two right angles and that two triangles on a sphere are congruent if the pairs of corresponding angles are congruent. The second book was on astronomy, and the third book dealt with spherical trigonometry (Kline 1972, 120). The first proposition of the third book is known as **Menelaus's Theorem**. It holds true in both two and three dimensions. The proof of the three-dimensional case can be found in Heath (1921b, 266–268). However, the *Sphaerica* does not include the plane version of the theorem. Heath conjectured that Menelaus had already proven it in one of his lost writings. We give the plane version proof here (Anglin and Lambek 1995, 104-105).

■ **THEOREM 9.3.3 [Menelaus]**

In $\triangle ABC$ let D be on the line through B and C, E on the line through A and C, and F on the line through A and B. When either exactly one or all three of the points D, E, F are not on the triangle, these three points are collinear if and only if $BF \cdot AE \cdot CD = AF \cdot CE \cdot BD$.

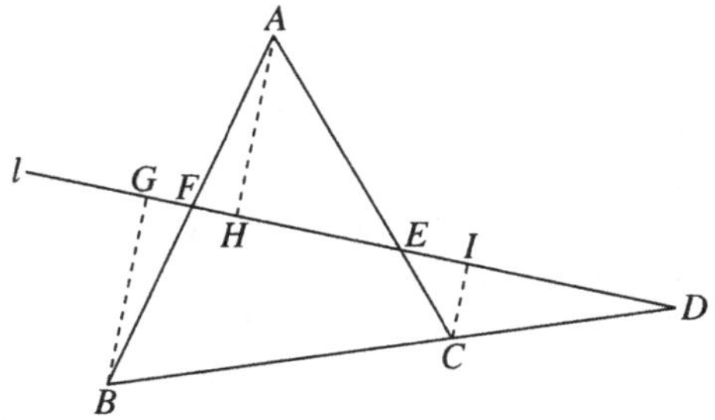

PROOF

We leave the proof for the case when all three points are off the triangle to Exercise 3. For now suppose that F is on $\overline{AB}$, E on $\overline{AC}$, but D on $\overline{BC}$ produced beyond C. First assume that D, E, and F are all on transversal l. If l passes through one of the vertices of $\triangle ABC$, then E and F must be vertices themselves. Hence, $BF = CE = 0$, and the equation holds. Suppose then that l does not include any of the vertices of the triangle. Find points G, H, and I on l such that $\overline{BG}$, $\overline{AH}$, and $\overline{CI}$ are perpendicular to l. We have $\measuredangle GFB \cong \measuredangle HFA$ and $\measuredangle AEH \cong \measuredangle CEI$ because they are pairs of vertical angles, so

$$\triangle BGF \sim \triangle AHF \text{ and } \triangle AHE \sim \triangle CIE.$$

Furthermore,

$$\triangle CID \sim \triangle BGD.$$

Therefore,

$$\frac{BF}{AF} = \frac{BG}{AH}, \frac{AE}{CE} = \frac{AH}{CI}, \text{ and } \frac{CD}{BD} = \frac{CI}{BG}.$$

When we multiply these equations together, we find that

$$\frac{BF}{AF} \cdot \frac{AE}{CE} \cdot \frac{CD}{BD} = \frac{BG}{AH} \cdot \frac{AH}{CI} \cdot \frac{CI}{BG} = 1.$$

Hence,

$$BF \cdot AE \cdot CD = AF \cdot CE \cdot BD.$$

Now suppose that the equation holds and note that we can work the above equations backward and find that $\triangle AHE \sim \triangle CIE$. By the definition of similar triangles, $\measuredangle AEH \cong \measuredangle CEI$, and so $\measuredangle AEH$ and $\measuredangle AEI$ are supplementary. Therefore, D, E, and F are collinear by Theorem 5.3.8. ■

The theorem due to Menelaus had been lost since antiquity until rediscovered in 1678 by Giovanni Ceva. He published it with a theorem that now bears his name. The proof is in the spirit of the proof of Menelaus's Theorem, so we leave it to Exercise 6 (Burton 1985, 436; Martin 1975, 464–468). A line that connects a vertex of a triangle to a nonvertex point on the opposite side is now known as a **cevian line**.

■ **THEOREM 9.3.4 [Ceva]**

In $\triangle ABC$ let D be on the line through B and C, E on the line through A and C, and F on the line through A and B. The cevian lines $\overline{AD}$, $\overline{BE}$, and $\overline{CF}$ are concurrent if and only if

$$AF \cdot BD \cdot CE = FB \cdot DC \cdot EA.$$

We did not include the possibility of the lines being parallel in the statement of Ceva's Theorem because in projective geometry if lines are parallel in a Euclidean sense, they are concurrent at an ideal point.

Another proposition that we can use Menelaus's Theorem to prove is due to the son of Etienne and Antoinette Pascal. His name was Blaise and he was born on June 19, 1623, the third of five children. When he was only 3, his mother died, and although he benefited from the services of a nanny, it was his father who played the central role in his life. In addition to being a prominent lawyer and high-ranking official in the government, Etienne was a reputable mathematician. Schooling his children himself, he soon realized that Blaise was exceptionally bright. It was claimed by family members that at the age of 11 the young Pascal wrote a small paper on the nature of sound, and a year later discovered the Pythagorean Theorem without help from his father, his sisters, or Euclid (Rogers 2003, 5–6; Houston 1989, 23).

At some point in his life Etienne realized that the life of a lawyer was not what he wanted. Instead, he desired to concentrate on research, so he resigned his position and moved his family to Paris. There he met Mersenne and joined his scientific group. He quickly became a leading member, and so would Blaise. Etienne had planned to introduce his son to mathematics around the age of 15 or 16, but Blaise showed such interest that Etienne had no choice but to begin to educate him in the field. The young Pascal soon found himself attending meetings and even contributing to discussions. It was here that Blaise met Desargues. He was working on his *Brouillon Project* and suggested to Pascal that he study conic sections using projective methods. In 1639, the same year that Desargues finished the *Brouillon Project*, Pascal wrote his *Essay on Conics* when he was 16 years old (Rogers 2003, 5–6; Kline 1972, 297). In this small tract he honored his new friend:

> We can also demonstrate a property stated below, due to M. Desargues of Lyons, one of the great geniuses of this time and well versed in mathematics, particularly in conics, whose writing on the subject although few in number give abundant proof of his knowledge to those who seek for information. I should like to say that I owe the little that I have found on this subject to his writings, and that I have tried to imitate his method. . . . (Smith 1959, 329)

The *Essay* was a single-page publication giving some new but fundamental results on conic sections but without proof. In it Pascal included the next result that is sometimes known as the **mystic hexagram theorem**.

■ **THEOREM 9.3.5 [Pascal]**

If a hexagon is inscribed within a circle, the points of intersection of its opposite sides are collinear.

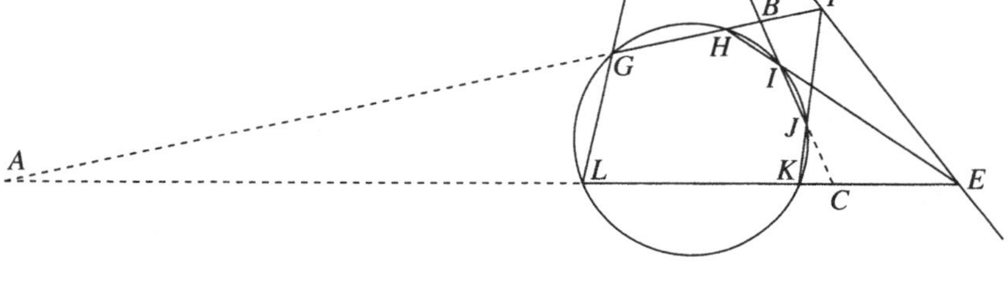

PROOF

Let hexagon $GHIJKL$ be inscribed within a circle. Let D be the intersection of $\overline{LG}$ and $\overline{JI}$ produced, F the intersection of the $\overline{GH}$ and $\overline{KJ}$ produced, and E the intersection of the $\overline{HI}$ and $\overline{LK}$ produced. Also, let A be the intersection of $\overline{LC}$ and $\overline{GB}$ and C be the intersection of $\overline{LE}$ and $\overline{DJ}$. We now apply Menelaus's Theorem three times to $\triangle ABC$:

- In the first case we have D on the line through B and C, G on $\overline{AB}$, and L on $\overline{AC}$ the Since L, G, and D are collinear, we may conclude that
$$LC \cdot AG \cdot BD = AL \cdot GB \cdot CD.$$
- Next we have F on the line through A and B, J on $\overline{BC}$, and K on $\overline{AC}$. Because the points K, J, and F are collinear,
$$AK \cdot JC \cdot BF = KC \cdot BJ \cdot AF.$$
- Finally, E is on the line through A and C, I on $\overline{BC}$, and H on $\overline{AB}$. We see that H, I, and E are collinear, so
$$AH \cdot BI \cdot CE = HB \cdot IC \cdot AE.$$

By the Secant Theorem (5.5.4), we also obtain the following equations:

$$\begin{aligned} AL \cdot AK &= AG \cdot AH, \\ HB \cdot GB &= BI \cdot BJ, \\ JC \cdot IC &= KC \cdot LC. \end{aligned}$$

Therefore, we may multiply the equations that resulted from Menelaus's Theorem, substitute into them using the equations obtained by the Secant Theorem, and cancel what we can from both sides to get

$$BD \cdot AF \cdot CE = CD \cdot BF \cdot AE.$$

When we apply Menelaus's Theorem to this last equation, we find that D, F, and E lie on the same line. ■

Exercises

1. Prove Theorem 9.3.2.

2. Two triangles are **Desarguean** if they satisfy the equivalent conditions of Desargues' Theorem. Show that the following pairs of triangles in a plane are Desarguean.
   (a) Triangles that share a common side.
   (b) Triangles that have parallel corresponding sides.
   (c) Any two triangles that each have two ideal points for vertices.

3. Prove the plane version of Menelaus's Theorem (9.3.3) for the case when all three points D, E, and F are not on the triangle.

4. In $\triangle ABC$, let D be on the line through B and C, E on the line through A and C, and F on the line through A and B. When either exactly one or all three of the points D, E, F are not on the triangle, these three points are collinear if and only if

$$\sin \measuredangle BAD \cdot \sin \measuredangle CBE \cdot \sin \measuredangle ACF = \sin \measuredangle DAC \cdot \sin \measuredangle EBA \cdot \sin \measuredangle FCB.$$

This is the trigonometric version of Menelaus's Theorem.

5. Prove that the interior angle bisectors of two angles of a triangle and the exterior angle bisector of the third angle intersect the opposite sides in three collinear points. (*Hint:* Use the trigonometric version of Menelaus's Theorem.)

6. Prove Ceva's Theorem (9.3.4).

7. Prove that Ceva's Theorem implies Menelaus's Theorem.

8. On quadrilateral $ABCD$, let A' be on $\overline{AB}$, B' on $\overline{BC}$, C' on $\overline{CD}$, and D' on $\overline{D'}$ such that

$$AA' \cdot BB' \cdot CC' \cdot DD' = A'B \cdot B'C \cdot C'D \cdot D'A.$$

Prove that $\overline{A'B'}$ and $\overline{C'D'}$ intersect on $\overline{AC}$ and $\overline{A'D'}$ and $\overline{B'C'}$ intersect on $\overline{BD}$.

9. Prove the following propositions using Ceva's Theorem.
   (a) The medians of a triangle are concurrent.
   (b) The bisectors of the internal angles of a triangle are concurrent.
   (c) The altitudes of a triangle are concurrent.

10. Let a circle contain vertices B and C of $\triangle ABC$ and intersect $\overline{AB}$ at P and $\overline{AC}$ at R. Prove that if $\overline{PR}$ meets $\overline{BC}$ (produced if necessary) at Q, then

$$\frac{QC}{QB} = \frac{RC \cdot AC}{PB \cdot AB}.$$

11. If the six vertices of a hexagon lie alternately on two lines, the three pairs of opposite sides meet in collinear points. This is known as Pappus's Theorem. Prove it and explain how it is a generalization of Pascal's Theorem.

9.4 INVOLUTIONS

In the projective geometry of Desargues, concepts such as distance and angle measure are less important than in Euclidean geometry because these properties are not preserved under projection. However, projections do preserve some properties, and Desargues wrote about one in his *Brouillon Project*. He took some of the ideas of Apollonius and Pappus that he had learned from the *Mathematical Collection* and made the following definition. It is the only one of Desargues' terms that we continue to use today (Kline 1972, 292; Coxeter 2003, 45).

DEFINITION 9.4.1

The three pairs of collinear points A, B; C, D; and E, F are said to be in **involution** if

$$\frac{AC \cdot BC}{AD \cdot BD} = \frac{EC \cdot FC}{ED \cdot FD}.$$

Denote this by $(AB)(CD)(EF)$.

Certainly, if $(AB)(CD)(EF)$, then $(BA)(DC)(FE)$. It should be noted further that three pairs of points being in involution is independent of the order in which the pairs are taken (Exercise 1).

Desargues proved that the involution is a property that is **preserved** under projection. That is to say, whenever we have a section cutting a projection in three pairs of points that form an involution, every section that cuts the projection form an involution. Desargues' proof of this theorem is translated from the *Brouillon Project* by Field and Gray (1987, 50–52, 94) and relies heavily on Menelaus's Theorem (9.3.3).

THEOREM 9.4.2

Involutions are preserved under projection.

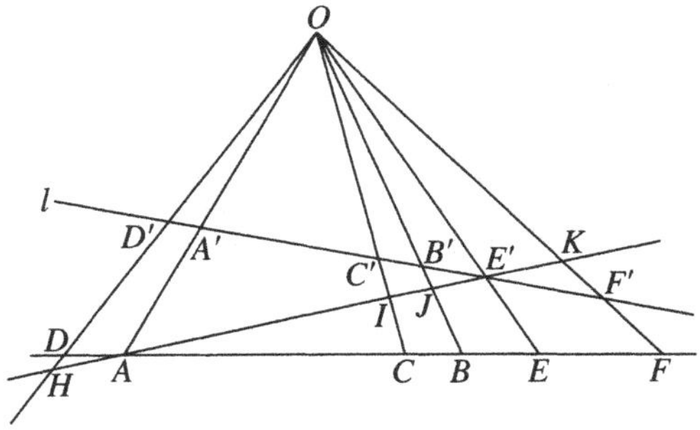

PROOF

Suppose that $(BC)(AE)(DF)$. This means that

$$\frac{AB \cdot AC}{EB \cdot EC} = \frac{AD \cdot AF}{ED \cdot EF}. \tag{9.1}$$

Further assume that A, B, C, D, E, and F are in a range intersecting a pencil through O. Take another line l intersecting the pencil and let A' be the intersection of l with $\overline{OA}$, B' be the intersection of l with $\overline{OB}$, and so on. We must show that A', B', C', D', E', and F' are in involution. To do this, begin by joining $\overline{AE'}$. Let $\overline{AE'}$ intersect $\overline{OD}$ at H, $\overline{OC}$ at I, $\overline{OB}$ at J, and $\overline{OF}$ at K. This will enable us to apply Menelaus's Theorem.

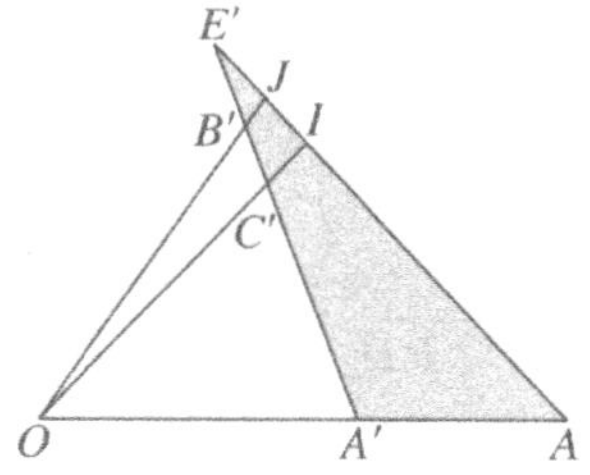

We first examine $\triangle AE'A'$. When we view this triangle with transversal $\overline{JO}$, we conclude by Menelaus's Theorem that

$$AJ \cdot E'B' \cdot A'O = E'J \cdot A'B' \cdot AO.$$

On the same triangle but with transversal $\overline{IO}$, we have

$$AI \cdot E'C' \cdot A'O = E'I \cdot A'C' \cdot AO.$$

Therefore,

$$\frac{A'B'}{E'B'} = \frac{A'O \cdot AJ}{AO \cdot E'J}$$

and

$$\frac{A'C'}{E'C'} = \frac{A'O \cdot AI}{AO \cdot E'I}.$$

Multiplying these last two equations yields

$$\frac{A'B' \cdot A'C'}{E'B' \cdot E'C'} = \frac{A'O^2 \cdot AJ \cdot AI}{AO^2 \cdot E'J \cdot E'I}. \tag{9.2}$$

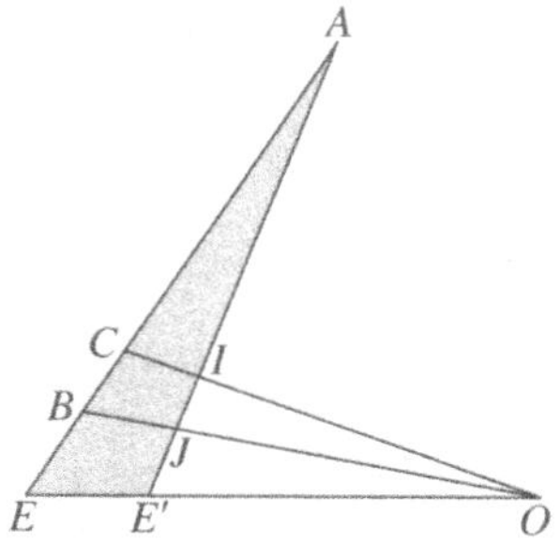

However, appealing to Menelaus's Theorem on $\triangle EAE'$ with transversal $\overline{BO}$ gives

$$EB \cdot AJ \cdot E'O = AB \cdot E'J \cdot EO$$

and with transversal $\overline{CO}$ gives

$$EC \cdot AI \cdot E'O = AC \cdot E'I \cdot EO.$$

From these equations we conclude that

$$\frac{AJ}{E'J} = \frac{AB \cdot EO}{EB \cdot E'O}$$

and

$$\frac{AI}{E'I} = \frac{AC \cdot EO}{EC \cdot E'O}.$$

Substituting into Equation 9.2 yields

$$\frac{A'B' \cdot A'C'}{E'B' \cdot E'C'} = \frac{A'O^2 \cdot EO^2 \cdot AB \cdot AC}{AO^2 \cdot E'O^2 \cdot EB \cdot EC}. \tag{9.3}$$

A similar sequence of proportions would also show that

$$\frac{A'D' \cdot A'F'}{E'D' \cdot E'F'} = \frac{A'O^2 \cdot EO^2 \cdot AD \cdot AF}{AO^2 \cdot E'O^2 \cdot ED \cdot EF}. \tag{9.4}$$

(See Exercise 5.) Hence by Equations 9.1, 9.3, and 9.4,

$$\frac{A'B' \cdot A'C'}{E'B' \cdot E'C'} = \frac{A'D' \cdot A'F'}{E'D' \cdot E'F'}.$$

This means that $(B'C')(A'E')(D'F')$. ■

The ratio found in Desargues' definition of involution defines a particular set of points that are collinear. Unlike Euclid's geometry, the projective geometry of Desargues is not concerned with lengths when it comes to describing its objects of study. Although distance did play a role in proofs, like the last one, it served only as a tool. The exact values of the distances are irrelevant.

Euclid defined a polygon as a collection of line segments joined at their endpoints in a particular order. When we view the same situation as a collection of points and lines instead of points and line segments, we have what is called a **configuration**. An example is the following, which was studied initially by Pappus (Boyer and Merzbach 1991, 360).

■ **DEFINITION 9.4.3**

Let A, B, C, and D be four points in a plane, no three of which are collinear. The configuration that consists of these four points and the six lines determined by all possible pairs of the points is called the **complete quadrangle** $ABCD$.

The points of the complete quadrangle $ABCD$ are called its **vertices**, and the lines are its **sides** (Figure 9.9). The six sides of this complete quadrangle are $\overline{AB}$, $\overline{BC}$, $\overline{CD}$, $\overline{AD}$, $\overline{AC}$, and $\overline{BD}$. Two sides are said to be **opposite** if they do not share a common vertex. There are three pairs of opposite sides. They intersect in three points called **diagonal points**. Join these points to form the **diagonal triangle** of $ABCD$. The diagonal triangle of the complete quadrangle in Figure 9.9 is $\triangle EFG$. Notice that if E and G are ideal points, then $\overline{AB}$ is parallel to $\overline{CD}$, and $\overline{AD}$ is parallel to $\overline{BC}$. In this case the complete quadrangle is a parallelogram.

Now to relate the complete quadrangle to the involution. We do it by cutting the complete quadrangle with a line and study the resulting section of points.

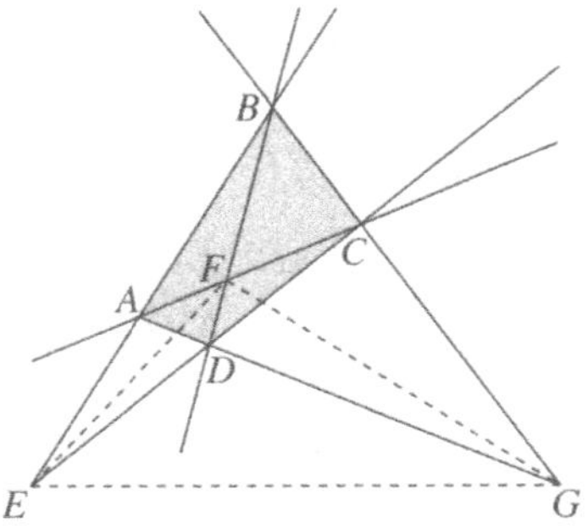

Figure 9.9 The complete quadrangle $ABCD$ with diagonal triangle EFG.

■ **DEFINITION 9.4.4**

A **quadrangular set** is the range of a complete quadrangle formed by a line that does not lie on a vertex of the complete quadrangle.

Following Veblen and Young (1910, 49), we notice that in the complete quadrangle of Figure 9.10 three of the sides meet in a single vertex and the other three form a triangle. We say that the first three form a **point triple of lines** and the latter three form a **triangle triple of lines**. Let l intersect the sides of the complete quadrangle to form the quadrangular set A, B, C, D, E, and F. The points A, B, and C that lie on a point triple of lines are called a **point triple of points**, and the points D, E, and F that lie on a triangle triple of lines are called a **triangle triple of points**. We denote this quadrangular set by

$$\mathsf{Q}(ABC,\ DEF),$$

where the first set of three points is the point triple of points and the second set is the triangle triple of points.

The quadrangular set will also be preserved by projection. Desargues proved this by first showing that a quadrangular set forms an involution. This is the first result of a two-part theorem from the *Brouillon Project* (Field and Gray 1987, 54–55, 107).

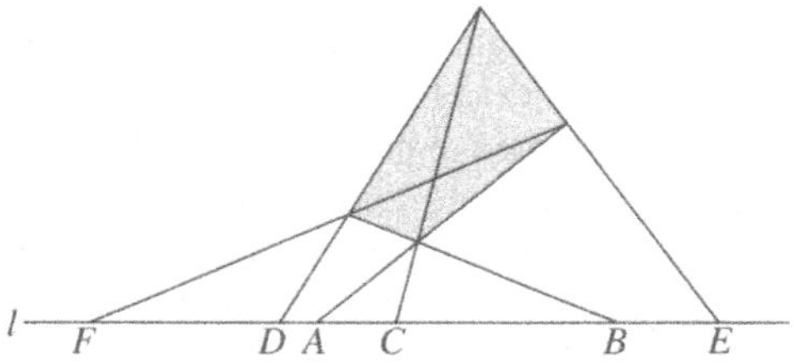

Figure 9.10 A quadrangular set on l.

THEOREM 9.4.5 [Desargues' Involution Theorem]

If Q$(ABC,\ DEF)$, then $(AD)(BE)(CF)$.

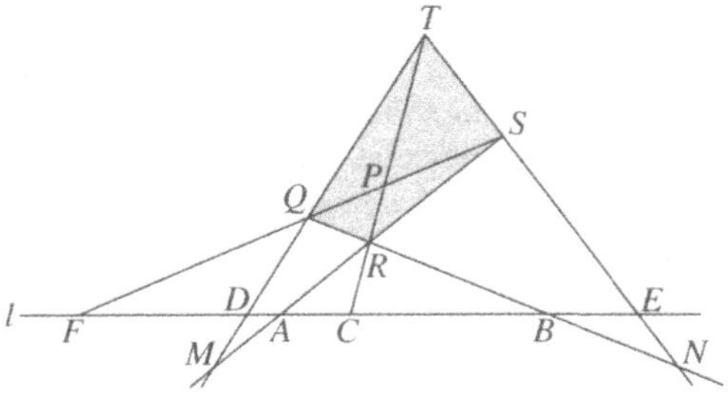

PROOF

Let $QRST$ be a complete quadrangle such that $\overline{QT}$ and $\overline{RS}$ meet at M, $\overline{QS}$ and $\overline{RT}$ at P, and $\overline{QR}$ and $\overline{TS}$ at N. Let l be a line that does not contain Q, R, S or T but does intersect the six sides of the complete quadrangular at the points F, D, A, C, B, and E as in the diagram. We see that Q$(ABC,\ DEF)$. We will now show that these three points are in involution. Consider $\triangle MDA$. Menelaus's Theorem with transversal $\overline{QRB}$ gives

$$DQ \cdot MR \cdot AB = MQ \cdot AR \cdot DB. \tag{9.5}$$

Using transversal $\overline{TSE}$, we have

$$DT \cdot MS \cdot AE = MT \cdot AS \cdot DE. \tag{9.6}$$

Finally, when we use transversals $\overline{FQS}$ and $\overline{CRT}$, we find that

$$AS \cdot MQ \cdot DF = MS \cdot DQ \cdot AF \tag{9.7}$$

and

$$DT \cdot MR \cdot AC = MT \cdot AR \cdot DC. \tag{9.8}$$

Therefore,

$$\frac{AB \cdot AE}{DB \cdot DE} = \frac{MQ \cdot AR}{DQ \cdot MR} \cdot \frac{MT \cdot AS}{DT \cdot MS} = \frac{AF \cdot AC}{DF \cdot DC}, \tag{9.9}$$

where the first equality follows from Equations 9.5 and 9.6 while the second follows from Equations 9.7 and 9.8. We then conclude that

$$(BE)(AD)(CF),$$

from which follows

$$(AD)(BE)(CF)$$

by Exercise 1. ■

Because the converse of Theorem 9.4.5 also holds (Coxeter 2003, 46), we have the following.

COROLLARY 9.4.6

Quadrangular sets are preserved under projection.

The second part of the Involution Theorem is the beginning of Desargues' investigation into the projective properties of conic sections. Due to the discussion on page 295, Desargues realized that in a projective plane there really is only one kind of conic. This is because any property that is preserved by projection will be shared among all conic sections. Therefore, if we can show that a property holds in a circle and that the property is preserved under projection, the property holds for all conics. Desargues used this strategy in the next theorem (Field and Gray 1987, 53; Coxeter 2003, 72).

THEOREM 9.4.7 [Desargues' Involution Theorem]

If a complete quadrangle is circumscribed by a conic section, any line not passing through a vertex of the complete quadrangle intersects it and the conic in six points in involution.

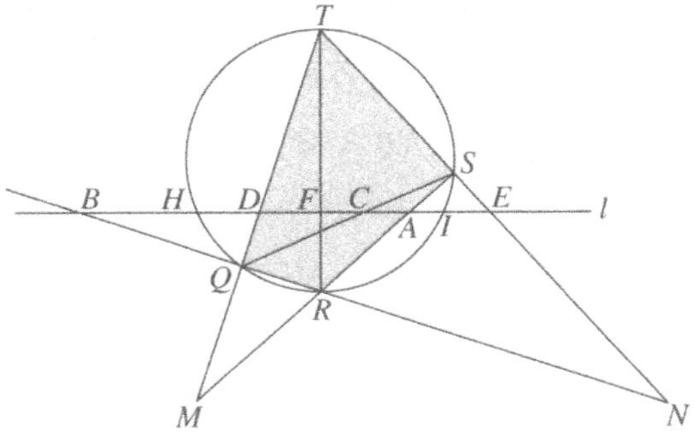

PROOF

Let complete quadrangle $QRST$ be circumscribed by a conic section. Assume first that the conic is a circle. Take a line l and suppose that it intersects the circle at H and I and intersects the sides of $QRST$ at points B, D, F, C, A, and E as in the diagram. Notice that this diagram has the same labeling as the diagram for the proof of Theorem 9.4.5. Since

$$AH \cdot AI = AS \cdot AR$$

and

$$DH \cdot DI = DQ \cdot DT$$

by the Chord Theorem (5.5.3), we have

$$\frac{AH \cdot AI}{DH \cdot DI} = \frac{AS \cdot AR}{DQ \cdot DT}.$$

Since the Secant Theorem (5.5.4) yields

$$MQ \cdot MT = MR \cdot MS,$$

Equation 9.9 of the proof of Theorem 9.4.5 yields

$$\frac{AH \cdot AI}{DH \cdot DI} = \frac{AB \cdot AE}{DB \cdot DE},$$

which means $(HI)(AD)(BE)$.

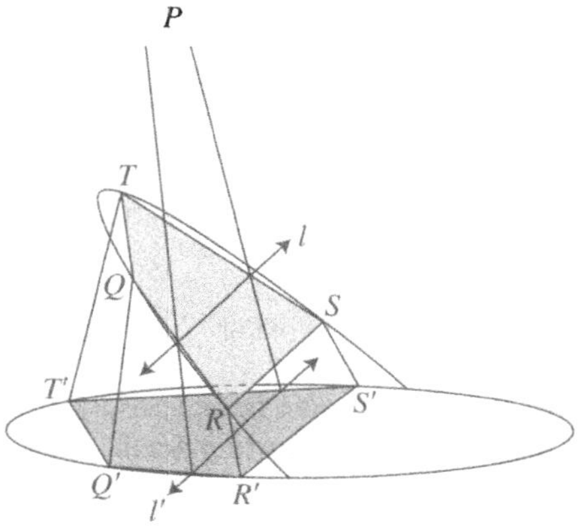

If the conic is not a circle, it can still be mapped by a projection onto a circle. For example, suppose that the complete quadrangle $QRST$ is inscribed within a parabola. Since this parabola is a section of a cone, $QRST$ is projected onto another complete quadrangle $Q'R'S'T'$ that is inscribed within a circle. Let l intersect the parabola and $QRST$. The preceding paragraph shows that this line is then projected onto a line l' that intersects the circle and $Q'R'S'T'$ in six points of involution. These points were projected from the six points of intersection of l with the parabola and $QRST$, which by Theorem 9.4.2 are now known to be in involution. ■

Exercises

1. In the *Brouillon Project*, Desargues introduced two terms (Field and Gray 1987, 47–49). Let A, B; C, D; and E, F be three pairs of collinear points. They form a **tree** if there exists a point O called a **stump** such that

$$OA \cdot OB = OC \cdot OD = OE \cdot OF.$$

Denote this tree by $\mathsf{T}(AB)(CD)(EF)$. Show that $\mathsf{T}(AB)(CD)(EF)$ with stump O implies the following:

(a) $OD/OF = OE/OC = DE/CF$ and $OF/OC = OD/OE = DF/CE$.

(b) $OD/OC = (DA \cdot DB)/(CA \cdot CB) = (DE \cdot DF)/(CE \cdot CF)$.

2. Prove that if $(AB)(CD)(EF)$, then $\mathsf{T}(AB)(CD)(EF)$.

3. Prove: If $(AB)(CD)(EF)$, then $(AB)(EF)(CD)$ and $(EF)(CD)(AB)$.

4. Suppose that $(AB)(CD)(EF)$ but $A = B$ and $E = F$. This means that the proportion from the definition of involution becomes

$$\frac{AC}{AD} = \frac{EC}{ED}.$$

Desargues called C, D; A, E **four points in involution**. Show that for these four points in involution, if M is the midpoint of $\overline{AE}$, then $EM \cdot AD = AC \cdot DM$.

5. Prove Equation 9.4 in the proof of Theorem 9.4.2.

6. Explain why four points in involution are preserved under projection.

7. Draw a complete quadrangle that has one of its vertices an ideal point. Draw another that has two ideal vertices.

8. Draw two complete quadrangles that intersect l in the same quadrangular set such that

   (a) both quadrangles have the same triangle triple of points. Such complete quadrangles are said to be **similarly placed**.
   (b) the point triple of points of one quadrangle is the triangle triple of points of the other. These complete quadrangles are called **oppositely placed**.
   (c) For each pair of complete quadrangles drawn, confirm that the points of intersection of the corresponding sides of the diagonal triangles are collinear.

9. Let A, B, C, D, and E be collinear points. Find F such that Q(ABC, DEF). Is F unique?

10. Let Q(ABC, DEF). Show that we have Q(AEF, DBC), Q(DBF, AEC), and Q(DEC, ABF).

11. Construct a complete quadrangle with a given diagonal triangle. Is the construction unique?

12. Find the vertices of a complete quadrangle given one of the vertices and the diagonal triangle.

PART IV

A STRANGE NEW WORLD

CHAPTER 10

GIOVANNI SACCHERI

Giovanni Girolamo Saccheri was born on September 5, 1667 in San Remo, Italy. He became a member of the Society of Jesus in 1685, and five years later he left for Milan to study theology and philosophy at the Jesuit College. While at the college he soon began teaching mathematics. It was at this time that Saccheri met Tommaso Ceva. Although the Jesuit Ceva was a notable poet and loved spending time writing, he was primarily a mathematician. He published a mathematical work on gravity called *Opuscula Mathematica*, and one day Saccheri became his student. This relationship would prove fruitful for Saccheri in that he would eventually meet some of the more notable mathematicians of his day. These included Giovanni Ceva, the brother of Tommaso, who proved the geometric theorem that is now named after him (Theorem 9.3.4), and Vincenzo Viviani, a one-time assistant to Galileo who later went on to study geometry and to calculate a more accurate value for the speed of sound. Under the tutelage of Tommaso Ceva, Saccheri wrote his first work, titled *Geometric Investigation*, in 1693. In it Saccheri solved numerous problems from elementary and analytic geometry.

Saccheri was ordained in 1694 and transferred to Turin, where he would teach philosophy and theology for three years. Although he did not lose his interest in mathematics, during this time his focus was on logic. These studies resulted in his

Revolutions of Geometry. By Michael O'Leary

Logica Demonstrativa of 1697. The work was significant in his day but was lost and not rediscovered until around the turn of the twentieth century. In the *Logica Demonstrativa* Saccheri treated logic as Euclid treated geometry by viewing the subject as a system of results based on definitions and postulates (Burton 1985, 529; Saccheri 1986, xv, xvii).

In the *Logica Demonstrativa*, Saccheri made three significant contributions. First he analyzed a type of flawed reasoning that he called *fallaciae duplicis hypothesis*. It consists of drawing a conclusion from a set of premises from which the negation of at least one of them can be deduced from the others (Saccheri 1986, xx–xxi). Such a premise set is called **inconsistent**. A set of premises is **consistent** if no contradiction can be deduced from it. Although an argument from inconsistent hypotheses may be deductively valid, it gives no reason to believe the conclusion. In fact, one can prove any proposition from contradictory hypotheses. To see this, suppose that q and *not* q both follow from a set of premises and let p be a random proposition. From q and *not* q we may conclude that (q *and not* q) *or* p. However, q *and not* q is a contradiction which means we that we must conclude p by the Disjunctive Syllogism (2.2.4).

The second contribution of the *Logica Demonstrativa* was a method to check the independence of premises (Saccheri 1986, xxi). We say that the proposition p is **independent** of the given propositions $q_1, q_2, \ldots, q_n$ if p cannot be proven or disproven from $q_1, q_2, \ldots, q_n$. To prove independence, Saccheri proposed doing the following:

- Find an interpretation of $q_1, q_2, \ldots, q_n$ and p that made the premises q_i true but the conclusion p false.
- Find another interpretation in which p and every q_i was true.

These interpretations were simply scenarios in which the premises were true and p was true in one case yet false in the other. This showed that neither p nor its negation necessarily followed from $q_1, q_2, \ldots, q_n$.

More significant to his geometric work was his third important contribution. Its original source was the *Elements*, where Euclid proved that a certain magnitude x measured y by assuming that it did not and then concluding that it did (Euclid 1925, IX.12, 397–399). This may seem odd at first, but it is a type of *reductio ad absurdum* argument. Suppose that p is the desired proposition. We can prove it by proving *if not* p *then* p, because these this conditional is logically equivalent to p. Saccheri may not have been the first to use such reasoning, but he was the first to include it in a systematic study of logic (Saccheri 1986, xxiv–xxvi).

Saccheri's interest in logic, and in particular this last strategy, combined with his expertise in mathematics led him to a line of research that had been a concern of many geometers since the time of Euclid.

10.1 THE QUESTION OF PARALLELS

Soon after Euclid gave his formulation of the Parallel Postulate (5.1.1.5), there were Greeks who had their doubts about it. The doubts were fueled by the fact that the

Parallel Postulate is fundamentally different from Euclid's other four postulates. The first three deal with constructions. We can connect two points with a straight line. We can identify a segment on a line. We can draw a circle. The fourth postulate states something equally obvious: that all right angles are congruent. We realize that when we see perpendicular lines, they look the same as any other pair of perpendicular lines that we have already seen. In summary the first four postulates reflect our experience. The Parallel Postulate is not like this. It claims that when the interior angles on the same side of a transversal add to less than two right angles, the lines will meet. If the two angles add to 25°, we can easily draw this situation and see that the lines intersect. However, what happens when they add to 179.9999°? If the transversal is perpendicular to one of the lines and the length of the transversal between the lines is 1 foot, the two triangles meet at a distance of about 110 miles. It is within our experience to draw such angles, but it is not within our experience to draw such a triangle. Maybe it is impossible to do so.

We could solve matters if we could prove the Parallel Postulate. This is what some of the ancient Greek geometers attempted to do. There is a lost work of Archimedes that was titled *On Parallel Lines*. It was included in the list of mathematical works in the *Book of Bibliography of the Sciences* by Abū l-Faraj Muhammad ibn al-Nadīm. It is conjectured by some that in it Archimedes replaced Euclid's Parallel Postulate with one stating that two lines are parallel if and only if the distance between them is constant. The Syrian mathematician Posidonius attempted a proof of the Parallel Postulate based on such an assumption. Ptolemy also tried to prove the Postulate. He did not use the equidistant assumption, but he assumed that if two parallel lines are cut by a transversal, the interior angles on the same side of the transversal are supplementary (Theorem 5.4.2) (Rosenfeld 1988, 40–41).

Another attempt to prove the Parallel Postulate was made by a Greek scholar by the name of Proclus. He was born in Byzantium in A.D. 411, and at a young age he traveled to Alexandria to begin his studies to become a lawyer. We are told by his student Marinus that on one of his trips back to Byzantium to visit his parents, he experienced a divine revelation. He was not to study law but philosophy. So, upon his return to Alexandria he began attending lectures on the works of Aristotle and on mathematics. He was soon dissatisfied with his education in Egypt and moved to Athens to learn at the Academy. There Proclus studied *De Anima* by Aristotle and then the *Phaedo* of Plato. His teachers encouraged him to take good notes at the lectures and afterward write his own comments on what he had learned. So began Proclus's career as a commentator on the great works of antiquity. His primary interest was philosophy, and he wrote numerous commentaries on the works of Plato, including the *Timaeus*. His other interests included mathematics, history, theology, poetry, and literary criticism. What made him successful was that not only did he work hard to understand the material, but he was also careful to fully examine the background that led to the creation of the works on which he wrote. Along with the *Mathematical Collection* of Pappus, the *Commentary on the First Book of Euclid's Elements* by Proclus serves as a crucial historic source for ancient Greek geometry in addition to providing valuable mathematical notes on Euclid's work (Burton 1985, 254–255; Heath 1921b, 530).

In his *Commentary* Proclus made an attempt to prove the Parallel Postulate. Like the others he believed that the Postulate should be considered a theorem. In his proof Proclus allowed himself to use any of the theorems of Euclid that were proven without the use of the Parallel Postulate. These are the results before *Elements* I.29 (Theorem 5.4.2). However, Proclus would not try to prove Euclid's formulation of the Parallel Postulate. Instead, he tried to prove a proposition that is now named after John Playfair, who had replaced Euclid's Parallel Postulate with the following statement in his successful *Elements of Geometry* of 1795 (Proclus 1970, 364; Burton 1985, 526).

■ POSTULATE 10.1.1 [Playfair]

Given a line l and a point P not on l, there exists exactly one line through P that is parallel to l.

The reason that Playfair was able to substitute his postulate for Euclid's is because the statements are logically equivalent (Burton 1985, 526–527).

■ THEOREM 10.1.2

Playfair's Postulate is equivalent to the Parallel Postulate.

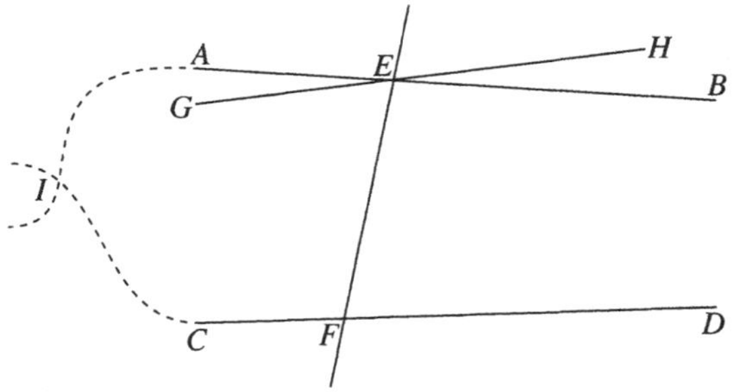

PROOF

The proof that the Parallel Postulate implies Playfair's Postulate is left to Exercise 2. We will show the converse here. Assume Playfair's Postulate. Let $\overline{AB}$ and $\overline{CD}$ be cut by transversal $\overline{EF}$ with E on $\overline{AB}$ and F on $\overline{CD}$. Let

$$m\measuredangle FEB + m\measuredangle EFD < \text{two right angles.}$$

We need to show that the $\overline{AB}$ and $\overline{CD}$ produced meet on the right side of $\overline{EF}$. Start by noting that

$$m\measuredangle EFC + m\measuredangle EFD = \text{two right angles.}$$

Thus, we have $m\measuredangle FEB + m\measuredangle EFD < m\measuredangle EFC + m\measuredangle EFD$, so then

$$m\measuredangle FEB < m\measuredangle EFC.$$

This means that we can construct $\overline{GH}$ through E such that

$$\measuredangle EFC \cong \measuredangle FEH$$

and $\overline{GE}$ is on the interior of $\measuredangle AEF$. By Theorem 5.4.1, $\overline{GH}$ is parallel to $\overline{CD}$, and by Playfair's Postulate this is the only line through E that is parallel to $\overline{CD}$. At this point we need to note that $\overline{AB}$ and $\overline{GH}$ are different lines. This is because $\measuredangle FEB$ is not congruent to $\measuredangle FEH$. Therefore, $\overline{AB}$ and $\overline{CD}$ intersect. Suppose that they meet at I on the left side of $\overline{EF}$. This would mean that we have ΔEIF with $\measuredangle FEB$ as an exterior angle to it. Theorem 5.3.9 tells us that the measure of an exterior angle to a triangle is greater than the measures of its opposite interior angles, but $m\measuredangle IFE > m\measuredangle FEB$. We then conclude that $\overline{AB}$ and $\overline{CD}$ must instead meet on the right-hand side of $\overline{EF}$. ■

Now on to Proclus's attempted proof. He started with an assumption, which he was comfortable making because he thought that Aristotle had proven it previously.

■ **POSTULATE 10.1.3 [Aristotle]**

If l and m intersect, for any magnitude r, there exists A on l and B on m such that $\overline{AB}$ is perpendicular to l and $AB > r$.

The proofs of the next two theorems are due to Proclus (1970, 371–373). In them he assumed properties concerning the continuity of lines. Proclus also assumed that parallel lines looked a particular way.

■ **THEOREM 10.1.4**

Aristotle's Postulate implies that if a line intersects one of two parallel lines, it intersects the other.

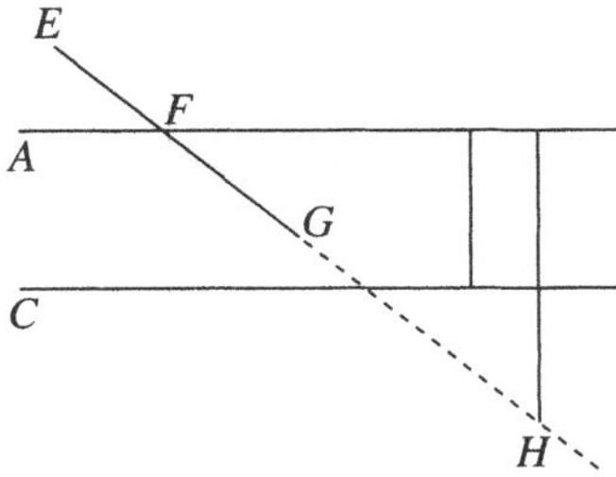

PROOF

Assume that Aristotle's Postulate holds. Let $\overline{AB}$ be parallel to $\overline{CD}$ and let $\overline{EG}$ meet $\overline{AB}$ at F. Therefore, by Aristotle's Postulate, there exists a point H on the $\overline{FG}$ produced so that the distance between $\overline{AB}$ and $\overline{FG}$ at H is greater than the distance between $\overline{AB}$ and $\overline{CD}$. Therefore, H is on the opposite side of $\overline{CD}$ than $\overline{AB}$, and we conclude that as $\overline{EG}$ is produced to H, $\overline{EG}$ and $\overline{CD}$ must intersect. ■

THEOREM 10.1.5

Aristotle's Postulate implies the Parallel Postulate.

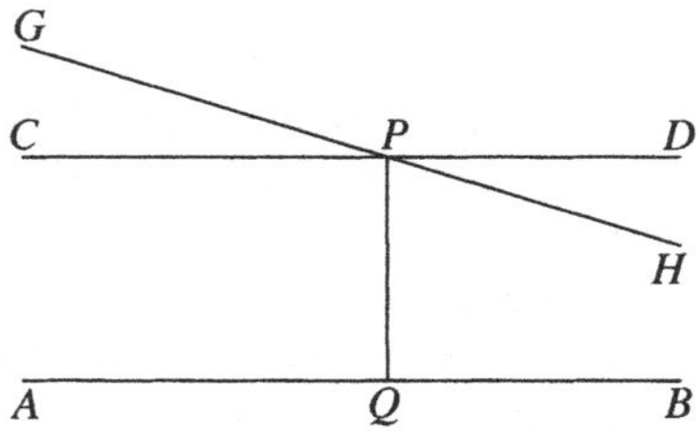

PROOF

Assume Aristotle's Postulate. Let P be a point not on $\overleftrightarrow{AB}$. Find Q on $\overleftrightarrow{AB}$ such that $\overleftrightarrow{PQ}$ is perpendicular to $\overleftrightarrow{AB}$. When we construct $\overleftrightarrow{CD}$ perpendicular to $\overleftrightarrow{PQ}$ at P, we see that $\overleftrightarrow{AB}$ and $\overleftrightarrow{CD}$ are parallel (Theorem 5.4.1). Let $\overleftrightarrow{GH}$ be another line containing P. By Theorem 10.1.4, $\overleftrightarrow{GH}$ must intersect $\overleftrightarrow{AB}$. Therefore, there is exactly one line through P parallel to $\overleftrightarrow{AB}$, which demonstrates Playfair's Postulate. Hence, the Parallel Postulate holds by Theorem 10.1.2. ■

Unfortunately for Proclus there is a hidden assumption in his proof of Theorem 10.1.4. Like Posidonius, he assumed that parallel lines are **equidistant**, that they are everywhere the same distance apart. However, such an assumption is equivalent to the Parallel Postulate.

THEOREM 10.1.6

The Parallel Postulate holds if and only if parallel lines are equidistant.

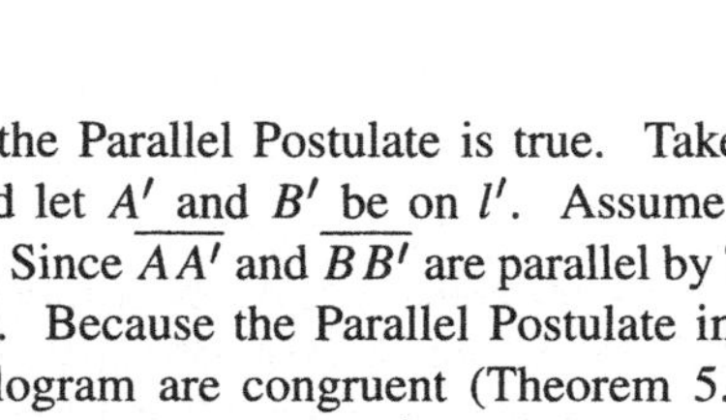

PROOF

First assume that the Parallel Postulate is true. Take lines l and l'. Let A and B be on l and let A' and B' be on l'. Assume that $\overline{AA'}$ and $\overline{BB'}$ are perpendicular to l. Since $\overline{AA'}$ and $\overline{BB'}$ are parallel by Theorem 5.4.1, $ABB'A$ is a parallelogram. Because the Parallel Postulate implies that the opposite sides of a parallelogram are congruent (Theorem 5.4.6), we conclude that $\overline{AA'} \cong \overline{BB'}$, which means that l and l' are equidistant.

To prove the converse, suppose that parallel lines are equidistant. Let A' not be on l. Construct $\overline{AA'}$ perpendicular to l and construct l' through A' perpendicular to $\overline{AA'}$. We know that l and l' are parallel. Let $\overleftrightarrow{A'C}$ be another line through A' that is parallel to l. Without loss of generality, let C be between l and l'. Drop a perpendicular through C to l at B and produce $\overline{BC}$ to intersect l' at B'. By hypothesis, $\overline{AA'} \cong \overline{BB'}$. However, $BB' = BC + CB'$, so $AA' > BC$ even though $\overline{A'C}$ is parallel to l. This is a contradiction. Therefore, l' is the only line through A' parallel to l, from which the Parallel Postulate follows by Theorem 10.1.2. ■

It is acceptable to view Theorem 10.1.4 as a proven result, but we must remember that it is a proposition that follows from all five of Euclid's postulates because of its reliance on parallel lines being equidistant. This means that what we have shown above can be summarized in the next theorem. Since the Parallel Postulate can also be used to prove Aristotle's Postulate (Exercise 2), we include it in the list.

■ **THEOREM 10.1.7**

The following are equivalent.

- The Parallel Postulate.
- Playfair's Postulate.
- Aristotle's Postulate.
- Parallel lines are equidistant.

The problem with all of these proofs, from that of Posidonius and Ptolemy to that of Proclus, is that the assumptions, which they thought were more self-evident than the Parallel Postulate, are in fact equivalent to it. This means that these geometers unknowingly assumed what they wanted to prove. They begged the question.

Exercises

1. Following the strategy of Posidonius, define *parallel* to mean *equidistant* and replace the Parallel Postulate with

> for every line l and point P not on l, there exists a line m on P such that m is equidistant to l.

Using the above replacements, the first four postulates of Euclid, and the theorems from the *Elements* that can be proved without the Parallel Postulate (those prior to *Elements* I.29, which is Theorem 5.4.2), prove the following:

(a) Two lines that are both perpendicular to a third line are equidistant.
(b) When two lines are cut by a transversal, the lines are equidistant if and only if corresponding angles are congruent.
(c) The sum of the internal angles of a triangle equals two right angles.

2. Prove that the Parallel Postulate implies both Playfair's Postulate and Aristotle's Postulate.

3. Replace the Parallel Postulate with Playfair's and prove that if two parallel lines are cut by a transversal, alternate interior angles are congruent. In addition to Playfair's Postulate, only use theorems prior to *Elements* I.29 (Theorem 5.4.2).

4. Repeat Exercise 3 replacing the Parallel Postulate with Aristotle's Postulate.

5. Proclus made the following argument concerning the Parallel Postulate (Euclid 1925, vol. 1, 206–207): Let $\overline{AB}$ and $\overline{CD}$ be chosen so that $\sphericalangle BAC$ and $\sphericalangle ACD$ add to less than two right angles. Let E be the midpoint of $\overline{AC}$ and find points F and G on $\overline{AB}$ and $\overline{CD}$, respectively, such that $\overline{AE} \cong \overline{AF} \cong \overline{CG}$. Let H be the midpoint of $\overline{FG}$ and again find points I and J on $\overline{FB}$ and $\overline{GD}$, respectively, such that $\overline{FH} \cong \overline{FI} \cong \overline{GJ}$. Continue this process indefinitely.
   (a) Why must F and G be different points? What about the points I and J or any other pair constructed in this manner?
   (b) Proclus argues that $\overline{AB}$ and $\overline{CD}$, when produced, must intersect, thus proving the Parallel Postulate. How must he have done this, and what mistake did he make?

6. Prove that Theorem 5.4.3 is equivalent to the Parallel Postulate.

7. Show that if parallel lines are not equidistant, a quadrilateral with three right angles is not necessarily a rectangle.

10.2 THE THREE HYPOTHESES

There were many other attempts to prove the Parallel Postulate. They included the Persians Umar Khayyām and Nasir al-Dīn al-Tūsī, who lived during the Middle Ages, and the Englishman John Wallis, who was a contemporary of Descartes. They followed the same general strategy as that of Proclus, but they were more explicit in stating their assumptions. Saccheri, for his part, criticized both Wallis' proof (Exercise 13) and the proof attributed to al-Tūsī, but he was influenced by the solution given by Khayyām. It involved a quadrilateral and three hypotheses that al-Tūsī had used. These would serve as the foundation for Saccheri's own attempt in his work entitled *Euclides ab omni naevo vindicatus* (*Euclid Freed of All Blemish*) (Rosenfeld 1988, 64–71, 74–85, 97–98).

Saccheri's work (Saccheri 1986, 19f) was completed shortly before his death on October 25, 1733 while he held the chair in mathematics at Pavia. His goal was to demonstrate that the Parallel Postulate necessarily follows from Euclid's other four postulates. Since Euclid did not appeal to the Parallel Postulate in any of his proofs until Proposition I.29 (Theorem 5.4.2), Saccheri allowed himself the use of the first 28 theorems found at the beginning of the *Elements*. The set of all propositions that can be proved from these theorems that exclude the Parallel Postulate is called **absolute geometry** (Bólyai 1987, 18, 21).

Saccheri's proof centered around one particular figure. Take $\overline{AB}$ and erect perpendiculars $\overline{AD}$ and $\overline{BC}$ such that $\overline{AD} \cong \overline{BC}$. Join $\overline{DC}$. The quadrilateral described is known as a **Saccheri quadrilateral** (see the figure for Theorem 10.2.1). The right

$\measuredangle DAB$ and $\measuredangle CBA$ are known as **base angles**, and $\overline{AB}$ is the **base**. The side opposite the base is the **summit**, and $\measuredangle ADC$ and $\measuredangle BCD$ are called **summit angles**. When identifying a Saccheri quadrilateral, the first two vertices will be those of the base angles and the second two those of the summit angles.

Saccheri wanted to know if the summit angles are right angles. If so, he would be able to prove the Parallel Postulate. He began with our first two propositions.

■ **THEOREM 10.2.1 [*Euclides Vindicatus* I.1]**

The summit angles of a Saccheri quadrilateral are congruent.

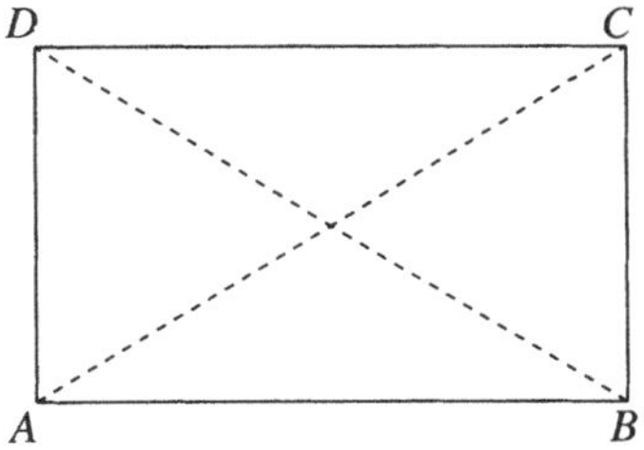

PROOF

In quadrilateral $ABCD$, let $\measuredangle A$ and $\measuredangle B$ be right angles and $\overline{AD} \cong \overline{BC}$. Since $\overline{AD} \cong \overline{BC}$, $\measuredangle A \cong \measuredangle B$, and $\overline{AB}$ is congruent to itself, we may use SAS to conclude that $\triangle ABD \cong \triangle BAC$. Therefore, $\overline{AC} \cong \overline{DB}$, so $\triangle ACD \cong \triangle BDC$ by SSS. Hence, $\measuredangle D \cong \measuredangle C$. ■

■ **THEOREM 10.2.2 [*Euclides Vindicatus* I.2]**

In a Saccheri quadrilateral the segment joining the midpoints of the summit and the base is perpendicular to both the summit and the base.

PROOF

Let $ABCD$ be a Saccheri quadrilateral, M be the midpoint of $\overline{AB}$, and N be the midpoint of $\overline{DC}$. Then $\overline{AD} \cong \overline{BC}$ and $\measuredangle A$ and $\measuredangle B$ are both right. Since M is the midpoint of $\overline{AB}$, we have $\overline{AM} \cong \overline{MB}$. Therefore, $\triangle AMD \cong \triangle BMC$ by SAS, and we see that $\overline{DM} \cong \overline{CM}$. Hence, $\triangle DMN \cong \triangle CMN$ by SSS because N is the midpoint of $\overline{DC}$. This yields $\measuredangle DNM \cong \measuredangle CNM$, which means that these are right angles. Similarly, we can show that $\measuredangle AMN$ and $\measuredangle BMN$ are right angles. ■

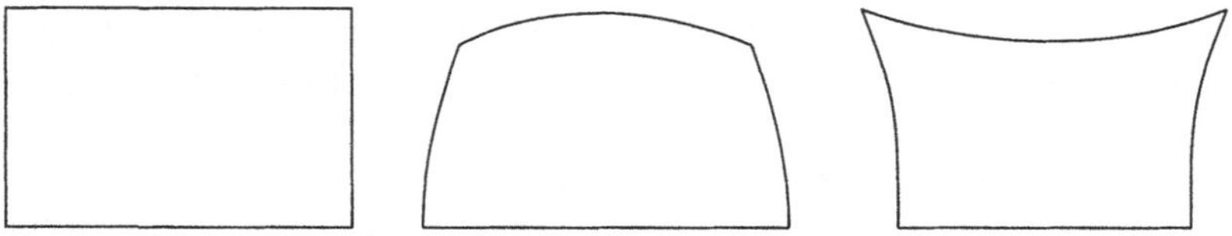

(a) Right summit angles (b) Obtuse summit angles (c) Acute summit angles

Figure 10.1 The three options for a Saccheri quadrilateral.

There are three possibilities for the measures of the summit angles of any given Saccheri quadrilateral. It was Saccheri's strategy to eliminate two of them, leaving the one that will yield the Parallel Postulate.

■ DEFINITION 10.2.3

- **HR**$(ABCD)$ is the sentence

  the summit angles of Saccheri quadrilateral ABCD are both right.

- **HO**$(ABCD)$ is the sentence

  the summit angles of Saccheri quadrilateral ABCD are both obtuse.

- **HA**$(ABCD)$ is the sentence

  the summit angles of Saccheri quadrilateral ABCD are both acute.

We should note that the Parallel Postulate suffices to prove proposition **HR**$(ABCD)$ (Exercise 9).

The resulting polygons that these three propositions describe are in Figure 10.1. Three of the sides in the cases when the summit angles are not right are represented by curved lines even though the lines themselves are "straight" and not curves. The length of the summit relative to the base is as indicated in the figure and is dependent upon which of the three hypotheses is true.

■ THEOREM 10.2.4 [*Euclides Vindicatus* I.3, 4]

- **HR**$(ABCD)$ if and only if the length of the summit of $ABCD$ is equal to the length of its base.

- **HO**$(ABCD)$ if and only if the length of the summit of $ABCD$ is less than the length of its base.

- **HA**$(ABCD)$ if and only if the length of the summit $ABCD$ is greater than the length of its base.

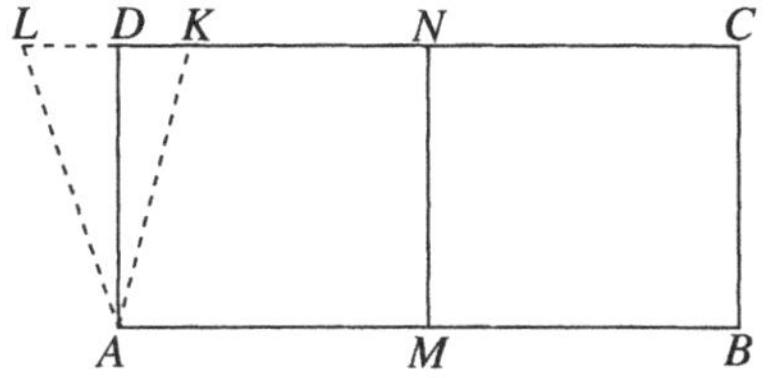

PROOF

We begin by proving sufficiency. Let $ABCD$ be a Saccheri quadrilateral with base $\overline{AB}$. Let M be the midpoint of $\overline{AB}$ and N the midpoint of $\overline{DC}$. Join $\overline{NM}$. We know that $\overline{AM}$ and $\overline{DN}$ are perpendicular to $\overline{NM}$ by Theorem 10.2.2. Now check the three cases depending on the hypothesis. We begin by proving sufficiency.

- When the summit angles are right is left for Exercise 10.

- Assume **HO**$(ABCD)$. First suppose that $\overline{AM} \cong \overline{DN}$. Then $MNDA$ is a Saccheri quadrilateral with $\sphericalangle A$ and $\sphericalangle D$ being its summit angles. Therefore, $\sphericalangle A \cong \sphericalangle D$ by Theorem 10.2.1, which means that $\sphericalangle D$ is a right angle, a contradiction. Next assume that $DN > AM$, and find K on $\overline{DN}$ such that $\overline{KN} \cong \overline{AM}$. We have Saccheri quadrilateral $MNKA$, so $\sphericalangle AKN \cong \sphericalangle KAM$. However, $\sphericalangle KAM$ is acute and $m\sphericalangle AKN$ is greater than $m\sphericalangle NDA$ because the latter angle is interior and opposite to $\sphericalangle AKN$ in ΔKDA (Theorem 5.3.9). This is impossible because $\sphericalangle NDA$ is obtuse. We may then conclude that $DN < AM$, and since $CD = 2DN$ and $AB = 2AM$, we have $CD < AB$.

- When **HA**$(ABCD)$ is true, we argue as in the preceding case. We see that $AM \neq DN$, so suppose that $DN < AM$. This means that we may extend $\overline{DN}$ to $\overline{LN}$ such that $\overline{LN} \cong \overline{AM}$. Join $\overline{LA}$ to form Saccheri quadrilateral $MNLA$. Since $\sphericalangle LAM$ is obtuse, $\sphericalangle NLA$ is obtuse. However, $\sphericalangle NLA$ is opposite and internal in ΔDLA with respect to $\sphericalangle NDA$. Therefore, $m\sphericalangle DLA$ is less than $m\sphericalangle NDA$, which is impossible since $\sphericalangle NDA$ is acute. We conclude that $CD > AB$.

For necessity, if $CD = AB$ but not **HR**$(ABCD)$, then either **HO**$(ABCD)$ or **HA**$(ABCD)$ must hold. In either case, by the previous arguments, $\overline{CD}$ is not congruent to $\overline{AB}$, which is a contradiction. Similarly, $CD < AB$ implies **HO**$(ABCD)$, and $CD > AB$ implies **HA**$(ABCD)$. ■

Saccheri now shows that the three hypotheses are incompatible with each other in the sense that if one of them is satisfied by a Saccheri quadrilateral, that same hypothesis is satisfied by them all. We start with the right-angle hypothesis and prove that if there exists Saccheri quadrilateral $ABCD$ such that **HR**$(ABCD)$, we can construct a Saccheri quadrilateral of any given dimension. We do this in stages by proving three propositions.

LEMMA 10.2.5

If there exists Saccheri quadrilateral $ABCD$ such that $\mathbf{HR}(ABCD)$, for all positive integers n, a Saccheri quadrilateral with base of length AB and height of nAD has right summit angles.

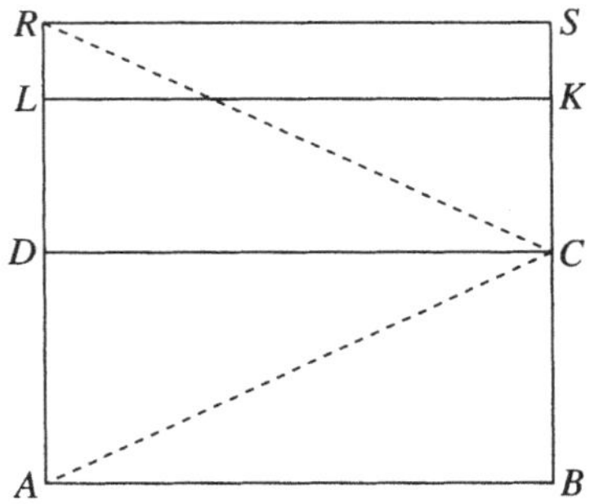

PROOF

Suppose that we have a Saccheri quadrilateral $ABCD$ such that $\sphericalangle C$ and $\sphericalangle D$ are right angles. Now find points R and S extending $\overline{AD}$ and $\overline{BC}$, respectively, such that $\overline{AD} \cong \overline{DR}$ and $\overline{BC} \cong \overline{CS}$. Notice that $ABSR$ is a Saccheri quadrilateral, so $\sphericalangle R \cong \sphericalangle S$. Moreover, we have $\triangle CDR \cong \triangle CDA$ by SAS. Thus, $\overline{RC} \cong \overline{AC}$ and $\sphericalangle RCD \cong \sphericalangle ACD$. Therefore, $\sphericalangle SCR \cong \sphericalangle BCA$, yielding $\triangle SCR \cong \triangle BCA$ by SAS. Since $\sphericalangle B$ is a right angle, we conclude that $\sphericalangle S$ is a right angle, and then $\sphericalangle R$ is also a right angle. This argument can be generalized to any positive integer multiple of AD. ■

The last lemma can be modified to give a Saccheri quadrilateral of an arbitrary height.

LEMMA 10.2.6

If there exists Saccheri quadrilateral $ABCD$ such that $\mathbf{HR}(ABCD)$, every Saccheri quadrilateral with base of length AB has right summit angles.

PROOF

Using the diagram above, take L on $\overline{AR}$ and K on $\overline{BS}$ such that $\overline{AL} \cong \overline{BK}$. Then $ABKL$ is a Saccheri quadrilateral. Furthermore, $\overline{LR} \cong \overline{KS}$, so $SRLK$ is also a Saccheri quadrilateral because $\sphericalangle R$ and $\sphericalangle S$ are right angles. To obtain a contradiction, first assume that $\sphericalangle ALK$ and $\sphericalangle BKL$ are acute. We conclude that $LK > AB$ (Theorem 10.2.4), and hence $LK > RS$. However, $\sphericalangle RLK$ and $\sphericalangle SKL$ are obtuse because $\sphericalangle ALK$ and $\sphericalangle BKL$ are acute. Therefore, $LK < RS$. This is absurd, so we conclude that the original summit angles are not acute. A similar argument will show that they are not obtuse, so we have shown that the summit angles of $ABKL$ are right angles. ■

The generalization Lemma 10.2.6 to an arbitrary base is the next theorem. Its proof is left to Exercise 11.

■ **THEOREM 10.2.7** [*Euclides Vindicatus* **I.5**]

If there exists Saccheri quadrilateral $ABCD$ such that **HR**$(ABCD)$, every Saccheri quadrilateral has right summit angles.

The same strategy will show that the proposition holds when the summit angles are obtuse. We state the theorem but leave its proof for Exercise 11.

■ **THEOREM 10.2.8** [*Euclides Vindicatus* **I.6**]

If there exists Saccheri quadrilateral $ABCD$ such that **HO**$(ABCD)$, every Saccheri quadrilateral has obtuse summit angles.

If there is Saccheri quadrilateral $ABCD$ that satisfies either **HR**$(ABCD)$ or **HO**$(ABCD)$, no Saccheri quadrilateral will have acute summit angles. Hence, if there exists Saccheri quadrilateral $ABCD$ that satisfies **HA**$(ABCD)$, none can have right or obtuse summit angles. Since these are the only alternatives to an angle being acute, we conclude the next result.

■ **THEOREM 10.2.9** [*Euclides Vindicatus* **I.7**]

If there exists Saccheri quadrilateral $ABCD$ such that **HA**$(ABCD)$, every Saccheri quadrilateral has acute summit angles.

Because Theorems 10.2.7, 10.2.8, and 10.2.9 state that all Saccheri quadrilaterals will share the same property regarding their summit angles, we should restate the three hypotheses. Therefore, we make the following definition and summarize the results with an obvious theorem.

■ **DEFINITION 10.2.10**

- **HR** is the sentence

  the summit angles of every Saccheri quadrilateral are both right.

- **HO** is the sentence

  the summit angles of every Saccheri quadrilateral are both obtuse.

- **HA** is the sentence

  the summit angles of every Saccheri quadrilateral are both acute.

We call **HR** the **Hypothesis of the Right Angle**, **HO** the **Hypothesis of the Obtuse Angle**, and **HA** the **Hypothesis of the Acute Angle**.

■ **THEOREM 10.2.11**

Exactly one of the following hypotheses is true: **HR**, **HO**, or **HA**.

Exercises

1. Let $ABCD$ be a Saccheri quadrilateral with base $AB = 10$ and side $BC = 6$. Answer the following if possible.
   (a) Find DC and AD.
   (b) Find $m\sphericalangle ADC$ and $m\sphericalangle BCD$.
   (c) Find $m\sphericalangle ADC$ given that $m\sphericalangle BCD = \pi/4$.
   (d) Let M be the midpoint of $\overline{AB}$. Find DM.
   (e) Let N be the midpoint of $\overline{DC}$. Find AN.

2. Use a straightedge and compass to construct a Saccheri quadrilateral. What appears true about its summit angles, and what do you think that this implies about the geometry of the paper, straightedge, and compass?

3. Prove the following propositions concerning Saccheri quadrilaterals.
   (a) The diagonals are congruent.
   (b) The base produced will not intersect the summit.
   (c) The segment joining the midpoints of the summit and the base divides a Saccheri quadrilateral into two congruent quadrilaterals.
   (d) The segment joining the midpoints of the sides will intersect neither the base nor the summit if produced.

4. Show that the length of the segment joining the midpoints of the summit and base of a Saccheri quadrilateral are shorter than the perpendiculars to the base if and only if **HA**. What can be said about this segment under the other two hypotheses?

5. Prove that $ABCD$ is a parallelogram with four right angles (that is to say, a rectangle) if and only if $ABCD$ can be viewed both as a Saccheri quadrilateral with base $\overline{AB}$ and as a Saccheri quadrilateral with base $\overline{AD}$.

6. Must the segment joining the sides of a Saccheri quadrilateral be perpendicular to its sides? Explain.

7. Let M and N be the midpoints of the two sides of Saccheri quadrilateral $ABCD$. Join $\overline{MC}$ and $\overline{DN}$ and let P be their intersection. Prove:

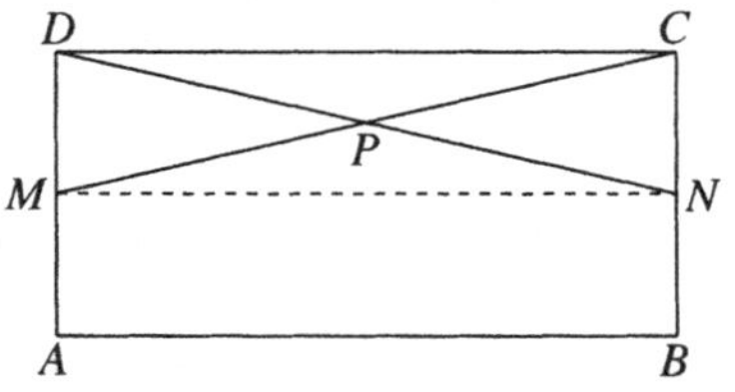

(a) $\overline{MC} \cong \overline{DN}$.
(b) $\triangle DCP$ is isosceles.
(c) $\triangle MDN \cong \triangle NCM$.
(d) $\triangle DMP \cong \triangle CNP$.

8. In the figure for Exercise 7, must $\triangle DMP$ be isosceles? Explain.

9. Prove that the Parallel Postulate implies $\mathbf{HR}(ABCD)$.

10. Demonstrate the first case of the proof of Theorem 10.2.4.

11. Prove Theorems 10.2.7 and 10.2.8.

12. Let $ABCD$ and $ABEF$ be two Saccheri quadrilaterals with $\overline{CD}$ and $\overline{EF}$ on opposite sides of $\overline{AB}$. Is $DCEF$ a parallelogram? Under what conditions will $DCEF$ be a Saccheri quadrilateral or a rectangle?

13. In 1693, Wallis published his *Treatise on the Quadrilateral.* In it he proved that the statement (now known as **Wallis's Postulate**) *to every triangle there exists a similar triangle of arbitrary magnitude* implies the Parallel Postulate (Kline 1972, 864–865; Euclid 1925, 210-211).

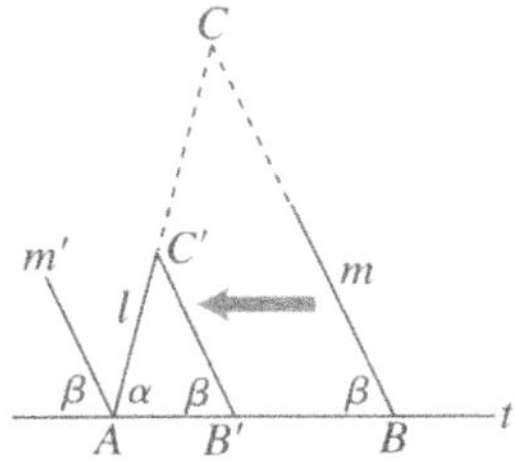

(a) Assume Wallis's Postulate. Let l and m intersect transversal t at points A and B, respectively. Denote the measure of the interior angle that l makes with t by α and let β be the measure of the interior angle that m makes with t on the same side as α. Assume that $\alpha + \beta$ is less than two right angles. Construct an angle with measure β with A as its vertex, one side on t, and the other side m' in the interior of the angle adjacent to α. We will now slide m towards m' stopping at B' when m intersects l at C'. Finish the proof to show that the Parallel Postulate follows.

(b) Prove that the Parallel Postulate implies Wallis' Postulate.

14. In his *Commentaries on the Difficulties in the Premises of Euclid's Book*, Umar Khayyām explicitly replaced Euclid's Parallel Postulate with Aristotle's Postulate. Doing the same, answer the following:

(a) Prove that two lines perpendicular to the same line do not intersect.
(b) What would it mean for the two perpendiculars to *diverge*?
(c) Prove that two perpendiculars cannot diverge.

10.3 CONCLUSIONS FOR TWO HYPOTHESES

Saccheri will soon have enough machinery to prove that the Parallel Postulate follows from the hypothesis of the right angle. He needs to know that the hypotheses not only determine the angles of a Saccheri quadrilateral, but that they also determine the angle sum of triangles and quadrilaterals. We state the results without proof for now. (The notation **[HR]**, **[HO]**, and **[HA]** means that the hypothesis indicated is assumed true for the proof of the labeled proposition.)

■ **THEOREM 10.3.1 [*Euclides Vindicatus* I.8, 9, 15]**

The sum of the internal angles of any triangle

- **[HR]** adds to two right angles.
- **[HO]** are together greater than two right angles.
- **[HA]** are together less than two right angles.

■ **THEOREM 10.3.2 [*Euclides Vindicatus* I.16]**

The sum of the internal angles of any quadrilateral

- **[HR]** adds to four right angles.
- **[HO]** are together greater than four right angles.
- **[HA]** are together less than four right angles.

We now need a technical result. It states that if two lines are cut by a transversal that is perpendicular to one of the lines but makes an acute angle with the other, the original lines must meet. Saccheri will use this to derive the Parallel Postulate from the Hypothesis of the Right Angle.

■ **LEMMA 10.3.3 [*Euclides Vindicatus* I.11]**

[HR] If $\overline{AD}$ and $\overline{BX}$ are cut by transversal $\overline{AB}$ such that $\overline{AB}$ is perpendicular to $\overline{BX}$ and $\sphericalangle DAB$ is acute, the $\overline{AD}$ and $\overline{BX}$ produced will intersect on the same side of $\overline{AB}$ as D.

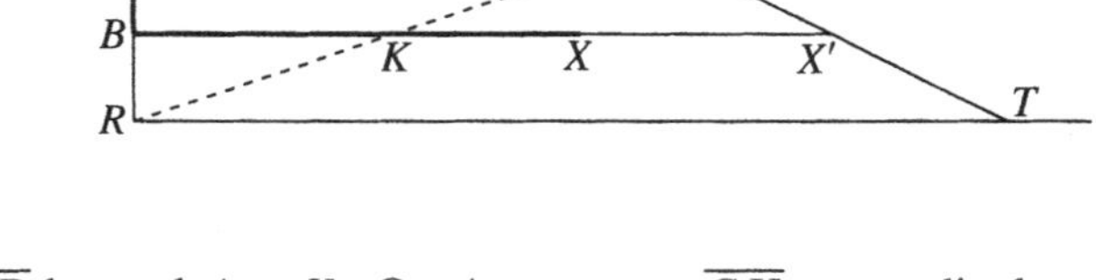

PROOF

Extend $\overline{AD}$ beyond A to Y. On A, construct $\overline{CH}$ perpendicular to $\overline{AB}$ such that C is on the interior of $\sphericalangle YAB$. Produce $\overline{AD}$ to F so that $\overline{AD} \cong \overline{DF}$. From D and F, drop perpendiculars to $\overline{AB}$ at E and M, respectively. Join $\overline{MD}$. Then

$$\overline{MD} \cong \overline{AD} \cong \overline{DF}$$

by Exercise 5. Thus, $AF = 2AD$ and $AM = 2AE$. By Archimedes' Postulate (6.2.1), there exists a positive integer k such that $kAE > AB$. Let R be on the $\overline{AB}$ produced so that $kAE = AR$. Now find point T on the $\overline{AD}$ produced so that $\overline{RT}$ is perpendicular to $\overline{AB}$. Observe that the point T must lie on the opposite side of $\overline{BX}$ as A. If not, suppose it is on the same side at point T' instead. Then perpendicular $\overline{RT'}$ must cut $\overline{BX}$ at some K. In this case $\triangle KBR$ consists of two right angles, which contradicts Theorem 10.3.1. Thus, $\overline{BX}$ enters $\triangle ART$ through side $\overline{AR}$, and since it cannot exit through $\overline{RT}$ by Theorem 10.3.1, it must intersect $\overline{AT}$ at some X'. This is because if a line enters through one side of a triangle it must exit through another side. ■

■ **THEOREM 10.3.4** [*Euclides Vindicatus* **I.13**]

HR is equivalent the Parallel Postulate.

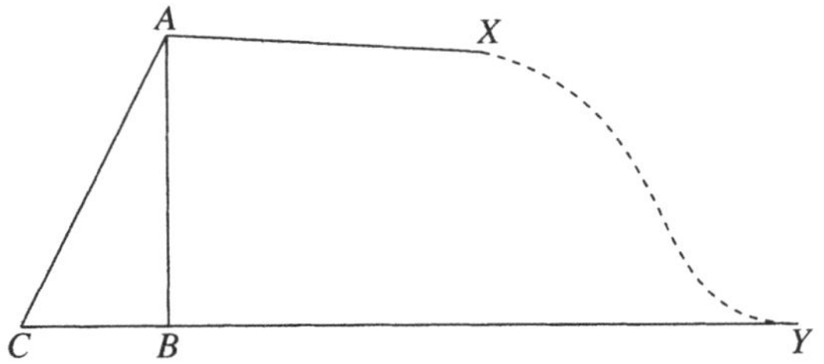

PROOF

We know by Exercise 9 of Section 10.2 that the Parallel Postulate implies **HR**, so to prove the converse assume **HR** and let $\overline{AX}$ and $\overline{CY}$ be cut by transversal $\overline{AC}$ such that

$$\text{m}\sphericalangle CAX + \text{m}\sphericalangle ACY < \text{two right angles.}$$

To show that the Parallel Postulate holds, we must prove that $\overline{AX}$ and $\overline{CY}$ meet when produced on the right side of $\overline{AC}$. First note that either m$\sphericalangle CAX$ or m$\sphericalangle ACY$ must be acute. Say that it is $\sphericalangle ACY$. From A drop a perpendicular to $\overline{CY}$, meeting it in B. Because of Theorem 10.3.1, $\overline{AB}$ must lie on the interior of $\sphericalangle CAX$. Also, by Theorem 10.3.1 we see that

$$\text{m}\sphericalangle BAC + \text{m}\sphericalangle BCA = \text{a right angle,}$$

but

$$\text{m}\sphericalangle BAC + \text{m}\sphericalangle BCA + \text{m}\sphericalangle BAX < \text{two right angles.}$$

Therefore, $\sphericalangle BAX$ is acute, and then by Lemma 10.3.3 the lines meet. ■

We could also have given Saccheri's proof that **HO** implies the Parallel Postulate. It is essentially the same as that of Theorem 10.3.4. We omit it because we cannot have **HO** under Saccheri's assumptions. This is because **HO** implies that the angle sum of a triangle is greater than two right angles (Theorem 10.3.1) but the Parallel Postulate implies that the angle sum is equal to two right angles (Theorem 5.4.3).

■ **THEOREM 10.3.5** [*Euclides Vindicatus* **I.14**]

HO is inconsistent with absolute geometry.

This quickly leads to a conclusion about the summit angles of a Saccheri quadrilateral.

■ **COROLLARY 10.3.6**

The summit angles of a Saccheri quadrilateral are not obtuse.

If Saccheri can also eliminate **HA** as inconsistent with absolute geometry, he will have proven the Parallel Postulate. This is his next goal. To accomplish this, Saccheri will work under the assumption that **HA** is true and seek a contradiction. If he can find one, he can reject **HA** along with **HO** as possibilities in absolute geometry. Then only **HR** will remain, from which will follow the Parallel Postulate. Since in absolute geometry **HA** is equivalent to the negation of the Postulate (Exercise 1), Saccheri's strategy amounts to proving the Parallel Postulate by assuming its negation.

Exercises

1. Prove that in absolute geometry **HA** is equivalent to the negation of the Parallel Postulate.

2. Draw examples of two triangles, one that has an internal angle sum of less than two right angles and another with angle sum of more than two right angles.

3. Draw examples of quadrilaterals, one assuming **HR** and the other assuming **HA**.

4. Show that Theorem 10.3.1 implies Theorem 10.3.2.

5. Show $\overline{MD} \cong \overline{DF}$ in the proof of Theorem 10.3.3.

6. Reprove Lemma 10.3.3 under the assumption **HO**.

7. Use Exercise 6 to show that **HO** implies the Parallel Postulate.

8. Let D be on $\overline{BC}$ such that $\overline{AD}$ is perpendicular to $\overline{BC}$ in $\triangle ABC$. Prove that $AC > AB$ if and only if $DC > BD$. Does this result require any of the three hypotheses, or is it a theorem of absolute geometry?

10.4 PROPERTIES OF PARALLEL LINES

There is little doubt that at this point Saccheri believed that the argument that he was developing in the *Euclides Vindicatus* was close to proving the Parallel Postulate. All that he needed to accomplish was to show that the Hypothesis of the Acute Angle is impossible in absolute geometry. As the odd results about straight lines under this hypothesis continued to mount, we can almost imagine Saccheri's excitement as he approached the proof that mathematicians since the days of ancient Greece were not able to find. As we read his results, we must remember that all lines are straight, despite the pictures. These are not curves.

An Acute-Angled Parallel

Saccheri begins with what will become a familiar construction. Start with two lines l and m cut by a transversal. Assume that the transversal is perpendicular to l. It is possible that m makes an acute angle with the transversal and seemingly heads towards l without ever intersecting it. There will be other lines, however, that also intersect the transversal in the same point that m does, make an acute angle with the transversal, but will intersect l. Between these two sets of lines, the intersecting and nonintersecting ones, will be a line that serves as a border between the two sets. Identifying this line is Saccheri's goal. He believed that it is the source of his sought-after contradiction.

■ **THEOREM 10.4.1 [*Euclides Vindicatus* I.17]**

[HA] If $\overline{AB}$ is perpendicular to $\overline{BY}$, there exists $\overline{AX}$ such that $\measuredangle BAX$ is acute but $\overline{AX}$ and $\overline{BY}$ produced do not intersect.

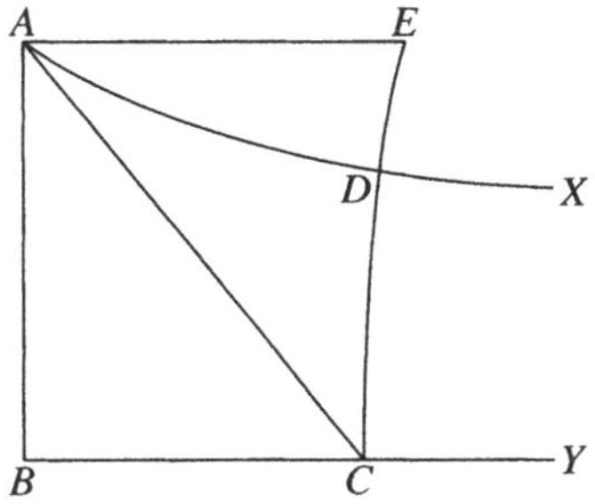

PROOF

Let $\overline{AE}$ and $\overline{BY}$ be perpendicular to $\overline{AB}$. Find C on $\overline{BY}$, extended if necessary, so that $\overline{CE}$ is perpendicular to $\overline{BY}$. Join $\overline{AC}$. Since it lies on the interior of a right angle, $\measuredangle ACE$ is acute. By Theorem 10.3.2 we see that $\measuredangle CEA$ is also acute. Now draw a perpendicular to $\overline{CE}$ through A. Call it $\overline{AX}$ and let D be the point of intersections between $\overline{AX}$ and $\overline{CE}$. Notice that D must fall below E, else there would be a triangle with two angles combined greater than two right angles in violation of Theorem 10.3.1. All of this means that $\overline{AX}$ is the line segment that we are after. This is because if $\overline{AX}$ and $\overline{BY}$ produced did meet, we would have a triangle with base $\overline{CD}$ that does not have an internal angle sum of less than two right angles. ■

Common Perpendiculars

By Theorem 10.2.2 we already know that the base and summit of a Saccheri quadrilateral have a **common perpendicular**, a segment that forms right angles with two different lines. Saccheri now shows that this is the case even when the "Saccheri quadrilateral" does not have congruent sides. His proof uses a continuity argument.

■ **THEOREM 10.4.2 [*Euclides Vindicatus* I.22]**

[HA] Let $\sphericalangle ADC$ and $\sphericalangle DCB$ be acute in quadrilateral $ABCD$. If $\overline{AD}$ and $\overline{BC}$ are perpendicular to $\overline{AB}$, there exists a common perpendicular to $\overline{AB}$ and $\overline{DC}$ between $\overline{AD}$ and $\overline{BC}$.

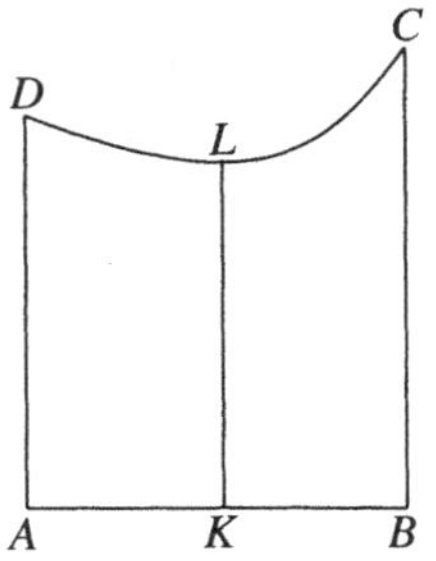

PROOF

Let $\overline{AD}$ and $\overline{BC}$ both be perpendicular to $\overline{AB}$. If $\overline{AD} \cong \overline{BC}$, we can join the midpoints of $\overline{AB}$ and $\overline{DC}$ to find the common perpendicular since $ABCD$ would be a Saccheri quadrilateral (Theorem 10.2.2). Otherwise, take any point L on $\overline{DC}$ and drop a perpendicular to $\overline{AB}$, intersecting it at K. As we have seen before, K must be between A and B to be consistent with Theorem 10.3.1. If $\sphericalangle CLK$ is a right angle, we are done, so suppose that it is obtuse. View $\overline{LK}$ as moving toward $\overline{AD}$ in a manner that L is always on $\overline{DC}$ and $\overline{LK}$ is always perpendicular to $\overline{AB}$. As $\overline{LK}$ approaches $\overline{AD}$, we see that $\sphericalangle CLK$ approaches $\sphericalangle ADC$, which means that $\sphericalangle CLK$ cannot remain obtuse. At the instant when the angle changes from obtuse to acute, we will have the common perpendicular. ■

The Defect of a Triangle

Assuming the Hypothesis of the Acute Angle, we know that the sum of the interior angles of a triangle is less than two right angles. How much that sum differs from a right angle will become very important, so we make the following definition.

■ **DEFINITION 10.4.3**

The difference between two right angles and the angle sum of a triangle is called the **(angle) defect** of the triangle. The defect of $\triangle ABC$ is denoted by $\mathsf{d}(ABC)$.

We will always consider the defect to be a nonnegative number. When we write $\mathsf{d}(ABC) = 0$, this means that the angle sum of $\triangle ABC$ is equal to two right angles. This happens under **HR**. In the case of **HO**, the difference between the angle sum of a triangle and two right angles is negative, so we turn the subtraction around and speak of the **(angle) excess** of the triangle. With this terminology, we may appeal to Theorem 10.3.1 and write the following:

THEOREM 10.4.4

- **[HR]** The angle sum of every triangle equals two right angles.
- **[HO]** All triangles have positive angle excess.
- **[HA]** All triangles have positive angle defect.

Up to this point we have compared the sum of the measure of the internal angles of a triangle to two right angles. To make our calculations easier to read, we now use

$$\pi = 4\left(1 - \frac{1}{3} + \frac{1}{5} - \frac{1}{7} + \cdots\right) \tag{10.1}$$

to represent the measure of two right angles and $\pi/2$ for the measure of one right angle. Although under Euclid's postulates this value is equal to the ratio of the circumference to the diameter of a circle, this will not be the case when we assume **HA** or any other hypothesis that implies negation of the Parallel Postulate. As of now, π is simply the number given in Equation 10.1. With this understanding we write

$$\mathsf{d}(ABC) = \pi - \mathrm{m}\measuredangle A - \mathrm{m}\measuredangle B - \mathrm{m}\measuredangle C.$$

We use this notation in the proof of the next theorem (Euclid 1925, Vol. I, 217–219).

THEOREM 10.4.5

If two given triangles are added together to form a third triangle, the defect of the third triangle equals the sum of the defects of the given triangles.

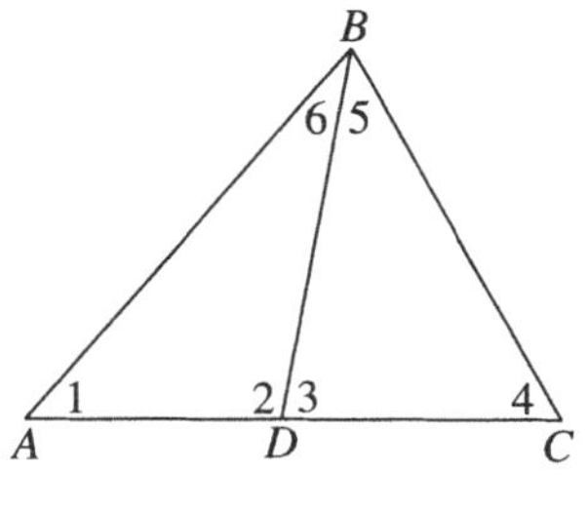

PROOF

Given $\triangle ABC$ and a point D that is not a vertex but lies on $\overline{AC}$, join $\overline{DB}$ forming $\triangle ABD$ and $\triangle DBC$. Letting the angles be numbered as in the diagram, we see that

$$\mathsf{d}(ABC) = \pi - \mathrm{m}\measuredangle 1 - \mathrm{m}\measuredangle 4 - \mathrm{m}\measuredangle 5 - \mathrm{m}\measuredangle 6,$$

$$\mathsf{d}(ADB) = \pi - \mathrm{m}\measuredangle 1 - \mathrm{m}\measuredangle 2 - \mathrm{m}\measuredangle 6,$$

and

$$\mathsf{d}(DBC) = \pi - \mathrm{m}\measuredangle 3 - \mathrm{m}\measuredangle 4 - \mathrm{m}\measuredangle 5.$$

Therefore, $\mathrm{d}(ADB) + \mathrm{d}(DBC)$ equals

$$2\pi - \mathrm{m}\sphericalangle 1 - \mathrm{m}\sphericalangle 2 - \mathrm{m}\sphericalangle 6 - \mathrm{m}\sphericalangle 3 - \mathrm{m}\sphericalangle 4 - \mathrm{m}\sphericalangle 5.$$

Since $\sphericalangle 2$ and $\sphericalangle 3$ are supplementary,

$$\mathrm{d}(ADB) + \mathrm{d}(DBC) = \pi - \mathrm{m}\sphericalangle 1 - \mathrm{m}\sphericalangle 6 - \mathrm{m}\sphericalangle 4 - \mathrm{m}\sphericalangle 5.$$

Therefore, $\mathrm{d}(ABC) = \mathrm{d}(ADB) + \mathrm{d}(DBC)$. ■

If we take a convex polygon with n sides, we can define its angle defect as the difference between $n - 2$ right angles and the sum of its interior angles. This is because a convex n-gon is the union of $n - 2$ triangles. We can now give the following corollary to Theorem 10.4.5. Its proof is Exercise 6(b).

■ **COROLLARY 10.4.6**

If two convex polygons are added together to form a third convex polygon, the defect of the third polygon is the sum of the defects of the given polygons.

Asymptotic Behavior

From the above we know that it is possible that two lines have a common perpendicular. However, what happens if they do not? Certainly, one possibility is that the lines intersect. Due to the hypothesis **HA**, it is also possible that there are two straight lines that do not intersect but instead approach each other in such a way that the distance between the straight lines continually diminishes.

■ **LEMMA 10.4.7 [*Euclides Vindicatus* I.23]**

[HA] Given $\overline{AX}$ and $\overline{BY}$ so that $\overline{AB}$ is perpendicular to $\overline{BY}$ and $\sphericalangle BAX$ is acute, exactly one of the following holds:

- $\overline{AX}$ and $\overline{BY}$ have a common perpendicular, or
- the distance between $\overline{AX}$ and $\overline{BY}$ decreases the further away from $\overline{AB}$.

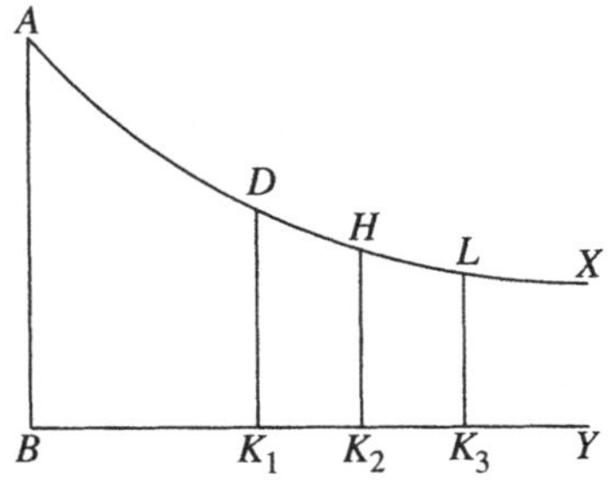

PROOF

Let $\measuredangle ABY$ be a right angle and let $\measuredangle BAX$ be acute. Take points D, H, and L in sequence on $\overline{AX}$ and drop perpendiculars to $\overline{BY}$. Let K_1, K_2, and K_3 be the points of intersection of the perpendiculars and $\overline{BY}$. If any of $\measuredangle ADK_1$, $\measuredangle AHK_2$, or $\measuredangle ALK_3$ are acute, then by Theorem 10.4.2 there is a common perpendicular between $\overline{AX}$ and $\overline{BY}$. If any of these angles are right, then the corresponding segment is the common perpendicular, so suppose that they are all obtuse. This means that their supplementary angles are acute. In quadrilateral K_1DHK_2 the angles at K_1 and K_2 are right but the angle at D is acute. Therefore, we have $DK_1 > HK_2$ (Exercise 7). Similarly, $HK_2 > LK_3$ and we find that the distance between the lines is strictly decreasing. ■

The next result further describes the parallel lines under **HA**. If there are two parallel lines cut by a transversal forming a right angle with one line and an acute angle with the other, the line that appears quite curved near the transversal will more closely resemble what we would consider a straight line as we move farther from the transversal. Saccheri measures this curvature by using perpendicular segments.

■ **LEMMA 10.4.8 [*Euclides Vindicatus* I.24]**

[HA] Take $\overline{AX}$ and $\overline{BY}$ so that $\overline{AB}$ is perpendicular to $\overline{BY}$. Assume that $\overline{AX}$ and $\overline{BY}$ neither intersect nor have a common perpendicular. Let K_1, K_2, and K_3 be points on $\overline{BY}$ in sequence and equally spaced with K_1 nearest to B. If $\overline{DK_1}$, $\overline{HK_2}$, and $\overline{LK_3}$ are constructed perpendicular to $\overline{BY}$ with points D, H, and L on $\overline{AX}$, then $\mathsf{d}(K_1K_2HD) > \mathsf{d}(K_2K_3LH)$.

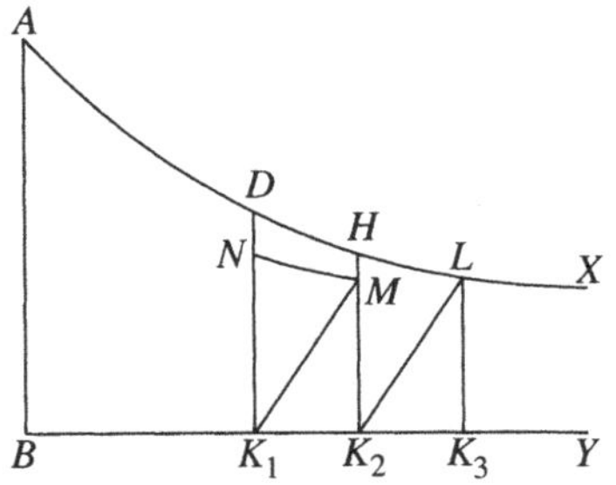

PROOF

Construct a diagram as in the previous proposition, and suppose that $\overline{K_1K_2}$ is congruent to $\overline{K_2K_3}$. We know by Lemma 10.4.7 that

$$DK_1 > HK_2 > LK_3.$$

Find M on $\overline{HK_2}$ such that $\overline{MK_2} \cong \overline{LK_3}$ and find another point N on $\overline{DK_1}$ so that $\overline{NK_1} \cong \overline{HK_2}$. Join $\overline{MN}$, $\overline{MK_1}$, and $\overline{LK_2}$. By SAS, $\triangle K_3K_2L$ is congruent to $\triangle K_2K_1M$. Thus, $\overline{LK_2} \cong \overline{MK_1}$, $\measuredangle LK_2K_3 \cong \measuredangle MK_1K_2$, and

$$\measuredangle K_2LK_3 \cong \measuredangle K_1MK_2. \tag{10.2}$$

Since $\measuredangle HK_2L$ is complementary with $\measuredangle LK_2K_3$ and $\measuredangle NK_1M$ is complementary with $\measuredangle MK_1K_2$, we have that $\measuredangle HK_2L \cong \measuredangle NK_1M$. Therefore, $\triangle NK_1M \cong \triangle HK_2L$ by SAS, so by definition $\overline{NM} \cong \overline{HL}$ and $\measuredangle K_1MN \cong \measuredangle K_2LH$. Then angle addition with Congruence 10.2 yields

$$\measuredangle NMK_2 \cong \measuredangle HLK_3. \tag{10.3}$$

We now make two observations, the details being checked in Exercise 9:

- The angle sum of K_2K_3LH equals the angle sum of K_1K_2MN.
- By **HA** and Theorem 10.3.2,

  $$\text{m}\measuredangle MNK_1 + \text{m}\measuredangle NMK_2 > \text{m}\measuredangle HDN + \text{m}\measuredangle DHM.$$

  Therefore, the angle sum of K_1K_2MN is greater than the angle sum of K_1K_2HD.

Therefore, the angle sum of K_2K_3LH is greater than the angle sum of K_1K_2HD, and so we have $\text{d}(K_1K_2HD) > \text{d}(K_2K_3LH)$. ■

■ **LEMMA 10.4.9 [*Euclides Vindicatus* I.25]**

Let $\overline{AB}$ be perpendicular to $\overline{BY}$, $\measuredangle XAB$ be acute, and $\overline{AX}$ approach $\overline{BY}$ such that the segments produced are always greater than some fixed distance apart. If $\overline{AB}$ and $\overline{BY}$ do not have a common perpendicular, **HA** is false.

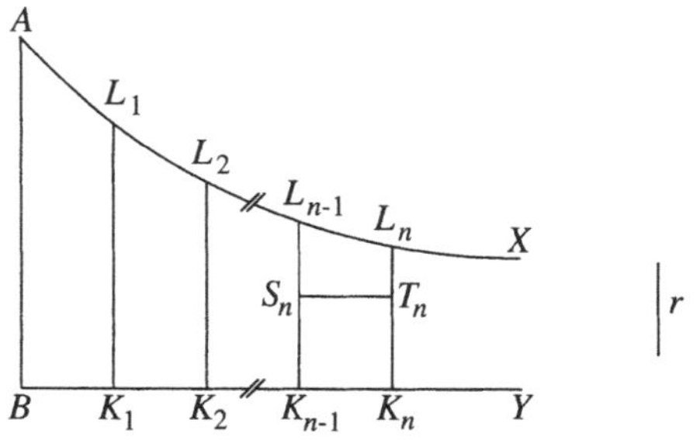

PROOF

Assume **HA** and that $\overline{AB}$ and $\overline{BY}$ do not have a common perpendicular. Let r be a magnitude such that $\overline{AX}$ and $\overline{BY}$ are always at least r units apart. Find points $K_1, K_2, \ldots, K_n, \ldots$ on $\overline{BY}$ such that $BK_1 = r$ and $K_{n-1}K_n = r$ for all integers $n \geqslant 2$. On each K_n erect a perpendicular to $\overline{BY}$ and let L_n be the intersection of this perpendicular with $\overline{AX}$. Now, we know by Lemma 10.4.8 that $\text{d}(K_{n-1}K_nL_nL_{n-1})$ strictly decreases as n increases. In fact, we may be more precise and claim that

$$\text{d}(K_{n-1}K_nL_nL_{n-1}) < \frac{\pi}{n}. \tag{10.4}$$

To see this, first notice that

$$\mathsf{d}(K_1K_2L_2L_1) = \mathsf{d}(BK_2L_2A) - \mathsf{d}(BK_1L_1A)$$

by Corollary 10.4.6. Therefore,

$$\mathsf{d}(K_1K_2L_2L_1) < \mathsf{d}(BK_2L_2A) - \mathsf{d}(K_1K_2L_2L_1),$$

and from this we conclude that

$$\mathsf{d}(K_1K_2L_2L_1) < \frac{1}{2}\mathsf{d}(BK_2L_2A) < \frac{\pi}{2}.$$

Similarly, since

$$\mathsf{d}(K_2K_3L_3L_2) = \mathsf{d}(BK_3L_3A) - \mathsf{d}(BK_2L_2A),$$

we have by Exercise 8 that

$$\mathsf{d}(K_2K_3L_3L_2) < \frac{1}{3}\mathsf{d}(BK_3L_3A) < \frac{\pi}{3}. \tag{10.5}$$

Generalizing this argument proves Inequality 10.4, and this in turn implies that

$$\lim_{n\to\infty} \mathsf{d}(K_{n-1}K_nL_nL_{n-1}) = 0. \tag{10.6}$$

Because $L_nK_n > r$ for all n, we can find points S_n on $\overline{L_{n-1}K_{n-1}}$ and T_n on $\overline{L_nK_n}$ such that

$$S_nK_{n-1} = T_nK_n = r.$$

By Theorem 10.3.2,

$$\mathrm{m}\sphericalangle K_nT_nS_n + \mathrm{m}\sphericalangle K_{n-1}S_nT_n > \mathrm{m}\sphericalangle T_nL_nL_{n-1} + \mathrm{m}\sphericalangle S_nL_{n-1}L_n.$$

Therefore, the sum of the internal angles of $K_{n-1}K_nT_nS_n$ is greater than the sum of the internal angles of $K_{n-1}K_nL_nL_{n-1}$. Therefore, there is a quadrilateral $KK'ST$ that is congruent to $K_nK_{n-1}S_nT_n$ and has the property that

$$\mathsf{d}(KK'TS) < \mathsf{d}(K_{n-1}K_nL_nL_{n-1})$$

for all n. By Equation 10.6 we may conclude that $\mathsf{d}(KK'ST) \leqslant 0$. Therefore, either **HR** or **HO** holds, which contradicts our assumption of **HA**. ■

The contrapositive of the theorem combined Lemma 10.4.7 give us the next proposition. We say that two lines approach each other **asymptotically** if the distance between them is decreasing and approaches but is never equal to zero.

■ THEOREM 10.4.10

[HA] Given $\overleftrightarrow{AX}$ and $\overleftrightarrow{BY}$ so that $\overline{AB}$ is perpendicular to $\overleftrightarrow{BY}$ and $\sphericalangle BAX$ is acute, exactly one of the following holds:

- $\overleftrightarrow{AX}$ and $\overleftrightarrow{BY}$ have a common perpendicular,
- $\overleftrightarrow{AX}$ and $\overleftrightarrow{BY}$ intersect, or
- $\overleftrightarrow{AX}$ and $\overleftrightarrow{BY}$ approach each other asymptotically.

Exercises

1. Prove that the following statement is equivalent to Playfair's Postulate:

> If $\overline{AB}$ is perpendicular to $\overline{BY}$ with X and Y on the same side of $\overline{AB}$, then $\overline{AX}$ is parallel to $\overline{BY}$ if and only if $\angle BAX$ is right.

2. Using the figure and hypotheses from Theorem 10.4.1, show that if $\overline{AZ}$ is on the interior of $\angle EAX$, then $\overline{AZ}$ produced will not intersect $\overline{BY}$ produced.

3. Exercise 2 suggests that there are infinitely many lines through A that do not intersect $\overline{BY}$. Where in the figure can we find infinitely many lines through A that do intersect $\overline{BY}$?

4. Prove that if a given triangle is cut by a line forming a smaller triangle and a quadrilateral, the defect of the given triangle is the sum of the defects of the smaller triangle and the quadrilateral.

5. Let $\overline{AD}$ be a cevian line (page 309) of $\triangle ABC$ partitioning it into $\triangle ABD$ and $\triangle ACD$. Prove that $\mathsf{d}(ABC) = \mathsf{d}(ABD) + \mathsf{d}(ACD)$.

6. Let P be a convex n-gon.
   (a) Partition P into $n - 2$ triangles and show that the defect of P equals the sum of the defects of the triangles.
   (b) Use part (a) to prove Corollary 10.4.6.

7. Assume **HA**. Let A be on $\overline{YX}$ and B on $\overline{Y'X'}$ so that $\overline{AB}$ is perpendicular to $\overline{YX}$ and $\angle ABX'$ is acute. Further assume that $\overline{BX'}$ does not intersect $\overline{AX}$ but, instead, approaches it asymptotically. Let E and C be on $\overline{AY}$ and D and F be on $\overline{BY'}$ such that $\overline{FE}$ and $\overline{DC}$ are perpendicular to $\overline{YX}$. Prove that if C is between E and A, then $FE > DC$.

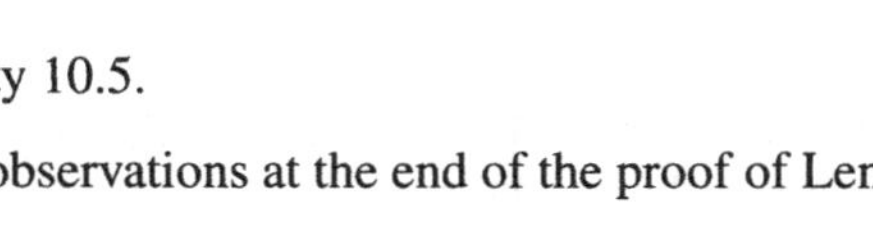

8. Confirm Inequality 10.5.

9. Confirm the two observations at the end of the proof of Lemma 10.4.8.

10. Prove or show false: Aristotle's Postulate (10.1.3) follows from **HA**.

11. In Euclid's geometry, straight lines do not approach each other asymptotically. They are either parallel or intersecting. Only a curve can approach a line asymptotically. Describe a method to construct an asymptote to a line in Euclid's geometry. Can it be done with only a straightedge and compass?

12. Let ABC be a semicircle with diameter $\overline{AB}$. Prove that **HR** if $\sphericalangle ACB$ is right, **HA** if $\sphericalangle ACB$ is acute, and **HO** if $\sphericalangle ACB$ is obtuse.

13. The proof of Theorem 10.4.2 used a continuity argument to find the location of the common perpendicular. Continuity of what magnitude did Saccheri assume? What was the range of the magnitude?

14. Why is π defined as in Equation 10.1? (*Hint:* This is a calculus question.)

10.5 PARALLELISM REDEFINED

With the properties obtained from Section 10.4, we can now describe completely the situation about parallel lines under the Hypothesis of the Acute Angle. Crucial to the characterization is the first theorem.

■ **LEMMA 10.5.1 [*Euclides Vindicatus* I.30]**

[HA] Let $\overline{AX}$ and $\overline{BY}$ be cut by transversal $\overline{AB}$ such that $\overline{BY}$ and $\overline{AB}$ are perpendicular.

- If $\sphericalangle BAX$ is obtuse, the $\overline{AX}$ produced does not share a common perpendicular with $\overline{BY}$.
- If $\sphericalangle BAX$ is acute and $\overline{AX}$ has a common perpendicular with $\overline{BY}$, there exists $\overline{AS}$ on the interior of $\sphericalangle BAX$ such that $\overline{AS}$ shares a common perpendicular with $\overline{BX}$.

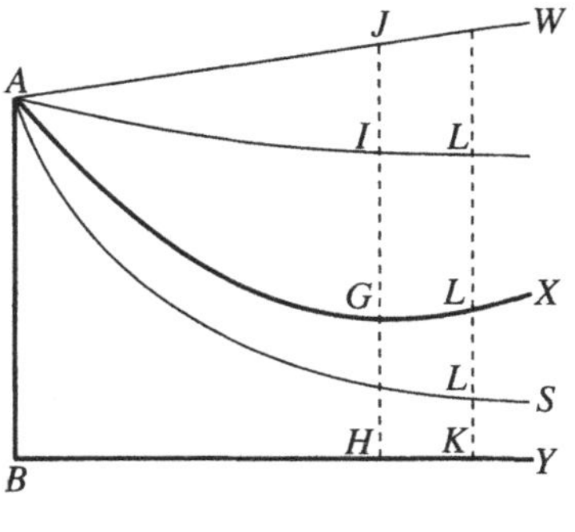

PROOF

Let $\overline{AB}$ be perpendicular to $\overline{BY}$. To prove the first part, construct $\overline{AW}$ on the same side of $\overline{AB}$ as X and Y so that $\sphericalangle WAB$ is obtuse. If there exists J on $\overline{AW}$ and H on $\overline{BY}$ such that $\overline{HJ}$ is perpendicular to both $\overline{AW}$ and $\overline{BY}$, the sum of the internal angles of quadrilateral $BHJA$ would be greater than four right angles contradicting **HA**.

We prove the second part indirectly. Let $\sphericalangle BAX$ be acute and suppose that it is the least angle such that $\overline{AX}$ has a common perpendicular with $\overline{BY}$. Let

$\overline{GH}$ be that perpendicular with G on $\overline{AX}$ and H on $\overline{BY}$. Find K on $\overline{BY}$ such that H is between K and B and $\overline{LK}$ is perpendicular to $\overline{BY}$. Join $\overline{AL}$. By Theorem 5.2.12, we may assume that $\overline{AL}$ is perpendicular to $\overline{LK}$. We will show that all locations of L lead to a contradiction.

- If $\overline{AL}$ lies on the interior of $\sphericalangle BAX$, then $\sphericalangle BAX$ is not the least angle so that $\overline{AX}$ and $\overline{BY}$ have a common perpendicular because $\overline{LK}$ would be a common perpendicular to $\overline{AL}$ and $\overline{BY}$.
- L cannot be on $\overline{AX}$ because $HKLG$ would have 4 right angles.
- Let $\overline{AL}$ intersect the $\overline{HG}$ produced at some point I on the exterior of $\sphericalangle BAX$. By Theorem 5.3.9, we see that $\sphericalangle AIG$ is acute since $\sphericalangle AGH$ is a right angle external to $\triangle AIG$. Hence, $\sphericalangle GIL$ is obtuse, and $HKLI$ has an angle sum greater than four right angles, another contradiction.

Therefore, if there is an acute $\sphericalangle BAX$ with the property that it has a common perpendicular with $\overline{BY}$, there necessarily is a smaller angle with the same vertex and sharing $\overline{AB}$ that has the same property. ■

Now where does Saccheri find himself? He proves this:

■ **THEOREM 10.5.2 [*Euclides Vindicatus* I.32]**

[HA] If $\overline{AB}$ is perpendicular to $\overline{BY}$, there exists X such that $\sphericalangle BAX$ is acute, $\overline{AX}$ approaches $\overline{BY}$ asymptotically, and for any point L on the same side of $\overline{AB}$ as X:

- If m$\sphericalangle BAL$ is less than m$\sphericalangle BAX$, the $\overline{AL}$ produced will intersect $\overline{BY}$.
- If m$\sphericalangle BAL$ is greater than m$\sphericalangle BAX$ but less than a right angle, the $\overline{AZ}$ and $\overline{BX}$ produced share a common perpendicular.

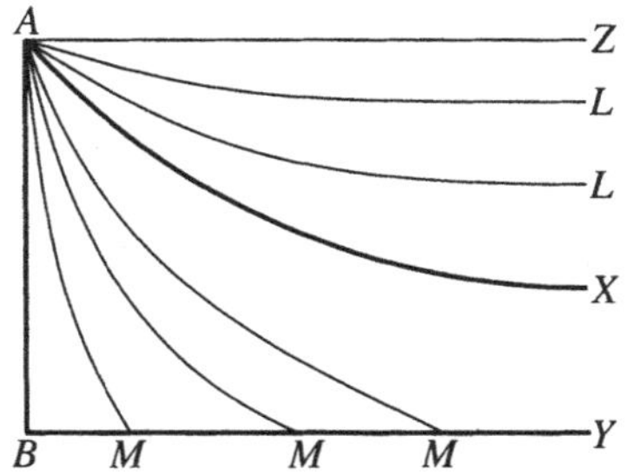

PROOF

Let $\overline{AB}$ be a common perpendicular to $\overline{AZ}$ and $\overline{BY}$. By Lemma 10.5.1, there exists a sequence of points L such that m$\sphericalangle BAL$ is acute and $\overline{AL}$ has a common perpendicular with $\overline{BY}$. Notice that none of the $\overline{AL}$ intersect $\overline{BY}$ and a line through A must have a common perpendicular with $\overline{BY}$ if it is between two lines, both containing A, that have a common perpendicular with $\overline{BY}$ (Exercise 9). Furthermore, for all points M on $\overline{BY}$, we can join $\overline{AM}$ to form a sequence of lines that intersect $\overline{BY}$ with the property that any line between any two lines of the sequence must intersect $\overline{BY}$ (Exercise 9). Therefore, there exists $a > 0$ such that:

- If $0 \leqslant m\sphericalangle BAL < a$, then $\overline{AL}$ intersects $\overline{BY}$.
- If $a < m\sphericalangle BAL \leqslant \pi/2$, then $\overline{AL}$ has a common perpendicular with $\overline{BY}$.

Let X be on the same side of $\overline{AB}$ as Y such that $m\sphericalangle BAX = a$. Notice that $\overline{AX}$ cannot have a common perpendicular with $\overline{BY}$ because then a would be the minimum angle measure that does not exist by Lemma 10.5.1. Similarly, $\overline{AX}$ does not intersect $\overline{BY}$ because there is no maximum measure for $\sphericalangle BAL$ such that $\overline{AL}$ intersects $\overline{BY}$. Hence, by Theorem 10.4.10, $\overline{AX}$ must approach $\overline{BY}$ asymptotically but never intersect it. ■

Saccheri called $\overline{AX}$ from the figure of Theorem 10.5.2 a **limit** because it can be viewed as the last line that does not intersect $\overline{BY}$. So that we will be consistent with subsequent terminology, we will say that $\overline{AX}$ is **parallel** to $\overline{BY}$, and it is the only line through A on that side of $\overline{AB}$ that is. We will simply say that the lines $\overline{AL}$ that share a common perpendicular with $\overline{BY}$ are **nonintersecting** with respect to $\overline{BY}$. We will not call them parallel. The lines $\overline{AM}$ that meet $\overline{BY}$ will be called **intersecting** with respect to $\overline{BY}$. The acute $\sphericalangle BAX$ is called the **angle of parallelism**.

At this point Saccheri thought that he was done and wrote, "The hypothesis of [the] acute angle is absolutely false because [it is] repugnant to the nature of the straight line." His proof?

> From the foregoing theorem may be established, that at length the hypothesis of acute angle inimical to the Euclidean geometry has as outcome that we must recognize two straights AX and $[BY]$, existing in the same plane, which produced *in infinitum* toward the parts of the points X [and Y] must run together at length into one and the same straight line, truly receiving, at one and the same infinitely distant point a common perpendicular in the same plane with them.

He viewed both $\overline{AX}$ and $\overline{AY}$ as intersecting at a point at infinity, and at that point the lines would align, allowing a common perpendicular. Saccheri thought that he had a contradiction because he had just proven that these were two mutually exclusive events. The problem is that he confused points at infinity with ordinary points. The phrases "being parallel" and "meeting in an ideal point" mean the same thing because there are no ideal points in his geometry (Saccheri 1986, 173, 251).

Evidently in the very next paragraph, Saccheri has a change of heart. Realizing that his argument relies more on intuition than firm logic, Saccheri proceeds to "diligently" prove **HA** false. Unfortunately for him, he continues on for another 69 pages searching for a contradiction, but none will come. He is so led astray by his preconceived notions about how a straight line must behave that he did not realize the mathematical gem that he had uncovered. It will be left for others to overcome the predisposition toward Euclid and draw the right conclusions. Saccheri, for his part, ended the *Euclides Vindicatus* with the words, "But this now is enough."

Exercises

Problems that require the Hypothesis of the Acute Angle are denoted **[HA]**.
The term parallel *refers to the new definition of parallel lines.*

1. Let l be a line and P a point not on l. Draw pictures that represent the parallel and nonintersecting lines to l through P that exist under **HR** and **HA**. If **HO** were allowed in Saccheri's geometry, what would the parallel lines look like?

2. **[HA]** Prove that if $\overline{AB}$ is a common perpendicular to $\overline{AX}$ and $\overline{BY}$, there is no magnitude r such that all common perpendiculars between $\overline{BY}$ and a line drawn from A have length of at least r.

3. **[HA]** Let l and m be cut by transversal t so that the interior angles on the same side of t are supplementary. Explain why l is not parallel to m. What is their relationship?

4. **[HA]** Let $\overline{AB}$ be perpendicular to $\overline{BY}$ and $\overline{AX}$ parallel to $\overline{BY}$. Let $\overline{A'X'}$ and $\overline{B'Y'}$ be two other segments. Prove that if $\sphericalangle BAX \cong \sphericalangle B'A'X'$, $\sphericalangle ABY \cong \sphericalangle A'B'Y'$, and $\overline{AB} \cong \overline{A'B'}$, then $\overline{A'X'}$ is parallel to $\overline{B'Y'}$.

5. **[HA]** Repeat Exercise 4 with the perpendicular condition removed.

6. Prove that two lines either have zero, exactly one, or infinitely many common perpendiculars.

7. **[HA]** Suppose that $\overline{AB}$ is perpendicular to $\overline{BY}$ and that both $\sphericalangle BAX$ and $\sphericalangle BAX'$ are acute, where X and X' are on the same side of $\overline{AB}$ as Y. Let H be on $\overline{AX}$, H' on $\overline{AX'}$, and G and G' on $\overline{BY}$ such that $\overline{GH}$ is the common perpendicular between $\overline{AX}$ and $\overline{BY}$ and $\overline{G'H'}$ is the common perpendicular between $\overline{A'X'}$ and $\overline{BY}$. Prove that if $\overline{AX'}$ is in the interior of $\sphericalangle BAX$, then G is between B and G'.

8. Prove that two parallel lines have a common perpendicular if and only if the Parallel Postulate holds.

9. **[HA]** Let $\overline{AB}$ be perpendicular to $\overline{BY}$ and let $\sphericalangle BAX$ is acute. Assume that L is on the same side of $\overline{AB}$ as X and Y. Prove:
   (a) If $\overline{AL}$ has a common perpendicular with $\overline{BY}$, then $\overline{AL}$ and $\overline{BY}$ do not intersect.
   (b) If $\overline{AL}$ is between two lines containing A that have a common perpendicular with $\overline{BY}$, then $\overline{AL}$ has a common perpendicular with $\overline{BY}$.
   (c) If $\overline{AL}$ is between two lines containing A that intersect $\overline{BY}$, then $\overline{AL}$ intersects $\overline{BY}$.

10. Prove that the angle of parallelism equals $\pi/2$ if **HR** is assumed.

CHAPTER 11

JOHANN LAMBERT

Johann Heinrich Lambert came from humble origins. He was born in August 1728 in Mulhouse in eastern France near the German and Swiss borders. His father, Lukas, was a tailor. His mother's name was Elisabeth. The Lambert side of the family were Calvinists and had fled persecution in the French region of Lorraine and arrived in Mulhouse in 1635. The family's finances were tight, and Lambert was required to leave school at an early age to help his father. This situation did not prevent him from continuing his studies, however. Whenever he could, Lambert taught himself and was always interested in learning new things. This served him well later in life when he earned a living as a private tutor, teaching his students mathematics as well as languages, geography, history, and religion. In 1756 he took some of his students on a journey through Europe. The destinations were the continent's great centers of learning. During this journey Lambert made contact with many notable mathematicians and scientists. Among these was the mathematician and historian A. G. Kästner. The trip lasted for two years (Scriba 1970, 595–596).

Upon his return, Lambert sought to gain a permanent university position. His top choice was the University of Göttingen, but his attempts to secure a position were unsuccessful. Lambert then moved to Zurich and worked as an astronomer and later moved to Augsburg and then Munich. In Munich he was chosen to organize the newly formed Bavarian Academy of Sciences, but differences arose and his time there was short lived.

Finally, on January 10, 1765 he found the stability for which he had so longed. Four years earlier, Lambert's name was proposed for a position at the Prussian Academy of Sciences in Berlin, but the appointment never came. After years of wandering, Lambert arrived in Berlin at the beginning of 1764. He had just been assigned a position at the Academy by Frederick the Great. The appointment also included a seat on a commission on economics with Leonard Euler. However, Lambert was a bit eccentric, and this delayed his installation. Among other things he dressed in very colorful clothes and preferred to stand at right angles to anyone with whom he was conversing. This made the king uncomfortable. He was not interested in such a "blockhead" serving at the Academy. Over time, though, Lambert would win him over with his excellent work in physics and astronomy. His contributions to astronomy included trigonometric tables that aided in the mundane and time-consuming calculations that astronomers often had to perform, and his work in physics included the study of measurement, particularly that of the intensity of light, humidity, and high temperatures. In addition, he was influential in leading an effort to bring about more collaboration between astronomers and promoted the publication of astronomical journals. (Scriba 1970, 596–598).

Besides his scientific interests, Lambert was also a philosopher contributing two works of note to the field: the *Neues Organon* of 1764 and the *Anlage zur Architectonik* of 1771. He sought to establish a logical system for all knowledge that would be based on a few fundamental axioms. The resulting system was to be **complete**, able to answer any question posed within it, and consistent. The system, however, did not rely solely on reason. Lambert believed that reason needed to be supplemented with experiment and observation. He based this on the fact that humans are imperfect, so their thought is subject to error. Any evidence obtained can be used to check the truth of the conclusion of an argument, and in addition, the act of collecting the evidence can provide a structure to guide one's thoughts. Even here, Lambert was careful. The data collected should be quantitative as much as possible, and we must always realize that our conclusions can only be as precise as our initial measurements. By carefully combining these two roads to discovery, Lambert developed a balanced search for truth (Gray and Tilling 1978, 15–17).

Finally, Lambert was a mathematician, and as varied as his general interests were, so were his mathematical pursuits. He studied number theory, trigonometry, hyperbolic functions, and geometry. Among his accomplishments are the first proofs that π and e are irrational. His proof concerning π began with the demonstration of a continued fraction representation for tangent:

$$\tan x = \cfrac{x}{1 - \cfrac{x^2}{3 - \cfrac{x^2}{5 - \cfrac{x^2}{\ddots}}}}.$$

From this he deduced that if x is rational, $\tan x$ is irrational. He concluded that $\pi/4$ is not rational since $\tan \pi/4 = 1$, so π is not rational. Lambert further conjectured that both π and e are not **algebraic**. Instead, they are **transcendental**: namely, there is no polynomial with integer coefficients that has π or e as a root. The confirmation of this would have to wait for Charles Hermite's proof for e in 1873 and Ferdinand von Lindemann's proof for π in 1882 (Laczkovich 1997, 439; Boyer and Merzbach 1991, 463, 573; Scriba 1970, 599).

Lambert had two significant geometric interests. The first was perspective drawing. In his *Freye Perspektive* he detailed how to draw accurately using perspective. This included the construction of the parallel lines that we see in our daily lives. Lambert noted that such constructions can be made with minimal data. His tools were a straightedge with either no compass or only one fixed circle. He then provided examples of constructions using only these tools. They included finding points on an ellipse given four points, the construction of right angles, the construction of parallel lines, and the bisection of a given angle (Gray and Tilling 1978, 30–31).

His second line of geometric research involved the Parallel Postulate. At a time when few mathematicians knew of Saccheri's work, Lambert's studies may have been influenced by the *Euclides Vindicatus*. He learned about it through the work of Georg Klügel. This mathematician lived from 1739 to 1812 and was a professor at the University of Helmstädt. In his doctoral dissertation, Klügel enumerated many of the attempts to decide the question of parallels. His survey included the work of Saccheri. As a result of his dissertation, Klügel concluded that the Postulate was not self-evident but should instead be accepted or rejected based on one's experience of the world. Since our experience is limited to this small corner of the universe, Klügel doubted that the question would ever be settled.

We should note that Klügel is also famous for his involvement in a controversy during the late eighteenth century concerning whether the best way to study geometry was analytic or synthetic. The English were supporters of the synthetic, whereas Klügel and his mentor, A. G. Kästner, fell on the side of the analytic. They believed that analysis was superior because it provided the reader with an understanding of the motivation behind a proof. Klügel and Kästner also charged the English, as Descartes had with the Greeks, with choosing to publish their results synthetically to show off their keen mathematical intellects. (Kline 1980, 81; Boyer and Merzbach 1991, 536).

11.1 THE THREE HYPOTHESES REVISITED

Since Klügel did not delve deeply into the *Euclides Vindicatus*, Lambert's knowledge of Saccheri's arguments was probably superficial. Nonetheless, it may have been the inspiration for his own study, which culminated in his work entitled the *Theory of Parallel Lines*, finished in 1766. The work begins with a discussion regarding past attempts to prove the Parallel Postulate, and then Lambert tries it himself (Bonola 1912, 44–46). Following a strategy similar to Saccheri's, Lambert constructs a quadrilateral $ABCD$ by starting with the **base** $\overline{AB}$. At the endpoints of the base he erects perpendiculars $\overline{AD}$ and $\overline{BC}$. At D he constructs a perpendicular to $\overline{AD}$

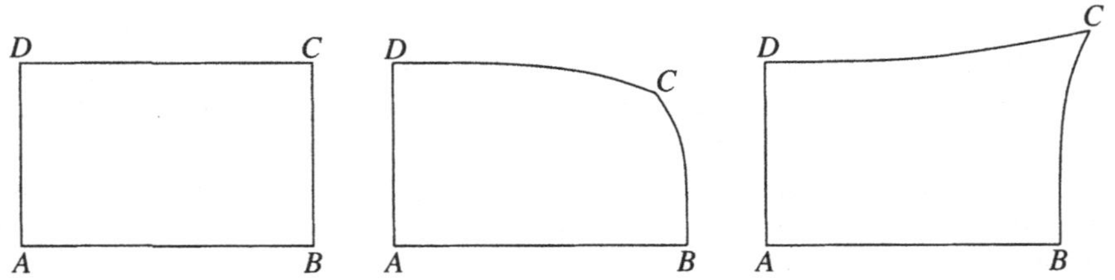

(a) Right fourth angle (b) Obtuse fourth angle (c) Acute fourth angle

Figure 11.1 The three options for a Lambert quadrilateral.

that meets $\overline{BC}$ at C. This quadrilateral with three right angles is called a **Lambert quadrilateral**, and we will call $\sphericalangle DCB$ its **fourth angle**. This fourth angle can either be right, obtuse, or acute, as illustrated in Figure 11.1, and its size will affect the shape of the quadrilateral. The possibilities are enumerated in the next theorem. Its proof relies heavily on Theorem 10.2.4 and is left to Exercise 2.

THEOREM 11.1.1

In every Lambert quadrilateral:

- If the fourth angle is also right, opposite sides will be congruent because it will then be a rectangle.
- If the fourth angle is obtuse, the sides adjacent to the obtuse angle have lesser length than their opposite sides.
- If the fourth angle is acute, the sides adjacent to the acute angle have greater length than their opposite sides.

The question concerning parallels reduces to what can be proven about the fourth angle. We begin by modifying Saccheri's three hypotheses for Lambert quadrilaterals. We again have three options.

DEFINITION 11.1.2

- **LHR**$(ABCD)$ is the sentence

  the fourth angle of Lambert quadrilateral ABCD is right.

- **LHO**$(ABCD)$ is the sentence

  the fourth angle of Lambert quadrilateral ABCD is obtuse.

- **LHA**$(ABCD)$ is the sentence

  the fourth angle of Lambert quadrilateral ABCD is acute.

As with Saccheri, Lambert is able to prove that if a Lambert quadrilateral satisfies one of the three hypotheses, all Lambert quadrilaterals will satisfy that same hypothesis. However, he did not rely on Saccheri's results to accomplish this. Since we do have them readily available, we will. We begin with a lemma.

■ LEMMA 11.1.3

Let $ABCD$ be a Lambert quadrilateral with base $\overline{AB}$.

- If **LHR**$(ABCD)$, then **HR**.
- If **LHO**$(ABCD)$, then **HO**.
- If **LHA**$(ABCD)$, then **HA**.

PROOF

We have three cases to prove.

- Assume **LHR**$(ABCD)$. This means that $ABCD$ is a rectangle and that the opposite sides of every rectangle are congruent. Therefore, $ABCD$ is a Saccheri quadrilateral with right summit angles. By Theorem 10.2.7 we conclude **HR**.

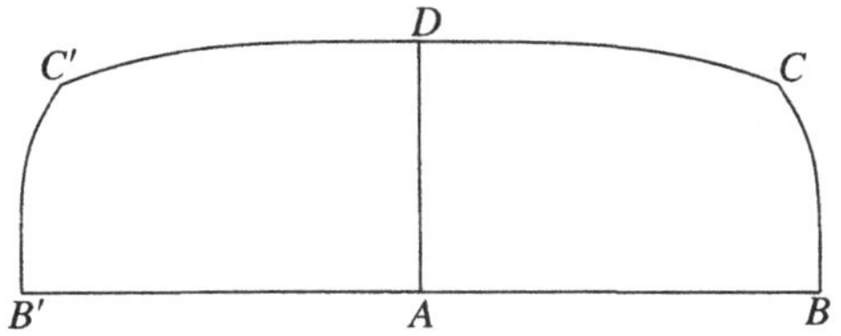

- Now suppose that the fourth angle of Lambert quadrilateral $ABCD$ is obtuse. Construct another Lambert quadrilateral $AB'C'D$ with base $\overline{B'A}$ such that $AB'C'D \cong ABCD$ (Exercise 1). In other words, construct a copy of the original Lambert quadrilateral on $\overline{AD}$ that is a reflection of the original. Hence, $\overline{B'C'}$ and $\overline{BC}$ are perpendicular to $\overline{B'B}$, and $\overline{B'C'} \cong \overline{BC}$. This means that $B'BCC'$ is a Saccheri quadrilateral with obtuse summit angles. Therefore, **HO**.
- If **LHA**$(ABCD)$, repeat the previous construction. This time the result is a Saccheri quadrilateral with acute summit angles, which gives us **HA**.

■

We can now Lemma 11.1.3 to show that once a Lambert quadrilateral satisfies one of the three new hypotheses, all Lambert quadrilaterals must satisfy that same hypothesis.

THEOREM 11.1.4

Let $ABCD$ be a Lambert quadrilateral with base $\overline{AB}$.

- If **LHR**$(ABCD)$, the fourth angle of every Lambert quadrilateral is right.
- If **LHO**$(ABCD)$, the fourth angle of every Lambert quadrilateral is obtuse.
- If **LHA**$(ABCD)$, the fourth angle of every Lambert quadrilateral is acute.

PROOF

Suppose **LHR**$(ABCD)$. By Lemma 11.1.3 we conclude **HR**. If there exists a Lambert quadrilateral with either an obtuse or acute fourth angle, we may conclude either **HO** or **HA**. This would contradict Theorem 10.2.7. Therefore, there is no Lambert quadrilateral that does not have a right fourth angle. The proof of the other two cases follows along these lines. ■

We now have everything set up to show that the three hypotheses of Saccheri are equivalent to the three hypotheses of Lambert.

THEOREM 11.1.5

- **HR** if and only if the fourth angle of every Lambert quadrilateral is right.
- **HO** if and only if the fourth angle of every Lambert quadrilateral is obtuse.
- **HA** if and only if the fourth angle of every Lambert quadrilateral is acute.

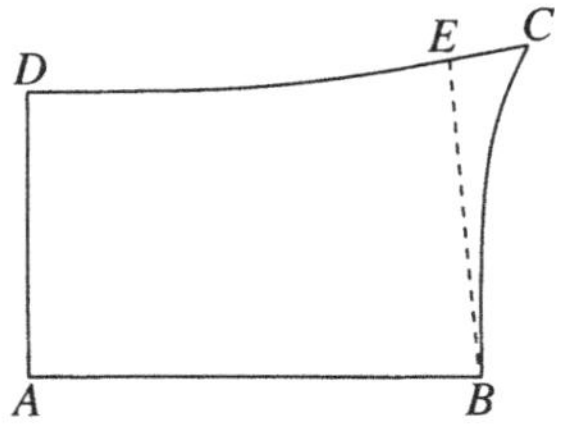

PROOF

We will prove the first equivalence. The other two are left for Exercise 9. First note that if **HR** is false, by Lemma 11.1.3 there can be no Lambert quadrilateral with a right fourth angle. So assume **HR** and let $ABCD$ be a Lambert quadrilateral with $\sphericalangle C$ being the fourth angle. If $\sphericalangle C$ is acute, then by Theorem 11.1.1 we have $DC > AB$. Find E on $\overline{DC}$ such that $\overline{DE} \cong \overline{AB}$. Join $\overline{EB}$ to create Saccheri quadrilateral $DABE$ with base $\overline{DA}$. Since $\overline{EB}$ is on the interior of right $\sphericalangle ABC$, it is impossible that $\sphericalangle ABE$ is a right angle, but it is by **HR**. Thus, $\sphericalangle C$ cannot be acute. A similar argument shows that $\sphericalangle C$ is not obtuse. Therefore, **LHR**$(ABCD)$, and we may conclude by Theorem 11.1.4 that all Lambert quadrilaterals have a right fourth angle. ■

Because of the equivalence between the two hypotheses, we redefine Saccheri's three hypotheses to include information about Lambert quadrilaterals.

■ **DEFINITION 11.1.6**

- **HR** is the sentence

  the summit angles of every Saccheri quadrilateral are both right, and the fourth angle of every Lambert quadrilateral is right.

- **HO** is the sentence

  the summit angles of every Saccheri quadrilateral are both obtuse, and the fourth angle of every Lambert quadrilateral is obtuse.

- **HA** is the sentence

  the summit angles of every Saccheri quadrilateral are both acute, and the fourth angle of every Lambert quadrilateral is acute.

Due to Theorems 11.1.4 and 11.1.5, we conclude that Theorem 10.2.11 still holds under the redefinition. We also know that Saccheri's **HR** is equivalent to the Parallel Postulate (Theorem 10.3.4 and Exercise 9 of Section 10.2). We restate both results.

■ **THEOREM 11.1.7**

Exactly one of the following hypotheses is true: **HR**, **HO**, or **HA**.

■ **THEOREM 11.1.8**

The Parallel Postulate is true if and only if **HR**.

PROOF

We already know that **HR** implies the Parallel Postulate (Theorem 10.3.4), so suppose that the Parallel Postulate holds and let $ABCD$ be a Lambert quadrilateral. By Theorem 5.4.4, the angle sum of $ABCD$ is four right angles. Since the Lambert quadrilateral already has three right angles, its fourth angle also must be right. Therefore, **HR**. ■

Like Saccheri, Lambert planned on proving that his version of **HR** was a theorem in absolute geometry, and like Saccheri, he did not succeed. He did show that **HO** can be eliminated as a possibility, but he could not do the same for **HA**. Unlike Saccheri, Lambert realized that he had failed to meet his goal. He also realized that the geometry that he was studying had many strange results but appeared to be a consistent mathematical system. He thought this because he would later prove that the theorems that followed from **HA** could be expressed using formulas from spherical trigonometry. The problem was that the formulas required imaginary values for the radii of the spheres. Lambert concluded that the geometry that followed from **HA** was the geometry on the surface of a sphere with imaginary radius, whatever that meant. Since spherical trigonometry was a well-established mathematical system, Lambert thought that it gave a foundation to his system, but he never did find a proof that completely settled the question. It is apparently for this reason that Lambert never published his results (Bonola 1912, 45, 49–50; Kline 1980, 81).

Exercises

1. Construct a Lambert quadrilateral and then construct another that is congruent and adjacent to the original so that combined, they form a Saccheri quadrilateral.

2. Prove Theorem 11.1.1.

3. Let $ABCD$ be a Lambert quadrilateral with the fourth angle being $\sphericalangle DCB$. Prove the following (near) converses to Theorem 11.1.1:

 (a) If opposite sides are congruent, then **HR**.
 (b) If $DC < AB$, then **HO** and $BC < AD$.
 (c) If $DC > AB$, then **HA** and $BC > AD$.

4. Why is the existence of a Lambert quadrilateral with an obtuse fourth angle incompatible with absolute geometry?

5. Identify the error in the following proof: Assume **HA**. Let $ABCD$ be a Lambert quadrilateral with $\overline{AD} \cong \overline{AB}$. Join $\overline{AC}$ to form $\triangle ABC$ and $\triangle ADC$. Since $\sphericalangle ABC$ and $\sphericalangle ADC$ are right, we may appeal to HL (Exercise 5.6.4) to conclude that $\triangle ABC \cong \triangle ADC$.

6. Explain why if both $ABCD$ and $BADC$ are Lambert quadrilaterals, $ABCD$ is a rectangle.

7. Let $ABCD$ be a Saccheri quadrilateral. Prove that if M is the midpoint of $\overline{AB}$ and N is the midpoint of $\overline{DC}$, then $MADN$ and $MBCN$ are Lambert quadrilaterals.

8. Given Lambert quadrilateral $MADN$, prove that Saccheri quadrilateral $ABCD$ can be constructed such that M is the midpoint of $\overline{AB}$ and N is the midpoint of $\overline{DC}$.

9. Prove the last two equivalences of Theorem 11.1.5.

10. Prove the following theorem that is due to Giordano Vitale: If three points of a line l are equidistant from three points on a line m, then l and m are equidistant (Rosenfeld 1988, 248).

11.2 POLYGONS

The three hypotheses determine the measure of some of the angles of Saccheri and Lambert quadrilaterals. The hypotheses also determine the sum of the internal angles of triangles and quadrilaterals, as we saw in Theorems 10.3.1 and 10.3.2. In this section we elaborate on this relationship.

Rectangles

The quadrilaterals of Saccheri and Lambert are nearly rectangles. The summit angles of a Saccheri quadrilateral may not be right angles, and a Lambert quadrilateral may be missing one. In fact, we should note that according to Theorem 5.4.1, a Lambert quadrilateral is a parallelogram in the sense that opposite sides will not intersect. But if the fourth angle is not right, we do not want to say that it is a rectangle even though it appears to satisfy the definition on page 151. That definition is legitimate only under the Parallel Postulate. Let us then slightly modify the definition of a rectangle so that we can use it whether or not we have the Parallel Postulate.

■ **DEFINITION 11.2.1**

A quadrilateral with four right angles is called a **rectangle**.

If we assume the Parallel Postulate, the figures that are considered rectangles under the previous definition are exactly those figures that are rectangles under the new definition (Exercise 2). Furthermore, based on our discussion it should come as no surprise that there is a close relationship between the Parallel Postulate and rectangles.

■ **THEOREM 11.2.2**

The Parallel Postulate is true if and only if there exists a rectangle.

PROOF

Suppose that the Parallel Postulate is true. By following the description of a Lambert quadrilateral on page 356, we know that we can construct a quadrilateral with three right angles. By Theorem 11.1.8, the fourth angle must be right, so this quadrilateral is a rectangle.

Suppose that we have constructed a rectangle. This means that we have a Lambert quadrilateral with a right fourth angle. Thus, **HR** holds, and we conclude that the Parallel Postulate is true by Theorem 11.1.8. ■

The existence of a rectangle does more than prove the Parallel Postulate. It also allows use to construct a rectangle of any dimension that we want. We will prove this in stages. We first construct a rectangle with one side that is greater in length than a given magnitude. Next we construct a rectangle with all sides greater than a given magnitude. Once this is done we will be able to construct our rectangle of arbitrary size. To begin we need a theorem of absolute geometry.

■ **THEOREM 11.2.3**

The fourth angle of a Lambert quadrilateral is not obtuse.

PROOF

Since **HO** is incompatible with absolute geometry (Theorem 10.3.5), only **HR** and **HA** are possible in that system. Therefore, by Theorem 11.1.5, the fourth angle of a Lambert quadrilateral is either right or acute. ■

Now for our first rectangle proposition.

■ LEMMA 11.2.4

For any magnitude r, if a rectangle exists, there exists a rectangle with a side of length greater than r.

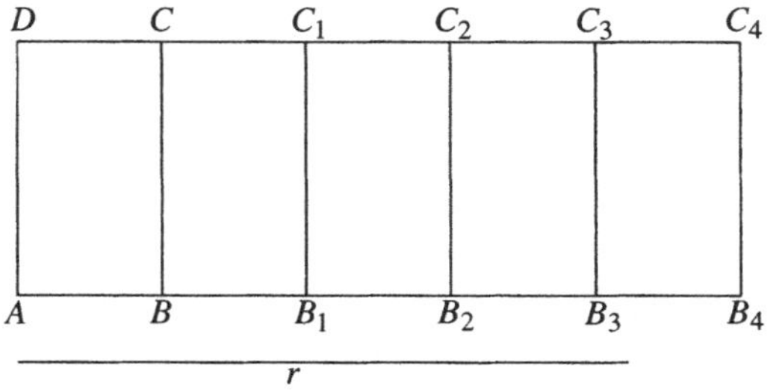

PROOF

Let $r > 0$ and let $ABCD$ be a rectangle. By Archimedes' Postulate (6.2.1) we can find k sufficiently large so that $kAB > r$ and $k > 2$. Notice that in the diagram we assume that $k = 5$. On $\overline{BC}$, construct BB_1C_1C to be a copy of the original rectangle. Then on $\overline{B_1C_1}$ construct $B_1B_2C_2C_1$ to be another copy of $ABCD$. Continue to copy rectangles until we have $k - 1$ copies, with the last one being $B_{k-2}B_{k-1}C_{k-1}B_{k-2}$. This means that we have constructed rectangle $AB_{k-1}C_{k-1}D$, where $AB_{k-1} > r$. ■

Since the proof of the next lemma simply takes the method of the preceding proof and applies it twice, once in each direction, we state it without proof.

■ LEMMA 11.2.5

For magnitudes r and s, if a rectangle exists, there exists a rectangle with one side of length greater than r and another side of length greater than s.

We can now prove our theorem. Be careful to note that it does not state that rectangles exist. It simply says that if one exists, we have them in all possible sizes.

■ THEOREM 11.2.6

If a rectangle exists, there exists a rectangle with sides of any given length.

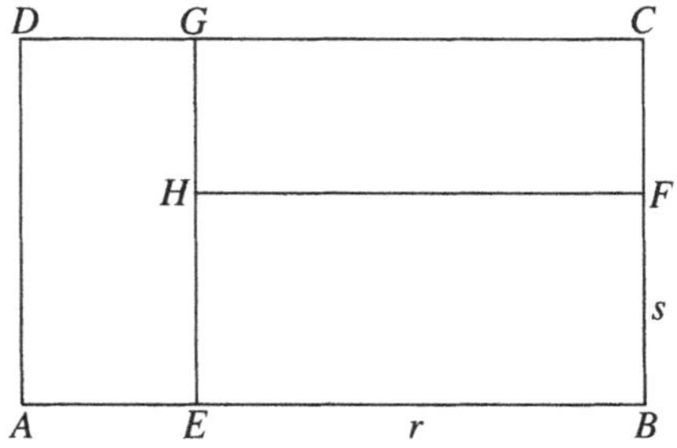

PROOF

Let $r > 0$ and $s > 0$. Suppose that we know that a rectangle exists. We will construct an $r \times s$ dimension rectangle. Let $ABCD$ be the rectangle that is guaranteed by Lemma 11.2.5 with $AB > r$ and $CD > s$. Find E on $\overline{AB}$ such that $EB = r$ and find F on $\overline{BC}$ so that $BF = s$. Find point G on $\overline{DC}$ such that $\overline{EG}$ is perpendicular to $\overline{AB}$ and find H on $\overline{EG}$ so that $\overline{HF}$ is perpendicular to $\overline{BC}$. This means that $BEGC$, $AEGD$, and $BFHE$ are Lambert quadrilaterals. By Theorem 11.2.3 we see that neither $\measuredangle DGE$ nor $\measuredangle CGE$ is obtuse. Since they cannot both be acute, they must both be right angles. From this we conclude that $CFHG$ is a Lambert quadrilateral. If $\measuredangle EHF$ is acute, $\measuredangle GHF$ is obtuse, contradicting Theorem 11.2.3. Therefore, $\measuredangle EHF$ is right, and we see that $EBFH$ is a rectangle of the required dimensions. ■

Legendre

The following result was known to Saccheri. It follows from Theorem 10.3.1 with the understanding that **HO** is inconsistent with absolute geometry (Theorem 10.3.5). The proof given here is due to Adrien-Marie Legendre (Bonola 1912, 55–56).

■ THEOREM 11.2.7 [Saccheri-Legendre]

The sum of the angle measures of a triangle is at most two right angles.

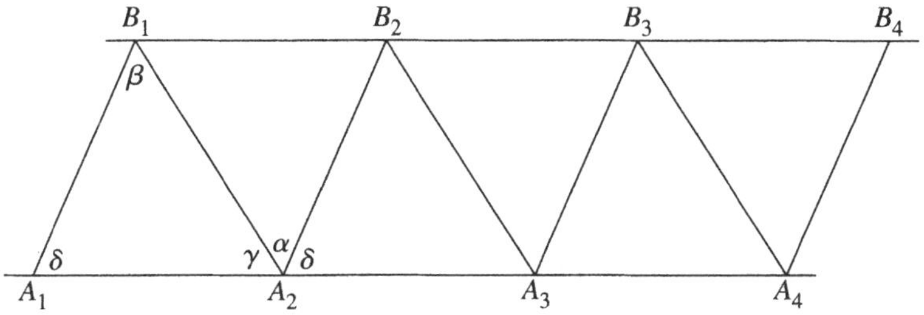

PROOF

Let $\overline{A_1A_2}$, $\overline{A_2A_3}$, ..., $\overline{A_nA_{n+1}}$ be congruent, consecutive segments on a line. Construct congruent triangles

$$\triangle A_1A_2B_1,\ \triangle A_2A_3B_2,\ \ldots,\ \triangle A_nA_{n+1}B_n.$$

Observe that

$$\triangle B_1B_2A_2,\ \triangle B_2B_3A_3,\ \ldots,\ \triangle B_nB_{n+1}A_{n+1}$$

are also all congruent (SAS). Let $\alpha = m\measuredangle B_1A_2B_2$ and $\beta = m\measuredangle A_1B_1A_2$. To obtain a contradiction, suppose that $\beta > \alpha$. This yields $A_1A_2 > B_1B_2$ by Exercise 6. Since

$$A_1B_1 + B_1B_{n+1} + A_{n+1}B_{n+1} > A_1A_{n+1},$$

we may conclude that

$$A_1B_1 + nB_1B_2 + A_{n+1}B_{n+1} > nA_1A_2.$$

In other words, for all positive integers n,

$$2A_1B_1 > n(A_1A_2 - B_1B_2), \tag{11.1}$$

but then by Archimedes' Postulate (6.2.1) we may choose n sufficiently large so that the opposite inequality to (11.1) holds. This is a contradiction . Therefore, $\beta \leqslant \alpha$. To finish the proof, let $\delta = \text{m}\measuredangle B_1A_1A_2$ and $\gamma = \text{m}\measuredangle A_1A_2B_1$. We may now conclude that

$$\beta + \delta + \gamma \leqslant \alpha + \delta + \gamma = \pi. \blacksquare$$

The Saccheri-Legendre Theorem generalizes quickly to give us a bound on the angle sum of quadrilaterals.

■ **THEOREM 11.2.8**

The sum of the angle measures of a quadrilateral is at most four right angles.

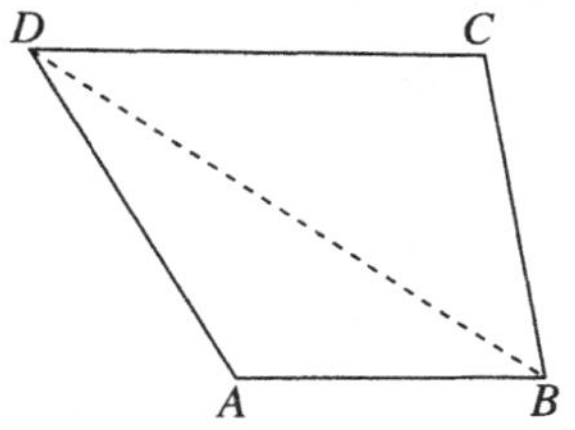

PROOF

Let $ABCD$ be quadrilateral. Join $\overline{DB}$ to form $\triangle ABD$ and $\triangle CBD$. If the quadrilateral is not convex, label the vertices so that the diagonal $\overline{DB}$ is in the interior of the quadrilateral. By the Saccheri-Legendre Theorem,

$$\text{m}\measuredangle ADB + \text{m}\measuredangle DBA + \text{m}\measuredangle DAB \leqslant \pi$$

and

$$\text{m}\measuredangle CDB + \text{m}\measuredangle DBC + \text{m}\measuredangle DCB \leqslant \pi.$$

Since

$$\text{m}\measuredangle ADB + \text{m}\measuredangle CDB = \text{m}\measuredangle ADC$$

and

$$\text{m}\measuredangle DBA + \text{m}\measuredangle DBC = \text{m}\measuredangle ABC,$$

we conclude that $\text{m}\measuredangle ABC + \text{m}\measuredangle DAB + \text{m}\measuredangle ADC + \text{m}\measuredangle DCB \leqslant 2\pi. \blacksquare$

Lambert completely generalized this result to any convex polygon (Bonola 1912, 46). See Exercise 9.

The mathematician that shares credit for the Saccheri-Legendre Theorem was born in Paris on September 18, 1752. Legendre was from a wealthy family, and this afforded him luxuries that other mathematicians did not always enjoy. He attended school at the Collège Mazarin where he studied mathematics and the sciences. He graduated in 1770 when he was 18, and due to his financial situation he dedicated full time to research. Legendre showed his promise when in 1782 he won the annual prize awarded by the Berlin Academy. He wrote a paper on the curve determined by the flight of a cannonball, which included the effects of air resistance, different initial velocities, and different angles of projection (Itard 1970, 135).

Legendre's family fortune did not prevent him from being willing to work. From 1775 to 1780 he taught mathematics at the École Militaire in Paris, and three years later when a position at the Académie des Sciences opened, he was elected to fill it. There Legendre continued his research, and his reputation as a mathematician continued to grow. His favorite subjects were number theory, celestial mechanics, and elliptic functions. In 1785 he advanced to the position of associate professor at the Académie, but when the school closed in 1793, Legendre was overlooked when many of his colleagues received invitations to help overhaul the French education system. However, a year later, in 1794, Legendre took a leadership position that was charged by the government with the oversight of weights and measures and the general promotion of science (Itard 1970, 135–136; Boyer and Merzbach 1991, 486; Burton 1985, 534).

Despite his significant mathematical work, Legendre was again slighted by the French government when in 1795 some 48 academics from around France were asked to be the initial faculty of the newly formed Institut National des Sciences et des Arts. His colleagues, though, would have none of this, and elected Legendre to be a resident member of the mathematics department. It is true that his mathematical work did not reach the level of the top French mathematicians at the time, such as Joseph-Louis Lagrange and Pierre-Simon Laplace, but it is also true that his contributions were significant. It is very possible that Legendre was overlooked by many in power because he was a man of conscience and not afraid to speak his mind. In France at that time such convictions were not always profitable. In 1824 Legendre refused to vote in support of the government's candidate for the institute. Because of this he lost his pension, and when he died on January 10, 1833, he was living in poverty (Burton 1985, 534; Gray 2007, 3).

Besides the theorem that bears his name, Legendre appears here because in 1794, the year of the Reign of Terror, Legendre published the first edition of his *Éléments de géométrie*. Although his main fields of interests did not include geometry, he was motivated to write by the poor state of mathematics education in France at the time, particularly in geometry. The *Éléments* became very influential and went through 20 editions. The work was similar to Euclid's but with a slightly different approach. Legendre tried to strike the right balance between the rigor that geometry requires and the clarity of explanation that both students and their teachers needed. It included his own proof that π is irrational and his own attempt at proving the Parallel Postulate by examining the angle sum of triangles. The book's influence even reached across the Atlantic, where the American edition, known as *Geometry and Trigonometry, from the Works of A. M. Legendre* (Davies 1858), was the top geometry textbook in the United States until 1885 (Boyer and Merzbach 1991, 485–486; Itard 1970, 136; Gray 2004, 31–32).

Triangles

Following the lead of Legendre, we now demonstrate that the angle sum of a triangle equals two right angles if and only if **HR**. Because of the Saccheri-Legendre Theorem (11.2.7), this then yields a related result for **HA**. We begin with a theorem that Legendre included in his *Éléments* (Euclid 1925, Vol. I, 217).

■ **THEOREM 11.2.9**

If the sum of the angles of a given triangle equals two right angles, the sum of the angles of a triangle formed by joining a vertex of the given triangle to its opposite sides also equals two right angles.

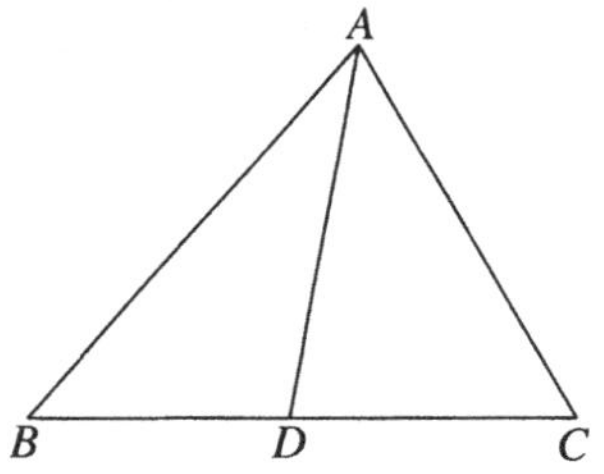

PROOF

Let $\triangle ABC$ have zero defect. Choose D on $\overline{BC}$ and join $\overline{AD}$. Suppose that $\mathsf{d}(ABD) > 0$. Since $\mathsf{d}(ABC) = \mathsf{d}(ABD) + \mathsf{d}(ADC)$ by Theorem 10.4.5, we conclude that $\mathsf{d}(ADC) < 0$. This means that the sum of the internal angles of $\triangle ADC$ is greater than two right angles, contradicting the Saccheri-Legendre Theorem (11.2.7). ■

Theorem 11.2.9 allows us to prove the next. It will connect our work on triangles with our work on rectangles.

■ **THEOREM 11.2.10**

If there exists a triangle with angle sum equal to two right angles, there exists a rectangle.

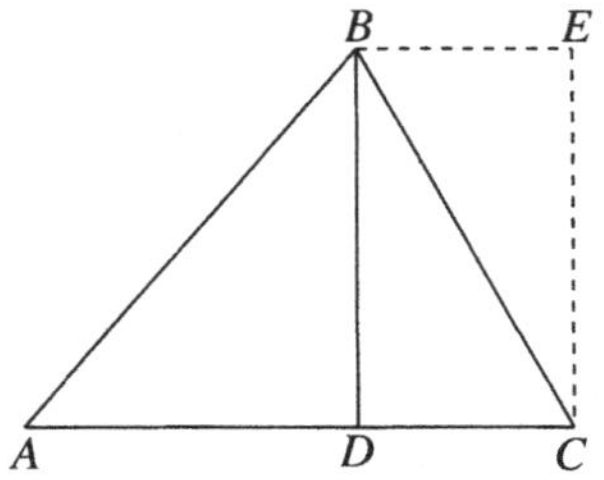

PROOF

Let $\triangle ABC$ have zero defect. Find D on $\overline{AC}$ such that $\overline{BD}$ is perpendicular to $\overline{AC}$. If this is not possible, choose another side. By Theorem 11.2.9, the defect of $\triangle DBC$ must be zero. Find the point E such that

$$\measuredangle DCB \cong \measuredangle EBC$$

and

$$\measuredangle DBC \cong \measuredangle ECB.$$

By ASA we see that $\triangle DBC \cong \triangle ECB$. Therefore, $\measuredangle BEC$ is right. We also have

$$m\measuredangle DBC + m\measuredangle DCB = \pi/2.$$

Therefore, we obtain

$$m\measuredangle DBC + m\measuredangle EBC = \pi/2$$

and

$$m\measuredangle ECB + m\measuredangle DCB = \pi/2.$$

This means that both $\measuredangle DBE$ and $\measuredangle DCE$ are right angles, so $BECD$ is a rectangle. ■

We know that the existence of a rectangle is equivalent to the Parallel Postulate (Theorem 11.2.2). Therefore,

> if there exists a triangle with angle sum equal to two right angles, the Parallel Postulate holds.

Moreover, if the Parallel Postulate holds, *all* triangles have angle sum equal to two right angles (Theorem 5.4.3). Hence:

■ **COROLLARY 11.2.11**

If there exists a triangle with angle sum equal to two right angles, all triangles have angle sum equal to two right angles.

■ **COROLLARY 11.2.12**

The following are equivalent in absolute geometry:

- The Parallel Postulate.
- All triangles have angle sum equal to two right angles.
- All quadrilaterals have angle sum equal to four right angles.

Finally, if there is a triangle with angle sum less than two right angles, all triangles must have angle sum less than two right angles. This follows from the Saccheri-Legendre Theorem (11.2.7) and Corollary 11.2.11.

■ COROLLARY 11.2.13

If there exists a triangle with angle sum less than two right angles, all triangles have angle sum less than two right angles.

We finish our sequence of corollaries by observing:

■ COROLLARY 11.2.14

The following are equivalent in absolute geometry:

- **HA**.
- All triangles have angle sum less than two right angles.
- All quadrilaterals have angle sum less than four right angles.

As triangles on a sphere bend outward, Lambert must have imagined that the triangles under the Hypothesis of the Acute Angle existed on some kind of surface that curved in such a way that the angles would bend inward. This is the idea behind his sphere of imaginary radius. So triangles with positive defect at least have some frame of reference with which to understand them. It is the next result, however, that will lead to a result that must have made Lambert realize that a world in which **HA** is true would be fundamentally different from the world of Euclid's geometry and from the immediate world that we seem to see around us (Bonola 1912, 46–50; Wolfe 1945, 83).

■ THEOREM 11.2.15 [AAA]

[HA] In a pair of triangles, if corresponding angles are congruent, the triangles are congruent.

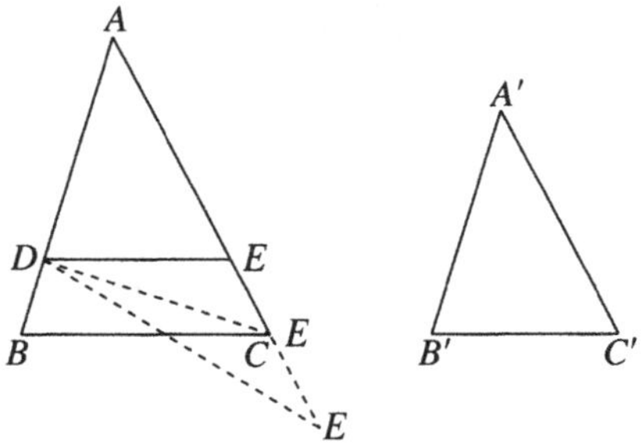

PROOF

Given $\triangle ABC$ and $\triangle A'B'C'$, assume that corresponding angles are congruent. By SAS we need only one pair of corresponding sides to be congruent to prove that the triangles are congruent. To accomplish this, we will show that $\overline{AB} \cong \overline{A'B'}$ indirectly. First suppose that $AB > A'B'$ and then find D on $\overline{AB}$

such that $\overline{AD} \cong \overline{A'B'}$. Also, find E on $\overline{AC}$ (produced if necessary) such that $\overline{AE} \cong \overline{A'C'}$. This means that $\triangle ADE \cong \triangle A'B'C'$ by SAS. Therefore,

$$\measuredangle ADE \cong \measuredangle ABC \text{ and } \measuredangle AED \cong \measuredangle ACB.$$

Notice that this implies that E is between A and C. This is because if C and E are the same point, $\overline{DE}$ is on the interior of $\measuredangle ACB$, which is congruent to $\measuredangle AED$, and if E is beyond C, we have an angle that is congruent to one of its opposite and internal angles, contradicting Theorem 5.3.9. Therefore, $DECB$ is a convex quadrilateral, and since $\measuredangle ADE$ and $\measuredangle EDB$ are supplementary and $\measuredangle AED$ and $\measuredangle DEC$ are also supplementary,

$$\text{m}\measuredangle ADE + \text{m}\measuredangle EDB + \text{m}\measuredangle AED + \text{m}\measuredangle DEC = 2\pi.$$

So we have

$$\text{m}\measuredangle ABC + \text{m}\measuredangle EDB + \text{m}\measuredangle ACB + \text{m}\measuredangle DEC = 2\pi.$$

This contradicts Corollary 11.2.14. We have a similar problem when we assume that $AB < A'B'$. Hence, $\overline{AB} \cong \overline{A'B'}$, and the triangles are congruent. ■

The theorem's corollary is quickly proven from the theorem. See Exercise 8.

■ COROLLARY 11.2.16

The Parallel Postulate is true if and only if there exists a pair of similar but noncongruent triangles.

Consider what this means. In Euclid's geometry, measurements of angles are based on magnitudes inherit to the system. We begin by constructing a right angle. That such a construction is possible follows logically from the postulates. From this we assign measures. The right angle is assigned the measure of $\pi/2$ radians or 90 degrees. A straight angle has either π radians or 180 degrees. The measure of half of a right angle is $\pi/4$ radians or 45 degrees. Both units of angle measure start with the right angle and proceed from there.

Lengths of line segments are different, at least in the Euclidean geometry with which we are familiar. If we take a foot-long segment, we can measure it as having a length of 1 foot, 12 inches, or one third of a yard. Which one is completely up to us. Unlike the right angle, there is no natural line segment that can be constructed from Euclid's postulates that would serve as a reference for all others. Since the distance between two points is represented by the length of the line segment connecting them, this discussion carries over to distances. The units used to measure the distance between two points can be feet, inches, yards, or any other arbitrarily chosen length.

■ DEFINITION 11.2.17

If a unit of measure can be defined in terms of an object that exists because it follows logically from the postulates, the unit is called an **absolute measure**. A unit is a **relative measure** if it cannot be based on such an object.

Degrees and radians are examples of absolute measures because they are defined in terms of the right angle. In Euclidean geometry measures of length are relative measures since they can only be chosen arbitrarily. Any measure for area or volume will also be relative in Euclid's system because they are based on length. However, things change when we assume **HA**. Take $\overline{AB}$ and construct an equilateral triangle with $\overline{AB}$ as one of its sides (Construction 5.2.6). We know that every equilateral triangle is equiangular and that this fact can be proven without the use of the Parallel Postulate (Exercise 14). Let θ represent the measure of the angles of the triangle. Let $\overline{CD}$ be another segment such that $AB \neq CD$ and construct an equilateral triangle on it. Let ψ be the measure of each angle in the new triangle. If $\psi = \theta$, the triangles will be similar and thus congruent (Theorem 11.2.15). This would mean that $AB = CD$, contradicting our hypothesis. Therefore, θ is the unique angle measure associated with $\overline{AB}$ and is dependent on its length. We may then define the length of $\overline{AB}$ to be θ. Hence, length is an absolute measure when **HA** is assumed (Gray 2007, 84).

Without providing the details for the converse, the previous argument results in the next theorem.

■ **THEOREM 11.2.18**

There exists an absolute measure for length if and only if the Parallel Postulate is false.

Including the work of Wallis (Exercise 13 in Section 10.2), Saccheri, Lambert, and Legendre, we have encountered various propositions that are equivalent to the Parallel Postulate. It is time to summarize them.

■ **THEOREM 11.2.19**

The following are equivalent in absolute geometry:

- The Parallel Postulate.
- The Hypothesis of the Right Angle.
- Playfair's Postulate.
- Aristotle's Postulate.
- Parallel lines are equidistant.
- Wallis's Postulate.
- The angle sum of every triangle equals two right angles.
- There exists a rectangle.
- There is no absolute unit of length.
- The Pythagorean Theorem.

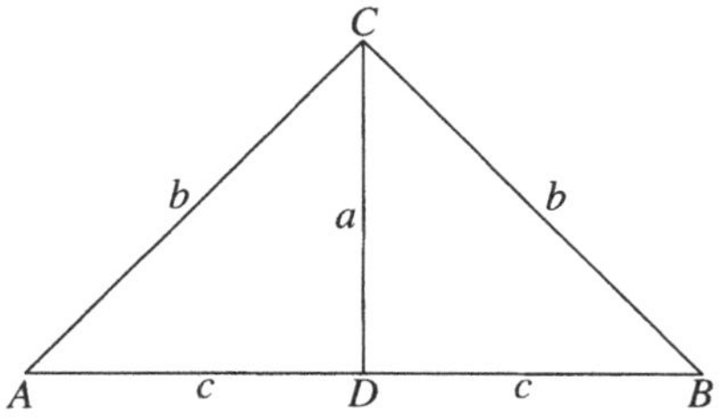

PROOF

Because this section is dedicated to triangles, we will prove that the Pythagorean Theorem (5.6.1) is equivalent to the Parallel Postulate. An examination of the proof on page 168 shows that the Parallel Postulate (plus the other theorems of absolute geometry) implies the Pythagorean Theorem. To prove the converse, assume that the Pythagorean Theorem holds. We must be careful not to use any result of the Parallel Postulate to deduce it. Let $\triangle ABC$ be an isosceles right triangle with $\sphericalangle C$ being the right angle. If we can prove that the angle sum of this triangle equals two right angles, by Corollary 11.2.12 we will have shown the Parallel Postulate. We begin by dropping a perpendicular from C to $\overline{AB}$, meeting the segment at D. By AAS,

$$\triangle ADC \cong \triangle BDC,$$

so $\overline{AD} \cong \overline{DB}$ and $\sphericalangle ACD \cong \sphericalangle BCD$. To ease our calculations, let $a = CD$, $b = AC = CB$, $c = AD = DB$. By the Pythagorean Theorem,

$$2b^2 = 4c^2 \text{ and } a^2 + c^2 = b^2.$$

Therefore, $b^2 = 2c^2$ and $a^2 + c^2 = 2c^2$. From this we conclude that $a = c$. Hence, $\triangle ADC$ and $\triangle BDC$ are isosceles triangles. We can now summarize the measures of the angles in the figure.

- Since $\sphericalangle ACD \cong \sphericalangle BCD$ and $\sphericalangle ACB$ is a right angle,
$$m\sphericalangle ACD = m\sphericalangle BCD = \pi/4.$$
- By Theorem 5.3.3 we see that $\sphericalangle A \cong \sphericalangle ACD$ and $\sphericalangle B \cong \sphericalangle BCD$, so
$$m\sphericalangle A = m\sphericalangle B = \pi/4.$$

Therefore, $m\sphericalangle A + m\sphericalangle B + m\sphericalangle ACB = \pi/4 + \pi/4 + \pi/2 = \pi$. ■

Exercises

1. List the properties that a rectangle has in common with a Lambert quadrilateral and list the properties they do not share. Repeat this for a Saccheri quadrilateral.

2. Assume the Parallel Postulate. Prove that $ABCD$ is a rectangle according to the definition on page 151 if and only if $ABCD$ is a rectangle according to Definition 11.2.1.

3. Prove that if $ABCD$ is both a Saccheri and Lambert quadrilateral, $ABCD$ is a rectangle.

4. Construct a quadrilateral with four congruent angles. Can this be done if **HA** is assumed?

5. Legendre gave multiple proofs of the Saccheri-Legendre Theorem (11.2.7). This exercise is based on one of them (Burton 1985, 540–541).

   (a) Given $\triangle ABC$, show that there exists $\triangle DEF$ such that the two triangles have the same angle sum but $\mathrm{m}\sphericalangle DEF \leqslant \frac{1}{2}\mathrm{m}\sphericalangle ABC$.
   (b) Use this to give an alternate proof of the Saccheri-Legendre Theorem.
   (c) In the proof for Exercise 5(b), was the assumption that lines have infinite length used?

6. Show that $\beta > \alpha$ implies that $A_1A_2 > B_1B_2$ in the proof of the Saccheri-Legendre Theorem (11.2.7).

7. Prove in absolute geometry that an exterior angle of a triangle has measure greater than or equal to the sum of the measures of the two opposite interior angles.

8. Prove Corollary 11.2.16.

9. Prove that for all integers $n > 2$, the sum of the angle measure of a convex n-gon is at most $2(n-2)$ right angles.

10. Given $\triangle ABC$ and $\triangle DEF$, prove that if $\mathsf{d}(ABC) > \mathsf{d}(DEF)$, the angle sum of $\triangle ABC$ is less than the angle sum of $\triangle DEF$.

11. Restate Theorem 11.2.10 in terms of angle defect.

12. Prove the following without appealing to the rectangle theorems of this section. If there exists a triangle with angle sum equal to two right angles, then for any magnitude r:

   (a) There exists a square with sides of length greater than r.
   (b) There exists an isosceles right triangle with legs of length greater than r.

13. Prove in absolute geometry: All triangles have angle sum less than two right angles if and only if **HA**.

14. Check that the proof of the equivalence of a triangle being equiangular and equilateral is a theorem of absolute geometry.

15. Suppose that there exists a number n such that the angle sum of every triangle equals n. Prove that the Parallel Postulate holds.

16. Prove that the following are equivalent to the Parallel Postulate.

   (a) If l and m are parallel and n is perpendicular to l, n is perpendicular to m.
   (b) A line perpendicular to one side of an acute angle intersects the other side.
   (c) The perpendicular bisectors of the sides of a triangle are concurrent.

17. Find propositions that can be added to the list of Theorem 11.2.19.

11.3 OMEGA TRIANGLES

Johann Friedrich Carl Gauss was born on April 30, 1777 in Brunswick in the German Duchy of Brunswick-Wolfenbüttel. He was born into a poor family. His paternal grandparents were originally small farmers but moved to Brunswick around 1740 in search of a better life. Gauss' parents, Gebhard and Dorothea, were hardworking but not well educated. Gebhard found that the local guilds did not welcome outsiders, so he struggled to find work. He would move from job to job, sometimes working as a gardener, a canal worker, a butcher, and sometimes as an accountant. This brought the family steady income, but acquiring wealth was very slow. This meant that the family goal of owning a home would take many years, but once achieved it would be a significant step up in social status (Bühler 1981, 5–6, 8; Burton 1985, 509–510).

After surviving a near drowning when he was around the age of 3, Gauss entered elementary school in 1784. In a class of about 50 students of various ages and abilities, Gauss stood out. He already knew how to read and write upon enrolling, abilities that he had taught himself. Supposedly, when he was 3 he could count and perform simple calculations, even to the point of correcting his father's ledgers. His masters at the school, J. G. Büttner and his assistant Martin Bartels, soon realized that they had an exceptional student on their hands. The details of the story differ among the various reports, but the essentials are that one day Gauss's class was given the task of adding the numbers from 1 to 100, either as simple busywork or as a punishment. Within moments Gauss had the answer, and it was the only correct one in the class! He had simply recognized the pattern in the sequence

$$1, 2, 3, \ldots, 98, 99, 100.$$

Since there are 50 pairs of numbers that add to 101, the sum must be 5050. Fortunately for the young Gauss, both Büttner and Bartels were good teachers, and they encouraged him in his studies. It was Büttner who aided Gauss in enrolling in secondary school in 1788. Gauss definitely benefited from his new surroundings. Not only was his academic life more structured, but he greatly improved his language skills. Up until this point, Gauss spoke only the local dialect of Brunswick. In secondary school he not only learned High German but also Latin (Bühler 1981, 7; Burton 1985, 510).

In 1791 an event happened to Gauss that would open more academic doors. Because of his excellent work in school, he was introduced to Duke Ferdinand of Brunswick. The duke was impressed with Gauss's achievements in school. As he had done with other promising students, Ferdinand granted Gauss the financial backing that would enable him to continue his education. This support was in the form of a stipend and would continue for many years. It eventually enabled Gauss to earn a doctorate and gave him the freedom to pursue some of his early mathematical research. Such backing was crucial for someone from his social standing at this time in Germany (Bühler 1981, 8; Burton 1985, 510).

The next stop in Gauss's academic career was Caroline College in 1792. He studied there until he was 18. This was a relatively new school, just 10 years old. Its curriculum was focused on the sciences, and it had a very good library. In it

Gauss found most of the great and contemporary works in mathematics. He studied the works of Newton and Archimedes, his two favorite mathematicians. It may have been in this library where Gauss began his work on the Parallel Postulate. As with many before him, he attempted to prove it indirectly, and as with many before him, his results were surprising but not contradictory. He would pursue this subject occasionally throughout the rest of his life (Bühler 1981, 8–9; Bonola 1912, 65).

Thus far, Gauss's education was within Brunswick-Wolfenbüttel. Ferdinand always appreciated it when those he supported attended his schools, but if they decided otherwise, he would still stand behind his students. This is what Gauss did in 1795. He packed up his bags and traveled 65 miles south to Göttingen to attend the university there. It had a solid reputation in the sciences and a good library. As with other institutions of higher learning at the time in Germany, the students were on their own. There were no mandatory classes or exams. The students were completely responsible for their education. Gauss took full advantage of this. He never did receive a Göttingen degree, but by the time that he left in 1798, he had laid the foundation for the mathematical research that would occupy much of his time over the next 25 years (Bühler 1981, 15, 17).

During his Göttingen years Gauss did not make many new friends. An exception was the Hungarian mathematician Farkas Bolyai, who he met in 1795. Bolyai was born on February 9, 1775 in the city of Bolya in Hungary. While he was a boy he attended the Lutheran and Calvinist College of Nagyenyed and showed a talent for mathematics. He was also skilled in music, drawing, and acting (Prékopa 2006, 11). While at the university, Gauss and Bolyai discovered that they shared an interest in the study of parallels. In June 1799, Bolyai left Göttingen to return to his home in Hungary, and for the next 54 years the two wrote frequently. In an example of their early correspondence from December 1799, Gauss discusses his attempts to prove the Parallel Postulate. He told Bolyai that he had tried to prove it by *reductio ad absurdum* arguments and had arrived at results that many would consider contradictory, but they were not. He wrote that because of this he had considered replacing the Parallel Postulate with another more intuitive one. He told Bolyai that "if it could be proved that a rectilinear triangle is possible with an area exceeding any given area," the Parallel Postulate would follow (Wolfe 1945, 128), but Gauss regarded such an assumption as no more likely to be true than the Postulate itself. As he had no experience with indefinitely producing two apparently parallel lines to see if they meet, Gauss had no experience with constructing arbitrarily large triangles. If the Parallel Postulate was to be replaced, any possible candidate must have at least some direct evidence in support of its truth (Bühler 1981, 16–17, 100–101; Bonola 1912, 65–66).

Gauss had wide and varying interests in addition to his mathematical studies. He was particularly fascinated with astronomy since his days at Caroline College. He enjoyed both the time at the telescope and the time crunching numbers, and he was good at both. He had the patience and attention to detail required of the discipline and the amazing ability for calculation that would bring him acclaim. The turn of the nineteenth century was an important time for astronomy. Uranus was discovered in 1781, and it would not be long until astronomers found Neptune in 1846. The instrumentation improved significantly, and with the aid that the newly discovered planets lent to orbital calculations, reliable sky maps were produced.

It was an event on January 1, 1801 that would set in motion a chain of events of which Gauss would take full advantage. The theory of the day stated that there should be a planet somewhere in the gap between Mars and Jupiter. On New Year's Day, Giuseppe Piazzi was observing the gap and thought that he had found the planet. He named it Ceres. We know it today as one of the larger minor planets in the asteroid belt. Piazzi successfully mapped approximately nine degrees of its orbit before it disappeared behind the "shadow of the sun." It was expected to reappear in late 1801 or early 1802, but since Piazzi had recorded only a little orbital data, where it would reappear was uncertain. Numerous astronomers made predications including Franz Xaver von Zach, one of Germany's top astronomers and the editor of the astronomical journal *Monatliche Correspondenz*. Gauss also made a prediction. When Ceres finally was sighted on December 7, 1801, it was Zach who found it. He published the results in the February issue of the *Monatliche Correspondenz*, proclaiming to the astronomical community that it was the mathematician Gauss whose prediction was closest. In 1809 Gauss published the full details of his technique in his famous *Theory of the Motion of the Heavenly Bodies Moving about the Sun in Conic Sections* (Pannekoek 1961, 352–353; Bühler 1981, 43–44, 46).

Gauss was well on his way to becoming the greatest mathematician of his day and possibly of all time. With the Ceres calculation he earned the distinction of being among the top astronomers. Such a reputation had its benefits. In 1802, he began discussions with the Brunswick government about building a new observatory. In that same year he received an invitation from the academy at St. Petersburg to become the director of their observatory. When the astronomer Wilhelm Obers, who had become a good friend of Gauss, heard this, he contacted the university at Göttingen in an attempt to convince them to offer Gauss the directorship of their newly planned observatory. He did not want Gauss to leave Germany. Although initial reaction was restrained, the administration did express interest, so negotiations began with Obers taking the lead. While all of this was happening, Gauss continued to pursue his scientific interests. Then in 1805, Obers finally finished the deal with Göttingen. The university offered to appoint Gauss the head of the observatory. If he accepted he would be very independent from the university, with little classroom time and nearly no administrative duties. The Brunswick observatory never materialized, and the offer was too good to refuse. So, in 1807, Gauss left Brunswick and returned to Göttingen (Bühler 1981, 45–46, 50–51).

Gauss began work again on the parallels problem in 1813. He began to realize that the Parallel Postulate could not be proven from the other axioms. He accepted what Saccheri could not: that there might be a geometry different from Euclid's. This realization must have made all of his strange discoveries make sense. His previous attempts were within a framework of looking for a contradiction. Now he understood that there was no contradiction. The results were strange only because they were viewed from a Euclidean point of view. They needed to be viewed as part of a completely different geometry. Gauss initially called this new system **anti-Euclidean geometry**, but finally settled on calling his system **non-Euclidean**

geometry, the term we use today for any geometry in which the Parallel Postulate does not hold (Bonola 1912, 66–67).

Gauss was very busy both as a mathematician and an astronomer. He had worked hard all of his life, and he valued his reputation. He would not risk it harm by publishing anything that might set him up for ridicule. For this reason he did not want his discoveries concerning the new geometry made public. He did not even trust the mathematical community to be ready to accept such results. In any age, new ideas that go against the mainstream must struggle for acceptance, and in the early nineteenth century Euclid was still king in mathematics and even philosophy. Not wanting to deal with such matters, Gauss only wrote about his research to a few select confidants. One of his few trusted friends was the mathematician and astronomer Friedrich Bessel, to whom in 1829 he wrote:

> It may take a very long time before I make public my investigations on the issue. In fact, it may not happen during my lifetime, since I fear the scream of the Boeotians were I to completely express my views.

The Boeotians were an ancient Greek tribe who were not known for their intelligence. Despite his concern, in 1831, Gauss began to compose his *Meditations*, in which he planned to summarize his discoveries concerning non-Euclidean geometry. He did not wish for his research on the theory of parallels to die with him (Burton 1985, 551; Bonola 1912, 67).

The Area of a Triangle

Although Gauss was involved with many different types of projects during the course of his career, he would occasionally return to the question of parallels. One such time was on March 6, 1832. Bolyai had again written on the question, and Gauss responded. In his letter he included among other things a proof that the defect of a triangle is proportional to its area (Wolfe 1945, 128–130). Gauss had previously written to Bolyai in 1799 about the fact that the Parallel Postulate follows if it could be shown that there is no greatest triangle with respect to area. However, Gauss could not disprove the existence of such a triangle, so instead he assumed that there was one to see what would happen. He realized that the maximum triangle would be the limit of a sequence of non-Euclidean triangles found by letting the triangles increase in area as one progresses through the sequence. As this happens, the sides of the triangle pinch in and the angles decrease in measure until the vertices become ideal points (Section 9.2) and the measures of the angles become zero (Figure 11.4). In these triangles with three ideal vertices, all sides are parallel to each other.

■ **DEFINITION 11.3.1**

A triangle in which at least one of its vertices is an ideal point is called an **omega triangle**. An **ordinary triangle** has no ideal vertices.

Now assume **HA** and let M represent the greatest possible area of a triangle. Take $\overline{AB}$ and two lines $\overline{CD}$ and $\overline{EF}$ which are both parallel to $\overline{AB}$ and meet at a point P as in Figure 11.2. (Be careful to note that this is now Saccheri's definition of parallel;

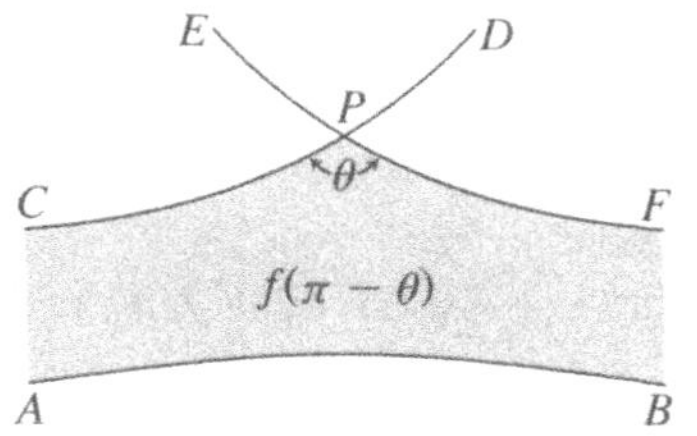

Figure 11.2 The area bounded by parallel lines.

see page 351.) Suppose that m$\sphericalangle CPF = \theta$. Define a function

$$f : [0, \pi] \to [0, M]$$

such that the area of the region bounded by these three lines equals $f(\pi - \theta)$. By the Saccheri-Legendre Theorem (11.2.7) we know that if $\theta = \pi$, the region can have no area. Therefore,

$$f(0) = 0.$$

If we take $\overline{CD}$ to be parallel to $\overline{EF}$ so that P is a point at infinity, then $\theta = 0$, and we have

$$f(\pi) = M.$$

Now imagine that the area transitions smoothly from 0 to M as θ ranges from π to 0. We see that it is not unreasonable to assume that f is not only continuous on $[0, \pi]$ but also differentiable. Furthermore, we infer from Figure 11.3 that

$$f(\pi - \theta) + f(\theta) = M$$

and

$$f(\theta) + f(\psi) + f(\pi - \theta - \psi) = M.$$

This last equation holds because

$$f(\pi - [\pi - \theta]) = f(\theta)$$

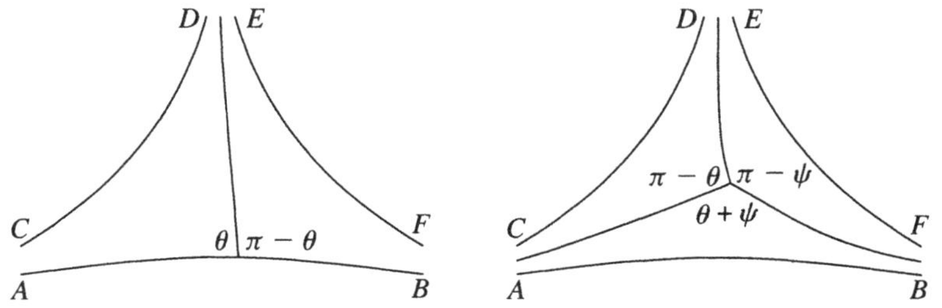

Figure 11.3 Omega triangles.

and

$$f(\pi - [\pi - \psi]) = f(\psi).$$

From this we conclude that

$$f(\psi) + f(\pi - \theta - \psi) = M - f(\theta) = f(\pi - \theta). \tag{11.2}$$

Next let $\gamma \geqslant 0$ be chosen such that $0 \leqslant \gamma \leqslant \pi$. Take $h \geqslant 0$, so that $\gamma + h \leqslant \pi$. Then there exists $\delta \geqslant 0$ such that $h = \pi - \delta - \gamma$. Notice that since $\delta = \pi - \gamma - h$, we have $0 \leqslant \delta \leqslant \pi$ and

$$\gamma + h = \gamma + \pi - \delta - \gamma = \pi - \delta. \tag{11.3}$$

From this we conclude that

$$f(\gamma + h) = f(\pi - \delta) = f(\gamma) + f(\pi - \delta - \gamma) = f(\gamma) + f(h),$$

where the first equation follows by Equation 11.3, the second by Equation 11.2, and the last by the definition of h. Hence for all α and β in the domain of f and $\alpha + h,\ \beta + h \leqslant \pi$,

$$\begin{aligned} f(\alpha + h) - f(\alpha) &= f(\alpha) + f(h) - f(\alpha) \\ &= f(\beta) + f(h) - f(\beta) = f(\beta + h) - f(\beta). \end{aligned}$$

This means that $f'(\alpha) = f'(\beta)$ for all α and β. It follows that there exists a constant $k > 0$ such that $f'(\theta) = k$, and since $f(0) = 0$ we see that for all θ, $f(\theta) = k\theta$. Because $f(\pi) = M$, we conclude that $M = k\pi$.

We are now ready to prove the result. Take a triangle with angles of measure α, β, and δ. Produce each of the sides as in Figure 11.4 and draw lines parallel to each of these extended sides producing a triangle with maximum area. Then

$$\text{area}(\triangle ABC) + f(\alpha) + f(\beta) + f(\delta) = M,$$

so

$$\text{area}(\triangle ABC) + k\alpha + k\beta + k\delta = k\pi.$$

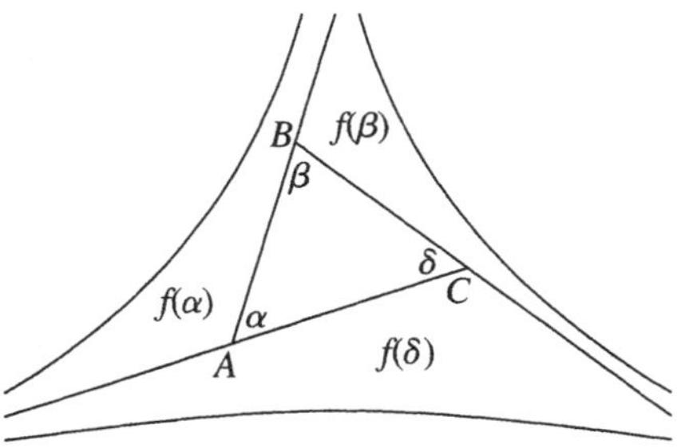

Figure 11.4 A triangle within a limiting maximum triangle.

From this we conclude that

$$\text{area}(\triangle ABC) = k(\pi - \alpha - \beta - \delta),$$

where $\pi - \alpha - \beta - \delta$ is the defect of the triangle. This shows that the area of a triangle is proportional to its defect assuming **HA**. This was a result also known to Lambert (Bonola 1912, 50).

Triangles with One Ideal Vertex

The omega triangle of Figure 11.4 has all of its vertices as ideal points. The triangles that we examine in this section will have only one ideal point as a vertex (see the figure for Theorem 11.3.4). These triangles share some of the same properties as regular triangles under **HA** (Wolfe 1945, 71–75).

Our first two results rely on an assumption that we have used before but now state explicitly. The postulate is named for Moritz Pasch, who in the late nineteenth century made significant contributions to the understanding of the logic behind geometry (Kline 1972, 1008). For these results, we say that a line **enters** a triangle if it contains a point on the interior of the triangle and either intersects a side or contains a vertex.

■ **POSTULATE 11.3.2 [Pasch]**

If a line enters an ordinary triangle through a side and does not contain one of its vertices, the line intersects exactly one of the other sides of the triangle.

The proof of the corollary is left for Exercise 2.

■ **COROLLARY 11.3.3 [Pasch]**

If a line enters an ordinary triangle through a given vertex, the line intersects the side of the triangle that is opposite the given vertex.

The next two theorems generalize these two propositions to omega triangles. We prove the first one here, but leave the second to Exercise 3.

■ **THEOREM 11.3.4**

[HA] If a line enters an omega triangle through a given vertex, the line will intersect the side opposite the given vertex.

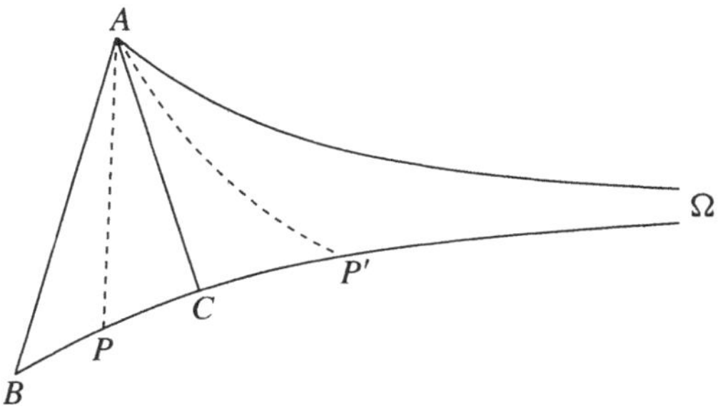

PROOF

Let $\overline{A\Omega}$ be parallel to $\overline{B\Omega}$. Find C on $\overline{B\Omega}$ such that $\overline{AC}$ is perpendicular to $\overline{B\Omega}$. If a line enters $\triangle AB\Omega$ through A, there are three options. The line could be on $\overline{AC}$. In this case it intersects $\overline{B\Omega}$ at C. It could lie on a line on the interior of $\measuredangle BAC$, but by Corollary 11.3.3 this line must intersect $\overline{BC}$ at some point P. Finally, if the line is on the interior of $\measuredangle CA\Omega$, it is below the parallel $\overline{A\Omega}$. Hence, it must intersect $\overline{C\Omega}$ at some P' by the definition of parallel line (Theorem 10.5.2 and page 351). ■

■ THEOREM 11.3.5

[HA] If a line enters an ideal triangle through a side and does not contain one of its vertices, the line intersects exactly one of the other sides of the triangle.

An important theorem regarding ordinary triangles states that an exterior angle is greater than either of its opposite interior angles (Theorem 5.3.9). The next theorem states that this also holds for triangles with an ideal vertex.

■ THEOREM 11.3.6

[HA] The exterior angles of $\triangle AB\Omega$ at A and B are greater than their respective opposite interior angles.

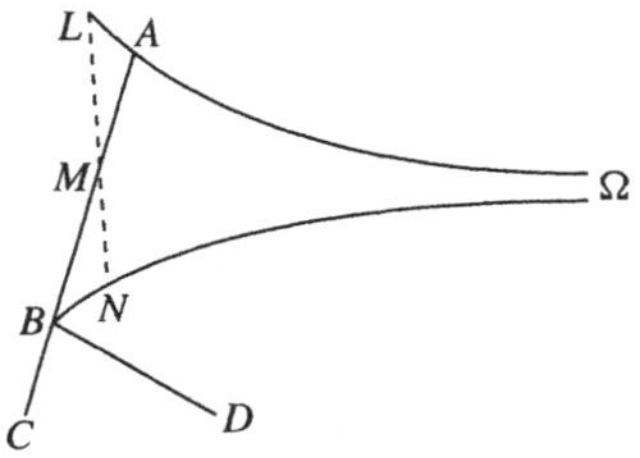

PROOF

In $\triangle AB\Omega$ produce $\overline{AB}$ to C. We will prove $m\measuredangle CB\Omega > m\measuredangle BA\Omega$. Copy $\measuredangle BA\Omega$ at B, producing $\measuredangle CBD$. First notice that $\overline{BD}$ will not intersect $\overline{A\Omega}$, else we would have a triangle in which one of its external angles is congruent to one of the opposite internal angles, and this would contradict Theorem 5.3.9. Therefore, $\overline{BD}$ is on the interior of $\measuredangle CB\Omega$ or D is on $\overline{B\Omega}$. Suppose the latter. This means that $\measuredangle CBD$ and $\measuredangle CB\Omega$ are the same angle. Let M be the midpoint of $\overline{AB}$ and find N on $\overline{B\Omega}$ such that $\overline{MN}$ is perpendicular to $\overline{B\Omega}$. Produce $\overline{A\Omega}$ to a point L through A such that $\overline{LA} \cong \overline{BN}$. Since our assumption on D implies that $\measuredangle CB\Omega \cong \measuredangle BA\Omega$, we have $\measuredangle LAM \cong \measuredangle NBM$. Thus, $\triangle MAL \cong \triangle MBN$ by SAS, and we may conclude that L, M, and N are collinear because $\measuredangle LMA \cong \measuredangle BMN$ (Theorem 5.3.8). From this we conclude that $\overline{LN}$ is perpendicular to both $\overline{A\Omega}$ and $\overline{B\Omega}$, but these two lines are parallel. This is a contradiction of **HA** (specifically, Theorem 10.5.2). Therefore, $\overline{BD}$ must lie on the interior of $\measuredangle CB\Omega$, which means that $m\measuredangle CB\Omega$ is greater than $m\measuredangle BA\Omega$. ■

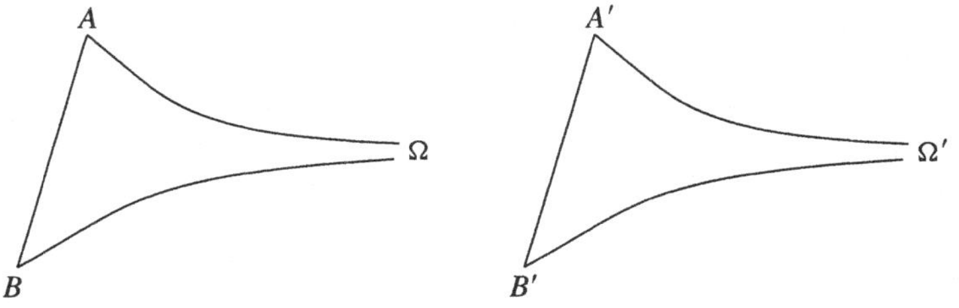

Figure 11.5 $\triangle AB\Omega \cong \triangle A'B'\Omega'$.

We finish the section by looking at what it means for two omega triangles with exactly one ideal vertex each to be congruent. Since any segment that has an ideal point as an endpoint is considered to be infinite in length, and since any angle that has an ideal point as a vertex is considered to have zero measure, we need to check only three conditions to show that these types of triangles are congruent (Figure 11.5).

■ **DEFINITION 11.3.7**

We say that $\triangle AB\Omega$ is **congruent** to $\triangle A'B'\Omega'$ if

- $\overline{AB} \cong \overline{A'B'}$
- $\measuredangle AB\Omega \cong \measuredangle A'B'\Omega'$.
- $\measuredangle BA\Omega \cong \measuredangle B'A'\Omega'$.

In this case we write $\triangle AB\Omega \cong \triangle A'B'\Omega'$.

We need to confirm only two of the three conditions to guarantee that a pair of such omega triangles are congruent.

■ **THEOREM 11.3.8 [SA]**

[HA] If $\overline{AB} \cong \overline{A'B'}$ and $\measuredangle BA\Omega \cong \measuredangle B'A'\Omega'$, then $\triangle AB\Omega \cong \triangle A'B'\Omega'$.

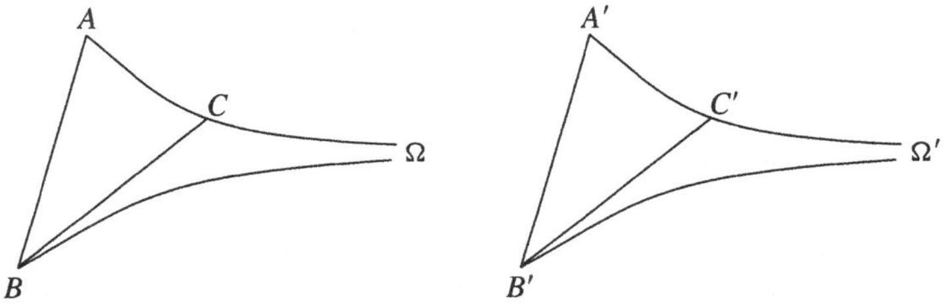

PROOF

Let $\overline{AB} \cong \overline{A'B'}$ and $\measuredangle BA\Omega \cong \measuredangle B'A'\Omega'$. We need to show that $\measuredangle AB\Omega$ is congruent to $\measuredangle A'B'\Omega'$. Suppose instead that $m\measuredangle AB\Omega > m\measuredangle A'B'\Omega'$. This means that there is a line on the interior of $\sphericalangle AB\Omega$ that forms an angle congruent to $\sphericalangle A'B'\Omega'$. This line must intersect $\overline{A\Omega}$ at some point C (Theorem 11.3.4). We may now write $\sphericalangle ABC \cong \sphericalangle A'B'\Omega'$. Next find C' on $\overline{A'\Omega'}$ such that $\overline{AC} \cong \overline{A'C'}$. We conclude that $\triangle BAC \cong \triangle B'A'C'$ by SAS. Therefore, $\sphericalangle ABC \cong \sphericalangle A'B'C'$. This is a contradiction because $\sphericalangle A'B'C' \cong \sphericalangle A'B'\Omega'$ by transitivity, but $\overline{B'C'}$ is on the interior of $\sphericalangle A'B'\Omega'$. The case when $m\sphericalangle AB\Omega$ is less than $m\sphericalangle A'B'\Omega'$ leads to a similar contradiction. ■

Our last theorem has a familiar name but an unfamiliar result. Compare this with Theorem 11.2.15.

■ THEOREM 11.3.9 [AA]

[HA] If $\sphericalangle BA\Omega \cong \sphericalangle B'A'\Omega'$ and $\sphericalangle AB\Omega \cong \sphericalangle A'B'\Omega$, then $\triangle AB\Omega$ is congruent to $\triangle A'B'\Omega'$.

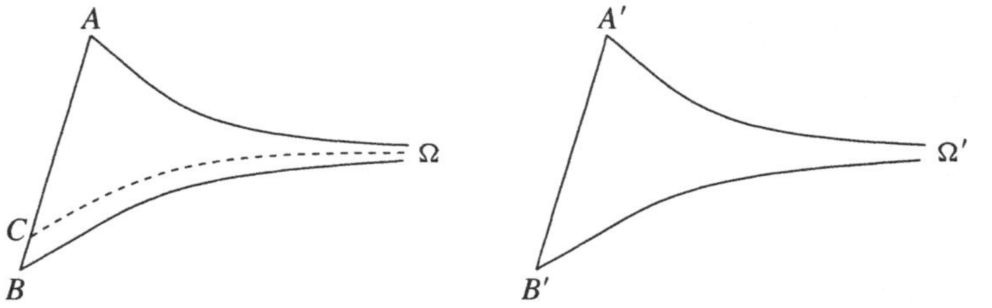

PROOF

Suppose that $\sphericalangle AB\Omega \cong \sphericalangle A'B'\Omega'$ and $\sphericalangle BA\Omega \cong \sphericalangle B'A'\Omega'$ in $\triangle AB\Omega$ and $\triangle A'B'\Omega'$, but $\overline{AB}$ is not congruent to $\overline{A'B'}$. This means either $AB > A'B'$ or $AB < A'B'$. It is enough to show that the first inequality leads to a contradiction, so suppose that we can find C on $\overline{AB}$ such that $\overline{AC} \cong \overline{A'B'}$. Thus, by SA, $\triangle AC\Omega \cong \triangle A'B'\Omega'$, which gives $\sphericalangle AC\Omega \cong \sphericalangle A'B'\Omega'$. Therefore, $\sphericalangle AC\Omega \cong \sphericalangle AB\Omega$ by transitivity. However, we know by Theorem 11.3.6 that $m\sphericalangle AC\Omega > m\sphericalangle AB\Omega$ because $\sphericalangle AC\Omega$ is exterior to $\triangle CB\Omega$. Therefore, $\overline{AB} \cong \overline{A'B'}$ and we conclude that $\triangle AB\Omega \cong \triangle A'B'\Omega'$. ■

Exercises

All problems except the first assume the Hypothesis of the Acute Angle.

1. Explain why Pasch's Postulate (11.3.2) is implied by the following statement: *If a line enters an ordinary triangle through a side and does not contain one of its vertices, the line intersects one of the other sides of the triangle.*

2. Prove Corollary 11.3.3.

3. Prove Theorem 11.3.5.

4. Let l intersect side $\overline{AB}$ of convex quadrilateral $ABCD$ and contain a point in the interior of $ABCD$. Prove that exactly one of the following holds: C is on l, D is on l, l intersects $\overline{BC}$, l intersects $\overline{CD}$, or l intersects $\overline{AD}$.

5. What proposition must be true in order for the area of a quadrilateral to be proportional to its defect? Assuming this proposition to be true, prove that the area of a quadrilateral is proportional to its defect.

6. Show that in $\triangle AB\Omega$ the sum of $m\sphericalangle AB\Omega$ and $m\sphericalangle BA\Omega$ is less than two right angles.

7. Let $\triangle AB\Omega$ be an omega triangle such that $\sphericalangle AB\Omega \cong \sphericalangle BA\Omega$. Let M be the midpoint of $\overline{AB}$ and let $\overline{M\Omega}$ be the line through M that is parallel to both $\overline{A\Omega}$ and $\overline{B\Omega}$. Prove that $\overline{M\Omega}$ is perpendicular to $\overline{AB}$.

8. Let A, B, and C be collinear with B between A and C. Let $\overline{AC}$ be perpendicular to $\overline{C\Omega}$. Demonstrate that $m\sphericalangle CA\Omega < m\sphericalangle CB\Omega$.

9. Show that the A in SA cannot be dropped by drawing two noncongruent omega triangles that have congruent corresponding sides.

10. Show that one of the A's in AA cannot be dropped by drawing two noncongruent omega triangles that have one pair of congruent corresponding angles

11. Let B be on $\overline{A\Omega}$ and C on $\overline{A\Omega'}$. Assume that $\overline{A\Omega}$ is perpendicular to $\overline{A\Omega'}$. Prove:

(a) $\overline{B\Omega'}$ intersects $\overline{C\Omega}$ in a point D.

(b) If $\overline{AB} \cong \overline{AC}$, then $\triangle BD\Omega \cong \triangle CD\Omega'$.

12. Let Ω and Ω' be on opposite sides of $\overline{AB}$. Suppose that $\overline{A\Omega}$ and $\overline{B\Omega'}$ are perpendicular to $\overline{AB}$. Prove that $\sphericalangle AB\Omega \cong \sphericalangle BA\Omega'$.

13. Does the Pythagorean Theorem hold when **HA** is assumed? Explain.

14. In $\triangle AB\Omega$ and $\triangle A'B'\Omega'$ we have $\sphericalangle AB\Omega \cong \sphericalangle A'B'\Omega'$ and $AB > A'B'$. Show that $m\sphericalangle BA\Omega < m\sphericalangle B'A'\Omega'$.

11.4 PURE REASON

Immanuel Kant was born in Königsberg on April 22, 1724 to Johann Georg and Ann Regina Kant. Immanuel was born the fourth child, but only an older sister was still alive at the time of his birth. Immanuel's father was a master harness maker and his mother was the daughter of a harness maker. Due to his father's expertise in his trade, the family lived well but not extravagantly and was very dependent on the local economy. As trade decreased in Königsberg, so did the family's fortunes, but they never reached a level that we would consider poor. Those early years with his family taught Kant the virtues of hard work, self-reliance, and honesty. Like his father before him, Kant determined that he would be a self-made man (Kuehn 2001, 26–31).

Kant's parents were very religious and greatly influenced by the Pietist movement within the Lutheran church. Pietism appeared in the late seventeenth century and was prominent in the first half of the eighteenth century. As a reaction to a church whose members appeared simply to be going through the motions, the Pietists attempted to bring back a living Christianity. Kant's parents wanted their son to receive a sound eduction, so they sent him to the Pietist *Collegium Fridericianum.* Little is known about the details of Kant's religious beliefs, but it is known that Kant hated the school. Even though it provided Kant with a good education and a chance for social advancement, he despised the school's attempts to convert him and the suffocating moralism of the faculty (Kuehn 2001, 34–35, 39–40, 45, 51–52, 54).

Not only did the Pietists run the *Collegium Fridericianum,* but they grew in influence at the University of Königsberg, where Kant began to study in 1740, when he was 17. The city of Königsberg was an important but remote city in northeastern Prussia near the Russian border. This remoteness had its disadvantages. It was difficult to recruit faculty, and the 44 full professors who were on staff earned low salaries. The remaining faculty were not paid by the university but instead were ***privatdozents***, faculty who were allowed to charge their students directly for attending their lectures. Kant would have attended both the free lectures given by those on the university's payroll and the paid ones. From these courses he learned subjects which included general philosophy, metaphysics, logic, ethics, natural law, physics, systematic theology, rhetoric, history, poetry, psychology, and mathematics. While at the *Collegium Fridericianum,* Kant became particularly interested in the classics, and he continued those studies at the university (Kuehn 2001, 45, 56, 63, 73, 75, 76).

A turning point came in 1744. Under the influence of faculty from both philosophy and physics, Kant turned to philosophy and began work on his first major philosophical piece. It was titled *Thoughts on the True Estimation of the Living Forces.* The work is a critique of both Newton and Leibnitz. Some also surmise that it contains a veiled attack on some of the Königsberg faculty. Eventually, this made Kant's life at Königsberg uncomfortable. 1744 also saw tragedy in the Kant family. Kant had lost his mother seven years earlier, and this year his father suffered a stroke that would lead to his death in 1746. This, combined with the turmoil at school and the fact that he had finished his formal studies, led to Kant's decision to leave Königsberg in 1748 and move to the surrounding countryside (Kuehn 2001, 80, 81, 86, 89–95).

There Kant became a private tutor. The position did not hold a high social status. It was something akin to a servant, and many who did not succeed at the university found themselves teaching the children of those wealthy enough to afford it. From the time he left his home to 1754, Kant worked for various families in Judtschea, Arnsdorf, and Rautenburg. He never did consider himself very good at teaching children, but his students and their parents thought otherwise and kept in contact with him long after his work was done (Kuehn 2001, 95–97).

It was apparently Kant's goal not to stay away from Königsberg forever. While working as a tutor he continued his research. In 1749 he published his *True Estimation of Living Forces*, and in 1754 he returned to Königsberg. On April 7 of the following year he turned in his dissertation. It was titled "Succinct Meditations on Fire," and in

June he made a public defense, the final step to his doctorate. At this point, however, Kant did not have the right to teach at the university. To earn this he would have to present another dissertation, called a ***Habilitationsschrift***, and then give a public lecture, the ***Habilitationsvortrag***. This occurred in September 1755 when he gave a talk entitled "A New Exposition of the First Principles of Metaphysics." For the next 15 years Kant taught students, conducted research, sought to be appointed to the position of full professor, and wrote. Some of his titles include *General History and Theory of the Heavens* (1755), *The Only Possible Argument in Support of a Demonstration of the Existence of God* (1763), *Observations on the Feeling of the Beautiful and Sublime* (1764), and his "prize essay" for the Prussian Academy of Sciences in Berlin entitled *Inquiry Concerning the Distinctness of the Principles of Natural Theology and Morality* (1764) (Kuehn 2001, xv–xvii, 98, 100).

Then in 1765, Kant received a letter that would begin a correspondence that would last for 12 years. It was from Lambert, who had read Kant's *The Only Possible Argument in Support of a Demonstration of the Existence of God* and found thoughts very similar to his own. Lambert was confident that if Kant would read his *Neues Organon* and *Anlage zur Architectonik*, Kant would share in this opinion. In writing, Lambert sought to exchange letters on subjects ranging from mathematics to metaphysics, with his main interest being Kant's views regarding space and time. Lambert wanted to prove that Euclidean geometry was true but had been unsuccessful. This caused him to recognize the logical possibility of an alternative geometry. Although Kant would reject Lambert's view of space, their letters greatly aided Kant's research and helped direct Kant's writing of his *Critique of Pure Reason* (Kant 1999, 77–79, 81–82; Kuehn 2001, xvii, 181, 196, 231).

The creation of the *Critique* began when the University of Königsberg appointed Kant to be professor of logic and metaphysics in 1770. Again a paper and a talk were required. Kant defended this inaugural dissertation on August 21. He sent a copy to Lambert a month later and wrote to him that he expected to quickly produce a major work from it by making only minor modifications. Kant later claimed that it took him four to five months of work to finish the *Critique*, but it was the result of 11 years of contemplation. In May 1781 the *Critique* was published, followed by a second edition in 1787. The fundamental question that Kant sought to answer was: "What can I know?" He concluded that traditional philosophy erred when it taught that one can discover truths about the world independent of experience. Instead, all of our knowledge of the world is obtained through the senses, but the structure that we put on those perceptions exists in our minds independent of any sensory input. The structure that we put onto space, for example, is described by Euclid's geometry. Metaphysics then becomes the study of the presuppositions of experience. We can only know the world as it appears to us, never as it actually is. Kant called his position **transcendental idealism** (Kuehn 2001, 231, 240–243, 246).

Kant intended to dedicate the *Critique of Pure Reason* to his good colleague Johann Lambert, but Lambert died on September 27, 1777 shortly before its publication. Lambert never saw the *Critique* in its final form, nor did he see the publication of his own work on geometry, entitled *Theory of Parallel Lines*. It was published in 1786 by Johann Bernoulli along with much of Lambert's correspondence, including his letters to Kant (Kuehn 2001, 231, 254; Kant 1999, 185–187; Boyer and Merzbach 1991, 462).

Describing Knowledge

Saccheri believed that the Parallel Postulate was true, so he went looking for a proof. He found one that he trusted, so he believed that the Postulate held. However, we know that his proof was not valid. We conclude that regardless of the truth of the Parallel Postulate, Saccheri did not know its truth value. This is because his reason for his belief was not sound. We base this conclusion on the definition of what it means to know that a statement is true.

■ **DEFINITION 11.4.1**

For a person to have **knowledge** that a proposition is true, three conditions must be satisfied:

- The person believes that the proposition is true.
- The statement is in fact true.
- The person has good reason to believe that the proposition is true.

The definition of knowledge and the discussion of this section is adapted from Halverson (1981, 25–63).

To clarify what it means to know that a proposition is true, let us examine a sentence that is simpler than the Parallel Postulate. Let p represent the claim that

there exist unicorns on Mars.

What would it mean for us to know p? With respect to the three conditions, it means:

- To know proposition p, we must first affirm it to be true. This is the bare minimum. If we believe that there are no unicorns on Mars, then whether there are unicorns or not, how could it be said that we know that they exist?
- Of course, simply believing p does not make it so. For us to know p, it must also be the case that p is true. If we believe that there are unicorns on Mars and there are none, we do not know that there are unicorns on Mars. Instead we believe a proposition that is false.
- Suppose that it is indeed the case that unicorns live on the red planet and we believe it. This is still not enough for us to be able to claim knowledge of this state of affairs. We must add to this a good reason to give credence to our belief. If we do not, we have no reason to have any confidence that our belief is accurate even though in reality it might be.

To have knowledge that p is true, we need the three listed conditions to hold, and it is crucial that we do not overlook the need for an adequate justification of our beliefs.

So what would be adequate justification for believing the Parallel Postulate? It depends on the type of proposition it is. Let us first compare it to p. Any justification of p requires information about the unicorns that is not included in the sentence. We need a photograph, a reliable eye witness, or some other kind of evidence that the words of the sentence do not provide. Nothing from the definition of a unicorn suggests that they live on Mars. Hence, the sentence p claims to give new information about its subject. Does the Parallel Postulate also claim to give new information about lines? The Postulate states that

> *if a straight line falling on two straight lines make the interior angles on the same side less than two right angles, the two straight lines, if produced indefinitely, meet on that side on which the angles are less than the two right angles.*

If there is something that the Parallel Postulate tells us about lines that is not included in the meaning of the word *line*, it is a sentence like p and requires evidence to determine its truth.

However, consider the assertion that

all bachelors are unmarried men.

For this statement no additional information is required to determine its truth. The definition of *bachelor* gives it to us. A bachelor is an unmarried man, and this is what the sentence states. It tells us nothing new about its subject. If the Parallel Postulate more closely resembles this type of assertion, observation is not needed to determine its truth. We need only to analyze the meanings of the words and possibly apply some logic. This and the previous case exhaust the possibilities, so we can make the following definitions.

■ **DEFINITION 11.4.2**

A **synthetic** proposition is one that gives new information about its subject.

■ **DEFINITION 11.4.3**

An **analytic** proposition is one that is not synthetic. It does not give new information about its subject.

The sentence *there exists unicorns on Mars* is synthetic. Either direct or indirect experience of the unicorns is required to prove it. If such evidence is found, we believe it, and if the sentence turns out to be true, we have gained knowledge of a particular type.

■ **DEFINITION 11.4.4**

Knowledge that is based on experience is called ***a posteriori***.

The term *a posteriori* refers to the knowledge coming *after* having experienced the state of affairs described by the sentence. If knowledge comes *before* experience,

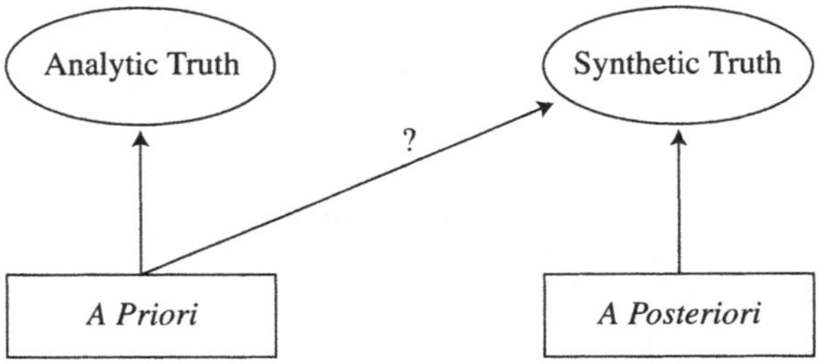

Figure 11.6 *A priori* and *a posteriori* knowledge.

such as that gained from the sentence *a bachelor is an unmarried man*, then it is not *a posteriori*. We do not need to meet a bachelor or have any other type of experience involving bachelors to know that the sentence is true. This represents the second type of knowledge.

■ **DEFINITION 11.4.5**

Knowledge that is not based on experience is called ***a priori***.

Based on our examples we may summarize by stating that we have *a posteriori* knowledge of some synthetic truths and we have *a priori* knowledge of some analytic truths. Can we push this further? Certainly all analytic propositions are known *a priori*, and all *a posteriori* knowledge is synthetic (Figure 11.6). However, is it possible to know a synthetic proposition *a priori*? Our answer to this question will put us into one of two mutually exclusive camps.

■ **DEFINITION 11.4.6**

A **rationalist** believes that there exists a true proposition about the world that can be known without either direct or indirect experience of it.

A rationalist does not deny that many propositions that describe the world accurately are known by observation. The claim is that there is at least one proposition that says something true about the world, yet we determine its truth simply by analyzing the terms of the sentence.

■ **DEFINITION 11.4.7**

An **empiricist** believes that the only way to know a true proposition about the world is by either direct or indirect experience of it.

The empiricist believes that there is no sentence that gives us new information about the world, yet we know it *a priori*. Only experiment or some other type of observation can lead to knowledge of the universe.

It is not uncommon for a scientist to speak as if the only way to learn about the universe is through the scientific method. All other modes of discovery are either analytic or superstition. It appears that scientists are empiricists. The rationalist will counter that this is a hasty generalization. There are synthetic *a priori* truths, and examples can be found in mathematics. This is what Kant believed.

Intuition

Mathematics served as Kant's main example of knowledge that is both *a priori* and synthetic. For example, consider Euclid's theorem regarding the sum of the interior angles of a triangle being equal to two right angles. Kant believed that this proposition is synthetic. Simply analyzing the meanings of the terms of the sentence will not yield whether the statement is true or not. We understand what *interior*, *angle*, and *triangle* mean, but none of their meanings involve *sum*, *two*, or *right*. We must arrive at the conclusion by some other way (Shabel 2006, 98).

The way to arrive at this conclusion is a method that Kant believed is unique to mathematics. Consider first this nonmathematical example. When we look at a tree, the light rays reflect off its surface and hit our eyes, which send signals along our optic nerves to our brains. We then carry the resulting image of the tree with us mentally. Kant had a name for this.

■ **DEFINITION 11.4.8**

An **intuition** is the immediate representation of an object in one's mind. An intuition that is caused by an object acting upon us in some way is called an **empirical intuition**, and one that is independent of any sensation is called a **pure intuition**.

The mental picture of the tree is an example of an empirical intuition. If we read the definition of a triangle, we see in our mind what it looks like based on the definition without needing a picture of the triangle. This is a pure intuition. From these intuitions we derive concepts.

■ **DEFINITION 11.4.9**

A **concept** is an abstract generalization of an intuition without the influence of the associated object.

We take our mental picture of the tree or the triangle and from that picture alone we create categories into which we place our intuitions. When we see other trees, new intuitions form in our minds, and the concepts that we have generalized from other intuitions allow us to group like intuitions under the same categories (Kant 1998, A19/B33, 155; cf. 36). This is illustrated in Figures 11.7 and 11.8.

Kant pulls this together to describe the very unique job of mathematicians. They arrive at truth by way of construction. According to Kant (1998, A713/B741, 630), "to **construct** a concept means to exhibit *a priori* the intuition corresponding to it." This means that if we take a mathematical concept, that of a triangle for instance, that

Figure 11.7 An empirical intuition leads to a concept.

is represented in our minds as an intuition, to construct that concept we must explain, illustrate, or describe the intuition in some way. The result is called a **construction**. Think of it as a picture that represents the concept. It can either be in one's mind or on a piece of paper. Either way, a construction is a generalization derived from the concept. For example, the construction of a triangle on paper is drawn in such a way that the lengths of its sides and the measures of its angles are irrelevant. What is important in the picture are the general properties of the triangle (Kant 1998, A 713–A 714/B 741–B 742, 630).

It is such constructions that make mathematical propositions synthetic. We see this in Kant's comparison between the philosopher and the mathematician. Consider the geometric proposition which states that the angle sum of a triangle equals two right angles. Philosophers will proceed by analysis. They will examine the meanings of *angle* and *side* and *vertex* and attempt to find within these terms a nugget of truth that will enable them to conclude the result about the angle sum. They will not find it. The mathematicians will find it because they will look beyond the definition. Their mental picture tells them that if through a vertex they draw a line parallel to the opposite side, they will see the angles of the triangle lying along the parallel at the chosen vertex. The truth about the angle sum is a synthetic one. The sentence gives us information about the triangle that goes beyond its definition, and it is the construction that shows this. Kant argued that this not only works for geometry but also for arithmetic. For example, the sentence *one plus one equals two* is synthetic because it tells us information about *one*, *two*, *plus*, and *equals* that go beyond their

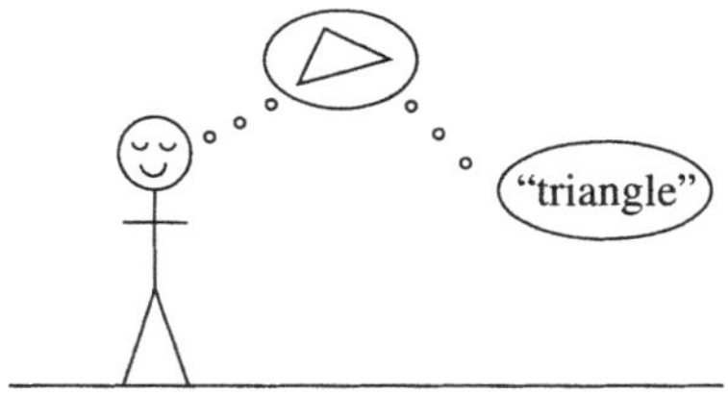

Figure 11.8 Pure intuition leads to a concept.

meanings. We then generalize to basic rules based on our intuition of numbers and addition. Because all of mathematics is either the study of the continuous (geometry) or the discrete (arithmetic), all mathematical propositions are synthetic (Shabel 2006, 101, 106–107; cf. Kant 1998, A 716/B 744, 631–632).

Now Kant asserts that mathematical knowledge is *a priori*. The example will again be the angle sum theorem for triangles. He claims that the proposition has two properties that make it clear that it entails *a priori* knowledge. They are the statement's necessity and universality. Kant claims that the angle sum result must be true because of its proof. The geometric demonstration does not depend on conclusions that are based on experimental evidence that later could be proven wrong. Instead, the proof transcends time and gives the geometer complete certainty of the proposition's truth. Necessity is not a quality that is determined *a posteriori*. We cannot determine whether a statement must be true by experiment alone. As for the fact that the sentence holds for all triangles at all times follows from the very act of construction. Indeed, the triangle constructed is a particular instance of the term, so the construction is particular, not universal. The algorithm that leads to the construction flows from the concept of triangle, and it is the same each time an arbitrary triangle is needed. This universality is also something that our senses cannot give us. When we try to generalize based on multiple observations, there is always at least some doubt that the next observation will not uphold our conclusion. Both necessity and universality are characteristics of *a priori* knowledge and not *a posteriori*. Because arithmetic propositions also have these two characteristics, Kant concluded that all of the true propositions of mathematics are instances of synthetic *a priori* knowledge (Shabel 2006, 107–109).

We may now define a **mathematical concept** as a synthetic definition that gives us the rule for describing the associated intuition. The definition of a triangle is an example. It tells us that to construct a triangle we need three noncollinear line segments joined at their endpoints. We now know how to construct it. This is an example of a specific type of mathematical concept called a **geometric concept**. Kant taught that geometric concepts give us a pattern for constructing space (Shabel 2006, 99). This means that since geometry is *a priori* and necessary, the same can be said about space. It is a pure intuition. It is not something that exists outside ourselves but originates in our minds as a fundamental framework to put meaning onto the input from our senses. Kant held this position against two prevailing views in which space exists in an objective sense either as the final container of objects or as an abstraction from the relations among the objects. He regarded the first view as untenable because it would make space into some kind of substance that was infinite in extent, and only God can be said to have such a quality. Moreover, Kant viewed God as existing outside space and time, and this appears to be impossible if the universe is infinite in size. Kant rejected the second view because it was not consistent with geometry being both necessarily true and an accurate description of the world. If space is abstracted from what we see, the structure of space is discovered by empirical means. This would mean that geometry is not *a priori* as Kant claimed.

To support his view, Kant notes that for us space is assumed by our minds when we use our senses. The standard example is that of a pair of hands. When we examine them, they are from our point of view essentially identical. To our eyes they look the same. If we measure the lengths of the fingers, the distance around the wrists, and so on, we find that they have the same dimensions. How, then, do we tell the left hand from the right? We are able to do this, according to Kant, because our spatial intuition comes preloaded with such concepts as up, down, in front, behind, and to the side. These *a priori* properties of space enable us to make the distinction. That this ability is universal in humans is further support for Kant's view. Since space is neither an object that exists independent of ourselves nor an abstraction based on observation, there are no competing realities when it comes to space. Everyone has the same intuition about its nature, and that intuition is described by Euclid's geometry. The mathematics that describe the universe is common to all people because everyone has the same intuition about space (Hatfield 2006, 73–78).

CHAPTER 12

NICOLAI LOBACHEVSKI AND JÁNOS BOLYAI

After he left Göttingen in 1799, Farkas Bolyai moved to the Hungarian city of Kolozsvár, where he began work as a tutor. He met Zsuzsanna Árkosi Benkő, and they soon were wed. The couple moved to Domáld but returned to Kolozsvár in the fall of 1802 expecting their first child. The baby was a boy, and they named him János (Prékopa 2006, 11).

János was a child prodigy. When he was 6, he taught himself to read. When he was 7 he learned German and how to play the violin. His father began to teach him mathematics when he was 9. He began formal schooling when he was 12, skipping the first three grades. János could easily solve calculus problems by the time he was 14. Passing his final exam in 1817 when he was 15 years old, he was ready to attend a university. The problem was that there were none in the local vicinity, and the closest ones, the University of Pest and the University of Vienna, did not have the caliber of mathematics faculty necessary to further János' education. The obvious solution was to send him to Gauss at Göttingen, so Farkas wrote his friend. At such a young age János would need proper supervision, so he inquired of Gauss whether his son might stay with him during his time at the university. Unfortunately, the letter could have been more tactfully written, for Farkas wanted to confirm that the Gauss household provided the proper environment for his son. It must be that Gauss was not "grumbling," his wife "exceptional," and his daughter not "dangerous." It is not surprising that Gauss did not answer the letter. It is a testament, however, to their friendship that this incident did not prevent future amicable correspondence (Prékopa 2006, 14–15).

So instead János attended the Vienna Academy of Military Engineers. Due to some financial problems, he enrolled a year later, in 1818. He excelled in his studies, and János was ranked first in his class by his professors. There was also a ranking based on student input in which he ranked second in his class. János did not enjoy drawing and his performance reflected this. This was evidently enough to keep him from first place. However, what he did enjoy was mathematics. Like his father he excelled at it, and like his father he was drawn to the study of parallel lines. Beginning in 1820 János researched the field thoroughly. His goal was to prove the Parallel Postulate once and for all. Learning of his son's intentions, Farkas wrote to him in a letter dated April 4, 1820 (Prékopa 2006, 15–16):

> You must not attempt this approach to parallels: I know this way to its very end. I have traversed this bottomless night, which extinguished all light and joy of my life. For God's sake! I entreat you leave parallels alone, abhor them like indecent talk, they may deprive you [just like me] from your time, health, tranquility and the happiness of your life. That bottomless darkness may devour a thousand tall towers of Newton and it will never bright up in the earth. ... I thought I would sacrifice myself for the sake of the truth. I was ready to become a martyr who would remove the flaw from geometry and return it purified to mankind. I accomplished monstrous, enormous labors; my creations are far better than those of others and yet I have not achieved complete satisfaction. ... I turned back when I saw that no man can reach the bottom of this night. I turned back unconsoled, pitying myself and all mankind. ... I have traveled past all reefs of this infernal Dead Sea and have always come back with broken mast and torn sail. The ruin of my disposition and my fall date back to this time. I thoughtlessly risked my life and happiness — *aut Caesar aut nihil.*

Evidently, Farkas's letter inspired János. If the Parallel Postulate could not be proven, then what? It should be replaced by something else, not with a statement that is equivalent to it because that would simply lead to all of the same problems that his father had encountered. Instead, the Parallel Postulate should be replaced by an assumption that would imply the Postulate's negation. He would continue his research. In 1822 János finished his studies at the Academy, but because he was one of their top two students, the school allowed him to stay an extra year. When he left Vienna after his extension, János was advanced to the position of sub-lieutenant and was assigned the position of Directorate of Fortification of Temesvár. It was there where things fell together. On November 3, 1823 he wrote to his father:

> I am resolved to publish a work on parallels as soon as I can complete and arrange the material, and the opportunity arises. At the moment I still do not clearly see my way through, but the path which I have followed is almost certain to lead me to my goal, provided it is at all possible. I have not quite reached it, but I have discovered things so wonderful that I was astounded and it would be an everlasting pity if these things were lost. When you, my dear father, see them, you will understand. All I can say at present is that out of nothing I have created a strange new world. All that I have sent you previously is like a house of cards in comparison with a tower. (Burton 1985, 554)

Possibly reflecting on the discovery of analytic geometry by Descartes and Fermat and that of the calculus by Newton and Leibnitz, Farkas wrote to János:

> It seems to me advisable, if you have actually succeeded in obtaining a solution of the problem, that, for a two-fold reason its publication be hastened: first, because ideas easily pass from one to another who, in that case, can publish them; secondly, because it seems to be true that many things have, as it were, an epoch in which they are discovered in several places simultaneously, just as the violets appear on all sides in the springtime. (Burton 1985, 554)

In 1825 János sent an abstract to his father, and then finally in 1829 he sent a copy of the work. Farkas was pleased but concerned about one point. An indeterminant constant appeared in his formula, and he did not understand why his son could not eliminate it. Nonetheless, the two came to agreement, and János's work would be appended to Farkas's *Essay on the Elements of Mathematics for Studious Youths*. The full title of János's text in Latin is *Appendix scientiam spatii absolute veram exhibens: a veritate aut falsitate Axiomatis XI. Euclidei, a priori haud unquam decidenda, independentem: adjecta ad casum falsitatis quadratura circuli geometrica*. This is shortened in English to *The Science of Absolute Space*, and it is often referenced simply as the *Appendix*. It was was sent to Gauss in 1831, but it was lost in transit. The *Appendix* was sent a second time in January 1832 (Bonola 1912, 99–100).

Dated March 6, 1832, the letter that Gauss sent in reply to Farkas is one of the most famous ever penned in the history of mathematics. As we have seen, Gauss studied the parallels question, and he came to many of the same results as others had. The difference was that he had kept his results private. Thus, we are not surprised by what Gauss wrote to his old colleague from Göttingen.

> If I commenced by saying that I *am unable to praise this work* (by Johann), you would certainly be surprised for a moment. But I cannot say otherwise. To praise it, would be to praise myself. Indeed the whole contents of the work, the path taken by your son, the results to which he is led, coincide almost entirely with my meditations, which have occupied my mind partly for the last thirty or thirty-five years. So I remained quite stupefied. So far as my own work is concerned, of which up till now I have put little on paper, my intention was not to let it be published during my lifetime. Indeed the majority of people have not clear ideas upon the questions of which we are speaking, and I have found very few people who could regard with any special interest what I communicated to them on this subject. To be able to take such an interest it is first of all necessary to have devoted careful thought to the real nature of what is wanted and upon this matter almost all are most uncertain. On the other hand it was my idea to write down all this later so that at least it should not perish with me. It is therefore a pleasant surprise for me that I am spared this trouble, and I am very glad that it is just the son of my old friend, who take the precedence of me in such a remarkable manner. (Bonola 1912, 100)

Farkas was pleased with the letter and viewed it as an honor. János, however, did not. He was devastated. He initially suspected that his father had secretly informed Gauss of his work and now Gauss was using that communication to upstage him. This could not have been further from the truth, because Farkas was very proud of his son and wanted only the best for him. This we have seen in Farkas's letters.

Nonetheless, the event strained the relationship between the two, and it never fully healed. It took nearly 10 years before János eventually accepted the truth of the situation, but he would never come to fully trust Gauss. Scholars dispute whether this distrust was justified. Gauss did work on the question of parallels, but whether his results rose to the level of János' work is left to various interpretations (Bonola 1912, 101; Gray 2004, 54–55).

János's acceptance of the facts did not mean that the letter did not have lasting effects. He would never publish again, although he did continue mathematical research. Making matters worse was the fact that the *Appendix* was ignored by the mathematical community, and then in 1848 János discovered a German edition of a treatise written by a Russian mathematician. Again he initially believed that the work must have been stolen since the results were so similar. They both used Saccheri's definition of parallel and proved the same basic results. They both examined spherical geometry, discovered the equation for the angle of parallelism, and constructed the same geometrical objects. There were some differences. The Russian tended to be more analytical, while Bolyai tended to focus on making clear the distinction between those theorems that followed from the negation of the Parallel Postulate and those that can be proven without any kind of parallel assumption (Burton 1985, 555; Prékopa 2006, 19).

The Russian's name was Nicolai Ivanovich Lobachevski. He was born on December 1, 1792 in the Russian city of Nizhny Novgorod. He was the son of Ivan Maksimovich Lobachevski, a government clerk, and his wife, Praskovia Aleksandrovna Lobachevskaya. Ivan died when the boy was only 7, so without an income, Praskovia moved to Kazan. The city was remote, near the Siberian border, and it had a large Muslim population. The benefit to moving to the city was that the Russian government supported Kazan both financially and educationally in an effort to spread Russian culture to the population. With the aid of a government scholarship, Lobachevski attended school. He was a bright student but not always interested in his studies. His teachers recognized in him an ability for mathematics, so they worked with him and helped him graduate.

With a goal of studying medicine, Lobachevski entered Kazan University in 1807. The university also benefited from the government's policies. They recruited professors to Kazan from Europe with the goal of creating a top center of Russian learning. One of these new faculty members was Martin Bartels from Caroline College. Lobachevski came into contact with him and soon abandoned medicine for mathematics. In 1811 he received his master's degree in mathematics and physics, and two years later he joined the university faculty teaching courses in his master's fields and astronomy. In 1816, at the age of 33, Lobachevski was promoted to the rank of full professor (Burton 1985, 557; Boyer and Merzbach 1991, 520).

Bartels and the other mathematics faculty at the University of Kazan educated Lobachevski in the German tradition. This meant that geometry played an important role in his education. He carried this influence into his professional life, often teaching the subject. One day in 1823 he wrote an outline for his students. He noted for his class that no one to date had found a proof of the Parallel Postulate. Such a note provides evidence that he was thinking about the subject, and as with every other interested mathematician, he must have thought about proving it. Three years later he presented a paper at his university in which he elaborated many of the results that follow when the Parallel Postulate is assumed to be false. Then in 1829 he did something that Gauss would not do. Convinced that it was impossible to prove the Parallel Postulate, he published his research on parallels as a new system of geometry with the Hypothesis of the Acute Angle as a postulate. It appeared in the *Kazan Messenger*, and it was titled "On the Principles of Geometry." He realized that the results were not in line with the mathematical community. Nonetheless, he believed that the theorems that flowed from the new postulate set formed a consistent logical system. Lobachevski called his new system **imaginary geometry** (Boyer and Merzbach 1991, 520–521).

Gauss appears to have been correct concerning the Boeotians. When Lobachevski submitted his 1829 paper, the referee, the person assigned to read the work and comment on its value and correctness, did not understand Lobachevski's ideas and commented negatively. Nonetheless, the paper still went to publication. Reviews were not favorable. One mathematician suggested that Lobachevski had titled his piece incorrectly. Instead of *On the Principles of Geometry*, it should have been named either *A Satire on Geometry* or *A Caricature of Geometry*. Lobachevski was not discouraged and continued to publish his research on his new imaginary geometry in 1835 with the *New Foundations of Geometry*, written in Russian. In 1840 he completed his *Geometrical Investigations on the Theory of Parallels*, written in German (Rosenfeld 1988, 208–209; Boyer and Merzbach 1991, 521). In it Lobachevski opens by writing:

> In geometry I find certain imperfections which I hold to be the reason why this science, apart from transition into analytics, can as yet make no advance from that state in which it has come to us from Euclid.
>
> As belonging to these imperfections, I consider the obscurity in the fundamental concepts of the geometrical magnitudes and in the manner and method of representing the measuring of these magnitudes, and finally the momentous gap in the theory of parallels, to fill which all efforts of mathematics have been so far in vain. (Lobachevski 1914, 11)

All of the efforts of the past had not solved the problem of parallels. Some were close, but none were able to take the last step. Lobachevski would be the first mathematician to publicly accept and detail results that would rival Euclid and revolutionize geometry.

12.1 PARALLEL FUNDAMENTALS

To begin our survey of the geometric work of Lobachevski and Bolyai, we start by proving some basic facts about parallel lines. Using Lobachevski's diagram (Figure 12.1), let $\overline{AN}$ be parallel to $\overline{BY}$ and let $\overline{AB}$ be perpendicular to $\overline{BY}$. By Theorem 10.5.2 acute $\sphericalangle BAN$ has the property that if $\overline{AK}$ is in the interior of $\sphericalangle BAN$, it intersects $\overline{BY}$, but if $\overline{AK}$ is in the interior of $\sphericalangle NAM$, it will not meet $\overline{BY}$. The point A is said to have the **characteristic of parallelism** for $\overline{AN}$ and $\overline{BY}$ because $\sphericalangle BAN$ has these properties. We should note that there is a mirror image of this

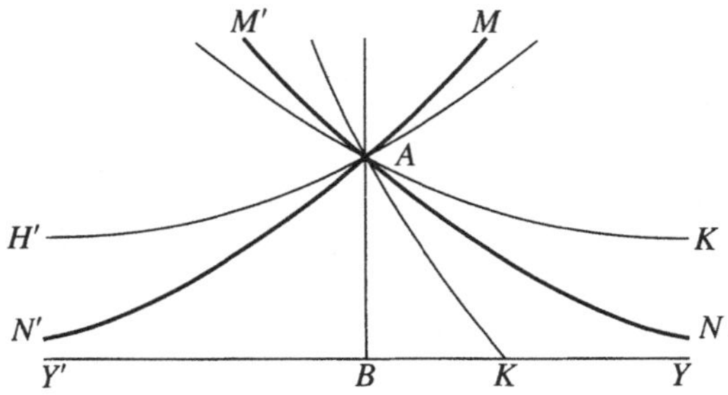

Figure 12.1 The parallel lines of Lobachevski.

situation on the other side of $\overline{AB}$ and A is not the only point on $\overline{AN}$ that has the characteristic of parallelism.

■ **THEOREM 12.1.1 [*Theory of Parallels* 17, *Appendix* 2]**

[HA] If $\overline{AB}$ is parallel to $\overline{CD}$, every point on $\overline{AB}$ has the characteristic of parallelism.

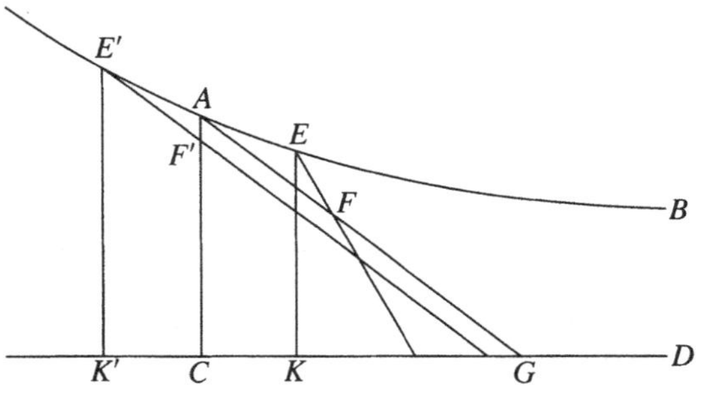

PROOF

Let $\overline{AB}$ be parallel to $\overline{CD}$ and let A have the characteristic of parallelism for $\overline{AB}$ and $\overline{CD}$. Take E to the right of A in the diagram making $\sphericalangle CAE$ acute. Find K on $\overline{CD}$ so that $\overline{EK}$ is perpendicular to $\overline{CD}$. Pick F in the interior of $\sphericalangle BEK$ and join both $\overline{EF}$ and $\overline{AF}$. Since $\overline{AF}$ is in the interior of $m\sphericalangle CAB$, we may conclude that $\overline{AF}$ will intersect $\overline{CD}$ at some G when extended. This forms $\triangle ACG$, so by Pasch's Postulate (11.3.2), $\overline{EF}$ produced must meet $\overline{CD}$.

Next, let E' be on $\overline{AB}$ but on the other side of $\overline{AC}$ and let K' be on the $\overline{CD}$ produced such that $\overline{E'K'}$ is perpendicular to $\overline{CD}$. Choose F' on $\overline{AC}$. Since A has the characteristic of parallelism, we can find G on $\overline{CD}$ such that $\sphericalangle BAG \cong \sphericalangle AE'F'$. We now have $\triangle ACG$, in which $\overline{E'F'}$ enters through $\overline{AC}$. If $\overline{E'F'}$ intersects $\overline{AG}$, we would have a triangle with angle sum greater than two right angles because $\sphericalangle E'AG$ and $\sphericalangle AE'F'$ are supplements. Hence, it must be the case that $\overline{E'F'}$ can be extended to meet $\overline{CD}$. ■

In the next two theorems we prove that parallel lines under the new definition share some common characteristics with Euclid's parallel lines. The first result states that lines are parallel is a symmetric relation. Lobachevski's proof relies on a continuity argument.

■ **THEOREM 12.1.2 [*Theory of Parallels* 18, *Appendix* 6]**

[HA] If $\overline{AB}$ is parallel to $\overline{CD}$, then $\overline{CD}$ is parallel to $\overline{AB}$.

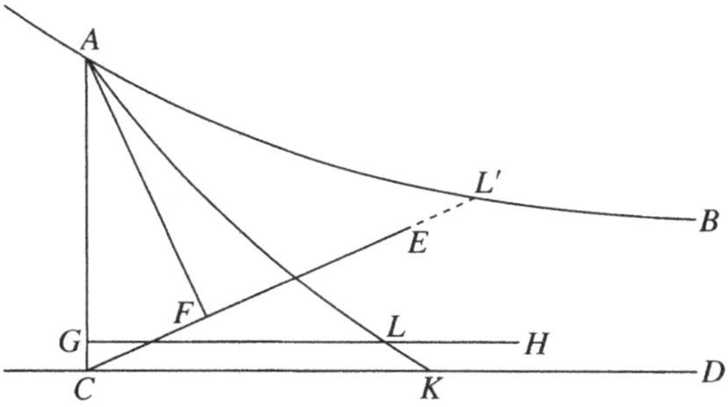

PROOF

Suppose that $\overline{AB}$ is parallel to $\overline{CD}$ with $\overline{AC}$ perpendicular to $\overline{CD}$. Because $\overline{AB}$ and $\overline{CD}$ do not intersect, to prove that $\overline{CD}$ is parallel to $\overline{AC}$, it suffices to show that any line in the interior of $\sphericalangle ACD$ must intersect $\overline{AB}$. To do this, let $\overline{CE}$ form the acute $\sphericalangle ACE$ in the interior of $\sphericalangle ACD$. Construct a perpendicular to $\overline{CE}$ through A. Let F be its point of intersection with $\overline{CE}$. The result is right $\triangle AFC$ with $AC > AF$. Thus, we can find G on $\overline{AC}$ so that $\overline{AG} \cong \overline{AF}$. Now construct $\overline{GH}$ perpendicular to $\overline{AC}$ and rotate $BAFE$ with pivot A so that $\overline{AF}$ and $\overline{AG}$ line up. This is equivalent to taking K on $\overline{CD}$ such that

$$\sphericalangle GAK \cong \sphericalangle FAB.$$

We can find such a K since $\overline{AB}$ is parallel to $\overline{CD}$. Let L be the intersection of $\overline{AK}$ and $\overline{GH}$. Find L' on $\overline{CE}$ (produced if necessary) such that $\overline{FL'} \cong \overline{GL}$. Because $\overline{AG} \cong \overline{AF}$, $\sphericalangle AGL \cong \sphericalangle AFL'$, and $\overline{GL} \cong \overline{FL'}$, we conclude by SAS that

$$\triangle AGL \cong \triangle AFL'.$$

Hence, $\sphericalangle FAL' \cong \sphericalangle GAL$. Since $\sphericalangle GAL$ and $\sphericalangle GAK$ are the same angle, we conclude that $\sphericalangle FAL' \cong \sphericalangle FAB$. The only way for this to happen is if L' is on $\overline{AB}$. Therefore, $\overline{CD}$ is parallel to $\overline{AB}$. ■

Being parallel is not transitive, as noted in Exercise 2. The next theorem is as close as we can get.

■ **THEOREM 12.1.3 [*Theory of Parallels* 25]**

[HA] If two distinct lines are parallel to a third, they are parallel to each other.

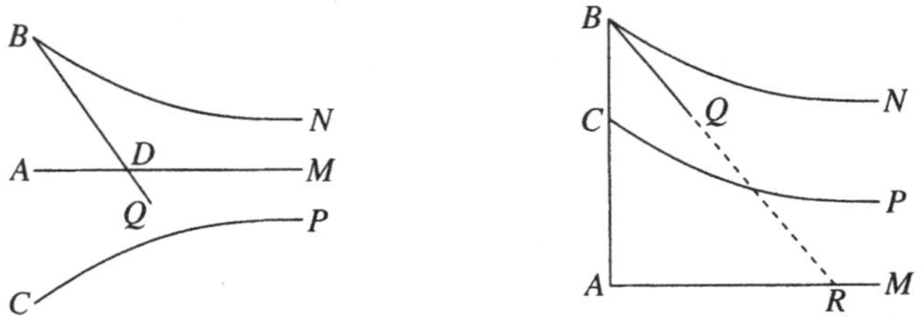

PROOF

Let both $\overline{BN}$ and $\overline{CP}$ be parallel to $\overline{AM}$ with C not on $\overline{BN}$. By Exercise 4, $\overline{BN}$ and $\overline{CP}$ do not intersect. We now have two cases to consider. First let $\overline{AM}$ be between $\overline{BN}$ and $\overline{CP}$. Take point Q and form $\measuredangle NBQ$ toward $\overline{AM}$, so $\overline{BQ}$ and $\overline{AM}$ must intersect at some point D. Since $\overline{AM}$ is parallel to $\overline{CP}$ (Theorem 12.1.2), $\overline{DQ}$ will meet $\overline{CP}$ since all of the points of $\overline{AM}$ have the characteristic of parallelism (Theorem 12.1.1). Thus, $\overline{BN}$ is parallel to $\overline{CP}$.

Next let $\overline{AM}$ not be between the other two. Without loss of generalization we may assume that $\overline{CP}$ is between $\overline{BN}$ and $\overline{AM}$ and for clarification in the diagram assume that B, C, and A are collinear. Let $\overline{BQ}$ form an angle within $\measuredangle NBA$. Since $\overline{BN}$ is parallel to $\overline{AM}$, we can produce $\overline{BQ}$ to intersect $\overline{AM}$ at some R. Now notice that $\overline{CP}$, $\overline{CA}$, and $\overline{AM}$ form an omega triangle (Definition 11.3.1). Since $\overline{BR}$ already intersects one of its sides, by Theorem 11.3.5, we conclude that $\overline{BQ}$ must intersect $\overline{CP}$. ■

The Angle of Parallelism

Let $\overline{AB}$ be perpendicular to $\overline{BY}$ and assume that $\overline{AN}$ is parallel to $\overline{BY}$ (Figure 12.2). Let $p = AB$. Since the measure of the angle of parallelism (page 351) is dependent only on the value of p, we will define

$$\Pi : [0, \infty) \to [0, \pi/2]$$

by

$$\Pi(p) = \begin{cases} \mathrm{m}\measuredangle BAN, & p > 0 \\ \pi/2, & p = 0, \end{cases}$$

where we define $\Pi(0) = \pi/2$ because

$$\lim_{p \to 0+} \Pi(p) = \pi/2 \tag{12.1}$$

by Exercise 7. Furthermore, if we let H be a point and set $\alpha = \mathrm{m}\measuredangle BAH$, we have:

- If $0 \leqslant \alpha < \Pi(p)$, the $\overline{AH}$ produced intersects $\overline{BY}$.
- If $\alpha = \Pi(p)$, then $\overline{AH}$ is parallel to $\overline{BY}$.
- If $\Pi(p) < \alpha \leqslant \pi/2$, then $\overline{AH}$ will not intersect $\overline{BY}$.

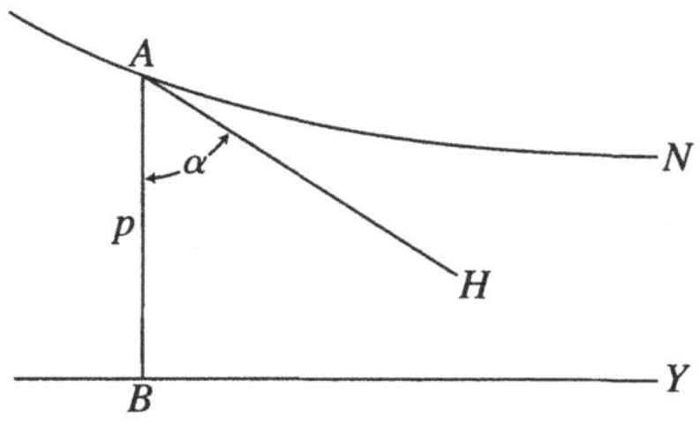

Figure 12.2 $\measuredangle BAN$ is the angle of parallelism.

If $\alpha > \pi/2$, view the line as on the opposite side of $\overline{AB}$ and treat accordingly (Lobachevski 1914, 13–14). It is important to note that **HR** implies that $\Pi(p) = \pi/2$ for all p. However, because the work of Lobachevski and Bolyai focused only on the **HA** case, we assume this hypothesis for the rest of the chapter.

The function Π has many nice properties that we now prove. The proof of the first result uses omega triangles.

■ THEOREM 12.1.4

[**HA**] Π is a strictly decreasing function.

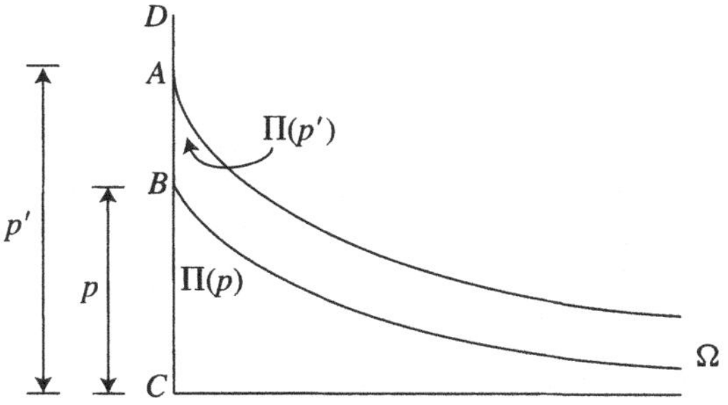

PROOF

Let $\overline{CD}$ be perpendicular to $\overline{C\Omega}$. Take A and B on $\overline{CD}$ and write $p = BC$ and $p' = AC$. To see that Π is strictly decreasing, suppose that $p < p'$. Remember that $\overline{A\Omega}$, $\overline{B\Omega}$, and $\overline{C\Omega}$ are all parallel to each other. Then $m\measuredangle CB\Omega = \Pi(p)$ and $m\measuredangle CA\Omega = \Pi(p')$. We observe that $\measuredangle DA\Omega$ is exterior to $\triangle AB\Omega$. Therefore, by Theorem 11.3.6,

$$m\measuredangle AB\Omega < m\measuredangle DA\Omega.$$

This gives

$$\pi - m\measuredangle AB\Omega > \pi - m\measuredangle DA\Omega,$$

from which we see that $\Pi(p) > \Pi(p')$, so Π is strictly decreasing. ■

COROLLARY 12.1.5

[**HA**] Π is one-to-one.

By Theorem 10.5.2 we know that to every magnitude there is associated an angle of parallelism. The next theorem states that the converse is also true, so we now know that Π is a one-to-one correspondence.

THEOREM 12.1.6 [*Theory of Parallels* 23]

[**HA**] Π is onto.

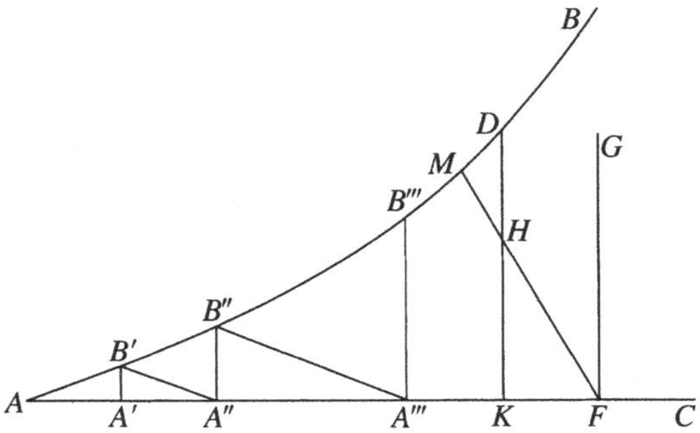

PROOF

Take α such that $0 \leqslant \alpha \leqslant \pi/2$. We must find p so that $\Pi(p) = \alpha$. Let $m\sphericalangle BAC = \alpha$. Take an arbitrary point B' on $\overline{AB}$ and drop a perpendicular to $\overline{AC}$, meeting it in A'. Find A'' on $\overline{A'C}$ such that $\overline{A'A''} \cong \overline{AA'}$. At A'' erect a perpendicular that intersects $\overline{AB}$ at B''. Continuing this process, we will find $\overline{FG}$ that is perpendicular to $\overline{AC}$ but does not intersect the $\overline{AB}$ produced. We will prove this fact indirectly, so assume that no such perpendicular exists. Appealing to Theorem 10.4.5, we know that

$$\mathsf{d}(AB'A'') = \mathsf{d}(AB'A') + \mathsf{d}(A''B'A').$$

Because $\triangle AB'A' \cong \triangle A''B'A'$ by SAS, $\mathsf{d}(AB'A'') = 2\mathsf{d}(AB'A')$. Therefore,

$$\mathsf{d}(AB''A'') > 2\mathsf{d}(AB'A')$$

by Theorem 10.4.5. Next find A''' on $\overline{AC}$ and B''' on $\overline{AB}$ such that $\overline{A''A'''}$ is congruent to $\overline{AA''}$ and $\overline{A'''B'''}$ is perpendicular to $\overline{AC}$. As before we have $\triangle AB''A'' \cong \triangle A'''B''A''$, and we conclude that

$$\mathsf{d}(AB'''A''') > 2^2\mathsf{d}(AB'A').$$

We can repeat this construction and find for each positive integer n an ordinary triangle with defect greater than $2^n\mathsf{d}(AA'B')$. That is, we can find a triangle with angle sum less than $\pi - 2^n\mathsf{d}(AA'B')$ for each n, which means that we can find a triangle with angle sum less than or equal to zero, which is impossible.

Without loss of generality we may assume that $\overline{FG}$ is the first perpendicular that does not intersect $\overline{AB}$. Its existence is guaranteed by an argument similar to that for the existence of Saccheri's limit (Theorem 10.5.2 and page 351). To prove that $\overline{AB}$ is parallel to $\overline{FG}$, it suffices to show that $\overline{FG}$ is parallel to $\overline{AB}$ (Theorem 12.1.2). To do this take H in the interior of $\sphericalangle AFG$. Choose K on $\overline{AC}$ so that $\overline{HK}$ is perpendicular to $\overline{AC}$. Since $\overline{FG}$ is the first not to intersect $\overline{AB}$, there exists D on $\overline{AB}$ where the $\overline{KH}$ produced intersects $\overline{AB}$. We obtain $\triangle AKD$. Since $\overline{FH}$ enters this triangle at $\overline{DK}$, it must exit through $\overline{AD}$ at some point M by Pasch's Postulate (11.3.2). Thus, $\overline{FG}$ is a parallel to $\overline{AB}$, and we may set $p = AF$. ■

All of this allows us to conclude that Π is also continuous (Wolfe 1945, 76–77; Lobachevski 1914, 21). Assuming **HR**, Π is a constant function, so it is continuous. To prove the result under **HA**, suppose that Π is not continuous at p_0. This means that

$$\Pi(p_0) \neq \lim_{p \to p_0} \Pi(p),$$

and this means that one of the one-sided limits does not equal $\Pi(p_0)$. Suppose that it is $p \to p_0^-$. Then there exists $\epsilon > 0$ such that for all x where $0 < x < p_0$,

$$\Pi(x) - \Pi(p_0) > \epsilon.$$

From this we conclude that no angle measure θ such that $\Pi(p_0) < \theta < \Pi(p_0) + \epsilon$ corresponds to a segment length since Π is strictly decreasing. This contradicts Theorem 12.1.6. Therefore, we have:

■ **THEOREM 12.1.7**

Π is continuous.

Exercises

1. Let $\overline{AX}$ be parallel to $\overline{BY}$ and $\overline{BY}$ be parallel to $\overline{CZ}$. Is it possible for $\overline{AD}$ to not intersect $\overline{BY}$ but intersect $\overline{CZ}$. Explain.

2. Is it possible for l to be parallel to m, m parallel to n, but l not parallel to n?

3. Use omega triangles to prove Theorem 12.1.2 and reprove the first case of the proof for Theorem 12.1.3.

4. Prove that $\overline{BN}$ and $\overline{CP}$ do not intersect in the proof of Theorem 12.1.3.

5. Generalize the definition of two lines being parallel to the three dimensions.

6. Let l, m, and n be distinct lines in three dimensions. Assume that l and m are parallel to n in space. Using the definition from Exercise 5, demonstrate that n is parallel to both l and m and l is parallel to m.

7. Show $\lim_{p \to 0^+} \Pi(p) = \pi/2$.

8. Explain why AA (Theorem 11.3.8) implies that Π is not a constant function.

9. True or false: There exists a positive real number p such that $\Pi(p) = \pi/2$ if and only if **HR**. Explain your reasoning.

10. Let l be a line and P a point off l. True or false: There exists exactly one line n through P that is parallel to l if and only if **HR**. Explain your reasoning.

11. Let A not be on l. Let $\overline{AN}$ and $\overline{AN'}$ both be parallel to l as in Figure 12.1. Prove that the angle bisector of $\sphericalangle N'AN$ is perpendicular to l.

12. Suppose that X and Y are on the same side of $\overline{AB}$. Let $\sphericalangle BAX$ and $\sphericalangle ABY$ be supplementary. Prove that $\overline{AX}$ is not parallel to $\overline{BY}$ and that the two lines do not intersect.

12.2 HOROCYCLES

In 1804, Farkas Bolyai sent Gauss what he thought was a proof of the Parallel Postulate. Gauss showed him his error in a later letter, but Farkas was nonetheless still hopeful of finding a proof and continued to study the subject (Kline 1972, 878). One of his attempts was his proof that the Parallel Postulate follows from the assumption

$$\text{three points that are not collinear lie on a circle.} \tag{12.2}$$

To prove this, take $\overline{AB}$ and points A' and B' on the same side of $\overline{AB}$ such that $\overline{BB'}$ and $\overline{AB}$ are perpendicular and $\sphericalangle BAA'$ is acute (Figure 12.3). We must show that $\overline{AA'}$ and $\overline{BB'}$ intersect on the same side of $\overline{AB}$ as both A' and B'. Take any point M on $\overline{AB}$. Find M' on the opposite side of $\overline{AA'}$ such that $\overline{AA'}$ is the perpendicular bisector of $\overline{MM'}$. Similarly, let M'' be on $\overline{AB}$ produced so that $\overline{BB'}$ is the perpendicular bisector of $\overline{MM''}$. We observe that M, M', and M'' are not collinear. Hence, they lie on a circle by Proposition 12.2. Both $\overline{AA'}$ and $\overline{BB'}$ are concurrent with radii of the circle, so they must intersect at the circle's center (Bonola 1912, 61).

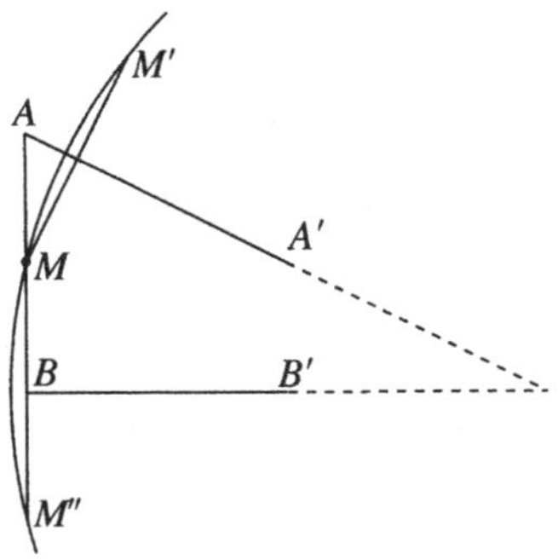

Figure 12.3 Farkas Bolyai's proof of the Parallel Postulate.

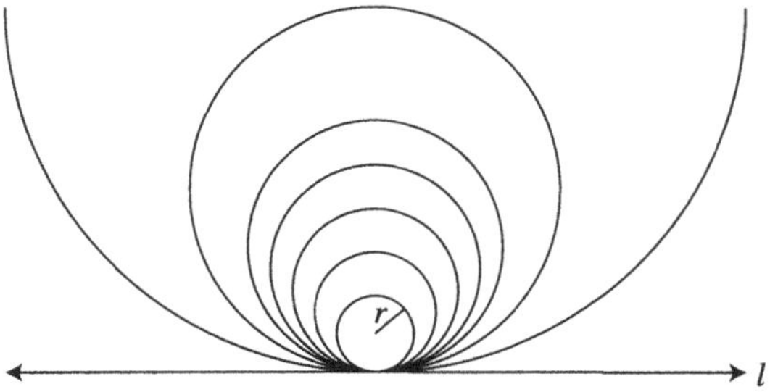

Figure 12.4 A circle with radius approaching infinity.

The problem with Farkas Bolyai's approach is that Proposition 12.2 is equivalent to the Parallel Postulate (Exercise 1). However, this result did inspire a student of Gauss by the name of Friedrich Ludwig Wachter. He also attempted to prove the Parallel Postulate. Despite believing otherwise, Wachter was not successful. In a letter written in 1816 to Gauss he described his efforts. He wrote concerning what he believed would happen when the radius of a sphere was allowed to approach infinity. Take a circle with radius r and let $r \to \infty$ as illustrated in Figure 12.4. The limit of the circles l is called the **circle with infinite radius**. If **HR** is assumed, l is a line; otherwise, l is a nonstraight curve. Now rotate the original circle about its diameter, forming a sphere. As $r \to \infty$ the sphere approaches the surface formed by revolving l about a point. When **HR** is assumed, the resulting surface is a plane, but under **HA** it will be a **sphere with infinite radius**. Wachter observed that even when working under the Hypothesis of the Acute Angle, **HR** held on the surface of rotation (Bonola 1912, 62–63). In this section we define a curve that is Wachter's circle with infinite radius. Later we consider his infinite sphere. Both of these played an important role in the work of Lobachevski and Bolyai.

Fundamentals

We begin with some required terminology. Take two parallel lines $\overleftrightarrow{A\Omega}$ and $\overleftrightarrow{B\Omega}$, where Ω is an ideal point (Figure 12.5). Join $\overline{AB}$. Construct the bisectors to $\sphericalangle BA\Omega$ and $\sphericalangle AB\Omega$. These two bisectors meet in a point E (Exercise 11). If we drop a perpendicular from E to $\overleftrightarrow{AB}$ intersecting it at F, then

$$\triangle AFE \cong \triangle ACE \text{ and } \triangle BFE \cong \triangle BDE$$

by AAS. Thus, $\overline{CE} \cong \overline{DE}$ because $\overline{CE} \cong \overline{FE}$ and $\overline{FE} \cong \overline{DE}$. Let $\overleftrightarrow{E\Omega}$ be parallel to both $\overleftrightarrow{A\Omega}$ and $\overleftrightarrow{B\Omega}$ (Exercise 10). This line is called the **line of symmetry** between the parallels $\overleftrightarrow{A\Omega}$ and $\overleftrightarrow{B\Omega}$. We note the following (Exercise 13):

- The line of symmetry bisects $\sphericalangle DEC$.
- The line of symmetry is equidistant between its two parallel lines.

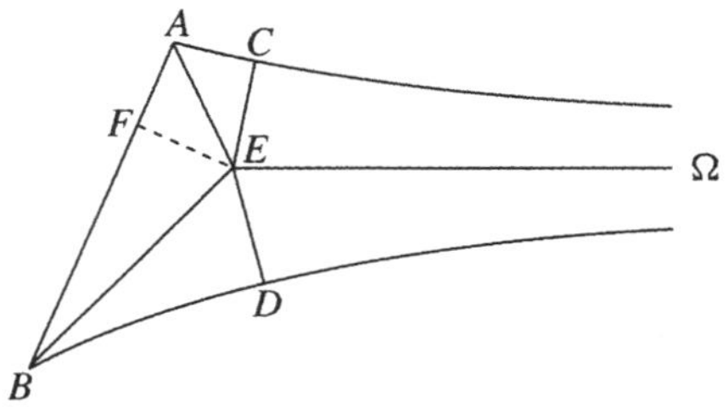

Figure 12.5 $\overline{E\Omega}$ is the line of symmetry between $\overline{A\Omega}$ and $\overline{B\Omega}$.

Furthermore, if we join C and D, the angles that $\overline{CD}$ makes with $\overline{A\Omega}$ and $\overline{B\Omega}$ are congruent (Exercise 12). We have a name for this (Wolfe 1945, 110–113).

■ **DEFINITION 12.2.1 [*Appendix* 8]**

If $\overline{C\Omega}$ is parallel to $\overline{D\Omega}$ and $\sphericalangle CD\Omega$ is congruent to $\sphericalangle DC\Omega$, the point C **corresponds** to the point D.

We now prove two basic results about corresponding points. The first involves finding the line of symmetry. The second states that given a point on one of two parallel lines, there is a unique point on the second parallel that corresponds to the given point.

■ **THEOREM 12.2.2**

[HA] If two points on different parallel lines correspond, the line of symmetry between the parallel lines is the perpendicular bisector of the segment joining the two points.

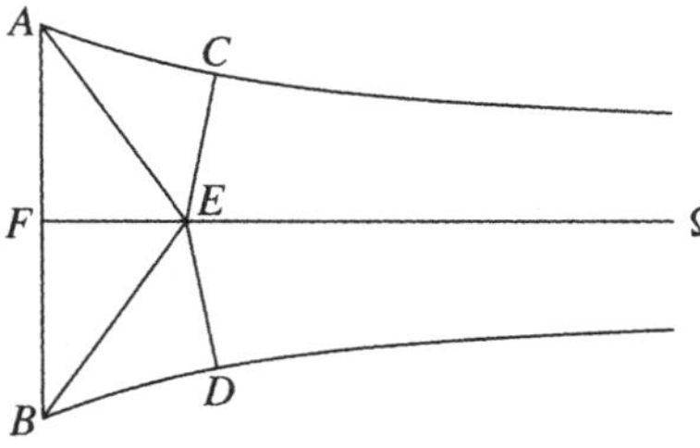

PROOF

Let $\overline{A\Omega}$ be parallel to $\overline{B\Omega}$ and suppose that A corresponds to B. Join $\overline{AB}$. Bisect $\sphericalangle BA\Omega$ and $\sphericalangle AB\Omega$ and let E be the intersection of the bisectors. Then $\overline{E\Omega}$ is the line of symmetry between $A\Omega$ and $B\Omega$. Extend it to intersect $\overline{AB}$ at F. Since A corresponds to B, $\sphericalangle BAE \cong \sphericalangle ABE$, and then $\overline{AE} \cong \overline{BE}$. Since $\overline{E\Omega}$ is the line of symmetry for the parallels, $\sphericalangle CE\Omega \cong \sphericalangle DE\Omega$. Thus, $\sphericalangle AEF \cong \sphericalangle BEF$. We now have $\Delta AFE \cong \Delta BFE$ by ASA. Therefore, $\overline{AF} \cong \overline{FB}$ and $\overline{F\Omega}$ is perpendicular to $\overline{AB}$ because $\sphericalangle AFE \cong \sphericalangle BFE$. ■

■ **THEOREM 12.2.3**

[HA] Given two parallel lines and any point on one of them, there exists exactly one point on the other line that corresponds to the first point.

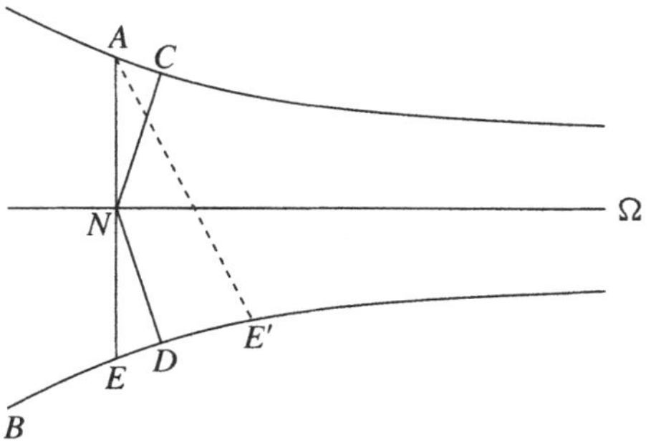

PROOF

Let $\overline{A\Omega}$ and $\overline{B\Omega}$ have line of symmetry $\overline{N\Omega}$. Let N be chosen so that $\overline{AN}$ is perpendicular to $\overline{N\Omega}$. From N construct $\overline{NC}$ perpendicular to $\overline{A\Omega}$ at C and $\overline{ND}$ perpendicular to $\overline{B\Omega}$ at D. Because $\overline{N\Omega}$ is the line of symmetry for $\overline{A\Omega}$ and $\overline{B\Omega}$, we have $\overline{NC} \cong \overline{ND}$. Finally, find E on $\overline{B\Omega}$ so that E is between B and D and $\overline{AC} \cong \overline{ED}$. Since $\triangle ACN \cong \triangle EDN$ by SAS, we see that A, N, and E are collinear (Theorem 5.3.8) and

$$\sphericalangle EAC \cong \sphericalangle AED.$$

That is, A corresponds to E.

To see that the point E is unique, suppose that A also corresponds to the point E' on $\overline{B\Omega}$. This means that

$$\sphericalangle AE'\Omega \cong \sphericalangle E'A\Omega.$$

We now examine the angle sum of $\triangle AEE'$. Since

$$\mathrm{m}\sphericalangle AE'E = \pi - \mathrm{m}\sphericalangle AE'\Omega$$

and

$$\mathrm{m}\sphericalangle AEE' = \mathrm{m}\sphericalangle EAE' + \mathrm{m}\sphericalangle E'A\Omega,$$

we find that

$$\mathrm{m}\sphericalangle EAE' + \mathrm{m}\sphericalangle AEE' + \mathrm{m}\sphericalangle AE'E = 2\mathrm{m}\sphericalangle EAE' + \pi.$$

This contradicts the Saccheri-Legendre Theorem (11.2.7). ■

We are now ready to formally define the sphere of infinite radius. Bolyai did not give this curve a name, so we use Lobachevski's terminology.

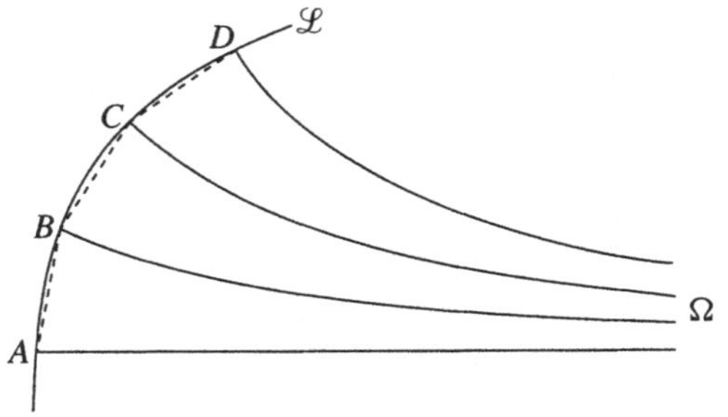

Figure 12.6 The horocycle $\mathcal{L}$ with corresponding points A, B, C, and D.

■ **DEFINITION 12.2.4 [*Theory of Parallels* 31, *Appendix* 11]**

Fix $\overline{A\Omega}$. The locus of all points B such that $\overline{B\Omega}$ is parallel to $\overline{A\Omega}$ and B corresponds to A is called a **horocycle** (or **boundary line**). The segment $\overline{A\Omega}$ is called the **axis** of the horocycle.

If we assume **HR**, the horocycle is a line. This is because when A and B are corresponding points on two parallel lines, they form interior angles on the same side that are congruent and supplementary. However, when we assume **HA**, the horocycle becomes a curve (Figure 12.6). This is the consequence of the next theorem.

■ **THEOREM 12.2.5**

[HA] If three points A, B, and C lie one on each of three mutually parallel lines such that A corresponds to B and B corresponds to C, the three points are not collinear.

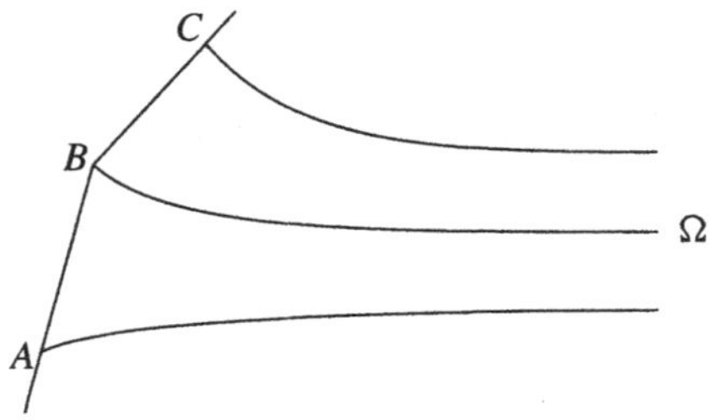

PROOF

Let A, B, and C be on distinct parallel lines through ideal point Ω. Let $\angle BA\Omega \cong \angle AB\Omega$ and $\angle CB\Omega \cong \angle BC\Omega$. Assume that A, B, and C are collinear. Then, $\angle AB\Omega$ and $\angle CB\Omega$ are supplementary. Hence, $\angle BA\Omega$ and $\angle BC\Omega$ are supplementary. This is impossible because we would have the angle sum of $\triangle AC\Omega$ at least as great as two right angles, so we conclude that A, B, and C cannot be collinear. ■

Not only is a horocycle $\mathcal{L}$ curved under **HA**, but any two points on $\mathcal{L}$ must correspond. This means that if a line drawn at a point on $\mathcal{L}$ is parallel to the defining axis of the horocycle, that line can be treated as the axis of the horocycle. For this reason we call every such line an **axis**. All of this follows from the next proposition.

■ **THEOREM 12.2.6 [*Theory of Parallels* 30, *Appendix* 10]**

[HA] If three points A, B, and C lie one on each of three parallel lines, where A corresponds to C and B corresponds to C, A must correspond to B.

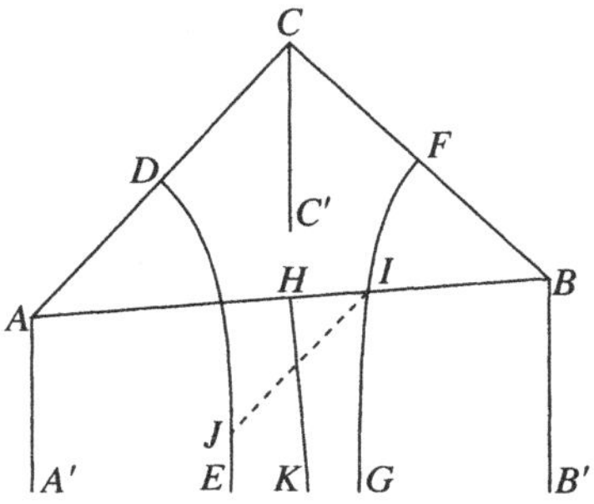

PROOF

Suppose that $\overleftrightarrow{AA'}$, $\overleftrightarrow{BB'}$, and $\overleftrightarrow{CC'}$ are parallel to each other with A and B corresponding with point C. Let $\overleftrightarrow{DE}$ be the line of symmetry between $\overleftrightarrow{AA'}$ and $\overleftrightarrow{CC'}$ and $\overleftrightarrow{FG}$ the line of symmetry between $\overleftrightarrow{CC'}$ and $\overleftrightarrow{BB'}$. Let $\overleftrightarrow{HK}$ be the perpendicular bisector of $\overline{AB}$ with H on $\overline{AB}$. We know that the two lines of symmetry are parallel to themselves and the three given lines (Theorem 12.1.3). Moreover, $\overleftrightarrow{HK}$ is parallel to $\overleftrightarrow{FG}$. To see this, let I be the intersection of $\overleftrightarrow{FG}$ and $\overline{AB}$ and let $\overrightarrow{IJ}$ be in the interior of $\sphericalangle HIG$. Since $\overleftrightarrow{DE}$ and $\overleftrightarrow{FG}$ are parallel, we may take J on $\overleftrightarrow{DE}$. Therefore, by Pasch's Postulate, $\overleftrightarrow{HK}$ must intersect $\overrightarrow{IJ}$. From this we may conclude that $\overleftrightarrow{HK}$ is parallel to both $\overleftrightarrow{AA'}$ and $\overleftrightarrow{BB'}$. Since $\overleftrightarrow{HK}$ is the perpendicular bisector to $\overline{AB}$, we have that $\sphericalangle A'AH$ is congruent to $m\sphericalangle B'BH$ because both have measure $\Pi(AH)$. Hence, the points A and B correspond to each other. ■

Concentric Horocycles

Let A be on horocycle $\mathcal{L}$ with axis $\overrightarrow{A\Omega}$ in a plane Λ. Let B be any other point on $\mathcal{L}$. By Definition 12.2.4, A corresponds with B on $\mathcal{L}$. Taking M to be the midpoint of $\overline{AB}$ implies that $\overrightarrow{M\Omega}$ is the line of symmetry between $\overrightarrow{A\Omega}$ and $\overrightarrow{B\Omega}$ and meets $\overline{AB}$ in a right angle (Theorem 12.2.2). Therefore, $m\sphericalangle MB\Omega = \Pi(AB/2)$, so we may describe $\mathcal{L}$ by

$$\mathcal{L} = \{X \in \Lambda : m\sphericalangle AX\Omega = \Pi(AX/2)\}. \tag{12.3}$$

Furthermore, since all of the points on a horocycle correspond to each other, any of the parallels at a given point along the curve could have been chosen as the axis upon which to define $\mathcal{L}$. This means that the curvature of the horocycle is constant at every point. In fact, since the horocycle is defined with Π, all horocycles have the same curvature, allowing us to conclude that any two are congruent. Therefore, if two horocycles in a plane have a common axis, they will share all of their axes and the two horocycles will be equidistant as measured along them. Such a pair of horocycles are called **concentric**. Suppose that $\mathscr{L}$ and $\mathscr{L}'$ are concentric with A on $\mathscr{L}$ and B on $\mathscr{L}'$. If B is between A and the ideal point Ω on a common axis, we say that $\mathscr{L}$ is the **outer horocycle** with respect to $\mathscr{L}'$.

In concentric horocycles a pair of axes will intercept arcs that have lengths that depend only on the distance that separates the horocycles. To prove this, we need two lemmas.

■ **LEMMA 12.2.7**

Let $\mathscr{L}$ and $\mathscr{L}'$ be concentric horocycles. Take A, B, and C on $\mathscr{L}$ and A', B', and C' on $\mathscr{L}'$ such that $\overline{AA'}$, $\overline{BB'}$, and $\overline{CC'}$ are common axes. If $\widehat{AB}$ and $\widehat{AC}$ are commensurable, then

$$\frac{\mathrm{m}\widehat{A'C'}}{\mathrm{m}\widehat{AC}} = \frac{\mathrm{m}\widehat{A'B'}}{\mathrm{m}\widehat{AB}}.$$

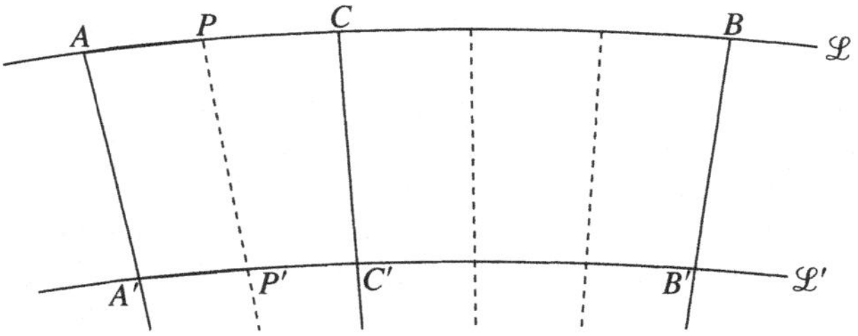

PROOF

We assume that the lengths of arcs of a horocycle are additive and we measure the arcs by measuring their intercepted line segments. Let r and s be positive integers such that

$$\mathrm{m}\widehat{AC} = r(\mathrm{m}\widehat{AP}) \text{ and } \mathrm{m}\widehat{AB} = s(\mathrm{m}\widehat{AP}),$$

where P is some point on $\mathscr{L}$. In the diagram, $r = 2$ and $s = 5$. Without loss of generality, P is on $\widehat{AC}$. Now join $\overline{AC}$ and $\overline{A'C'}$. Notice that P is the midpoint of $\widehat{AC}$. Draw the axis to $\mathscr{L}$ at P and let it intersect $\mathscr{L}'$ at P'. Therefore, $\overline{PP'}$ is perpendicular to $\overline{AC}$. Since A corresponds to C and A' corresponds to C', $\sphericalangle A'AC \cong \sphericalangle C'CA$ and $\sphericalangle AA'C' \cong \sphericalangle CC'A'$. Finally, $\overline{AA'} \cong \overline{CC'}$ since the horocycles are concentric. Hence, $APP'A' \cong CPP'C'$. From this we conclude that $\overline{A'P'} \cong \overline{P'C'}$, so $\mathrm{m}\widehat{A'P'} = \mathrm{m}\widehat{P'C'}$ and

$$\mathrm{m}\widehat{A'C'} = r(\mathrm{m}\widehat{A'P'}).$$

Similarly, $\mathrm{m}\widehat{A'B'} = s(\mathrm{m}\widehat{A'P'})$. Since this argument generalizes to all positive integers r and s, we conclude that $\mathrm{m}\widehat{A'C'}/\mathrm{m}\widehat{AC} = \mathrm{m}\widehat{A'B'}/\mathrm{m}\widehat{AB}$. ■

The second lemma is needed here and in a future application.

■ LEMMA 12.2.8

Let f be a continuous, positive-valued function such that $f(1) < 1$ and

$$f(nx) = [f(x)]^n$$

for every $x > 0$ and positive integer n. Then there exists $k > 0$ such that $f(x) = e^{-kx}$ for all $x > 0$.

PROOF

Let $x > 0$ and take positive integers r and s. Observe that

$$f\left(\frac{r}{s}x\right) = f\left(r\frac{x}{s}\right) = \left[f\left(\frac{x}{s}\right)\right]^r$$

and

$$\left[f\left(\frac{x}{s}\right)\right]^s = f\left(s\frac{x}{s}\right) = f(x).$$

Therefore,

$$f\left(\frac{r}{s}x\right) = \left[f\left(\frac{x}{s}\right)\right]^r = \left[f\left(\frac{x}{s}\right)^s\right]^{r/s} = [f(x)]^{r/s}. \tag{12.4}$$

Now let $a > 0$ be a real number. There exist positive rational numbers $q_1, q_2, q_3, \ldots$ such that

$$\lim_{i\to\infty} q_i = a.$$

By Equation 12.4 we know that

$$f(q_i x) = [f(x)]^{q_i}$$

for all $i = 1, 2, 3, \ldots$. Therefore, since f is continuous,

$$f(ax) = \lim_{i\to\infty} f(q_i x) = \lim_{i\to\infty} [f(x)]^{q_i} = [f(x)]^a.$$

Finally, choose $k > 0$ so that $f(1) = e^{-k}$. Then

$$f(x) = [f(1)]^x = e^{-kx}. \blacksquare$$

We can now prove the main result.

■ THEOREM 12.2.9 [*Theory of Parallels* 33]

Given concentric horocycles $\mathcal{L}$ and $\mathcal{L}'$ such that $\mathcal{L}$ is outer with respect to $\mathcal{L}'$, there exists $k > 0$ such that for all $\overline{AA'}$ and $\overline{BB'}$ that are shared axes of $\mathcal{L}$ and $\mathcal{L}'$ with A and B on $\mathcal{L}$ and A' and B' on $\mathcal{L}'$,

$$m\widehat{A'B'} = m\widehat{AB} \cdot e^{-k(AA')}.$$

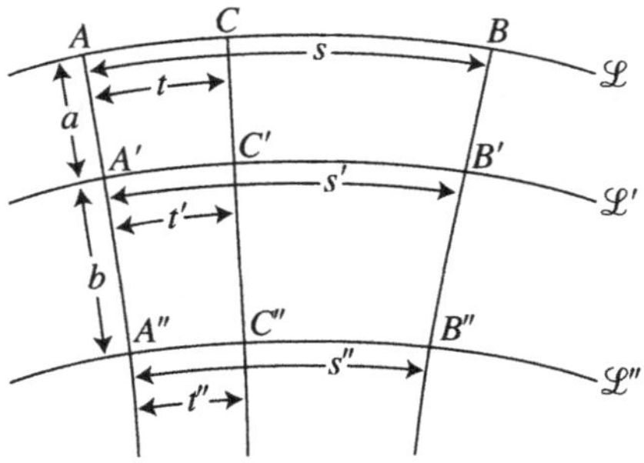

PROOF

Let $\mathcal{L}''$ be a third horocycle concentric with $\mathcal{L}$ and $\mathcal{L}''$. Take any C on $\widehat{AB}$ and let the axis through C intersect $\mathcal{L}'$ at C' and $\mathcal{L}''$ at C''. Make the following assignments:

$$s = \mathrm{m}\widehat{AB},\ s' = \mathrm{m}\widehat{A'B'},\ s'' = \mathrm{m}\widehat{A''B''}$$

and

$$t = \mathrm{m}\widehat{AC},\ t' = \mathrm{m}\widehat{A'C'},\ t'' = \mathrm{m}\widehat{A''C''}.$$

Assume that s and t are commensurable. Since the horocycles are concentric, we know by Lemma 12.2.7 that

$$\frac{t'}{t} = \frac{s'}{s},\ \frac{t''}{t'} = \frac{s''}{s'},\ \text{and}\ \frac{t''}{t} = \frac{s''}{s}.$$

Since these ratios remain constant, provided that the distance between the horocycles remains constant, we conclude that we can define a function $f(x)$ that equals the value of the ratio of the smaller arc to the larger one when the distance between the horocycles is x. Based on the above, we have

$$f(a) = \frac{s'}{s},\ f(b) = \frac{s''}{s'},\ \text{and}\ f(a+b) = \frac{s''}{s}.$$

Therefore,

$$f(a)f(b) = \frac{s'}{s}\frac{s''}{s'} = \frac{s''}{s} = f(a+b).$$

Since a and b were chosen arbitrarily, we conclude that for all $x > 0$ and positive integers n,

$$\begin{aligned} f(nx) &= f([n-1]x + x) \\ &= f([n-1]x)f(x) \\ &= f([n-2]x)[f(x)]^2 \\ &\ \vdots \\ &= [f(x)]^n. \end{aligned}$$

Since f is certainly a continuous function and $f(x) < 1$, Lemma 12.2.8 allows us to conclude that for some $k > 0$, $f(x) = e^{-kx}$ for every $x > 0$. Hence,

$$e^{-k(AA')} = f(AA') = \mathrm{m}\widehat{A'B'}/\mathrm{m}\widehat{AB}. \blacksquare$$

Exercises

1. Prove that the Parallel Postulate implies that three points determine a circle.

2. Show that Theorem 12.2.3 also holds for **HR**.

3. Prove: If $\overline{AB}$ is parallel to $\overline{CD}$, $\sphericalangle CAB \cong \sphericalangle ACD$, and $\overline{EF}$ is the perpendicular bisector of $\overline{AC}$, then $\overline{EF}$ is parallel to both $\overline{AB}$ and $\overline{CD}$.

4. Prove:
   (a) If perpendiculars are drawn at the midpoints of the sides of a triangle and two of them are parallel, all three are parallel.
   (b) If perpendiculars are drawn at the midpoints of the sides of a triangle such that no two are parallel but they do not intersect each other, they are all perpendicular to the same line.

5. In a plane, how many horocycles pass through two given points? Explain.

6. Prove that a horocycle is determined uniquely by three distinct points.

7. Describe a mechanical device that could be used to draw a horocycle.

8. Let $\overline{A\Omega}$ be an axis of horocycle $\mathscr{L}$. Show that if $\overline{AT}$ is perpendicular to $\overline{A\Omega}$, then $\overline{AT}$ is tangent to $\mathscr{L}$. Is the converse true?

9. If a horocycle is revolved about any two of its axes, explain why the resulting surfaces are congruent. Are the two surfaces the same surface?

10. Let $\overline{A\Omega}$ be parallel to $\overline{B\Omega}$. Let C be a point not on either of these two parallel lines. Prove that there exists a line through C parallel to both $\overline{A\Omega}$ and $\overline{B\Omega}$.

11. In Figure 12.5, prove that the bisectors of $\sphericalangle BA\Omega$ and $\sphericalangle AB\Omega$ intersect in E.

12. Show that C and D are corresponding points in Figure 12.5.

13. Let n be the line of symmetry between l and m. There are two ways to interpret that n is equidistant between l and m.
    (a) Let A be on l and find B on n such that $\overline{AB}$ is perpendicular to l. Next find C on m such that $\overline{BC}$ is perpendicular to m. Show that $\overline{AB} \cong \overline{BC}$.
    (b) Let B be on n. Find A on l and C on m such that $\overline{AB}$ and $\overline{BC}$ are perpendicular to n. Prove that $\overline{AB} \cong \overline{BC}$.

14. Prove that the line of symmetry between two parallel lines is unique.

15. Let $\mathscr{L}$ and $\mathscr{L}'$ be concentric horocycles. Let A and B be on $\mathscr{L}$ and A' and B' be on $\mathscr{L}'$ such that $\overline{AA'}$ and $\overline{BB'}$ are common axes. Prove that if $\mathscr{L}$ is the outer horocycle with respect to $\mathscr{L}'$, then $AB > A'B'$.

16. Show that the area enclosed by two horocycles and two common axes is proportional to the distance between the horocycles. (*Hint:* First determine how to measure this area.)

17. Assume **HR**. Let $\widehat{AB}$, $\widehat{A'B'}$, and $\widehat{A''B''}$ be on concentric circles such that $\overline{AA'A''}$ and $\overline{BB'B''}$ are common radii and $\overline{AA'} \cong \overline{A'A''}$. Show that

$$\frac{m\widehat{AB}}{m\widehat{A'B'}} \neq \frac{m\widehat{A'B'}}{m\widehat{A''B''}}.$$

18. Let f and g be real-valued, continuous functions. Answer the following and justify your responses.

(a) If $f(n) = g(n)$ for all integers n, does $f = g$?
(b) If $f(x) = g(x)$ for all rational numbers x, does $f = g$?
(c) If $f(x) = g(x)$ for all irrational numbers x, does $f = g$?

19. Given $\overline{AB}$ and $k > 0$. The set of all points P such that the perpendicular distance from P to $\overline{AB}$ equals k is called an **equidistant curve**. $\overline{AB}$ is called the **baseline**, and any segment between the curve and the baseline that is perpendicular to the baseline is an **axis**.

(a) Explain why each equidistant curve has two **branches**, one "below" the baseline and one "above."
(b) Draw an equidistant curve assuming **HR** and then assuming **HA**.
(c) Assuming **HA**, draw two equidistant curves to $\overline{AB}$, one with distance k_1 and the other with distance $k_2 > k_1$.

20. Prove the following about equidistant curves:

(a) A line intersects an equidistant curve at most twice.
(b) A line that intersects an equidistant curve at P and is perpendicular to the axis at P is tangent to the curve.

21. Let l be a line and $\mathscr{C}$ an equidistant curve to l. Let $\overline{AB}$ be an axis with A on $\mathscr{C}$ and B on l. Prove that for any two distinct points C and D on $\mathscr{C}$ such that $\overline{CA} \cong \overline{AD}$, there exist C' and D' on l such that $\measuredangle C'CA \cong \measuredangle D'DA$ and $\measuredangle CC'B \cong \measuredangle DD'B$. Can the construction be redone so that $\measuredangle C'CA \cong \measuredangle CC'B$?

22. Conjecture why the horocycle is viewed as a circle of infinite radius.

12.3 THE SURFACE OF A SPHERE

Thus far our introduction to non-Euclidean geometry has involved references to spheres. Lambert conjectured that his new geometry could be interpreted as existing on the surface of a sphere with an imaginary radius. Wachter wrote to Gauss about his sphere of infinite radius. In each of these cases, the surface of the sphere is what was important, so in this section we survey some of the results concerning this surface in Euclidean space. These results will later be generalized to help us understand the geometry of the Hypothesis of the Acute Angle.

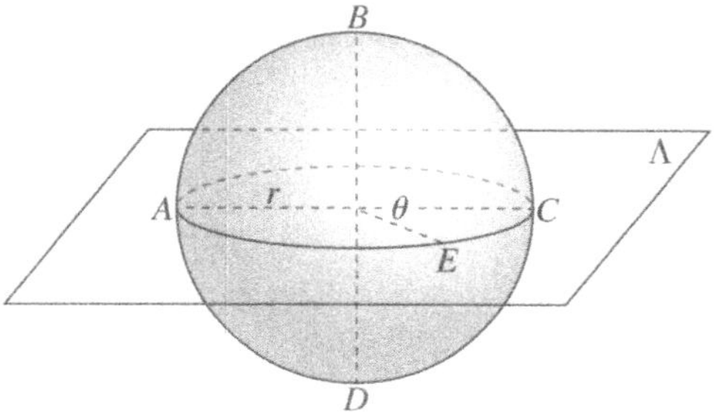

Figure 12.7 The sphere $ABCD$ cut by plane Λ forming a great circle.

The Area of a Triangle

Fix a sphere of radius r. The points of the geometry will be standard points on the surface of this sphere. The lines will be **great circles**. These are the circles formed by the intersection of the sphere with a plane that contains the sphere's center, as in Figure 12.7. A line segment in this geometry will be an arc of a great circle. Hence, the length of a segment is given by

$$\text{arc length} = \text{sphere's radius} \times \text{arc's central angle}, \qquad (12.5)$$

where all angles are measured in radians. For example, in Figure 12.7 the length of $\widehat{EC}$ is $r\theta$.

As lines in the plane can be used to construct angles and triangles, the same is true for great circles on the surface of the sphere. Let Λ and K be two planes that cut the sphere, forming two great circles, as in Figure 12.8. Suppose that B is a point of intersection of the two great circles and assume that the plane A_1BC_1 is tangent to the sphere at B. We will denote the **spherical angle** formed by points A, B, and C by our usual angle notation, and its measure will be defined as the measure of $\measuredangle A_1BC_1$. In some cases three great circles will meet to form a triangle. Such a figure is called a

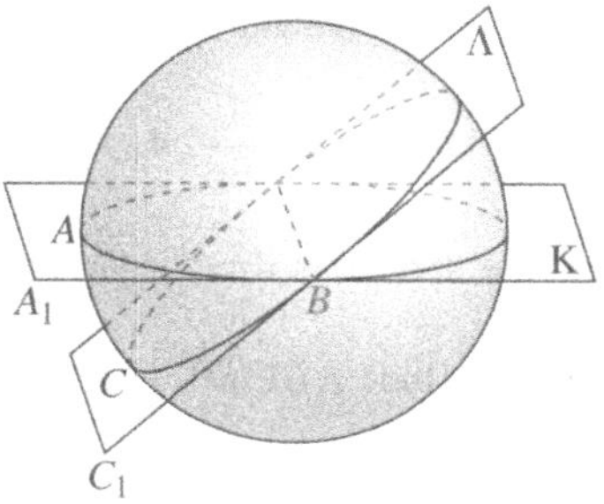

Figure 12.8 The measure of $\measuredangle ABC$ equals the measure of $\measuredangle A_1BC_1$.

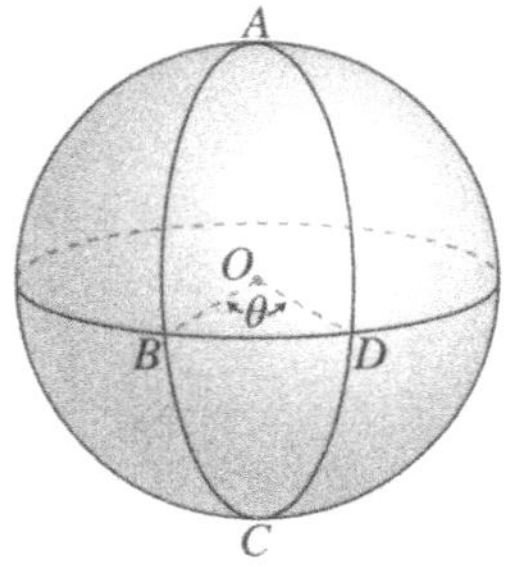

Figure 12.9 The lune $ABCD$.

spherical triangle, and again we will not change notation. The type of triangle will be clear from context. As opposed to the situation with triangles constructed under **HA**, triangles on the surface of a sphere have a positive angle excess. This means that for spherical $\triangle ABC$,

$$\text{angle excess} = (\text{m}\sphericalangle A + \text{m}\sphericalangle B + \text{m}\sphericalangle C) - \pi > 0.$$

The proposition that allows us to conclude this is called **Girard's Theorem**. Albert Girard was Flemish and lived from 1590 to 1633. He is noted for his early promotion of algebraic symbolism, and he was the first to formulate the Fundamental Theorem of Algebra (Boyer and Merzbach 1991, 303, 319; Burton 1985, 512). His work in trigonometry led him to prove the theorem that bears his name. The proof given below, however, is due to Euler (Rosenfeld 1988, 31–32). He proved it using lunes. A **lune** is the region on the surface of a sphere that lies between two great circles. (Compare this definition with that of Exercise 3 in Section 4.3.) The great circles intersect at the endpoints of a diameter of the sphere. These two points are called **antipodal**.

Let us find the area of a lune on a sphere with center O and radius r. Suppose that A, B, and C lie on a great circle and A, D, and C lie on another great circle (Figure 12.9). Let $\theta = \text{m}\sphericalangle BOD$. The ratio of the area of the lune swept out by θ to the surface area of the sphere is $\theta/2\pi$. Thus,

$$\text{area(lune } ABCD) = \frac{\theta}{2\pi} \cdot 4\pi r^2 = 2\theta r^2$$

because the surface area of the sphere is $4\pi r^2$. On the unit sphere the equation simplifies to

$$\text{area(lune } ABCD) = 2\theta.$$

This is what is needed in Euler's proof.

■ THEOREM 12.3.1 [Girard]

The area of a spherical triangle equals its angle excess.

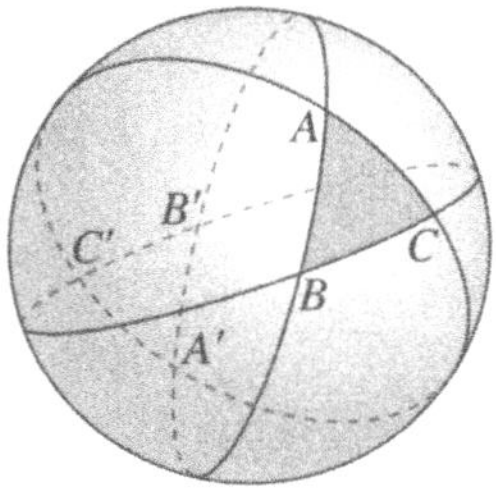

PROOF

Begin by choosing a sphere with unit radius. Let $\triangle ABC$ be on the sphere's surface. Let A', B', and C' be antipodal, respectively, to A, B, and C. Write

$$\alpha = \mathrm{m}\sphericalangle BAC, \beta = \mathrm{m}\sphericalangle ABC, \text{ and } \delta = \mathrm{m}\sphericalangle ACB.$$

Then

$$\text{area}(ABA'C) = 2\alpha, \text{ area}(BCB'A) = 2\beta, \text{ and area}(CAC'B) = 2\delta.$$

Notice that $S = \text{area}(\triangle ABC)$ is the area of the intersection of all three lunes. Therefore, the areas of the lunes without $\triangle ABC$ are $2\alpha - S$, $2\beta - S$, and $2\delta - S$. Observe that there are four pairs of congruent triangles on the sphere:

$$\begin{aligned} \triangle ABC &\cong \triangle A'B'C', \\ \triangle AB'C' &\cong \triangle A'BC, \\ \triangle BA'C' &\cong \triangle B'AC, \\ \triangle CA'B' &\cong \triangle C'AB. \end{aligned}$$

Moreover, these triangles cover the sphere, so their total sum must equal 4π. Hence,

$$2S + 4(\alpha + \beta + \delta) - 6S = 4\pi,$$

and then

$$S = (\alpha + \beta + \delta) - \pi.$$

The generalization to an arbitrary sphere is left to Exercise 6. ■

Identities

We shall use the convention that in $\triangle ABC$, the lowercase letters a, b, and c represent the lengths of the sides opposite the angles A, B, and C, respectively, and both McCleary (1994, 3–9) and Granville (1909, 193f) were used as sources. We begin with some fundamental identities.

■ **THEOREM 12.3.2**

If $\triangle ABC$ is on a sphere of radius r and $\sphericalangle C$ is the only right angle, then

- $\sin a/r = \sin c/r \sin A$.
- $\sin b/r = \sin c/r \sin B$.
- $\cos A = \cos a/r \sin B$.
- $\cos B = \cos b/r \sin A$.
- $\cos c/r = \cos a/r \cos b/r$.

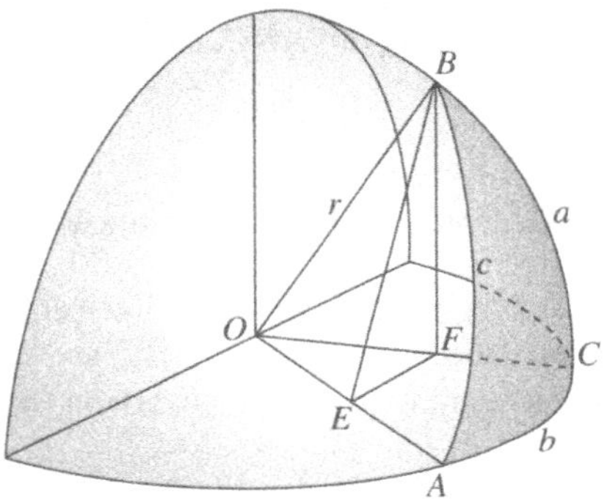

PROOF

We will prove the first and the last, leaving the others to Exercise 2. Let O be the center of the sphere and from B drop a perpendicular to $\overline{OC}$ at F and then find E on $\overline{OA}$ so that $\overline{EF}$ is perpendicular to $\overline{OA}$. First notice that

$$\text{m}\sphericalangle BOF = \frac{a}{r}, \sphericalangle EOF = \frac{b}{r}, \text{and m}\sphericalangle BOE = \frac{c}{r}.$$

Then

$$\sin\frac{a}{r} = \frac{BF}{OB}, \sin\frac{c}{r} = \frac{BE}{OB}, \text{ and } \sin A = \sin\sphericalangle BEF = \frac{BF}{BE}.$$

Hence, $\sin a/r = \sin c/r \sin A$. We also have

$$\cos\frac{c}{r} = \frac{OE}{OB}, \cos\frac{a}{r} = \frac{OF}{OB}, \text{ and } \cos\frac{b}{r} = \frac{OE}{OF}.$$

Thus, $\cos c/r = \cos a/r \cos b/r$. ■

The remaining theorems that we will prove concerning spherical triangles all have corresponding plane versions. We begin with the spherical analog to the Pythagorean Theorem.

■ **THEOREM 12.3.3 [Spherical Pythagorean Theorem]**

If $\measuredangle C$ is the only right angle in $\triangle ABC$ on a sphere of radius r, then

$$\cos\frac{c}{r} = \cos\frac{a}{r}\cos\frac{b}{r}.$$

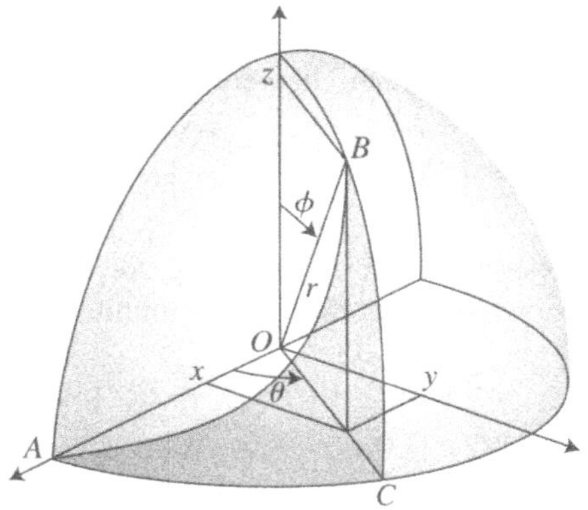

PROOF

We may assume by rotating the sphere, if necessary, that the coordinates of A are $(r,\ 0,\ 0)$ and C lies in the xy-plane. Recall that any point B on the sphere can be identified by the point $(r,\ \phi,\ \theta)$ in spherical coordinates. The point B can then be converted to Cartesian coordinates using the equations

$$x = r\sin\phi\cos\theta,\ y = r\sin\phi\sin\theta, \text{ and } z = r\cos\phi,$$

where $\phi = \pi/2 - a/r$ and $\theta = b/r$. Therefore,

$$\begin{aligned} B &= \left(r\sin\left[\frac{\pi}{2}-\frac{a}{r}\right]\cos\frac{b}{r},\ r\sin\left[\frac{\pi}{2}-\frac{a}{r}\right]\sin\frac{b}{r},\ r\cos\left[\frac{\pi}{2}-\frac{a}{r}\right]\right) \\ &= \left(r\cos\frac{a}{r}\cos\frac{b}{r},\ r\cos\frac{a}{r}\sin\frac{b}{r},\ r\sin\frac{a}{r}\right) \end{aligned}$$

and

$$C = \left(r\cos\frac{b}{r},\ r\sin\frac{b}{r},\ 0\right).$$

Now apply some linear algebra and view $A(r,\ 0,\ 0)$ and $B(x,\ y,\ z)$ as vectors originating from the origin. Since the angle between these two vectors is $\measuredangle AOB$ and the measures of this angle is c/r, we can apply the dot product to write

$$A \bullet B = \|A\|\,\|B\|\cos\frac{c}{r}.$$

Thus,

$$\cos\frac{c}{r} = \frac{A \bullet B}{\|A\|\,\|B\|} = \frac{r^2\cos a/r\cos b/r}{r^2} = \cos\frac{a}{r}\cos\frac{b}{r}. \ ■$$

The previous theorem is called the Spherical Pythagorean Theorem for two reasons. First, it gives a relationship between the sides of a spherical right triangle. Secondly, and more important, it is the Pythagorean Theorem when examined correctly. Using the Maclaurin series,

$$\cos x = \sum_{n=0}^{\infty} (-1)^n \frac{x^{2n}}{(2n)!},$$

we see that the Spherical Pythagorean Theorem becomes

$$\sum_{n=0}^{\infty} (-1)^n \frac{(c/r)^{2n}}{(2n)!} = \left(\sum_{n=0}^{\infty} (-1)^n \frac{(a/r)^{2n}}{(2n)!} \right) \cdot \left(\sum_{n=0}^{\infty} (-1)^n \frac{(b/r)^{2n}}{(2n)!} \right).$$

By multiplying the series on the right, subtracting one from both sides, and multiplying through by $-2r^2$, we find that there exists infinite series $S(r)$ and $T(r)$ such that

$$c^2 + S(r) = a^2 + b^2 + T(r), \tag{12.6}$$

where both $S(r) \to 0$ and $T(r) \to 0$ as $r \to \infty$. We conclude that as the radius of the sphere increases without bound, the equation of the Spherical Pythagorean Theorem approaches that of the standard Pythagorean Theorem. In other words, as the volume of the sphere increases, the geometry on its surface becomes more Euclidean.

The Law of Sines for plane trigonometry states that

$$\frac{a}{\sin A} = \frac{b}{\sin B} = \frac{c}{\sin C}.$$

The following is the analogous theorem for the sphere.

■ THEOREM 12.3.4 [Spherical Law of Sines]

In $\triangle ABC$ on a sphere of radius r,

$$\frac{\sin(a/r)}{\sin A} = \frac{\sin(b/r)}{\sin B} = \frac{\sin(c/r)}{\sin C}.$$

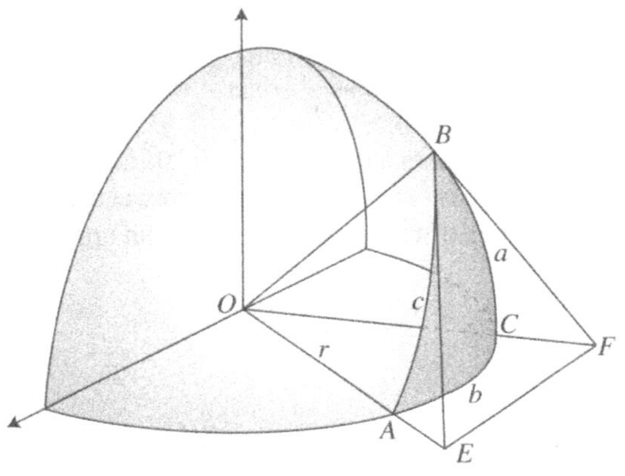

PROOF

We will prove the theorem for right triangles first and then generalize. Let $\measuredangle C$ in $\triangle ABC$ be right on the surface of a sphere with center O and radius r. Let plane EBF be tangent to the sphere at B. Assume that $\overline{OA}$ is produced to E and $\overline{OC}$ produced to F. Since $\overline{EB}$ and $\overline{BF}$ are tangent to the sphere, $\measuredangle OBE$ and $\measuredangle OBF$ are right angles. Hence, appealing to Pythagoras we see that

$$OE^2 = BE^2 + OB^2$$

and

$$OF^2 = BF^2 + OB^2.$$

Observe that planes EBF and OBF are perpendicular, so $\measuredangle OFE$ is right. We then conclude that

$$OE^2 = EF^2 + OF^2.$$

Therefore,

$$EF^2 = OE^2 - OF^2 = (BE^2 + OB^2) - (BF^2 + OB^2),$$

and then

$$BE^2 = EF^2 + BF^2.$$

Therefore, $\measuredangle EFB$ is right, and we then have

$$\sin \measuredangle AOC = \frac{EF}{OE} = \frac{EF}{BE} \cdot \frac{BE}{OE} = \sin B \sin \measuredangle AOB.$$

Since $b/r = \mathrm{m}\measuredangle AOC$,

$$\frac{\sin(b/r)}{\sin B} = \sin \measuredangle AOB.$$

Similarly, we could redraw the plane parallel at A instead of at B and find that

$$\frac{\sin(a/r)}{\sin A} = \sin \measuredangle AOB.$$

Because $\sin C = 1$ and $c/r = \mathrm{m}\measuredangle AOB$, we have

$$\frac{\sin(a/r)}{\sin A} = \frac{\sin(b/r)}{\sin B} = \frac{\sin(c/r)}{\sin C}.$$

Now take $\triangle ABC$ to be an arbitrary spherical triangle. Drop $\overline{BD}$ perpendicular to $\overline{AC}$. By examining $\triangle ABD$ and $\triangle CBD$, the equation for right triangles yields

$$\sin \frac{a}{r} \sin C = \sin \frac{BD}{r} = \sin \frac{c}{r} \sin A. \qquad (12.7)$$

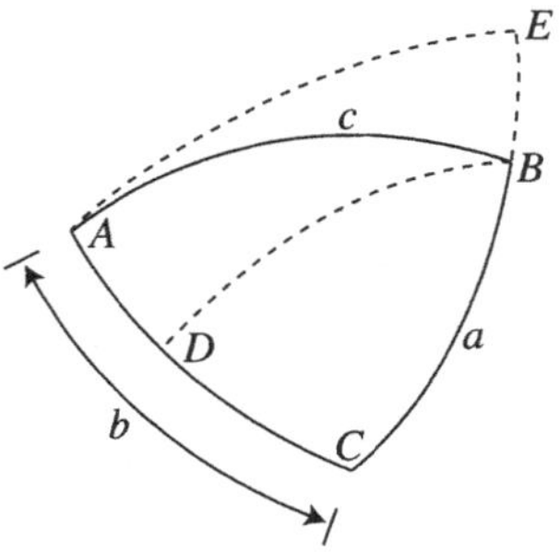

Extend $\overline{CB}$ to E such that $\overline{AE}$ joined is perpendicular to $\overline{EC}$. When we examine $\triangle ACE$ and $\triangle ABE$, we find that

$$\sin\frac{b}{r}\sin C = \sin\frac{AE}{r} = \sin\frac{c}{r}\sin(\pi - B) = \sin\frac{c}{r}\sin B. \qquad (12.8)$$

These equations complete the proof. ■

The last theorem of this section is the spherical version of the Law of Cosines. Recall that the plane version states that in any $\triangle ABC$,

$$c^2 = a^2 + b^2 - 2ab\cos C.$$

When we examine the following result when r is small, we will find that it approximates the plane version (Exercise 11).

■ **THEOREM 12.3.5 [Spherical Law of Cosines]**

In $\triangle ABC$ on a sphere of radius r,

$$\cos\frac{a}{r} = \cos\frac{b}{r}\cos\frac{c}{r} + \sin\frac{b}{r}\sin\frac{c}{r}\cos A.$$

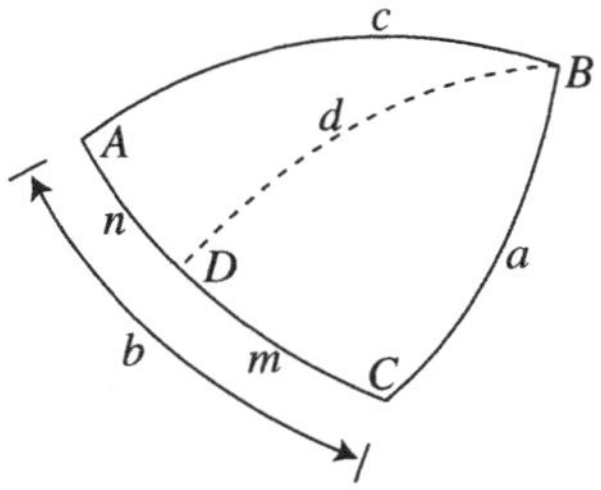

PROOF

Leaving the generalization to Exercise 10, we will prove the case when $r = 1$. Take spherical $\triangle ABC$ and drop an altitude from B to D on $\overline{AC}$. When we apply the Spherical Pythagorean Theorem (12.3.3) to $\triangle BCD$, letting $d = DB$, $m = DC$, and $n = AD$, we find that

$$\cos a = \cos d \cos m = \cos d \cos(b - n) = \cos d(\cos b \cos n + \sin b \sin n).$$

Therefore,

$$\cos a = \cos d \cos b \cos n + \cos d \sin b \sin n. \tag{12.9}$$

When we apply Theorem 12.3.3 to $\triangle BAD$, we get $\cos c = \cos d \cos n$, and substituting this into Equation 12.9 we see that

$$\cos a = \cos b \cos c + \cos d \sin b \sin n. \tag{12.10}$$

Furthermore, by Theorem 12.3.2 we have

$$\cos d = \frac{\cos c}{\cos n}.$$

Multiplying both sides of this equation by $\sin n$ we obtain

$$\cos d \sin n = \cos c \tan n. \tag{12.11}$$

However,

$$\cos A = \tan n \cot c \tag{12.12}$$

by Exercise 9, so $\tan n = \tan c \cos A$. We now substitute this into Equation 12.11 to find

$$\cos d \sin n = \cos c \tan c \cos A = \sin c \cos A,$$

and we may complete the proof by substituting into Equation 12.10. ■

Exercises

1. What would a triangle on the sphere look like if it had two right angles? Is it possible for a spherical triangle to have three right angles? Explain.

2. Let $\triangle ABC$ be on sphere of radius r with $\sphericalangle C$ the only right angle. Show the following identities.
   (a) $\sin b/r = \sin c/r \sin B$.
   (b) $\cos A = \cos a/r \sin B$.
   (c) $\cos B = \cos b/r \sin A$.

3. Let $\triangle ABC$ be on the unit sphere with $\sphericalangle C$ as the only right. Prove:
   (a) $\cos A = \tan b \cot c$.
   (b) $\sin b = \tan a \cot A$.
   (c) $\cos c = \cot A \cot B$
   (d) $\sec a = \sec A \sin B$

4. Find the remaining sides and angles (in degrees) of the spherical triangles given the following. Assume that C is the only right angle of the triangle.

(a) $c = 110°, A = 80°$.
(b) $c = 100°, b = 115°$.
(c) $c = 120°, a = 20°$.
(d) $B = 150°, b = 160°$.
(e) $B = 150°, a = 110°$.

5. Prove that the following geometric results hold for $\triangle ABC$ on the unit sphere.

(a) $a < b + c$.
(b) If $a \neq b$, then $A \neq B$.
(c) $a < b$ if and only if $A < B$.
(d) $a + b + c < 2\pi$.
(e) $\pi < A + B + C < 3\pi$.

6. Given (r, ϕ, θ) in spherical coordinates, show that the corresponding point in Cartesian coordinates is given by

$$x = r \sin \phi \cos \theta,$$
$$y = r \sin \phi \sin \theta,$$
$$z = r \cos \phi.$$

7. Prove Equation 12.12 in the proof of Theorem 12.3.5.

8. Generalize the proof of Theorem 12.3.5 to a sphere with any radius.

9. Use the Maclaurin series for sine and cosine to show that the Spherical Law of Cosines (12.3.5) yields an equation that approximates the plane Law of Cosines for small spherical triangles.

10. Prove Girard's Theorem (12.3.1) for a sphere of any radius.

11. Confirm Equations 12.7 and 12.8 from the proof of Theorem 12.3.4.

12. Verify for $\triangle ABC$ on the unit sphere:

(a) $\cos A = \cos a \sin B \sin C - \cos B \cos C$.
(b) $\cot A \sin C + \cos b \cos C = \cot a \sin b$.

12.4 HOROSPHERES

Both Bolyai and Lobachevski defined a three-dimensional analog to the horocycle. Bolyai did it by selecting a point A in space and an ideal point Ω (*Appendix* 11). He defined

$$\mathcal{F} = \{X : \sphericalangle AX\Omega \cong \sphericalangle XA\Omega\}.$$

The line $\overleftrightarrow{A\Omega}$ we will call the **axis** of $\mathcal{F}$. Like horocycles, at each point on $\mathcal{F}$ there is a line that is parallel to the axis. Each of these lines is parallel to each other (Exercise 6 in Section 12.1), so we can also view them as axes.

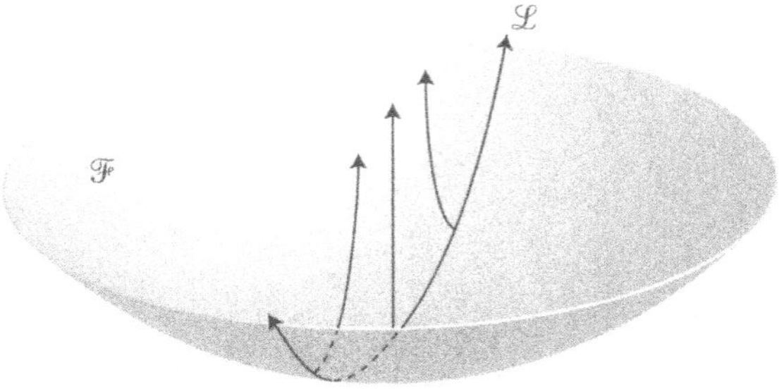

Figure 12.10 The horosphere $\mathscr{F}$ with generating horocycle $\mathscr{L}$.

If we let $B \in \mathscr{F}$, the plane Λ that contains A, B, and Ω intersects $\mathscr{F}$, resulting in a horocycle $\mathscr{L}$ as described in Equation 12.3. If we revolve Λ about $\overline{A\Omega}$, then we see that Bolyai's definition is equivalent to Lobachevski's.

■ **DEFINITION 12.4.1 [*Theory of Parallels* 34]**

A **horosphere** (or **boundary surface**) is the surface $\mathscr{F}$ formed by revolving a horocycle about one of its axes.

Notice that due to the constant curvature of horocycles, if Λ instead makes an angle with $\overline{A\Omega}$, it will determine a circle on the surface of $\mathscr{F}$ (Exercise 3). Since a horocycle under **HR** is a line, the horosphere is a plane in Euclidean geometry. Under **HA** it looks more like Figure 12.10.

Induced Euclidean Geometry

Despite assuming the Hypothesis of the Acute Angle, there is a natural interpretation of the term *line* that yields a Euclidean geometry on the surface of a horosphere. We first need to prove a theorem.

■ **THEOREM 12.4.2 [*Theory of Parallels* 28]**

[HA] If three planes intersect each other in parallel lines, the sum of the angles that the planes make with each other equals two right angles.

PROOF

Suppose that three planes intersect in $\overline{AA'}$, $\overline{BB'}$, and $\overline{CC'}$ and that the three lines are parallel. These planes make angles with each other of measure x at $\overline{AA'}$, y at $\overline{BB'}$, and z at $\overline{CC'}$ (not identified in the diagram on page 426). Take D on $\overline{BB'}$ and pass through it a plane containing A and C. This fourth plane intersects $ABB'A'$ in $\overline{AD}$ and $CBB'C'$ in $\overline{CD}$. Denote the angle of inclination of plane ACD with $ACC'A'$ by w. We now construct three spheres, one each

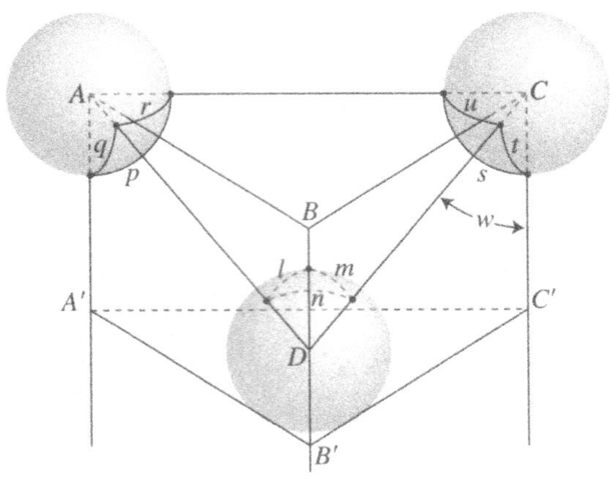

centered at A, C, and D. This results in three spherical triangles:

- The vertices of the spherical triangle at A are the intersections of $\overline{AA'}$, $\overline{AD}$, and $\overline{AC}$ with the sphere. Let its area be α and its sides have length p, q, and r. Observe that the measure of the angle opposite q is w, the angle opposite r is x, and by Girard's Theorem (12.3.1) the angle opposite p has measure $\alpha - w - x + \pi$. This is summarized in the following table:

length	opposite angle
p	$\alpha - w - x + \pi$
q	w
r	x

- Similarly, the sphere at C intersects $\overline{AC}$, $\overline{CD}$, and $\overline{CC'}$, forming a spherical triangle of area β:

length	opposite angle
s	$\beta - w - z + \pi$
t	w
u	z

- Let δ be the area of the spherical triangle formed by the sphere at D. We note that the angle opposite l in the diagram is supplementary to the angle opposite s in the sphere at C. This is because they both have one side in the plane ACD and their other side in the plane $BCC'B'$, but these sides are on opposite sides of ACD. Similarly, the angle opposite m and the angle opposite p are also supplementary. Therefore, we describe the spherical triangle at D by a table:

length	opposite angle
l	$w + z - \beta$
m	$w + x - \alpha$
n	y

Therefore,

$$\delta = (x + y + z - \pi) + 2w - \beta - \alpha$$

by Girard's Theorem. Now let $w \to 0$. We see that both $q \to 0$ and $t \to 0$, so $\alpha \to 0$ and $\beta \to 0$. Also, the sides l and m approach zero, which means that $\delta \to 0$. But the angles x, y, and z remain fixed. Hence, $x + y + z = \pi$. ■

On a horosphere $\mathcal{F}$ let the word *point* refer to a regular point and let the word *line* refer to a horocycle. These choices impart a geometric system upon $\mathcal{F}$ called the **induced Euclidean geometry**. Let A, B, and C be points on $\mathcal{F}$ but not all on the same horocycle. Call the three horocycles determined by these points AB, BC, and AC. Take three planes that each contain an axis of $\mathcal{F}$ and also intersect the horosphere in AB, BC, and AC, respectively. Define the angle between any two of these horocycles to be the angle between their corresponding planes. (This is similar to how we defined the measure of angles on a sphere.) By Theorem 12.4.2, the sum of the internal angles of $\triangle ABC$ on $\mathcal{F}$ is equal to two right angles. This means that the Parallel Postulate holds true in the induced geometry by Corollary 11.2.12.

■ **THEOREM 12.4.3 [*Theory of Parallels* 34, *Appendix* 21]**

The induced geometry on a horosphere is Euclidean.

The last result means that inside a geometry that satisfies **HA**, there is a surface that can naturally be interpreted as a Euclidean plane. The German mathematician David Hilbert showed that there is no such copy of **HA** in Euclidean geometry (Bonola 1912, 145).

Spherical Trigonometry and HA

As we study the sphere with infinite radius under **HA**, it is natural to ask about the status of a sphere with finite radius under the same hypothesis. The answer will involve two technical results, the second requires two horospheres.

Let $\triangle ABC$ be such that $m\sphericalangle A = \Pi(\alpha)$, $m\sphericalangle B = \Pi(\beta)$, and $\sphericalangle C$ is a right angle (Figure 12.11). Find A' in space so that $\overline{AA'}$ is perpendicular to the plane ABC. In addition, find B' and C' so that $\overline{BB'}$ and $\overline{CC'}$ are parallel to $\overline{AA'}$. These lines are also mutually parallel and are the intersections of three planes that make angles with each other with the following measures:

$$\Pi(\alpha) \text{ at } \overline{AA'},\ \Pi(\alpha') \text{ at } \overline{BB'}, \text{ and } \pi/2 \text{ at } \overline{CC'},$$

where α' is chosen so that

$$\Pi(\alpha) + \Pi(\alpha') = \pi/2$$

by Theorem 12.4.2. At B construct a unit sphere. Let B_1, B_2, and B_3 be the intersections of the sphere with $\overline{BC}$, $\overline{BA}$, and $\overline{BB'}$, respectively. Then the measurements of spherical $\triangle B_1B_2B_3$ are

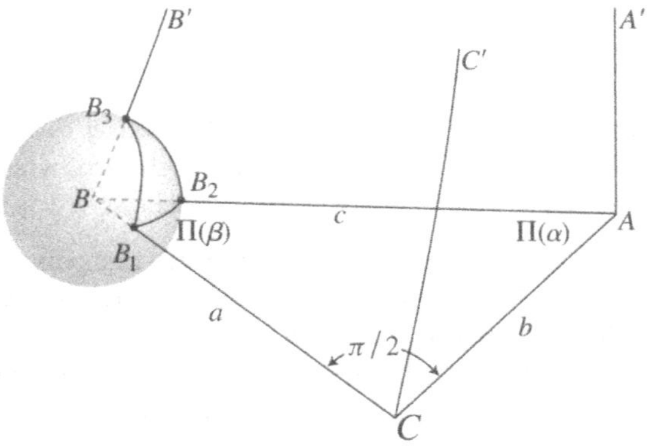

Figure 12.11 Deriving a spherical triangle from a plane triangle.

vertex	measure	opposite side
B_1	$\Pi(b)$	$\Pi(c)$
B_2	$\pi/2$	$\Pi(a)$
B_3	$\Pi(\alpha')$	$\Pi(\beta)$

We will return to this shortly, but first we prove the following.

■ THEOREM 12.4.4 [*Theory of Parallels* 35]

[HA] If $m\sphericalangle A = \Pi(\alpha)$, $m\sphericalangle B = \Pi(\beta)$, and $\sphericalangle C$ is right in $\triangle ABC$, then:

- $\sin \Pi(c) = \sin \Pi(a) \sin \Pi(b)$.
- $\sin \Pi(\beta) = \cos \Pi(\alpha) \sin \Pi(a)$.
- $\sin \Pi(\alpha) = \cos \Pi(\beta) \sin \Pi(b)$.
- $\cos \Pi(b) = \cos \Pi(c) \cos \Pi(\alpha)$.
- $\cos \Pi(a) = \cos \Pi(c) \cos \Pi(\beta)$.

PROOF

Let $\triangle ABC$ be as described in the hypothesis of the theorem. Let $\overline{AA'}$ be perpendicular to ABC at A and let $\overline{BB'}$ and $\overline{CC'}$ be parallel to $\overline{AA'}$. With axis $\overline{AA'}$ describe a horosphere that contains A and intersects $\overline{BB'}$ at B'' and $\overline{CC'}$ at C''. These three points are joined by horocycles. Each is the intersection of the horosphere with a plane that contains an axis of the horosphere. Since the angle between two of these horocycles is defined to be the angle between the two planes and letting $p = B''C''$, $q = AC''$, and $r = AB''$, we see that $\triangle AB''C''$ on the surface of the horosphere can be described by

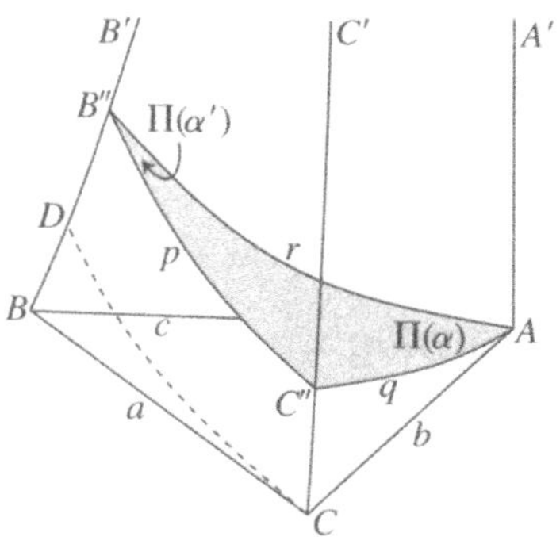

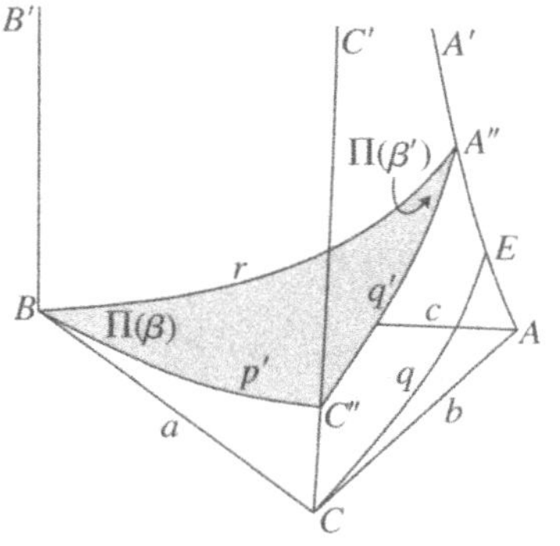

vertex	measure	opposite side
A	$\Pi(\alpha)$	p
B''	$\Pi(\alpha')$	q
C''	$\pi/2$	r

We may apply plane trigonometry because the induced geometry on the horosphere is Euclidean (Theorem 12.4.3) and write

$$q = r \cos \Pi(\alpha). \tag{12.13}$$

Now reverse the setup by erecting the perpendicular at B, but still let $\overline{AA'}$, $\overline{BB'}$, and $\overline{CC'}$ be parallel. Place a horosphere with axis $\overline{BB'}$ at B and let it intersect $\overline{AA'}$ at A'' and $\overline{CC'}$ at C''. Let $q' = C''A''$ and $p' = BC''$. Since all horocycles have the same curvature, $BA'' = r$. Place a horocycle at C so that it is concentric with $C''A''$. Its length is q. Since all horocycles are congruent, there exists a function f that measures the distance from the plane ABC to a horocycle along an axis. That is, $f(b) = AE$, $f(a) = CC'' = EA''$, and we may conclude that

$$f(c) = AA'' = AE + EA'' = f(a) + f(b). \tag{12.14}$$

Then by Theorem 12.2.9, $q = q'e^{kf(a)}$ for some $k > 0$, and again since the induced geometry on a horosphere is Euclidean, we have $q' = r \sin \Pi(\beta)$, so that

$$q = re^{kf(a)} \sin \Pi(\beta). \tag{12.15}$$

Therefore, setting Equation 12.13 equal to Equation 12.15 gives

$$\cos \Pi(\alpha) = e^{kf(a)} \sin \Pi(\beta). \tag{12.16}$$

A similar construction of $\triangle ABC$ with the two nonright angles having measures $\Pi(b')$ and $\Pi(c)$ will yield

$$\sin \Pi(b) = e^{kf(a)} \sin \Pi(c). \tag{12.17}$$

because $\cos \Pi(b') = \sin \Pi(b)$ (Exercise 5). Therefore, by multiplying both sides of Equation 12.17 by $e^{kf(b)}$, we have

$$e^{kf(b)} \sin \Pi(b) = e^{kf(c)} \sin \Pi(c)$$

by Equation 12.14. An argument similar to the above will yield

$$e^{kf(a)} \sin \Pi(a) = e^{kf(b)} \sin \Pi(b).$$

Finally, we note that the lengths of a and b may be chosen independently of each other. Hence, since $f(0) = 0$ and $\Pi(0) = \pi/2$, let $b = 0$ and conclude that

$$\sin \Pi(a) = e^{-kf(a)}.$$

Therefore, by Equation 12.17,

$$\sin \Pi(c) = \sin \Pi(a) \sin \Pi(b),$$

and then by Equation 12.16,

$$\sin \Pi(\beta) = \cos \Pi(\alpha) \sin \Pi(a).$$

The remaining three identities follow by a "mutation of the letters." ■

We close the section with the remarkable result that the geometry on the surface of a sphere where *line* is interpreted as *great circle* is independent of the hypothesis chosen. We accomplish this by examining the sphere described in Figure 12.11 and the corresponding table at the top of page 36.

■ **COROLLARY 12.4.5 [*Theory of Parallels* 35, *Appendix* 26]**

The geometry on the surface of a sphere is the same whether **HR** or **HA** is assumed.

PROOF

If we assign $a = \Pi(c)$, $b = \Pi(\beta)$, $c = \Pi(a)$, $A = \Pi(b)$, and $B = \Pi(\alpha')$ for the dimension of a triangle on a unit sphere, the identities of Theorem 12.4.4 are seen to be identical to the identities of Theorem 12.3.2. ■

Exercises

1. Let A, B, and C be points on a horosphere $\mathcal{F}$ but not all on the same horocycle. The three horocycles through these points form a triangle on $\mathcal{F}$. Show that if $\measuredangle C$ is right, then $AB^2 = AC^2 + BC^2$.

2. Let $\mathcal{L}$ be a horocycle on the surface of a horosphere $\mathcal{F}$ with axis $\overline{A\Omega}$. Let B be on $\mathcal{F}$ but not on $\mathcal{L}$. Prove:
   (a) B corresponds with A.
   (b) B corresponds with every other point on $\mathcal{F}$.

3. Take A, B, and C to be points on the surface of a horosphere but not all on the same horocycle. Show that A, B, and C lie on a circle on the horosphere. (*Hint:* Find the intersection of the lines that are the perpendicular bisectors of the sides of the triangle.)

4. Redo Exercise 3 using another method.

5. Let $\triangle ABC$ be as in the proof of Theorem 12.4.4 but such that $m\sphericalangle A = \Pi(b')$, $m\sphericalangle B = \Pi(c)$, $m\sphericalangle C = \pi/2$, $BC = a$, $AC = \alpha$, and $AB = \beta$.
   (a) Show $\cos \Pi(b') = \sin \Pi(b)$.
   (b) Use $\triangle ABC$ to demonstrate Equation 12.17.

6. Suppose that spherical $\triangle B_1B_2B_3$ in Figure 12.11 was, instead, constructed at A. Provided that $\overline{AA'}$ is still perpendicular to plane ABC, find the measures of the angles and sides of $\triangle B_1B_2B_3$, where B_3 is the intersection of $\overline{AA'}$ with the sphere.

7. Let f be the function defined in the proof of Theorem 12.4.4. In the first diagram of that theorem, draw a horocycle through C that is concentric to the horocycle $B''C''$ and intersects $\overline{BB''}$ at D. Using B, B'', and D, show that $f(c) = f(a) + f(b)$.

8. Give the details for the "mutation of letters" from the proof of Theorem 12.4.4:
   (a) $\sin \Pi(\alpha) = \cos \Pi(\beta) \sin \Pi(b)$.
   (b) $\cos \Pi(b) = \cos \Pi(c) \cos \Pi(\alpha)$.
   (c) $\cos \Pi(a) = \cos \Pi(c) \cos \Pi(\beta)$.

9. Provide the details for the steps labeled as "similar" in the proof of Theorem 12.4.4.

12.5 EVALUATING THE PI FUNCTION

We have seen that Lobachevski's Π function has some very nice properties. In this section we derive a formula for it. The proof given is by Lobachevski with some modification by Manning (1901, 45–46, 58).

LEMMA 12.5.1

If n is a positive integer and

$$f(x) = \frac{f(2x)}{f(x)} = \frac{f(3x)}{f(2x)} = \cdots = \frac{f(nx)}{f([n-1]x)},$$

then $f(nx) = [f(x)]^n$.

PROOF

By hypothesis we conclude that

$$f(nx) = \frac{[f([n-1]x)]^2}{f([n-2]x)}, \quad f([n-1]x) = \frac{[f([n-2]x)]^2}{f([n-3]x)}, \quad \ldots,$$

$$f(3x) = \frac{[f(2x)]^2}{f(x)}, \quad f(2x) = [f(x)]^2.$$

Hence,

$$f(nx) = \frac{[f([n-1]x)]^2}{f([n-2]x)} = f([n-1]x)\frac{f([n-1]x)}{f([n-2]x)} = f([n-1]x)f(x),$$

with the last equality holding by the hypothesis. By a similar argument,

$$f([n-1]x) = f([n-2]x)f(x),$$

so

$$f(nx) = f([n-2]x)[f(x)]^2.$$

Continuing, we find that $f(nx) = [f(x)]^n$. ■

The next theorem contains the exact relationship between the length of a segment and its corresponding angle of parallelism. Lobachevski's formula did not include the constant k. Instead, he modified the units along the x-axis so that the constant could be taken to equal 1. Bolyai did not do this, so the constant remained an important part of his equation. His father was very much troubled that his son could never find a way to remove the constant. Farkas believed that the final description of the new geometry should be able to determine the value of k. János was content to leave it undetermined (Bonola 1912, 99).

■ **THEOREM 12.5.2 [*Theory of Parallels* 36, *Appendix* 29]**

[HA] There exists $k > 0$ such that for all $x \geqslant 0$,

$$\tan \frac{1}{2}\Pi(x) = e^{-kx}.$$

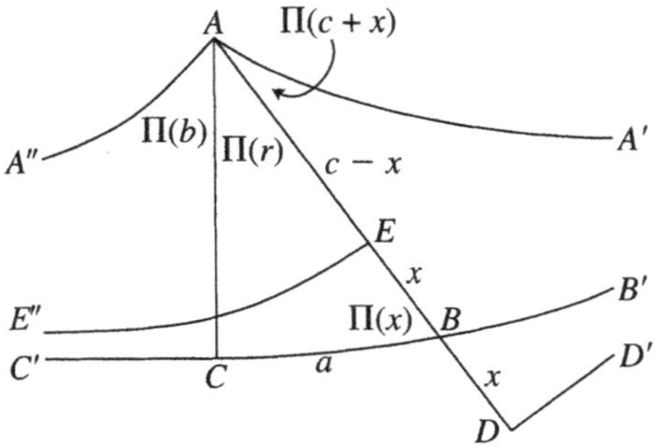

PROOF

We first note that if $x = 0$,

$$\tan \frac{1}{2}\Pi(0) = \tan \pi/4 = 1$$

because $\Pi(0) = \pi/2$, so the equation holds. So, let $x > 0$ and form $\triangle ABC$ with $m\sphericalangle ABC = \Pi(x)$ and $\sphericalangle ACB$ right. Further, since Π is onto (Theorem 12.1.6), there exists $r > 0$ such that $m\sphericalangle BAC = \Pi(r)$. To ease our calculations later, let $a = BC$, $b = AC$, and $c = AB$. Notice that the diagram assumes that $c > x$. Next extend $\overline{AB}$ to point D, where $BD = x$. Find D' so that $\overline{DD'}$ is perpendicular to $\overline{AD}$. Producing $\overline{CB}$ to B', we see that $\overline{BB'}$ is parallel to $\overline{DD'}$ because $m\sphericalangle B'BD = \Pi(x)$. Finally, draw $\overline{AA'}$ so it is parallel to $\overline{DD'}$. This gives us $m\sphericalangle A'AD = \Pi(c + x)$. Thus, by Theorem 12.1.1 and Theorem 12.1.3, $\overline{AA'}$ is also parallel to $\overline{CB'}$ and $m\sphericalangle A'AC = \Pi(b)$. Hence,

$$\Pi(b) = \Pi(r) + \Pi(c + x). \tag{12.18}$$

There are now two cases to consider. Either $c > x$ or it is not. Suppose that $c > x$ and find E on $\overline{AB}$ such that $AE = c - x$ and $EB = x$. At E construct $\overline{EE'}$ so that it meets $\overline{AB}$ in right angles and produce $\overline{BC}$ to C'. We see that $\overline{BC'}$ and $\overline{EE'}$ are parallel since $m\sphericalangle ABC = \Pi(x)$. Draw $\overline{AA''}$ parallel to $\overline{CC'}$. This means that we have $m\sphericalangle CAA' = \Pi(b)$. Therefore, $\overline{AA'}$ is parallel to $\overline{EE'}$ by Theorem 12.1.3. Thus, $m\sphericalangle EAA' = \Pi(c - x)$, and we conclude that

$$\Pi(b) = \Pi(c - x) - \Pi(r). \tag{12.19}$$

That this equation holds when $c \leqslant x$ is left to Exercise 2.

We may now combine Equations 12.18 and 12.19 to find that

$$2\Pi(b) = \Pi(c - x) + \Pi(c + x),$$
$$2\Pi(r) = \Pi(c - x) - \Pi(c + x).$$

Therefore,

$$\frac{\cos \Pi(b)}{\cos \Pi(r)} = \frac{\cos \frac{1}{2}[\Pi(c - x) + \Pi(c + x)]}{\cos \frac{1}{2}[\Pi(c - x) - \Pi(c + x)]}.$$

By Theorem 12.4.4,

$$\cos \Pi(c) = \frac{\cos \Pi(b)}{\cos \Pi(r)},$$

and if we let $\theta = \frac{1}{2}\Pi(c - x)$ and $\phi = \frac{1}{2}\Pi(c + x)$, we find

$$\cos \Pi(c) = \frac{\cos(\theta + \phi)}{\cos(\theta - \phi)},$$

from which some trigonometric identities give

$$2\cos^2 \tfrac{1}{2}\Pi(c) = \frac{\cos(\theta + \phi) + \cos(\theta - \phi)}{\cos(\theta - \phi)},$$

and then

$$\sec^2 \tfrac{1}{2}\Pi(c) - 1 = \frac{\sin\theta \sin\phi}{\cos\theta \cos\phi}.$$

Therefore,

$$\tan^2 \tfrac{1}{2}\Pi(c) = \tan \tfrac{1}{2}\Pi(c - x) \tan \tfrac{1}{2}\Pi(c + x).$$

For all positive integers n, let $y = c - nx$ so that

$$\tan^2 \tfrac{1}{2}\Pi(y + nx) = \tan \tfrac{1}{2}\Pi(y + (n-1)x) \tan \tfrac{1}{2}\Pi(y + (n+1)x),$$

and then

$$\frac{\tan \frac{1}{2}\Pi(y + x)}{\tan \frac{1}{2}\Pi(y)} = \frac{\tan \frac{1}{2}\Pi(y + 2x)}{\tan \frac{1}{2}\Pi(y + x)} = \cdots = \frac{\tan \frac{1}{2}\Pi(y + nx)}{\tan \frac{1}{2}\Pi(y + (n-1)x)}.$$

Because $\tan \frac{1}{2}\Pi(0) = 1$, letting $y = 0$ yields

$$\tan \tfrac{1}{2}\Pi(x) = \frac{\tan \frac{1}{2}\Pi(2x)}{\tan \frac{1}{2}\Pi(x)} = \cdots = \frac{\tan \frac{1}{2}\Pi(nx)}{\tan \frac{1}{2}\Pi([n-1]x)}.$$

Hence, by Lemma 12.5.1,

$$\tan \tfrac{1}{2}\Pi(nx) = \tan^n \tfrac{1}{2}\Pi(x)$$

for all positive integers n. Since $\Pi(x) < \pi/2$, we have $\tan \frac{1}{2}\Pi(x) < 1$. Therefore, by Lemma 12.2.8 there exists $k > 0$ such that

$$\tan \tfrac{1}{2}\Pi(x) = e^{-kx}. \blacksquare$$

Since Π is undefined for negative numbers, Lobachevski extended the domain of Π to all real numbers by defining

$$\Pi(x) + \Pi(-x) = \pi. \qquad (12.20)$$

With this definition we can show that $\tan \frac{1}{2}\Pi(x) = e^{-kx}$ for all real numbers x (Exercise 3).

The Graph of Pi

The Theorem 12.5.2 allows us to compute values for Π, for it follows quickly that

$$\Pi(x) = 2\tan^{-1} e^{-kx}.$$

The graph of this function is given in Figure 12.12 for $k = 1$. Notice that this confirms the basic properties of Π listed in Section 12.1.

We should now observe that there is a relationship between Π and the **hyperbolic functions**. For any real number x, define **hyperbolic cosine** and **hyperbolic sine** as

$$\cosh x = \frac{e^x + e^{-x}}{2},$$

$$\sinh x = \frac{e^x - e^{-x}}{2}.$$

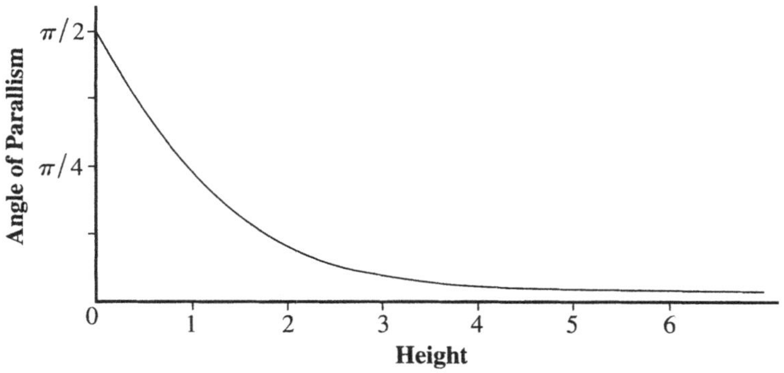

Figure 12.12 The graph of $\Pi(x) = 2\tan^{-1} e^{-x}$.

The hyperbolic versions of tangent, cotangent, secant, and cosecant are defined as their trigonometric counterparts. In addition, there are hyperbolic identities such as

$$1 - \tanh^2 x = \operatorname{sech}^2 x. \tag{12.21}$$

The hyperbolic functions can be used to help evaluate expressions that involve the composition of standard trigonometric functions and Π. For example, to evaluate $\cos \Pi(x)$, first note that

$$e^{-2kx} = \tan^2 \tfrac{1}{2}\Pi(x) = \frac{1 - \cos \Pi(x)}{1 + \cos \Pi(x)}.$$

Therefore,

$$\cos \Pi(x) + e^{-2kx} \cos \Pi(x) = 1 - e^{-2kx},$$

and then

$$\cos \Pi(x) = \frac{1 - e^{-2kx}}{1 + e^{-2kx}} = \frac{e^{kx} - e^{-kx}}{e^{kx} + e^{-kx}} = \frac{\sinh kx}{\cosh kx},$$

and we may conclude that

$$\cos \Pi(x) = \tanh kx. \tag{12.22}$$

We use this last equation to evaluate $\sin \Pi(x)$. From

$$\cos^2 \Pi(x) + \sin^2 \Pi(x) = 1$$

we may substitute and use Equation 12.21 to find

$$\sin^2 \Pi(x) = 1 - \cos^2 \Pi(x) = 1 - \tanh^2 kx = \operatorname{sech}^2 kx.$$

Therefore,

$$\sin \Pi(x) = \operatorname{sech} kx. \tag{12.23}$$

Pi in Space

Beginning in 1818 there was a large effort to map the entire Kingdom of Hanover. The interest in geodesy at the time was partly a theoretical one to use the methods of calculus to determine the shape of the globe and partly a practical one to create accurate maps to the benefit of commerce and the military. It was Gauss who was asked to lead the initial effort. He had previously made some geodetic studies after his return to Brunswick, so he was interested when he was offered the leadership position. Over a period of three summers, Gauss spent most of his time traveling. He led a small team of military men who took the measurements with a device called a heliotrope, which Gauss had invented (Bühler 1981, 95–98).

The work allowed Gauss an opportunity to conduct an experiment that would hopefully determine whether it was Euclidean or non-Euclidean geometry that was true. He decided that a check of the angle sum of a large triangle would provide a perfect test, so Gauss chose a triangle near Göttingen. Its vertices were mountain tops that had been surveyed previously, and its sides were approximately 40 miles long. Measured in degrees, the calculations yielded an angle sum for the triangle that was less than 2 seconds short of two right angles. This was well within the margin of error of the instruments that Gauss used, so he concluded that as far as measurement was concerned, it was safe to assume that the geometry near earth was Euclidean. However, since the defect did not calculate to zero, there was still some question (Burton 1985, 552).

Although not influenced directly by Gauss, Lobachevski shared his belief that empirical methods should determine the truth of the Parallel Postulate. Like Gauss, he rejected the prevalent Kantian philosophy of his day and believed that whether Euclid's geometry was true or not must be determined by experiment. Our senses must be allowed to inform our beliefs regarding mathematics. Our intuition and any notions that we may have pre-wired in our brains must not be trusted. To find his evidence, Lobachevski decided that he would need to examine large triangles. He did not use mountains; instead, Lobachevski found his large triangles in space (Rosenfeld 1988, 208).

Take the vertices of the triangle to be the points A, B, and C in Figure 12.13. Locate them at the centers of the earth, the sun, and a distant star, respectively. Assume that $\overline{AB}$ is perpendicular to $\overline{BC}$. With this setup Lobachevski used the

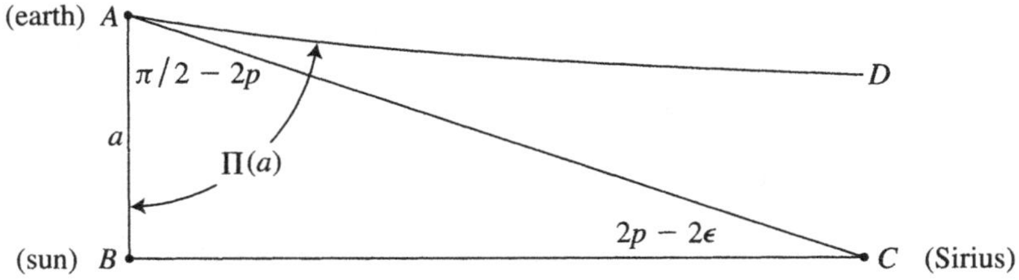

Figure 12.13 Lobachevski testing the geometry of space.

parallax angle, which is $\sphericalangle BCA$ in Figure 12.13, to measure how close to being Euclidean is space. Lobachevski chose Sirius for his calculations. Let $a = AB$,

$$\text{m}\sphericalangle BAC = \frac{\pi}{2} - 2p, \text{ and } \text{m}\sphericalangle BCA = 2p - 2\epsilon,$$

where $\epsilon \geqslant 0$ to allow for a possible positive defect in the triangle. Choose D so that $\sphericalangle BAD$ is the angle of parallelism. We conclude that

$$\Pi(a) > \frac{\pi}{2} - 2p,$$

and then

$$e^{-ka} = \tan\frac{1}{2}\Pi(a) > \tan\left(\frac{\pi}{4} - p\right) = \frac{1 - \tan p}{1 + \tan p}$$

for some constant k. Therefore, we conclude that

$$e^{ka} < \frac{1 + \tan p}{1 - \tan p}.$$

We know from calculus that if $|x| < 1$, then

$$\ln\frac{1+x}{1-x} = 2\left(x + \frac{x^3}{3} + \frac{x^5}{5} + \cdots + \frac{x^{2n+1}}{2n+1} + \cdots\right).$$

Since $0 < p < \pi/4$, we have $\tan p < 1$, and hence

$$ka < \ln\frac{1 + \tan p}{1 - \tan p} = 2\left(\tan p + \frac{\tan^3 p}{3} + \frac{\tan^5 p}{5} + \cdots + \frac{\tan^{2n+1} p}{2n+1} + \cdots\right).$$

Moreover,

$$2(\tan p + \tan^3 p + \tan^5 p + \cdots + \tan^{2n+1} p + \cdots) = \frac{2\tan p}{1 - \tan^2 p} = \tan 2p,$$

and we finally have that

$$ka < \tan 2p.$$

Lobachevski took the parallax of Sirius to be 1.24 arcseconds, but we know that it is closer to .375 arcseconds. Using the value of 93 million miles for the radius of earth's orbit, a little calculation yields

$$k < 1.95 \times 10^{-14}.$$

Therefore, Lobachevski drew the same conclusion as Gauss. Since $k = 0$ means that $\Pi(x) = \pi/2$ for all x, the very small k tells us that if the universe is not Euclidean, it is nearly so (Bonola 1912, 94–96).

The Imaginary

Lambert believed that his geometry would hold true on a sphere of imaginary radius (page 360). He did not explain how to visualize this, but if we take Lobachevski's equations and replace radius r with radius ir, we obtain familiar formulas from spherical geometry. To see how this works, we must make two notes. First, since the length of an arc of a circle with real radius r and central angle θ is $r\theta$,

> the length of an arc of a great circle on a sphere of imaginary radius is $ri\theta$.

Second, we need to extend the domain of Π to include any complex number. We already know that $\Pi(x) = 2\tan^{-1} e^{-kx}$. Without going into the details, to calculate $\Pi(a+bi)$, use the power series expansion for $2\tan^{-1} e^{-kx}$ and then substitute $a+bi$ in for x.

We now apply trigonometric functions to $\Pi(xi)$. We prove the first formula and leave the other two to Exercise 9. Using the power series for hyperbolic cosine, we have

$$\begin{aligned}\cosh xi &= 1 + \frac{(xi)^2}{2!} + \frac{(xi)^4}{4!} + \frac{(xi)^6}{6!} + \cdots \\ &= 1 - \frac{x^2}{2!} + \frac{x^4}{4!} - \frac{x^6}{6!} + \cdots \\ &= \cos x.\end{aligned}$$

Therefore, by Equation 12.23,

$$\sin \Pi(xi) = \sec kx.$$

The other two substitutions are

$$\begin{aligned}\cos \Pi(xi) &= i \tan kx, \\ \tan \Pi(xi) &= -i \csc kx.\end{aligned}$$

Lobachevski used these equations to translate his trigonometric equations into those of spherical trigonometry. He showed (*Theory of Parallels* 37) that

$$\sin A \tan \Pi(a) = \sin B \tan \Pi(b), \tag{12.24}$$

$$\cos A \cos \Pi(b) \cos \Pi(c) + \frac{\sin \Pi(b) \sin \Pi(c)}{\sin \Pi(a)} = 1, \tag{12.25}$$

$$\cot A \sin C \sin \Pi(b) + \cos C = \frac{\cos \Pi(b)}{\cos \Pi(a)}, \tag{12.26}$$

$$\cos A + \cos B \cos C = \frac{\sin B \sin C}{\sin \Pi(a)}. \tag{12.27}$$

Then on a sphere of imaginary radius the second equation becomes

$$\cos A \cos \Pi(bi) \cos \Pi(ci) + \frac{\sin \Pi(bi) \sin \Pi(ci)}{\sin \Pi(ai)} = 1.$$

Using the substitutions above we obtain

$$\cos A(i \tan kb)(i \tan kc) + \frac{\sec kb \sec kc}{\sec ka} = 1,$$

which gives

$$-\cos A \cdot \frac{\sin kb \sin kc}{\cos kb \cos kc} + \frac{\cos ka}{\cos kb \cos kc} = 1.$$

Multiplying through by $\cos kb \cos kc$ yields

$$-\cos A \sin kb \sin kc + \cos ka = \cos kb \cos kc.$$

Thus,

$$\cos ka = \cos kb \cos kc + \sin kb \sin kc \cos A,$$

and this is Theorem 12.3.5. It is left to Exercise 10 to show that similar work with Lobachevski's trigonometric formulas gives

$$\sin A \sin kb = \sin B \sin ka$$

and

$$\cos A = \cos ka \sin B \sin C - \cos B \cos C.$$

These equations are the spherical trigonometry identities found in Theorem 12.3.4 and Exercise 12 of Section 12.3. This is what it means for this new geometry to be found on the surface of a sphere of imaginary radius.

Exercises

1. Confirm the following using the equation for $\Pi(x)$ given by Theorem 12.5.2.
   (a) Equation 12.1
   (b) Π is a strictly decreasing continuous function.

2. Prove Equation 12.19 for the cases $c = x$ and $c < x$.

3. Use Lobachevski's extension of Π in Equation 12.20 to show that for all real numbers x, $\tan \frac{1}{2}\Pi(x) = e^{-kx}$.

4. Draw the following, paying careful attention to all measurements.
   (a) The graph of Π when $k = .1, k = .5, k = 2, k = 10$.
   (b) A $60 - 90$ omega triangle ($k = 1$).
   (c) $\triangle ABC$ where $m\sphericalangle C = 90°$, $m\sphericalangle B = 60°$, and $AC = 1$ ($k = 1$).

5. The area of $\triangle ABC$ is $k^2(\pi - A - B - C)$. Use this to explain why the maximum area of a triangle πk^2?

6. Find the area of $\triangle ABC$ where $m\sphericalangle C = 90°$, $m\sphericalangle B = 60°$, and $AC = 1$ $(k = 1)$.

7. Give a formula for the area of a quadrilateral $ABCD$ and find its maximum area.

8. Use the value of 93 million miles for the radius of earth's orbit about the sun.
   (a) Confirm that if the parallax of Sirius is .375 arcseconds, $k < 1.95 \times 10^{-14}$.
   (b) Find the maximum value of k using Lobachevski's value of 1.24 arcseconds for the parallax of Sirius.
   (c) Calculate $\Pi(1)$ using both of these k values.
   (d) Find the distance to Sirius assuming Euclidean geometry and assuming that $k = 1.95 \times 10^{-14}$. Is the difference significant?

9. Confirm that the following hold.
   (a) $\cos \Pi(xi) = i \tan kx$.
   (b) $\tan \Pi(xi) = i \csc kx$.

10. Show that for triangles on a sphere of imaginary radius,
   (a) $\sin A \sin kb = \sin B \sin ka$.
   (b) $\cos A = \cos ka \sin B \sin C - \cos B \cos C$.

11. The circumference of a circle with radius r is

$$C = \frac{2\pi}{k} \sinh kr,$$

and its area is

$$A = \frac{4\pi}{k^2} \sinh^2 2kr.$$

Confirm that these equations approach the Euclidean formulas as $k \to 0$.

12. For very small values of x,

$$\cos \Pi(x) \approx kx, \sin \Pi(x) \approx 1 - \frac{k^2x^2}{2}, \text{ and } \cot \Pi(x) \approx kx.$$

   (a) Confirm that this is the case.
   (b) Use these approximations to show that Lobachevski's trigonometric equations (12.24–12.27) are equivalent to

$$b \sin A = a \sin B,$$

$$a^2 = b^2 + c^2 - 2bc \cos A,$$

$$a \sin(A + C) = b \sin A,$$

$$\cos A + \cos(B + C) = 0.$$

   (c) Use these results to show that the angle sum of very small triangles is approximately equal to two right angles, showing that in the geometry of Lobachevski and Bolyai, very small regions of the plane are essentially Euclidean.

PART V

NEW DIRECTIONS

CHAPTER 13

BERNHARD RIEMANN

Riemann was born in Breselenz in Hanover on September 17, 1826. His full name was Georg Friedrich Bernhard Riemann, and he was the son of a Lutheran pastor. Following his father's wishes, he moved to Göttingen in 1846 to study theology and philosophy. After attending various mathematics courses, he realized that he had an aptitude for the subject, so after receiving permission from his father, Riemann switched fields. Unfortunately, he found himself unhappy at the university. Despite Gauss holding a chair in mathematics, Riemann was disappointed in the level of academic rigor of the courses. Because of this he moved to Berlin and studied mathematics with such notables as Dirichlet, Eisenstein, and Jacobi (Burton 1985, 563; Boyer and Merzbach 1991, 545).

In 1848, revolution embroiled Berlin, and Riemann found himself in the middle of something that was quite foreign to his nature. Riemann was a frail, unassuming person, but evidently motivated by a sense of duty to his country, he joined a group of students who were loyal to the crown. His activities for the group included a two-day stint in which he stood guard at the royal palace. This proved to be too much excitement for Riemann. He decided that it was time to leave, and in 1849 he returned to Göttingen to finish his doctoral degree. With Gauss as his dissertation advisor, he worked on a topic that was new and significant. It involved surfaces over the complex plane that we now call **Riemann surfaces**. The dissertation also included equations that were discovered independently by Augustin-Louis Cauchy. They are called the **Cauchy-Riemann equations**, and the proposition states that if f is a complex differentiable function of the form

$$f(x, y) = u(x, y) + iv(x, y),$$

then

$$\frac{\partial u}{\partial x} = \frac{\partial v}{\partial y} \text{ and } \frac{\partial v}{\partial x} = -\frac{\partial u}{\partial y}.$$

Gauss recognized in his student exceptional ability and was pleased with his results (Burton 1985, 563–564; Boyer and Merzbach 1991, 545).

Over the next few years Riemann continued his research. Besides his mathematical pursuits, he studied physics as an assistant to Wilhelm Weber. As a result of his work, Riemann published numerous papers. The topics included heat, light, gases, magnetism, gravitation, fluid dynamics, and acoustics. Riemann even conjectured that space was curved and that this was the reason that gravity works on objects at a distance (Kline 1972, 655–656; Gray 2007, 189).

Then in 1854 Riemann was ready to become a *privatdozent* at Göttingen. For his *Habilitationsschrift*, Riemann gave three topics to Gauss. The first involved using a trigonometric series to represent a function. The second dealt with quadratic equations. The last was titled *On the Hypotheses Which Lie at the Bases of Geometry*. Riemann was quoted as being surprised at Gauss going against the "usual convention" by picking his last suggestion, but Riemann should not have been surprised at the choice. The first two were too elementary for the presentation. Moreover, there is evidence that Riemann was both willing and ready for the topic (Laugwitz 1999, 219–220; Gray 2007, 189).

On June 10, 1854 Riemann gave his talk. Since mathematics was housed in the school of philosophy at Göttingen, the audience included both mathematicians and philosophers. For this reason his essay (Riemann 1854) and the accompanying presentation focused on big-picture issues and provided little detail. Despite the subject being geometry, Riemann included no diagrams and only two equations. Nonetheless, in this one presentation, Riemann gave a completely new definition of geometry. From the time of Euclid geometry had been the study of objects in space, and it is the choice of postulates that determines the behavior of these objects. Up until the time of Gauss, these postulates were viewed as self-evident truths. Gauss's work, along with that of Lobachevski and Bolyai, showed otherwise by proving that the Parallel Postulate may or may not describe space accurately. Its truth is now an empirical question, one Riemann, as Lobachevski before him (page 436), believed would be answered by astronomy (Burton 1985, 565).

The goal of Riemann's redefinition was to generalize the system of Euclid. He did not simply change the postulate set as did Lobachevski and Bolyai. In fact, it is doubtful that he knew their work. Instead, Riemann set his focus on the two different types of problems of the geometer. The first seeks to determine the intrinsic properties of a surface. An **intrinsic** property is one that requires no information about the encompassing space. Imagine little creatures living on a surface. They would not need to leave the surface to determine an intrinsic property. An example is the measure of angles. One of the creatures can easily take a little protractor and find the measure of any given angle without ever leaving home. The second type of geometric problem involves the examination of the various methods by which one surface can be mapped onto another. Sometimes these functions distort the surface by bending or warping it, but sometimes the maps simply superimpose one surface on top of the other. We do this, for example, when we study congruent figures. Congruence means that the figures look identical in size and shape so that one can be lined up on top of the other (Gray 2004, 85; Laugwitz 1999, 237).

Since geometry assumes that surfaces exist in three dimensions, Riemann needed a new geometric notion that could generalize the Euclidean concept of space. He accomplished this with what he called a **manifoldness** (Riemann 1854, I.1, 57). A manifoldness was a multidimensional space consisting of points. These points were identified by the use of coordinates, and the number of coordinates needed gave the number of dimensions of the space. Today we use the term *manifold* instead of *manifoldness*, and we define a **manifold** to be a topological space that has the property that every point is contained within a neighborhood that is essentially Euclidean. The primary example of a manifold is $\mathbb{R}^n$. Other examples include the surfaces of spheres, cylinders, and tori.

Riemann's manifoldness had the additional property that the length of any curve in the space was **rectifiable**; that is, its length could be measured. This was done using infinitesimally small rulers. Hence, instead of geometric postulates Riemann used the methods of calculus to study his manifolds. This field of mathematics that studies curves and surfaces using calculus is called **differential geometry** (Gray 2004, 85).

13.1 METRIC SPACES

Riemann taught that all that is needed for a geometry are its points and a function to compute distances. Since there are infinitely many such functions, there are infinitely many geometries. This implies the concept of a metric space (O'Leary 2002, 325–348; Marsden and Hoffman 1993; Simmons 1963). Although it came after Riemann's work, a metric space shares two main features with his manifoldness, namely, a set of points and a distance function.

■ DEFINITION 13.1.1

Let M be a nonempty set and $d : M \times M \to \mathbb{R}$ a function such that for all $x, y, z \in M$:

- $d[x, y] \geqslant 0$.
- $d[x, y] = 0$ if and only if $x = y$.
- $d[x, y] = d[y, x]$.
- $d[x, y] \leqslant d[x, z] + d[z, y]$.

The pair (M, d) is called a **metric space**. The function d is the distance function, and it is called a **metric**. The last condition is called the **Triangle Inequality**.

There are many examples of metric spaces. For example, $(\mathbb{R}, d)$ is a metric space where

$$d[x, y] = |x - y|.$$

The check of the first three conditions is left to Exercise 1. To see that the Triangle Inequality holds, notice that for all $a, b \in \mathbb{R}$,

$$-|a| \leqslant a \leqslant |a| \text{ and } -|b| \leqslant b \leqslant |b|.$$

Adding these inequalities yields

$$-(|a| + |b|) \leqslant a + b \leqslant |a| + |b|.$$

Therefore,

$$|a + b| \leqslant |a| + |b|. \tag{13.1}$$

Similarly, if we define the distance function d as

$$d[X, Y] = \sqrt{(x_2 - x_1)^2 + (y_2 - y_1)^2}$$

for all $X = (x_1, y_1)$ and $Y = (x_2, y_2)$, we see that $(\mathbb{R}^2, d)$ is a metric space (Exercise 1). This metric can be generalized to any dimension. For all

$$X = (x_1, x_2, \ldots, x_n),$$
$$Y = (y_1, y_2, \ldots, y_n),$$

define

$$d[X, Y] = \sqrt{(x_1 - y_1)^2 + (x_2 - y_2)^2 + \cdots + (x_n - y_n)^2}.$$

Then $(\mathbb{R}^n, d)$ is a metric space. These distance functions are known as the **standard metrics** on $\mathbb{R}^n$, and unless otherwise noted, they will be the assumed metrics for these spaces.

For another example of a metric space, let $C[a, b]$ be the set of all functions continuous on $[a, b]$ and define for all $f, g \in C[0, 1]$,

$$d[f, g] = \int_0^1 |f(x) - g(x)|\, dx.$$

This is a natural definition for the distance between two functions because the closer two functions are to each other, the smaller the area between them will be (Figure 13.1). For instance, to measure the distance between $f(x) = x^2$ and $g(x) = x$ over the interval $[0, 1]$, we compute

$$d[f, g] = \int_0^1 |x^2 - x|\, dx = 1/6.$$

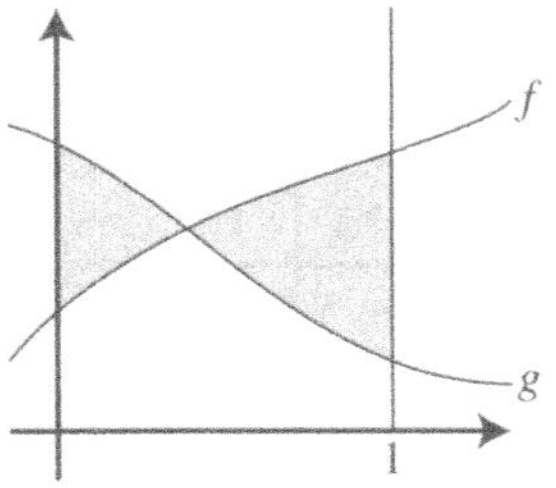

Figure 13.1 Measuring the distance between two functions using area.

Now check that $(C[0, 1], d)$ is a metric space. Since it is the most involved, we will check first that the Triangle Inequality works. Take $f, g, h \in C[0, 1]$ and calculate:

$$\begin{aligned} d[f, g] &= \int_0^1 |f(x) - g(x)|\, dx \\ &= \int_0^1 |f(x) - h(x) + h(x) - g(x)|\, dx \\ &\leqslant \int_0^1 (|f(x) - h(x)| + |h(x) - g(x)|)\, dx \\ &= \int_0^1 |f(x) - h(x)|\, dx + \int_0^1 |h(x) - g(x)|\, dx \\ &= d[f, h] + d[h, g]. \end{aligned}$$

The inequality holds due to Equation13.1. For the other three:

- Since $|f(x) - g(x)| \geqslant 0$ for all x,
$$d[f, g] = \int_0^1 |f(x) - g(x)|\, dx \geqslant 0.$$
- If $d[f, g] = 0$, this means that the area under the curve $y = |f(x) - g(x)|$ on the interval $[0, 1]$ equals 0. Since $|f(x) - g(x)|$ is never negative, the only way that this can happen is when $f = g$. Conversely, if $f = g$, then $|f(x) - g(x)| = 0$ for all x. Hence, $d[f, g] = 0$.
- Because $|f(x) - g(x)| = |g(x) - f(x)|$, we have that $d[f, g] = d[g, f]$.

Therefore, $(C[0, 1], d)$ is a metric space.

Open Sets

Manifolds have the property that the nearby region surrounding any point of the space resembles $\mathbb{R}^n$. In the one-dimensional case, this means that the region looks like an open interval. To generalize this to higher dimensions, we need our first definition.

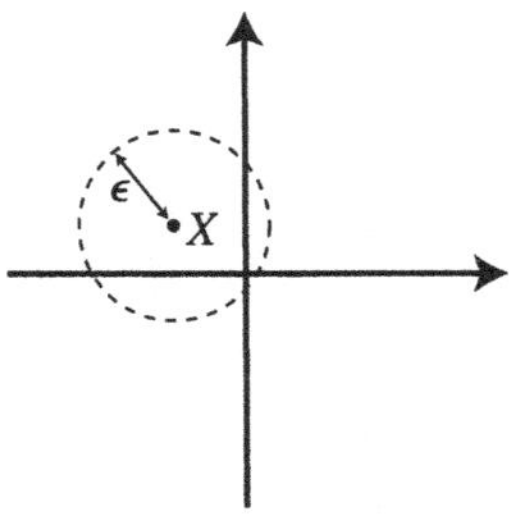

Figure 13.2 The open disk $D(x, \epsilon)$ in $\mathbb{R}^2$.

■ **DEFINITION 13.1.2**

Let (M, d) be a metric space and take $x \in M$. For all $\epsilon > 0$, define

$$D_M(x, \epsilon) = \{y \in M : d[x, y] < \epsilon\}.$$

This set is called an **open disk** about x in M.

If the metric space M is clear from context, we can write $D(x, \epsilon)$ for $D_M(x, \epsilon)$. For example, if $M = \mathbb{R}$ with the standard metric,

$$D(x, \epsilon) = (x - \epsilon, x + \epsilon),$$

and if $M = \mathbb{R}^2$ with the standard metric, $D(X, \epsilon)$ is the interior of a circle excluding its circumference. This is illustrated in Figure 13.2, where the dashed lines represent the exclusion of the circumference of the circle.

The open disk will allow us to make our main definition.

■ **DEFINITION 13.1.3**

Let M be a metric space. A subset U of M is called **open** in M if for all $x \in U$, there exists $\epsilon > 0$ such that $D_M(x, \epsilon) \subseteq U$.

We should note that for any metric space M, both M and $\varnothing$ are open. The is because $D_M(x, \epsilon)$ is a subset of M for any $x \in M$ and $\epsilon > 0$. As for the empty set, the proposition

$$\text{if } x \in \varnothing, \text{ then there exists } \epsilon > 0 \text{ such that } D_M(x, \epsilon) \subseteq \varnothing$$

is true for all x because $x \in \varnothing$ is always false (see page 38).

Here are some other examples of open sets.

- Let $a, b \in \mathbb{R}$. To prove that (a, b) is open, let $a < x < b$. We must isolate x inside an open interval within (a, b). To do this, find the distance from x to each endpoint and let ϵ be the minimum of the two distances. The notation that we will use for this is:

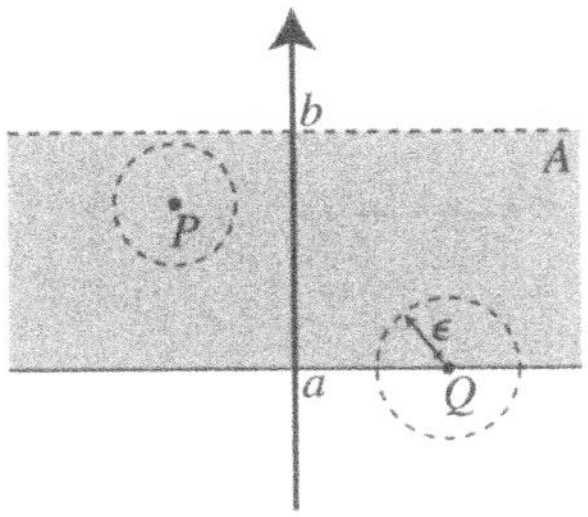

Figure 13.3 The set A is not open.

$$\epsilon = \min\{x - a,\ b - x\}.$$

Now let $y \in D(x, \epsilon)$. This means that $x - \epsilon < y < x + \epsilon$, and we may conclude that

$$a = x - (x - a) \leqslant x - \epsilon < y < x + \epsilon \leqslant x + (b - x) = b.$$

Therefore, $D(x, \epsilon) = (x - \epsilon, x + \epsilon) \subseteq (a, b)$.

- We can use the strategy of the previous example on other sets. Let A equal the set $(1, 2) \cup (4, 7)$. To show that A is open, pick a real number x such that $1 < x < 2$ or $4 < x < 7$. Define

$$\epsilon = \min\{x - 1,\ 2 - x,\ x - 4,\ 7 - x\}.$$

With this definition, $(x - \epsilon, x + \epsilon) \subseteq A$.

To show that a set A is not open in a metric space M, we must prove that there exists $x \in A$ such that $D(x, \epsilon)$ is not a subset of A for all $\epsilon > 0$. For this example, let $a, b \in \mathbb{R}$ so that $a < b$. The set

$$A = \{(x, y) \in \mathbb{R}^2 : a \leqslant y < b\}$$

is not open in $\mathbb{R}^2$ (Figure 13.3). Any point P that is on the interior of A can be surrounded by a disk that is a subset of A. However, this is not true for a point on the bottom border, so take Q to be $(3, a)$. Then for all $\epsilon > 0$,

$$D((3, a), \epsilon) \setminus A \neq \varnothing$$

because $(3, a - \epsilon/2)$ is a member of $D((a, 3), \epsilon)$ but not of A.

Now that we have seen examples of open and nonopen sets, we should prove that the disk is actually open and justify its name. The proof is a generalization of the fact that (a, b) is open in $\mathbb{R}$.

■ THEOREM 13.1.4

$D(x, \epsilon)$ is open for any x in a metric space M.

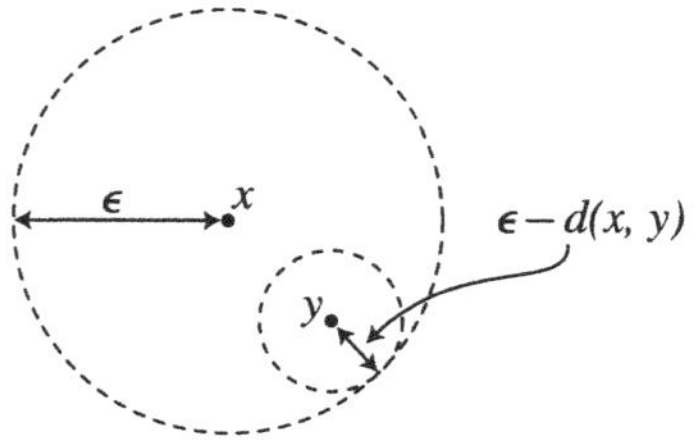

PROOF

We must show that we can isolate any element of $D(x, \epsilon)$ inside an open disk. So let $y \in D(x, \epsilon)$. The open disk about y that we want is

$$D(y, \epsilon - d[x, y]).$$

To show that this disk is included in $D(x, \epsilon)$, let $z \in D(y, \epsilon - d[x, y])$. This means that

$$d[y, z] < \epsilon - d[x, y],$$

and hence

$$d[x, z] \leqslant d[x, y] + d[y, z] < \epsilon$$

by the Triangle Inequality. We conclude that $z \in D(x, \epsilon)$. ■

An application of the definition is one way to prove that a set is open. Other ways include the next two theorems of the section.

■ **THEOREM 13.1.5**

Let M be a metric space.

- The union of a collection of open sets is open.
- The intersection of a finite collection of open sets is open.

PROOF

To prove the first part, let $\mathcal{F} = \{U_i : i \in I\}$ be a family of open sets. Take $x \in \bigcup \mathcal{F}$. This means that there exists $i \in I$ such that $x \in U_i$. Since U_i is open, we have $\epsilon > 0$, so that $D(x, \epsilon) \subseteq U_i$, but this implies that $D(x, \epsilon) \subseteq \bigcup \mathcal{F}$.

For the second part let $x \in U_1 \cap U_2 \cap \cdots \cap U_k$, where k is a positive integer ($k \in \mathbb{Z}^+$). By definition, x is an element of each U_i. Since each of these is open, there exists $\epsilon_i > 0$ so that $D(x, \epsilon_i)$ is a subset of U_i for $i = 1, 2, \ldots, k$. Define

$$\epsilon = \min\{\epsilon_1, \epsilon_2, \ldots, \epsilon_k\}.$$

Then $\epsilon \leqslant \epsilon_i$ for all i, and this means that $D(x, \epsilon) \subseteq D(x, \epsilon_i)$. Hence, $D(x, \epsilon)$ is a subset of the intersection. ■

To see that the finiteness condition is necessary for the intersection of open sets to be open, consider $\{(-1/n, 1/n) : n \in \mathbb{Z}^+\}$. This is a family of open sets in $\mathbb{R}$.

However,

$$\bigcap_{n=1}^{\infty}(-1/n,\ 1/n) = \{0\},$$

which is not open in $\mathbb{R}$.

The next definition relies on Theorem 13.1.5.

DEFINITION 13.1.6

If M is a metric space and A a subset of M, define

$$\text{int}(A) = \bigcup\{U : U \text{ is open in } M \text{ and } U \subseteq A\}.$$

This set is called the **interior** of A.

Since it is the union of open sets, the interior of a set A is open. Moreover, it must be the largest open subset of A (Exercise 14). As examples, $\text{int}([0,\ 1]) = (0,\ 1)$ and $\text{int}(\mathbb{Z}) = \varnothing$ in $\mathbb{R}$, because the empty set is the only open subset of $\mathbb{Z}$. In $\mathbb{R}^2$,

$$\text{int}([0,\ 1] \times [2,\ 4]) = (0,\ 1) \times (2,\ 4)$$

and

$$\text{int}(\{(x,\ y) \in \mathbb{R}^2 : xy \leqslant 1\}) = \{(x,\ y) \in \mathbb{R}^2 : xy < 1\}.$$

One way to identify the points in a set's interior is to use the next theorem.

THEOREM 13.1.7

If A is a subset of a metric space, then $x \in \text{int}(A)$ if and only if there exists $\epsilon > 0$ so that $D(x,\ \epsilon) \subseteq A$.

PROOF

First let $x \in \text{int}(A)$. This means that there exists an open subset U of A that contains x. So there exists $\epsilon > 0$ such that $D(x,\ \epsilon) \subseteq U \subseteq A$.

Conversely, if $D(x,\ \epsilon) \subseteq A$ for some $\epsilon > 0$, then x must be an element of the interior because $D(x,\ \epsilon)$ is open. ■

For instance, define $\mathbb{Z} = \{\ldots,\ -2,\ -1,\ 0,\ 1,\ 2,\ \ldots\}$ and view it as a metric space with the standard metric from $\mathbb{R}$. Let $A = \{2k : k \in \mathbb{Z}\}$, the set of even integers. The interior of A is A. To see this, let $n \in \mathbb{Z}$. Then

$$D_{\mathbb{Z}}(2n,\ 1) = \{m \in \mathbb{Z} : |2n - m| < 1\} = \{2n\}.$$

Therefore, $2n \in \text{int}(A)$, which implies that $A \subseteq \text{int}(A)$. Since we already know that $\text{int}(A) \subseteq A$, we conclude that $\text{int}(A) = A$. This means that A is open in $\mathbb{Z}$.

Closed Sets

The difference between an open interval and a closed one lies with the endpoints. Open intervals do not contain the endpoints. Instead, the endpoints belong to the complement of the open interval. A closed interval will contain its endpoints. We will take this idea and define what it means for a set to be closed in an arbitrary metric space and then prove some results.

■ DEFINITION 13.1.8

A subset A of a metric space M is **closed** in M if $M\backslash A$ is open in M.

Both M and $\varnothing$ are closed because $M\backslash M = \varnothing$ and $M\backslash\varnothing = M$ and both of these are open. Letting $A^{\mathrm{C}} = M\backslash A$, here are some examples of closed sets.

- $[a, b]$ is closed in $\mathbb{R}$ for all real numbers $a < b$.
- Since $D(x, \epsilon)$ is open, $D(x, \epsilon)^{\mathrm{C}}$ is closed.
- In $\mathbb{R}^2$ the set $\{(x, y) \in \mathbb{R}^2 : xy \geqslant 1\}$ is closed.

There are sets that are neither open nor closed. In $\mathbb{R}$, $(0, 1]$ provides a simple example. The set $A = \{(x, y) \in \mathbb{R}^2 : a \leqslant x < b\}$ is such an example in $\mathbb{R}^2$. Here is a more interesting one. Let $A = \{1/k : k \in \mathbb{Z}^+\}$.

- To see that A is not open, let n be a positive integer and consider
$$D(1/n, \epsilon) = (1/n - \epsilon, 1/n + \epsilon)$$
for any $\epsilon > 0$. Since there exists $m \in \mathbb{R}$ such that
$$\max\left\{\frac{1}{n+1}, \frac{1}{n} - \epsilon\right\} < m < \frac{1}{n},$$
$m \in D(1/n, \epsilon)$ but $m \notin A$. Hence, A is not open.
- To check that A is not closed, we show that A^{C} is not open. Suppose that $\epsilon > 0$. Then there exists a positive integer n such that $1/n < \epsilon$. Thus, $D(0, \epsilon) \cap A \neq \varnothing$, which means that $D(0, \epsilon)$ is not a subset of A^{C} even though $0 \in A^{\mathrm{C}}$.

We shall see shortly that there is another way to show that $\{1/n : n \in \mathbb{Z}^+\}$ is not closed.

To emphasize the point of the examples, remember that although the complement of an open set is closed and the complement of a closed set is open, the terms are not negations of each other. There are sets that are both open and closed, and there are sets that are neither. Hence, we cannot show that a set is closed by proving that it is not open. Similarly, we cannot show that a set is open by showing that it is not closed. However, for the closed case the next theorems can help.

■ THEOREM 13.1.9

In any metric space:

- The intersection of a collection of closed sets is closed.
- The union of a finite collection of closed sets is closed.

PROOF

Let $\{C_i : i \in I\}$ be a family of closed sets. To prove that $\bigcap_{i\in I} C_i$ is closed, we must show that its complement is open. Since each C_i is closed, C_i^c is open. Therefore,

$$\left(\bigcap_{i\in I} C_i\right)^c = \bigcup_{i\in I} C_i^c$$

is open by Theorem 13.1.5.

Similarly, if I is a finite index set equal to $\{1, 2, \ldots, k\}$, then

$$C_1 \cup C_2 \cup \cdots \cup C_k$$

is closed because its complement

$$C_1^c \cap C_2^c \cap \cdots \cap C_k^c$$

is open. ■

The largest open set within a set is its interior. Corresponding to this is the smallest closed set that contains a given set. This is our next definition.

■ DEFINITION 13.1.10

Let A be a subset of a metric space M. Define

$$\mathrm{cl}(A) = \bigcap\{F : F \text{ is closed in } M \text{ and } A \subseteq F\}$$

and call this set the **closure** of A.

In $\mathbb{R}$, $\mathrm{cl}((0, 1)) = [0, 1]$ and $\mathrm{cl}(\mathbb{Z}) = \mathbb{Z}$. In $\mathbb{R}^2$,

$$\mathrm{cl}((0, 1) \times (2, 4)) = [0, 1] \times [2, 4]$$

and

$$\mathrm{cl}(\{(x, y) \in \mathbb{R}^2 : xy < 1\}) = \{(x, y) \in \mathbb{R}^2 : xy \leqslant 1\}.$$

We note that the closure of a set is closed. To see this, examine the complement of $\mathrm{cl}(A)$:

$$\mathrm{cl}(A)^c = \left(\bigcap\{F : F \text{ is closed and } A \subseteq F\}\right)^c = \bigcup\{F^c : F \text{ is closed and } A \subseteq F\}.$$

Since each F is closed, each F^c is open. Therefore, the union is open (Theorem 13.1.5). We should also note that a set A is closed in a metric space M if and only if $\mathrm{cl}(A) = A$. The proof of this is left as Exercise 15.

The way to picture a closed set is to think of it like a closed interval. A closed set contains all of the points of its interior plus all of the points that form a boundary around its interior. We therefore need a definition that will name all of these points.

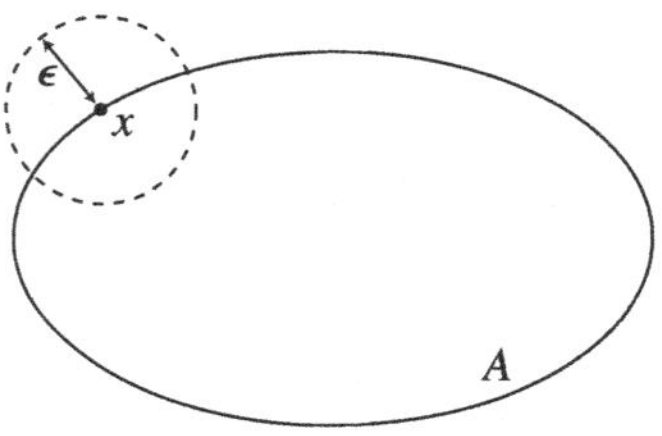

Figure 13.4 x is an accumulation point of A.

■ **DEFINITION 13.1.11**

Let A be a subset of a metric space M. An element $x \in M$ is an **accumulation point** of A if for all $\epsilon > 0$,

$$(D(x, \epsilon)\backslash\{x\}) \cap A \neq \varnothing.$$

Basically, an accumulation point of a set A has the property that there are points in A that are arbitrarily close to it (see Figure 13.4). We exclude the accumulation point from the criterion since it is always very close to itself!

Here is an example. Define $A = \{(x, y) \in \mathbb{R}^2 : 0 < x < 2\}$. The interior point $(1, 0)$ is an accumulation point. To prove that $(0, 0)$ is also an accumulation point, let $\epsilon > 0$ and take $a \in \mathbb{R}$ so that $0 < a < \min\{1, \epsilon\}$. Then,

$$(a, 0) \in [D((0, 0), \epsilon)\backslash\{(0, 0)\}] \cap A,$$

which means that the desired intersection is nonempty.

The last example shows that an accumulation point need not be an element of the set. In the metric space $\mathbb{R}$, we see that 0 is an accumulation point for both $(0, 1)$ and $[0, 1]$. For $\mathbb{R}^2$, $(1, 0)$ is an accumulation point for the open disk

$$\{(x, y) \in \mathbb{R}^2 : x^2 + y^2 < 1\}$$

and the line

$$\{(x, x - 1) : x \in \mathbb{R}\}.$$

On the other hand, a set need not have any accumulation points. Take the metric space $\mathbb{Z} \times \mathbb{Z}$ with the standard metric from $\mathbb{R}^2$. Then for all $a, b \in \mathbb{Z}$, (a, b) cannot be an accumulation point because

$$D_{\mathbb{Z}\times\mathbb{Z}}((a, b), 1/2)\backslash\{(a, b)\} = \varnothing.$$

We are now ready for our last test for closure.

■ **THEOREM 13.1.12**

A set is closed if and only if it contains its accumulation points.

PROOF

Let A be a subset of a metric space M. Assume that A is closed. Let x be an accumulation point of A. To obtain a contradiction, assume that $x \notin A$. This means that $x \in A^c$, but this set is open. Hence, there is an $\epsilon > 0$ with the property that $D(x, \epsilon) \subseteq A^c$. Therefore,

$$D(x, \epsilon) \cap A = \varnothing,$$

contradicting the fact that x is an accumulation point.

Now suppose that A contains all its accumulation points. We must show that A^c is open. Take $x \in A^c$. By hypothesis, x cannot be an accumulation point of A. This means that there exists $\epsilon > 0$ such that $D(x, \epsilon) \cap A = \varnothing$. Therefore, $D(x, \epsilon)$ is a subset of A^c. ■

Exercises

Let (M, d) be a metric space.

1. Prove that $\mathbb{R}$, $\mathbb{R}^2$, and $\mathbb{R}^n$ when paired with the standard metrics are metric spaces.

2. Prove that for all $x, y \in M$ such that $x \neq y$, there exists $\epsilon > 0$ such that $D(x, \epsilon) \cap D(y, \epsilon) = \varnothing$.

3. Suppose that $\epsilon_1 \leqslant \epsilon_2$. Prove for any $x \in M$, $D(x, \epsilon_1) \subseteq D(x, \epsilon_2)$.

4. Define

$$d'(x, y) = \frac{d[x, y]}{1 + d[x, y]}.$$

Show that d' is a metric on M.

5. For all $x, y \in M$, let $d[x, y] = 0$ if and only if $x = y$, else $d[x, y] = 1$. Show that d is a metric on M.

6. Define the following metric on $\mathbb{R}^2$: For all $(x, y), (x', y') \in \mathbb{R}^2$,

$$d[(x, y), (x', y')] = |x - x'| + |y - y'|.$$

Let $N = \{(x, y) : x, y \in \mathbb{Z} \text{ and } x, y \geqslant 0\}$. Show that (N, d) is a metric space. (The function d is called the **taxicab** metric on N because N is viewed as the grid of a downtown area like Chicago. If a taxi is boarded at intersection $(1, 3)$ and ridden to intersection $(4, 1)$, the total number of blocks traveled is $|1 - 4| + |3 - 1| = 5$.)

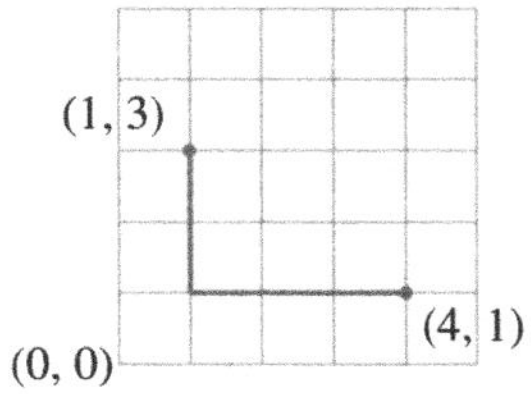

7. Sometimes a set will be open in one metric space but not open in another. Let A be the set of points that lie on the real line between 1 and 2, exclusive.

(a) Explain why A is open in $\mathbb{R}$.
(b) Describe A as a subset of $\mathbb{R}^2$.
(c) Explain why A is not open in $\mathbb{R}^2$.

8. Prove that the following are open in the metric spaces indicated. Assume the usual metric for each space.

(a) $\mathbb{R}\backslash\{0\}$ in $\mathbb{R}$
(b) $\mathbb{R}^2\backslash\{(0, 0)\}$ in $\mathbb{R}^2$
(c) $\mathbb{R}^2\backslash\{(x, y) \in \mathbb{R}^2 : y = 2x - 1\}$ in $\mathbb{R}^2$
(d) $\{(x, y) \in \mathbb{R}^2 : x + y < 1\}$ in $\mathbb{R}^2$
(e) $\{(x, y, z) \in \mathbb{R}^3 : x^2 + y^2 < 1\}$ in $\mathbb{R}^3$

9. Prove that $\{(x, y) \in \mathbb{R}^2 : a < x < b\}$ is open in $\mathbb{R}^2$.

10. Prove that every subset of M is open if and only if $\{a\}$ is open for all $a \in M$.

11. View $\mathbb{R}$ as a metric space with the standard metric. Verify Theorem 13.1.5 by directly proving the following propositions.

(a) $\displaystyle\bigcup_{n=1}^{\infty} D(0, 1/n)$ is open.
(b) $\displaystyle\bigcap_{n=1}^{k} D(0, 1/n)$ is open for all $k \in \mathbb{Z}^+$.
(c) $\displaystyle\bigcap_{n=1}^{\infty} D(0, 1/n)$ is not open.

12. Describe the interior of each of the following subsets of $\mathbb{R}$: $\{0\}$, $(0, 2]$, $[0, 2]$, and $(0, 2) \cup \{-1, 3\}$.

13. Describe the interior of each of the given sets in $\mathbb{R}^2$:

(a) $\{(x, y) \in \mathbb{R}^2 : x^2 + y^2 < 1\}$
(b) $\{(x, y) \in \mathbb{R}^2 : x^2 + y^2 \leqslant 1\}$
(c) $\{(x, y) \in \mathbb{R}^2 : y \leqslant 1/x\}$

14. Let $A \subseteq M$. Show that $\text{int}(A)$ is the largest open subset of A by proving that if $U \subseteq A$ is open, then $U \subseteq \text{int}(A)$.

15. Let M be a metric space. Prove that $A \subseteq M$ is closed in M if and only if $\text{cl}(A) = A$.

16. Let $x \in M$. Prove $\{x\}$ is closed in M.

17. Show that the following are closed in the given metric spaces.

(a) $\{0\}$ in $\mathbb{R}$
(b) $\{(0, 0)\}$ in $\mathbb{R}^2$

(c) $\{(x, y) \in \mathbb{R}^2 : y = 2x - 1\}$ in $\mathbb{R}^2$
(d) $\{(x, y) \in \mathbb{R}^2 : 5 \leqslant x \leqslant 7\}$ in $\mathbb{R}^2$
(e) $\{(x, y) \in \mathbb{R}^2 : x + y \leqslant 1\}$ in $\mathbb{R}^2$
(f) $\{(x, y, 1) : x, y \in \mathbb{R}\}$ in $\mathbb{R}^3$

18. Let $A \subseteq M$. Show that $A \setminus \text{int}(A)$ is closed.

19. Take $a \in M$. Prove that $\{b \in M : d[a, b] \leqslant 1\}$ is closed using both the definition and Theorem 13.1.12.

20. Let $A \subseteq B \subseteq M$. Prove that if a is an accumulation point of A, it is an accumulation point of B.

21. Suppose that A and B are subsets of M. Prove the following about the closure:
(a) $A \subseteq \text{cl}(A)$.
(b) If $A \subseteq B$, then $\text{cl}(A) \subseteq \text{cl}(B)$.
(c) $\text{cl}(\varnothing) = \varnothing$.
(d) $\text{cl}(A) = \text{cl}(\text{cl}(A))$.
(e) $\text{cl}(A)^{\text{c}} = \text{int}(A^{\text{c}})$.

22. Describe the set of accumulation points of each of the following as subsets of the given metric space:
(a) $(0, 2)$ in $\mathbb{R}$
(b) $(0, 2]$ in $\mathbb{R}$
(c) $(0, 2) \cup \{-1, 3\}$ in $\mathbb{R}$
(d) $\{(x, y) \in \mathbb{R}^2\ x^2 + y^2 < 1\}$ in $\mathbb{R}^2$

23. Let f be a continuous function on $[0, 1]$. Prove that $\{f \in C[0, 1] : \|f\| < 1\}$ is open in $C[0, 1]$ with the metric defined on page 446, and the **norm** $\|f\|$ defined by

$$\|f\| = \int_0^1 |f(x)|\, dx.$$

13.2 TOPOLOGICAL SPACES

Let us begin by recalling the definition of a continuous function from calculus. Suppose that $f : \mathbb{R} \to \mathbb{R}$ is continuous at $x = x_0$. This means that for all real numbers x and $\epsilon > 0$, there exists $\delta > 0$ such that

$$\text{if } |x - x_0| < \delta, \text{ then } |f(x) - f(x_0)| < \epsilon. \tag{13.2}$$

Let d be the standard metric on $\mathbb{R}$ and observe that we may write Sentence 13.2 as

$$\text{if } d[x, x_0] < \delta, \text{ then } d[f(x), f(x_0)] < \epsilon. \tag{13.3}$$

However, the x values that satisfy $d[x, x_0] < \delta$ are exactly the elements of $D(x_0, \delta)$. Therefore, we may rewrite Sentence 13.3 as

$$\text{if } x \in D(x_0, \delta), \text{ then } f(x) \in D(f(x_0), \epsilon). \tag{13.4}$$

These three sentences all describe continuity based on the standard metric on $\mathbb{R}$. To generalize the definition of continuity to other metric spaces, all that is needed is to drop the assumption that d is the standard metric. Sentences 13.3 and 13.4 are then definitions for continuity in any metric space. However, can we generalize further and define what it means for a function to be continuous without using a metric? Consider the following. Let $f : M \to N$ be a function where (M, d) and (N, d') are metric spaces. Assume that for all subsets V of N,

$$\text{if } V \text{ is an open in } N\text{, then } f^{-1}[V] \text{ is open in } M. \tag{13.5}$$

The set $f^{-1}[V] = \{x \in M : f(x) \in V\}$ is called the **inverse image** of V under f. We will now show that Sentence 13.5 is equivalent to Sentence 13.4.

- Assume Sentence 13.4 holds for all elements of M. Let V be open in N. To show that $f^{-1}[V]$ is open in M, take $x_0 \in f^{-1}[V]$. This means that $y_0 = f(x_0)$ is an element of V. Since V is open in N, there exists $\epsilon > 0$ such that $D_N(y_0, \epsilon) \subseteq V$. By hypothesis there exists $\delta > 0$ so that for all x,

  $$\text{if } x \in D_M(x_0, \delta)\text{, then } f(x) \in D_N(y_0, \epsilon).$$

  However, letting a be an element of $D_M(x_0, \delta)$, we find that $f(a)$ is an element of $D_N(y_0, \epsilon)$, which is a subset of V. Thus, $a \in f^{-1}[V]$, and this shows that we have $D_M(x_0, \delta) \subseteq f^{-1}[V]$.

- Now suppose that Sentence 13.5 is true. Let $x_0 \in M$ and $\epsilon > 0$. Write $y_0 = f(x_0)$. Since $D_N(y_0, \epsilon)$ is open in N, we know that $f^{-1}[D_N(y_0, \epsilon)]$ is open in M. Thus, we have $D_M(x_0, \delta) \subseteq f^{-1}[D_N(y_0, \epsilon)]$ for some $\delta > 0$. This is the δ that we want because if $x \in D_M(x_0, \delta)$, then $x \in f^{-1}[D_N(y_0, \epsilon)]$, and this means that $f(x) \in D_N(y_0, \epsilon)$.

Since the metric definition of continuity is equivalent to the definition given in Sentence 13.5, we may use the new sentence instead. Furthermore, notice that the new definition only requires an examination involving the open sets of the space. Therefore, a complete generalization of the concept of continuity will involve excluding the metric from the space in favor of an identification of the space's open sets. For this reason we make the following definition.

DEFINITION 13.2.1

Let X be a nonempty set. A collection $\mathcal{T}$ of subsets of X is called a **topology** on X if:

- $\varnothing$ and X are both elements of $\mathcal{T}$.
- the union of every collection of sets from $\mathcal{T}$ is an element of $\mathcal{T}$.
- the intersection of every finite collection of sets from $\mathcal{T}$ is again in $\mathcal{T}$.

The pair $(X, \mathcal{T})$ is called a **topological space**. The subsets found in $\mathcal{T}$ are defined to be the **open** sets of X, and the elements of X are called **points**.

The definition is chosen so that the metric space (M, d) can be viewed naturally as a topological space. Define $\mathcal{T}$ to be the open sets of M in the sense of Definition 13.1.3. Then by Theorem 13.1.5, $\mathcal{T}$ is a topology on M. A topology defined on a metric space M in this manner is called the **induced topology** on M.

Another example is the **discrete topology**. Let X be a set and let $\mathcal{T}$ be the set of subsets of X. Since the union and intersection of subsets of X is a subset of X, $\mathcal{T}$ is a topology on X. On the other extreme, if $\mathcal{T}' = \{\varnothing, X\}$, then $\mathcal{T}'$ is the smallest possible topology on X and is called the **trivial topology**. A more general example is the next theorem.

■ THEOREM 13.2.2

The intersection of two topologies on a set X is a topology on X.

PROOF

Let X be a set and $\mathcal{T}$ and $\mathcal{T}'$ be topologies on X. To show that $\mathcal{T} \cap \mathcal{T}'$ is a topology on X, we must check the three conditions.

- $X, \varnothing \in \mathcal{T} \cap \mathcal{T}'$ since $X, \varnothing \in \mathcal{T}$ and $X, \varnothing \in \mathcal{T}'$.
- Let $\mathcal{F} \subseteq \mathcal{T} \cap \mathcal{T}'$. Then, $\mathcal{F} \subseteq \mathcal{T}$ and $\mathcal{F} \subseteq \mathcal{T}'$. Therefore, $\bigcup \mathcal{F} \in \mathcal{T}$ and $\bigcup \mathcal{F} \in \mathcal{T}'$, and we conclude that
$$\bigcup \mathcal{F} \in \mathcal{T} \cap \mathcal{T}'.$$
- Let $\{U_1, U_2, \ldots, U_k\} \subseteq \mathcal{T} \cap \mathcal{T}'$. From this $U_i \subseteq \mathcal{T}$ and $U_i \subseteq \mathcal{T}'$ for all i. Therefore,
$$U_1 \cap U_2 \cap \cdots \cap U_k \in \mathcal{T},$$
and
$$U_1 \cap U_2 \cap \cdots \cap U_k \in \mathcal{T}'.$$
Hence,
$$U_1 \cap U_2 \cap \cdots \cap U_k \in \mathcal{T} \cap \mathcal{T}'. \blacksquare$$

Our generalization of continuity is complete with the next definition.

■ DEFINITION 13.2.3

Let X and Y be topological spaces. A function $f : X \to Y$ is called **continuous** when $f^{-1}[V]$ is open in X for all open subsets V of Y.

From our work above, any function between metric spaces that is continuous under the metric definition is continuous under the topological definition provided that the set has the induced topology.

For another example, let X be a set and let $f : X \to X$ be a function. View X as a topological space using the discrete topology. Let V be a subset of X. Under the discrete topology, since every subset of X is open, both V and $f^{-1}[V]$ are open. Therefore, in this case every function on X is continuous. This is also true when X

is given the trivial topology. Since only $\varnothing$ and X are the only open sets in X using this topology, $f^{-1}[X] = X$ and $f^{-1}[\varnothing] = \varnothing$ show that f is continuous.

We know from calculus that the composition of two continuous functions is continuous. The same result holds for the topological definition.

■ **THEOREM 13.2.4**

The composition of continuous functions is a continuous function.

PROOF

Let X, Y, and Z be topological spaces. Assume that $f : X \to Y$ and $g : Y \to Z$ are continuous functions. Let $V \subseteq Z$. We first observe that

$$(g \circ f)^{-1}[V] = f^{-1}[g^{-1}[V]].$$

To prove this, let $x \in (g \circ f)^{-1}[V]$. This means that there exists $z \in V$ such that $g(f(x)) = z$. Let $y = f(x)$ and write $z = g(y)$. Therefore, $y \in g^{-1}[V]$, and we conclude that $x \in f^{-1}[g^{-1}[V]]$. The other inclusion is left for Exercise 1.

To prove that the composition is continuous, let $V \subseteq Z$ be open. Since g is continuous, $g^{-1}[V]$ is open in Y, and since f is continuous, $f^{-1}[g^{-1}[V]]$ is open in X, so $(g \circ f)^{-1}[V]$ is open in X. Therefore, the composition is continuous. ■

We need one more test for continuity. We start by observing that the open disks of a metric space serve as the building block for all of the open sets of the metric space. To understand what this means in general, let X be a metric space and $\mathcal{B}$ the set of open disks in X. Let U be an open subset of X. For all $x \in X$, there exists $\epsilon_x > 0$ such that $D(x, \epsilon_x) \subseteq U$. Then

$$U = \bigcup_{x \in U} D(x, \epsilon_x). \tag{13.6}$$

Therefore, every open subset of X is the union of sets in $\mathcal{B}$. We now generalize this idea to topological sets.

■ **DEFINITION 13.2.5**

Let $(X, \mathcal{T})$ be a topological space. An **open base** for X is a set $\mathcal{B}$ with the property that every open set in $\mathcal{T}$ is the union of sets of $\mathcal{B}$. The sets of $\mathcal{B}$ are called **basic open sets**.

The reason that we need an open base is that it will make showing that a function is continuous more straightforward.

■ **THEOREM 13.2.6**

Let X and Y be topological spaces and $f : X \to Y$ a function. If $\mathcal{B}$ is an open base for Y, f is continuous if and only if the inverse image of each basic open set in $\mathcal{B}$ is open in X.

PROOF

Since sufficiency is clear, we shall let $\mathcal{B}$ be an open base for Y such that for all $B \in \mathcal{B}$, $f^{-1}[B]$ is open in X. To prove that f is continuous, take V to be an open subset of Y. Then there exists $\{B_i : i \in I\} \subseteq \mathcal{B}$ such that $V = \bigcup_{i \in I} B_i$. Thus,

$$f^{-1}[V] = f^{-1}\left[\bigcup_{i \in I} B_i\right] = \bigcup_{i \in I} f^{-1}[B_i].$$

Since the union of opens sets is open, $f^{-1}[V]$ is open. ■

Homeomorphisms

We should note that simply because $f : X \to Y$ is continuous and a one-to-one correspondence, it need not be the case that f^{-1} is continuous. If it is, by definition $f[U]$ is open for every open subset U of X. This is because $(f^{-1})^{-1} = f$. In this case we say that f **preserves** open sets and call f an **open mapping**.

■ **DEFINITION 13.2.7**

A **homeomorphism** is a continuous, one-to-one correspondence between two topological spaces that preserves open sets.

If there exists a homeomorphism between the topological spaces $(X, \mathcal{T})$ and $(Y, \mathcal{T}')$, we say that X and Y are **homeomorphic**. What this means is that the sets are the same size, or have the same **cardinality**, because there is a one-to-one correspondence between them, and there is a correspondence between the two spaces that associates the open sets of the one space with the open sets of the other. Essentially the two spaces look the same as topological spaces. For example, the function

$$f : (0, 1) \to \mathbb{R}$$

defined by $f(x) = \cot \pi x$ is a homeomorphism. We also have the following, whose proof we leave to Exercise 10.

■ **THEOREM 13.2.8**

The inverse of a homeomorphism is a homeomorphism. The composition of homeomorphisms is a homeomorphism.

We need an observation for our last definition. Let $(X, \mathcal{T})$ be a topological space and $Y \subseteq X$. Define

$$\mathcal{T}' = \{Y \cap U : U \in \mathcal{T}\}.$$

This set is a topology on Y. We will check the details:

- Since $Y = Y \cap X$ and $\varnothing = Y \cap \varnothing$, both Y and $\varnothing$ are elements of $\mathcal{T}'$.

- Let $\{U_i : i \in I\}$ be a collection of open sets from $\mathcal{T}$. Then because

$$\bigcup_{i \in I}(Y \cap U_i) = Y \cap \bigcup_{i \in I} U_i$$

and $\bigcup_{i \in I} U_i \in \mathcal{T}$, we have $\bigcup_{i \in I}(Y \cap U_i) \in \mathcal{T}'$.

- Now let $U_1, U_2, \ldots, U_k$ be sets in $\mathcal{T}$. Because
$$(Y \cap U_1) \cap (Y \cap U_2) \cap \cdots \cap (Y \cap U_k) = Y \cap (U_1 \cap U_2 \cap \cdots \cap U_k)$$
and $U_1 \cap U_2 \cap \cdots \cap U_k \in \mathcal{T}$, we may conclude that
$$(Y \cap U_1) \cap (Y \cap U_2) \cap \cdots \cap (Y \cap U_k) \in \mathcal{T}'.$$

$\mathcal{T}'$ is called the **relative topology** on Y, and Y is said to be a **subspace** of X. Therefore, every subset of a topological space can itself be viewed as a topological space with the relative topology. Also, if $f : X \to Z$ is continuous, then $g : Y \to Z$ is also continuous when we define $g(x) = f(x)$ for all $x \in Y$. However, there is no reason to expect g to be a homeomorphism when f is.

We need the concept of a homeomorphism to make the last definition of the section.

■ **DEFINITION 13.2.9**

A topological space X is **Euclidean** if it is homeomorphic to a subspace of $\mathbb{R}^n$.

We may then define a topological space X to be **locally Euclidean** if for all $x \in X$, there exists a Euclidean open set containing x. A **manifold** is a locally Euclidean topological space. Examples of spaces that are manifolds include the conic sections and surfaces of spheres, tori, conoids, and spheroids.

Exercises

Unless otherwise noted, $(X, \mathcal{T})$ *and* $(Y, \mathcal{T}')$ *are topological spaces and* $\mathbb{R}^n$ *has the induced topology.*

1. Let X, Y, and Z be sets. If $f : X \to Y$ and $f : Y \to Z$ are functions, then for all subsets V of Z,
$$f^{-1}[g^{-1}[V]] \subseteq (g \circ f)^{-1}[V].$$

2. Let $\mathcal{T}$ and $\mathcal{T}'$ be topologies on X. Give an example which shows that $\mathcal{T} \cup \mathcal{T}'$ may not be a topology on X.

3. Take X to be $\{a, b, c\}$ and $\mathcal{T} = \{\varnothing, \{a\}, \{a, b\}, X\}$. Check that $(X, \mathcal{T})$ is a topological space and find all other topologies on X.

4. Let $Z \subseteq Y \subseteq X$. Let $\mathcal{S}$ be the topology of Z relative to X and $\mathcal{S}'$ the topology of Z relative to Y. Show that $\mathcal{S} = \mathcal{S}'$.

5. Confirm that $\mathcal{T} = \{U : \mathbb{R} \setminus U \text{ is finite}\} \cup \{\varnothing\}$ is a topology on $\mathbb{R}$.

6. Let $f, g : \mathbb{R} \to \mathbb{R}$ be two continuous functions. Prove that $f + g$ and fg are continuous using the metric definition of continuity. Can the topological definition be used for this problem?

7. Suppose that $f, g : \mathbb{R} \to \mathbb{R}$ are continuous functions. Prove using the metric definition of continuity that $f \circ g$ is continuous.

8. Prove Equation 13.6.

9. Let $\mathcal{B} \subseteq \mathbb{R}^2$ be the set of open squares, so

$$\mathcal{B} = \{(a, b) \times (c, d) : a < b, c < d\},$$

where $\times$ denotes the Cartesian product. Prove that $\mathcal{B}$ is an open base for $\mathbb{R}^2$.

10. Prove Theorem 13.2.8.

11. Find an example of a continuous one-to-one correspondence between topological spaces that is not a homeomorphism.

12. Let $\mathcal{B}$ be an open base for X. Prove that $f : X \to Y$ is an open mapping if and only if $f[U]$ is open for all $U \in \mathcal{B}$.

13. Suppose that $\mathcal{B} = \{U_i : i \in I\}$ and $\mathcal{B}' = \{U_i' : i \in I\}$ are open bases for X and Y, respectively. Prove that if f is a one-to-one correspondence such that $f[U_i] = U_i'$, then f is a homeomorphism.

14. Show that $\{(x, x^2) : x \in \mathbb{R}\}$ is homeomorphic to $\mathbb{R}$ by finding two bases and a homeomorphism as described in the previous problem.

15. Show that $(0, 1)$ is homeomorphic to both $(5, 6)$ and $\mathbb{R}$.

16. Let $\mathcal{C} = \{(x, y, z) \in \mathbb{R}^3 : x^2 + y^2 = 1\}$ be a subset of $\mathbb{R}^3$.
   (a) Give a geometric description for $\mathcal{C}$.
   (b) Show that $\mathcal{C} \setminus \{(1, 0, z) : z \in \mathbb{R}\}$ is homeomorphic to $(-1, 1) \times \mathbb{R}$.
   (c) Is $\mathcal{C}$ homeomorphic to $(-1, 1) \times \mathbb{R}$ or even $\mathbb{R}^2$? Explain.

17. Define a subset C of X to be **closed** in X if $X \setminus C$ is open. Prove:
   (a) X and $\varnothing$ are closed.
   (b) The intersection of closed sets is closed.
   (c) The union of finitely many closed sets is closed.

18. Let $f : X \to Y$ be a homeomorphism. Show:
   (a) If C is closed in X, then $f[C]$ is closed in Y.
   (b) If C is closed in Y, then $f^{-1}[C]$ is closed in X.

19. The **closure** of a subset A of X is the intersection of all closed subsets of X that contain A. This is the same definition as for metric spaces (Definition 13.1.10). Prove that A is closed if and only if $\mathrm{cl}(A) = A$.

20. Let $f : X \to Y$ be a function. Prove that the following are equivalent:
   (a) f is continuous.
   (b) $f^{-1}[F]$ is closed in X for all F closed in Y.
   (c) $f[\mathrm{cl}(A)] \subseteq \mathrm{cl}(f[A])$ for each $A \subseteq X$.
   (d) $\mathrm{cl}(f^{-1}[B]) \subseteq f^{-1}[\mathrm{cl}(B)]$ for each $B \subseteq Y$.

13.3 STEREOGRAPHIC PROJECTION

Because of Exercise 16(b) of Section 13.2, we know that the surface of a cylinder is locally Euclidean. Therefore, it is a manifold. Like the cylinder, the surface of a sphere is also a locally Euclidean topological space, making it a manifold.

Take a sphere with center O and radius $\overline{ON}$. Construct $\overline{OU}$ and $\overline{OV}$ such that $\overline{ON}$, $\overline{OU}$, and $\overline{OV}$ are perpendicular to each other, setting up a coordinate system with O as the origin, $\overline{OU}$ as the x-axis, $\overline{OV}$ as the y-axis, and $\overline{ON}$ as the z-axis, as in Figure 13.5. With these axes the equation for the sphere is

$$x^2 + y^2 + z^2 = r^2, \tag{13.7}$$

where $r = ON$. Let $\mathcal{S}$ be the surface of the sphere without N and let $\mathcal{P}$ be the xy-plane. So we have

$$\mathcal{S} = \left\{ \left(x,\ y,\ \pm\sqrt{r^2 - x^2 - y^2} \right) : x \neq 0 \text{ or } y \neq 0 \right\} \cup \{(0,\ 0,\ -r)\}$$

and

$$\mathcal{P} = \{(x,\ y,\ 0) : x,\ y \in \mathbb{R}\}.$$

We wish to define a one-to-one correspondence between the points on the sphere (excluding N) and the points on the plane.

■ **DEFINITION 13.3.1**

A **stereographic projection** of a sphere $\mathcal{S}$ with **center of projection** N and **plane of projection** $\mathcal{P}$ is a function $\phi : \mathcal{S} \to \mathcal{P}$ that associates every point P on $\mathcal{S}$ with the point P' on $\mathcal{P}$ such that N, P, and P' are collinear.

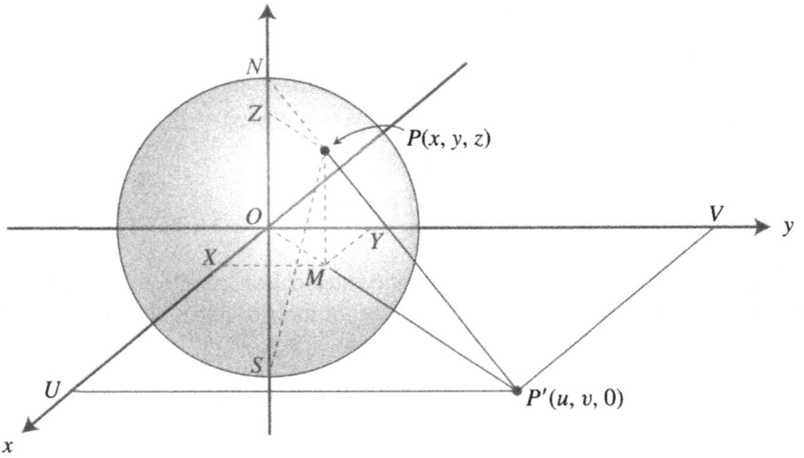

Figure 13.5 The stereographic projection maps P to P'.

Transformation Equations

We will follow Sommerville (1958, 172–174). Let N have coordinates $(0, 0, r)$ and S have coordinates $(0, 0, -r)$ using the labels of Figure 13.5. Let $\phi(P) = P'$, where P has coordinates (x, y, z) and P' has $(u, v, 0)$. Because $\triangle OUP'$ is similar to $\triangle OXM$, we have

$$\frac{u}{x} = \frac{v}{y} = \frac{OP'}{OM},$$

and if we take Z on $\overline{NS}$ such that $\overline{ZP}$ is parallel to $\overline{OM}$,

$$\frac{OP'}{OM} = \frac{OP'}{ZP} = \frac{NP'}{NP} = \frac{NO}{NZ} = \frac{r}{r-z}$$

because $\triangle NOP' \sim \triangle NZP$ and $\overline{OM} \cong \overline{ZP}$. Observe that since $\sphericalangle NPS$ and $\sphericalangle NOP'$ are right angles, $\triangle NPS$ is similar to $\triangle NOP'$. Therefore,

$$NP \cdot NP' = NO \cdot NS = 2r^2.$$

Solving for NP' and substituting, we find that

$$\frac{NP'}{NP} = \frac{2r^2}{NP^2} = \frac{2r^2}{OM^2 + NZ^2} = \frac{2r^2}{x^2 + y^2 + (r-z)^2}.$$

Similarly, when we solve for NP we conclude that

$$\frac{NP'}{NP} = \frac{(NP')^2}{2r^2} = \frac{(OP')^2 + ON^2}{2r^2} = \frac{u^2 + v^2 + r^2}{2r^2}.$$

Therefore,

$$\frac{u}{x} = \frac{v}{y} = \frac{2r^2}{x^2 + y^2 + (r-z)^2} \tag{13.8}$$

and

$$\frac{u}{x} = \frac{v}{y} = \frac{r}{r-z} = \frac{u^2 + v^2 + r^2}{2r^2}, \tag{13.9}$$

These two equations can be used to give us parametric equations that will allow us to transform a point on the sphere to the projected point on the plane, and conversely. In particular, if (x, y, z) is any point on the sphere other than N and $(u, v, 0)$ is its projection, then by solving for x, y, and z in Equation 13.9, we obtain

$$x = \frac{2r^2 u}{u^2 + v^2 + r^2}, \tag{13.10}$$

$$y = \frac{2r^2 v}{u^2 + v^2 + r^2}, \tag{13.11}$$

$$z = \frac{r(u^2 + v^2 + r^2) - 2r^3}{u^2 + v^2 + r^2}, \tag{13.12}$$

and by using Equation 13.8 we see that the stereographic projection ϕ is given by the equation

$$\phi(x, y, z) = \left(\frac{2r^2 x}{x^2 + y^2 + (r - z)^2}, \frac{2r^2 y}{x^2 + y^2 + (r - z)^2}, 0 \right),$$

but since $x^2 + y^2 + (r - z)^2 = 2r^2 - 2rz$, we may simplify this formulation to

$$\phi(x, y, z) = \left(\frac{rx}{r - z}, \frac{ry}{r - z}, 0 \right). \tag{13.13}$$

Images

We will find the image of circles on the surface of the sphere under the stereographic projection given by Equation 13.13. We want to do this for two reasons. First, the great circles of a sphere are considered to be the lines in the geometry found on its surface (page 415), so we would like to know their images. Second, we want to show that the sphere is locally Euclidean.

THEOREM 13.3.2

The stereographic projection maps a circle on the surface of the sphere onto

- a circle in the plane of projection if the circle does not contain the center of projection.
- a line in the plane of projection if the circle contains the center of projection.

PROOF

Circles on the sphere can be identified by slicing the surface with a plane. We will take the resulting circle, find its image in the plane of projection, and examine the equation that describes the image. Let

$$ax + by + cz = d$$

be a plane that intersects the sphere forming a circle and assume that (x, y, z) is on the surface of the sphere. Using the transformation Equations 13.10, 13.11, and 13.12, we conclude that d equals

$$a \left(\frac{2r^2 u}{u^2 + v^2 + r^2} \right) + b \left(\frac{2r^2 v}{u^2 + v^2 + r^2} \right) + c \left(\frac{r(u^2 + v^2 + r^2) - 2r^3}{u^2 + v^2 + r^2} \right).$$

Factoring out some common terms yields

$$(2r^2)(au + bv - cr) + cr(u^2 + v^2 + r^2) = d(u^2 + v^2 + r^2),$$

and from here we put the equation into the form of a general conic section with indeterminants u and v as

$$(cr - d)(u^2 + v^2) + (2r^2)(au + bv) - dr^2 - cr^3 = 0. \qquad (13.14)$$

If $cr - d \neq 0$, the coefficients on the two square terms are equal, and this conic is a circle. Its center and radius are found in Exercise 1. If $cr - d = 0$, then $(0, 0, r)$ is on the circle. In this case $(cr - d)(u^2 + v^2)$ drops out, and we are left with the equation for a line. ■

The function ϕ is a one-to-one correspondence because it has an inverse (Equations 13.10, 13.11, and 13.12). It is continuous because $f(x, z) = rx/(r - z)$ is continuous for $r \neq 0$. Since ϕ sends open disks to open disks as a result of Theorem 13.3.2, and since the set of open disks on the sphere serves as a base for its standard topology, ϕ is an open map (Theorem 13.2.6). We may then conclude the next two corollaries.

■ **COROLLARY 13.3.3**

The sphere without a point is homeomorphic to $\mathbb{R}^2$.

■ **COROLLARY 13.3.4**

The sphere is locally Euclidean.

Conformal Functions

A function that preserves angle measure is called **conformal.** We now show that the stereographic projection is a conformal map. What this means for the stereographic projection takes some explanation. Take, for instance, the planes Λ and K, and let them intersect the sphere in two great circles (Figure 13.6). To simplify our diagram, assume that the great circle that contains points A and B lies in the plane of projection K, so that it is projected onto itself. The great circle that contains C and D in Λ intersects the first great circle in B and E and is projected onto K, making the circle that contains C', D', F', and G'. The two projected circles meet in points B and E. We have two choices for the angle between the two great circles. We may choose

$$\theta = \text{m}\measuredangle ABC = \text{m}\measuredangle A_1B_1C_1$$

or its supplement ψ. Since both θ and ψ can be used to measure the the angle between planes Λ and K, they also measure the angle formed by the normal vectors of these two planes.

With this setup we can now examine the angles formed by the projected circles. Looking down from above N (Figure 13.7) we see two circles meeting in points B and E. Their centers are O and Q. We have two good options on how to measure the angle at B.

- Use $\measuredangle OBQ$. This is equivalent to finding the angle between the vector normal to $\odot O$ at B and the vector normal to $\odot Q$ at B.

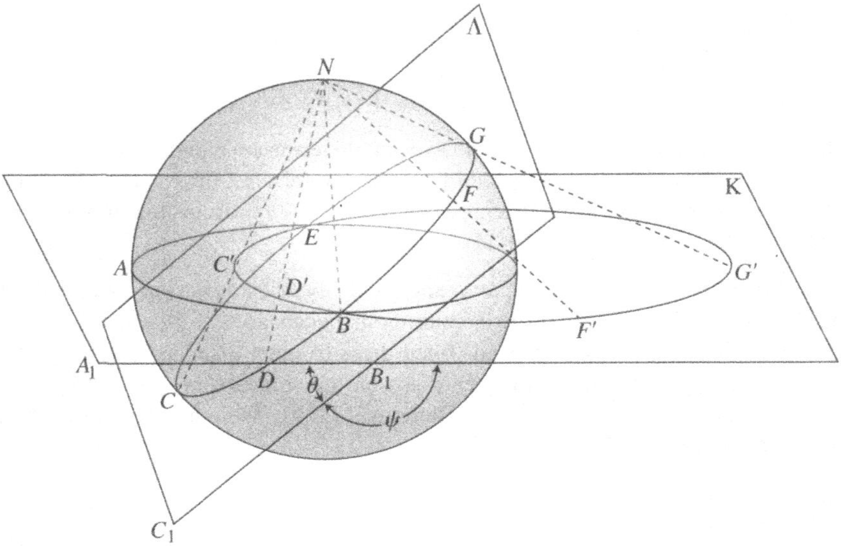

Figure 13.6 The stereographic projection of great circles onto the plane.

- Use $\measuredangle PBR$. This is the angle formed by the tangent to $\odot O$ at B and the tangent to $\odot Q$ at B. This angle is the supplement to $\measuredangle OBQ$ (Exercise 6).

The next theorem shows that $\theta = \text{m}\measuredangle OBQ$ and $\psi = \text{m}\measuredangle PBR$.

■ **THEOREM 13.3.5**

The stereographic projection maps angles between great circles on the surface of the sphere onto angles of equal measure in the plane of projection.

PROOF

To simplify our calculations, we will assume that the sphere has radius $r = 1$ and neither circle contains N. Take two planes that contain the center of the sphere in Figure 13.6. Assuming that the center also serves as the origin of a coordinate system, let the equations of the planes be

$$ax + by + cz = 0$$

and

$$dx + ey + fz = 0$$

and suppose that

$$a^2 + b^2 + c^2 = 1 \text{ and } d^2 + e^2 + f^2 = 1. \tag{13.15}$$

By Equations 13.15 these planes cut two great circles that form an angle of measure

$$\theta = \cos^{-1}(ad + be + cf).$$

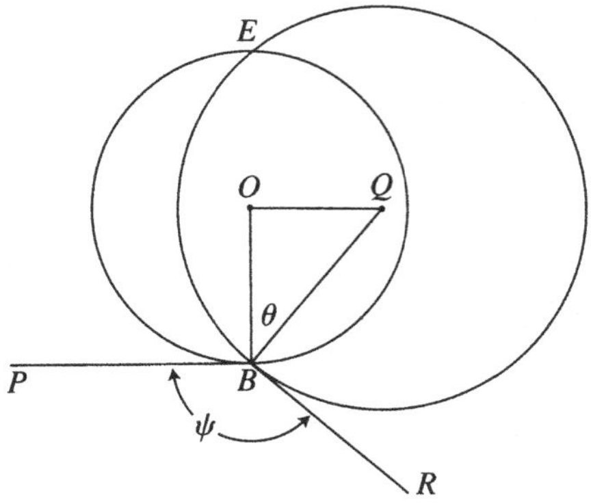

Figure 13.7 The top view of the images in the plane of two great circles.

By Equation 13.14 the projection of the two great circles onto the plane are the circles $\odot O$ and $\odot Q$ with

$$cu^2 + cv^2 + 2au + 2bv - c = 0 \tag{13.16}$$

and

$$fu^2 + fv^2 + 2du + 2ev - f = 0 \tag{13.17}$$

as their equations. Again due to Equations 13.15, we can complete the square on both of these last two equations to find

$$\left(u + \frac{a}{c}\right)^2 + \left(v + \frac{b}{c}\right)^2 = \frac{1}{c^2}$$

and

$$\left(u + \frac{d}{f}\right)^2 + \left(v + \frac{e}{f}\right)^2 = \frac{1}{f^2}.$$

Therefore, $OB = 1/c$ and $QB = 1/f$. We will obtain $m\measuredangle OBQ$ by using the law of cosines. We first need to compute the distance between the two centers of the circles. It is

$$OQ = \sqrt{\left(\frac{a}{c} - \frac{d}{f}\right)^2 + \left(\frac{b}{c} - \frac{e}{f}\right)^2}.$$

By the law of cosines,

$$\left(\frac{a}{c} - \frac{d}{f}\right)^2 + \left(\frac{b}{c} - \frac{e}{f}\right)^2 = \frac{1}{c^2} + \frac{1}{f^2} - \frac{2\cos B}{cf}.$$

Square the binomials to find that

$$\frac{a^2}{c^2} - \frac{2ad}{cf} + \frac{d^2}{f^2} + \frac{b^2}{c^2} - \frac{2be}{cf} + \frac{e^2}{f^2} = \frac{1}{c^2} + \frac{1}{f^2} - \frac{2\cos B}{cf}.$$

When we gather terms on the left we see that

$$-\frac{2ad}{cf} - \frac{2be}{cf} + \frac{a^2 + b^2 - 1}{c^2} + \frac{d^2 + e^2 - 1}{f^2} = -\frac{2\cos B}{cf}.$$

Because of Equations 13.15, we obtain

$$-\frac{2ad}{cf} - \frac{2be}{cf} - 2 = \frac{-2\cos B}{cf},$$

which finally yields

$$ad + be + cf = \cos B.$$

Therefore, $m\sphericalangle OBQ = \theta$, and since $\sphericalangle OBQ$ and $\sphericalangle PBR$ are supplements, $m\sphericalangle PBR = \psi$. ■

Exercises

1. Find the center and radius of the circle given in Equation 13.14.

2. Suppose that the sphere in Figure 13.5 has unit radius. Find equations for the images of the given circles under ϕ.
   (a) $\{(x, 0, z) : x^2 + z^2 = 1\}$
   (b) $\{(x, y, 0) : x^2 + y^2 = 1\}$
   (c) $\{(x, y, 1/2) : x^2 + y^2 = 3/4\}$
   (d) $\{(x, y, -1/2) : x^2 + y^2 = 3/4\}$
   (e) The circle through $(2/3, 2/3, 1/3)$, $(-2/3, 2/3, 1/3)$, and $(0, \sqrt{3}/2, 1/2)$
   (f) The intersection of the plane $x + y + z = 0$ and the sphere

3. The points $A(2/3, 2/3, 1/3)$, $B(-2/3, 2/3, 1/3)$, and $C(0, \sqrt{3}/2, 1/2)$ also describe an angle on the surface of the unit sphere centered at the origin. Let A be the vertex of this angle.
   (a) Find the equations for the planes whose intersections with the unit sphere are the great circles that contain $\overset{\frown}{AB}$ and $\overset{\frown}{AC}$.
   (b) Find the lengths of $\overset{\frown}{AB}$ and $\overset{\frown}{AC}$.
   (c) Find the measure of spherical $\sphericalangle BAC$.
   (d) Find measure of the angle in $\mathbb{R}^2$ formed by $\phi[\overset{\frown}{AB}]$ and $\phi[\overset{\frown}{AC}]$.

4. What on the surface of the sphere projects onto the y-axis under the stereographic projection? What projects onto the circle $x^2 + y^2 = 4$?

5. What is the image of the upper hemisphere under ϕ, and what is the image of the front hemisphere?

6. Show that $\sphericalangle OBQ$ and $\sphericalangle PBR$ in Figure 13.7 are supplementary.

7. Find the angle formed by the circles $x^2 + y^2 - 2y = 0$ and $x^2 + y^2 - 2y - 5 = 0$.

8. Define ϕ' to be the stereographic projection of a sphere onto a plane that is tangent to the sphere at its south pole. Find parametric equations that describe ϕ'.

9. Reprove Theorem 13.3.5 assuming that at least one of the great circles contains N.

13.4 CONSISTENCY OF NON-EUCLIDEAN GEOMETRY

During the last five years of Gauss's life, his health deteriorated due to severe heart problems, yet there were periods when he fought his illness successfully and showed increased stamina. One such time was upon publication of Riemann's *Habilitationsschrift*. Although he expressed excitement about the results, his failing memory and trouble concentrating left him unable and unwilling for deep mathematical conversation. When able, he did involve himself with some simple problems, but he seems to have preferred reading newspapers and other popular literature. During this time, Gauss expressed a belief in God. He was not a Christian but would be better described as a deist who believed in an afterlife. He thought that when one died, one would perceive more clearly the geometric nature of the universe. Gauss evidently hoped that it would be non-Euclidean, since he liked the concept of an absolute measure of length. On February 23, 1855, Gauss died in his home. It is said that his watch stopped moments after his death (Bühler 1981, 152–155).

Two years later, in 1857, Riemann joined the regular faculty at Göttingen, and after another two years he was promoted to the chair of mathematics that had been held by Gauss. Unfortunately, Riemann had contacted tuberculosis. His doctor recommended a warmer climate, so he moved to Italy. He died there near Lake Maggiore on July 20, 1866 at the age of 39. Dying at such a young age, Riemann did not live to see what just a few years later would follow from his work (Gray 2007, 189; Bühler 1981, 152–155; Burton 1985, 564).

Beltrami Models

Lobachevski wrote his geometry using algebra, trigonometry, and calculus, so he believed that his new system was without contradiction. However, he did not provide a proof. The first to put the new geometry on a sound logical footing was Eugenio Beltrami. He was born in Cremona, Italy on November 16, 1835. Many of his relatives were talented in various forms of art, including the engraving of cameos and the painting of miniatures. Beltrami himself enjoyed music and would eventually study its relationship to mathematics. Beltrami was a student from 1853 to 1856 at the University of Pavia, but financial problems would take him away from his academic life to work as a secretary for the railroad in Verona. His new occupation would later take him to Milan, where in his free time he resumed his mathematical pursuits. His early interests were greatly influenced by both Gauss and Riemann and included differential geometry. Beltrami gained a solid reputation for mathematics and would eventually hold various positions in algebra, geometry, and mechanics at universities throughout Italy, including Bologna, Pisa, and Rome. No biography of Beltrami is complete, even one as short as this, without mentioning that it was he who was responsible for rediscovering Saccheri's *Euclides Vindicatus* (Struik 1970a, 599-600; Gray 2006, 79; Stillwell 1996, 3).

Building on Riemann's work, Beltrami was the first to prove that the Parallel Postulate was not provable. He did this by allowing the terms *point* and *line* to be reinterpreted, such as when great circles are considered to be the lines on a sphere. Any example that satisfies the postulates under the new interpretation will be called a **model** of the postulates. If we find such a model, we conclude that the postulates are consistent (page 324). Finding a model yields consistency because we believe that neither mathematics nor the physical world are self-contradictory, so any set of statements interpreted within either of these contexts could not lead to a contradiction.

Beltrami's first consistency proof is found in his "Essay on the Interpretation of Noneuclidean Geometry" (*Saggio di Interpretazione della Geometria Non-euclidea*). In this paper, Beltrami modeled the new geometry using geodesics on a surface of constant negative curvature. The notion of curvature was developed by Gauss in his *Disquisitiones Circa Superficies Curvas* of 1827. It is an intrinsic property of a surface and is measured with real numbers. Spheres have constant positive curvature, while planes and cylinders are considered flat, having zero curvature. The advantage of having any kind of constant curvature is that shapes can be moved on the surface without deforming. An example of a surface with constant negative curvature was studied by Gauss's student Ferdinand Minding in 1840. All of the trigonometric formulas of Lobachevski and Bolyai held true on this surface, so it was a logical candidate for interpreting non-Euclidean geometry (Stillwell 1996, 1–2; Boyer and Merzbach 1991, 505–506).

Minding's surface is brought about by revolving a particular curve about the y-axis. Attach a rope to a relatively heavy object sitting on the ground. While holding the rope, walk away from it until the rope is taught. Now start to pull the object by turning right and walking along the straight line that is perpendicular to the rope. As you do this, the object will be dragged along a path known as a **tractrix** (Figure 13.8). This curve was originally described by Leibnitz. Some have used the German word *hundkurve* to describe it because we can imagine the curve representing the path that a dog would take as it goes on a walk with a leash (Steinhaus 1999, 250–251).

To derive the equation for the tractrix, place the object along the x-axis at $x = a$. Let O be the origin and P an arbitrary point along the curve. Let T represent the point of intersection between the y-axis and the tangent to the tractrix at P. As the object is pulled (or the dog led), the rope is represented by that tangent, so $PT = a$. If the coordinates of P are (x, y) and $x \leqslant a$, then

$$OT = y + \sqrt{a^2 - x^2}$$

and

$$\frac{dy}{dx} = -\frac{\sqrt{a^2 - x^2}}{x}.$$

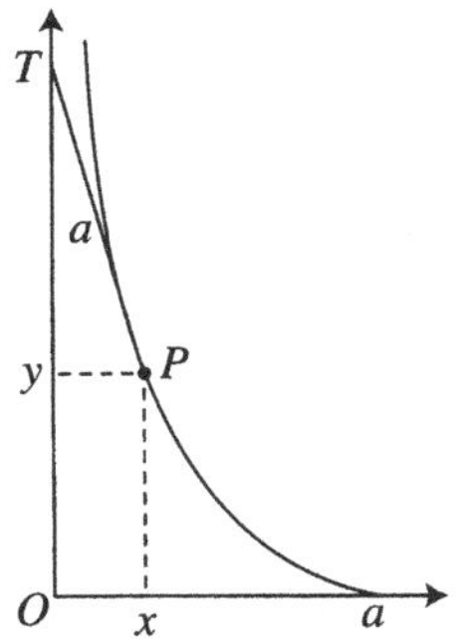

Figure 13.8 The tractrix.

To solve this differential equation we write

$$\int dy = -\int \frac{\sqrt{a^2 - x^2}}{x}\, dx.$$

Focusing our work on the right-hand side of the equation, we substitute $x = a \sin\theta$:

$$\int \frac{\sqrt{a^2 - x^2}}{x}\, dx = a \int \frac{\cos^2\theta}{\sin\theta}\, d\theta = a \int (\csc\theta - \sin\theta)\, d\theta.$$

Because

$$\int (\csc\theta - \sin\theta)\, d\theta = -\ln\left|\frac{a}{x} + \frac{\sqrt{a^2 - x^2}}{x}\right| + \frac{\sqrt{a^2 - x^2}}{a} + C$$

and $y = 0$ when $x = a$, we conclude that

$$y = a \ln\left(\frac{a + \sqrt{a^2 - x^2}}{x}\right) - \sqrt{a^2 - x^2}. \qquad (13.18)$$

The original derivation of the equation for the tractrix is due to James Bernoulli (Kline 1972, 473–474).

Now take the tractrix and rotate it about its asymptote. The result is a solid called a **pseudosphere** (Figure 13.9). This surface is the simplest that has constant negative curvature. However, since the pseudosphere has a single point at infinity, it cannot represent the entire Lobachevskian plane. Instead, it models a portion that is bounded by two parallel lines and an arc of a horocycle. The surface that Beltrami used to show the consistency of **HA** was essentially a covering that wrapped around the pseudosphere infinitely many times. With minor modification this can be made to represent an arbitrarily large portion of the non-Euclidean plane (Stillwell 1996, 2; Sommerville 1958, 169–170).

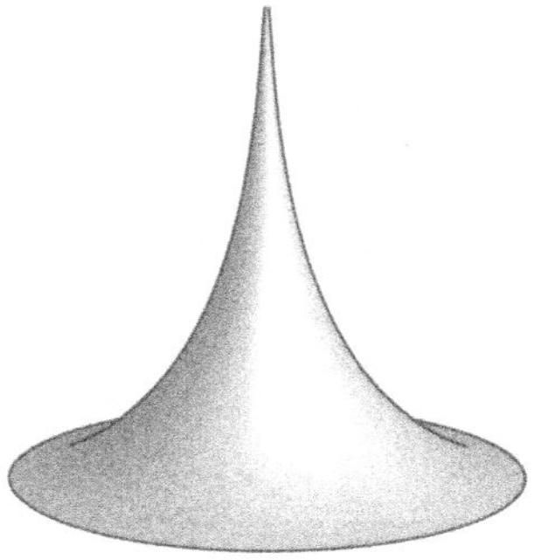

Figure 13.9 The pseudosphere.

In another of his papers, entitled "Fundamental Theory of Spaces of Constant Curvature" (*Teoria fondamentale degli spazii di curvatura constante*), Beltrami generalized his results to n dimensions. He also gave early versions of models that are now named after Henri Poincaré and Felix Klein and one that we will call Beltrami's **hemisphere model** (Figure 13.10). Take the hemisphere in $\mathbb{R}^3$ given by the equation

$$x^2 + y^2 + z^2 = r^2$$

and let $z > 0$. The surface of the hemisphere is interpreted to be the plane, *points* are regular points on the surface, and *lines* are the geodesics formed by slicing the model with vertical planes. The great circle $z = 0$ represents the line ω at infinity. To find the parallel lines of this model, let l be an arc formed by cutting the hemisphere with perpendicular plane Λ and let P be a point on the surface but not on l. Turn Λ until it intersects P by using the perpendicular to one of the endpoints of l as an axis. The resulting arc contains P and intersects l at a point on ω. This arc is considered parallel to l. If Λ had been turned using the other endpoint of l, the result would be the other parallel to l through P (Stillwell 1996, 35–37).

Both of Beltrami's models show that the geometries of Lobachevski and Bolyai are consistent by transferring the question to one within Euclidean geometry. If

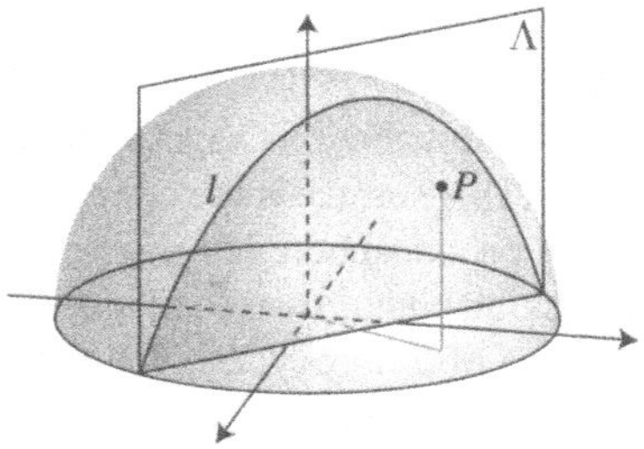

Figure 13.10 The hemisphere model.

there is an interpretation of non-Euclidean geometry that satisfies the negation of the Parallel Postulate, then non-Euclidean geometry is without contradiction if the system in which it is being interpreted has no contradictions. Since mathematicians were confident in the consistency of Euclid's system, they should now have confidence in the new geometry. This concept is known as **relative consistency**. All consistency proofs are actually proofs of relative consistency.

Poincaré Models

There are two models associated with Poincaré that can be obtained from Beltrami's hemisphere model. Obtain the first by stereographically projecting the hemisphere onto the plane $z = 0$ by completing the bottom half of the sphere and using the south pole as the center of projection (Definition 13.3.1). Let $\mathscr{C}$ be the circle with equation

$$x^2 + y^2 = r^2.$$

The images of points from the hemisphere lie within $\mathscr{C}$ and are the *points* of the model. The images of the arcs of the hemisphere will be projected onto either segments within $\mathscr{C}$ or half-circles within $\mathscr{C}$ that intersect the circle at right angles (Theorem 13.3.2). These are the *lines* of the model. Two lines are said to be parallel in this model if they intersect at $\mathscr{C}$. This means that $\mathscr{C}$ is interpreted as the line at infinity. The image of the hemisphere within the circle $\mathscr{C}$ is called the **Poincaré disk model** (Figure 13.11). Poincaré published this model in 1882 (Stillwell 1996, 35–36).

The second model is known as the **Poincaré half-plane model**. Take a plane Λ that is tangent to the Beltrami hemisphere at Q but perpendicular to the plane $z = 0$ (Figure 13.12). Let P be on the hemisphere opposite Q so that $\overline{PQ}$ is perpendicular to Λ. The image on the upper half of Λ of the hemisphere under the stereographic projection with center P is Poincaré's model. Points and lines are interpreted as in the Poincaré disk, and the intersection of Λ and $z = 0$ is the line at infinity (Stillwell 1996, 36–37).

The mathematician responsible for these two models earned a reputation that approached that of Gauss. Poincaré was born in Nancy, France on April 29, 1854.

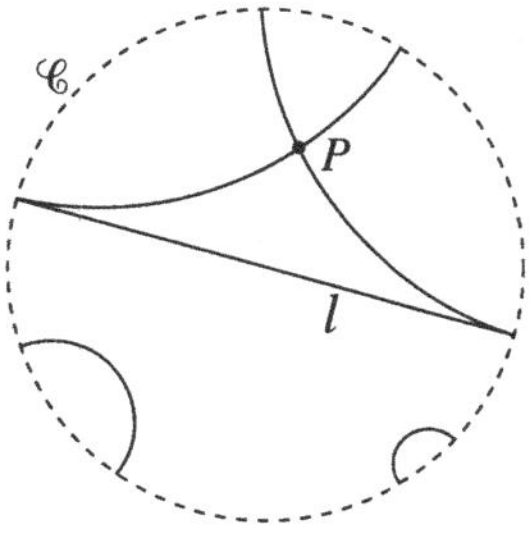

Figure 13.11 The Poincaré disk.

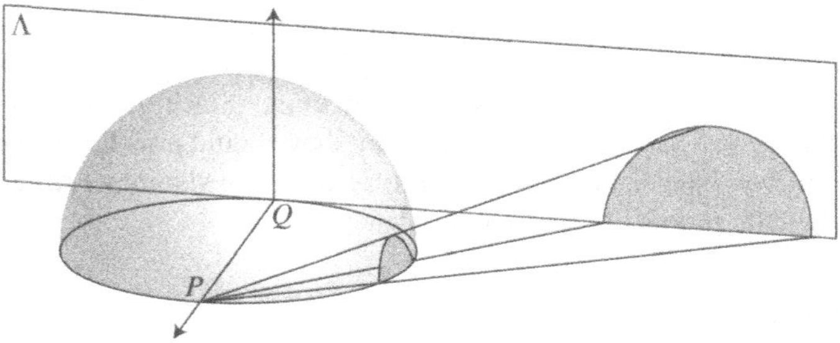

Figure 13.12 Stereographic projection of the hemisphere onto the Poincaré half-plane.

He was from an upper-middle-class family, and his relatives included physicians, professors, and politicians. He attended the École Polytechnique, and after graduating Poincaré worked at engineering while he studied for his doctorate degree in mathematics. He finished in 1879. Afterward he taught at the University of Caen and became a professor at the University of Paris in 1881. At the young age of 33, he was elected to the Académie des Sciences. He had wide mathematical interests that included function theory, number theory, algebra, and the foundations of mathematics. In addition to mathematics, Poincaré made important contributions to astronomy and physics (Dieudonné 1970, 51–61).

Klein Models

Felix Klein was another leading nineteenth century mathematician. He was born in Düsseldorf on April 25, 1849. His father worked in the government finance office and had very traditional Prussian views. This resulted in Klein attending the gymnasium in his hometown where he studied classical languages but very little mathematics. After graduation he attended the University of Bonn in 1866, where he met Julius Plücker. Klein was soon recruited by Plücker and became his assistant with the goal being to make Klein into a physicist. Klein had the talent for it, and it might have happened except that Plücker died in 1868. At the time Plücker had been working on various papers, including one on geometry. As his assistant, Klein prepared the unfinished works for publication and was particularly inspired by Plücker's *Neue Geometrie des Raumes*. By the end of the year, Klein finished his doctorate degree, moved to Göttingen for a few months, and then moved on to Berlin. This was followed by a trip to Paris that was interrupted by the Franco-Prussian War, and then a return to Göttingen, where he taught mathematics as a lecturer in 1871 (Yaglom 1988, 25–27; Burau and Schoenberg 1970, 396).

It was during this year that Klein's "On the So-called Noneuclidean Geometry" was published. It included a model of the geometry of Lobachevski and Bolyai. If Beltrami's hemisphere model is orthogonally projected onto the plane, the result is an open disk with the geodesics of the hemisphere mapped to chords of the circle. Viewing the plane as a copy of $\mathbb{R}^2$, the equation for the disk is $x^2 + y^2 < r^2$, where r

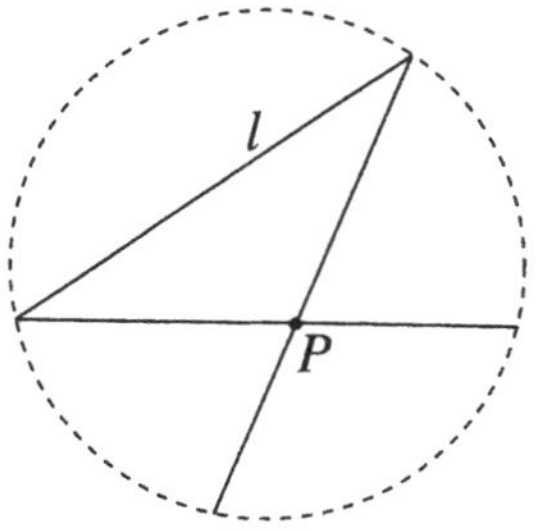

Figure 13.13 The Klein disk.

is the radius of the hemisphere. This is called the **Klein disk model** (Figure 13.13). The *points* of the model are regular points, and the *lines* are chords of the circle without the endpoints. Two lines are parallel if they meet at a point on the circle, and they are nonintersecting if the do not meet anywhere within or on the circle.

Klein also provided models that showed that the Hypothesis of the Obtuse Angle was possibly true. His motivation for this can be traced to Riemann. From Saccheri to Lobachevski and Bolyai, geometers did not accept geometries in which **HO** held. In such systems, Theorem 5.3.9 was false. Since this proposition was part of their data from Euclid that was not based on the Parallel Postulate, they consistently rejected **HO** as a possibility. This is surprising since it was a known fact that the angle sum of a triangle on the surface of a sphere had angle sum greater than two right angles. It was left to Riemann to accept that this surface yielded its own geometric system (Bonola 1912, 141).

Riemann observed that there are essentially three types of two-dimensional geometries, based on the type of surface on which it lies:

- A flat surface such as a plane or cylinder
- A surface of constant negative curvature such as a pseudosphere
- A surface of constant positive curvature

The first of these is Euclidean geometry and the second is Lobachevski's. The third can be interpreted as existing on the surface of a sphere, but to do this we must rethink our understanding of lines. According to Riemann:

> In the extension of space-construction to the infinitely great, we must distinguish between *unboundedness* and *infinite extent*; the former belongs to the extent relations, the latter to the measure-relations. That space is an unbounded three-fold manifoldness, is an assumption which is developed by every conception of the outer world; according to which every instant the region of real perception is completed and the possible positions of a sought object are constructed, and which by these applications is forever confirming itself. The unboundedness of space possesses in this way a greater empirical certainty than any external experience. But its infinite extent by no means follows from this; on the other hand if we assume independence of bodies from position, and therefore ascribe

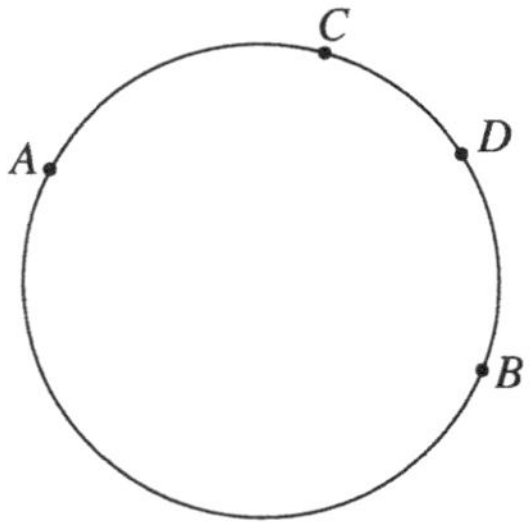

Figure 13.14 A finite, closed line.

> to space constant curvature, it must necessarily be finite provided this curvature has ever so small a positive value. If we prolong all the geodesics starting in a given surface-element, we should obtain an unbounded surface of constant curvature, *i.e.*, a surface which in a *flat* manifoldness of three dimensions would take the form of a sphere, and consequently be finite. (Riemann 1854, III.2, 67–68)

The result of Riemann's comments is two definitions that had previously not been applied to lines. We say that a line is **finite** if its length is finite. A line l is **closed** if given any two points A and B on l, there exist points C and D on l such that C and D are between A and B but B is between C and D (Figure 13.14).

This change requires a change in the postulates of Euclid. Since the days of the Greeks, geometers nearly universally assumed that lines were infinite. Even many of Euclid's proofs assumed this, despite this property not being found among the postulates. Therefore, this new geometry requires that Euclid's first and last postulates be replaced with the following two.

■ **POSTULATES 13.4.1 [Riemann]**

1. All lines are closed and finite.

2. All lines intersect.

In the development of this **Riemannian geometry** we may keep certain of Euclid's theorems up through *Elements* I.28 (Theorem 5.4.1), but others will have to be jettisoned. This includes the previously mentioned Theorem 5.3.9. We also cannot keep Theorem 5.3.12. The proofs of both of these propositions relied on lines being infinite, as indicated in Exercise 8.

Since the next theorem implies the existence of a triangle that has an angle sum greater than two right angles, we conclude by Theorem 10.3.1 that **HO** holds in this geometry.

■ **THEOREM 13.4.2 [Riemann]**

All lines perpendicular to a given line are concurrent.

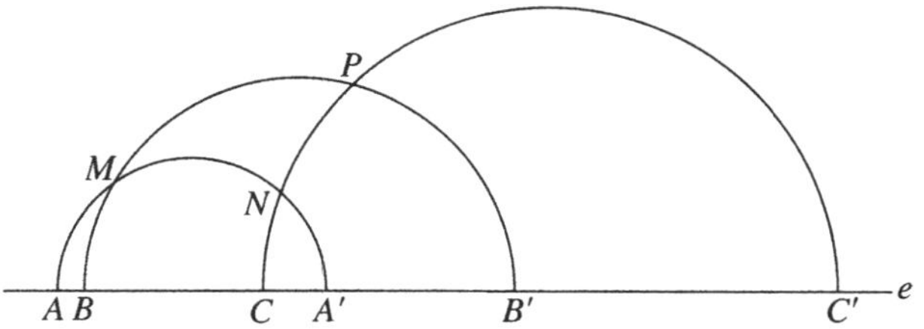

PROOF

Let e be a line and A a point on e. Let A' be chosen so that $\overline{AA'}$ is perpendicular to e. Since $\overline{AA'}$ is closed, it must intersect e again. The second time $\overline{AA'}$ intersects e is either back at A by looping back or by intersecting in some new point. Suppose it is the latter and that this new point is A'. Draw two more lines perpendicular to e at B and C and let the respective lines intersect e for a second time at B' and C'. By Postulate 13.4.1.2, the three pairs of lines intersect. Let M be the intersection of $\overline{AA'}$ and $\overline{BB'}$, N the intersection of $\overline{AA'}$ and $\overline{CC'}$, and P the intersection of $\overline{BB'}$ and $\overline{CC'}$. Since $\sphericalangle MAB$ and $\sphericalangle MBA$ are both right, $\triangle ABM$ is an isosceles triangle. Hence, $\overline{AM} \cong \overline{BM}$. Similarly, $\triangle BA'M$ is isosceles, so we conclude that $\overline{BM} \cong \overline{MA'}$. Thus, M is the midpoint of $\overline{AA'}$. A similar argument will show that N is also the midpoint of $\overline{AA'}$, from which we conclude that $M = N$. Since M and P are midpoints of $\overline{BB'}$, we see that they are in fact the same point. Hence, the three lines are concurrent. ■

The point N that exists on one side of the line e in the proof of Theorem 13.4.2 on which all perpendiculars to the line fall is called the **absolute pole** of e. The line e is the **absolute polar** of N. For any point P on e, $\overline{NP}$ is called a **quadrant**, and we say that N is **orthogonal** to P. All quadrants on the same side of e are congruent (Exercise 14).

Certainly on the other side of e we may repeat the construction above and find a point S that is contained by all perpendiculars. The question now is whether the points N and S are different. It turns out that both options are consistent with the postulates for this geometry. If N and S are different, the points are called **antipodal** and the geometry is called **double elliptic**. If this is the case, lines meet in two points, so Euclid's first postulate cannot hold. If, instead, the points are the same, we regain Euclid's assumption that two points determine a line. This geometry is called **single elliptic**. This terminology is due to Klein but from a different context. It has now become common to use the terms **elliptic** for geometries that satisfy **HO** and **hyperbolic** for the geometry of Lobachevski and Bolyai (Bonola 1912, 164; Wussing 1984, 177).

The two elliptic geometries were proven consistent by models given by Klein (Burton 1985, 568). He realized them on the surface of a sphere. Double elliptic is illustrated in Figure 13.15(a). *Lines* in the model are great circles of the sphere, and *points* are regular points on its surface. There is no line at infinity because there are no parallels. This model satisfies Postulates 13.4.1 and the three remaining postulates of Euclid.

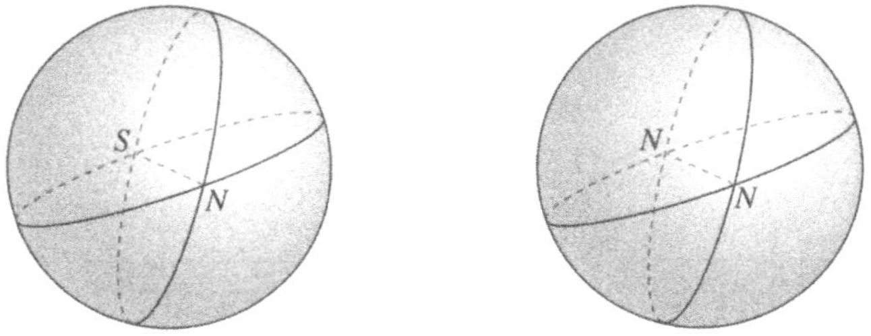

(a) Double elliptic (b) Single elliptic

Figure 13.15 Klein's models of elliptic geometry.

Single elliptic geometry is modeled in Figure 13.15(b). Here, *lines* are as in double elliptic; however, *point* is understood to refer to pairs of antipodal points on the surface of the sphere. Postulates 13.4.1 hold true in this model as well as all of Euclid's postulates except for the last postulate concerning parallels.

Exercises

1. Locate the point at infinity for the pseudosphere. Use this point to identify which objects on the surface of the pseudosphere should represent lines, parallel lines, and nonintersecting lines.

2. Assume $a = 1$ in the equation for the tractrix (Equation 13.18). This makes the equation equivalent to $y = \text{sech}^{-1}x - \sqrt{1 - x^2}$.
   (a) Confirm this result.
   (b) Show that the surface area of the pseudosphere (the surface area of revolution of the tractrix) can be approximated by 12.56. What is the significance of this result?
   (c) Show that the volume of the pseudosphere can be approximated by 2.09. What is the significance of this result?

3. A **pseudosquare** is a quadrilateral that is equilateral and equiangular. How can a pseudosquare be constructed in hyperbolic geometry?

4. Draw a picture representing parallel lines in Beltrami's hemisphere model.

5. Create a diagram for each of the Poincaré models and the Klein model that includes examples of points, lines, parallel lines, nonintersecting lines, and triangles.

6. The Poincaré models and the Klein model are intended to represent the non-Euclidean plane of Lobachevski and Bolyai. If this is the case, what is missing from the descriptions of these models in the section?

7. Draw horocycles in the Poincaré and Kline models.

8. Show that Theorems 5.3.9 and 5.3.12 are false under **HO**.

9. Let the open square $(0, 1) \times (0, 1)$ be the plane, *points* be regular points, and *lines* be line segments with endpoints on the border of the square. Is this geometry Euclidean? What is the angle sum of triangles? How do you harmonize the answers to these questions?

10. Prove that rectangles do not exist in elliptic geometry.

11. Explain why in single elliptic geometry a line does not separate the plane into two half planes. With this in mind, give an example of a "flat" model of double elliptic geometry.

12. Do lines divide the plane into two half planes in double elliptic geometry? Explain.

13. Does Pasch's Postulate (11.3.2) hold in either of the elliptic geometries?

14. Prove that all quadrants on the same side of e are congruent in Riemann's elliptic geometry.

15. What is the maximum measure of the third angle of a triangle with two right angles?

CHAPTER 14

JEAN-VICTOR PONCELET

Gaspard Monge was a significant figure in late eighteenth century French mathematics. In 1775, at the age of 29, he was appointed the Royal Professor of Mathematics and Physics at the École Militaire de Mézièrs. There he developed a method of designing forts that effectively defended the soldiers within from gunfire that originated from multiple directions. This led to Monge's creation of a field of mathematics known as **descriptive geometry**. Monge left Mézièrs in 1784 to become the head of naval education in France, and 10 years later he was appointed to another influential education position, one that led to the creation of École Polytechnique in June 1795. This was a two-year school designed to prepare students for the study of engineering. Its curriculum, which included both descriptive and differential geometry, was designed by mathematicians who believed that a study of pure mathematics was the best way to prepare their students for the various options they would have in their future studies. Besides his duties at École Polytechnique, Monge remained professionally active outside of the school. This included working to revive the French professional organizations, which had been neglected after the revolution, and accompanying Napoleon to Egypt, where he led the scientific team. When Monge returned a year later, he continued teaching at École Polytechnique until 1809. During his tenure there, Monge was a champion of synthetic geometry, and he conveyed this passion to

Revolutions of Geometry. By Michael O'Leary

one of his students: Jean-Victor Poncelet (Gray 2007, 4–6, 7; Gray 2004, 30; Boyer and Merzbach 1991, 468, 478).

The son of wealthy and politically influential parents, Poncelet was born on July 1, 1788 in the French town of Metz, a city that remained dear to him his entire life. At a very early age, Poncelet was sent away for his preliminary schooling and then returned to his family in 1804. Three years later he enrolled at École Polytechnique but fell behind due to poor health and took an extra year to complete his studies. After completing his work at École Polytechnique, Poncelet joined the corps of engineers in September 1810. A year and a half later he was assigned to the Dutch island of Walcheren, and in June 1812 he took part in the French invasion of Russia. Despite his role as an engineer and not as a fighting soldier, Poncelet was taken prisoner on November 18, 1812 during the retreat. He was held as a prisoner of war at Saratov and not released until June 1814. While in Russia, Poncelet spent his time writing. The result was a text on analytic geometry based on his studies at École Polytechnique. He would eventually have it published but not for 50 years. It was intended to be an introduction to his second work, the *Traité des propriétés projectives des figures*. This work would be published in 1822. Begun during the days of his imprisonment, in the *Traité* Poncelet aimed to write a geometry treatise using synthetic methods, yet with the generality of analytic geometry (Taton 1970, 76; Boyer and Merzbach 1991, 535).

The methods Poncelet used in the *Traité* were projective. For example, he used ideal points and introduced a new concept called **imaginary points**. These were defined in such a way that any line and circle would always intersect at one of these points (Boyer and Merzbach 1991, 535). Unfortunately, Poncelet had little familiarity with the work of Desargues and Pascal. The *Brouillon Project* of Desargues was lost soon after he wrote it, and it remained lost until 1847, when the handwritten copy made by Philippe de Lahire was discovered in a Paris library. This was 33 years after Poncelet's time in Russia. The result was that Poncelet spent unnecessary time reproving many of the early projective theorems. Nonetheless, Poncelet made further advancements in the subject, including developments regarding involutions and conic sections (Kline 1972, 840, 842; Boyer and Merzbach 1991, 361).

Poncelet viewed his projective geometry as a new branch of mathematics with its own distinct methods. One example of this is Poncelet's **principle of continuity**. Let us illustrate how it works by considering two theorems regarding secants and tangents to a circle. Let A and D be on a circle and C outside the circle such that they form secants $\overline{AC}$ and $\overline{DC}$ that intersect the circle in points B and D', as in Figure 14.1. We know by the Secant Theorem (5.5.4) that

$$AC \cdot BC = DC \cdot D'C.$$

Now let D move along the circumference of the circle so that it approaches D'. As this happens, the secant $\overline{DC}$ will eventually intersect the circle in only one point, so that it becomes a tangent, and this happens when D and D' coincide. Therefore, the equation obtained from the Secant Theorem becomes

$$AC \cdot BC = DC^2,$$

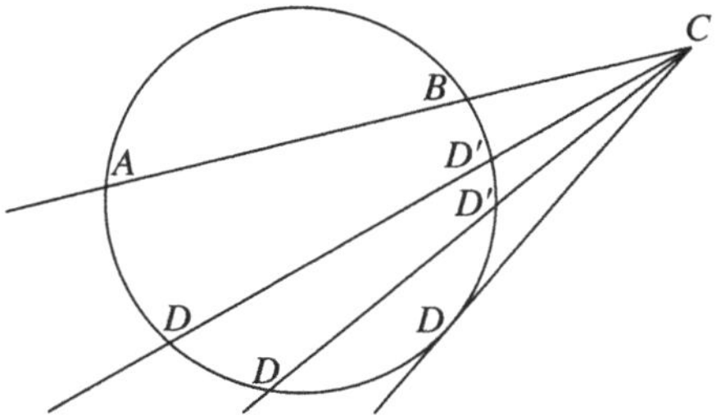

Figure 14.1 The Tangent-Secant Theorem and the principle of continuity.

which is what follows from the Tangent-Secant Theorem (Theorem 5.5.5). Poncelet's strategy is familiar to us. We have seen such continuity arguments used at various times. However, during the nineteenth century mathematicians were becoming more aware of the unstated assumptions that they had been using in the past. A champion of these new standards was Cauchy. He is known for his foundational work on the definitions of limit and continuity in calculus. He believed that it was important that these ideas have a rigorous foundation and was highly critical of Poncelet's principle (Burton 1985, 535; Kline 1972, 843, 950–953).

Poncelet took Cauchy's criticisms personally. He did not consider his principle of continuity a less rigorous form of mathematics. Quite the contrary, Poncelet viewed his principle as a postulate that allowed him to argue as he did. He rejected the notion that his projective geometry needed the analytic foundations that Cauchy apparently desired. Despite Cauchy's formidable reputation as one of the leading mathematicians of the time, Poncelet held his ground and set out to expand and clarify his geometric innovations. The result was an additional opening section to the *Traité*. This addition would make the *Traité* the major work of nineteenth-century projective geometry (Taton 1970, 77).

Partly due to the issues he had with Cauchy, and partly due to other controversies with which he found himself embroiled, Poncelet would eventually all but abandon geometry in favor of applied mechanics. In 1824 he became professor at École d'Application de l'Artillerie et du Génie at Metz. His appreciation of both the theoretical aspect of mechanics and its experimental side made him an excellent teacher. Eventually, the course on machine mechanics that he developed became very popular and widely used at other institutions. Poncelet's interest in machinery was not limited to the classroom, however. His fascination with mechanical devices led to his 1825 tour of France, Belgium, and Germany to visit various factories and study their machinery. Also in that year, Poncelet won a prize from the Académie des Sciences for his design of a more efficient undershot waterwheel. Its secret was in its curved paddles (Taton 1970, 77, 78).

As with many with a military background, Poncelet acquired an interest in politics. In 1830 he became a member of the city council of Metz. Seemingly destined to live the rest of his life in his hometown, plans changed in 1834, when Poncelet accepted an appointment to the school of mechanics at the Académie des Sciences in Paris. There he continued teaching, and later, in 1837, he created a course for the Faculty of Sciences of Paris. He married in 1842, and at this point he was no longer doing original research. He was happy to spend his time teaching and continuing his technical investigations. This all changed with the revolution of 1848. Poncelet was appointed brigadier general and assigned to be the commandant of École Polytechnique. While there, Poncelet was seemingly involved in every aspect of the institution: from reforming the school's curriculum to maintaining student order during the riots of mid-1848 (Taton 1970, 78–79).

Poncelet retired in 1850 and decided to edit and publish his collected works. He completed only four volumes. Two volumes were a second edition of the *Traité des propriétés projectives des figures*, and the other two were titled *Applications d'analyse et de géométrie*. This was the work on analytic geometry that he had written while a prisoner in Russia. Before he could complete the volumes on mechanics, Poncelet died in Paris, on December 22, 1867 (Taton 1970, 76, 78–79).

14.1 THE PROJECTIVE PLANE

We defined the projective plane in Section 9.2 when we studied Desargues. We defined it by adding points and a line at infinity to a Euclidean plane. The modern viewpoint turns this around and first defines the projective plane and then by removing elements constructs a Euclidean one.

Projective Axioms

The work in this section is based on Bennett (1995, 41–49) and the classic Veblen and Young (1910, 15–20). Let $\mathcal{P}$ and $\mathcal{L}$ be two nonempty sets. The elements of $\mathcal{P}$ will be called **points**, and the elements of $\mathcal{L}$ will be called **lines**. Assume that the points are fundamental and that a line is a set of points. We will use geometric terminology when we write sentences about the points and lines, but these should be interpreted with our new understanding. If we say that a point P **lies on** or is **incident** with a line l, this means that $P \in l$. Similarly, the line l **contains** or is **incident** with a point P if $P \in l$. Two points P and Q in $\mathcal{P}$ are **collinear** if there exists a line in $\mathcal{L}$ that contains both P and Q as elements, and two lines m and l are **concurrent** if there exists a point that is an element of both lines. Both collinear and concurrent can be generalized to any number of points and lines. Finally, if points P and Q chosen from $\mathcal{P}$ are the same, write $P = Q$; otherwise, they are **distinct** and we write $P \neq Q$. The pair $(\mathcal{P}, \mathcal{L})$ is called a **projective plane** when it follows these rules:

■ **POSTULATES 14.1.1**

1. For all distinct $P, Q \in \mathcal{P}$, there exists a unique line l in $\mathcal{L}$ such that P and Q are incident with l.

2. For all distinct $l, m \in \mathcal{L}$, there is a point incident with both l and m.

3. Every line contains at least three points.

4. There are at least two lines.

The geometric theory that satisfies these postulates is called a **projective geometry**. Notice that the second postulate states that projective geometry is elliptic.

Different authors will choose different sets of postulates. These are from Bennett. Veblen and Young use a different collection, which replaces some of the propositions of Postulates 14.1.1. These will serve as our first theorem, but its proof is left to Exercise 10.

THEOREM 14.1.2

Let $(\mathcal{P}, \mathcal{L})$ be a projective plane.

- There exists $P, Q \in \mathcal{P}$ such that $P \neq Q$ and P and Q are not collinear.
- For all $P, Q, R \in \mathcal{P}$, if P, Q, and R are not collinear, there exist points S, T, and U in $\mathcal{P}$ with $P \neq S$ such that P, R, and T are collinear, P, Q, and U are collinear, and S, T, and U are collinear.

Consistency

We now show that the projective geometry described by Postulates 14.1.1 is consistent by finding a model.

THEOREM 14.1.3

If $\mathcal{P} = \{0, 1, 2, 3, 4, 5, 6\}$ and $\mathcal{L}$ is the set

$$\{\{0, 1, 3\}, \{1, 2, 4\}, \{2, 3, 5\}, \{3, 4, 6\}, \{4, 5, 0\}, \{5, 6, 1\}, \{6, 0, 2\}\},$$

$(\mathcal{P}, \mathcal{L})$ is a projective plane.

PROOF

To prove this, we must systematically check that the four postulates hold under this definition of $\mathcal{P}$ and $\mathcal{L}$.

1. The table enumerates all possible pairs of distinct points and the unique line to which the pair belongs, so the first postulate is seen to be satisfied.

Points		Line	Points		Line	Points		Line
0	1	{0, 1, 3}	1	3	{0, 1, 3}	2	6	{6, 0, 2}
0	2	{6, 0, 2}	1	4	{1, 2, 4}	3	4	{3, 4, 6}
0	3	{0, 1, 3}	1	5	{5, 6, 1}	3	5	{2, 3, 5}
0	4	{4, 5, 0}	1	6	{5, 6, 1}	3	6	{3, 4, 6}
0	5	{4, 5, 0}	2	3	{2, 3, 5}	4	5	{4, 5, 0}
0	6	{6, 0, 2}	2	4	{1, 2, 4}	4	6	{3, 4, 6}
1	2	{1, 2, 4}	2	5	{2, 3, 5}	5	6	{5, 6, 1}

2. The next table shows that all pairs of distinct lines contain a common point.

Lines		Point	Lines		Point
{0, 1, 3}	{1, 2, 4}	1	{2, 3, 5}	{3, 4, 6}	3
{0, 1, 3}	{2, 3, 5}	3	{2, 3, 5}	{4, 5, 0}	5
{0, 1, 3}	{3, 4, 6}	3	{2, 3, 5}	{5, 6, 1}	5
{0, 1, 3}	{4, 5, 0}	0	{2, 3, 5}	{6, 0, 2}	2
{0, 1, 3}	{5, 6, 1}	1	{3, 4, 6}	{4, 5, 0}	4
{0, 1, 3}	{6, 0, 2}	0	{3, 4, 6}	{5, 6, 1}	6
{1, 2, 4}	{2, 3, 5}	2	{3, 4, 6}	{6, 0, 2}	6
{1, 2, 4}	{3, 4, 6}	4	{4, 5, 0}	{5, 6, 1}	5
{1, 2, 4}	{4, 5, 0}	4	{4, 5, 0}	{6, 0, 2}	0
{1, 2, 4}	{5, 6, 1}	1	{5, 6, 1}	{6, 0, 2}	6
{1, 2, 4}	{6, 0, 2}	2			

3. Every line has at least three points by definition.

4. There are clearly at least two lines. ■

There are **infinite models** of projective geometry. The Euclidean projective plane of Desargues (Section 9.2) is an example. Other models include those for single elliptic geometry and the **Riemann sphere**, the complex numbers joined with a point at infinity. The Riemann sphere is denoted by $\mathbb{C}^*$ or $\mathbb{C} \cup \{\infty\}$.

Isomorphic Models

Although one model is sufficient to show the consistency of a set of postulates, there are many different models for projective geometry. Our first was an example of a **finite model**. Here is another. It is named after Gino Fano (1871–1952). Fano was a respected projective and algebraic geometer who was a professor at the University of Turin until he lost his position under the Fascist laws of 1938. During World War II he taught Italian students at Lausanne, and after the war he continued his research and lecturing in both Italy and the United States (Struik 1970b, 522–523).

■ **THEOREM 14.1.4 [Fano Plane]**

If $\mathcal{P}_F$ is the set of dots illustrated by Figure 14.2 and

$$\mathcal{L}_F = \{\overline{EBA}, \overline{ADG}, \overline{EFG}, \overline{BCG}, \overline{ECD}, \overline{ACF}, \odot BDF\},$$

then $(\mathcal{P}_F, \mathcal{L}_F)$ is a projective plane.

We leave it to Exercise 8 to check that the interpretation satisfies Postulates 14.1.1 and that it is the smallest model to do so.

It should be noted that there is actually little difference between Theorem 14.1.3 and Theorem 14.1.4. The only variation between the two are the names of the elements. To see this, let $\mathcal{P}$ be from Theorem 14.1.3 and define $\phi : \mathcal{P}_F \to \mathcal{P}$ by

$$\phi(A) = 0, \phi(B) = 1, \phi(C) = 2, \phi(D) = 5, \phi(E) = 3, \phi(F) = 6, \text{ and } \phi(G) = 4.$$

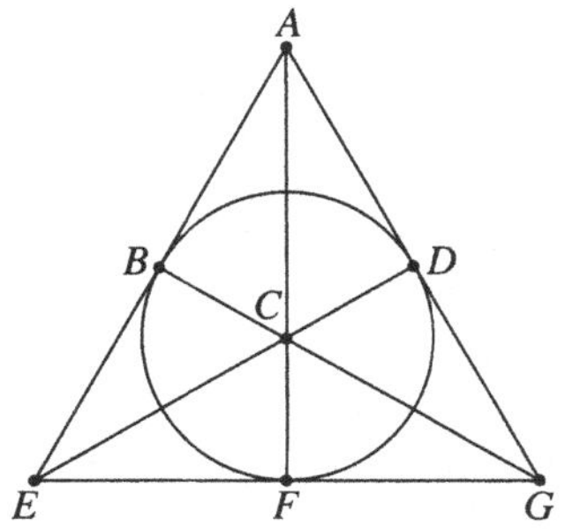

Figure 14.2 The Fano plane.

This function associates pairs of elements in such a way that if three points are on a line in one model, the corresponding points would be on a line in the other model. For instance, $\overline{EBA} = \{E, B, A\}$, $\phi(E) = 3$, $\phi(B) = 1$, and $\phi(A) = 0$. Thus,

$$\phi[\{E, B, A\}] = \{0, 1, 3\},$$

and for the rest of the lines we have

$$\begin{aligned}\phi[\{A, D, G\}] &= \{4, 5, 0\},\\ \phi[\{E, F, G\}] &= \{3, 4, 6\},\\ \phi[\{B, C, G\}] &= \{1, 2, 4\},\\ \phi[\{E, C, D\}] &= \{2, 3, 5\},\\ \phi[\{A, C, F\}] &= \{6, 0, 2\},\\ \phi[\{B, D, F\}] &= \{5, 6, 1\}.\end{aligned}$$

We now generalize this result.

DEFINITION 14.1.5

Let $(\mathcal{P}, \mathcal{L})$ and $(\mathcal{P}', \mathcal{L}')$ be projective planes. If $\phi : \mathcal{P} \to \mathcal{P}'$ is a one-to-one correspondence such that for all $P, Q, R \in \mathcal{P}$,

$$R \in \overline{PQ} \text{ if and only if } \phi(R) \in \overline{\phi(P)\phi(Q)},$$

then ϕ is a **collineation**.

If there is a collineation between two projective planes, the planes look identical except for the names put on their elements.

DEFINITION 14.1.6

The projective planes $(\mathcal{P}, \mathcal{L})$ and $(\mathcal{P}', \mathcal{L}')$ are **isomorphic** if there exists a collineation $\phi : \mathcal{P} \to \mathcal{P}'$.

Independence

Because of the work by Beltrami, Klein, and Poincaré in Section 13.4, we know that the Parallel Postulate is independent (page 324) of Euclid's other postulates. This means that the Parallel Postulate can neither be proven true nor proven false from Euclid's first four assumptions. It is important for any proposition in a set of postulates to have this property because we never want to assume a proposition that could have been proven from the others. We say that a set of propositions is **independent** if every proposition of the set is independent of the others. To prove that a set of propositions $\{p_1, p_2, \ldots, p_k\}$ is independent, for each i we must show that both

$$\{p_1, p_2, \ldots, p_i, \ldots, p_k\}$$

and

$$\{p_1, p_2, \ldots, \textit{not } p_i, \ldots, p_k\}$$

are consistent. In this way, neither *not* p_i nor p_i necessarily follow from the other postulates.

To prove that the set of projective postulates is independent, we first observe that the Fano plane shows that the postulates are consistent. It remains to show that the negation of any of the projective postulates plus the remaining three is a consistent set of propositions. The diagram of Figure 14.3(a) shows that it is possible that the first postulate is false while the others are true, and Figure 14.3(b) shows that it is possible for the third postulate to be false with the others true. The Euclidean plane is a model in which Postulates 1, 3, and 4 are satisfied but Postulate 2 is not because there are parallel lines in the Euclidean plane. The empty set is a model in which the last postulate is false because that postulate entails the existence of objects in the geometry. The others are **vacuously true** on the empty set. This refers to the fact that a universal sentence can be written as a conditional. For example, Postulate 14.1.1.1 can be stated as

if P and Q are distinct points in $\mathcal{P}$, there exists a unique line l in $\mathcal{L}$ such that P and Q are incident with l.

If $\mathcal{P} = \varnothing$, the antecedent of the conditional is false, and this implies that the conditional is true (page 38).

(a) Negation of Postulate 1 (b) Negation of Postulate 3

Figure 14.3 Independence of the projective postulates.

Exercises

1. Given the following set of postulates for **incidence geometry**:

   Postulate 1. There exist at least three points.
   Postulate 2. Two distinct points are on exactly one line.
   Postulate 3. There are at least two points on every line.
   Postulate 4. For every line l, there is a point not on l.

   (a) Show that this is a consistent set of postulates using a finite model.
   (b) Find an infinite model for this geometry.
   (c) Show that this is an independent set of postulates.
   (d) Determine whether incidence geometry is a projective geometry.

2. Given the following set of postulates for **three-point geometry**:

   Postulate 1. There exist exactly three points.
   Postulate 2. Two distinct points are on exactly one line.
   Postulate 3. Every two distinct lines are on at least one point.
   Postulate 4. For every line l, there is a point not on l.

   (a) Show that this is a consistent set of postulates.
   (b) Show that this is an independent set of postulates.
   (c) Determine whether three-point geometry is a projective geometry.

3. Prove the following theorems in three-point geometry.
   (a) Two distinct lines are on exactly one point.
   (b) There exist exactly three lines.
   (c) Lines cannot contain three distinct points.
   (d) There exists a set of two lines containing all of the points.
   (e) There exists a set of two points such that each line lies on at least one of the points.

4. Given the following set of postulates for **four-line geometry**:

   Postulate 1. There exist exactly four lines.
   Postulate 2. Two distinct lines intersect on exactly one point.
   Postulate 3. Every point is on exactly two lines.

   (a) Show that this is a consistent set of postulates.
   (b) Show that this is an independent set of postulates.
   (c) Determine whether four-line geometry is a projective geometry.

5. Given the following set of postulates for **Fano's geometry**:

   Postulate 1. There exists at least one line.
   Postulate 2. Every line contains exactly three points.
   Postulate 3. Not all points are on the same line.
   Postulate 4. Two distinct points are on exactly one line.
   Postulate 5. Two distinct lines intersect in at least one point.

(a) Show that this is a consistent set of postulates.
(b) Show that this is an independent set of postulates.
(c) Determine whether Fano's geometry is a projective geometry.

6. Prove the following theorems in Fano's geometry:
   (a) Every two lines have exactly one point in common.
   (b) There exist exactly seven points and seven lines.
   (c) Each point lies on exactly three lines.
   (d) Lines through any given point contain all of the points.
   (e) For any pair of points, there is exactly two lines not containing either point.
   (f) Given a set of three lines not all containing the same point, there exists exactly one point not on any of the three lines.

7. **Young's geometry** shares the first four postulates of Fano's geometry, but Postulate 5 is replaced with

   Postulate 5. If a point does not lie on a given line, there exists exactly one line on the point that does not intersect the given line.

   (a) Show that this is a consistent set of postulates.
   (b) Show that this is an independent set of postulates.
   (c) Determine whether Young's geometry is a projective geometry.

8. Show that the Fano plane is the smallest model of projective geometry.

9. Show that two lines in a projective plane intersect in exactly one point.

10. Prove Theorem 14.1.2.

11. Let $(\mathcal{P}, \mathcal{L})$, $(\mathcal{P}', \mathcal{L}')$, and $(\mathcal{P}'', \mathcal{L}'')$ be projective planes. Suppose that

$$\phi : \mathcal{P} \to \mathcal{P}'$$

and

$$\psi : \mathcal{P}' \to \mathcal{P}''$$

are collineations. Show that following:
   (a) ϕ^{-1} is a collineation.
   (b) $\psi \circ \phi$ is a collineation.

12. Do any of the five geometries defined in this exercise set have two nonisomorphic models? If so, which ones?

14.2 DUALITY

Throughout this section we assume that all points and lines are from the projective plane $(\mathcal{P}, \mathcal{L})$. We begin this section by proving a sequence of four theorems that follow quickly from the definition of a projective plane.

THEOREM 14.2.1

For all distinct l, $m \in \mathcal{L}$, there is a unique point $P \in \mathcal{P}$ so that P is on l and m.

PROOF

Let l, $m \in \mathcal{L}$. By Postulate 14.1.1.2 we know that there exists a point P that is incident with l and m. We must show that P is unique, so suppose that there exists another point Q incident with l and m that is different from P. This means that we have two distinct points on two distinct lines, which contradicts Postulate 14.1.1.1. ∎

THEOREM 14.2.2

For all distinct P, $Q \in \mathcal{P}$, there is a line incident with both P and Q.

PROOF

This is just Postulate 14.1.1.1 without the uniqueness condition. ∎

THEOREM 14.2.3

Every point is incident with at least three lines.

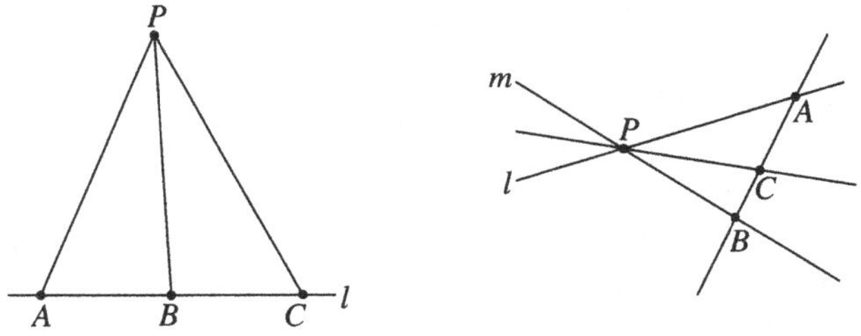

PROOF

Let P be a point and let l and m be the two lines guaranteed by Postulate 14.1.1.4. We have two cases to consider.

- Suppose that P is not on l. We know by Postulate 14.1.1.3 that l has three points on it. Call them A, B, and C. By Postulate 14.1.1.1 we may join $\overline{PA}$, $\overline{PB}$, and $\overline{PC}$. None of these lines equals l because they contain P but l does not. We also see that $\overline{PA}$, $\overline{PB}$, and $\overline{PC}$ are all distinct because if we suppose that any two of these lines are the same, there would be a pair of points on two distinct lines contradicting Postulate 14.1.1.1. For example, if $\overline{PA}$ is the same line as $\overline{PB}$, we conclude that A and B are on both $\overline{PA}$ and l.

- Now let P be on l. If P is not on m, we may construct three lines as we did in the first part, so suppose that P is also on m. We know that there is a point A on l and a B on m that are both different from P. Join $\overline{AB}$ by Postulate 14.1.1.1, and by Postulate 14.1.1.3 find another point C on

$\overline{AB}$. Join $\overline{PC}$. This is the third line through P because if $\overline{PC}$ were either equal to l or m, we would have two distinct points on two distinct lines. ■

■ **THEOREM 14.2.4**

There are at least two points.

PROOF

Postulate 14.1.1.4 tells us that there is a line, and by Postulate 14.1.1.3 there are at least three points on that line. Therefore, there are at least two points. ■

When we compare the four theorems with Postulates 14.1.1, we see that they are basically the same except that the words *point* and *line* have been interchanged. What this means is that every theorem about the projective plane has a corresponding theorem called its **dual** that is obtained from the original by exchanging every concept related to the word *point* with the corresponding concept related to the word *line*. This dual follows automatically without further proof once the original theorem is proven. This is a very important property of projective geometry called the **principle of duality**.

The principle of duality was discovered by Poncelet but not without controversy. In 1810 an artillery officer by the name of Joseph-Diez Gergonne published the first privately published journal of mathematics. He titled it the *Annales de Mathématiques Pures et Appliquées*. Gergonne was a fine mathematician and a strong supporter of analytic methods. Poncelet, on the other hand, was a staunch defender of the synthetic. These two soon became point men in the synthetic–analytic debate that had previously embroiled Klügel and Kästner with the English (page 355). At first their competition was civil, with both publishing papers espousing their points of view in Gergonne's *Annales* in 1818. Needless to say, the controversy was not settled, and years later, in 1826, matters heated up when both men claimed priority in the discovery of the principle of duality. Gergonne claimed that he had proven it using analytic methods, while Poncelet insisted that he had discovered duality during a study of the conic sections. The reality of the situation was that although the term *dual* was due to Gergonne, it was Poncelet who was first (Boyer and Merzbach 1991, 506–507, 536; Kline 1972, 845).

The Complete Quadrilateral

To illustrate the principle of duality, we consider two pairs of geometric figures. The first pair is the **triangle**,

three noncollinear points that join to form three sides,

and the **trilateral**,

three nonconcurrent lines that meet to form three vertices.

These definitions are duals of each other. One sentence is formed from the other by making the replacement:

point	$\longleftrightarrow$	line
join	$\longleftrightarrow$	meet
vertex	$\longleftrightarrow$	side
collinear	$\longleftrightarrow$	concurrent

When we draw both of these figures, we find that they are the same. We say that the figures are **self-dual**.

The second pair involves the complete quadrangle. We repeat its definition:

■ **DEFINITION 14.2.5**

Let A, B, C, and D be four points in a plane, no three of which are collinear. The configuration that consists of these four points and the six lines determined by all possible pairs of the points is called the **complete quadrangle** $ABCD$.

Since being *concurrent* is the dual of being *collinear*, we may define the dual of the complete quadrangle.

■ **DEFINITION 14.2.6**

Let a, b, c, and d be four lines in a plane, no three of which are concurrent. The configuration that consists of these four lines and the six points determined by all possible pairs of the lines is called the **complete quadrilateral** $abcd$.

Compare the following with the content on page 315. The lines of the complete quadrilateral $abcd$ are called its **sides**, and the points of $abcd$ are its **vertices**. The complete quadrilateral $abcd$ is illustrated in Figure 14.4. It consists of the four lines a, b, c, and d and the six vertices

$$ab,\ bc,\ cd,\ ad,\ ac,\ bd,$$

where ab represents the intersection of lines a and b. Two vertices are said to be **opposite** if they do not share a common side. There are three pairs of opposite

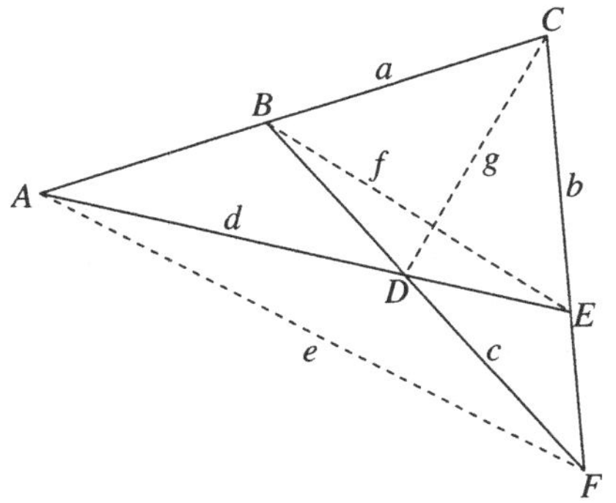

Figure 14.4 A complete quadrilateral $abcd$.

vertices. They lie on three lines called **diagonal lines**. The intersection of these lines form the **diagonal trilateral** of $abcd$. The diagonal trilateral of the quadrilateral in Figure 14.4 is Δefg.

Harmonic Sets

In Books III and IV of the *Conic Sections*, Apollonius determines necessary and sufficient conditions for a line through a point outside a conic section to be tangent to that section (Kline 1972, 96–97). Let A be a point external to a conic such that $\overline{AP}$ and $\overline{AQ}$ are tangent to the given conic section (Figure 14.5). Suppose that P and Q are the points of tangency. Join $\overline{PQ}$. This segment is called the **polar** of the **pole** A. Through any point C on $\overline{PQ}$, draw the line that joins A and C and let B and D be the points of intersection of this line with the conic. Apollonius showed that

$$\frac{AB}{AD} = \frac{CB}{CD}. \tag{14.1}$$

The phrase that is traditionally used to describe this proportion is:

A divides $\overline{BD}$ externally in the same ratio as C divides it internally.

Conversely, suppose that we have a point outside a conic and want to find the tangents to the conic that pass through the point. To do this, we again let A be off the conic and external to it. Draw a line through A that meets the conic at B and D. Draw another line through A that intersects the conic at B' and D' (Figure 14.5). On $\overline{BD}$ find C such that Equation 14.1 holds and find C' on $\overline{B'D'}$ so that

$$\frac{AB'}{AD'} = \frac{C'B'}{C'D'}.$$

Apollonius showed that the line through C and C' intersects the conic section at the points P and Q, respectively, so that $\overline{AP}$ and $\overline{AQ}$ are tangent lines.

Now let us change our perspective. Take quadrangular set $Q(ABC,\ EDF)$. By Theorem 9.4.5,

$$\frac{AB \cdot EB}{AD \cdot ED} = \frac{CB \cdot FB}{CD \cdot FD}.$$

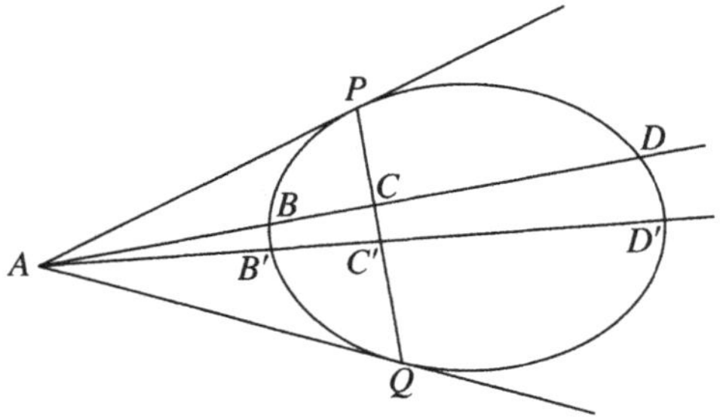

Figure 14.5 A pole and polar of a conic section.

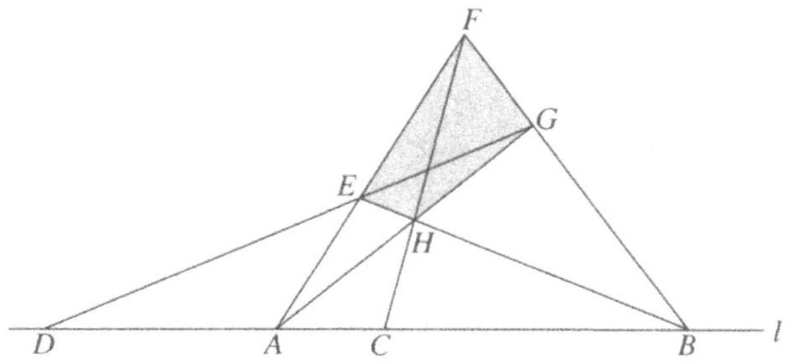

Figure 14.6 H(AB, CD).

Suppose that point A coincides with E and that C coincides with F. We then have

$$\frac{AB \cdot AB}{AD \cdot AD} = \frac{CB \cdot CB}{CD \cdot CD},$$

and this is equivalent to Equation 14.1. Desargues called the original points A, B, C, and D four points in involution (Exercise 4 of Section 9.4). We have a different name for this relation. It is defined in terms of a modified quadrangular set and illustrated in Figure 14.6.

■ DEFINITION 14.2.7

The four collinear points A, B, C, D form a **harmonic set of points** if there is a complete quadrangle such that points A and B are two of the vertices of its diagonal triangle and points C and D are on the sides through the third vertex of the diagonal triangle. The harmonic set of points is denoted by H(AB, CD).

Observe that

$$\text{H}(AB,\ CD) \text{ if and only if } \text{Q}(ABC,\ ABD).$$

Moreover, like the notation for involutions (page 313), there are permutations of the points of a harmonic set that immediately from the definition also form harmonic sets, but H(CD, AB) is not one of them. That H(AB, CD) implies H(CD, AB) requires proof.

■ THEOREM 14.2.8

For any four points A, B, C, D, if H(AB, CD), then H(CD, AB).

PROOF

Let A, B, C, and D be collinear such that H(AB, CD). This means there exists complete quadrangle $EFGH$ such that A is the intersection of $\overline{EF}$ and $\overline{HG}$, B is the intersection of $\overline{EH}$ and $\overline{FG}$, C is on $\overline{FH}$, and D is on $\overline{EG}$. To prove H(CD, AB) we must find a complete quadrangle such that C and D

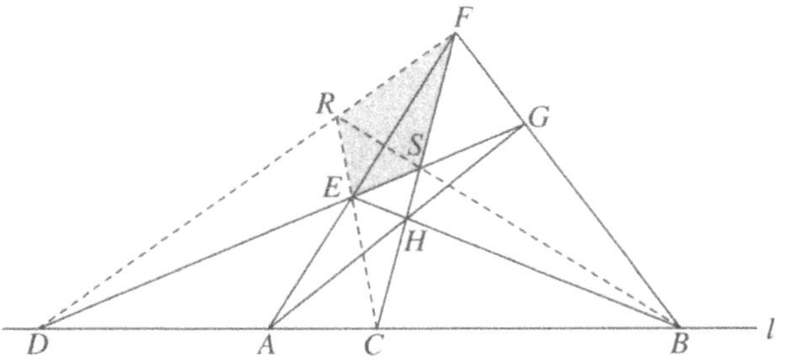

are vertices of its diagonal triangle and A and B are on lines through the third vertex of its diagonal triangle. To do this, join $\overline{DF}$ and $\overline{EC}$. Extend $\overline{EC}$ to meet $\overline{DF}$ at R. Let S be the intersection of $\overline{DG}$ and $\overline{CF}$. We want R, S, and B to be collinear. To obtain this, consider $\triangle REF$ and $\triangle SHG$. Notice that $\overline{RF}$ and $\overline{SG}$ meet at D, $\overline{EF}$ and $\overline{HG}$ meet at A, and $\overline{RE}$ and $\overline{SH}$ meet at C. This means that the points of intersection of pairs of corresponding sides of the two triangles are collinear. Therefore, the lines joining corresponding vertices are concurrent by Desargues' Theorem (9.3.2). Since we know that $\overline{EH}$ and $\overline{FG}$ already intersect at B, it must be the case that the $\overline{RS}$ produced will contain B. We conclude that $RFSE$ is the desired complete quadrangle because C and D are vertices of its diagonal triangle and A and B are on lines that intersect the diagonal triangle's third vertex. Therefore, $\mathsf{H}(CD,\ AB)$. ■

Choose three points A, B, and C on a line l. We can prove that there is a unique point D on l so that $\mathsf{H}(AB,\ CD)$. The point D is called the **harmonic conjugate** of C with respect to A and B.

■ THEOREM 14.2.9

Let A, B, and C be three points on a line l. There exists exactly one point D on l such that $\mathsf{H}(AB,\ CD)$.

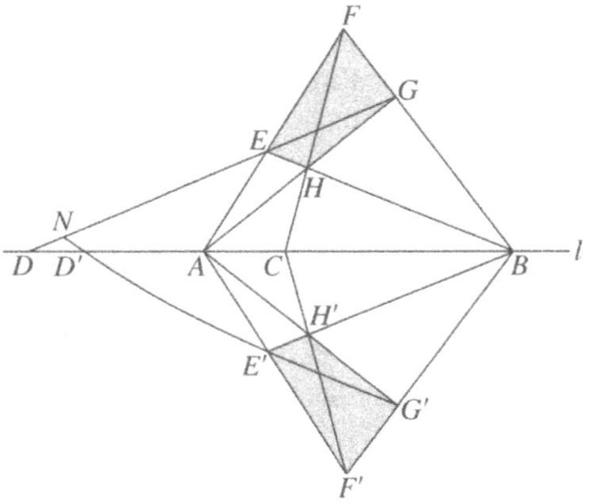

PROOF

Let l be a line and take A, B, and C to be points on l. We first show that the desired D exists. Take a point F not on l. After joining $\overline{FB}$, pick G on that segment. Join $\overline{AG}$ and $\overline{FC}$ and let these segments intersect at H (Theorem 9.2.8). Join $\overline{AF}$ and $\overline{BH}$, producing $\overline{BH}$ to meet $\overline{AF}$ at E. Take the point D to be the intersection of l and $\overline{GE}$ produced. Since $EFGH$ is a complete quadrangle, we see that $\mathsf{H}(AB, CD)$.

To show that D is unique, repeat the construction so that we have complete quadrangle $E'F'G'H'$ with $\overline{E'F'}$, $\overline{F'G'}$, and $\overline{H'F'}$ intersecting l at A, B, and C, respectively, but $\overline{E'G'}$ meeting l at D', which is different from D. By Theorem 9.2.8, there is a point N that is incident with both $\overline{GD}$ and $\overline{G'D'}$. First consider $\triangle EFH$ and $\triangle E'F'H'$. Since their corresponding sides intersect at A, C, and B on line l, we conclude by Desargues' Theorem (9.3.2) that $\overline{EE'}$, $\overline{HH'}$, and $\overline{FF'}$ are concurrent (not pictured). The same argument using $\triangle FGH$ and $\triangle F'G'H'$ allows us to conclude that $\overline{HH'}$, $\overline{FF'}$, and $\overline{GG'}$ are concurrent. Therefore, the lines joining corresponding vertices of $\triangle HEG$ and $\triangle H'E'G'$ are concurrent, so by another application of Desargues' Theorem (9.3.1), points N, A, and B are collinear. This means that N must be on l, so $D = D'$. ■

The dual of a harmonic set of points is a harmonic set of lines.

■ **DEFINITION 14.2.10**

The four concurrent lines a, b, c, d form a **harmonic set of lines** if there exists a complete quadrilateral such that a and b are two of the sides of its diagonal trilateral and c and d are on the points on the third side of the diagonal trilateral. When a, b, c, and d form a harmonic set in this fashion, we write $\mathsf{H}(ab, cd)$.

Exercises

1. Given ellipse $x^2 + 4y^2 = 16$, find
   (a) the polar of $(-7, 0)$.
   (b) the pole of the vertical polar at $x = 2$.
   (c) the pole of the segment that joins the points of intersection of $y = 2x + 3$ and the ellipse.
2. Write the duals of each of the following sentences.
   (a) *There exist at least two lines.*
   (b) *There are at least two lines on every point.*
   (c) *Not all points lie on the same line.*
   (d) *Two distinct lines intersect in a unique point.*
   (e) *The line intersects one of the sides of a complete quadrangle.*
   (f) *The line joining two points contains the third point.*
   (g) *All points lie on either the first or the second line.*

3. Brianchon's Theorem is the dual of Pascal's Theorem (9.3.5). State Brianchon's Theorem.

4. Let $\mathsf{H}(AB, CD)$. Write seven other orders of points A, B, C, and D that form harmonic sets.

5. Construct a complete quadrangle given its diagonal triangle and one vertex.

6. Let A, B, C, and D be collinear points. Find P and Q such that $\mathsf{H}(AB, PQ)$ and $\mathsf{H}(CD, PQ)$.

7. The **cross ratio** of the four collinear points A, B, C, and D is

$$(AB, CD) = \frac{AC \cdot BD}{AD \cdot BC}.$$

Show that $\mathsf{H}(AB, CD)$ if and only if $(AB, CD) = 1$.

8. Let A, B, C, and D be collinear. Show the following:
   (a) $(AB, CD) = (CD, AB)$.
   (b) $(AB, DC) = 1/(AB, CD)$.
   (c) Use these two results to confirm the findings of Exercise 4.

9. Assume that for any ordinary points A and B and ideal point Ω, $A\Omega = B\Omega$ because both distances are infinite. With this in mind, prove that if $\mathsf{H}(AB, CD)$ and C is the midpoint of $\overline{AB}$, then D is an ideal point.

10. Let A, B, C, D, E be collinear. Show that:
   (a) $(AB, CD) = (AB, CE) \cdot (AB, ED)$.
   (b) $(AB, CD) = (AE, CD) \cdot (EB, CD)$.
   (c) $(AB, DE) = (AB, CD)/(AB, CE)$

11. Let O, A, B, C, A', B', and C' be collinear and

$$OA \cdot OA' = OB \cdot OB' = OC \cdot OC'.$$

Prove that $\mathsf{H}(AB', BC)$ if and only if $\mathsf{H}(A'B, B'C')$.

12. Prove that $\mathsf{H}(AB, CD)$ if and only if

$$\frac{1}{AB} = \frac{1}{2}\left(\frac{1}{AC} + \frac{1}{AD}\right).$$

13. For the given sets of three points, find the Cartesian coordinates of the harmonic conjugate of C with respect to A and B.
   (a) $A(0, 0)$, $B(3, 0)$, $C(8, 0)$
   (b) $A(-1, 3)$, $B(1, 2)$, $C(5, 0)$

14. Write the duals for Theorems 14.2.8 and 14.2.9.

15. Let a, b, and c be concurrent lines. Explain the procedure for finding the line d that is the harmonic conjugate of the given lines.

16. Prove that given concurrent lines a, b, and c, there exists a unique line d such that $\mathsf{H}(ab,\ cd)$.

17. Suppose that l intersects concurrent lines a, b, c, and d in points A, B, C, and D. Prove that $\mathsf{H}(AB,\ CD)$ if and only if $\mathsf{H}(ab,\ cd)$.

14.3 PERSPECTIVITY

Poncelet made new contributions to the projection of Desargues. Let $(\mathcal{P},\ \mathcal{L})$ be a projective plane, and take $l,\ m \in \mathcal{L}$, $O \in \mathcal{P}$ but O is concurrent with neither l nor m. Define $T : l \to m$ such that $T(A) = A'$ if and only if A and A' are collinear with O (Figure 14.7). The function T is called a **central perspectivity** and O is the **center of perspectivity**. We represent $T(A) = A'$ by

$$A \overset{O}{\barwedge} A'.$$

Read this as "A and A' are in **perspective** with respect to O."

The central perspectivity is the projection of the earlier projective geometers (Section 9.1), and it is an invertible function.

■ THEOREM 14.3.1

A central perspectivity is a one-to-one correspondence.

PROOF

Let $T : l \to m$ be a central perspectivity with center O. Let A and B be on l and assume that

$$T(A) = T(B) = A'.$$

If $A \neq B$, we conclude that there are two distinct lines that pass through O and A'. This contradicts Postulate 14.1.1.1. Hence, $A = B$, and we conclude that T is one-to-one. To prove that T is onto, let A' be incident with m and choose A to be the intersection of $\overline{OA'}$ with l. This means that $T(A) = A'$. ■

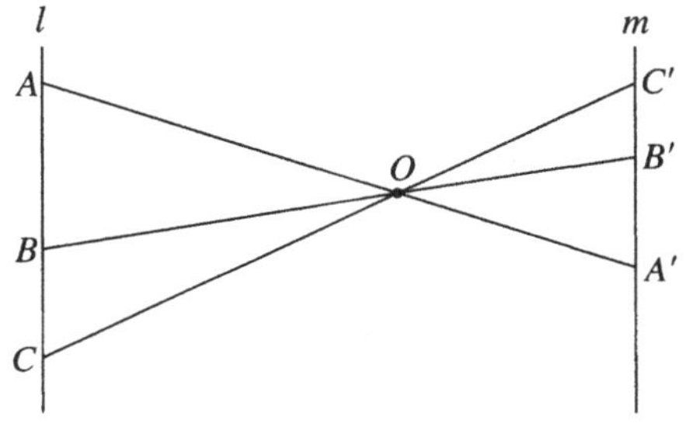

Figure 14.7 A central perspectivity.

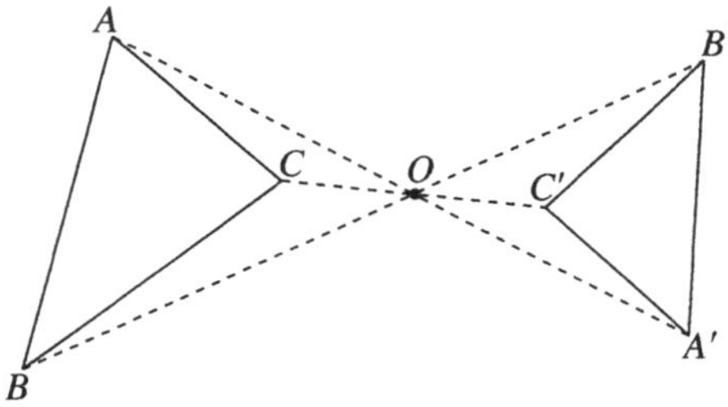

(a) $\triangle ABC$ and $\triangle A'B'C'$ are perspective from P

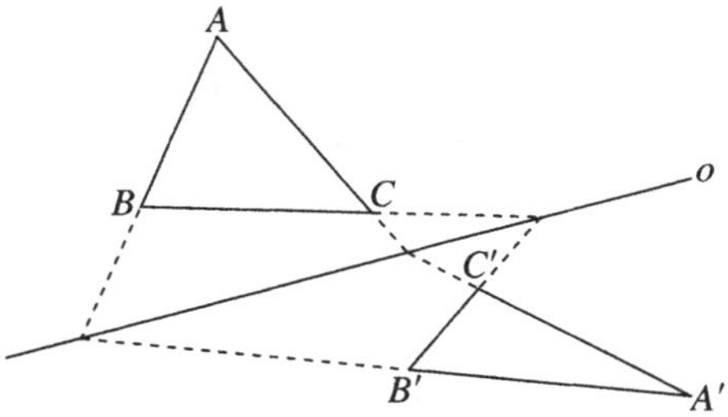

(b) $\triangle ABC$ and $\triangle A'B'C'$ are perspective from p

Figure 14.8 Desargues' Theorem.

Although a perspectivity is a function between two lines, there is a natural way to generalize it to plane figures (Figure 14.8).

■ **DEFINITION 14.3.2**

Two figures in the plane are **perspective from a point** if there exists a point O in the plane such that any line joining corresponding points from the figures contain O.

The dual of two figures perspective from a point is perspective from a line.

■ **DEFINITION 14.3.3**

Two figures in the plane are **perspective from a line** if there exists a line o in the plane such that all corresponding lines from the figures intersect at o. The line o is called the **axis of perspectivity**.

The new terminology allows us to restate Desargues' Theorem (9.3.1 and 9.3.2). Remember that it was required in the proof to pass to the third dimension. Since we are limiting ourselves to the projective plane, a proof would require an additional postulate justifying a key step (Coxeter 2003, 18–20). We will, nonetheless, still label it as a theorem.

■ **THEOREM 14.3.4 [Desargues]**

Two triangles are perspective from a point if and only if they are perspective from a line.

Projectivities

Since perspectivities are functions, we can compose them under the right conditions. This amounts to taking a sequence of points $O_1, O_2, \ldots, O_k$ and lines $l_1, l_2, \ldots, l_k, l_{k+1}$ such that O_i is not on l_i or l_{i+1} and defining perspectivities $T_i : l_i \to l_{i+1}$ with center of perspectivity O_i. The resulting function plays a central role in the study of projective geometry.

■ **DEFINITION 14.3.5**

The composition of central perspectivities is called a **projectivity**.

Since the composition of one-to-one and onto functions is one-to-one and onto, we conclude that projectivities are one-to-one correspondences and thus invertible.

As an example, let l and m be two lines with $A, B, C \in l$ and $A', B', C' \in m$ (Figure 14.9). We claim that we can find a projectivity T between l and m such that

$$T(A) = A', T(B) = B', T(C) = C',$$

and T is the composition of two central perspectivities. To do this, find C'' not on l or m. Join $\overline{AA'}$ and $\overline{CC''}$ and let O be their intersection. Let B'' be the intersection of $\overline{BO}$ and $\overline{C''A'}$. Now join $\overline{C''C'}$ and $\overline{B''B'}$ and let their intersection by O'. We now have two central perspectivities, one with center O and the other with center O'. Their composition is our desired T because

$$A \overset{O}{\barwedge} A' \overset{O'}{\barwedge} A', \ B \overset{O}{\barwedge} B'' \overset{O'}{\barwedge} B', \text{ and } C \overset{O}{\barwedge} C'' \overset{O'}{\barwedge} C'.$$

For a second example we write the identity map on a line l as the composition of three central perspectivities. Let A, B, and C be points on l (Figure 14.10). Take

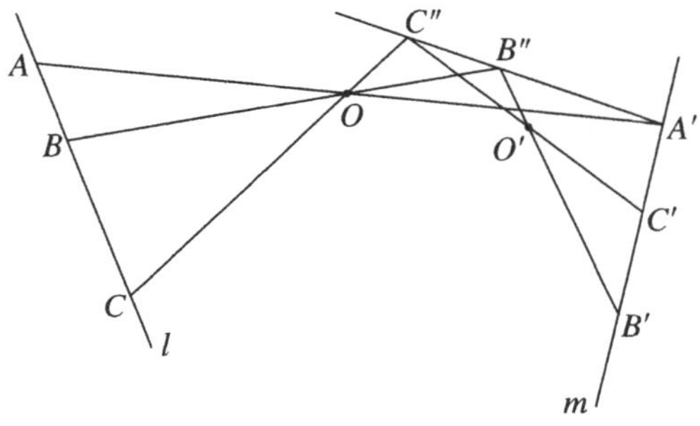

Figure 14.9 A projectivity that is the composition of two central perspectivities.

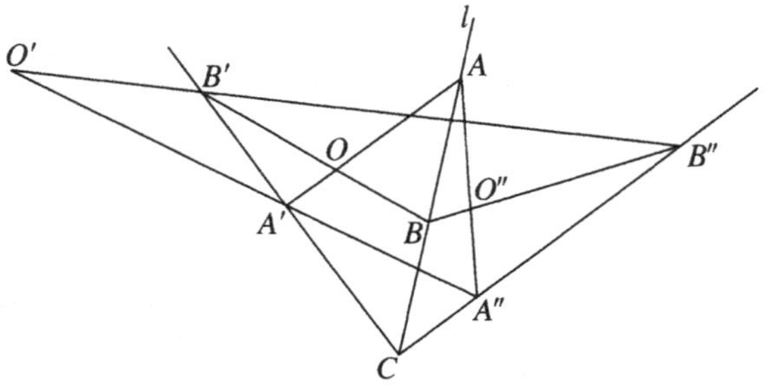

Figure 14.10 The identity map is the composition of three perspectivities.

B' and B'' not on l. Let A' be on $\overline{B'C}$ but not equal to either B' or C. Let O be the intersection of $\overline{A'A}$ with $\overline{B'B}$. Find a new point A'' on $\overline{CB''}$ and let O'' be the intersection of $\overline{AA''}$ and $\overline{BB''}$. Finally, let O' be the intersection of $\overline{A'A''}$ and $\overline{B'B''}$. This defines three perspectivities. Their centers are O, O', and O'', so we may write

$$A \overset{O}{\barwedge} A' \overset{O'}{\barwedge} A'' \overset{O''}{\barwedge} A$$

and similarly for B and C. These three points are invariants.

■ **DEFINITION 14.3.6**

A point that is mapped to itself under a projectivity is called **invariant**.

The question now is whether the composition of these three perspectivities yields the identity on l. If we pick a few random points from l and apply the projectivity (Exercise 14), we discover that the checked points are invariant. This, of course, is far from a proof.

Harmonic Nets

As Desargues was interested in what properties were preserved by projections, modern projective geometry asks what properties are preserved by projectivities. These are essentially the same question, so we should expect the next theorem. Compare this with Theorem 9.4.2. We state it without proof.

■ **THEOREM 14.3.7**

Let T be a projectivity with A, B, C, D in the domain of T. If $\mathsf{H}(AB, CD)$, then $\mathsf{H}(T(A)T(B), T(C)T(D))$.

Suppose that T is a projectivity with the line l as its domain. Take A, B, $C \in l$. By Theorem 14.2.9, there exists a unique $D \in l$ such that $\mathsf{H}(AB, CD)$. Therefore, by Theorems 14.3.7 and 14.2.9, $T(D)$ is the harmonic conjugate of $T(C)$ with respect to $T(A)$ and $T(B)$, and $T(D)$ could equal no other point. Therefore,

T is determined uniquely on $\{A, B, C, D\}$
because it is determined on $\{A, B, C\}$.

There also are unique points E and F in the domain of T and not among $\{A, B, C, D\}$ such that $\mathsf{H}(AC, BE)$ and $\mathsf{H}(BC, AF)$ (Exercise 15). This means that $T(E)$ and $T(F)$ are determined uniquely because projectivities preserve harmonic sets. Hence,

T is determined uniquely on $\{A, B, C, D, E, F\}$
because it is determined on $\{A, B, C\}$.

We may continue in this manner and either exhaust all of the points of l or find a countably infinite collection of points of l on which T is determined uniquely. Call the resulting set a **harmonic net** of l. All of this is restated as the next theorem.

■ THEOREM 14.3.8

There exists a unique projectivity that maps three distinct points of a harmonic net onto three distinct points in a given order.

The identity map I on l is a projectivity (Exercise 8) that leaves all points on a harmonic net invariant. If a projectivity T with domain l leaves three points invariant on the same harmonic net, we know that T and I must agree on all points of the harmonic net. This proves the corollary.

■ COROLLARY 14.3.9

A projectivity that leaves three points invariant on a harmonic net of points leaves all points invariant on the same harmonic net.

The corollary also implies Theorem 14.3.8 (Exercise 18), so the two statements are equivalent. Our ultimate goal is to generalize these equivalent statements so that they hold on the entire domain of a projectivity. If the domain is finite, the domain is a harmonic net, and the generalization follows easily. Otherwise, we are left in an impossible situation. It turns out that any attempt at generalizing Theorem 14.3.8 or its corollary requires knowing that the harmonic net is distributed on the domain of a projectivity in much the same way that the rational numbers are spread out on the real line. This seems to require the concepts of distance and convergence that have been deliberately left out of the projective postulates. Therefore, since we think both generalizations are reasonable, we will choose one of them as an assumption and prove that the other follows.

■ POSTULATE 14.3.10

If a projectivity leaves three points invariant, it is the identity.

With this we may prove the central theorem of Poncelet's new geometry.

■ THEOREM 14.3.11 [Fundamental Theorem of Projective Geometry]

A projectivity is determined uniquely by three points.

PROOF

Let T and T' be projectivities with domain l. Assume that $T(A) = T'(A) = A'$, $T(B) = T'(B) = B'$, and $T(C) = T'(C) = C'$, but $T(D) = D'$ and $T'(D) = D''$ with $D' \neq D''$. We know that T^{-1} exists (Theorem 14.3.1), so we check:

$$(T' \circ T^{-1})(A') = T'(T^{-1}(A')) = T'(A) = A'.$$

Similarly, B' and C' are invariant under $T' \circ T^{-1}$. But $(T' \circ T^{-1})(D') = D''$. This contradicts Postulate 14.3.10 because $T' \circ T^{-1}$ is a projectivity that holds three points invariant but is not the identity. ■

The Fundamental Theorem answers the question posed by the example on page 504 in the affirmative because the composition defined there and the identity are projectivities that agree on three points. The Fundamental Theorem also shows that the projectivity found in the first example on page 503 is unique (Exercise 17).

Exercises

1. Show that a central perspectivity has exactly one center if the perspectivity's domain and range are different.

2. Is it possible for a central perspectivity to have two centers if its domain and range share a point?

3. Prove that a central perspectivity is the identity map if the domain equals the range.

4. Let l and m be lines with $A,\ B \in l$ and $A',\ B' \in m$. Find a central perspectivity with center O such that

$$A \overset{O}{\barwedge} A' \text{ and } B \overset{O}{\barwedge} B'.$$

5. Show that a central perspectivity is determined uniquely by two points.

6. The dual of a central perspectivity is an **axial perspectivity**.
   (a) Define an axial perspectivity.
   (b) Without simply writing the dual of Theorem 14.3.1, show that an axial perspectivity is a one-to-one correspondence.
   (c) What is the dual of a projectivity?

7. Define central perspectivity and axial perspectivity using the terms *pencil* and *range*.

8. Show that the identity map on a line is a projectivity.

9. Prove that a projectivity $T : l \to l$ can be written as the composition of three central perspectivities.

10. Let T be a projectivity that leaves exactly one point invariant. Prove that T is a perspectivity.

11. Define $(ABCD) \barwedge (A'B'C'D')$ to mean that there exists a projectivity T such that $T(A) = A'$, $T(B) = B'$, $T(C) = C'$, and $T(D) = D'$. Show for any collinear A, B, C, and D, we have $(ABCD) \barwedge (BADC) \barwedge (CDAB) \barwedge (DCBA)$.

12. Prove that $(AECF) \barwedge (BDCF)$ if and only if $\mathsf{Q}(ABC,\ DEF)$.

13. Draw an example of a projectivity that leaves exactly two points invariant.

14. Choose a point on l in Figure 14.10 and confirm that it is invariant under the composition of the three perspectivities.

15. Assume $\mathsf{H}(AB,\ CD)$, where A, B, C, and D are on a line l. Show that there exists E and F on l and not among $\{A,\ B,\ C,\ D\}$ such that $\mathsf{H}(AC,\ BE)$ and $\mathsf{H}(BC,\ AF)$.

16. Prove that a harmonic net is determined uniquely by three points.

17. Draw a line l containing the points A, B, and C and another line containing A', B', and C'. Let T and T' be projectivities such that

$$T(A) = T'(A) = A',\ T(B) = T'(B) = B',\ \text{and } T(C) = T'(C) = C'.$$

Draw T as the composition of two central perspectivities with centers O and O' as on page 503. Also, draw T' as the composition of two central perspectivities with centers P and P'. Pick a point D on l and confirm that $T(D) = T'(D)$.

18. Prove that Corollary 14.3.9 implies Theorem 14.3.8.

14.4 HOMOGENEOUS COORDINATES

To assign points using Cartesian coordinates, we fix a reference point called the origin, draw two perpendicular lines through it, and then by measuring distances horizontally and vertically along a grid of parallel lines, a point is labeled using an ordered pair consisting of those distances. But how do you assign coordinates in a system that does not allow for angle measure, has no parallel lines, and has no concept of distance? We use triangulation and the methods of linear algebra.

Points and Lines

Let E_1, E_2, and E_3 be points in a projective plane. Label E_1 as $(1,\ 0,\ 0)$, E_2 as $(0,\ 1,\ 0)$, and E_3 as $(0,\ 0,\ 1)$, as in Figure 14.11. Join $\overline{E_1E_2}$, $\overline{E_2E_3}$, and $\overline{E_1E_3}$ to form the **reference triangle**. Finding new points on these three lines requires the introduction of a fourth point E that we will label $(1,\ 1,\ 1)$ and place in the middle of the reference triangle. We call E the **unit point**. Joining $\overline{EE_1}$ results in a new point A on $\overline{E_2E_3}$ because any two lines in a projective plane intersect in a point (Postulate 14.1.1.2). Similarly, $\overline{EE_2}$ and $\overline{EE_3}$ result in points B and C on $\overline{E_1E_3}$ and $\overline{E_1E_2}$, respectively.

We know that given three points on a line, we can construct another point so that the four points are a harmonic set. Since the harmonic conjugate is unique, we have

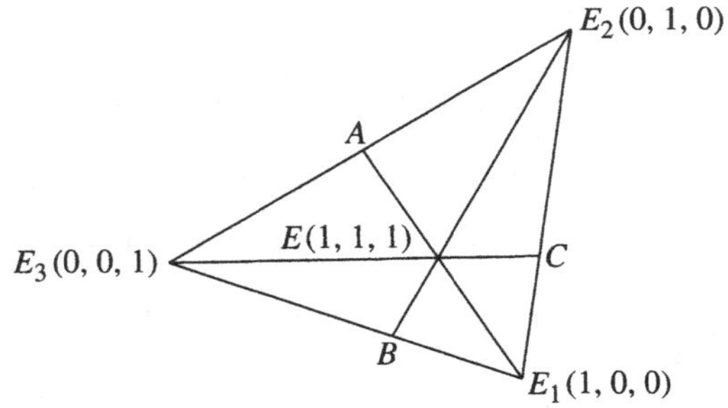

Figure 14.11 The reference triangle and the unit point.

a method, without a metric for either segments or angles, to generate points uniquely in a projective plane. It works like this. Start with three collinear points X, X_0, and X_1. If we choose X_2 such that $\mathsf{H}(XX_1, X_0X_2)$, X_2 will be between X_1 and X (Figure 14.12). Similarly, we may find X_3 between X_2 and X if $\mathsf{H}(XX_2, X_1X_3)$. We thus find a sequence

$$X_1, X_2, X_3, \ldots, X_n, \ldots$$

continuing indefinitely toward X provided that

$$\mathsf{H}(XX_{n-1}, X_{n-2}X_n)$$

for $n > 2$. To find a point $X_{(m+n)/2}$ between a pair of points X_m and X_n, use

$$\mathsf{H}(XX_{(m+n)/2}, X_mX_n).$$

If the process continues indefinitely, it will generate a harmonic net, which fills the line as the rational numbers fill $\mathbb{R}$. For this reason a harmonic net is often called a **net of rationality**.

Now that we know how to find points on a line in the projective plane, we define an analytic geometry so that we can write equations to identify the points and lines. For every point P in the plane, assign a set of ordered triples of the form

$$(kx_1, kx_2, kx_3),$$

where kx_1, kx_2, kx_3 are real numbers, not all zero. For example, the ordered triples $(1, 2, 3)$ and $(4, 8, 12)$ identify the same point. Because all nonzero multiples of (x_1, x_2, x_3) represent the same point, we say that (kx_1, kx_2, kx_3) is written using **homogeneous coordinates**. Lines in this system will be identified by equations of the form

$$a_1x_1 + a_2x_2 + a_3x_3 = 0,$$

where a_1, a_2, and a_3 are not all zero. This resembles the equation for a line using Cartesian coordinates except that the third term is not a constant.

To see how this works, let us find the equation for the line joining $(1, 2, 3)$ and $(2, 4, 1)$. To accomplish this, we must find real numbers a_1, a_2, and a_3 so that

$$\begin{array}{rcrcrcl} a_1 & + & 2a_2 & + & 3a_3 & = & 0, \\ 2a_1 & + & 4a_2 & + & a_3 & = & 0. \end{array}$$

This is equivalent to forming the matrix

$$\begin{bmatrix} 1 & 2 & 3 \\ 2 & 4 & 1 \end{bmatrix},$$

and putting it into row-reduced echelon form,

$$\begin{bmatrix} 1 & 2 & 0 \\ 0 & 0 & 1 \end{bmatrix}.$$

From this we find that the coefficients for the equation of the line are

$$a_1 = -2k, a_2 = k, \text{ and } a_3 = 0$$

for all nonzero k. We arbitrarily choose $k = -1$, so the equation for the line between $(1, 2, 3)$ and $(2, 4, 1)$ can be written as

$$2x_1 - x_2 = 0.$$

Any point (x_1, x_2, x_3) that satisfies this equation will lie on the line. Because of the principle of duality, this line should receive a name like the points. The natural choice is to use the coefficients, so call the line $[2, -1, 0]$. Because of the parameter k, each triple $[2k, -k, 0]$ names the same line for all possible values of $k \neq 0$. Evidently, the sides of the reference triangle are $[1, 0, 0]$, $[0, 1, 0]$, and $[0, 0, 1]$ [Exercise 3(b)].

Since every line has an equation, by duality every point also should have an equation. Take lines $[1, 2, 3]$ and $[2, 4, 1]$. The equations for these lines are

$$\begin{array}{rcrcrcl} x_1 & + & 2x_2 & + & 3x_3 & = & 0, \\ 2x_1 & + & 4x_2 & + & x_3 & = & 0. \end{array}$$

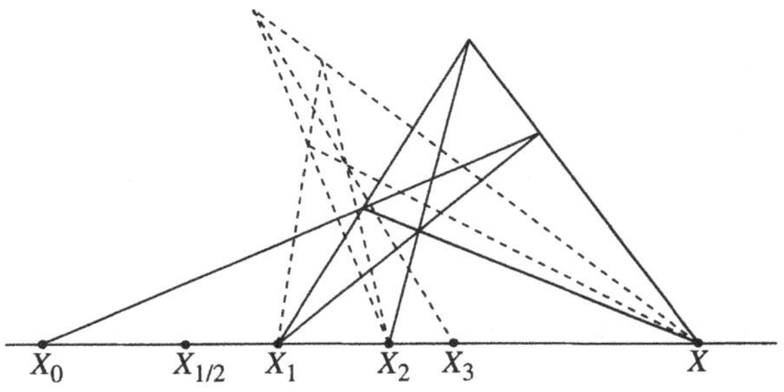

Figure 14.12 The points on a line form a harmonic net.

The solution to this linear system of equations is

$$x_1 = -2k,\ x_2 = k, \text{ and } x_3 = 0.$$

Therefore, the point of intersection can be taken to be $(2,\ -1,\ 0)$, and its equation is

$$2a_1 - a_2 = 0.$$

Any line $[a_1,\ a_2,\ a_3]$ that satisfies this equation will contain the point $(-2k,\ k,\ 0)$, provided that $k \neq 0$.

We may then combine these two concepts in the next definition.

DEFINITION 14.4.1

$(x_1,\ x_2,\ x_3)$ is **incident** with $[a_1,\ a_2,\ a_3]$ or $[a_1,\ a_2,\ a_3]$ **contains** $(x_1,\ x_2,\ x_3)$ if and only if

$$a_1x_1 + a_2x_2 + a_3x_3 = 0.$$

The definition makes clear why the ordered triple $(0,\ 0,\ 0)$ is excluded from our system of homogenous coordinates. Because $(0,\ 0,\ 0)$ satisfies the equation

$$a_1x_1 + a_2x_2 + a_3x_3 = 0$$

for all real numbers a_1, a_2, and a_3, the point that $(0,\ 0,\ 0)$ represents would be incident with every line. This is impossible in the projective plane (Exercise 1).

Incidence Results

Let $\mathcal{P} = \{P_1,\ P_2,\ \ldots,\ P_n\}$ be a set of points written using homogeneous coordinates. We define the set $\mathcal{P}$ to be **linearly independent** if it is linearly independent when viewed as a set of vectors. If $\mathcal{P}$ is not linearly independent, it is **linearly dependent**. Since $(x_1,\ x_2,\ \ldots,\ x_n)$ and $(kx_1,\ kx_2,\ \ldots,\ kx_n)$ with $k \neq 0$ represent the same point, two points in homogeneous coordinates are the same point if and only if they form a linearly dependent set. For three points, we have the following.

THEOREM 14.4.2

The set of points $\{(x_1,\ x_2,\ x_3),\ (y_1,\ y_2,\ y_3),\ (z_1,\ z_2,\ z_3)\}$ written in homogeneous coordinates is linearly dependent if and only if the points are collinear.

PROOF

Suppose that the three points are collinear. This means that there exists a line $[a_1,\ a_2,\ a_3]$ that is a solution to the equations for each of the three points:

$$\begin{aligned} x_1a_1 &+ x_2a_2 + x_3a_3 = 0, \\ y_1a_1 &+ y_2a_2 + y_3a_3 = 0, \\ z_1a_1 &+ z_2a_2 + z_3a_3 = 0. \end{aligned}$$

The solution must be nontrivial in order for it to be a line, since $[0, 0, 0]$ does not represent a line. This can happen only if

$$\{(x_1, x_2, x_3), (y_1, y_2, y_3), (z_1, z_2, z_3)\}$$

is linearly dependent. Since these steps are reversible, we have proven the theorem. ■

A set of three vectors in $\mathbb{R}^3$ is linearly dependent if and only if the determinant of the matrix formed by the vectors of the set is zero.

■ **COROLLARY 14.4.3**

If (x_1, x_2, x_3), (y_1, y_2, y_3), (z_1, z_2, z_3) are in homogeneous coordinates,

$$\begin{vmatrix} x_1 & x_2 & x_3 \\ y_1 & y_2 & y_3 \\ z_1 & z_2 & z_3 \end{vmatrix} = 0$$

if and only if the points are collinear.

Theorem 14.4.2 and its corollary also hold for lines in the projective plane, by the principle of duality.

Equivalence Classes

Let $\sim$ be an equivalence relation on $\mathscr{A}$ (page 51). If $a \in \mathscr{A}$, the **equivalence class** of a is defined as the set

$$[a] = \{x \in \mathscr{A} : x \sim a\}.$$

The set of all equivalence classes,

$$\mathscr{A}/\!\sim\; = \{[a] : a \in \mathscr{A}\},$$

is called a **quotient set**, and we read $\mathscr{A}/\!\sim$ as "A modulo tilde." It is always the case that

$$\mathscr{A} = \bigcup_{a \in \mathscr{A}} [a].$$

The equivalence relation associates elements of the set that are considered the same type of element with respect to the equivalence relation. This has the effect of dividing the set up into cells that do not overlap, yet exhaust the entire set.

■ **DEFINITION 14.4.4**

Let $\mathscr{A} \neq \varnothing$ be a set and $\mathscr{P}$ be a set of subsets of $\mathscr{A}$. The set $\mathscr{P}$ is a **partition** of $\mathscr{A}$ if and only if

- $\bigcup \mathscr{P} = \mathscr{A}$.
- For all $\mathscr{B}$, $\mathscr{C} \in \mathscr{P}$, either $\mathscr{B} = \mathscr{C}$ or $\mathscr{B} \cap \mathscr{C} = \varnothing$.

The second condition of the definition means that $\mathcal{P}$ is **pairwise disjoint**. To illustrate this definition, let $\mathcal{A} = \{1, 2, 3, 4, 5, 6, 7\}$ and define $\mathcal{A}_1 = \{1, 2, 5\}$, $\mathcal{A}_2 = \{3\}$, and $\mathcal{A}_3 = \{4, 6, 7\}$. The family $\mathcal{P} = \{\mathcal{A}_1, \mathcal{A}_2, \mathcal{A}_3\}$ is pairwise disjoint, and $\mathcal{A} = \mathcal{A}_1 \cup \mathcal{A}_2 \cup \mathcal{A}_3$. Hence, $\mathcal{P}$ is a partition of $\mathcal{A}$. Similarly, the quotient set of an arbitrary set under an equivalence relation will be a partition for the set.

■ THEOREM 14.4.5

If $\sim$ is an equivalence relation on $\mathcal{A}$, then $\mathcal{A}/{\sim}$ is a partition of $\mathcal{A}$.

PROOF

Take a set $\mathcal{A}$ with an equivalence relation $\sim$. We first prove that $\bigcup(\mathcal{A}/{\sim}) = \mathcal{A}$. Suppose that $x \in \bigcup(\mathcal{A}/{\sim})$. This means that there exists $a \in \mathcal{A}$ such that $x \in [a]$. Since $[a] \subseteq \mathcal{A}$, x is an element of $\mathcal{A}$. To prove the converse, let $x \in \mathcal{A}$. Because $x \sim x$, x is an element of $[x]$. This implies that $x \in \bigcup(\mathcal{A}/{\sim})$ since $[x] \in \mathcal{A}/{\sim}$.

To show that $\mathcal{A}/{\sim}$ is pairwise disjoint, let $[a]$, $[b] \in \mathcal{A}/{\sim}$ and assume that $[a] \cap [b] \neq \varnothing$. We conclude that there exists $y \in [a] \cap [b]$. In other words, $y \sim a$ and $y \sim b$. Now take $x \in [a]$. This means that $x \sim a$. Since $a \sim y$ by symmetry, $x \sim y$ and then $x \sim b$, both by transitivity. Thus, $x \in [b]$, which shows that $[a] \subseteq [b]$. Similarly, $[b] \subseteq [a]$. Therefore, $[a] = [b]$. ■

We now define an equivalence relation on a set of homogeneous coordinates and examine the resulting partition. Let

$$\mathcal{H} = \{(x_1, x_2, x_3) : x_1, x_2, x_3 \in \mathbb{R} \text{ but not all zero}\}.$$

Since all multiples of a given ordered triple identify the same point, define

$$(x_1, x_2, x_3) \sim (y_1, y_2, y_3)$$

if and only if there exists $k \in \mathbb{R}$ such that

$$x_1 = ky_1,$$
$$x_2 = ky_2,$$
$$x_3 = ky_3.$$

To show that this is an equivalence relation, take (x_1, x_2, x_3), (y_1, y_2, y_3), and (z_1, z_2, z_3) in $\mathcal{H}$.

- $(x_1, x_2, x_3) \sim (x_1, x_2, x_3)$ because $x_1 = 1x_1$, $x_2 = 1x_2$, and $x_3 = 1x_3$.
- Assume that $(x_1, x_2, x_3) \sim (y_1, y_2, y_3)$. From this we conclude that there exists $k \in \mathbb{R}$ such that $x_1 = ky_1$, $x_2 = ky_2$, and $x_3 = ky_3$. Since not all of the coordinates are zero, k must be nonzero. Therefore,

  $$y_1 = (1/k)x_1,\ y_2 = (1/k)x_2, \text{ and } y_3 = (1/k)x_3,$$

  which means that $(y_1, y_2, y_3) \sim (x_1, x_2, x_3)$

- Let $(x_1, x_2, x_3) \sim (y_1, y_2, y_3)$ and $(y_1, y_2, y_3) \sim (z_1, z_2, z_3)$. Then there exists $j, k \in \mathbb{R}$ so that

  $$x_1 = jy_1, x_2 = jy_2, \text{ and } x_3 = jy_3$$

  and

  $$y_1 = kz_1, y_2 = kz_2, \text{ and } y_3 = kz_3.$$

  Hence,

  $$x_1 = (jk)z_1, x_2 = (jk)z_2, \text{ and } x_3 = (jk)z_3.$$

  Therefore, $(x_1, x_2, x_3) \sim (z_1, z_2, z_3)$.

Since $\sim$ is an equivalence relation on $\mathcal{H}$, the quotient set

$$\mathbb{P}^2 = \mathcal{H}/\sim$$

is a partition of $\mathcal{H}$. If we take $(x_1, x_2, x_3) \in \mathcal{H}$, the points in $\mathcal{H}$ that are equivalent to it are exactly those ordered triples that are nonzero multiples of (x_1, x_2, x_3). We conclude that

$$[(x_1, x_2, x_3)] = \{(kx_1, kx_2, kx_3) : k \neq 0\}.$$

Although there is not a unique ordered triple for every point in the projective plane, there is a unique equivalence class. When we work with these classes, we typically choose a particular element of the class, called a **representative**, and label the point with it. The quotient set $\mathbb{P}^2$ is called the **real projective plane**. It is left to Exercise 2 to confirm that $\mathbb{P}^2$ satisfies the postulates of projective geometry.

Of course, everything just written about points holds also for lines, so we have developed a complete system to label the objects in a projective plane. Although homogeneous coordinates were created independently by several mathematicians, including August Möbius, the full system is due to the German mathematician Julius Plücker. He was born in Elberfeld in 1801 and lived for 67 years. He studied at various universities throughout Europe, earning his doctorate from the University of Marburg in 1824. His mathematical specialty was analytic geometry, which he then applied to the projective plane. Since most methods in projective geometry at the time were synthetic, as appears to be the nature of the two camps, controversy again arose over which of the two methods was best. Plücker sided with the analysts, a decision which at one point cost him a residency position at Berlin. Despite the controversy, Plücker became a top German mathematician with a reputation for elegant work. His other interests included physics, on which he had a lasting correspondence with Michael Faraday (Burau and Schoenberg 1970, 44, 46; Boyer and Merzbach 1991, 540–543).

Collineations

The functions that we have examined so far both synthetically and analytically in a projective plane have been functions between ranges (e.g., central perspectivities) or between pencils (e.g., axial perspectivities in Exercise 6 of Section 14.3). We now want to define a function whose domain is an entire projective plane but still resembles the other functions. We will do our work on $\mathbb{P}^2$, and the function will be a collineation (Definition 14.1.5).

Suppose that $\phi : \mathbb{P}^2 \to \mathbb{P}^2$ is a collineation and write

$$\begin{aligned}\phi(1, 0, 0) &= (x_1, y_1, z_1),\\ \phi(0, 1, 0) &= (x_2, y_2, z_2),\\ \phi(0, 0, 1) &= (x_3, y_3, z_3),\\ \phi(1, 1, 1) &= (x_4, y_4, z_4).\end{aligned}$$

This function has the property that it maps all points on a line to all of the points of another line, so ϕ must be a linear transformation. Therefore, there exists a 3×3 matrix A such that

$$\phi(X) = AX,$$

where X is a point from $\mathbb{P}^2$. Since ϕ is a one-to-one function, the determinant of A is not equal to zero.

As an example, define $\phi : \mathbb{P}^2 \to \mathbb{P}^2$ by

$$\begin{aligned}\phi(1, 0, 0) &= (1, -2, 3),\\ \phi(0, 1, 0) &= (5, 6, 5),\\ \phi(0, 0, 1) &= (1, 0, 4),\\ \phi(1, 1, 1) &= (3, 2, 1).\end{aligned}$$

Observe that

$$\phi(1, 1, 1) = \phi(1, 0, 0) + \frac{2}{3}\phi(0, 1, 0) - \frac{4}{3}\phi(0, 0, 1)$$

or

$$3\phi(1, 1, 1) = 3\phi(1, 0, 0) + 2\phi(0, 1, 0) - 4\phi(0, 0, 1).$$

Therefore,

$$3\begin{bmatrix} x_1' \\ x_2' \\ x_3' \end{bmatrix} = \begin{bmatrix} 3 & 10 & -4 \\ -6 & 12 & 0 \\ 9 & 10 & -16 \end{bmatrix}\begin{bmatrix} x_1 \\ x_2 \\ x_3 \end{bmatrix},$$

which gives rise to this system of linear equations:

$$\begin{aligned}3x_1' &= 3x_1 + 10x_2 - 4x_3,\\ 3x_2' &= -6x_1 + 12x_2,\\ 3x_3' &= 9x_1 + 10x_2 - 16x_3.\end{aligned}$$

We see that we can generalize this to the next proposition.

THEOREM 14.4.6

The collineation $\phi : \mathbb{P}^2 \to \mathbb{P}^2$, where $\phi(x_1, x_2, x_3) = (x_1', x_2', x_3')$, is defined by the equations

$$\begin{aligned} kx_1' &= a_{11}x_1 + a_{12}x_2 + a_{13}x_3, \\ kx_2' &= a_{21}x_1 + a_{22}x_2 + a_{23}x_3, \\ kx_3' &= a_{31}x_1 + a_{32}x_2 + a_{33}x_3 \end{aligned}$$

such that $k \neq 0$ and

$$\begin{vmatrix} a_{11} & a_{12} & a_{13} \\ a_{21} & a_{22} & a_{23} \\ a_{31} & a_{32} & a_{33} \end{vmatrix} \neq 0.$$

Finally, let ϕ be a collineation on $\mathbb{P}^2$. Observe that if we take lines l and m in $\mathbb{P}^2$, their images $\phi[l]$ and $\phi[m]$ are also lines in $\mathbb{P}^2$. Let O be a point not on l or m, so $\phi(O)$ is not on $\phi[l]$ or on $\phi[m]$. If A is on m and B on m such that A, O, and B are collinear, $\phi(A)$ is on $\phi[l]$, $\phi(B)$ is on $\phi[m]$, and $\phi(A)$, $\phi(O)$, and $\phi(B)$ are collinear. All of this follows because ϕ is a collineation. Therefore, if

$$T : l \to m$$

is a central perspectivity,

$$\phi \circ T \circ \phi^{-1} : \phi[l] \to \phi[m]$$

is a central perspectivity. It is then reasonable to say that perspectivities are preserved under collineations. Similarly, since a projectivity is the composition of perspectivities, we can also say that projectivities are preserved under collineations. For this reason, a collineation on $\mathbb{P}^2$ is usually called a **projective collineation**.

Exercises

1. Explain why it is not possible that a point exists in the projective plane that is on every line.
2. Confirm that $\mathbb{P}^2$ is a projective plane.
3. Using Figure 14.11:
   (a) Find the coordinates of points A, B, and C.
   (b) Show that the coordinates for $\overline{E_2E_3}$, $\overline{E_1E_3}$, and $\overline{E_1E_2}$ are, respectively, $[1, 0, 0]$, $[0, 1, 0]$, and $[0, 0, 1]$.
   (c) Find the coordinates for $\overline{AE_1}$, $\overline{BE_2}$, and $\overline{CE_3}$.
   (d) Find two more points between B and E_1 and two more between E_3 and B.
   (e) Find the intersection of $\overline{AC}$ and $\overline{BE_2}$.
4. Find the homogeneous coordinates for the line between $(3, 0, 6)$ and $(0, 4, 1)$.
5. Find the equation for the intersection of $[6, -4, 2]$ and $[1, 1, 8]$.

6. Determine whether the following points are collinear.
   (a) $(3, 0, 0), (3, -4, 0), (1, 1, 1)$
   (b) $(9, 8, -5), (2, -8, 9), (2, -6, 9), (2, -8, 4)$

7. Find ordered triples (a_1, a_2, a_3), (b_1, b_2, b_3), and (c_1, c_2, c_3) in homogeneous coordinates that are multiples of the given points and such that

$$(a_1, a_2, a_3) = (b_1, b_2, b_3) + (c_1, c_2, c_3).$$

   (a) $(0, 0, 5), (1, 1, 1), (7, 7, 0)$
   (b) $(-2, 0, 1), (1, -1, 2), (4, -2, 3)$
   (c) $(1, 2, 3), (0, 1, 2), (-1, 0, 1)$

8. Using Figure 14.11:
   (a) Find the lines of the complete quadrangle with vertices E, C, E_1, and B.
   (b) Confirm that the diagonal points are not collinear.
   (c) Generalize this to show that the diagonal points of any complete quadrangle are not collinear.

9. For all nonzero real numbers a and b, let $a \sim b$ if and only if $ab > 0$.
   (a) Show that $\sim$ is an equivalence relation on $\mathbb{R}\backslash\{0\}$.
   (b) Find the equivalence classes $[1]$ and $[-3]$.

10. For all $a, b \in \mathbb{Z}$, let $a \sim b$ if and only if $|a| = |b|$.
   (a) Prove that $\sim$ is an equivalence relation on $\mathbb{Z}$.
   (b) Describe $\mathbb{Z}/\sim$.

11. For all $(a, b), (c, d) \in \mathbb{Z}\backslash\{0\} \times \mathbb{Z}\backslash\{0\}$, define $(a, b) \sim (c, d)$ if and only if $ad = bc$.
   (a) Show that $\sim$ is an equivalence relation on $\mathbb{Z}\backslash\{0\} \times \mathbb{Z}\backslash\{0\}$.
   (b) What is the equivalence class of $(1, 2)$?
   (c) Describe $(\mathbb{Z}\backslash\{0\} \times \mathbb{Z}\backslash\{0\})/\sim$.

12. Prove that the following are partitions of $\mathbb{R}^2$.
   (a) $\mathscr{E} = \{\{(a, b)\} : a, b \in \mathbb{R}\}$.
   (b) $\mathscr{F} = \{V_r : r \in \mathbb{R}\}$, where $V_r = \{(r, y) : y \in \mathbb{R}\}$ for all real numbers r.
   (c) $\mathscr{H} = \{H_n : n \in \mathbb{Z}\}$, where $H_n = \{\mathbb{R} \times (n, n+1] : n \in \mathbb{Z}\}$ for all $n \in \mathbb{Z}$.

13. Is $\{[n, n+1] \times (n, n+1) : n \in \mathbb{Z}\}$ a partition of $\mathbb{R}^2$? Explain.

14. Define the collineation $\phi : \mathbb{P}^2 \to \mathbb{P}^2$ by $\phi(X) = AX$, where

$$A = \begin{bmatrix} 1 & 1 & 1 \\ 0 & 2 & 1 \\ 1 & 0 & 1 \end{bmatrix}.$$

   (a) Find $\phi(1, 2, 3)$ and $\phi[4, 5, 6]$.
   (b) Write the image of the unit point as a linear combination of the images of the vertices of the reference triangle.

(c) Find $\phi^{-1}(4, 0, 1)$.

(d) Find the intersection of $\phi^{-1}[-1, 4, 6]$ and $\phi^{-1}[0, 0, 2]$.

15. Let $\psi : \mathbb{P}^2 \to \mathbb{P}^2$ be a collineation such that $\psi(1, 0, 0) = (2, -2, 8)$, $\psi(0, 1, 0) = (2, 5, 1)$, $\psi(0, 0, 1) = (2, 2, 4)$, and $\psi(1, 1, 1) = (0, 1, -1)$.

(a) Find the matrix for ψ.

(b) Write ψ as a system of linear equations.

16. Let $\phi : \mathbb{P}^2 \to \mathbb{P}^2$ be a collineation such that $\phi(2, 1, 1) = (3, 1, -5)$, $\phi(2, -1, 1) = (1, 1, -3)$, $\phi(1, 2, 1) = (-1, 0, 2)$, $\phi(5, 2, 3) = (6, 4, -12)$.

(a) Find the matrix for $\psi_1 : \mathbb{P}^2 \to \mathbb{P}^2$ such that $\psi(1, 0, 0) = (2, 1, 1)$, $\psi(0, 1, 0) = (2, -1, 1)$, $\psi(0, 0, 1) = (1, 2, 1)$, $\psi(1, 1, 1) = (5, 2, 3)$.

(b) Find the matrix for $\psi_2 : \mathbb{P}^2 \to \mathbb{P}^2$ such that we have $\psi(1, 0, 0) = (3, 1, -5)$, $\psi(0, 1, 0) = (1, 1, -3)$, $\psi(0, 0, 1) = (-1, 0, 2)$, and $\psi(1, 1, 1) = (6, 4, -12)$.

(c) Use the matrices for ψ_1 and ψ_2 to find the matrix for ϕ.

(d) Write ϕ as a system of linear equations.

17. Let $E_1 = (1, 0, 0)$, $E_2 = (0, 1, 0)$, $E_3 = (0, 0, 1)$, and $E = (1, 1, 1)$. The point O at the intersection of $\overline{EE_3}$ and $\overline{E_1E_2}$ defines a central perspectivity between $\overline{E_1E_3}$ and $\overline{E_2E_3}$. After finding the homogeneous coordinates for A and B (Exercise 3), find an equation relating the homogeneous coordinates of A with those of B that must be satisfied in order for $A \overset{O}{\barwedge} B$ through O.

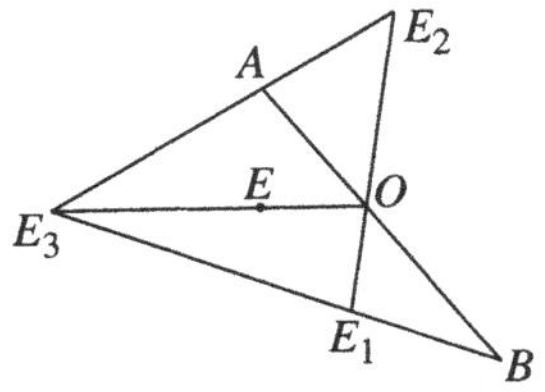

CHAPTER 15

FELIX KLEIN

After lecturing at Göttingen for a year, Klein joined the faculty at the University of Erlangen in 1872. The public lecture that earned him this position he titled *A Comparative Review of Recent Researches in Geometry* (Klein 1893). In it he introduced a new means of classifying geometries that became known as the **Erlangen Program**. It characterized a geometry based on the properties that were preserved under certain sets of functions. This is similar to what Poncelet had done in his *Traite*. There Poncelet looked for properties that were preserved under projectivities. One such property was that of being a harmonic set. Klein's scheme would begin with the projective transformations, and then by placing restrictions on these functions, find other collections that would yield their own types of invariants on other geometries. Each of the sets of functions were required to satisfy certain conditions, making them into what is called a group. Klein had learned of this algebra concept from Camille Jordan when he traveled to Paris with Sophus Lie (Yaglom 1988, 128–129; Burau and Schoenberg 1970, 396–397; Wussing 1984, 27).

Lie was from Norway and was seven years older than Klein. Lie was an interesting combination of outdoorsman and overall success in all of the academic subjects. When it came time for college, he had his choice of fields. Lie's father was a Lutheran pastor, so Lie considered studying theology, but decided instead on mathematics and the sciences at Christiania University. It was there where he read the works of Poncelet and Plücker. These two mathematicians had a lasting impact on Lie, and their work would inspire Lie's long research program on geometric transformations. Lie went to Berlin in 1870 to continue his mathematical education. It was in Berlin where Klein and Lie first met. They would become lifelong friends and work on many projects together (Yaglom 1988, 22; Burau and Schoenberg 1970, 396).

One of their first joint undertakings was to go to Paris and learn about current French mathematics. They arrived in late 1870. In Paris Klein and Lie studied geometry and learned about group theory. While there it was their intention to meet Camille Jordan, who had recently published *Commentaire sur le mémorie de Galois* in 1865 and *Commentaire sur Galois* in 1869. These were expositions on the work of Evariste Galois, a gifted mathematician who began a study of the subject when he was 15, developed an approach to deal with equations solvable by radicals, spent time in jail as a political prisoner, and was killed in a dual at the age of 21. Klein was particularly impressed with Jordan's *Traité des substitutions et des équations algébriques*. Inspired by what they learned, Klein and Lie coauthored two papers, one published in June 1870 and the other in March 1871. These would become the preliminary work for Klein's Erlangen lecture (Wussing 1984, 135–136, 180–184; Yaglom 1988, 22; Kline 1972, 755–756).

Klein and Lie left Berlin in July 1870, at the start of the Franco-Prussian War. Being German, Klein needed to leave the country. When he arrived home, he planned to join the Prussian army, but he caught typhus. When he recovered, Klein returned to Göttingen. All the while Lie remained in France to hike the French countryside, see the Alps, and then to see Italy. Although it was only for a month, he did spend some time in a French prison under suspicion of spying for Germany (Wussing 1984, 179; Yaglom 1988, 24, 27).

15.1 GROUP THEORY

So what is a group? Start with the set of integers $\mathbb{Z}$. If we add any two integers together, the result will be an integer, and this result will be unique to the original two integers. For example, the sum of 5 and 3 is 8 and always will be 8. The same thing goes for subtraction and multiplication. Division is a problem, however. We cannot divide by zero, and an integer divided by an integer is not guaranteed to be an integer. On the set of integers, addition, subtraction, and multiplication are examples of binary operations, whereas division is not.

■ **DEFINITION 15.1.1**

A **binary operation** on a set $\mathcal{S}$ is a function $* : \mathcal{S} \times \mathcal{S} \to \mathcal{S}$.

Since a binary operation is a generalization of the more concrete operations of addition and multiplication, we write

$$a * b$$

instead of $*(a, b)$. The following are examples of binary operations.

- $a * b = |ab|$ on $\mathbb{R}$.
- Addition, subtraction, and multiplication on $\mathbb{R}$.
- Division on $(-\infty, 0) \cup (0, \infty)$.
- $a * b = \sqrt{ab}$ on $[0, \infty)$.

To show that a rule on a set $\mathcal{S}$ is a binary operation, we must show that is satisfies two properties. First, it must be well defined as a function, and second, it must be true that $a * b \in \mathcal{S}$ for all $a, b \in \mathcal{S}$. When this second condition is satisfied, we say that $\mathcal{S}$ is **closed under** $*$.

Consider the following example. Denoting the set of positive integers by $\mathbb{Z}^+$, for all $n \in \mathbb{Z}^+$ and $a \in \mathbb{Z}$, define

$$[a]_n = \{a + kn : k \in \mathbb{Z}\}.$$

This is called the **congruence class** of a **modulo** n (see Exercise 12 from Section 2.3 and also page 511). For instance,

$$\begin{aligned}
[0]_4 &= \{\ldots, -4, 0, 4, 8, \ldots\}, \\
[1]_4 &= \{\ldots, -3, 1, 5, 9, \ldots\}, \\
[2]_4 &= \{\ldots, -2, 2, 6, 10, \ldots\}, \\
[3]_4 &= \{\ldots, -1, 3, 7, 11, \ldots\}.
\end{aligned}$$

The collection cycles because $[4]_4 = [0]_4$. We now define

$$\mathbb{Z}_n = \{[a]_n : a \in \mathbb{Z}\},$$

and on this set specify $+$ by

$$[a]_n + [b]_n = [a + b]_n.$$

Be careful to note that the symbol $+$ is being used in two ways. The $+$ on the left is the new definition, but the $+$ on the right is standard addition.

We first show that this addition is well defined. This is particularly important to prove this in this case since each congruence class can be represented using infinitely many different integers. Take $[a]_n, [b]_n, [a']_n, [b']_n \in \mathbb{Z}_n$ and suppose that

$$[a]_n = [a']_n \text{ and } [b]_n = [b']_n.$$

This does *not* mean that $a = a'$ and $b = b'$. Instead, we conclude that

$$a = a' + jn \text{ and } b = b' + kn$$

for some $j, k \in \mathbb{Z}$. To show that the result of operating these two equal pairs of elements together yields the same answer, we must show that

$$[a + b]_n = [a' + b']_n.$$

To do this, let $x \in [a+b]_n$. This means that $x = a + b + ln$ for some $l \in \mathbb{Z}$. Therefore,

$$x = a' + jn + b' + kn + ln = a' + b' + (j + k + l)n,$$

from which follows that $x \in [a' + b']_n$ because $j + k + l \in \mathbb{Z}$. Hence,

$$[a+b]_n \subseteq [a' + b']_n.$$

The other inclusion follows similarly.

It is a simple matter to see that $\mathbb{Z}_n$ is closed under $+$. Let $[a]_n, [b]_n \in \mathbb{Z}_n$. This means that a and b are integers. Since we know that $\mathbb{Z}$ is closed under standard addition, $a + b \in \mathbb{Z}$. Therefore, $[a]_n + [b]_n = [a+b]_n \in \mathbb{Z}_n$.

Groups

Now consider the properties of addition as viewed as a binary operation on the set of integers. We know that addition is commutative and associative (page 32). There is a particular element of $\mathbb{Z}$ that is denoted by 0 that has the property

$$\text{for all } a \in \mathbb{Z}, 0 + a = a + 0 = a.$$

The number 0 is the additive identity of $\mathbb{Z}$. Finally, every integer a has an additive inverse. This inverse is denoted by $-a$ and has the property that

$$a + (-a) = (-a) + a = 0.$$

In other words, an integer added to its inverse yields the identity. Ignoring the commutative condition for the moment, we make our definition.

DEFINITION 15.1.2

Let $\mathcal{G}$ be a set and $*$ a binary operation with the following properties:

- $a * (b * c) = (a * b) * c$ for all $a, b, c \in \mathcal{G}$.
- There exists $e \in \mathcal{G}$ such that $e * a = a * e = a$ for all $a \in \mathcal{G}$. The element e is called the **identity** of $\mathcal{G}$.
- For all $a \in \mathcal{G}$, there exists $a' \in \mathcal{G}$ such that $a * a' = a' * a = e$. The element a' is called the **inverse** of a' in $\mathcal{G}$.

The pair $(\mathcal{G}, *)$ is called a **group**. When the group's binary operation is clear from context, the set $\mathcal{G}$ can be referred to as the group.

The term *group* was first used by Galois (Wussing 1984, 17). A group with a commutative binary operation (that $a * b = b * a$ for all a and b in the group) is known as an **abelian group.** It is named after the Norwegian mathematician Niels Henrik Abel.

+	[0]	[1]	[2]	[3]
[0]	[0]	[1]	[2]	[3]
[1]	[1]	[2]	[3]	[0]
[2]	[2]	[3]	[0]	[1]
[3]	[3]	[0]	[1]	[2]

(a) $\mathbb{Z}_4$

$*$	e	a	b	c
e	e	a	b	c
a	a	e	c	b
b	b	c	e	a
c	c	b	a	e

(b) Klein-4 group

Figure 15.1 The groups of order four.

The pairs $(\mathbb{Z}^+, +)$, $(\mathbb{Z}, \cdot)$, and $(\mathbb{R}, \cdot)$ are not groups, but $(\mathbb{Z}, +)$, $(\mathbb{Q}, +)$, $(\mathbb{R}, +)$, $(\mathbb{Q}\setminus\{0\}, \cdot)$, and $(\mathbb{R}\setminus\{0\}, \cdot)$ are. These are examples of infinite groups. When the group's set is finite, we use the term **order** to refer to its cardinality. The group $(\{e\}, *)$, where $*$ is any binary operation, is the only group of order 1. This means that any other group with one element, such as $(\{0\}, +)$ or $(\{1\}, \cdot)$, has the same structure as $(\{e\}, *)$. We say that these three groups are **isomorphic**. Any group of order 2 will be isomorphic to $\mathbb{Z}_2$, and any group of order 3 will be isomorphic to $\mathbb{Z}_3$. There are essentially two groups of order 4: those that are isomorphic to $\mathbb{Z}_4$ and those that are isomorphic to the **Klein-4** group. Their tables are given in Figure 15.1.

All of the examples given so far have been abelian. Define $\text{GL}(n, \mathbb{R})$ to be the pair consisting of the set of all $n \times n$ invertible matrices with real entries and matrix multiplication. This is called the **general linear group**. We know that matrix multiplication is associative, that its identity is the identity matrix, which is invertible, and that every matrix in the set has an inverse by definition. However, we know that even for invertible matrices, multiplication is not commutative. Therefore, $\text{GL}(n, \mathbb{R})$ is a nonabelian group.

The group definition states only that in a group there is an identity and there are inverses. Based on what we know about the integers, we should be able to prove more about these elements.

■ THEOREM 15.1.3

In a group there is exactly one identity and every element has a unique inverse.

PROOF

The uniqueness of the identity is left to Exercise 8. To show the uniqueness of inverses, let $\mathcal{G}$ be a group, and take $a \in \mathcal{G}$. Suppose that both a' and a'' are inverses of a in $\mathcal{G}$. Then

$$a' = a' * e = a' * (a * a'') = (a' * a) * a'' = e * a'' = a''. \blacksquare$$

The purpose behind a group is that it is the most basic algebraic structure in which simple equations can be solved. For example, to solve $x + 4 = 10$, we need the additive inverse of 4 to add to both sides of the equation. After this we need to be able to regroup so that we may add the inverse to 4 to obtain the identity. As a consequence of these three properties, we will discover that the solution found is the unique one for the equation. All of this is summarized in the next theorem.

THEOREM 15.1.4 [Cancellation Law]

Let $\mathcal{G}$ be a group and a' be the inverse of a. For all $a,\ b \in \mathcal{G}$:

- The unique solution to $a * x = b$ is $x = a' * b$.
- The unique solution to $x * a = b$ is $x = b * a'$.

PROOF

That $a' * b$ is a solution of $a * x = b$ simply requires a check:

$$a * (a' * b) = (a * a') * b = e * b = b.$$

To see that the solution is unique, suppose that both u and v are solutions to the equation. Again we calculate:

$$\begin{aligned} a * u &= a * v, \\ a' * (a * u) &= a' * (a * v), \\ (a' * a) * u &= (a' * a) * v, \\ e * u &= e * v, \\ u &= v. \end{aligned}$$

Therefore, $a' * b$ is the only solution to $a * x = b$. That the other equation has a unique solution as indicated is proven similarly. ■

Subgroups

Let a be an element of a group $\mathcal{G}$. For all positive integers n, define a^n to be the result of operating a with itself n times. That is,

$$a^1 = a,\ a^2 = a * a,\ a^3 = a * a * a,\ \ldots$$

and

$$a^m * a^n = a^{m+n}.$$

Further define $a^0 = e$ and a^{-1} to be the inverse of a. With this notation we observe that

$$(a * b)^{-1} = b^{-1} * a^{-1}$$

and

$$a^{-n} = (a^n)^{-1} = (a^{-1})^n.$$

We then gather all of these elements into a set,

$$\langle a \rangle = \{a^n : n \in \mathbb{Z}\},$$

and define the following.

■ **DEFINITION 15.1.5**

A group $\mathcal{G}$ is **cyclic** if there exists $a \in \mathcal{G}$ such that $\mathcal{G} = \langle a \rangle$. The element a is called a **generator** of $\mathcal{G}$.

For example, $(\mathbb{Z}, +)$ is a cyclic group. Both 1 and -1 are generators. However, $\mathbb{Q}$ and $\mathbb{R}$ are not cyclic groups under addition. As for finite groups, each $\mathbb{Z}_n$ is cyclic, generated by $[1]_n$, but the Klein-4 group is not cyclic because $a^2 = e$ for all a in the group.

An element a of a group may not generate the entire group, but since $e \in \langle a \rangle$ and both a^n and a^{-n} are elements of $\langle a \rangle$, the set generated by a is a group using the operation from $\mathcal{G}$.

■ **DEFINITION 15.1.6**

Let $\mathcal{G}$ be a group with binary operation $*$ and $\mathcal{H} \subseteq \mathcal{G}$. We say that $\mathcal{H}$ is a **subgroup** of $\mathcal{G}$ if $(\mathcal{H}, *)$ is a group.

Every group with at least two elements has at least two subgroups, itself (the **improper subgroup**) and $\{e\}$ (the **trivial subgroup**). A group that has as most these two subgroups is called **simple**. For example, $\mathbb{Z}_2$ and $\mathbb{Z}_3$ are simple groups, but $\mathbb{Z}_4$ is not. It has $\{[0]_4, [2]_4\}$ as a subgroup. Other examples of nonsimple groups are $(\mathbb{R}, +)$ because $(\mathbb{Z}, +)$ is one of its subgroups and $(\langle 2 \rangle, +)$ because $(\langle 6 \rangle, +)$ is one of its subgroups.

Since every element of a cyclic group can be written as a generator to a power, we can write equations for each of the elements of a cyclic group. This simplifies our work when we deal with these types of groups. The cyclic groups then pass this property to their subgroups. To prove this we need the following proposition, known as the **Division Algorithm**, that was presupposed by Euclid in his method for finding the greatest common divisor of two positive integers (Euclid 1925, VII.2). We omit the proof of the Division Algorithm, but it can be found in any introductory number theory text such as Rosen (1993, 37–38).

■ **THEOREM 15.1.7 [Division Algorithm]**

If $m > n > 0$ are integers, there exists unique integers q and r such that $0 \leqslant r < n$ and

$$m = nq + r.$$

The integer q is called the **quotient** and r is called the **remainder**. We need the theorem to prove the following.

■ **THEOREM 15.1.8**

A subgroup of a cyclic group is cyclic.

PROOF

Let $\mathcal{G}$ be a cyclic group with generator a. Let $\mathcal{H}$ be a subgroup of $\mathcal{G}$. If $\mathcal{H} = \{e\}$, the subgroup is cyclic with generator e. So suppose that $\mathcal{H}$ is not the trivial subgroup. Because $\mathcal{G}$ is cyclic, there exists a minimum positive integer n such that $a^n \in \mathcal{H}$. Suppose that this element is not a generator of $\mathcal{H}$. This means that there exists an integer $m > n$ such that $a^m \in \mathcal{H}$ but $a^m \notin \langle a^n \rangle$. This combined with the Division Algorithm yields unique integers q and r such that $m = nq + r$ and $0 < r < n$. Therefore,

$$a^m = a^{nq+r} = a^{nq} * a^r,$$

and from this we conclude that

$$a^r = a^m * a^{-nq}.$$

Since both a^m and a^{-nq} are elements of $\mathcal{H}$, a^r is an element of $\mathcal{H}$. This contradicts the minimality of n because $r < n$. Thus, a^n is a generator of $\mathcal{H}$. ■

To show that a set is a subgroup of a group, we do not need to show directly that the set satisfies the three group properties. We may check a different set of conditions instead.

■ **THEOREM 15.1.9**

Let $(\mathcal{G}, *)$ be a group with identity e. A subset $\mathcal{H}$ of $\mathcal{G}$ is a subgroup of $\mathcal{G}$ if and only if:

- $\mathcal{H}$ is closed under $*$.
- $e \in \mathcal{H}$.
- $a^{-1} \in \mathcal{H}$ for all $a \in \mathcal{H}$.

PROOF

Suppose that the three hypotheses of the theorem hold. We conclude the following:

- $*$ is a binary operation on $\mathcal{H}$ because $\mathcal{H}$ is closed under $*$.
- Since $*$ is associative on $\mathcal{G}$, it must be associative on $\mathcal{H}$.
- $e \in \mathcal{H}$, and every element of $\mathcal{H}$ has its inverse in $\mathcal{H}$.

Therefore, $(\mathcal{H}, *)$ is a group. Conversely, if $\mathcal{H}$ is a subgroup of $\mathcal{G}$, the three conditions clearly hold. ■

To illustrate the theorem, take a group $\mathcal{G}$ and a family of subgroups $\{\mathcal{H}_i : i \in I\}$. We claim that $\bigcap_{i \in I} \mathcal{H}_i$ is a subgroup of $\mathcal{G}$.

- Let $a, b \in \bigcap_{i \in I} \mathcal{H}_i$. This means that $a, b \in \mathcal{H}_i$ for all $i \in I$. Since each $\mathcal{H}_i$ is closed under the operation of $\mathcal{G}$, $a * b \in \mathcal{H}_i$ for all i. Hence,

$$a * b \in \bigcap_{i \in I} \mathcal{H}_i.$$

- Since e in each $\mathcal{H}_i$ for every $i \in I$, we must have $e \in \bigcap_{i \in I} \mathcal{H}_i$.
- Take $a \in \bigcap_{i \in I} \mathcal{H}_i$. Then because $a \in \mathcal{H}_i$ for all $i \in I$, we have $a^{-1} \in \mathcal{H}_i$, so $a^{-1} \in \bigcap_{i \in I} \mathcal{H}_i$.

We do not obtain the same result for union (Exercise 15).

Exercises

1. Which of the following are binary operations on the given sets?
    (a) Subtraction on $[0, \infty)$
    (b) Multiplication on $\mathbb{R}$
    (c) Division on $\mathbb{Z}^+$
    (d) $a * b = a^b$ on $\mathbb{Z}$
    (e) $a * b = \max\{a, b\}$ on $\mathbb{Z}$
    (f) $a * b = \sqrt{ab}$ on $\mathbb{R}$

2. What assumptions about $\mathbb{Z}$ must be made to show that $x * y = 2x - y$ is a binary operation on $\mathbb{Z}$?

3. Let $\mathcal{S}$ be a set and $P(\mathcal{S})$ the set of subsets of $\mathcal{S}$. The family of sets $P(\mathcal{S})$ is called the **power set** of $\mathcal{S}$.
    (a) Define $*$ such that for all $\mathcal{A}, \mathcal{B} \in \mathcal{S}$, $\mathcal{A} * \mathcal{B} = \mathcal{A} \cap \mathcal{B}$. Prove that $*$ is a binary operation on $\mathcal{S}$.
    (b) Is $(\mathcal{S}, *)$ a group? Explain.

4. Explain why $(\mathbb{Z}^+, +)$, $(\mathbb{Z}, \cdot)$, and $(\mathbb{R}, \cdot)$ are not groups.

5. Suppose that $*$ is defined by $x * y = x + y + 2$.
    (a) Identify the identity and the inverses for this group.
    (b) Prove that $(\mathbb{Z}, *)$ is a group.
    (c) Solve $8 * x = 10$ in $(\mathbb{Z}, *)$.

6. Let $\mathcal{F}$ be the set of **real-valued** functions. That is, each element of $\mathcal{F}$ is a function $f : \mathbb{R} \to \mathbb{R}$.
    (a) Prove that $(\mathcal{F}, +)$ is a group where

$$(f + g)(x) = f(x) + g(x).$$

    (b) Is $(\mathcal{F}, /)$ a group where / is function division?
    (c) Is $(\mathcal{F}, \circ)$ a group where $\circ$ is composition of functions? Explain.

7. Let n be a positive integer and define

$$M_n = \left\{ \begin{bmatrix} a_{1,1} & a_{1,2} & \cdots & a_{1,n} \\ a_{2,1} & a_{2,2} & \cdots & a_{2,n} \\ \vdots & \vdots & & \vdots \\ a_{n,1} & a_{n,2} & \cdots & a_{n,n} \end{bmatrix} : a, b, c, d \in \mathbb{R} \right\}.$$

(a) Show that M_n under addition is a group.

(b) Solve the equation in M_2:

$$\begin{bmatrix} 1 & 4 \\ -3 & 0 \end{bmatrix} + \begin{bmatrix} a & b \\ c & d \end{bmatrix} = \begin{bmatrix} -3 & 8 \\ 0 & -6 \end{bmatrix}.$$

(c) Show that M_n under multiplication is not a group.

8. Prove that the identity of a group is unique.

9. Let $\mathscr{G}$ be a group. Define $\phi : \mathscr{G} \to \mathscr{G}$ by $\phi(a) = a^{-1}$. Prove that ϕ is a one-to-one correspondence.

10. Let $(\mathscr{G}, *)$ and $(\mathscr{G}', *')$ be two groups. Define the **Cartesian product** of $\mathscr{G}$ and $\mathscr{G}'$ as

$$\mathscr{G} \times \mathscr{G}' = \{(a, a') : a \in \mathscr{G} \text{ and } a' \in \mathscr{G}'\}.$$

Define $(a, a') \cdot (b, b') = (a * b, a' *' b')$.

(a) Confirm that $\cdot$ is a binary operation on $\mathscr{G} \times \mathscr{G}'$.

(b) Show that $(\mathscr{G} \times \mathscr{G}', \cdot)$ is a group. Prove that it is abelian if and only if both $\mathscr{G}$ and $\mathscr{G}'$ are abelian.

11. Let $\mathscr{G}$ be a group so that $(a * b)^2 = a^2 * b^2$ for all $a, b \in \mathscr{G}$. Show that $\mathscr{G}$ is abelian.

12. Identify all of the subgroups of the following groups.

(a) the Klein-4 group

(b) $\mathbb{Z}_5$

(c) $\mathbb{Z}_8$

(d) $\mathbb{Z}_2 \times \mathbb{Z}_3$

(e) $\mathbb{Z}_2 \times \mathbb{Z}_6$

13. Let $\mathscr{F}$ be the set of real-valued functions (Exercise 6). Show that

$$\{f \in \mathscr{F} : f(0) = 0\}$$

is a subgroup of $(\mathscr{F}, +)$.

14. Define $\mathscr{H} = \{A \in M_n : a_{1,1} + a_{2,2} + \cdots + a_{n,n} = 0\}$. Prove that $\mathscr{H}$ is a subgroup of $(M_n, +)$.

15. Let $\mathscr{G}$ be a group and $\{\mathscr{H}_i : i \in I\}$ a family of subgroups of $\mathscr{G}$.

(a) Give an example to show that $\bigcup_{i \in I} \mathscr{H}_i$ is not necessarily a subgroup of $\mathscr{G}$.

(b) Assume that for all $i, j \in I$, either $\mathscr{H}_i \subseteq \mathscr{H}_j$ or $\mathscr{H}_j \subseteq \mathscr{H}_i$. Show that $\bigcup_{i \in I} \mathscr{H}_i$ is a subgroup of $\mathscr{G}$.

16. Let $\mathscr{H}$ and $\mathscr{K}$ be two subgroups of an abelian group $\mathscr{G}$. Define

$$\mathscr{H}\mathscr{K} = \{a * b : a \in \mathscr{H} \text{ and } b \in \mathscr{K}\}.$$

Prove that $\mathscr{H}\mathscr{K}$ is a subgroup of $\mathscr{G}$. Is it abelian?

17. For any group $\mathcal{G}$, let $\mathcal{S} \subseteq \mathcal{G}$. Prove that

$$\mathcal{H} = \{a \in \mathcal{G} : a * b = b * a \text{ for all } b \in \mathcal{S}\}$$

is a subgroup of $\mathcal{G}$.

18. Let $s \in \mathcal{G}$. Define $\mathcal{G}_s = \{a \in \mathcal{G} : a * s = s\}$. Show that this set, called the **stabilizer** of s, is a subgroup of $\mathcal{G}$.

19. Let $\mathcal{H}$ and $\mathcal{H}'$ be subgroups of groups $\mathcal{G}$ and $\mathcal{G}'$, respectively. Show that $\mathcal{H} \times \mathcal{H}'$ is a subgroup of $\mathcal{G} \times \mathcal{G}'$ under the operation defined in Exercise 10.

20. Write four elements of each of the following cyclic subgroups.
   (a) $\langle 1/2 \rangle$ in $(\mathbb{Q}, +)$
   (b) $\langle 1/2 \rangle$ in $(\mathbb{R}\backslash\{0\}, \cdot)$
   (c) $\langle [3] \rangle$ in $\mathbb{Z}_8$
   (d) $\langle x^2 + 1 \rangle$ the group of real-valued functions under addition

21. Take a to be an element of a group with binary operation $*$. Let k be a positive integer. Prove that $\{a^{nk} : n \in \mathbb{Z}\}$ is a subgroup of $\langle a \rangle$.

22. Demonstrate that simple groups are cyclic.

15.2 TRANSFORMATION GROUPS

In this section we introduce the type of group that was central to Klein's Erlangen Program. We begin with an arbitrary set $\mathcal{S}$ and define a **transformation** of $\mathcal{S}$ to be a function $\phi : \mathcal{S} \to \mathcal{S}$ that is a one-to-one correspondence. Let Trans($\mathcal{S}$) be the set of all transformations of $\mathcal{S}$. We make four observations.

- If $\phi, \psi \in$ Trans($\mathcal{S}$), then $\phi \circ \psi \in$ Trans($\mathcal{S}$) since the composition of two one-to-one correspondences on $\mathcal{S}$ is a one-to-one correspondence on $\mathcal{S}$.

- Composition is associative, namely for all $\phi, \psi, \sigma \in$ Trans($\mathcal{S}$),

$$(\phi \circ \psi) \circ \sigma = \phi \circ (\psi \circ \sigma).$$

- The identity map is a one-to-one correspondence, so it is an element of Trans($\mathcal{S}$).

- If $\phi \in$ Trans($\mathcal{S}$), then $\phi^{-1} \in$ Trans($\mathcal{S}$) because the inverse of any one-to-one correspondence on $\mathcal{S}$ is a one-to-one correspondence on $\mathcal{S}$.

Therefore, (Trans($\mathcal{S}$), $\circ$) is a group. We will call any subgroup of Trans($\mathcal{S}$) a **transformation group**. An example is the set of collineations on $\mathbb{P}^2$. They form a subgroup **PR** of Trans($\mathbb{P}^2$) (Exercise 10). Harmonic sets are preserved by every element of **PR**, so we say that harmonic sets are an **invariant** of **PR**. On the other hand, distance and angle measure are not invariants for **PR**.

Klein used **PR** to categorize projective geometry, and he planned to use other groups, including subgroups of **PR**, to characterize other geometries by identifying their invariants under a given group. In this section we examine one such example. The texts of Bennett (1995, 18–33) and Eves (1965, 149–162), serve as general sources.

Affine Planes

Remember that Desargues defined the projective plane from the Euclidean plane by adding a line and points at infinity (Section 9.2). Let us now reverse that process and recoup the plane from $\mathbb{P}^2$. Let $\triangle E_1E_2E_3$ be the reference triangle (Figure 15.2). Let the ideal line ω be the line with equation $x_3 = 0$ and let all points on $x_3 = 0$ be the ideal points. These will be removed from the space but still used to identify parallel lines. Define lines l and m to be **parallel** if and only if l and m are different lines and there exists a point Ω on ω such that l and m both contain Ω. In Figure 15.2, both $\overline{AE_1}$ and $\overline{BE_1}$ are interpreted as being parallel to $\overline{E_1E_3}$.

When we remove the points with $x_3 = 0$, the remaining ordered triples in homogenous coordinates are of the form (kx_1, kx_2, kx_3), with both k and x_3 nonzero. However, since $x_3 \neq 0$, any of these points can be written as

$$(x_1/x_3,\ x_2/x_3,\ 1).$$

This means that we can uniquely identify each of these points by an ordered pair (x, y), where $x = x_1/x_3$ and $y = x_2/x_3$. These are called **nonhomogeneous coordinates**. Similarly, if we take a line that is represented using homogeneous coordinates $[a, b, c]$, its equation in homogeneous coordinates is

$$ax_1 + bx_2 + cx_3 = 0,$$

and letting $x = x_1/x_3$ and $y = x_2/x_3$, its equation in nonhomogeneous coordinates is

$$ax + by + c = 0. \tag{15.1}$$

Notice that if this line is parallel to

$$dx + ey + f = 0,$$

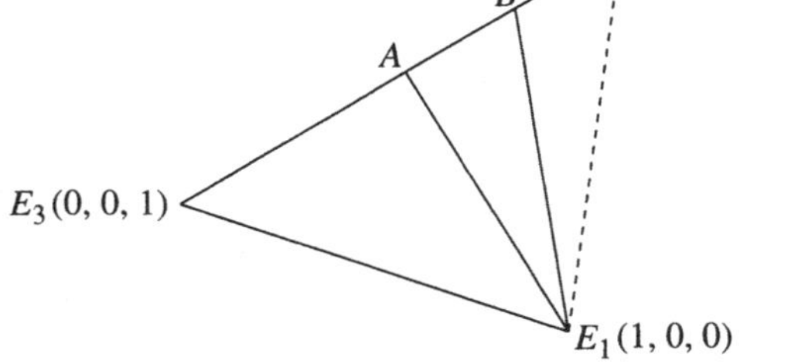

Figure 15.2 Interpreting $\overline{AE_1}$ and $\overline{AE_2}$ as parallel to $\overline{E_1E_2}$.

then

$$\begin{vmatrix} a & b \\ d & e \end{vmatrix} = 0.$$

If we convert these back to homogeneous coordinates, we obtain the system

$$\begin{array}{ccccccc} ax_1 & + & bx_2 & + & cx_3 & = & 0 \\ dx_1 & + & ex_2 & + & fx_3 & = & 0, \end{array}$$

which when put into a matrix and transform into row-reduced echelon form will look like either

$$\begin{bmatrix} 1 & * & * \\ 0 & 0 & 0 \end{bmatrix} \text{ or } \begin{bmatrix} 1 & * & 0 \\ 0 & 0 & 1 \end{bmatrix}.$$

The first form tells us that the lines are the same, while the second means that the point in common lies on the line $x_3 = 0$. This confirms that in our new geometry parallel lines meet at the line at infinity.

We now define new planes using nonhomogeneous coordinates. Let $(\mathcal{P}, \mathcal{L})$ be a projective plane with its points and lines identified by homogeneous coordinates. Define

$$\mathbb{A}(\mathcal{P}) = \{(x, y) : (x, y, 1) \text{ represents a point in } \mathcal{P}\}$$

and

$$\mathbb{L}(\mathcal{L}) = \{\{(x, y) : ax + by + c = 0\} : [a, b, c] \text{ represents an ordinary line in } \mathcal{L}\}.$$

Notice that $[0, 0, 1]$ is not used to define $\mathbb{L}(\mathcal{L})$. When we examine the basic properties of these two sets, we find that they satisfy the following list of postulates.

■ POSTULATES 15.2.1

Let $\mathcal{P}_A$ be a set of objects that are called **points** and $\mathcal{L}_A$ be a set of objects that are called **lines**.

1. For all distinct $P, Q \in \mathcal{P}_A$, there exists a unique line l in $\mathcal{L}_A$ such that P and Q are incident with l.

2. Every line contains at least two points.

3. There exists at least three noncollinear points.

4. Given a line l and a point P not on l, there exists exactly one line through P that is parallel to l.

We call the pair $(\mathcal{P}_A, \mathcal{L}_A)$ an **affine plane** if it satisfies the four postulates. It is left to Exercise 1 to show that $(\mathbb{A}(\mathcal{P}), \mathbb{L}(\mathcal{L}))$ is an affine plane for all projective planes $(\mathcal{P}, \mathcal{L})$. Other examples of affine planes include $\mathbb{R}^2$ and $\mathbb{Q}^2$, where the points are ordered pairs and the lines are sets of points that satisfy equations of the form given in Equation 15.1. In fact, $\mathbb{R}^2$ is the affine plane obtained from $\mathbb{P}^2$. A finite example is

$$\mathcal{P}_A = \{A, B, C, D\}$$

and

$$\mathcal{L}_A = \{\{A, B\}, \{A, C\}, \{A, D\}, \{B, C\}, \{B, D\}, \{C, D\}\}.$$

This affine plane is obtained from the Fano projective plane (Exercise 4). That this is the smallest affine plane is the result of the next theorem.

■ **THEOREM 15.2.2**

An affine plane has at least four points and at least six lines.

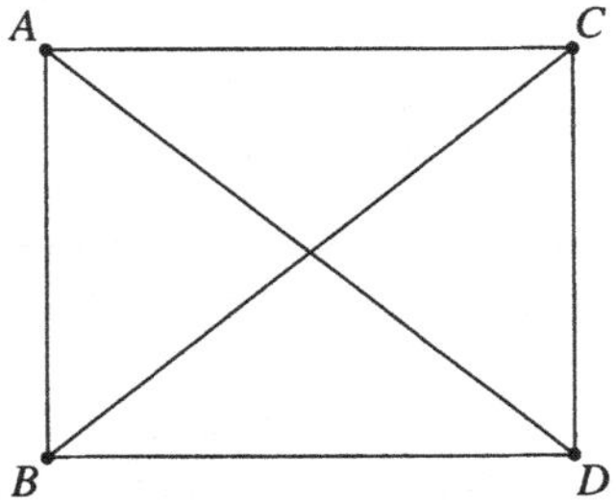

PROOF

By Postulate 3 of 15.2.1, there exists at least three noncollinear points in an affine plane. Let them be labeled A, B, and C. Join $\overline{AB}$, $\overline{AC}$, and $\overline{BC}$. These three lines are distinct. Since C is not on $\overline{AB}$, Postulate 4 guarantees a line on C that is parallel to $\overline{AB}$, and by Postulate 2 this line must contain a new point D. By the first postulate, join $\overline{BD}$ and $\overline{AD}$. The resulting diagram satisfies the postulates for an affine plane. ■

Notice that affine geometry feels a bit like projective geometry except that there are parallel lines. This is due not only to the construction of the affine plane from the projective plane but also because like projective geometry, affine geometry lacks a metric.

Affinities

A collineation on $\mathbb{R}^2$ is equivalent to a collineation on $\mathbb{P}^2$ that leaves the line $x_3 = 0$ fixed. No points can be mapped to or from it because it is not part of $\mathbb{R}^2$. By Theorem 14.4.6, the equations in homogeneous coordinates that describe these collineations are

$$\begin{bmatrix} kx_1' \\ kx_2' \\ kx_3' \end{bmatrix} = \begin{bmatrix} a_{11} & a_{12} & a_{13} \\ a_{21} & a_{22} & a_{23} \\ 0 & 0 & 1 \end{bmatrix} \begin{bmatrix} x_1 \\ x_2 \\ x_3 \end{bmatrix}, \tag{15.2}$$

where

$$\begin{vmatrix} a_{11} & a_{12} \\ a_{21} & a_{22} \end{vmatrix} \neq 0.$$

If we use nonhomogeneous coordinates, Equation 15.2 becomes

$$\begin{bmatrix} x' \\ y' \end{bmatrix} = \begin{bmatrix} a_{11} & a_{12} \\ a_{21} & a_{22} \end{bmatrix} \begin{bmatrix} x \\ y \end{bmatrix} + \begin{bmatrix} a_{13} \\ a_{23} \end{bmatrix}. \tag{15.3}$$

■ DEFINITION 15.2.3

The function $\phi : \mathbb{R}^2 \to \mathbb{R}^2$ is called an **affinity** if there exist an invertible 2×2 matrix A with real entries and vector $\mathbf{a}$ in $\mathbb{R}^2$ such that

$$\phi(\mathbf{x}) = A\mathbf{x} + \mathbf{a}.$$

Let **AF** denote the set of all affinities. We show that **AF** is a subgroup of **PR**. To do this, first note that by our work on the bottom of page 532, **AF** $\subseteq$ **PR**. Next take $\phi, \psi \in$ **AF**. This means that we can write ϕ as in Definition 15.2.3 and ψ as

$$\psi(\mathbf{x}) = B\mathbf{x} + \mathbf{b},$$

where $\det(B) \neq 0$ and $\mathbf{b} \in \mathbb{R}^2$. We check the three conditions of Theorem 15.1.9.

- **AF** is closed under composition because
$$(\phi \circ \psi)(\mathbf{x}) = \phi(B\mathbf{x} + \mathbf{b}) = AB\mathbf{x} + A\mathbf{b} + \mathbf{a},$$
which is an element of **AF** because AB is invertible and $A\mathbf{b} + \mathbf{a} \in \mathbb{R}^2$.
- Certainly, the identity map is an element of **AF**.
- Define $\psi(x) = A^{-1}\mathbf{x} - A^{-1}\mathbf{a}$ and calculate:
$$(\phi \circ \psi)(\mathbf{x}) = \phi(A^{-1}\mathbf{x} - A^{-1}\mathbf{a}) = A(A^{-1}\mathbf{x} - A^{-1}\mathbf{a}) + \mathbf{a} = \mathbf{x}$$
and
$$(\psi \circ \phi)(\mathbf{x}) = \psi(A\mathbf{x} + \mathbf{a}) = A^{-1}(A\mathbf{x} + \mathbf{a}) - A^{-1}\mathbf{a} = \mathbf{x}.$$
Therefore, $\psi = \phi^{-1}$.

Affinities preserve neither distance nor angle measure. For example, take the affinity ϕ defined by

$$\begin{bmatrix} x' \\ y' \end{bmatrix} = \begin{bmatrix} 1 & 2 \\ 4 & 5 \end{bmatrix} \begin{bmatrix} x \\ y \end{bmatrix} + \begin{bmatrix} 3 \\ 6 \end{bmatrix}. \tag{15.4}$$

Then, $\phi(0, 0) = (3, 6)$ and $\phi(1, 1) = (6, 15)$. We see that the distance between $(0, 0)$ and $(1, 1)$ is different from the distance between $(3, 6)$ and $(11, 29)$. Moreover, if we compare the angle of inclination α of the original segment with the angle of inclination β of the image, we compute that $\cos\alpha = 1/\sqrt{2}$ and $\cos\beta = 3/\sqrt{89}$, which imply that $\alpha \neq \beta$.

However, affinities do preserve parallel lines.

■ THEOREM 15.2.4

The image of two parallel lines under an affinity is two parallel lines.

PROOF

Let ϕ be an affinity of an affine plane, so we can view ϕ as a projective collineation that leaves the line at infinity fixed. Assume that lines l and m are parallel. We conclude that there exists an ideal point Ω that is incident with both l and m. Therefore, $\phi(\Omega)$ is an ideal point, and since ϕ is a collinearity, $\phi[l]$ and $\phi[m]$ must intersect at $\phi(\Omega)$. Hence, $\phi[l]$ and $\phi[m]$ are parallel. ■

Exercises

1. Prove that $(\mathbb{A}(\mathcal{P}),\ \mathbb{L}(\mathcal{L}))$ is an affine plane for all projective planes $(\mathcal{P},\ \mathcal{L})$.

2. Can $\mathbb{Z}^2$ be considered an affine plane?

3. Let $\mathcal{P}_A = \{A,\ B,\ C,\ D\}$ and

$$\mathcal{L}_A = \{\{A,\ B\},\ \{A,\ C\},\ \{A,\ D\},\ \{B,\ C\},\ \{B,\ D\},\ \{C,\ D\}\}.$$

   (a) Show that $(\mathcal{P}_A,\ \mathcal{L}_A)$ is an affine plane.
   (b) Find all affinities of $(\mathcal{P}_A,\ \mathcal{L}_A)$.

4. Can the Fano plane (Theorem 14.1.4) be considered an affine plane? Explain.

5. In Figure 14.2 of the Fano plane, let $A = (0,\ 1,\ 0)$, $E = (0,\ 0,\ 1)$, $G = (1,\ 0,\ 0)$, and $C = (1,\ 1,\ 1)$.
   (a) Find the homogeneous coordinates for B, D, and F.
   (b) Which line in Figure 14.2 should be considered the line at infinity?
   (c) If possible, find the nonhomogeneous coordinates for A, B, C, D, E, F, and G.
   (d) Let $\mathbb{A}$ represent the ordered pairs in nonhomogeneous coordinates from Exercise 5(c). Let $\mathbb{L}$ be the set of lines through the points of $\mathbb{A}$. Find $\mathbb{L}$.
   (e) Show that $(\mathbb{A},\ \mathbb{L})$ is an affine plane.
   (f) Find an affinity $\phi : \mathbb{A} \to \mathbb{A}$ such that $\phi(E) = E$, $\phi(B) = F$, and $\phi(F) = B$. Write the function in matrix form (Equation 15.3).
   (g) Two affine planes are **isomorphic** if there exists a collineation between them (compare Definition 14.1.6). Using the affine plane $(\mathcal{P}_A,\ \mathcal{L}_A)$ from Exercise 3, show that $(\mathbb{A},\ \mathbb{L})$ is isomorphic to $(\mathcal{P}_A,\ \mathcal{L}_A)$.

6. Let $(\mathcal{P}_A,\ \mathcal{L}_A)$ be an arbitrary affine plane. Prove the following.
   (a) Two distinct lines in $\mathcal{L}_A$ have at most one point in common.
   (b) For all $l,\ m,\ n \in \mathcal{L}_A$, if m and n intersect and l is parallel to m, then l intersects n.
   (c) If there exists a line in $\mathcal{L}_A$ that contains n points, all lines in $\mathcal{L}_A$ contains n points.
   (d) For all $l,\ m \in \mathcal{L}_A$, let $l \sim m$ if and only if $l = m$ or l is parallel to m. Show that $\sim$ is an equivalence relation.

7. Can an affine plane be constructed with 10 points? 16 points? Explain.

8. Which of Postulates 14.1.2 are implied by Postulates 15.2.1?

9. Prove that Postulates 15.2.1 are consistent and independent.

10. Confirm that the set of collineations form a subgroup of $\text{Trans}(\mathbb{P}^2)$.

11. Find the homogeneous coordinates for the point $(4,\ 10)$ given in nonhomogeneous coordinates, and give the nonhomogeneous coordinates for the homogeneous $(4,\ 10,\ 7)$.

12. Write the function given in Equation 15.4 using homogeneous coordinates.

13. Let $\phi : \mathbb{R}^2 \to \mathbb{R}^2$ be an affinity such that $\phi(1, 1) = (4, 0)$ and $\phi(3, 2) = (2, 6)$. Give an equation for ϕ in the form $\phi(\mathbf{x}) = A\mathbf{x} + \mathbf{a}$.

14. A **dilatation** is a collineation $\phi : \mathbb{R}^2 \to \mathbb{R}^2$ such that for all lines l, $\phi[l]$ is parallel to l.
   (a) Why is a dilatation an affinity?
   (b) Draw an example of a dilatation in $\mathbb{R}^2$ and give its equation in the form $\phi(\mathbf{x}) = A\mathbf{x} + \mathbf{a}$.
   (c) Is the set of dilatations a group under composition?
   (d) Prove that a dilatation that has two invariant points is the identity map.
   (e) A **stretch** is an affinity ψ so that $\psi(x, y) = (kx, ky)$ for some k. Show that a stretch is a dilatation.

15. An **affine translation** is the identity map or a dilatation with no fixed points. Show that the set of affine translations is a group under composition.

16. Let τ be an affine translation and ϕ an affinity. Prove that $\phi^{-1} \circ \tau \circ \phi$ is an affine translation.

17. An affinity ϕ such that $\phi(A) = B$ is a **shear** if it fixes every point on lines that are parallel but not equal to $\overline{AB}$.
   (a) Draw an example of a shear in $\mathbb{R}^2$ and give its equation in the form $\phi(\mathbf{x}) = A\mathbf{x} + \mathbf{a}$.
   (b) Does the set of shears form a group under composition? Explain.

18. An affinity ψ is an **affine reflection** if for some points A and B, $\psi(A) = B$, $\psi(B) = A$, and there exists a point C not on $\overline{AB}$ such that $\psi(C) = C$.
   (a) Draw an example of an affine reflection in $\mathbb{R}^2$ and give its equation in the form $\psi(\mathbf{x}) = A\mathbf{x} + \mathbf{a}$.
   (b) Does the set of affine reflections form a group under composition?
   (c) Let M be the midpoint of $\overline{AB}$. Show that ψ fixes every point on $\overline{CM}$.

15.3 THE PRINCIPAL GROUP

Jordan taught his students that transformation groups would one day play an important role in the development of geometry. This was realized by the work of Klein and Lie. Influenced by what they learned in Paris, Klein focused on discrete groups, while Lie made studying **continuous groups** a lifelong project. Examples of these uncountable groups include $(\mathbb{R}, +)$ and the groups of projective and affine transformations. Initially, however, Klein's and Lie's group theoretic interests coincided with the group of transformations of the Euclidean plane (Yaglom 1988, 95, 97).

As projective geometry can be defined as the study of the properties that remain invariant under the group of projectivities, Klein defined Euclidean geometry as the study of properties that are preserved under transformations of the Euclidean plane.

To understand the reason for his definition, we need to go back to Euclid. Although he used it reluctantly, Euclid applied superposition to prove SAS (page 138). He assumed that he could take a triangle and move it so that it aligned with any other triangle to which it was congruent. As the triangle moved, it did not deform. Its sides remained the same length and its angles kept their measure. Euclid also studied similar figures. Two figures are similar if there exists a transformation that changes only the size of the figure, not its angle measures. All of this means that Euclid indirectly applied transformations of the plane that preserved both distance and angle measure. With this in mind, Klein (1893, 218) made the following definition.

■ **DEFINITION 15.3.1**

The **principal group PG** is the transformation group on $\mathbb{R}^2$ that consists of the following functions:

- motions in the plane (translations, rotations)
- transformations into symmetrical figures (reflections)
- transformations into similar figures (dilations and contractions)
- any composition of the above

Isometries

All of the maps of the principal group are conformal, but the first two are also guaranteed to preserve distance. Throughout, let d represent the standard metric.

■ **DEFINITION 15.3.2**

A one-to-one correspondence $f : \mathbb{R}^2 \to \mathbb{R}^2$ is an **isometry** if

$$d[A, B] = d[f(A), f(B)]$$

for all $A, B \in \mathbb{R}^2$. In this case we say that f **preserves distance**.

We begin by proving three basic properties of isometries.

■ **THEOREM 15.3.3**

An isometry is a collineation.

PROOF

Let ϕ be an isometry and assume that $\phi(A) = A'$ and $\phi(B) = B'$. Take P on $\overline{AB}$. To obtain a contradiction, suppose that $\phi(P) = P'$ is not on $\overline{A'B'}$. By the Triangle Inequality (5.3.11),

$$A'B' < A'P' + P'B'.$$

However, this is impossible because

$$A'B' = AB = AP + PB = A'P' + P'B'.$$

Therefore, P' is on $\overline{A'B'}$. ■

■ THEOREM 15.3.4

Isometries are conformal.

PROOF

Let $f : \mathbb{R}^2 \to \mathbb{R}^2$ be an isometry and choose three points A, B, C in $\mathbb{R}^2$. Then

$$\begin{aligned} d[A, B] &= d[\phi(A), \phi(B)], \\ d[A, C] &= d[\phi(A), \phi(C)], \\ d[B, C] &= d[\phi(B), \phi(C)]. \end{aligned}$$

Thus, by SSS, $\triangle ABC \cong \triangle f(A)f(B)f(C)$, from which we conclude that $\sphericalangle ABC \cong \sphericalangle f(A)f(B)f(C)$. ■

■ THEOREM 15.3.5

An isometry is uniquely determined by three noncollinear points.

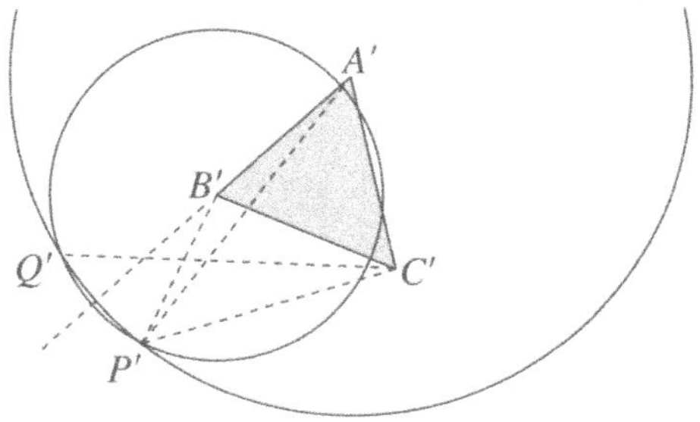

PROOF

Let A, B, and C be distinct noncollinear points. Let ϕ and ψ be two isometries and assume that

$$\phi(A) = \psi(A) = A',\ \phi(B) = \psi(B) = B', \text{ and } \phi(C) = \psi(C) = C'.$$

We need to show that $\phi = \psi$. To do this, take any point P in the plane and let $\phi(P) = P'$. We will show that P' is uniquely determined by its distance to each of the points A', B', and C', thus showing that $\phi(P) = \psi(P)$. Suppose that there is another point Q' that has the same distances from A', B', and C' as P'. We can then draw $\odot A'$ with radius $\overline{A'P'}$ and $\odot B'$ with radius $\overline{B'P'}$ such that Q' is also on the intersection of the two circles. Join $\overline{C'Q'}$ and $\overline{C'P'}$. The only way that these two segments can be congruent is for C' to lie on the perpendicular bisector of $\overline{Q'P'}$. However, this is not the case because the $\overline{A'B'}$ produced is that perpendicular bisector and C' is not collinear with A' and B'. Therefore, Q' does not exist, and P' is described uniquely by its distances to A', B', and C'. Since the distances from $\phi(A)$, $\phi(B)$, and $\phi(C)$ to P', respectively, equal the distances from $\psi(A)$, $\psi(B)$, and $\psi(C)$ to P', it must be the case that $\psi(P) = P'$ since ψ preserves distances. ■

The fact that three noncollinear points determine an isometry uniquely will enable us to show that all isometries are affinities.

■ **THEOREM 15.3.6**

A function $\phi : \mathbb{R}^2 \to \mathbb{R}^2$ is an isometry if and only if there exists a 2×2 matrix A, a vector $\mathbf{b} \in \mathbb{R}^2$, and an angle θ such that either

$$A = \begin{bmatrix} \cos\theta & -\sin\theta \\ \sin\theta & \cos\theta \end{bmatrix} \text{ or } A = \begin{bmatrix} \cos\theta & \sin\theta \\ \sin\theta & -\cos\theta \end{bmatrix}$$

and $\phi(\mathbf{x}) = A\mathbf{x} + \mathbf{b}$.

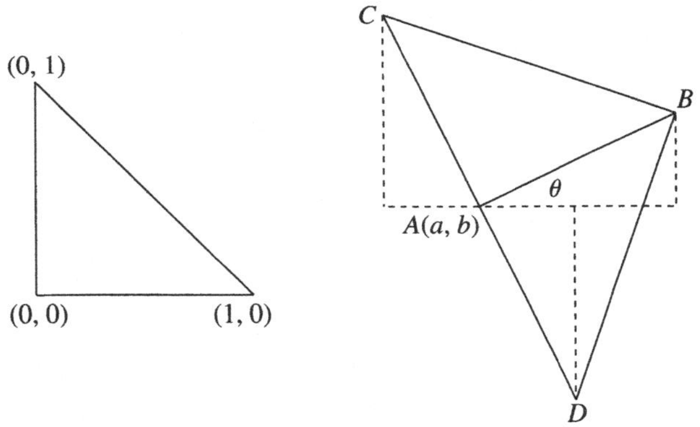

PROOF

We leave to Exercise 4 the proof that $\phi(\mathbf{x}) = A\mathbf{x} + \mathbf{b}$ with A and $\mathbf{b}$ as described is an isometry. So to prove the converse, let ϕ be an isometry. First suppose that $\phi(0, 0) = (a, b) = A$. We have two cases to check, depending on whether ϕ preserves the orientation of the vertices of the triangle. Assume that it does. This means that $\phi(1, 0) = B$ and $\phi(0, 1) = C$, as in the diagram. Let θ be the angle of inclination of $\overline{AB}$. Then, since $AB = 1$,

$$\phi(1, 0) = (a + \cos\theta, b + \sin\theta),$$
$$\phi(0, 1) = (a - \sin\theta, b + \cos\theta).$$

Since ϕ is an isometry, we may conclude by Theorem 15.3.5 that

$$\phi(x, y) = \begin{bmatrix} \cos\theta & -\sin\theta \\ \sin\theta & \cos\theta \end{bmatrix} \begin{bmatrix} x \\ y \end{bmatrix} + \begin{bmatrix} a \\ b \end{bmatrix}$$

because a function of this form is an isometry (Exercise 4). If the isometry does not preserve the order of the vertices, then $\phi(0, 1) = D$ and

$$\phi(0, 1) = (a + \cos(\theta - \pi/2), b + \sin(\theta - \pi/2)) = (a + \sin\theta, b - \cos\theta).$$

Therefore,

$$\phi(x, y) = \begin{bmatrix} \cos\theta & \sin\theta \\ \sin\theta & -\cos\theta \end{bmatrix} \begin{bmatrix} x \\ y \end{bmatrix} + \begin{bmatrix} a \\ b \end{bmatrix},$$

again due to Theorem 15.3.5. ■

Since both versions of A in the theorem have a nonzero determinant, we may conclude the following result.

■ COROLLARY 15.3.7

An isometry of $\mathbb{R}^2$ is an affinity.

Now to confirm the group structure of the set of isometries. We leave the proof of the first part of the lemma to Exercise 7.

■ LEMMA 15.3.8

- The composition of isometries is an isometry.
- The inverse of an isometry is an isometry.

PROOF

Let $f : \mathbb{R}^2 \to \mathbb{R}^2$ be an isometry. Take $X', Y' \in \mathbb{R}^2$ and let X and Y be in the plane so that $f^{-1}(X') = X$ and $f^{-1}(Y') = Y$, which means that $f(X) = X'$ and $f(Y) = Y'$. Therefore,

$$d[X', Y'] = d[f(X), f(Y)] = d[X, Y] = d[f^{-1}(X'), f^{-1}(Y')]. \blacksquare$$

Lemma 15.3.8 combined with the fact that the identity is an isometry yield the next theorem.

■ THEOREM 15.3.9

The set of isometries of $\mathbb{R}^2$ is a subgroup of the principal group.

Similarities

Isometries are conformal and preserve distances. If we require a transformation on $\mathbb{R}^2$ to only be conformal, we obtain another group of functions of the plane.

■ DEFINITION 15.3.10

The transformation $\sigma : \mathbb{R}^2 \to \mathbb{R}^2$ is a **similarity** if there exists $r > 0$ such that for all $A, B \in \mathbb{R}^2$, if $A' = \sigma(A)$ and $B' = \sigma(B)$, then $A'B' = rAB$. The real number r is called the **ratio** of the similarity.

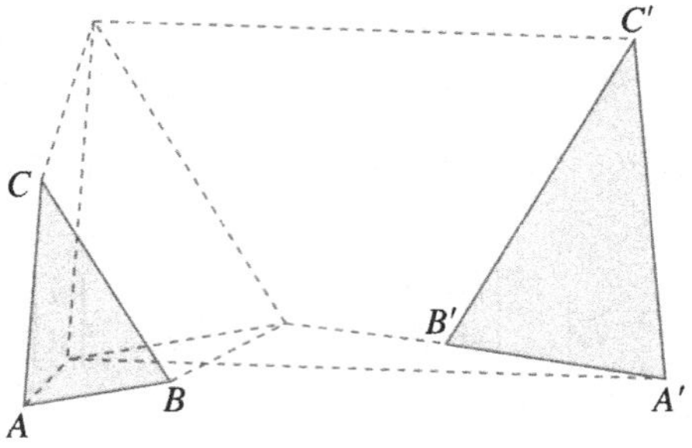

Figure 15.3 $\triangle A'B'C'$ is the image of $\triangle ABC$ under a similarity.

The similarity σ_0 defined by $\sigma_0(x, y) = (rx, ry)$ is an example of a **homothety** (Exercise 12). It leaves the origin fixed and either dilates or contracts all other points, depending on the value of r. To obtain formulas for the other similarities, take an isometry ϕ and compose it with σ_0 (Figure 15.3). That the resulting composition is a similarity is seen by

$$\|(\phi \circ \sigma_0)(x, y)\| = \|\phi(rx, ry)\| = \|(rx, ry)\| = r\|(x, y)\|,$$

where the second equality holds because ϕ is an isometry. This result can be used to prove the following characterization of similarities (Exercise 11).

■ THEOREM 15.3.11

If σ is a similarity with ratio r, there exists a 2×2 invertible matrix A and a vector $\mathbf{b} \in \mathbb{R}^2$ such that

$$\sigma(\mathbf{x}) = rA\mathbf{x} + \mathbf{b},$$

where

$$\phi(\mathbf{x}) = A\mathbf{x} + \mathbf{b}$$

is an isometry.

■ COROLLARY 15.3.12

A similarity is an affinity.

Similarities do share some properties with isometries. The following is an example. Its proof is a modification of Theorems 15.3.3 and 15.3.4 (Exercise 16).

■ THEOREM 15.3.13

A similarity is a conformal collineation.

The similarities also form a subgroup of the principal group. To see this, we first show that the set of similarities is closed under composition.

LEMMA 15.3.14

The composition of similarities is a similarity.

PROOF

Let σ_1 and σ_2 be similarities with ratios r_1 and r_2. Then there are invertible matrices A_1 and A_2 with $\mathbf{b}_1,\ \mathbf{b}_2 \in \mathbb{R}^2$ so that $A_1\mathbf{x}+\mathbf{b}_1$ and $A_2\mathbf{x}+\mathbf{b}_2$ represent isometries and

$$\sigma_1(\mathbf{x}) = r_1A_1\mathbf{x} + \mathbf{b}_1 \text{ and } \sigma_2(\mathbf{x}) = r_2A_2\mathbf{x} + \mathbf{b}_2.$$

We now compute for all $\mathbf{x} \in \mathbb{R}^2$,

$$\begin{aligned}\|(\sigma_1 \circ \sigma_2)(\mathbf{x})\| &= \|\sigma_1(r_2A_2\mathbf{x} + \mathbf{b}_2)\| \\ &= \|r_1r_2A_1A_2\mathbf{x} + r_1A_1\mathbf{b}_2 + \mathbf{b}_1\| \\ &= r_1r_2\left\|A_1A_2\mathbf{x} + \frac{1}{r_2}A_1\mathbf{b}_2 + \frac{1}{r_1r_2}\mathbf{b}_1\right\| \\ &= r_1r_2\|\mathbf{x}\|.\end{aligned}$$

The last equality follows from the fact proven in Lemma 15.3.8 that the composition of isometries is an isometry (Exercise 8). Therefore, we conclude that $\sigma_1 \circ \sigma_2$ is a similarity. ■

The lemma, together with the facts that the identity map is a similarity with $r = 1$ and the inverse of a similarity with ratio r is a similarity with ratio $1/r$ (Exercise 13), gives us the next theorem.

THEOREM 15.3.15

- The set of similarities form a subgroup of the principal group.
- The set of isometries form a subgroup of the set of similarities.

Many more results concerning isometries and similarities, including applications, may be found in Dodge (1972, 48–130). That text served as the general reference for this section and the next.

Exercises

1. Define $f : \mathbb{R} \to \mathbb{R}$ by $f(x) = x^3 + 1$.
   (a) Prove that f is a transformation.
   (b) Define what it would mean for f to be an isometry on $\mathbb{R}$. Based on the definition chosen, show that f not an isometry.
   (c) Find an example of an isometry on $\mathbb{R}$. Must it be linear?
2. Let $g : \mathbb{R}^2 \to \mathbb{R}^2$ be defined by $g(x, y) = (x + 2, y^5 - 1)$.
   (a) Prove that g is a transformation.
   (b) Is g an isometry? Explain.

3. Prove directly from the definition that $f(x, y) = (x - 7, y + 5)$ is an isometry.

4. Prove that $\phi(\mathbf{x}) = A\mathbf{x} + \mathbf{b}$ is an isometry if either

$$A = \begin{bmatrix} \cos\theta & -\sin\theta \\ \sin\theta & \cos\theta \end{bmatrix} \text{ or } A = \begin{bmatrix} \cos\theta & \sin\theta \\ \sin\theta & -\cos\theta \end{bmatrix}.$$

5. Prove that isometries preserve the areas of rectangles and triangles. Do isometries preserve the areas of all plane figures?

6. Prove that the image of a circle under an isometry is a circle.

7. Prove that the composition of isometries is an isometry (Lemma 15.3.8).

8. Let ϕ be an isometry. Show that for all $\mathbf{b} \in \mathbb{R}^2$, $\psi(\mathbf{x}) = \phi(\mathbf{x}) + \mathbf{b}$ is an isometry.

9. Is an isometry that fixes three noncollinear points the identity map? Explain.

10. Let $g(x, y) = (4x + 2, 4y - 8)$. Prove that g is a similarity.

11. Prove Theorem 15.3.11.

12. Given a point C called the **center** and $r > 0$, a **homothety** η is a transformation of the plane such that for all points A,

$$d[\eta(A), C] = rd[A, C].$$

(a) Prove that a homothety has only one fixed point.
(b) Show that the composition of two homotheties with the same center is a homothety. Identify the composition's ratio and center.
(c) Let $\triangle ABC$ and $\triangle A'B'C'$ be similar but not congruent triangles such that corresponding sides are parallel. Find a homothety η such that the image of $\triangle ABC$ under η is $\triangle A'B'C'$.
(d) Prove or disprove: The set of homotheties form a subgroup of the set of similarities.

13. Prove that the inverse of a similarity with ratio r is a similarity with ratio $1/r$.

14. Is the group of similarities abelian? Explain.

15. Let σ be a similarity with ratio r. If $r \geqslant 1$, then σ is a **dilation**, and if $0 < r < 1$, then σ is a **contraction**. Is the set of dilations a subgroup of the group of similarities? What about the set of contractions? Explain.

16. Prove Theorem 15.3.13

17. Let σ be an similarity with ratio r. Let $\mathcal{F}$ be a convex figure in the plane. True or false: The area of $\sigma[\mathcal{F}]$ equals r^2 times the area of $\mathcal{F}$. Give an example.

18. Prove the following for transformations of $\mathbb{R}^2$:
(a) The composition of conformal maps is conformal.
(b) The inverse of a conformal map is conformal.

19. Define $\phi(\mathbf{x}) = rA\mathbf{x} + \mathbf{b}$. For each A and $\mathbf{b}$ given, find ϕ^{-1}, the image of the triangle with vertices (0, 0), (1, 0), and (0, 1), and whether ϕ is an isometry or a similarity.

(a) $A = \begin{bmatrix} 1 & 0 \\ 0 & -1 \end{bmatrix}$, $\mathbf{b} = \begin{bmatrix} -4 \\ 3 \end{bmatrix}$.

(b) $A = \begin{bmatrix} \sqrt{2}/2 & -\sqrt{2}/2 \\ \sqrt{2}/2 & \sqrt{2}/2 \end{bmatrix}$, $\mathbf{b} = \begin{bmatrix} 0 \\ 0 \end{bmatrix}$.

(c) $A = \begin{bmatrix} 2\sqrt{3} & -2 \\ 2 & 2\sqrt{3} \end{bmatrix}$, $\mathbf{b} = \begin{bmatrix} 1 \\ 2 \end{bmatrix}$.

15.4 ISOMETRIES OF THE PLANE

There are four different types of isometries of the principal group. In this section we examine each and then prove that there are no other types.

Reflections

We begin our introduction to the transformations of the principal group by defining the function that is used to determine the symmetries of a plane figure. These functions also have the property that they map figures to congruent figures.

DEFINITION 15.4.1

The transformation $\mu : \mathbb{R}^2 \to \mathbb{R}^2$ is a **reflection** through a line l if for all $A \in \mathbb{R}^2$, if A is not on l, the line l is the perpendicular bisector of the segment joining A and $\mu(A)$, otherwise $\mu(A) = A$. We call l the **mirror** of the reflection.

The definition is illustrated in Figure 15.4. Although a reflection is an isometry and every reflection is its own inverse (Exercise 6), the diagram makes clear that the

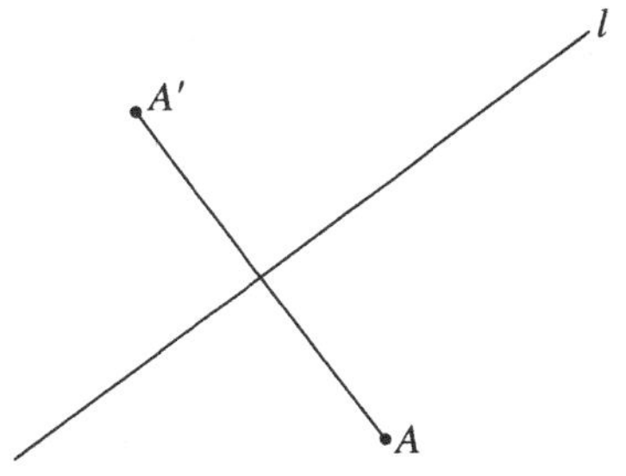

Figure 15.4 A reflection of A to image A' through l.

identity is not a reflection. Hence, the set of reflections do not form a subgroup of the principal group. Instead, they serve as the building blocks for the group of isometries.

THEOREM 15.4.2

Every isometry is the composition of at most three reflections.

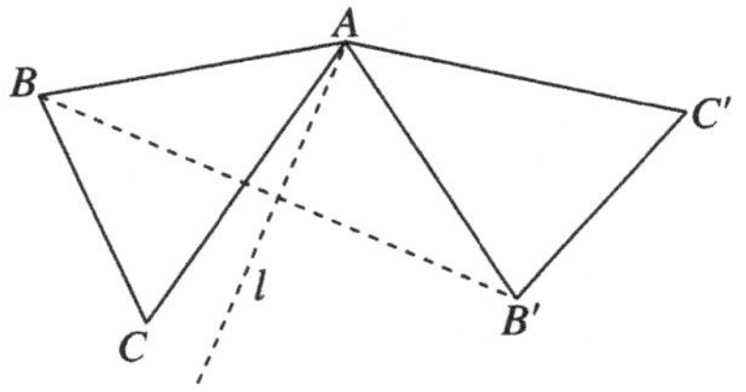

PROOF

Let ϕ be an isometry on $\mathbb{R}^2$. By Theorem 15.3.5 it suffices to examine ϕ on any three noncollinear points A, B, and C. We have four cases to consider.

- Let $\phi(A) = A$, $\phi(B) = B$, and $\phi(C) = C$. This means that ϕ is the identity, so let μ be any reflection. Then $\phi = \mu \circ \mu$.

- Let $\phi(A) = A$, $\phi(B) = B$, and $\phi(C) \neq C$. In this case ϕ is a reflection with mirror $\overline{AB}$ (Exercise 4).

- Let $\phi(A) = A$, $\phi(B) = B' \neq B$, and $\phi(C) = C' \neq C$. This is illustrated in the diagram for this theorem. Since $\triangle ABB'$ is isosceles, A is on the perpendicular bisector l of $\overline{BB'}$. Therefore, the reflection μ with mirror l has

$$\mu(A) = A \text{ and } \mu(B') = B,$$

and then

$$(\mu \circ \phi)(A) = A \text{ and } (\mu \circ \phi)(B) = \mu(B') = B.$$

By the previous two cases, $\mu \circ \phi$ is either the identity or a reflection. If $\mu \circ \phi$ is the identity, ϕ is a reflection. If $\mu \circ \phi$ is a reflection, ϕ is the composition of two reflections.

- The case where $\phi(A) \neq A$, $\phi(B) \neq B$, and $\phi(C) \neq C$ is also left to Exercise 4. ■

The natural question to ask next is whether all reflections are isometries. The next proposition gives the answer. Since it can be proven either along the lines of Theorem 15.3.6 or by using methods from later in this section, its proof is left to Exercise 5.

■ **THEOREM 15.4.3**

The equation for the reflection μ with a mirror that has an angle of inclination $\theta/2$ and passes through the point (a, b) is

$$\mu(x, y) = \begin{bmatrix} \cos\theta & \sin\theta \\ \sin\theta & -\cos\theta \end{bmatrix} \begin{bmatrix} x - a \\ y - b \end{bmatrix} + \begin{bmatrix} a \\ b \end{bmatrix}.$$

The consequence of the theorem is that reflections are isometries because they can be written in the form of Theorem 15.3.6.

■ **COROLLARY 15.4.4**

Reflections are isometries.

Translations

The most basic plane motion is when a figure is slid in the direction of a given vector.

■ **DEFINITION 15.4.5**

The transformation $\tau : \mathbb{R}^2 \to \mathbb{R}^2$ is a **translation** if there exists a vector $\langle a, b\rangle$ such that

$$\tau(x, y) = (x + a, y + b).$$

We say that $\langle a, b\rangle$ **describes** τ. This is illustrated in Figure 15.5.

Observe that the translation τ in the definition can be written in the form

$$\tau(x, y) = \begin{bmatrix} 1 & 0 \\ 0 & 1 \end{bmatrix} \begin{bmatrix} x \\ y \end{bmatrix} + \begin{bmatrix} a \\ b \end{bmatrix}. \tag{15.5}$$

Conversely, any function of the form given in Equation 15.5 is a translation. Thus, we have a proposition that quickly follows from the definition by Theorem 15.3.6.

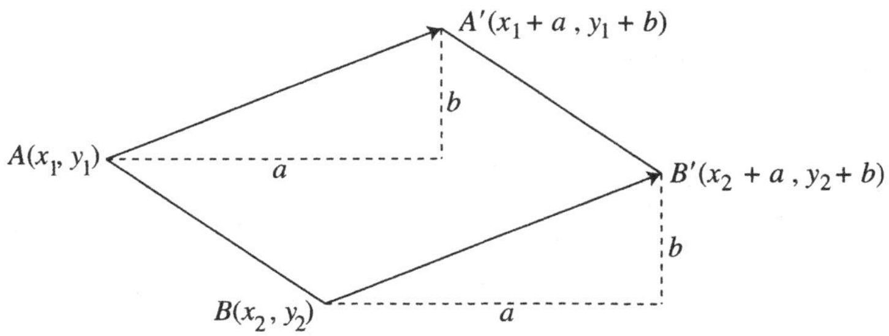

Figure 15.5 A translation of $\overline{AB}$ to image $\overline{A'B'}$.

■ THEOREM 15.4.6

A translation is an isometry.

As an isometry, a translation can be written as the composition of at most three reflections (Theorem 15.4.2). The next theorem improves on that.

■ THEOREM 15.4.7

A transformation is a translation if and only if it is the composition of two reflections with parallel mirrors.

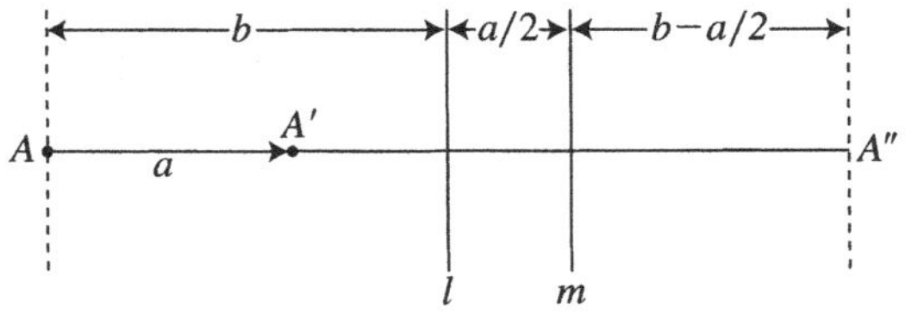

PROOF

We prove sufficiency. Take a vector of length a and let it describe a translation. Let l and m be two parallel lines that are perpendicular to the vector and are a distance $a/2$ apart. We claim that the translation is the composition of the reflection with mirror m and the reflection with mirror l. To prove this, let A be a point in the plane and let A' be its image under the translation. Let b be the distance from A to l. Take A'' to be the image of A by the reflection through l, so the distance from A to A'' is $2b$ and the distance from A'' to m is $b - a/2$, allowing for the possibility that this distance is signed. Now reflect A'' through m. Since $2b - 2(b - a/2) = a$, the image of this last reflection is A'. ■

We next want to show that the set of translations form a subgroup of the principal group under composition. For this we need some results.

■ LEMMA 15.4.8

The composition of translations is a translation.

PROOF

Let τ_1 and τ_2 be two translations of the plane. Let $\langle a_1, b_1\rangle$ be the vector that describes τ_1 and $\langle a_2, b_2\rangle$ be the vector that describes τ_2. When we compose the two functions, we find that

$$\begin{aligned}(\tau_1 \circ \tau_2)(x, y) &= \tau_1(\tau_2(x, y))\\ &= \tau_1(x + a_2, y + b_2)\\ &= ([x + a_2] + a_1, [x + b_2] + b_1)\\ &= (x + [a_2 + a_1], y + [b_2 + b_1]).\end{aligned}$$

The last ordered pair in this equation is that which defines the translation with vector $\langle a_2 + a_1, b_2 + b_1\rangle$. ■

Notice that although function composition is not generally commutative, it is in the case of translations. Namely, for any two translations τ_1 and τ_2,

$$\tau_1 \circ \tau_2 = \tau_2 \circ \tau_1.$$

This is because $\langle a_1 + a_2,\ b_1 + b_2\rangle = \langle a_2 + a_1,\ b_2 + b_1\rangle$.

■ **LEMMA 15.4.9**

If $\langle a,\ b\rangle$ describes the translation τ, $\langle -a,\ -b\rangle$ describes τ^{-1}.

PROOF

Let $\langle a,\ b\rangle$ describe τ and let $\langle -a,\ -b\rangle$ describe σ. We calculate:

$$(\tau \circ \sigma)(x,\ y) = \tau(x - a,\ y - b) = ([x - a] + a,\ [y - b] + b) = (x,\ y).$$

Therefore, $\tau \circ \sigma$ is the identity map. Since we know that composition is commutative for translations, we conclude that $\sigma \circ \tau$ is also the identity map. Thus, $\sigma = \tau^{-1}$. ■

The previous two lemmas, plus the fact that the identity map is a translation, allow us to draw our conclusion.

■ **THEOREM 15.4.10**

The set of translations of $\mathbb{R}^2$ is an abelian subgroup of the principal group.

Rotations

To rotate a point A around a center point O means to move A along the arc of a circle with center O and radius of length OA through a fixed angle measure θ. The result of the motion is A' in Figure 15.6. Algebraically, this means the following: Let $(x,\ y)$ be a point in $\mathbb{R}^2$. Let ρ be the transformation that rotates $(x,\ y)$ to $(x',\ y')$ along an arc of measure θ radians around the origin. Let r be the radius, so

$$r = \sqrt{x^2 + y^2},$$

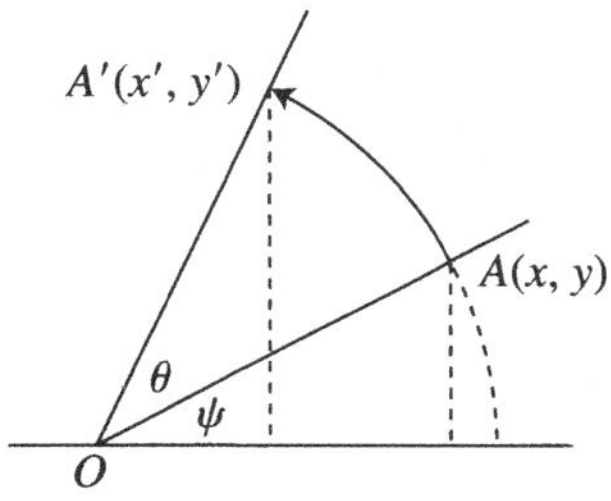

Figure 15.6 A rotation of A about the origin to image A'.

and let ψ be the angle that the line through the origin and (x, y) makes with the x-axis. This means that

$$x = r\cos\psi, \; y = r\sin\psi$$

and

$$x' = r\cos(\theta + \psi), \; y' = r\sin(\theta + \psi).$$

Using the addition formulas from trigonometry yields

$$x' = r\cos\theta\cos\psi - r\sin\theta\sin\psi = x\cos\theta - y\sin\theta$$

and

$$y' = r\sin\theta\cos\psi + r\cos\theta\sin\psi = x\sin\theta + y\cos\theta.$$

DEFINITION 15.4.11

The transformation $\rho : \mathbb{R}^2 \to \mathbb{R}^2$ is a **rotation** about the origin if there exists an angle measure θ such that

$$\rho(x, y) = (x\cos\theta - y\sin\theta, \; x\sin\theta + y\cos\theta).$$

We say that θ **describes** ρ.

We will not specifically define a function that represents a rotation about an arbitrary point. Instead, we compose translations with a rotation about the origin. Fix the point P with coordinates (a, b). Let τ be the translation described by $\langle -a, -b\rangle$ and ρ the rotation described by θ. The function

$$\tau^{-1} \circ \rho \circ \tau$$

is the rotation about P of measure θ (Figure 15.7). From this we can show that an arbitrary rotation with center (a, b) has the form

$$\rho(x, y) = \begin{bmatrix} \cos\theta & -\sin\theta \\ \sin\theta & \cos\theta \end{bmatrix} \begin{bmatrix} x - a \\ y - b \end{bmatrix} + \begin{bmatrix} a \\ b \end{bmatrix}. \qquad (15.6)$$

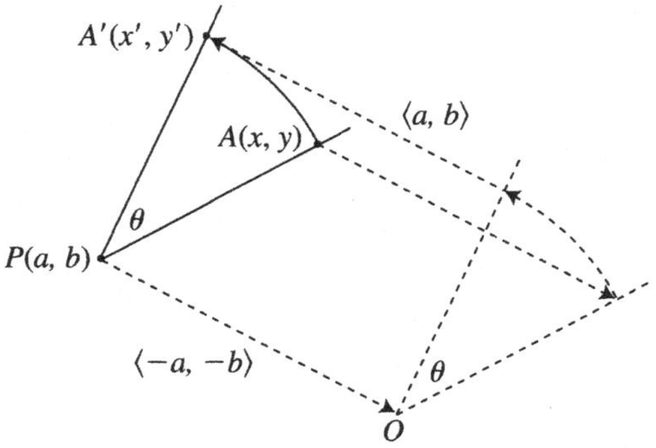

Figure 15.7 A rotation of A about P to image A'.

Moreover, any equation of the form given in Equation 15.6 is a rotation.

As with translations, we can write a rotation as a composition of two reflections.

■ THEOREM 15.4.12

A transformation is a rotation if and only if it is the composition of two reflections with intersecting mirrors.

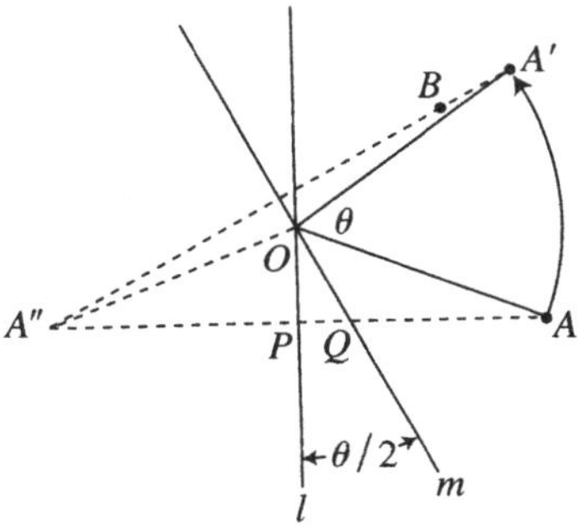

PROOF

We prove sufficiency. Let O be the center of the rotation and θ its angle. Let A be a point in the plane and A' its image under the rotation. Choose any line l containing O and let m also be through O, forming an angle of measure $\theta/2$ measured counterclockwise from l. Let A'' be the image of A under the reflection through mirror l and B the image of A'' through the mirror m. Then

$$\triangle A''PO \cong \triangle APO$$

by SAS. From this we conclude that $\overline{OA''} \cong \overline{OA}$, so A, A', and A'' lie on the same circle with center O. Moreover, letting Q be the intersection of m and $\overline{A''A}$, we see that $m\sphericalangle QA''B = \theta/2$ because this angle is the complement of $\sphericalangle OQP$. Therefore, the point B is actually the point A' because $\sphericalangle QA''A$ has the measure of an inscribed angle in $\odot O$ that intercepts $\widehat{AA'}$. Hence, A reflected through l and then through m results in A'. ■

■ COROLLARY 15.4.13

A rotation is an isometry.

We now identify the group structure of rotations. The first lemma involves the composition of two rotations. Supposing that they are both centered at the same point, the composition is another rotation about that point. Otherwise, we might obtain a translation.

■ LEMMA 15.4.14

The composition of two rotations is either a rotation or a translation.

PROOF

Let ρ_1 and ρ_2 be rotations about the points O and P, respectively. Let θ_1 be the angle the describes ρ_1 and θ_2 be the angle that describes ρ_2. Join $\overline{OP}$. As in the proof of Theorem 15.4.12, find l through O and a line m through P such that the measure of the angle from l to $\overline{OP}$ is $\theta_1/2$ and the measure of the angle from $\overline{OP}$ to m is $\theta_2/2$. That is,

$$\rho_1 = \mu_2 \circ \mu_1$$

and

$$\rho_2 = \mu_3 \circ \mu_2,$$

where μ_1 is the reflection with mirror l, μ_2 is the reflection with mirror $\overline{OP}$, and μ_3 is the reflection with mirror m. Thus,

$$\rho_2 \circ \rho_1 = \mu_3 \circ \mu_2 \circ \mu_2 \circ \mu_1 = \mu_3 \circ \mu_1.$$

If l and m are parallel, the composition is a translation (Theorem 15.4.7), otherwise it is a rotation (Theorem 15.4.12). ■

The next proposition states that the inverse of a rotation is the rotation in the opposite direction. Its proof is Exercise 14.

■ LEMMA 15.4.15

If θ describes the rotation ρ, then $-\theta$ describes ρ^{-1}.

We now have the pieces to prove that these transformations have a group structure under composition (Exercise 8).

■ THEOREM 15.4.16

The following are subgroups of the principal group. The first subgroup is abelian, but the second is not.

- The set of rotations about a given point.
- The set of translations and rotations.

There is one more transformation needed to complete our list of isometries. A **glide reflection** γ (Figure 15.8) is the composition of a reflection with mirror l and a translation described by a vector $\mathbf{v}$ that is parallel to l with angle of inclination θ. By Theorem 15.4.3 and Equation 15.5, the equation for a glide reflection with a mirror through a point (a, b) can be described by a function of the form

$$\gamma(x, y) = \begin{bmatrix} \cos\theta & \sin\theta \\ \sin\theta & -\cos\theta \end{bmatrix} \begin{bmatrix} x - a \\ y - b \end{bmatrix} + \begin{bmatrix} a + \|\mathbf{v}\| \cos\theta \\ b + \|\mathbf{v}\| \sin\theta \end{bmatrix} \tag{15.7}$$

Hence, when we examine the general form of an isometry given by Theorem 15.3.6 with Theorem 15.4.3 and Equations 15.5, 15.6, and 15.7, we can conclude that the translations of this section exhaust all possibilities for isometries in the plane. Therefore, we write the following result.

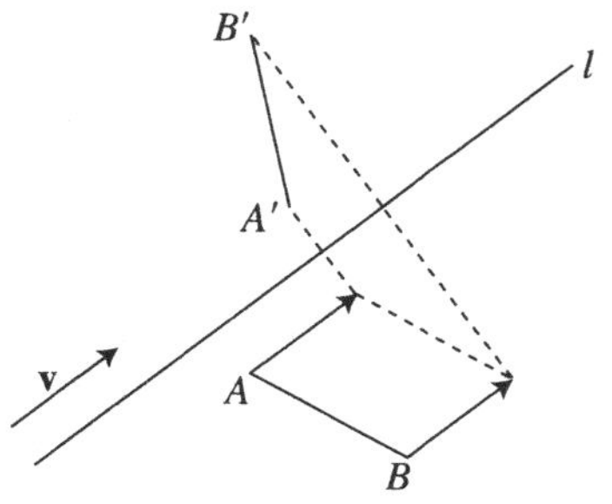

Figure 15.8 A glide reflection with vector **v** and mirror l.

THEOREM 15.4.17

The set of reflections, translations, rotations, and glide reflections of the plane is identical to the set of isometries.

Exercises

1. Find the image of the triangle with vertices (0, 0), (1, 0), and (0, 1) under the following isometries.
   (a) a reflection with mirror $y = -4$
   (b) a reflection with mirror $2x + 5y = 1$
   (c) a translation described by vector $\langle 2, -3\rangle$
   (d) a translation $4\sqrt{2}$ units along the line $y = x$
   (e) a rotation about the origin of measure $\pi/6$
   (f) a rotation about the point $(-4, -3)$ of measure $3\pi/4$
   (g) a glide reflection with mirror $x = 3$ and vector $\langle 0, 4\rangle$
   (h) a glide reflection with mirror $x - 2y = 10$ and vector $\langle 2, 1\rangle$

2. For the transformations of Exercise 1.
   (a) Write each as a reflection or product of reflections.
   (b) Write each in the form $\phi(\mathbf{x}) = A\mathbf{x}+\mathbf{b}$, where A is a 2×2 invertible matrix and $\mathbf{b}$ a vector in the plane.
   (c) Find the inverse of each transformation geometrically and then find it algebraically using the form $\phi(\mathbf{x}) = A\mathbf{x} + \mathbf{b}$.

3. Identify all isometries that can map the given figure to itself.
   (a) equilateral triangle
   (b) square
   (c) regular hexagon
   (d) parallelogram

4. Prove the second and fourth cases in the proof of Theorem 15.4.2.

5. Prove Theorem 15.4.3.

6. Prove that a reflection is an isometry and a reflection is its own inverse.

7. Prove necessity for Theorems 15.4.7 and 15.4.12.

8. Prove Theorem 15.4.16.

9. How can a reflection be used to show that the base angles of an isosceles triangle are congruent?

10. Let A and B be distinct points in the plane. Prove that there exists exactly one reflection μ so that $\mu(A) = B$.

11. Show that the composition of three reflections is a reflection or a glide reflection.

12. Define $\mathcal{G}$ to be the set of translations described by a vector with integer coordinates. Show that $\mathcal{G}$ is a subgroup of the set of translations.

13. Show that the composition of two homotheties is a translation or a homothety.

14. Prove that if θ describes the rotation ρ, then $-\theta$ describes ρ^{-1}.

15. Prove that the composition of two rotations about the same point is another rotation about that point.

16. Let $n \in \mathbb{Z}$ and define $\mathcal{G}_n$ to be the set of rotations about the origin of measure $2k\pi/n$ $(k \in \mathbb{Z})$. Prove that $\mathcal{G}_n$ is a subgroup of the set of all rotations about the origin.

17. Does the set of glide reflections form a subgroup of the principal group?

18. What is the inverse of a glide reflection?

19. Give the equation for the image of the line $2x - y = 0$ under the given isometries.
   (a) a reflection with mirror $x + 5y = 3$
   (b) a translation described by the vector $\langle -3, 7 \rangle$
   (c) a rotation of measure $\pi/6$ about the point $(1, 2)$
   (d) a glide reflection with mirror $x + y = 6$ and vector $\langle 3, -3 \rangle$

20. Let $a, b \in \mathbb{R}$. Show for all $(x_1, y_1), (x_2, y_2) \in \mathbb{R}^2$,

$$d[(x_1, y_1), (x_2, y_2)] = d[(x_1 + a, y_1 + b), (x_2 + a, y_2 + b)]$$

and

$$d[(x_1, y_1), (x_1 + a, y_1 + b)] = d[(x_2, y_2), (x_2 + a, y_2 + b)].$$

21. We say that a plane figure $\mathcal{F}$ is **symmetric** with respect to a line l if given the reflection μ with mirror l, $\phi[\mathcal{F}] = \mathcal{F}$. In this case, l is called a **line of symmetry** for $\mathcal{F}$. Find the lines of symmetry of the given figures.
   (a) equilateral triangle
   (b) regular hexagon
   (c) parallelogram

22. Using the definition given in Exercise 21, show the each figure is symmetric through the given lines.
   (a) parallelogram through a diagonal
   (b) a triangle through a median
   (c) circle through a diameter
   (d) a regular pentagon through the line connecting a vertex with the midpoint of the opposite side

15.5 CONSISTENCY OF EUCLIDEAN GEOMETRY

We have seen three transformation groups connected with Klein's Erlangen Program,

$$\mathbf{PG} \subseteq \mathbf{AF} \subseteq \mathbf{PR},$$

and each of these groups leaves different sets of geometric properties invariant (Figure 15.9). Furthermore, the "larger" the group, the more general is its geometry. For instance, since **PG** is a subgroup of **AF**, a model of Euclidean geometry can be interpreted as an affine model but not necessarily conversely. Also, both of these geometries are contained within projective geometry when the line at infinity is included in the plane. We should note that not all geometries fit into Klein's classification scheme. Differential geometry is an example. However, the Erlangen Program directed research for about 50 years and provided a framework in which the relationships between many geometries could be studied (Kline 1972, 920).

Leipzig

After Klein introduced his program in 1872, he remained at the University of Erlangen for four years. Because of his outstanding work, the *Technische Hochschule* in

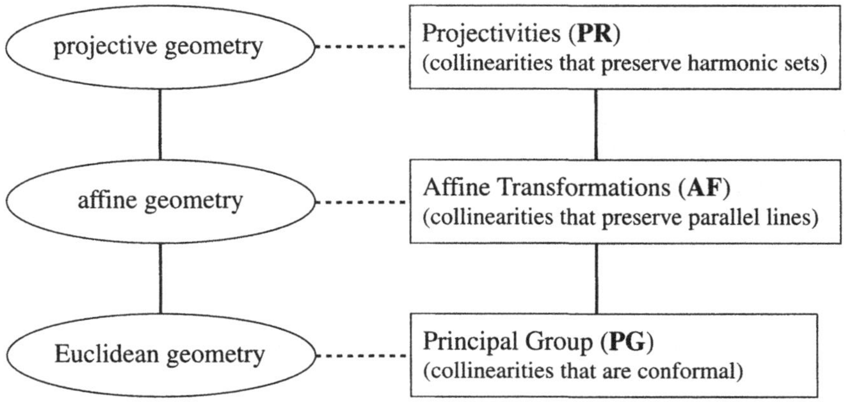

Figure 15.9 A hierarchy of geometries.

Munich offered him a position in the geometry department. As an institution with an outstanding scientific reputation, Klein was happy to accept. He spent several years there but moved on to Leipzig in 1880. At the university he began teaching geometry, set up a mathematical reading room, and taught students how to create mathematical drawings. However, his techniques were not well received by other faculty. Also during this time Klein began a correspondence with Poincaré concerning a mutual field of interest. In an attempt to beat him to results, Klein pushed himself to keep up with Poincaré. All of this stress led to a mental breakdown in 1882. Then in 1885, shortly after Klein had begun to recover from his rivalry with Poincaré, Klein met David Hilbert. At the time, Klein was 36 and Hilbert, 23 (Yaglom 1988, 125–126; Reid 1996, 18; Gray 2007, 225, 279).

Hilbert was born in Wehla near Kant's home town of Königsberg on January 23, 1862. His father was a judge and his mother had an interest in philosophy. Hilbert began school at the gymnasium when he was 8 years old. This was two years later than what was typical, probably because his mother home-schooled him for the first two years. Hilbert did not enjoy his school years because he thought that there was too much memorization. To him this was not interesting. What he did enjoy was mathematics. Hilbert thought the subject was easy because he did not have to memorize the answers. He could always simply figure it out. In the fall of 1880 he began his collegiate life at the University of Königsberg. Although his father wanted him to study law, Hilbert chose the only subject that truly interested him. It was traditional at the time in Germany to study at various schools, but Hilbert left Königsberg for only one semester. He much preferred to remain close to his home. After his coursework was completed, Lindemann, who was then at Königsberg, suggested Hilbert study invariants for his dissertation. He defended on December 11, 1884 (Reid 1996, 1, 3–4, 6–7, 9, 15).

After hearing Hilbert's first presentation at the *Technische Hochschule*, Klein realized that he would become a leading mathematician, so Klein suggested to Hilbert that he should travel to Paris as he had done 15 years earlier. Hilbert left in the early spring of 1886. While in Paris he met many of the leading French mathematicians and wrote back to Klein detailing his time there. He met Jordan, who he thought was a very pleasant person. He also met Poincaré; Hilbert's impression of the mathematician was that he was brilliant but quite shy. He also thought that Poincaré's tremendous output was due partly to his publishing even the smallest of results. Nonetheless, Hilbert benefited from the contacts he made in Paris and from the mathematics that he learned (Reid 1996, 19–20).

Göttingen

In 1886, Klein made the last move of his career. It was to the university of Gauss and Riemann at Göttingen. In June of that year Hilbert returned to Germany and made a stop at Klein's new home to make his last report in person. Hilbert did not stay long but returned to his home in Königsburg. By July he finished his *Habilitationsschrift*, which had been put on hold because of his travels. Hilbert was now a *privatdozent* at his hometown university, and he was pleased about his situation (Reid 1996, 25, 27).

Over the next several years the two mathematicians took different paths. Hilbert pushed ahead with his research. His main field of interest was invariant theory, a subject that Hilbert would eventually exhaust. Klein, on the other hand, had seen the last of his productive years as a mathematician. However, Klein was not the type to be inactive. He threw himself into other projects. For one, he was determined to make the university at Göttingen the leader in mathematical and scientific research. He began by changing the atmosphere of the university. He established a reading room on the third floor of the main hall as he had done in Leipzig. Students were encouraged to spend time there to discuss their work or freely take books from the shelves. In one of the hallways Klein displayed numerous mathematical models. This hall was famous among the students, so much so that they called it the "room of models," despite its being a simple hallway. To complete his plan, Klein then moved to recruit many of the top mathematicians to Göttingen. In 1894 a position opened and Klein made it clear that he wanted Hilbert to fill it. The search committee agreed and Hilbert accepted (Reid 1996, 45–46, 48; Yaglom 1988, 135).

At Göttingen Hilbert continued his mathematical pursuits and taught classes. It was here where the two brilliant mathematicians differentiated themselves from each other in a very public way. Hilbert was of the habit of preparing only broad outlines for his lectures and would then develop the mathematics in front of his students. He would often make errors when working the details, but he did not mind because he was showing his students how a mathematician worked. Hilbert's content was direct and to the point, and he would often repeat himself in an attempt to make sure that all of his students understood. Although some of his colleagues thought that his style was better suited for the gymnasium, most of his students appreciated the "beautiful insights" that they gleamed from the time in his classes.

Klein's approach was different. When it came to his lectures, Klein was a perfectionist. He would arrive at the room an hour early to mentally prepare his presentation. His assistant would have prepared for him a list of references. Klein would check these and use them all. By the time the lecture had started, he had prepared the entire lecture, including all of the formulas and diagrams that were required. By the end the chalkboard would be full of mathematics. Klein believed that he should present only an outline of the topic because he thought that students should spend a good four hours working on the mathematics outside his one-hour lecture (Reid 1996, 48–49).

New Postulates

During the 1898–1899 academic year, Hilbert surprised his students by giving a sequence of lectures on geometry. They had known him as an expert in invariant theory or his newfound interest of number theory, but not in geometry. Hilbert did have experience in the subject and had attended lectures on modern viewpoints that went beyond Lobachevski, Poncelet, and Riemann. Although inspired by their work, contemporary mathematicians such as Moritz Pasch had reduced the subject to pure logic, and the mathematician Giuseppe Peano took Pasch's work and translated it into his new symbolic notation. Essentially, in their hands geometry had become an abstract study not of points and lines but of sets of any kind that obeyed certain rules. Although Hilbert agreed with this new view, he realized that the new notation made it hard to grasp. Hilbert solved the problem by using the old Euclidean vocabulary when dealing with the new abstractions. The terms *point* and *line* simply became names for the objects and remain undefined. These objects could be anything as long as they could be interpreted to obey the postulates. Hilbert once clarified this viewpoint by stating that

> [o]ne must be able to say at all times — instead of points, straight lines, and planes — tables, chairs, and beer mugs.

Nonetheless, the familiar terms of *line* and *point* were preferred because they served as a mnemonic device for his students, making the new viewpoint easier to learn (Reid 1996, 57, 59, 61; Boyer and Merzbach 1991, 608).

Hilbert then took his geometric understanding to its logical conclusion. If the words *point* and *line* were going to be reexamined, why should not the same be done to Euclid's entire geometry? Despite his *Elements* remaining at the center of mathematics for over two millennia, Euclid's work was not without critiques. Gauss observed that Euclid had assumed betweenness properties in his proofs, and Leibnitz noted that he had assumed continuity. Neither of these are surprising when geometric proofs rely heavily on diagrams. Both betweenness and continuity are easily assumed from pictures. There is nothing wrong with these assumptions. The problem was that Euclid did not include them among his postulates.

To remedy matters, Hilbert introduced a new set of postulates and then proceeded to reprove the basic results of the first book of the *Elements*. As was standard, the number of postulates should be as few as possible, but they need not be self-evident because the terms *point* and *line* could represent anything. Nonetheless, the sentences chosen should use familiar terminology. More importantly, the postulate set must satisfy three criteria:

- The set must be independent.
- The set must be consistent.
- The set must be complete.

Although we saw the concepts of independence and consistency as early as Saccheri's work, Hilbert is responsible for making these criteria desirable for all modern axiom sets (Kline 1972, 102; Reid 1996, 63–64).

These were the ideas that Hilbert introduced to his students. His lectures became the basis for his *Foundations of Geometry*. It begins with these words,

> Geometry, like arithmetic, requires only a few and simple principles for its logical development. These principles are called the **axioms** of geometry. The establishment of the axioms of geometry and the investigation of their relationships is a problem which has been treated in many excellent works of the mathematical literature since the time of Euclid. This problem is equivalent to the logical analysis of our perception of space. (Hilbert 1992, 2)

Hilbert then begins to set up his system. He takes two types of objects. He called them **points** and **lines**. Denote points by uppercase Latin letters and lines by lowercase Latin letters. The points and lines exist in a **plane**.

When describing relationships among points and lines, it is common to use such words as **lie**, **between**, and **congruent**. For example,

- The point P *lies* on the line l.
- The point A is *between* the points B and C.
- The segments $\overline{AB}$ and $\overline{CD}$ are *congruent*.

These relationships will also be undefined, yet we will have some initial information about them because the axioms will codify our intuition concerning these relationships. The **incidence axioms** will explain *lie*, the **order axioms** will explain *between*, and the **congruence axioms** will explain *congruent*. There are two more sets of axioms: the singleton **parallel axiom**, which is a proposition equivalent to Euclid's Parallel Postulate, and the **continuity axioms**, which identify the principles that prevent Euclidean geometry from being discrete.

The incidence axioms determine what it means for one geometric object to lie on, pass through, intersect, or join another object (Hilbert 1992, 3–4). Compare them with the projective postulates of Sections 9.2 and 14.1.

■ **AXIOMS 15.5.1 [Incidence, *Foundations* §2]**

1. For every two points A and B, there exists a line l that contains each of the points A and B.
2. For every two points A and B, there exists no more than one line that contains each of the points A and B.
3. There exists at least two points on a line. There exist at least three points that do not lie on a line.

It is interesting that Hilbert stated the first two incident axioms as separate statements instead of combining them into one. However, it is reasonable to state the results as he did. Although we have not seen any geometry that does not satisfy the first, there is a model of the elliptic geometry of Riemann that does not satisfy the second (Section 13.4). However, both axioms together quickly yield the expected result.

■ **THEOREM 15.5.2**

For every two points A and B, there exists exactly one line that contains each of the points A and B.

Notice that $\mathbb{R}^2$ satisfies the Incidence Axioms, but there are other significantly smaller models that do the same thing. By Axiom 15.5.1.3 we arrive at a minimum of three points, A, B, and C. Axiom 15.5.1.1 guarantees that we have $\overleftrightarrow{AB}$, $\overleftrightarrow{AC}$, and $\overleftrightarrow{BC}$. If these are all the same line, then by Axiom 15.5.1.3, there exists at least three more points. However, it may be the case that $\overleftrightarrow{AB}$ and $\overleftrightarrow{AC}$ are different lines, so that the same axioms requires only two extra points, say D and E. Hence, only five points are required to satisfy Axioms 15.5.1.1 to 15.5.1.3. This is illustrated by the model

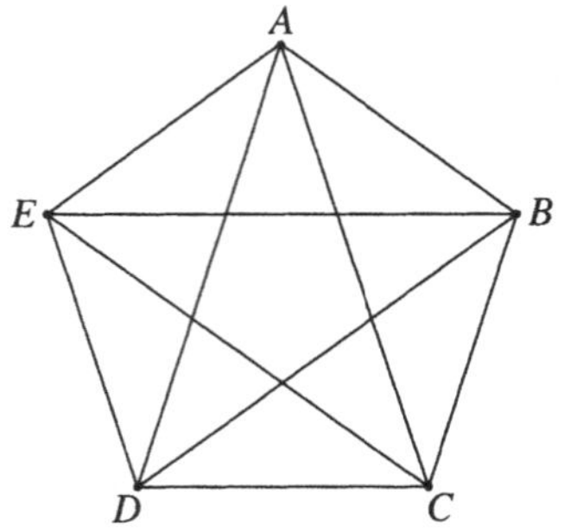

Figure 15.10 A finite model for the Incidence Axioms.

shown in Figure 15.10. Although the segments found inside the pentagon appear to intersect, they do not. Segments in this model intersect only at the points labeled.

The order axioms give meaning to the word *between*. We will write A–B–C to mean that B lies between A and C.

■ **AXIOMS 15.5.3 [Order, *Foundations* §3]**

1. If A–B–C, the points A, B, and C are three distinct points lying on a line, and C–B–A.
2. For two points A and B, there is a point C on $\overline{AB}$ produced such that A–B–C.
3. Of any three points on a line, there exists no more than one that lies between the other two.
4. **[Pasch]** In $\triangle ABC$, if a line intersects $\overline{AB}$ in a point D such that $A - D - B$, the line also intersects exactly one of the other sides in a point E such that either $A - E - C$ or $B - E - C$.

The order axioms allow us to conclude that points on lines behave as we expect

■ **THEOREM 15.5.4 [*Foundations* §4.3]**

For any two points A and C, there exists a point B on $\overline{AC}$ such that A–B–C.

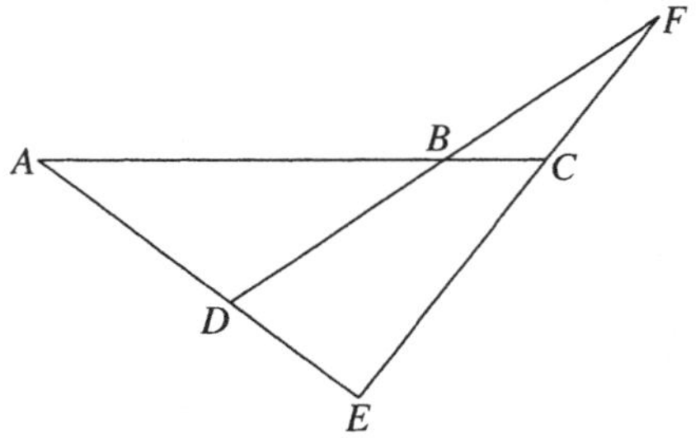

PROOF

Suppose that A and C are two points. Let D be a point not on $\overline{AC}$ by Axiom 15.5.1.3. Appealing to Axiom 15.5.3.2, we know that there is a point E such that A–D–E, and by the same axiom there is a point F such that E–C–F. If $\overline{DF}$ intersected $\overline{EC}$, Axiom 15.5.1.2 would be violated. Hence, $\overline{DF}$ intersects side $\overline{AC}$ at a point B by Pasch's Axiom (15.5.3.4). ■

If points A, B, and C are collinear, Axiom 15.5.3.3 prevents us from writing any two of A–B–C, B–A–C, or A–C–B at the same time, but none of the axioms state that one of these options must hold. Instead, this is a theorem. Its proof is important to read very carefully. Do not rely on results that seem to appear in the diagram but have yet to be proven.

■ **THEOREM 15.5.5 [*Foundations* §4.4]**

Of any three collinear points A, B, and C, there exists one that lies between the other two.

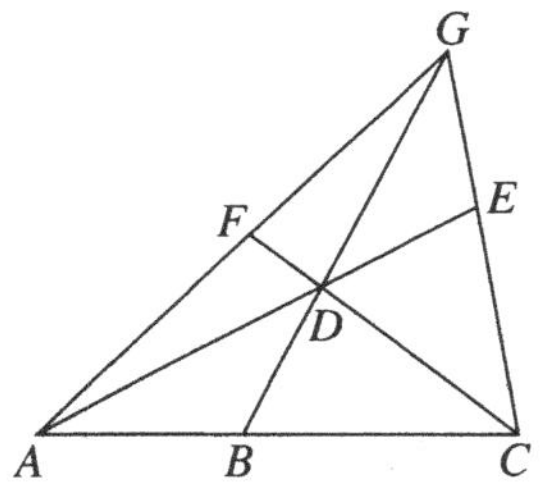

PROOF

Suppose that neither B–A–C nor A–C–B. Take D not on $\overline{AC}$ and find G such that B–D–G (Axioms 15.5.1.3 and 15.5.3.2). Using Pasch's Axiom (15.5.3.4) on $\triangle BCG$, it follows that $\overline{AD}$ intersects $\overline{GC}$ at E, where G–E–C. Notice that at the moment that we do not know if A–D–E. By a similar argument, $\overline{DC}$ intersects $\overline{AG}$ at F with A–F–G. Furthermore, it is still a question whether F–D–C.

Now apply Pasch's Axiom to $\triangle FGC$ using $\overline{AE}$. Since this line intersects $\overline{GC}$ at E and it is impossible for it to intersect $\overline{FG}$ (Axiom 15.5.1.2), we see that $\overline{AE}$ must intersect $\overline{FC}$. Since already $\overline{AE}$ meets $\overline{FC}$ at D, this point must be the point given by Pasch's Axiom. Hence, F–D–C. But, applying this same argument to $\triangle AFC$ with $\overline{DG}$ intersecting $\overline{FC}$ at D, we find that A–B–C. ■

Although the Incidence Axioms can be satisfied by a finite model, that is not the case for the Order Axioms. By a sequence of applications of Theorem 15.5.4, we may prove the next result.

■ **THEOREM 15.5.6 [*Foundations* §4.6]**

Between any two points on a line there exists a countably infinite number of points.

The next set of axioms deal with the familiar concept of congruence. Euclid wrote none of these as postulates in the *Elements*, but he used them all.

■ **AXIOMS 15.5.7 [Congruence, *Foundations* §5]**

1. If A and B are points on a line l, and A' is a point on the same or on another line l', there exists a point B' on a given side of the line that contains B' such that $\overline{AB} \cong \overline{A'B'}$.
2. If $\overline{A'B'} \cong \overline{AB}$ and $\overline{A''B''} \cong \overline{AB}$, then $\overline{A'B'} \cong \overline{A''B''}$.
3. On the line l, let $\overline{AB}$ and $\overline{BC}$ have no points in common other than B. On a line l' that may or may not be the same line as l, let $\overline{A'B'}$ and $\overline{B'C'}$ have no points in common other than B'. In this case, if $\overline{AB} \cong \overline{A'B'}$ and $\overline{BC} \cong \overline{B'C'}$, then $\overline{AC} \cong \overline{A'C'}$.
4. For any $\sphericalangle ABC$ and $\overline{B'A'}$, there exists a unique $\overline{B'C'}$ on a given side of $\overline{B'A'}$ such that $\sphericalangle ABC \cong \sphericalangle A'B'C'$. Moreover, every angle is congruent to itself.
5. In $\triangle ABC$ and $\triangle A'B'C'$, if
$$\overline{AB} \cong \overline{A'B'},\ \overline{AC} \cong \overline{A'C'}, \text{ and } \sphericalangle BAC \cong \sphericalangle B'A'C',$$
then $\sphericalangle ABC \cong \sphericalangle A'B'C'$.

Hilbert concluded his axiom set by including propositions that are required to make the Euclidean plane have the properties that we expect. That is, the plane should have parallel lines and be continuous. Hilbert accomplished this by including Archimedes' Postulate (6.2.1) and by returning to the definition of parallel used by Euclid. Instead of using the ancient version of the Parallel Postulate, Hilbert chose a modified version of Playfair's Postulate (10.1.1).

■ **DEFINITION 15.5.8**

Two lines are **parallel** if they do not intersect.

■ **AXIOM 15.5.9 [Euclid]**

If l is any line and P a point not on l, then there exists at most one line that passes through P and is parallel to l.

The Model

To show that Hilbert's axioms for Euclidean geometry are consistent, we must find a model in which the axioms can be interpreted as true statements. Because of Theorem 15.5.6, the model must be infinite. A suitable chosen set of rational numbers will suffice, but along with Hilbert we will choose the Cartesian plane $\mathbb{R}^2$.

We interpret the undefined terms *point*, *line*, *incident*, *between*, and *congruent* as follows:

- A *point* is an element of $\mathbb{R}^2$.
- A *line* is an element of $\mathcal{L} = \{[u, v, w] : u, v, w \in \mathbb{R}\}$.
- The point (x, y) is *incident* with line $[u, v, w]$ if and only if
$$ux + vy + w = 0.$$
- The point (x_2, y_2) is *between* (x_1, y_1) and (x_3, y_3) if and only if all three points are different and there exists a line on which all three points are incident and
$$\text{either } x_1 \leqslant x_2 \leqslant x_3 \text{ or } x_1 \geqslant x_2 \geqslant x_3, \text{ and}$$
$$\text{either } y_1 \leqslant y_2 \leqslant y_3 \text{ or } y_1 \geqslant y_2 \geqslant y_3.$$
- $\overline{AB}$ is *congruent* to $\overline{A'B'}$ if and only if there exists an isometry ϕ such that $\phi[\overline{AB}] = \overline{A'B'}$, where for every two points A and B,
$$\overline{AB} = \{C : A - C - B\}.$$
- $\measuredangle ABC$ is *congruent* to $\measuredangle A'B'C'$ if and only if there exists a similarity ψ such that $\psi[\measuredangle ABC] = \measuredangle A'B'C'$. An angle in the model is two line segments that share exactly one endpoint and are produced indefinitely in one direction.

Call $\mathbb{R}^2$ with these interpretations the **Cartesian model** of Euclidean geometry. To illustrate these interpretations, observe the following:

- $(0, 1)$ is incident with $[2, 4, -4]$ because $(0, 1)$ is a solution to $2x+4y-4 = 0$.
- $(0, 4)$ is between $(0, 1)$ and $(0, 5)$ because these points are on $[1, 0, 0]$, $0 \leqslant 0 \leqslant 0$, and $1 \leqslant 4 \leqslant 5$.
- The line segment with endpoints $(1, 2)$ and $(-3, 5)$ is congruent to the line segment with endpoints $(3, -1)$ and $(-1, 2)$ because the second segment is the image of the first under the translation described by $\langle 2, -3\rangle$.

We will check that some of the axioms hold true under this interpretation and leave the rest for Exercise 5. We begin with an incidence axiom.

> *For every two points A and B, there exists a line l that contains each of the points A and B.*

Let $A = (x_1, y_1)$ and $B = (x_2, y_2)$. Using some basic algebra, we find that
$$y - y_1 = \frac{y_2 - y_1}{x_2 - x_1}(x - x_1).$$
Therefore,
$$(y_1 - y_2)x + (x_2 - x_1)y + x_1y_2 - x_2y_1 = 0,$$
which means that both A and B are on $[y_1 - y_2, x_2 - x_1, x_1y_2 - x_2y_1]$.

Next we examine an axiom of order.

For two points A and B, there is a point C on $\overline{AB}$ produced such that A–B–C.

Let $A = (x_1, y_1)$ and $B = (x_2, y_2)$. We already know that these points are on $[y_1 - y_2, x_2 - x_1, x_1y_2 - x_2y_1]$. There are many cases to consider when finding C. For example, if $0 \leqslant x_1 \leqslant x_2$ and $0 \leqslant y_1 \leqslant y_2$, then

$$C = (2x_2 - x_1, 2y_2 - y_1)$$

is on the line and $A - B - C$ because

$$x_1 \leqslant x_2 \leqslant 2x_2 - x_1$$

and

$$y_1 \leqslant y_2 \leqslant 2y_2 - y_1.$$

Finally, we check a congruence axiom.

On the line l, let $\overline{AB}$ and $\overline{BC}$ have no points in common other than B. On a line l′ that may or may not be the same line as l, let $\overline{A'B'}$ and $\overline{B'C'}$ have no points in common other than B′. In this case, if $\overline{AB} \cong \overline{A'B'}$ and $\overline{BC} \cong \overline{B'C'}$, then $\overline{AC} \cong \overline{A'C'}$.

Let $A - B - C$ on l and $A' - B' - C'$ on l'. Assume that $\overline{AB} \cong \overline{A'B'}$ and $\overline{BC} \cong \overline{B'C'}$. Under our interpretation, this means that there exist isometries ϕ and ψ such that

$$\phi[\overline{AB}] = \overline{A'B'} \text{ and } \psi[\overline{BC}] = \overline{B'C'}.$$

Using these two functions, we define

$$\eta(X) = \begin{cases} \phi(X) & \text{if } A - X - B \text{ or } X = A \text{ or } X = B \\ \psi(X) & \text{if } B - X - C \text{ or } X = C. \end{cases}$$

This is the isometry desired.

The models of the hyperbolic and elliptic geometries show that these new systems are consistent if Euclidean geometry is consistent. By using the Cartesian plane as a model of his axioms, Hilbert showed that Euclidean geometry is consistent if arithmetic is consistent. Since mathematicians had little doubt concerning the consistency of arithmetic, Hilbert succeeded in putting all of the geometries on a solid logic footing.

Final Note

Lie, who had been a professor of mathematics at Christiania University, took Klein's place at Leipzig by Klein's recommendation. Three years later, in 1898, Lie received the first International Lobachevski Prize for his geometric work. The prize was established by the Kazan society in 1893 to celebrate the 100th anniversary of

Lobachevski's birth. Later winners would include Klein and Poincaré. In 1898, Lie returned to his home in Norway, and he died the following year on February 18, 1899 (Yaglom 1988, 126–130).

In 1872, Klein began assisting the editors of the journal *Mathematische Annalen* and became its chief editor in 1876. As was typical for Klein, he was instrumental in earning the journal its reputation as the leading mathematical journal. In 1898 he led the team whose task was to published Gauss's collected works. They finished in 1918. His other ventures included the publication of his own books and the lecture notes from the classes he taught at Göttingen (Yaglom 1988, 132–134).

Influenced by his childhood days at the Düsseldorf gymnasium, days that he did not remember with fondness, Klein led efforts to reform the teaching of mathematics. Himself known to be an excellent teacher, in 1898 he organized the International Commission on Mathematics Education. The commission recommended numerous changes that we see still today in the mathematics classroom:

- The use of visual aids
- The emphasis on the study of functions in algebra
- The application of geometric transformations in geometry
- A more unified approach to various mathematical subjects
- A closer connection between mathematics and science education

Such activities, combined with his academic accomplishments, led to his being the representative for Göttingen University in the upper house of the Prussian parliament (Yaglom 1988, 136–137).

Klein lived out his days in Göttingen and died on July 22, 1925 (Yaglom 1988, 137). Under Klein's leadership the university had risen from the mediocrity that Riemann encountered to being a preeminent institute for mathematics. He left the university in good hands. Not only did Hilbert remain to carry on the tradition, but there were other notable mathematicians, including Paul Bernays, Richard Courant, Edmund Landau, Emmy Noether, and Hermann Weyl, in addition to Max Born and James Franck in physics. In 1932, Hilbert celebrated his seventieth birthday, and the next year saw the beginning of the end of what Klein had established at Göttingen (Reid 1996, 191, 203–204).

Exercises

1. Show that the following axiom sets are independent.
   (a) Incidence axioms (15.5.1)
   (b) Order axioms (15.5.3)
   (c) Congruence axioms (15.5.7)
2. Which of Hilbert's axioms can be shown to be consistent with a finite model?
3. Identify all pairs of parallel lines in Figure 15.10.

4. Prove that Hilbert's geometry contains infinitely many lines.

5. Show that Hilbert's axioms hold in the Cartesian model.

6. Prove that congruence of line segments is an equivalence relation.

7. How would we show that a set of axioms is complete?

8. Use Hilbert's concept of betweenness to define the interior and exterior of an angle.

9. Define what it means for two points to be on opposite sides of a line segment. What would it mean for two points to be on the same side of a line segment?

10. Prove the following from Hilbert's *Foundations of Geometry*. Assume the standard definitions for all defined terms.

(a) For any two points A and B, there exists C collinear with A and B such that C is between A and B.

(b) Of any three distinct collinear points, there exists one the is between the other two.

(c) The base angles of an isosceles triangle are congruent.

(d) SAS

(e) ASA

(f) Let $\measuredangle ABC$ be the supplement of $\measuredangle CBD$ and $\measuredangle A'B'C'$ the supplement of $\measuredangle C'B'D'$. If $\measuredangle ABC \cong \measuredangle A'B'C'$, then $\measuredangle CBD \cong \measuredangle C'B'D'$.

(g) Vertical angles are congruent.

(h) Let $\overline{BD}$ be in the interior of $\measuredangle ABC$ and $\overline{B'D'}$ be in the interior of $\measuredangle A'B'C'$. If $\measuredangle DBC \cong \measuredangle D'B'C'$ and $\measuredangle ABC \cong \measuredangle A'B'C'$, then $\measuredangle ABD \cong \measuredangle A'B'D'$.

(i) If two points A and C are put on different sides of $\overline{BD}$ and if $\overline{AD} \cong \overline{CD}$ and $\overline{BA} \cong \overline{BC}$, then $\measuredangle ABD \cong \measuredangle CBD$.

(j) SSS

(k) If $\measuredangle A' \cong \measuredangle A$ and $\measuredangle A'' \cong \measuredangle A$, then $\measuredangle A' \cong \measuredangle A''$.

(l) All right angles are congruent to each other.

(m) The exterior angle of a triangle is greater than any interior angle that is not adjacent to it.

(n) In every triangle the greater angle lies opposite the greater side.

(o) A triangle with two congruent angles is isosceles.

(p) AAS

(q) Every segment can be bisected.

REFERENCES

Anglin, W. S. and J. Lambek (1995). *The Heritage of Thales*. New York: Springer-Verlag.

Annas, Julia (2000). Plato. In *Greek Thought: A Guide to Classical Knowledge*, ed. Jacques Brunschwig and Geoffrey E. R. Lloyd, pp. 672–692. Cambridge, MA: Belknap Press.

Anton, Howard, Irl Bivens, and Stephen Davis (2009). *Calculus: Early Transcendentals*, 9th ed. New York: Wiley.

Apollonius (1889). *Treatise on Conic Sections*, ed. T. L. Heath. Reprint, New York: Barnes & Noble, 1961.

Apostle, H. G. (1952). *Aristotle's Philosophy of Mathematics*. Chicago: University of Chicago Press.

Archimedes (1897). *The Works of Archimedes*, ed. Sir Thomas L. Heath. Reprint, New York: Dover, 2002.

Aristotle (1984). *The Complete Works of Aristotle*, ed. Jonathan Barnes. Princeton, NJ: Princeton University Press.

Artin, Michael (1991). *Algebra*. Englewood Cliffs, NJ: Prentice Hall.

Bennett, M. K. (1995). *Affine and Projective Geometry*. New York: Wiley.

Bibby, Neil (2006). *Tuning and Temperment: Closing the Spiral*. Music and Mathematics: From Pythagoras to Fractals, ed. John Fauvel, Raymond Flood, and Robin Wilson, pp. 13–27. Oxford, UK: Oxford University Press.

Boardman, John (1982). *The Cambridge Ancient History*, Vol. I, P. 2. London: Cambridge University Press.

Bocheński, I. M. (1968). *Ancient Formal Logic*. Studies in Logic and the Foundations of Mathematics, ed. L. E. J. Brouwer, E. W. Beth, and A. Heyting. Amsterdam: North-Holland.

Bodéüs, Richard (2000). The Statesman as Political Actor. In *Greek Thought: A Guide to Classical Knowledge*, ed. Jacques Brunschwig and Geoffrey E. R. Lloyd, trans. Elizabeth Rawlings and Jeannine Pucci, pp. 125–146. Cambridge, MA: Belknap Press.

Bólyai, János (1987). *Appendix, the Theory of Space*, ed. Ference Kárteszi and Barna Szénássy. Amsterdam: North-Holland.

Bonola, Roberto (1912). *Non-Euclidean Geometry: A Critical and Historical Study of Its Development*. Reprint, New York: Dover, 1955.

Borsuk, Karol and Wanda Szmielew (1960). *Foundations of Geometry*, trans. Erwin Marquit. Amsterdam: North-Holland.

Boyer, Carl B. (1956). *History of Analytic Geometry*. The Scripta Mathematica Studies, Nos. 6 and 7. New York: Scripta Mathematica.

Boyer, Carl B. and Uta C. Merzbach (1991). *A History of Mathematics*, 2nd ed. New York: Wiley.

Bühler, W. K. (1981). *Gauss: A Biographical Study*. Berlin: Springer-Verlag.

Bunt, Lucas N. H., Phillip S. Jones, and Jack D. Bedient (1976). *The Historical Roots of Elementary Mathematics*. Englewood Cliffs, NJ: Prentice Hall.

Burau, Werner and Bruno Schoenberg (1970). Klein, Christian Felix. In *Dictionary of Scientific Biography*, ed. Charles Coulston Gillispie, Vol. VII, pp. 396–400. New York: Scribner.

Burton, David M. (1985). *The History of Mathematics: An Introduction*. Newton, MA: Allyn and Bacon.

Cajori, Florian (1928). *A History of Mathematical Notations*. Chicago: Open Court.

Cherniss, Harold (1962). *Aristotle's Criticism of Plato and the Academy*. New York: Russell & Russell.

Cicero, Marcus Tullius (1951). The Basic Works of Cicero. In *The Modern Library of the World's Best Books*, Vol. 272. New York: Modern Library.

Clayton, Peter A. (1995). *Chronicle of the Pharaohs*. London: Thames & Hudson.

Coolidge, Julian Lowell (1963). *A History of Geometrical Methods*. Reprint, Mineola (N. Y.): Dover, 2003.

Copi, Irving M. and Carl Cohen (2002). *Introduction to Logic*, 11th ed. Upper Saddle River, NJ: Prentice Hall.

Coxeter, H. S. M. (2003). *Projective Geometry*, 2nd ed. New York: Springer.

Crawford, Harriet (2004). *Sumer and the Sumerians*, 2nd ed. Cambridge, UK: Cambridge University Press.

Davies, Charles (1858). *Geometry and Trigonometry, from the Works of A. M. Legendre*. New York: Barnes.

Descartes, René (1925). *The Geometry of Rene Descartes*, trans. David Eugene Smith and Marcia L. Latham. Reprint, New York: Dover, 1954.

Descartes, René (1985). *The Philosophical Writings of Descartes*, ed. John Cottingham, Robert Stoothoff, and Dugald Murdoch. Cambridge, UK: Cambridge University Press.

Dieudonné, Jean (1970). Poincaré, Jules Henri. In *Dictionary of Scientific Biography*, ed. Charles Coulston Gillispie, Vol. XI, pp. 51–61. New York: Scribner.

Dijksterhuis, E. J. (1987). *Archimedes*, trans. C. Dikshoorn. Princeton, NJ: Princeton University Press.

Dillon, John (2003). *The Heirs of Plato: A Study of the Old Academy (347–274 BC)*. Oxford, UK: Clarendon Press.

Diogenes (1966). *Lives of the Eminent Philosophers,* Vol. 1. Loeb Classic Library, trans. R. D. Hicks. Cambridge: Harvard University Press.

Dodge, Clayton W. (1972). *Euclidean Geometry and Transformations*. Reprint, Mineola, NY: Dover, 2004.

Euclid (1925). *The Elements*, ed. Sir Thomas Heath. Reprint, New York: Dover, 1956.

Evans, James (1998). *The History and Practice of Ancient Astronomy*. New York: Oxford University Press.

Eves, Howard (1965). *A Survey of Geometry*, Vol. 2. Boston: Allyn and Bacon.

Eves, Howard (1992). *Fundamentals of Modern Elementary Geometry*. Boston: Jones & Bartlett.

Ewald, Günter (1971). *Geometry: An Introduction*. Belmont, CA: Wadsworth.

Field, J. V. and J. J. Gray (1987). *The Geometrical Work of Girard Desargues*. New York: Springer-Verlag.

Fishback, W. T. (1962). *Projective and Euclidean Geometry*. New York: Wiley.

Fraleigh, John B. (1999). *A First Course in Abstract Algebra*, 6th ed. Reading, MA: Addison-Wesley.

Gaukroger, Stephen (1995). *Descartes: An Intellectual Biography*. Oxford, UK: Clarendon Press.

Gillings, Richard J. (1972). *Mathematics in the Time of the Pharaohs*. Reprint, Mineola, NY: Dover, 1982.

Gottlieb, Anthony (2000). *The Dream of Reason: A History of Philosophy from the Greeks to the Renaissance*. New York: W. W. Norton.

Grafton, Anthony, ed. (2000). *Leon Battista Alberti: Master Builder of the Italian Renaissance*. New York: Hill & Wang.

Granville, William Anthony (1909). *Plane and Spherical Trigonometry and Four-Place Tables of Logarithms*. Boston: Ginn.

Gray, Jeremy (2006). Gauss and Non-Euclidean Geometry. In *Non-Euclidean Geometries: János Bolyai Memorial Volume*, ed. András Prékopa and Emil Molnár, pp. 61–80. New York: Springer.

Gray, Jeremy (2007). *Worlds Out of Nothing: A Course in the History of Geometry in the 19th Century*. London: Springer.

Gray, Jeremy J. (2004). *János Bolyai, Non-Euclidean Geometry, and the Nature of Space*. Cambridge, MA: Burndy Library.

Gray, J. J. and L. Tilling (1978). Johann Heinrich Lambert, Mathematician and Scientist 1728–1777. *Historia Mathematica, 5*(1), 13–41.

Grimal, Nicolas (1992). *A History of Ancient Egypt*, trans. Ian Shaw. Oxford, UK: Blackwell.

Guthrie, Kenneth Sylvan (1988). *The Pythagorean Sourcebook and Libary*, ed. David Fideler. Grand Rapids, MI: Phanes.

Guthrie, W. K. C. (1962). *A History of Greek Philosophy*. London: Cambridge University Press.

Halverson, William H. (1981). *A Concise Introduction to Philosophy*, 4th ed. New York: Random House.

Hatfield, Gary (2006). Kant on the Perception of Space (and Time). In *The Cambridge Companion to Kant and Modern Philosophy*, ed. Paul Guyer, pp. 61–93. Cambridge, UK: Cambridge University Press.

Heath, Sir Thomas (1921a). *A History of Greek Mathematics,* Vol. 1; *From Thales to Euclid*. Reprint, New York: Dover, 1981.

Heath, Sir Thomas (1921b). *A History of Greek Mathematics,* Vol. 2; *From Aristarchus to Diophantus*. Reprint, New York: Dover, 1981.

Herodotus (1987). *The History*, trans. David Grene. Chicago: University of Chicago Press.

Hilbert, David (1992). *Foundations of Geometry*. La Salle, IL: Open Court.

Houston, James M., ed. (1989). *The Mind on Fire: An Anthology of the Writings of Blaise Pascal*. Portland, OR: Multnomah.

Hurley, Patrick J. (2000). *A Concise Introduction to Logic*, 7th ed. Belmont, CA: Wadsworth/Thomson.

Itard, Jean (1970). Legendre, Adrien-Marie. In *Dictionary of Scientific Biography*, ed. Charles Coulston Gillispie, Vol. VIII, pp. 135–143. New York: Scribner.

Jackson, Samuel Macauley, ed. (1953). *The New Schaff-Herzog Encyclopedia of Religious Knowledge*. Grand Rapids, MI: Baker.

Jaeger, Werner (1948). *Aristotle: Fundamentals of the History of His Development*, 2nd ed., trans. Richard Robinson. Oxford, UK: Clarendon Press.

Kadison, Lars and Matthias T. Kromann (1996). *Projective Geometry and Modern Algebra*. Boston: Birkhäuser.

Kahn, Charles H. (2001). *Pythagoras and the Pythagoreans: A Brief History*. Indianapolis, IN: Hackett.

Kant, Immanuel (1998). *Critique of Pure Reason*. The Cambridge Edition of the Works of Kant, ed. and trans. Paul Guyer and Allen W. Wood. Cambridge, UK: Cambridge University Press.

Kant, Immanuel (1999). *Correspondence*. Cambridge, UK: Cambridge University Press.

Klein, Felix (1893). A Comparative Review of Recent Researches in Geometry. *Bulletin of the New York Mathematical Society,* 2, 215–249.

Klein, Felix (1987). *Famous Problems of Elementary Geometry*. Reprint, New York: Cosimo Classics.

Kline, Morris (1953). *Mathematics in Western Culture*. London: Oxford University Press.

Kline, Morris (1972). *Mathematical Thought from Ancient to Modern Times*. New York: Oxford University Press.

Kline, Morris (1980). *Mathematics: The Loss of Certainty.* New York: Oxford University Press.

Kneale, William and Martha Kneale (1964). *The Development of Logic.* Oxford, UK: Clarendon Press.

Kuehn, Manfred (2001). *Kant: A Biography.* Cambridge, UK: Cambridge University Press.

Laczkovich, M. (1997). On Lambert's Proof of the Irrationality of π. *American Mathematical Monthly, 104*(5), 439–443.

Laugwitz, Detlef (1999). *Bernhard Riemann, 1826–1866: Turning Points in the Conception of Mathematics*, trans. Abe Shenitzer. Boston: Birkhäuser.

Leick, Gwendolyn (2003). *Historical Dictionary of Mesopotamia.* Lanham, MD: Scarecrow Press.

Lévy, Carlos (2000). The Academy. In *Greek Thought: A Guide to Classical Knowledge*, ed. Jacques Brunschwig and Geoffrey E. R. Lloyd, trans. Catherine Porter, pp. 799–820. Cambridge, MA: Belknap Press.

Livio, Mario (2002). *The Golden Ratio.* New York: Broadway Books.

Lobachevski, Nikolai Ivanovich (1914). *Geometrical Researches on the Theory of Parallels*, trans. George Bruce Halsted. LaSalle, IL: Open Court.

Mahoney, Michael Sean (1994). *The Mathematical Career of Pierre de Fermat: 1601–1665*, 2nd ed. Princeton, NJ: Princeton University Press.

Manning, Henry Parker (1901). *Non-Euclidean Geometry.* Reprint, New York: Dover, 1963.

Maor, Eli (2007). *The Pythagorean Theorem: A 4,000-Year History.* Princeton, NJ: Princeton University Press.

Marsden, Jerrold E. and Michael J. Hoffman (1993). *Elementary Classical Analysis.* New York: W.H. Freeman.

Martin, George E. (1975). *The Foundations of Geometry and the Non-Euclidean Plane.* New York: Springer.

McCleary, John (1994). *Geometry from a Differential Viewpoint.* Cambridge, UK: Cambridge University Press.

Moise, Edwin E. (1974). *Elementary Geometry from an Advanced Standpoint*, 2nd ed. Reading, MA: Addison-Wesley.

Moore, A. W. (1990). *The Infinite.* New York: Routledge, Chapman & Hall.

Netz, Reviel (2004). The Two Books on the Sphere and the Cylinder. In *The Works of Archimedes*, Vol. 4. Cambridge, UK: Cambridge University Press.

Netz, R. and W. Noel (2007). *The Archimedes Codex: How a Medieval Prayer Book Is Revealing the True Genius of Antiquity's Greatest Scientist.* Philadelphia: De Capo.

Nicomachus (1926). *Nicomachus of Gerasa: Introduction to Arithmetic*, ed. Frank Egleston Robbins and Louis Charles Karpinski, trans. Martin Luther D'Ooge. London: Macmillan.

O'Leary, Michael L. (2002). *The Structure of Proof: With Logic and Set Theory.* Upper Saddle River, NJ: Prentice Hall.

Pannekoek, A. (1961). *A History of Astronomy.* London: George Allen.

Pappus (1986). *Pappus of Alexandria: Book 7 of the Collection*, ed. Alexander Jones. Sources in the History of Mathematics and Physical Sciences, Vol. 8. New York: Springer.

Plato (1961). *The Collected Dialogues of Plato Including the Letters*, ed. Edith Hamilton and Huntington Cairns. Princeton, NJ: Princeton University Press.

Plutarch (1932). *The Lives of the Noble Grecians and Romans — Dryden Translation*, trans. John Dryden. New York: Modern Library.

Posamentier, Alfred A. and Charles T. Salkind (1988). *Challenging Problems in Geometry.* Palo Alto, CA: Dale Seymour.

Prékopa, András (2006). The Revolution of János Bolyai. In *Non-Euclidean Geometries: János Bolyai Memorial Volume*, ed. András Prékopa and Emil Molnár, pp. 3–60. New York: Springer.

Pressley, Andrew (2001). *Elementary Differential Geometry.* London: Springer.

Proclus (1970). *A Commentary on the First Book of Euclid's* Elements, trans. Glen R. Morrow. Princeton, NJ: Princeton University Press.

Reid, Constance (1996). *Hilbert.* New York: Copernicus.

Riemann, Bernhard (1854). On the Hypotheses Which Lie at the Bases of Geometry. In William Kingdon Clifford, *Mathematical Papers*, ed. Robert Tucker. London: Macmillan, 1882.

Roberts, J. A. G. (1999). *A Concise History of China.* Cambridge: Harvard University Press.

Rogers, Ben (2003). Pascal's Life and Times. In *The Cambridge Companion to Pascal*, ed. Nicholas Hammond, pp. 4–19. Cambridge, UK: Cambridge University Press.

Rosen, Kenneth H. (1993). *Elementary Number Theory and Its Applications*, 3rd ed. Reading, MA: Addison-Wesley.

Rosenfeld, B. A. (1988). *A History of Non-Euclidean Geometry: Evolution of the Concept of a Geometric Space.* Studies in the History of Mathematics and Physical Sciences, ed. Hardy Grant, trans. Abe Shenitzer, Vol. 12. New York: Springer-Verlag.

Ryan, Patrick J. (1986). *Euclidean and Non-Euclidean Geometry: An Analytic Approach.* Cambridge, UK: Cambridge University Press.

Saccheri, Girolamo (1986). *Euclides Vindicatus.* New York: Chelsea.

Schaff, Philip (1907). The Middle Ages from Gregory VII, 1049, to Boniface VIII, 1294. In *History of the Christian Church*, Vol. V. New York: Scribner.

Scriba, Christoph J. (1970). Lambert, Johann Heinrich. In *Dictionary of Scientific Biography*, ed. Charles Coulston Gillispie, Vol. VII, pp. 595–600. New York: Scribner.

Shabel, Lisa (2006). Kant's Philosophy of Mathematics. In *The Cambridge Companion to Kant and Modern Philosophy*, ed. Paul Guyer, pp. 94–122. Cambridge, UK: Cambridge University Press.

Shaughnessy, Edward L. (1999). Western Zhou History. In *The Cambridge History of Ancient China: From the Origins of Civilization to 222 B.C.*, ed. Michael Loewe and Edward L. Shaughnessy, pp. 292–361. Cambridge, UK: Cambridge University Press.

Shaw, Ian, ed. (2000). *The Oxford History of Ancient Egypt.* Oxford, UK: Oxford University Press.

Simmons, George F. (1963). *Topology and Modern Analysis.* New York: McGraw-Hill.

Smart, James R. (1998). *Modern Geometries*, 5th ed. Pacific Grove, CA: Brooks/Cole.

Smith, David Eugene, ed. (1959). *A Source Book in Mathematics.* Reprint, New York: Dover.

Sommerville, D. M. Y. (1958). *The Elements of Non-Euclidean Geometry*. Reprint, Mineola, NY: Dover, 2005.

Stein, Sherman K. (1999). *Archimedes: What Did He Do Besides Cry Eureka?* Washington, DC: Mathematical Association of America.

Steinhaus, Hugo (1999). *Mathematical Snapshots*, 3rd ed. New York: Dover.

Stillwell, John (1996). *Sources of Hyperbolic Geometry*. Providence, RI: American Mathematical Society.

Struik, D. J. (1970a). Beltrami, Eugenio. In *Dictionary of Scientific Biography*, ed. Charles Coulston Gillispie, Vol. I, pp. 599–600. New York: Scribner.

Struik, Dirk J. (1970b). Fano, Gino. In *Dictionary of Scientific Biography*, ed. Charles Coulston Gillispie, Vol. IV, pp. 522–523. New York: Scribner.

Swetz, Frank J. and T. I. Kao (1977). *Was Pythagoras Chinese?* University Park, PA: Pennsylvania State University Press.

Taton, René (1970). Poncelet, Jean Victor. In *Dictionary of Scientific Biography*, ed. Charles Coulston Gillispie, Vol. IX, pp. 76–82. New York: Scribner.

Tidman, Paul and Howard Kahane (2003). *Logic and Philosophy: A Modern Introduction*, 9th ed. Belmont, CA: Thomson.

Veblen, Oswald and John Wesley Young (1910). *Projective Geometry*, Vol. I. Boston: Ginn.

Venn, John (1894). *Symbolic Logic*, 2nd ed. London: Macmillan.

Viète, François (1983). *The Analytic Art: Nine Studies in Algebra, Geometry, and Trigonometry from the Opus Restitutae Mathematicae Analyseos, Seu, Algebrâ Novâ*, trans. T. R. Witmer. Kent, OH: Kent State University Press.

Vlastos, Gregory (1975). *Plato's Universe*. Seattle, WA: University of Washington Press.

Vrooman, Jack Rochford (1970). *René Descartes: A Biography*. New York: Putnam.

Waerden, B. L. van der (1971). *Science Awakening I: Egypt, Babylonian and Greek Mathematics*. Oxford, UK: Oxford University Press.

Wallace, Edward C. and Stephen F. West (1998). *Roads to Geometry*, 2nd ed. Upper Saddle River, NJ: Prentice Hall.

Walser, Hans (2001). *The Golden Section*, trans. Peter Hilton and Jean Pedersen. Washington, DC: Mathematical Association of America.

Watson, Richard A. (2002). *Cogito, Ergo Sum: The Life of René Descartes*. Boston: David R. Godine.

Wedberg, Anders (1955). *Plato's Philosophy of Mathematics*. Stockholm: Almqvist & Wiksell.

Wolfe, Harold (1945). *Introduction to Non-Euclidean Geometry*. New York: Henry Holt.

Wölfflin, Heinrich (1971). *The Art of Albrecht Dürer*, trans. Alastair Grieve and Heide Grieve. New York: Phaidon.

Wussing, Hans (1984). *The Genesis of the Abstract Group Concept*, ed. Hardy Grant, trans. Abe Shenitzer. Cambridge, MA: MIT Press.

Yaglom, I. M. (1988). *Felix Klein and Sophus Lie: Evolution of the Idea of Symmetry in the Nineteenth Century*, ed. Hardy Grant and Abe Shenitzer, trans. S. Sossinsky. Boston: Birkhäuser.

Yǎn, Lǐ and Dù Shírán (1987). *Chinese Mathematics: A Concise History*, trans. John N. Crossley and Anthony W.C. Lun. Oxford, UK: Clarendon Press.

INDEX

A posteriori, 387, 388, 391
A priori, 388, 391
AA, 156, 382
AAA, 156, 368
AAS, 141
Abel, Niels Henrik, 522
Abelian group, 522
Abscissa, 248, 259, 264
Absolute geometry, 330, 340, 359, 361, 363, 371
Absolute measure, 369, 370, 471
Absolute polar and pole, 479
Absolute value, 50
Académie des Sciences, 365, 476, 485, 486
Academy, 54, 103, 182, 325
Accident, 60
Accumulation point, 454
Achilles, 61
Acute, 131
Addition Property, 32, 47
Additive identity, 32, 47, 522
Additive inverse, 32, 33, 522
Adimantus, 57
Adjacent, 151
Aethalides, 89
Affine geometry, 532
Affine plane, 531, 532
Affine reflection, 535
Affine translation, 535
Affinity, 533, 535, 538, 539
Affirmative, 64
Affirming the consequent, 40
Ahmes, 6
Akkadians, 4, 13
Akousmatikoi, 98
Al-Nadīm, Abū l-Faraj Muhammad ibn, 325
Al-Tūsī, Nasir al-Dīn, 330
Alberti, Leone Battista, 294, 295
Alexander the Great, 55, 246
Alexandria, 124, 173, 197, 198, 308, 325
Algebra, 7, 112, 129, 229–231, 258, 274, 293, 416, 519, 523
Algebraic number, 355
Alternate angles, 148, 149
Alyattes, 28
Amenemhat II, 4
Amsterdam, 277
Amyclas, 123
Amyntas II, 54
Analysis, 230, 231, 260
Analytic art, 230–233, 236, 246, 257, 258
Analytic geometry, 264, 290, 323, 355

Analytic proposition, 387
Anaxagoras, 123
Angle, 131
Angle addition, 138
Angle bisector, 134
Angle of parallelism, 351, 400, 432
Angle subtraction, 138
Anlage zur Architectonik, 354, 385
Annales de Mathématiques Pures et Appliquées, 494
Anselm, 267
Antecedent, 37
Anthemius, 262
Anti-earth, 90
Anti-Euclidean geometry, 375
Antiphon, 183
Antipodal, 416, 479
Apex, 10
Apollo, 88
Apollonius, 246, 247, 250, 253, 257, 258, 264, 279, 293, 294, 312, 496
Apophis I, 6
Apothem, 17, 215, 216, 218, 221
Appendix, *see The Science of Absolute Space*
Applications d'analyse et de géométrie, 486
Aquinas, Thomas, 267
Archimedes, 173–174, 199, 236, 238, 257, 325, 374
Archimedes' Postulate, 184, 560
Archytas, 89, 102–104, 123, 182, 246
Area, 151
Argument, 29
Aristaeus, 246, 258, 279
Aristarchus, 180
Aristomache, 54
Aristotle, 54–55, 59–64, 67, 68, 74, 75, 77–82, 125, 194, 227, 267, 268, 280, 325, 327
Aristotle's Postulate, 327, 337, 370
Arithmetic, 61, 391, 562
Arithmetic mean, 94
Arithmetical Classic of the Gnomon and the Circular Paths of Heaven, 20, 22, 23
Arnsdorf, 384
ASA, 141
Asia Minor, 182
Associative, 32, 47
Astronomy, 20, 28, 54, 88, 102, 182, 229, 268, 298, 308, 354, 374, 396, 444, 476
Aswan, 198
Asymptote, asymptotically, 251, 347
Athenaeus, 123
Athens, 27, 53–55, 103, 123, 124, 182, 183, 325
Augsburg, 353
Augustine, 267
Axial perspectivity, 506, 513
Axial triangle, 247
Axiom, 32, 556, 557
Axiomatic system, 33
Axis
  cone, 247
  conic section, 248
  cylinder, 207
  equidistant curve, 414
  horocycle, 408
  horosphere, 424
  perspectivity, 502

Babylon, Babylonian, 4, 5, 13–18, 20, 21, 28, 29, 88, 97, 98, 182
Balance point, 195
Bartels, Martin, 373, 396
Barzizza, Gasparino, 294
Base
  isosceles triangle, 138
  Lambert quadrilateral, 355
  parallelogram, 154
  pyramid, 10
  Saccheri quadrilateral, 331
  trapezoid, 158
  triangle, 154
Base angles, 331
Baseline, 414
Basic open set, 460
Bavaria, 275
Bavarian Academy of Sciences, 354
Beaumont-de-Lomagne, 257
Beeckman, Isaac, 269, 271, 274
Beer mugs, 556
Beg the question, 74, 329
Behistun Cliff, 13
Beltrami hemisphere model, 474–476
Beltrami, Eugenio, 471–474
Benkő, Zsuzsanna Árkosi, 393
Berlin, 354, 443, 520
Berlin Academy, 365
Bernays, Paul, 563
Bernoulli, James, 473
Bernoulli, Johann, 385
Between, 557, 560
Bhaskara, 25
Bias, 27
Biconditional, 41
Biconditional proof, 45
Binary operation, 520

Bisector, 134
Boeotians, 376, 397
Bologna, 294, 472
Bolyai, Farkas, 374, 376, 393–395, 404, 405, 432
Bolyai, János, 393–396, 401, 404, 405, 407, 424, 432, 444, 477
Bonn, University of, 476
Book of Bibliography of the Sciences, 325
Book of Lemmas, 181
Bordeaux, 257
Born, Max, 563
Bosse, Abraham, 306
Boundary, 130
Boundary line, *see* horocycle
Boundary surface, *see* horosphere
Brahe, Tycho, 298, 301
Branch, 249, 414
Breda, 269, 274
Bride's Chair, The, 168
British Museum, 7, 13
Broken bamboo problem, 23
Brouillon Project, 294, 306, 310, 312, 313, 316, 319, 484
Brunelleschi, Filippo, 298
Brunswick-Wolfenbüttel, 373, 436
Büttner, J. G., 373
Byzantium, 325

Caen, University of, 476
Calandri, Philippi, 23
Calculus, 204, 228, 257, 393, 395, 436, 445, 471, 485
Calvin, John, 268
Cancellation Law, 524
Cardinality, 461
Caroline College, 373, 374, 396
Cartesian coordinates, 419, 507, 508
Cartesian model, 561
Cartesian product, 528
Cases, proof by, 49
Castres, 265
Categorical proposition, 64, 65
Categorical syllogism, 75
Categories, 63
Category, 64
Cauchy, Augustin-Louis, 444, 485
Cauchy-Riemann equations, 444
Cause, 62
Celestial mechanics, 365
Celestial spheres, 93
Center
  circle, 130
  homothety, 542
  hyperbola, 249
Center of gravity, 175, 194, 195, 197
Center of perspectivity, 501
Center of projection, 464
Central angle, 17, 415, 438
Central perspectivity, 501, 503, 513
Centroid, 167
Cephalus, 63
Ceres, 375
Ceva, Giovanni, 309, 323
Ceva, Tommaso, 323
Ceva's Theorem, 309
Cevian line, 309
Chalcis, 55
Chaldeans, 88
Characteristic of parallelism, 397
Charles IX, 229
Charmides, 53
Chemnitz, Martin, 268
Cheops, *see* Khufu
Chilon, 27
China, 5–6, 20–25
Chord, 159, 248
Chord Theorem, 161
Christiania University, 520
Christianity, 267, 384
Church, A. H., 114
Cicero, 223
Circle, 8, 130, 159, 258, 259, 289
  area, 8, 21, 186, 188
  circumference, 189
Circle with infinite radius, 405
Circular definition, 129
Circumcenter and circumcircle, 164
Circumference, 130
Circumscribe, 163, 164, 190
Cissoid, 239, 240, 258
Class, 64
Cleobulus, 27
Closed, 452, 454, 463, 478, 521
Closure, 32, 453, 463
Cnidus, 182
Cogent, 33
Collegium Fridericianum, 384
Collinear, 131, 486, 495
Collineation, 489, 514, 515, 532, 536, 540
Columbia University, 13, 14
Commensurable, 95, 101, 410, 412
Commensurable in square, 102
Commentaire sur Galois, 520
Commentaire sur le mémorie de Galois, 520
Commentaries on the Difficulties in the Premises of Euclid's Book, 337
Commentary, 23

Commentary on the First Book of Euclid's Elements, 325
Common notion, 129
Common perpendicular, 341
Commutative, 32
Comparative Review of Recent Researches in Geometry, A, 519
Compass, 56, 58, 106, 114, 130, 132, 236–238, 246, 258, 270, 281, 355
  collapsible, 59, 130, 133
Compendium of Calculation, 23
Complement, 68
Complete, 354, 556
Complete quadrangle, 315, 495
Complete quadrilateral, 495
Componendo, 94
Composite number, 47, 91
Composition, 460, 461, 503, 506, 515, 529, 536, 539, 541, 546, 549, 550
Concentric horocycles, 410
Concept, 389
Conchoid, 239, 241, 258
Conclusion, 30
Concurrent, 486, 495
Conditional, 37
Conditionals, 38
Cone, 11, 186, 209, 247, 295
Configuration, 315
Conformal, 467, 536, 537, 539, 540
Confucius, 6
Congruence axioms, 557, 560
Congruence class, 521
Congruent, 10, 44, 131, 133, 557, 560
Congruent modulo *n*, 51
Conic section, 246, 262, 288, 296, 299, 318, 496
Conic Sections, 246, 258, 259, 496
Conics, 246
Conjugate diameters, 256
Conjunction, 36
Connectives, 36
Conoids, 179
Conon, 173, 175
Consequent, 37
Consistent, 324, 474, 556
Constantinople, 202
Construct, 389
Constructed on, 176
Construction, 130, 281, 390
Contains a point, 486, 510
Continued fraction, 121, 354
Continued mean proportionals, 104, 105, 242, 270
Continuity axioms, 557
Continuity, continuous, 132, 278, 399, 403, 413, 446, 458–461
Continuity, principle of, 299, 484
Continuous group, 535
Contraction, 536, 542
Contradiction, 46
Contradictory, 69
Contrapositive, 44, 68
Contrary, 61, 69
Converse, 40, 67, 68, 80
Convertendo, 94
Convex, 118
Copernicus, Nicolaus, 228, 298, 300, 301
Cored rhombus, *see* rhombus, cored
Corollary, 33
Correspond, 58, 406, 407, 409
Corresponding angles, 148, 149
Council of Trent, 268
Counterexample, 67, 72
Courant, Richard, 563
Cremona, 471
Critias, 53
Critique of Pure Reason, 385
Cross ratio, 500
Croton, 88, 89, 101
Cube, 118, 301
Cuneiform, 4, 13
Curve, 125
Cutting plane, 247
Cyaxares, 28
Cyclic group, 525
Cylinder, 22, 186, 207, 445, 464
Cyrene, 197

D'Aubeterre, Antoinette, 228
Dali, Salvador, 116
Darius, 13
Data, 231
Daunian bear, 88
De Anima, 325
De Interpretatione, 63
De Prospettiva Pingendi, 298
Decad, 90
Deductive argument, 33
Defect, angle, 342, 344, 376
Definition, 32
Delian problem, 103, 104, 106, 123, 197, 236, 238, 241, 242, 246, 270
Delos, 88, 103
Delphi, 87
Demiurge, 118
Denmark, 301
Denying the antecedent, 40
Depression, 232

Desargues, Gérard, 277, 293–294, 310, 312, 313, 315, 316, 318, 484, 486, 488, 497, 501, 504, 530
Desargues' Involution Theorem, 317, 318
Desargues' Theorem, 306, 308, 498, 502
Descartes, Joachim, 268
Descartes, René, 116, 259, 267–281, 289, 294, 330, 355, 395
Descartes' Rule of Signs, 273
Describe, 545, 548
Descriptive geometry, 483
Determinant, 514, 539
Diagonal, 151
Diagonal line, 496
Diagonal point, 315
Diagonal triangle, 315, 498
Diagonal trilateral, 496, 499
Diairesis, 74
Dialectic, 63, 74, 118, 268
Dialogue on the Two Chief World Systems, 280
Diameter, 130, 248
Diana, feast of, 174
Differential geometry, 445, 472, 483, 553
Dilatation, 535
Dilation, 536, 542
Diocles, 239, 240
Diogenes Laertius, 87
Dion, 54, 103
Dionysius I, 54, 103
Dionysius II, 54, 103
Diophantus, 257
Dioptrics, 281
Diotima, 56
Direct proof, 43–45
Directrix, 262
Dirichlet, Lejeune, 443
Discourse on Method for Reasoning Well and for Seeking Truth in the Sciences, 281
Discrete topology, 459
Disjunction, 36
Disjunction proof, 48
Disjunctive Syllogism, 37
Disquisitiones Circa Superficies Curvas, 472
Distance, 146, 158, 239
Distinct, 486
Distributive, 32
Divide, 47, 49
Divine Proportion, The, 115
Division Algorithm, 525
Division, method of, 74
Djoser, 3
Dodecahedron, 118, 301
Domáld, 393
Donatello, 298
Doris, 54
Dositheus, 173, 177
Double elliptic geometry, 479
Double-napped cone, 247
Dual, 494
Duality, principle of, 494
Dulcarnon, 168
Duplication of the cube, *see* Delian problem
Dürer, Albrecht, 228, 298
Düsseldorf, 476, 563
Dyad, 90

e, 355
Earth measure, 6
Eastern Orthodox, 202
Eccentricity, 262
École d'Application de l'Artillerie et du Génie, 485
École Militaire, 365
École Militaire de Mézièrs, 483
École Polytechnique, 483, 484, 486
Edfu, 8
Efficient cause, 62
Egypt, Egyptians, 3–4, 6–12, 14, 18, 21, 28, 29, 88, 112, 173, 182, 197, 198, 246, 325, 483
Egyptian Mathematical Leather Scroll, 7
Eisenstein, Gotthold, 443
Elberfeld, 513
Elea, 60, 72
Elements of a plane, 302
Elements, 123–125, 128, 138, 140, 142, 147, 159, 163, 169, 194, 230, 302, 324, 326, 329, 330, 478, 556, 560
Éléments de géométrie, 365
Elements of Geometry, 326
Ellipse, 246, 247, 253, 254, 258, 259, 289
Ellipsoid of revolution, 179
Elliptic geometry, 479, 487, 557, 562
Empirical intuition, 389
Empiricist, 388
Endpoint, 125
Enter a triangle, 379
Epicureans, 136
Epicycle, 246
Equidistant, 328, 329
Equidistant curve, 414
Equilateral triangle, 132, 138, 370
Equilibrium, 194
Equivalence class, 511
Equivalence relation, 51, 512

Equivalent, logically, 41, 46, 68
Erasmus, Desiderius, 228
Eratosthenes, 173, 177, 197–199, 270
Erlangen Program, 519, 520, 529, 553
Erlangen, University of, 519, 553
Essay on Conics, 310
Essence, *see* form
Euclid, 124, 231, 246, 258, 262, 279, 315, 376, 444, 478, 525
Euclid Freed of All Blemish, *see* *Euclides Vindicatus*
Euclid's Axiom, 560
Euclidean geometry, 427, 535, 553, 557, 562
Euclidean topological space, 462
Euclides Vindicatus, 330, 340, 351, 355
Eudoxus, 123, 182, 186, 246
Euler, Leonard, 354
Euphorbus, 89
Eureka, 180
Eutocius, 219, 222
Even number, 45, 91
Excess, angle, 342, 416
Exclusive or, 36
Exegetics, 233
Exhaustion, method of, 186, 214, 217, 219
Existential import, 68, 69
Existential quantifier, 64
Exterior, 130
Exterior angle, 143, 150
Exterior angles on the same side, 148, 149
Extreme and mean ratio, 110
Extremes and means, 94
Extremity, 125

Fallaciae duplicis hypothesis, 324
Fallacy, 40
Fano plane, 488, 490, 532
Fano's geometry, 491
Fano, Gino, 488
Faraday, Michael, 513
Ferdinand of Brunswick, 373
Fermat, Pierre de, 257–259, 264, 265, 269, 289, 294, 395
Fibonacci, *see* Leonardo of Pisa
Fibonacci sequence, 112, 113
Fifth, perfect, 93
Figure, 77–82, 130
Final cause, 62
Finck, Thomas, 229
Finite, 478
Finite model, 488
First circle, 177
First conchoid, 239
Five Regular Solids, 115
Flat spheroid, 179
Florence, 294
Focus, 262, 298, 299
Focused system, 295
Form, 56–59
Formal cause, 62
Foundations of Geometry, 556
Four-line geometry, 491
Fourth angle, 356
Fourth, perfect, 93
Francesca, Piero della, 115, 298
Francisci turnica, 168
Franck, James, 563
Franco-Prussian War, 476, 520
Frederick the Great, 354
Freye Perspektive, 355
Frustum, 11, 211, 212, 215, 216
Fulcrum, 193
Fundamental Theorem of Projective Geometry, 505
"Fundamental Theory of Spaces of Constant Curvature", 474

Galileo, Galilei, 228, 280, 298, 323
Galois, Evariste, 520, 522
Gaojing, 5
Gauss, Johann Friedrich Carl, 373–376, 393, 395, 397, 404, 414, 436, 443, 444, 471, 475, 554, 556, 563
Gelon, 173
General History and Theory of the Heavens, 385
General linear group, 523
Generator of a cone or cylinder, 207, 209
Generator of a group, 525
Genoa, 294
Geodesic, 126, 472, 474
Geodesy, 436
Geometric concept, 391
Geometric Investigation, 323
Geometric mean, 95
Geometric sequence, 97
Geometric series, 177
Geometrical Investigations on the Theory of Parallels, 397
Geometry, 61, 391
Geometry, 271, 281, 283, 289
Geometry and Trigonometry, from the Works of A. M. Legendre, 366
Geometry by René Descartes, 283
Geometry, rules of, *see* rules of geometry
Gergonne, Joseph-Diez, 494
Girard's Theorem, 416
Girard, Albert, 416

Giza, 3
Glaucon, 57
Glide reflection, 550, 551
Gnomon, 20, 23
God, 267, 278–280, 391, 471
Golden ratio, 110, 112, 113, 115
Golden rectangle, 115
Golden triangle, 121
Golenischev Papyrus, *see* Moscow Papyrus
Golius, Jacobus, 279, 280
Göttingen, 353, 374, 375, 436, 443, 444, 471, 476, 519, 520, 554, 555, 563
Gōugǔ Theorem, 22–25
Grätz, 301
Great circle, 126, 415, 430
Great Pyramid, 3, 9, 28
Greece, 6
Group, 522

Habilitationsschrift, 385, 444, 471, 554
Habilitationsvortrag, 385
Hades, 89
Half-prism, 22
Hammurabi, 4
Han dynasty, 6, 20
Hanover, 436
Harmonic conjugate, 498
Harmonic mean, 95
Harmonic net, 505, 508
Harmonic set of lines, 499
Harmonic set of points, 497, 504, 505, 508
Heiberg, John L., 202
Hekademos, 54
Heliopolis, 182
Heliotrope, 436
Hemisphere model, *see* Beltrami hemisphere model
Henri III, 229
Henri IV, 229
Heptagon, 17
Hermes, 89
Hermite, Charles, 355
Hermodorus, 231
Hermotimus, 89, 123
Herodotus, 4, 6, 9, 28
Hertz, 97
Hestia, 90
Hexad, 92
Hexagon, 17
Hexahedron, 118
Hieratic, 7
Hieroglyphics, 6, 7, 13
Hieron II, 173–175, 179
Hilbert axioms, 557–560
Hilbert, David, 427, 554–556
Hippasus, 101, 110
Hippias, 123, 236
Hippocrates, 104, 123
Holland, 269, 277
Holy Sepulchre, 202
Homeomorphism, 461
Homogeneous coordinates, 508, 510–512, 532
Homogeneous terms, 230
Homothety, 540, 542
Horizontal suspension point, 194
Horocycle, 408, 409, 424, 427
Horocycle, outer, 410
Horosphere, 425, 427
Horus, temple of, 8
Hsuan-thu, 24, 25
Hundkurve, 472
Hydrostatics, 179
Hyperbola, 246, 247, 249–251, 258–261, 289
Hyperbolic functions, 354, 434
Hyperbolic geometry, 479, 562
Hyperboloid of revolution, 179
Hypothesis of the Acute Angle, 332, 335, 349, 359, 368, 397, 405
Hypothesis of the Obtuse Angle, 332, 335, 339, 359, 473, 477, 479
Hypothesis of the Right Angle, 332, 335, 338, 359, 370

Iamblichus, 87
Icosahedron, 118, 301
Idea, 57
Ideal plane, 305
Ideal point, 302, 304, 351, 376, 405, 424, 484, 530
Identity, 32, 48, 522
Identity map, 503, 529
If and only if, 41
If-then statement, 37
Image, 466
Imaginary geometry, 397
Imaginary point, 484
Imaginary radius, 360
Imhotep, 3, 6
Imperfect syllogism, 78
Implication, 37
Improper subgroup, 525
Incenter, 164
Incidence axioms, 557
Incidence geometry, 491
Incident, 486, 510, 560
Inclusive or, 36
Incommensurable, 101, 182
Inconsistent, 324

Independent, 324, 490, 556
Indeterminant, 246, 260, 289, 395, 467
India, 6, 23, 25
Indirect proof, 46–48
Induced Euclidean geometry, 427
Induced topology, 459
Inductive argument, 33
Infinite, 47, 60, 93, 125, 126, 236, 238, 273, 278, 391, 478, 523, 559, 560
Infinite model, 488
Infinite regression, 128
Infinite, actual, 60
Infinite, potential, 61
Infinitesimal, 185, 194, 445
Initial line, 177
Inquiry Concerning the Distinctness of the Principles of Natural Theology and Morality, 385
Inscribe, 164
Institut National des Sciences et des Arts, 365
Interior, 130, 451
Interior angle, 143, 150, 151
Interior angles on the same side, 148, 149
Intermediate forms, 58
International Commission on Mathematics Education, 563
International Lobachevski Prize, 562
Intersecting lines, 351
Intersection, 450, 452, 458, 526
Interval, 93
Introduction to Plane and Solid Loci, 258, 289
Introduction to the Analytic Art, 230, 257
Introduction to Vitello's Optics, 298
Intuition, 389
Invalid, 34
Invariant, 504, 505, 529, 553
Inverse, 32, 48, 522
Inverse function, 461, 547, 550
Inverse image, 458
Invertendo, 94
Involution, 313, 315
Involution, four points in, 320, 497
Ionia, 28, 88
Irrational number, 101
Isometry, 536–539, 541, 544, 545, 549–551
Isomorphic, 489, 523, 534
Isosceles trapezoid, 158
Isosceles triangle, 138, 139

Jacobi, Carl, 443
Jerusalem, 202
Jesuits, 268, 323
Jordan, Camille, 519, 520, 554
Judtschea, 384
Juel's pyramid, 13
Jupiter, 375
Justinian, 54

Kant, Immanuel, 383–386, 389, 391, 436, 554
Kästner, A. G., 353, 355, 494
Kazan, 396, 562
Kazan Messenger, 397
Kepler, Johannes, 298–301
Khayyām, Umar, 330, 337
Khufu, 3
Kish, 4
Klein disk model, 477
Klein, Felix, 474, 476, 519–520, 529, 535, 553–555, 562–563
Klein-4 group, 523, 525
Klügel, Georg, 355, 494
Knowledge, 386
Kolozsvár, 393
Königsberg, 383, 384, 554
Königsburg, 554

La Flèche, 268
La Haye, 267
Lagrange, Joseph-Louis, 365
Lahire, Philippe de, 294, 484
Lambert, Johann Heinrich, 353–355, 359, 368, 370, 379, 385, 414, 438
Lambert quadrilateral, 356–361
Lamina, 194
Landau, Edmund, 563
Laplace, Pierre-Simon, 365
Lateral surface area, 12, 207, 212
Latitudes, 227
Latus rectum, see parameter of the ordinates
Lausanne, 488
Law of Cosines, 172, 422
Law of Cosines, Spherical, 422
Law of Homogeneous Terms, 230, 232
Law of Sines, 420
Law of Sines, Spherical, 420
Law of the Lever, 175, 194
Leg, 158
Legendre, Adrien-Marie, 363, 365–366, 370
Leibnitz, Gottfried, 228, 556
Leiden, University of, 279
Leipzig, 554, 555, 562
Lemma, 33
Length, 44
Leodamas, 123
Leon, 123
Leonardo da Vinci, 115, 172, 228, 298
Leonardo of Pisa, 112, 227

Lever, 193
Liber abaci, 112
Lie, lies on, 486, 557
Lie, Sophus, 519, 520, 535
Limit, 351
Lindemann, Ferdinand von, 355, 554
Line, 125, 259, 474, 475, 479, 480, 494, 531, 556, 560
Line at infinity, 302, 475, 479, 486, 531, 553
Line of symmetry, 405, 552
Line segment, 44, 133
Linear loci, 258
Linear number, 91
Linearly dependent, 510
Linearly independent, 510
Lines, 486
Liu Bang, 5
Lobachevskaya, Praskovia Aleksandrovna, 396
Lobachevski, Ivan Maksimovich, 396
Lobachevski, Nicolai Ivanovich, 396–397, 401, 405, 407, 424, 431, 432, 436–438, 444, 471, 477, 555
Lobachevski Prize, 562
Locally Euclidean, 462, 464, 467
Locus, 108, 179, 232, 239, 240, 247, 248, 258, 260, 262, 279, 281, 288, 289, 295, 408
Logarithmic spiral, 116
Logica Demonstrativa, 324
Lombard, Peter, 267
Longitude, 227
Lorraine, 353
Louvre, 13
Lune, 108, 416
Luoyi, 5
Luther, Martin, 228, 268
Luxor, 6
Lyceum, 55, 63
Lydia, 28
Lyon, 293, 294

Macedonia, 54
Machiavelli, Niccolò, 228
Maggiore, Lake, 471
Magnitude, 61, 183
Major premises, 75
Major term, 75
Mamercus, 123
Manifold, 445, 447, 462, 464, 477
Manifoldness, *see* manifold
Marburg, University of, 513
Marcellus, 173
Marcus Claudius Marcellus, *see* Marcellus
Marinus, 325
Mars, 375
Masaccio, 298
Material cause, 62
Mathematical Collection, 230, 238, 262, 312, 325
Mathematical concept, 391
Mathematics, 62, 391
Mathêmatikoi, 98, 101
Mathematische Annalen, 563
Mathesis universalis, 274
Maulbronn, College of, 300
Maximum triangle, 376
Mazarin Collège, 365
Mean, 94
Mean proportional, 104, 110, 170, 206, 207, 212, 213, 241, 274
Means and extremes, 94
Measure, 131
Measurement of a Circle, 180, 189, 202
Mechanics, 56, 175, 197, 269, 365, 472, 485, 486
Media, 28
Median, 158, 167
Medici, 295
Meditations, 376
Memphis, 3
Menaechmus, 123, 246, 255
Menelaus, 308
Menelaus's Theorem, 308, 309, 312, 313
Menes, 3
Meno, 62, 63
Mersenne, Marin, 268, 277, 280, 293
Mesolabe, 270, 273
Mesopotamia, 4–5
Metapontum, 89, 91
Meteorology, 281
Method, The, 177, 197, 202, 204
Method, Descartes', 275
Metric, 446
Metric space, 446
Metz, 484, 485
Michelangelo, 228
Middle Ages, 228, 267, 294, 330
Middle term, 75
Midpoint, 134
Miletus, 28
Minding, Ferdinand, 472
Minim, 268
Minor premises, 75
Minor term, 75
Mirror, 182
Mirror of a reflection, 543
Mnesarchus, 87
Möbius, August, 513

Model, 472, 474, 476, 479, 487, 488, 490, 553, 557, 560
Modulo, 511, 521
Modus Ponens, 39
Modus Tolens, 39
Monad, 90
Monatliche Correspondenz, 375
Monge, Gaspard, 483
Monochord, 92
Mood, 78
More, Thomas, 228
Moscow Museum of Fine Arts, 7
Moscow Papyrus, 7, 10
Mulhouse, 353
Multilateral, 138
Multiplication Property, 32
Munich, 353, 554
Mursili, 5
Muses, 54
Mydorge, Claude, 268, 269, 277
Mysterium Cosmographicum, 301
Mystic hexagram theorem, 310

N-gon, 9
Nancy, France, 475
Napoleon, 4, 6, 483
Nebuchadnezzar II, 28
Necessary, 39, 46
Negation, 37
Negative, 64
Neoclides, 123
Neptune, 374
Net of rationality, 508
Neue Geometrie des Raumes, 476
Neues Organon, 354, 385
Neugebauer, Otto, 13, 14
Neusis, 238, 240
New Foundations of Geometry, 397
New York Historical Society, 7
Newton, Isaac, 183, 228, 374
Nicomedes, 238, 240
Nine Chapters on the Mathematical Art, 20
Nine Chapters on the Mathematical Art, 21
Nizhny Novgorod, 396
Noether, Emmy, 563
Non-Euclidean geometry, 376, 471
Nonhomogeneous coordinates, 530
Nonintersecting lines, 351
Norm, 457
Norway, 519, 522, 563
Number, 61
Number theory, 45, 183, 197, 257, 268, 354, 365, 476, 525, 555

Obers, Wilhelm, 375
Objects, 302
Oblique cone, 247
Oblique ordinates, 259
Oblong number, 96
Oblong spheroid, 179
Observations on the Feeling of the Beautiful and Sublime, 385
Obtuse, 131
Obtuse-angled conoid, 179
Octahedron, 118, 301
Octave, 93
Odd number, 45, 91
Oenopides, 123
Omega triangle, 376
On Burning-glasses, 239
On Centers of Gravity, 180
On Conoids and Spheroids, 179
On Floating Bodies, 202
On Levers, 180
On Means, 197
On Music, 102
On Painting, 295
On Plane Equilibriums, 174
On Speeds, 182
On Sphere-making, 180
On Spirals, 177, 202, 238, 258
On the Equilibrium of Planes, 194, 202
On the Hypotheses Which Lie at the Bases of Geometry, 444
On the Measurement of the Earth, 197
On the Principles of Geometry, 397
"On the So-called Noneuclidean Geometry", 476
On the Sphere and Cylinder, 177, 202, 204, 214
One, 90
One-to-one correspondence, 402, 461, 464, 467, 489, 501, 503, 529, 536
One-to-one function, 402
Only Possible Argument in Support of a Demonstration of the Existence of God, The, 385
Onto function, 402
Open, 448, 450, 458
Open base, 460
Open disk, 448
Open mapping, 461
Opposite, 151, 315, 495
Opposite interior angle, 143
Oppositely placed, 320
Opuscula Mathematica, 323
Order, 523
Order axioms, 557, 558

Ordinary, 302
Ordinary triangle, 376
Ordinate, 248, 259, 264
Oresme, Nicole, 227, 228
Organon, 63
Orléans, University of, 257
Orthogonal, 479
Outer horocycle, 410

$\mathbb{P}^2$, 513, 530
Pacioli, Luca, 115
Padua, 294
Pairwise disjoint, 512
Palimpsest, 202
Pappus, 172, 197, 230, 231, 238, 257, 258, 262, 279, 312, 315, 325
Pappus Problem, 279, 284, 285, 289
Parabola, 175, 246–249, 258, 259, 288
Parabolic section, 175
Paraboloid of revolution, 179
Parallax, 180, 436, 437, 440
Parallel, 128, 148, 149, 351, 530, 533, 560
Parallel axiom, 557
Parallel Postulate, 127, 149, 151, 154, 159, 169, 303, 324–326, 328, 330, 332, 337–340, 355, 359, 361, 365, 367, 369, 370, 374–376, 386, 394, 396, 404, 436, 444, 472, 477, 490, 557, 560
Parallelogram, 151
  height, 154
Parameter, 230
Parameter of the ordinates, 248
Paris, 269, 277, 293, 310, 365, 476, 484, 486, 519, 535, 554
Parmenides, 59, 96
Parmenides, 60, 72
Part, 125
Partake, 57
Parthenon, 116
Particular, 64
Partition, 511, 512
Pascal, Antoinette, 309
Pascal, Blaise, 294, 309, 310, 484
Pascal, Etienne, 309, 310
Pascal's Theorem, 310
Pasch's Postulate, 379, 558
Pasch, Moritz, 379, 555
Pavia, 330
Peano, Giuseppe, 555
Peloponnesian War, 53
Pencil, 301
Pennsylvania, University of, 13
Pentagon, 17, 116
Pentagonal number, 96
Perfect syllogism, 78
Perfection, 278
Perga, 246
Periander, 27
Permutando, 94
Perpendicular, 131, 135, 136
Perpendicular, common, 341
Persia, 88
Perspective, 295, 501
Perspective from a line, 502
Perspective from a point, 502
Phaedo, 325
Phenomena, 182
Phidias, 173
Phillippus, 123
Philolaus, 90, 101
Phoenicians, 88
Photius, 87, 103
Pi (π), 8, 180, 189, 192, 236, 343, 355
Pi function (Π), 400–403, 409, 427, 428, 431, 432, 434, 435, 438
Piazzi, Giuseppe, 375
Pietism, 384
Pisa, 112, 472
Pitch, 92
Pittacus, 27
Plane, 556
Plane angle, 131
Plane figure, 130
Plane Loci, 258
Plane loci, 258
Plane number, 92
Plane of projection, 464
Plane surface, 126
Planum, 232
Plato, 53–54, 56, 57, 59–62, 103, 106, 107, 118, 123, 130, 183, 194, 236, 325
Platonic solid, 118, 124, 301
Playfair, John, 303, 326
Playfair's Postulate, 326, 370, 560
Plimpton collection, 14
Plücker, Julius, 476, 513, 520
Poincaré, Henri, 474, 475, 554, 563
Poincaré disk model, 475
Poincaré half-plane model, 475
Point, 125, 458, 474, 475, 479, 480, 486, 494, 531, 556, 560
Point at infinity, 299, 530
Point of tangency, 160
Point triple of lines, 316
Point triple of points, 316
Poitiers, 228, 268
Polar, 496

Pole, 238, 240, 496
Polycrates, 88
Polygon, 9, 16, 17, 118, 124, 138, 151, 163, 187, 189, 190, 215, 216, 218, 315, 332
 circumscribed, 190, 205–207, 220, 221
 convex, 151, 170, 195, 344, 364
 inscribed, 183, 191, 192, 205–208, 220
 side, 9
 similar, 170
 standard, 214–220
 vertex, 9
Polyhedron, 118
Poncelet, Jean-Victor, 484–486, 494, 505, 519, 520, 555
Porisms, 258
Poristics, 231, 233
Porphyry, 87
Posidonius, 325, 329
Posterior Analytics, 63
Postulate, 31, 127
Postulate of Archimedes, 184
Power set, 527
Prague, 301
Predicate term, 64
Premise, 30
Preserve, 313
Preserve distance, 536
Preserve open sets, 461
Prime number, 47, 91, 197
Principal group, 536, 539, 540, 543, 544, 546, 547, 550
Principia Mathematica, 183
Principle of least time, 257
Prior Analytics, 63, 77
Prism, 10, 22, 186
 base, 10
 height, 10
Privatdozent, 384, 444, 554
Problematic analysis, 231
Proclus, 126, 136, 169, 239, 325, 329, 330
Projection, 295, 301
Projective collineation, 515
Projective geometry, 297, 487, 532, 535, 553
Projective plane, 302, 486
Projective space, 305
Projectivity, 503, 535
Proof, 29
Proof by cases, 49
Proof by contradiction, 46
Proportion, 93, 182
Proportional, 154
Proposition, 31
Proposition, categorical, 64
Prussia, 384, 563
Prussian Academy of Sciences, 354, 385
Pseudosphere, 473
Ptah, 3
Ptolemy, 325, 329
Ptolemy Evergetes, 197
Ptolemy's Theorem, 172
Pure intuition, 389
Pyramid, 9–12, 92, 186, 199, 295
 face, 10
 height, 10
 truncated, 11
Pyramid, Great, *see* Great Pyramid
Pyramid, Juel's, 13
Pyramid number, 96
Pyramid, visual, 296
Pyrrhus, 89
Pythagoras, 87–90, 92, 93, 98, 123, 168
Pythagorean scale, 95
Pythagorean Semicircle, 88
Pythagorean Theorem, 14, 16, 24, 98–100, 110, 118, 124, 138, 159, 167, 168, 170, 183, 310, 370, 371, 418
Pythagorean Theorem, Spherical, 419
Pythagorean triples, 16
Pythagoreans, 87, 89, 91, 94, 95, 98, 99, 101, 150, 182
Pythais, 87
Pythia, 88

Qin dynasty, 5, 20
Qin Shihuangdi, 5
Quadrangular set, 316, 317, 496, 497, 507
Quadrant, 479
Quadratic equation, 283, 444
Quadratic Formula, 14
Quadratrix, 236
Quadrature of the Parabola, 175, 258
Quadrilateral, 138, 151
 area, 8
Quadrivium, 102
Quality, 64
Quantifier, 64
Quantity, 61, 64
Quotient, 525
Quotient set, 511

$\mathbb{R}^n$, 445–447, 462, 476, 539, 560
Radius, 130
Rameses II, 4
Range, 302
Raphael, 228
Ratio, 93, 539

Rational number, 95
Rationalist, 388
Rautenburg, 384
Real projective plane, 513
Real-valued function, 527
Rectangle, 151, 361, 362, 370
Rectifiable, 445
Rectilineal angle, 131
Rectilineal figure, 138
Rectilinear number, 91
Reductio ad absurdum, 46, 123, 324, 374
Reduction, 232
Reference triangle, 530
Reflection, 536, 543–546, 549, 551
Reflexive, 44, 51
Regular, 9, 118
    polygon, 9, 18, 118, 190, 205–207, 219, 222, 236
    pyramid, 13
Reign of Terror, 365
Relative consistency, 475
Relative measure, 369
Relative topology, 462
Remainder, 525
Remo, 323
Renaissance, 228, 294, 298
Representative, 513
Republic, 57, 58
Retrograde motion, 246
Rheneia, 88
Rhind, A. Henry, 6, 7
Rhind Mathematical Papyrus, 7, 8, 12
Rhombus, cored, 211
Rhombus, solid, 209, 210
Riemann sphere, 488
Riemann surface, 444
Riemann, Georg Friedrich Bernhard, 443–444, 471, 477, 478, 554, 555, 557, 563
Riemannian geometry, 478
Right angle, 131
Right triangle, 138
Right-angled conoid, 179
Rome, 472
Roomen, Adriaan van, 229
Rosetta Stone, 6, 7, 13
Rotation, 536, 548–551
Rudolph II, 301
Ruler, 238
Rules of geometry, 31
Rules of logic, 35

SA, 381
Saccheri quadrilateral, 330–335, 357, 360, 361, 371
Saccheri, Giovanni Girolamo, 323–324, 330, 337, 350, 351, 355, 370, 396, 472, 477, 556
Saccheri-Legendre Theorem, 363, 364, 366, 367, 377
Sachs, Abraham, 14
Sacrament of the Last Supper, 116
Same parallels, in the, 152, 153
Samos, 87, 173, 180
Sand-reckoner, The, 180
Saqqara, 3
Saratov, 484
Sargon, 4
SAS, 138, 536
Scale, 93
Scalene triangle, 138
Scholasticism, 227, 267, 268, 279
Schooten, Frans van, 283
Science of Absolute Space, The, 395
Screw, Archimedean, 173
Secant, 159, 229
Secant Theorem, 161
Section, 295
Sectiones Conicae, 294
Segment addition, 139
Segment bisector, 134
Segment subtraction, 139
Seked, 12
Self-dual, 495
Self-evident, 31
Separando, 94
Sesostris, 4, 6
Sety I, 4
Seven Sages, 27
Sexagesimal, 15
Shakespeare, William, 228
Shang dynasty, 5
Shear, 535
Shih Huang-ti, 20
Sicily, 54, 89, 103, 173
Side of a complete quadrangle, 315
Side of a complete quadrilateral, 495
Sieve of Eratosthenes, 197
Similar, 154, 159, 256
Similarity, 539–541
Similarly placed, 320
Simple subgroup, 525
Single elliptic geometry, 479
Sirius, 437
Slant height, 209
Smith, Edwin, 7
Socrates, 53, 55–57, 62, 72, 183
Socrates, the younger, 72
Solid Loci, 246, 258

Solid loci, 258
Solid number, 92
Solid rhombus, *see* rhombus, solid
Solidum, 232
Solon, 27, 53
Sophist, 73
Sophist, 236
Sphaerica, 308
Sphere, 186, 430, 445, 464, 467
surface area, 223
volume, 22, 223
Sphere and Cylinder, 258
Sphere with imaginary radius, 360, 438, 439
Sphere with infinite radius, 405, 407, 414
Spherical angle, 415
Spherical Law of Cosines, 422
Spherical Law of Sines, 420
Spherical Pythagorean Theorem, 419, 420
Spherical triangle, 416
Spherical trigonometry, 359, 417, 438, 439
Spheroids, 179
Spiral, 258
Spiral of Archimedes, 177
Square, 151
Square number, 92
Square of opposition, 70, 71
Square pyramid, 10
Squaring the circle, 236
SSS, 140
St. Petersburg, 375
Stabilizer, 529
Standard form, 65, 75
Standard metric, 446
Statement, 31
Stereographic projection, 464, 468
Stobaeus, 124
Stomachion, 180
Straight line, *see* line
Straightedge, 56, 58, 106, 114, 236–238, 246, 258, 281, 355
Stretch, 535
Strictly decreasing, 401
Stump, 319
Subalternation, 70
Subcontrary, 69
Subgroup, 525, 526
Subject term, 64
Subspace, 462
Substance, 60
Sufficient, 39, 46
Sumerians, 4, 13, 15
Summit, 331
Summit angles, 331
Superimpose, 138, 445, 536
Supplement to Geometry, A, 240
Surface, 126
Surface Loci, 258, 262
Susa, 4, 16, 18
Syene, 198
Syllogism, 37
Symbolic logistic, 233
Symmetric, 44, 47, 51, 552
Symmetry, 536
Symposium, 56
Syncrisis, 233
Synthesis, 231, 260, 261
Synthetic geometry, 231, 293, 355, 483, 494, 513
Synthetic proposition, 387
Syracuse, 54, 103, 173, 174, 223
System, 194

Tangent, 159, 167, 229
Tangent-Secant Theorem, 161, 485
Tarentum, 88, 89, 102
Taxicab metric, 455
Technische Hochschule, 553, 554
Term, 64
Tetrad, 90
Tetrahedral number, 96
Tetrahedron, 22, 118, 301
Tetraktys, 90, 93
Thales, 27–29, 87, 89, 123, 138
Thales' Theorem, 29, 30, 43, 44, 106
Theaetetus, 72, 118, 123, 124, 183
Thebes, 6
Theodorus, 72, 123, 183
Theorem, 33
Theorem of the Bride, The, 168
Theorematic analysis, 231
Theory of Parallel Lines, 355, 385
Theory of the Motion of the Heavenly Bodies Moving about the Sun in Conic Sections, 375
Theudius, 123
Third Man Argument, 59
Third proportional, 159
Thirty Tyrants, 53
Thoughts on the True Estimation of the Living Forces, 384
Three and Four Line Locus Problem, *see* Pappus Problem
Three-point geometry, 491
Thureau-Dangin, Paul, 13
Timaeus, 118, 325
Tone, 92
Topics, 63
Topological space, 458

Topology, 458
Torus, 105, 445
Toulouse, 257, 265
Touraine, 267
Tractrix, 472
Traité des propriétés projectives des figures, 484–486, 519
Traité des substitutions et des équations algébriques, 520
Transcendental idealism, 385
Transcendental number, 355
Transformation, 529
Transformation group, 529
Transitive, 44, 47, 51, 399
Translation, 536, 545–547, 549–551
Transposition, 232
Transversal, 128, 147
Trapezoid, 152
Treasury of Analysis, 197, 230
Treatise on the Quadrilateral, 337
Tree, 294, 319
Triad, 90
Triangle, 138, 494
  equilateral, 132, 138, 370
  height, 154
  isosceles, 138, 139
  scalene, 138
Triangle Inequality, 136, 144, 446, 447, 536
Triangle triple of lines, 316
Triangle triple of points, 316
Triangular number, 91
Trident of Descartes, 291
Trigonometry, 198, 229, 308, 354, 359, 416, 438, 471
Trilateral, 138, 494
Trisection, 236, 241, 243, 270
Trisection compass, 270
Trivial subgroup, 525
Trivial topology, 459
True Estimation of Living Forces, 384
Truth value, 30
Tübingen, University of, 300
Turin, 323
Tyre, 87

Undefined term, 128
Undershot waterwheel, 485
Unicorn, 386, 387
Union, 450, 452, 458, 527
Unique, 47
United States, 366, 488
Universal, 64
Universal quantifier, 64
Ur, 4
Ur-Zababa, 4
Uranus, 374

Vacuously true, 490
Valid, 34, 74, 76, 77
Vanishing point, 298
Vector, 545, 550
Venice, 115
Venn diagram, 65, 66, 68, 69, 74–76, 78, 81
Vertex, 131, 248, 315, 495
Vertical angles, 146
Vertical suspension point, 195
Vienna Academy of Military Engineers, 394
Vienna, University of, 393
Viète, François Viète, 228–229, 231–233, 236, 257, 258, 281, 289
Visual pyramid, 295, 296
Viviani, Vincenzo, 323

Wachter, Friedrich Ludwig, 405, 414
Walcheren, 484
Wallis, John, 330, 337, 370
Wallis's Postulate, 337, 370
Weber, Wilhelm, 444
Wehla, 554
Weil, 300
Wen, 5
Weyl, Hermann, 563
World, or a Treatise on Light, The, 280
Wu, 5

Xenocrates, 55

Yale, 13
Young's geometry, 492

Zach, Franz Xaver von, 375
Zeno, 60, 72
Zetetics, 231–233
Zhou dynasty, 5, 20
Zurich, 353

PURE AND APPLIED MATHEMATICS

A Wiley-Interscience Series of Texts, Monographs, and Tracts

ADÁMEK, HERRLICH, and STRECKER—Abstract and Concrete Catetories
ADAMOWICZ and ZBIERSKI—Logic of Mathematics
AINSWORTH and ODEN—A Posteriori Error Estimation in Finite Element Analysis
AKIVIS and GOLDBERG—Conformal Differential Geometry and Its Generalizations
ALLEN and ISAACSON—Numerical Analysis for Applied Science
*ARTIN—Geometric Algebra
ATKINSON, HAN, and STEWART—Numerical Solution of Ordinary Differential Equations
AUBIN—Applied Functional Analysis, Second Edition
AZIZOV and IOKHVIDOV—Linear Operators in Spaces with an Indefinite Metric
BASENER—Topology and Its Applications
BERG—The Fourier-Analytic Proof of Quadratic Reciprocity
BERKOVITZ—Convexity and Optimization in $\mathbb{R}^n$
BERMAN, NEUMANN, and STERN—Nonnegative Matrices in Dynamic Systems
BOYARINTSEV—Methods of Solving Singular Systems of Ordinary Differential Equations
BRIDGER—Real Analysis: A Constructive Approach
BURK—Lebesgue Measure and Integration: An Introduction
*CARTER—Finite Groups of Lie Type
CASTILLO, COBO, JUBETE, and PRUNEDA—Orthogonal Sets and Polar Methods in Linear Algebra: Applications to Matrix Calculations, Systems of Equations, Inequalities, and Linear Programming
CASTILLO, CONEJO, PEDREGAL, GARCIÁ, and ALGUACIL—Building and Solving Mathematical Programming Models in Engineering and Science
CHATELIN—Eigenvalues of Matrices
CLARK—Mathematical Bioeconomics: The Optimal Management of Renewable Resources, Second Edition
COX—Galois Theory
†COX—Primes of the Form $x^2 + ny^2$: Fermat, Class Field Theory, and Complex Multiplication
*CURTIS and REINER—Representation Theory of Finite Groups and Associative Algebras
*CURTIS and REINER—Methods of Representation Theory: With Applications to Finite Groups and Orders, Volume I
CURTIS and REINER—Methods of Representation Theory: With Applications to Finite Groups and Orders, Volume II
DINCULEANU—Vector Integration and Stochastic Integration in Banach Spaces
*DUNFORD and SCHWARTZ—Linear Operators
Part 1—General Theory
Part 2—Spectral Theory, Self Adjoint Operators in Hilbert Space
Part 3—Spectral Operators

*Now available in a lower priced paperback edition in the Wiley Classics Library.
†Now available in paperback.

FARINA and RINALDI—Positive Linear Systems: Theory and Applications
FATICONI—The Mathematics of Infinity: A Guide to Great Ideas
FOLLAND—Real Analysis: Modern Techniques and Their Applications
FRÖLICHER and KRIEGL—Linear Spaces and Differentiation Theory
GARDINER—Teichmüller Theory and Quadratic Differentials
GILBERT and NICHOLSON—Modern Algebra with Applications, Second Edition
*GRIFFITHS and HARRIS—Principles of Algebraic Geometry
GRILLET—Algebra
GROVE—Groups and Characters
GUSTAFSSON, KREISS and OLIGER—Time Dependent Problems and Difference Methods
HANNA and ROWLAND—Fourier Series, Transforms, and Boundary Value Problems, Second Edition
*HENRICI—Applied and Computational Complex Analysis
Volume 1, Power Series—Integration—Conformal Mapping—Location of Zeros
Volume 2, Special Functions—Integral Transforms—Asymptotics—Continued Fractions
Volume 3, Discrete Fourier Analysis, Cauchy Integrals, Construction of Conformal Maps, Univalent Functions
*HILTON and WU—A Course in Modern Algebra
*HOCHSTADT—Integral Equations
JOST—Two-Dimensional Geometric Variational Procedures
KHAMSI and KIRK—An Introduction to Metric Spaces and Fixed Point Theory
*KOBAYASHI and NOMIZU—Foundations of Differential Geometry, Volume I
*KOBAYASHI and NOMIZU—Foundations of Differential Geometry, Volume II
KOSHY—Fibonacci and Lucas Numbers with Applications
LAX—Functional Analysis
LAX—Linear Algebra and Its Applications, Second Edition
LOGAN—An Introduction to Nonlinear Partial Differential Equations, Second Edition
LOGAN and WOLESENSKY—Mathematical Methods in Biology
MARKLEY—Principles of Differential Equations
MORRISON—Functional Analysis: An Introduction to Banach Space Theory
NAYFEH—Perturbation Methods
NAYFEH and MOOK—Nonlinear Oscillations
O'LEARY—Revolutions of Geometry
O'NEIL—Beginning Partial Differential Equations, Second Edition
PANDEY—The Hilbert Transform of Schwartz Distributions and Applications
PETKOV—Geometry of Reflecting Rays and Inverse Spectral Problems
*PRENTER—Splines and Variational Methods
PROMISLOW—A First Course in Functional Analysis
RAO—Measure Theory and Integration
RASSIAS and SIMSA—Finite Sums Decompositions in Mathematical Analysis
RENELT—Elliptic Systems and Quasiconformal Mappings
RIVLIN—Chebyshev Polynomials: From Approximation Theory to Algebra and Number Theory, Second Edition
ROCKAFELLAR—Network Flows and Monotropic Optimization
ROITMAN—Introduction to Modern Set Theory
ROSSI—Theorems, Corollaries, Lemmas, and Methods of Proof
*RUDIN—Fourier Analysis on Groups
SENDOV—The Averaged Moduli of Smoothness: Applications in Numerical Methods and Approximations

*Now available in a lower priced paperback edition in the Wiley Classics Library.
†Now available in paperback.

SENDOV and POPOV—The Averaged Moduli of Smoothness
SEWELL—The Numerical Solution of Ordinary and Partial Differential Equations, Second Edition
SEWELL—Computational Methods of Linear Algebra, Second Edition
SHICK—Topology: Point-Set and Geometric
*SIEGEL—Topics in Complex Function Theory
Volume 1—Elliptic Functions and Uniformization Theory
Volume 2—Automorphic Functions and Abelian Integrals
Volume 3—Abelian Functions and Modular Functions of Several Variables
SMITH and ROMANOWSKA—Post-Modern Algebra
ŠOLÍN–Partial Differential Equations and the Finite Element Method
STADE—Fourier Analysis
STAHL—Introduction to Topology and Geometry
STAKGOLD—Green's Functions and Boundary Value Problems, Second Editon
STANOYEVITCH—Introduction to Numerical Ordinary and Partial Differential Equations Using MATLAB®
*STOKER—Differential Geometry
*STOKER—Nonlinear Vibrations in Mechanical and Electrical Systems
*STOKER—Water Waves: The Mathematical Theory with Applications
WATKINS—Fundamentals of Matrix Computations, Second Edition
WESSELING—An Introduction to Multigrid Methods
†WHITHAM—Linear and Nonlinear Waves
ZAUDERER—Partial Differential Equations of Applied Mathematics, Third Edition

*Now available in a lower priced paperback edition in the Wiley Classics Library.
†Now available in paperback.

Wiley The manufacturer's authorized representative according to the EU General Product Safety Regulation is Wiley-VCH GmbH, Boschstr. 12, 69469 Weinheim, Germany, e-mail: Product_Safety@wiley.com.

Printed and bound by CPI Group (UK) Ltd, Croydon, CR0 4YY

19/09/2026

02219301-0003